MW00835275

MOVING THE EARTH

Excavation Equipment, Methods, Safety, and Cost

Robert L. Schmitt, P.E.

Clifford J. Schexnayder, P.E.

Aaron B. Cohen, CPC

Herbert L. Nichols, Jr.

David A. Day, P.E.

Seventh Edition

New York Chicago San Francisco Athens London
Madrid Mexico City Milan New Delhi
Singapore Sydney Toronto

Library of Congress Control Number: 2018959522

Moving the Earth: Excavation Equipment, Methods, Safety, and Cost, Seventh Edition

1 2 3 4 5 6 7 8 9 LCR 23 22 21 20 19 18

ISBN 978-1-260-01164-7
MHID 1-260-01164-X

Sponsoring Editor
Lauren Poplawski

Editing Supervisor
Stephen M. Smith

Production Supervisor
Pamela A. Pelton

Acquisitions Coordinator
Elizabeth M. Houde

Project Manager
Neha Bhargava,
Cenveo® Publisher Services

Copy Editor
Lisa McCoy

Proofreader
Upendra Prasad,
Cenveo Publisher Services

Indexer
Robert Swanson

Art Director, Cover
Jeff Weeks

Composition
Cenveo Publisher Services

CONTENTS

Chapter 16. Blasting 541

Chapter 17. Aggregate Processing 583

ABOUT THE AUTHORS

Robert L. Schmitt, P.E., is Professor of Civil Engineering at the University of Wisconsin, Platteville. He has worked as a consultant on major construction projects and served on technical committees for the American Society of Civil Engineers, the Associated General Contractors, and the Wisconsin DOT.

Clifford J. Schexnayder, P.E., has worked with major heavy/highway construction contractors as field engineer, estimator, and corporate chief engineer. He served as a consultant to the Autoridad del Canal de Panama and is a Member of the National Academy of Construction and the Beavers.

Aaron B. Cohen, CPC, is a product manager at InEight building estimating and project management software solutions and teaches courses on heavy construction methods at Arizona State University. He was previously president of Apollo Trenchless, Inc., specializing in the trenchless construction of underground utility projects.

Herbert L. Nichols, Jr. (deceased), was a private excavation consultant and author of numerous books and manuals on the topic.

David A. Day, P.E. (deceased), was a civil engineering and construction consultant. He was a Life Member of the National Society of Professional Engineers and the American Society of Civil Engineers, where he helped to start the Underground Technology Conference.

PREFACE TO THE SEVENTH EDITION

Construction techniques and equipment are constantly evolving. This seventh edition of *Moving the Earth* captures many of the machine advances and technologies showcased at the 2017 CONEXPO-CON/AGG, together with the experiences of the author team as we visited projects both in the United States and internationally since publication of the sixth edition. Most machines work in combination with others, such as an excavator loading trucks, so the text emphasizes the importance of thinking in terms of linked systems. Because machine technology continually evolves, a contractor must stay abreast of machine improvements to remain successful in determining the most economical grouping of machines.

This edition describes in detail how to investigate equipment productivity, understand linked system productivity, and perform step-by-step calculations to determine system productivity. Skill in work task analysis, together with familiarity with appropriate machine applications, is imperative to successfully completing an earthwork project. Machines are designed for specific tasks and are economical only when used in the proper manner.

Important changes have been made to this edition:

- Calculating machine ownership and operating costs is presented in detail.
- Earthwork quantity takeoff and development of earthwork estimates are consolidated in a single chapter.
- Soil and rock are explained in terms of their engineering properties.
- Machine power is described based on rolling and grade resistance to movement.
- Operational capabilities of machines by type characteristics are detailed, and methods for calculating productivity with sample calculations are presented. (Based on best professional practices, we have tried to present standard formats for analyzing production.)

To improve the organization and presentation of concepts, all chapters have undergone revision or consolidation. Numerous photographs are now used to illustrate the latest equipment and methods. Safety concerns are highlighted when discussing specific machines. Construction equipment is manufactured globally in a dynamic market setting, so we have searched worldwide for the latest ideas and trends in machine technology.

Many individuals and firms have supplied information and illustrations for this work, and we owe them a great debt of gratitude. Others have freely given of their time to explain important machine concepts; their kindness is sincerely appreciated. However, full responsibility for all content rests with the author team.

We solicit comments on this edition.

Robert L. Schmitt, P.E.
Clifford J. Schexnayder, P.E.
Aaron B. Cohen, CPC

CHAPTER 1
TOOLS AND TASKS—PRIDE AND REALITY

Men and machines, moving earth and rock, transform a project plan into reality. This book describes the fundamental concepts of machine utilization and how to economically match machine capability to specific earth-moving requirements. Because careful planning is of critical importance to a contractor's ability to survive in the marketplace, the second objective of this book is to explain how to plan the execution of earth-moving tasks.

REALITY

Reality or myth, the constructor who moves the earth with big machines is an illiterate character with muddy boots and a hard hat covering an even harder head. But there is also a glorious side: the constructor who moves the earth is a proud craftsman. The truth is somewhere between illiterate and pride, for an uncertain Mother Nature may bless you with sunshine or rain (Fig. 1.1), and those who work moving the earth are gamblers with guts and a developed skill for mastering the vagaries of nature. All successful earthmovers soon come to possess a humbleness and respect for the awe and unknown quantity of Mother Nature.

Before William S. Otis, working for the Philadelphia contracting firm of Carmichael & Fairbanks, built the first practical steam shovel excavating machine in 1837 (Fig. 1.2), all movement of the rock and soil was a challenge. Men many centuries before were dreaming of machines to move the earth. Giovanni Fontana in 1420 diagramed his dream of a mechanical dredging machine. The "Yankee Geologist," as Otis's machines were called, was his mechanical answer to moving the earth on an 1838 railroad project south of Boston, Massachusetts. Development of the steam shovel was driven by a demand for an economical mass excavation machine to support the era of railroad construction. Similarly the Cummins diesel engine was developed in the early 1900s as the road-building phase of earthwork construction began. In the short-term future, the basic machine frame will not change, but productivity, accuracy, and utility should improve because of enhancements. Machines will possibly evolve into mobile counterweights driven by an energy-efficient and environmentally friendly power plant. Machines may become mobile counterweights, simply a work platform for an array of hydraulic tools. Basic engineering tenets of leverage, torque, tension, and force will be boldly built into early three-dimensional machinery following the lead of William Otis.

FIGURE 1.1 A dozer in the throes of Mother Nature's mud.

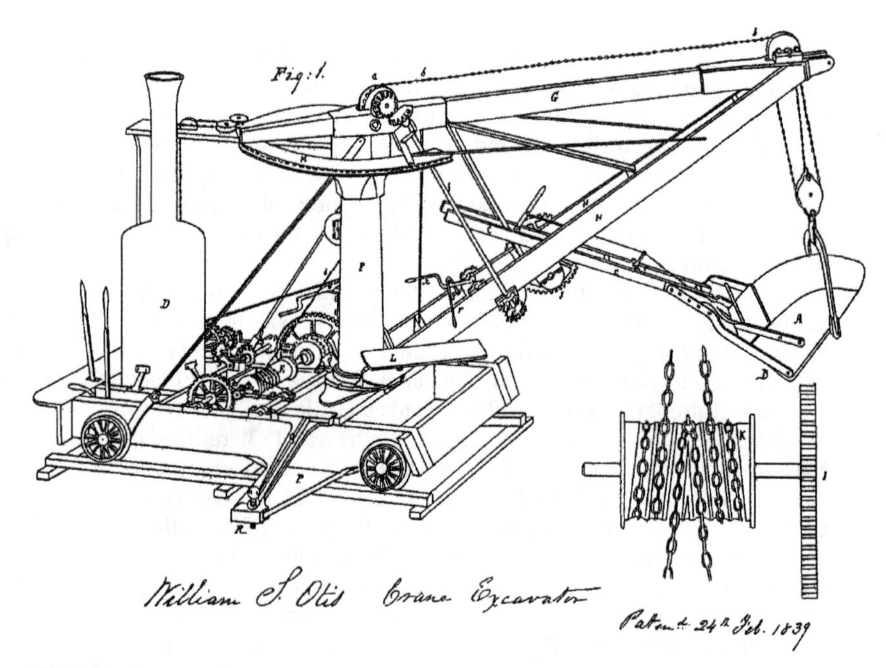

FIGURE 1.2 The William S. Otis steam shovel patent drawing.

Another part of folklore describes the proud earthmover as rolling in high profits and a consumer of large quantities of hard drink. And again the reality is very different. The high contractor failure rate quickly disproves the first notion, and for those who fail to carefully plan their earth-moving operations and monitor financial performance, maybe the drink is the result of despair. To overcome the vagaries of nature, all who succeed with a profit belong to a fraternity steeped in attention to careful planning and the deployment of superior cost analysis tools. Preparing accurate cost estimates is part of every successful earth-moving project won by competitive bidding. Still, for every contract awarded, a group of four or five other earth-moving contractors will say the price was too low! There may be mud on the boots, but the triumph of bringing a job in on time and budget goes to those who

- Understand machine capabilities
- Appreciate the physical characteristics of rock and dirt, both in its natural state and when manipulated
- Prepare detailed work plans
- Spend time studying contract language
- Carefully read cost reports

These are the subjects this text will scrutinize and explain in detail.

EARTH-MOVING EQUIPMENT

One of the most obvious problems in constructing a project is how to transport and manipulate rock and dirt. Machines provide the economical solution to this problem. The proof of how well the planner understands the work to be accomplished and selects appropriate machines is revealed by counting the money when the contract is completed. Did the company make a profit or sustain a loss?

From the time man first decided to build a simple structure for protection from the elements, to the inspiring construction achievements of the Pyramids, the Great Wall of China, the temples at Angkor Wat in Cambodia, and up until the middle of the nineteenth century, earthwork was accomplished by the muscle of man and beast. When New York State began excavating the Erie Canal on July 4, 1817, men bent to the task of digging the trench by the strength of their arms. The first mechanization—earth-moving machines—came with railroad construction in the late 1830s. The steam shovel transformed how projects were planned and executed.

The Bucyrus Foundry and Manufacturing Company came into being because of railroad building. Between 1885 and 1897 approximately 70,000 miles of railway were constructed in the United States. Later the Bucyrus shovels would be one of the key machines responsible for the Americans' success in completing the Panama Canal.

Internal Combustion Engines

By 1890 many companies had begun experimenting with gasoline engine–powered carriages using Nikolaus Otto's patented four-cycle gasoline engine. The Best Manufacturing Company (the predecessor to Caterpillar, Inc.) demonstrated a gasoline tractor in 1893. The first application of the internal combustion engine to excavating equipment occurred in 1910 when the Monighan Machine Company of Chicago shipped a dragline powered by an Otto engine to the Mulgrew-Boyce Company of Dubuque, Iowa. Henry Harnischfeger brought out a gasoline engine–powered shovel in 1914. C. L. "Clessie" Cummins, working out of an old cereal mill in Columbus, Indiana, developed his diesel engine in the early 1900s, and after World War I, the diesel engine began to appear in excavators. Warren A. Bechtel built a reputation as a successful railroad grading contractor. He pioneered the use of motorized trucks, tractors, and diesel-powered shovels in construction.

The Boulder Dam project (later named the Hoover Dam) was a colossal proving ground for construction equipment and techniques. With R. G. LeTourneau's development in the Nevada desert of welded equipment and cable-operated attachments, the use of bolted connections for joining machine pieces together came to an end. LeTourneau, fostering innovations in tractor/scraper design, made possible the machines later used to build airfields around the world during World War II. At the massive Hoover Dam project, sophisticated aggregate production plants were introduced, as well as the use of long-flight conveyor systems for material delivery.

Significant Developments

Torque converter A fluid-type coupling that enables an engine to be somewhat independent of the transmission.

After World War II, the interstate highway program caused a road building surge. Because of mandated design grades, there was a need to move very large quantities of rock and dirt. To support a road-building effort requiring large-scale movement of earth, scrapers increased in capacity from 10 to 30 cy. With the development of the **torque converter** and the power shift transmission, the front-end loader began to displace the old cable-operated "dipper" stick shovels. But the three most important developments were high-strength steels, nylon cord tires, and high-output diesel engines.

1. *High-strength steels*. Up to and through World War II, machine frames had been constructed with steels in the 30,000- to 35,000-psi yield range. After the war, steels in the 40,000- to 45,000-psi range with better fatigue properties were introduced. The new high-strength steel made possible the production of machines having greatly reduced overall weight.
2. *Nylon cord tires*. The utilization of nylon cord material in tire structures made larger tires with increased load capacity and heat resistance a practical reality. Nylon permitted the number of plies to be reduced with the same effective carcass strength.
3. *High-output diesel engines*. Manufacturers developed new ways to coax greater horsepower from a cubic inch of engine displacement. Compression ratios and engine speeds were raised, and the art of turbocharging was perfected.

BEING COMPETITIVE

This book introduces the engineering fundamentals for planning, selection, and utilization of equipment used for moving rock and dirt. It enables one to analyze operational problems and to arrive at practical solutions. It is about the application of engineering fundamentals, and the use of analysis to plan earth-moving activities and to make decisions based on economic comparisons of machine choices.

A contractor's ability to win contracts and to perform them at a profit is determined by two vital assets: people and equipment. To be competitive, the equipment must be competitive,

FIGURE 1.3 An equipment loading-hauling-compaction spread.

both mechanically and technologically. Old machines, which require costly repairs, cannot compete successfully with new equipment, having lower repair costs and higher production rates.

In most cases, a piece of equipment does not work as a stand-alone unit. Individual pieces of earthmoving equipment work together as a unit: a loader or excavator and dump trucks. When the truck dumps its load, the material is then spread by a dozer and densified by compaction equipment. This group of machines working together is commonly referred to as an *equipment spread* (Fig. 1.3). The operational productivity of each machine must match the capability of the other machines in the spread. For production efficiency, the machines must mesh together like gears.

Optimization in the management of an equipment spread is critical for a contractor, both in achieving a competitive pricing position and in accumulating the corporate operating capital required to finance the expansion of project performance capability. This book explains the fundamental concepts of machine utilization, which economically match machine capability to specific project requirements.

There are no unique solutions to the problem of selecting a machine to move material. All machine selection problems are influenced by external environmental conditions imposed by nature and the contract (a project must be completed on time or a late penalty is charged from the contractor). To appreciate how environmental conditions influence the utilization of heavy construction equipment, one must understand the mechanics of how the construction industry operates.

The success of earth-moving operations can set the tone for the entire project. Earth-moving is the first visible task, and without its success and profitability, an atmosphere of doom can loom over subsequent tasks. There may be added pressure for later tasks to make up for earth-moving losses. Owner relations can be strained because of early change orders. An understanding and intermingling of project management principles and machine capability is paramount.

THE CONSTRUCTION INDUSTRY

By the nature of the product, each earthwork project proceeds under a unique set of production conditions, and those conditions directly affect equipment selection. Whereas most manufacturing companies have a permanent factory where raw materials flow in and finished products flow out in a repetitive, assembly-line process, those who perform earthwork carry their factory with them from job to job. At each new site, the builder proceeds to set up and produce a one-of-a-kind project. If the work goes as planned, the job will be completed on time and with a profit.

Equipment-intensive projects present a great financial risk. Many projects involving earthwork are bid on a unit-price basis, and there can be large variations between estimated and actual quantities. Some projects require an equipment commitment greater than the contract amount. Such a situation forces a contractor into a continuing sequence of jobs to support long-term equipment payments.

Additional risk factors facing contractors in equipment-intensive work include their financing structure, construction activity levels (the amount of work being put out for bid), labor legislation and agreements, and safety regulations. Project size and weather-dependent outdoor work can contribute to long project durations. Projects requiring two or more years to complete are not uncommon in the industry. Subsurface conditions often differ from those anticipated. Utility interference always complicates earth-moving operations—consider the frequency of news reports where a construction crew has disturbed a gas main or fiber-optic line.

Government-initiated actions that seriously affect the operating environment of the construction contractor are labor legislation and safety regulations. In each of these areas, many regulations affect a contractor's operations. These actions can directly influence equipment decisions. Sound and emissions are issues receiving greater regulatory attention, and some owners are limiting machine noise levels through contract clauses. Newer emission requirements have forced equipment manufacturers to reorient their research and development programs.

SAFETY

The rate of personal injury and death resulting from construction work is too high. Of all major industry classifications, construction has one of the poorest safety records. The construction industry employs nearly 6.4 million people—roughly 6% of the American workforce. However, according to the National Safety Council, the industry has about 23% of the deaths and 10.3% of the injury accidents every year. Several of the major causes of deaths and injuries are earthwork equipment and task related: being struck by a machine, being caught in between equipment, and trench excavation cave-ins. The keys to better work site safety are first leadership, then programs, followed by the incentives to create a safe industry.

In 1970, Congress enacted the Williams-Steiger Act, more commonly referred to as the Occupational Safety and Health Act (OSHA). The act provides a comprehensive set of safety rules and regulations, inspection procedures, and safety record-keeping requirements. It also permits states to enact their own OSHA legislation as long as the state legislation is at least as stringent as the federal legislation. The OSHA rules and regulations are published in the *Federal Register*. Construction and Health Regulations, Code of Federal Regulations, Part 1926, pertains specifically to construction contractors and construction work. The act provides both civil and criminal penalties for violations of OSHA regulations. Contractors must maintain a current, up-to-date file of OSHA regulations and work proactively to comply with OSHA requirements.

THE CONTRACTING ENVIRONMENT

Construction contractors work within a unique market situation. The job plans and specifications supplied by the client dictate the sales conditions and product but not the price. Almost all work in the equipment-intensive earthwork fields of construction is awarded on a bid basis, through either open or selective tender procedures. Under the design–bid–build method of contracting, the contractor states a price after estimating the cost based on a completed design supplied by the owner. But there is often a risk component concerning payment: Is the work to be performed lump sum, all quantity risk on the builder, or will the quantities be measured and the work paid on a unit price basis? The offered price includes overhead, project risk contingency, and the desired profit.

Design–build contracts, where the contractor has control of the project design are becoming more popular. In the case of transportation projects with a total value under \$10M, 25% of the projects awarded in recent years were let design–build best value. With a design–build project, the contractor must state a guaranteed price before the design is completed. This adds another element of risk, because estimating the quantities of materials required to complete the project becomes very subjective. But the advantage to the contractor is that the design can be matched in the most advantageous way to the contractor's construction skills and equipment fleet.

Complicating this matter is the fact that the owner dictates the terms of the contract and the project delivery method, not the contractor. On most projects, the contractor is better suited to control the risk associated with this decision. This risk is inherently built into bids in the form of contingencies for each of the individual bid line items. Not infrequently, however, the range between the high and low bids is much greater than these factors would justify. A primary cause of variance in bid prices is a contractor's inability to estimate costs accurately.

Two quick measures—contract volume and contract turnover—indicate the financial strength of the firm. Contract volume refers to the total dollar value of awarded contracts (under contract) *at any given time.* Contract turnover measures the dollar value of work a firm completes during a specific *time interval.* Contract volume is a guide to the magnitude of resources committed at any one time, as well as to possible profit if the work is completed as estimated. But contract volume alone fails to answer any timing questions. A contractor, who, with the same contract volume as the competition, is able to achieve a more rapid project completion, and therefore a higher capital turnover rate while maintaining the revenue-to-expense ratio, will be able to increase the firm's profits. Contractors who finish work ahead of schedule usually make money. Overhead to support direct construction activities is typically time based. Simply put, the less time to build, the lower the overhead cost.

PLANNING EQUIPMENT UTILIZATION

Each piece of construction equipment is specifically designed to perform certain mechanical operations. The task of the earthwork builder is to match the right machine or combination of machines to the project task. Considering individual tasks, the quality of performance is measured by matching the equipment spread's production against its cost. Production is work done; it can be the volume or weight of material moved, the number of pieces of material cut, the distance traveled, or any similar measurement of progress. To estimate the equipment component of project cost, it is necessary to first determine machine *productivity.* Productivity is governed by engineering fundamentals. Chapter 5 covers the fundamentals controlling machine productivity. Each level of productivity has a corresponding cost associated with the effort expended. The expenses experienced through machine ownership and use, and the method of analyzing such costs are presented in Chap. 2.

Although each major type of equipment has different operational characteristics, it is not always obvious which machine is best for a particular project task. After studying the plans and specifications, visiting the project site, and performing a quantity take-off, the planner must visualize how to best employ specific pieces of equipment to accomplish the work. Is it less expensive to make an excavation with scrapers or to top-load trucks with a hoe or shovel? Both methods will yield the required end result, but which is the most economical for the given project conditions?

To answer this question, the planner develops an initial plan for employment of the scrapers and then calculates their production rate and the subsequent cost. The same process is followed for the top-load operation. The equipment spread with the lowest estimated total cost to move the material, including mobilization of the machines to the site, is selected for the job.

To perform such analyses, the planner must consider both machine capability and methods of employment. In developing suitable equipment employment techniques, the planner must have knowledge of the material quantities involved, which is the focus of Chap. 3. If it is determined that different equipment and methods will be used as an excavation progresses, then it is necessary to divide the quantity take-off in a manner compatible with the proposed equipment utilization. The person performing the quantity take-off must calculate the quantities in groups by material type or condition (dry earth, wet earth, rock). It is not just a question of estimating the total quantity of rock or the total quantity of material to be excavated. All factors that affect equipment performance and choice of construction method, such as location of the water table, clay or sand seams, site dimensions, depth of excavations, and compaction requirements, must be considered when making the quantity take-off.

The normal operating modes of the particular equipment types are discussed in Chaps. 6 to 11 and 13 to 15. No two projects are exactly alike; therefore, it is important to consider each project with a completely open mind and review all possible options. Moreover, machines are constantly being improved and new equipment being introduced.

Heavy equipment is usually classified or identified by one of two methods: functional identification or operational identification. A bulldozer used to push a stockpile of material could be identified as a support machine for an aggregate production plant, a grouping that could also include front-end loaders. The bulldozer could, however, be functionally classified as an excavator. In this book, combinations of functional and operational groupings are used. The basic purpose is to explain the critical performance characteristics of a particular piece of earth-moving equipment and then to describe the most common machine applications.

The efforts of builder and equipment manufacturers, daring to develop new ideas, constantly push machine capabilities forward. As equipment capabilities change, the importance of careful planning and execution of earthwork operations increases. New machines enable greater economies. The central focus of this book is to improve the ability of those engaged in "Moving the Earth" to match specific machines to individual project situations.

CHAPTER 2
COMPANY AND MACHINE COST

Cash flow is vital to any construction business. A contractor expends the client's funds, then, in turn, provides a valued service in the form of a constructed project. Contractors must be good stewards of this money and be able to track their expenses to build while simultaneously receiving and making payments. Payments for material invoices, weekly payroll, and subcontractor draws can complicate the accounting of money. Running a construction business with the associated buying and selling of machines requires solid accounting practices.

Two important accounting documents used to monitor a company's financial performance are the balance sheet and the income statement. The balance sheet is a snapshot of financial health at a particular moment in time. A company's income statement shows all of the money a company earned (revenues) and all of the money a company spent (expenses) during a specific period, usually the fiscal year. The income statement provides a map of how a company is changing and the pace of change.

A correct and thorough understanding of the costs resulting from equipment ownership and operation provides companies a market advantage that leads to greater profits. Ownership cost is the cumulative result of those cash flows an owner experiences, whether or not the machine is productively employed on a project. Operating cost is the sum of those expenses an owner experiences by using a machine on a project. There are three basic methods for securing a particular machine for project use: (1) buy, (2) rent, or (3) lease.

CONSTRUCTION COMPANY ACCOUNTING

Tracking the Money

> **Accounting**
> The practice of recording, classifying, reporting, and interpreting the financial data of an organization.

There are three types of **accounting**: (1) generally accepted accounting principles (GAAP accounting) for financial statements (lots and lots of rules), (2) management accounting (no rules) used in running the firm, and (3) tax accounting. This chapter will discuss the first two and leave the tax aspect to accounting professionals who are constantly trying to understand the tax laws that politicians and bureaucrats can dream up and continually change.

Accounting is the process of recording, classifying, reporting, and interpreting the financial data of an organization. One of the main purposes of accounting is to aid decision makers in choosing among alternative courses of action—management

Cost accounting The phase of accounting that deals with collecting and controlling the cost of producing a product or service.

accounting. Effective management of a construction company requires proper **cost accounting** and the availability of current and accurate cost information. For information to be useful, there must be consistent use of accounting principles from one period to another (e.g., a method of recording and reporting information).

Financial statements communicate useful monetary information and are the result of simplifying, condensing, and aggregating transactions. No one financial statement provides sufficient information by itself, and no one item or part of each statement can summarize the information. The most important accounting document used to monitor a company's earnings is the income statement. The balance sheet shows the company's overall financial health and potential for future growth. The income statement shows the sources of a company's production costs, other expenses, and income.

Income Statement

A company's income statement is a record of its earnings or losses for a given period. This can be thought of as a moving picture of company performance. It shows all of the revenue a company earned and all of the costs and expenses a company experienced during a specific period, usually the fiscal year. It also accounts for the effects of some basic accounting principles such as depreciation.

The income statement provides a record of a company's performance across time. Most importantly, the income statement tells if operations are yielding a profit. It provides a map (Fig. 2.1) of how a company is changing and the pace of change. It is essential for management to know how to analyze different elements of this important document.

Operating income (revenue) does not include interest earned. Nor does it include income generated outside the normal activities of the company, such as income on investments. Operating income is particularly important because it is a measure of profitability based on a company's operations. In other words, it assesses whether or not the core operations of a company are profitable. It ignores income or losses outside a company's normal domain. Earnings before taxes are the sum of operating and nonoperating income. Net earnings or net income (loss) after taxes is the proverbial bottom line. It is the amount of profit a company makes after all of its income and all of its expenses.

Retained Earnings. Retained earnings is the amount of equity from profits that a company has accumulated through all of its transactions since the beginning of the firm. If the company continually makes substantial profits, it is a stable company, or if it is like the Low Bid Corporation [see "Income (loss) from operations" in Fig. 2.1], there are major problems and the viability of the company is clearly threatened. In addition to gross profits that are way too thin, it appears that Low Bid is carrying way too much general and administrative expenses. These could include unnecessary travel, a company airplane, entertaining clients, or even unnecessary estimating cost caused by attempting to bid everything that comes along. Estimating can range from 0.1% to over 0.2% of the bid price, and work must be won to recover this cost.

LOW BID CONSTRUCTION CORPORATION
STATEMENT OF INCOME AND RETAINED EARNINGS

	For the Years Ended December 31,		
	2015	**2014**	**2013**
Revenue (Notes 10 and 16)	$174,063,148	$163,573,258	$210,002,272
Cost of revenue (Note 10)	169,404,787	158,935,103	200,070,826
Gross profit	4,658,361	4,638,155	9,931,446
General and administrative expenses (Notes 1 and 10)	8,062,623	6,777,840	6,671,035
Income (loss) from operations	(3,404,262)	(2,139,685)	3,260,411
Other income (expense):			
Interest income	319,797	646,480	668,928
Interest expense	(485,937)	(250,996)	(209,872)
Other income (expense)	377,840	(115,246)	211,119
	211,700	280,238	670,175
Income (loss) before income taxes	(3,192,562)	(1,859,447)	3,930,586
Income tax benefit (expense)	668,631	284,861	(1,590,480)
Net income (loss) (Note 17)	$ (2,523,931)	$ (1,574,586)	$2,340,106
BEGINNING RETAINED EARNINGS	$3,090,214	$4,664,800	$4,664,800
ENDING RETAINED EARNINGS	$566,283	$3,090,214	

Notes

10. Related Party Transactions:

Management believes that the fair value of the following transactions reflect current amounts that the Company could have consummated transactions with other third parties.

Revenue:

During the years ended December 31, 2015 and 2014, the Company provided construction materials to various related parties in the amounts of $108,112 and $26,556, respectively. Included in accounts receivable at December 31, 2015 and 2014 are amounts due from related parties, in the amounts of $27,337 and $15,132, respectively.

Professional Services:

During the years ended December 31, 2015, 2014, and 2013, a related party rendered professional services to the Company in the amounts of $14,573, $23,342, and $7,944, respectively. During the years ended December 31, 2015, 2014, and 2013, the Company paid $30,000, $30,000, and $5,000, respectively, to outside members of the board of directors.

Subcontractor/Supplier:

Various related parties provided materials and equipment used in the Company's construction business during the years ended December 31, 2015, 2014, and 2013, in the amounts of $4,114,319, $535,694, and $65,441, respectively. Included in accounts payable at December 31, 2015 and 2014 are amounts due to related parties, in the amounts of $1,046,908 and $154,861, respectively, related to supplies.

FIGURE 2.1 Sample income statement.

Royalties:

During the years ended December 31, 2015, 2014, and 2013, the Company paid a related party mining royalties in the amounts of $390,144, $328,310, and $182,061, respectively. Included in accrued liabilities December 31, 2015 and 2014, are amounts due to related parties, in the amounts of $30,464 and $49,983, respectively, related to royalties.

Commitments:

The Company leased office space in the state on a month-to-month basis, at a rental rate of $840 per month, from a related party of the Company. The lease terms also required the Company to pay common maintenance, taxes, insurance, and other costs. Rental expense under the lease for the years ended December 31, 2015, 2014, and 2013 amount to $9,240, $10,080, and $10,080, respectively.

16. Significant Customers:

For the years ended December 31, 2015, 2014, and 2013, the Company recognized a significant portion of its revenue from four Customers (shown as an approximate percentage of total revenue):

	For the Years Ended December 31,		
	2015	2014	2013
A	21.9%	17.5%	26.2%
B	12.5%	16.3%	28.7%
C	9.6%	23.0%	17.2%
D	14.7%	6.1%	5.8%

At December 31, 2015 and 2014, amounts due from the aforementioned Customers included in restricted cash and accounts receivables are as follows:

	For the Years Ended December 31,	
	2015	2014
A	$3,809,567	$2,968,786
B	$6,255,403	$1,855,666
C	$350,962	$1,124,196
D	$2,218,976	$762,181

17. Stock Option Plan:

In November 2008, the Company adopted a Stock Option Plan providing for the granting of both qualified incentive stock options and nonqualified stock options. The Company has reserved 1,200,000 shares of its common stock for issuance under the plan. Granting of the options is at the discretion of the Board of Directors and may be awarded to employees and consultants. Consultants may receive only nonqualified stock options. The maximum term of the stock options are 10 years and may be exercised as follows: 33.3% after one year of continuous service, 66.6% after two years of continuous service, and 100% after three years of continuous service. The exercise price of each option is equal to the market price of the Company's common stock on the date of grant.

FIGURE 2.1 *(Continued)*

Balance Sheet

A company's balance sheet (Figs. 2.2 and 2.3) describes its financial position at a specific date—a single point in time. This is like a snapshot picture of the financial condition. Because it is a single-point-in-time representation, it is sometimes referred to as a position statement. It is prepared at least once per year, but it may also be presented quarterly, semiannually, or even monthly. The balance sheet provides information on what the company owns (its assets), what it owes (its liabilities), and the value of the business to its owners (equity). The name *balance sheet* is derived from the fact that these accounts must always be in balance. Assets must always equal the sum of liabilities plus owner's equity. If liabilities exceed assets, the owner experiences a negative equity, or in simple terms, suffers a loss.

CONSOLIDATED BALANCE SHEET
(Thousands of dollars)

December 31,	2016	2015
ASSETS		
Current assets		
Cash	$ 11,330	$ 10,025
Short-term investments, at cost, which approximates market	8,620	14,738
Accounts receivable including retentions $44,943 and $65,064	208,007	182,319
Refundable federal income taxes	—	11,225
Costs and earnings in excess of billings on uncompleted contracts	192,727	164,232
Equity in joint ventures	96,655	82,311
Real property held for development and sale at the lower of cost or market	45,163	33,000
Other	11,462	7,790
Total current assets	573,964	505,640
Investments and other assets		
Equity in affiliated enterprises	9,815	16,349
Marketable securities, at cost, market $9,667 and $10,741	259	259
Other investments, at cost, which approximates market	41,318	15,896
Notes receivable and other assets	18,024	8,959
Total investments and other assets	69,416	41,463
Property and equipment, at cost		
Land	2,596	2,566
Buildings, ways, and wharves	164,322	148,900
Construction and other equipment	223,144	212,186
Total property and equipment	390,062	363,652
Less accumulated depreciation	(140,169)	(123,702)
Property and equipment—net	249,893	239,950
Total assets	$893,273	$787,053

The accompanying notes are an integral part of the financial statements.

FIGURE 2.2 Sample balance sheet showing the assets portion.

December 31,	2016	2015
LIABILITIES AND STOCKHOLDERS' EQUITY		
Current liabilities		
Short-term and current portion of long-term debt	$ 72,742	$ 29,350
Accounts payable and accrued expenses	228,115	200,761
Billings in excess of costs and earnings on uncompleted contracts	66,982	70,846
Advances from clients	19,687	32,076
Income taxes		
Currently payable	11,462	3,476
Deferred	6,047	46,866
Dividends payable	3,236	2,672
Total current liabilities	408,271	386,047
Non-current liabilities		
Deferred income taxes	43,666	20,611
Deferred income and compensation	15,513	15,602
Accrued workmen's compensation	27,973	23,189
Long-term debt	103,611	76,063
Total non-current liabilities	190,763	135,465
Commitments		
Stockholders' equity		
Common stock, par value $3.33⅓, authorized 20,000,000 shares, issued 10,616,391 shares	35,388	35,388
Capital in excess of par value	74,056	73,273
Retained earnings	196,764	170,421
	306,208	279,082
Less cost of treasury stock, 802,878 and 899,744 shares	(11,969)	(13,541
Total stockholders' equity	294,239	265,541
Total liabilities and stockholders' equity	$893,273	$787,053

FIGURE 2.3 Sample balance sheet showing liabilities and stockholders' equity part.

Assets. Assets are economic resources that are expected to produce economic benefits for their owners. They are presented first on the balance sheet (Fig. 2.2). The assets of a construction company are usually money (cash); receivables; equipment; and plant, property, and materials used to construct projects. A company's good name, or *goodwill,* is also considered an asset where the actual value is realized when the company is purchased, acquired, or merged by another.

Many construction companies have contractual obligations that significantly outweigh their physical assets. For example, a company may have $500 million in ongoing prime contracts and separate subcontract work, while their physical assets may only total $100 million. For this reason alone, the contractor must understand how the assets of the active contractual obligations are counted so that the company remains profitable.

Compared to other types of businesses, construction companies typically have less in the way of physical assets. A manufacturing business will have a large investment in buildings and machinery, and may even at times carry significant inventory of raw material with which to produce the product or an inventory of completed products that have not been sold. These are physical assets a lender relies on as collateral against a loan to the company.

A construction company, by comparison, conducts its business on property owned by someone else and only needs to actually possess a small office for corporate management. A heavy construction company might have a fleet of equipment, which represents substantial assets, but many times such equipment is leased or rented so it does not belong to the construction company. The one exception would be a construction company that produces concrete and asphalt. Many of these companies own their quarries from which they get the raw aggregate for their product. Quarries with their aggregate reserves represent significant asset holdings. In the case of contractors owning property for quarry mining, the reserve ore also represents an asset holding that is realized as the quarry is excavated and the product is sold. Therefore, because the value of actual physical assets possessed by a construction company is usually minimal, lenders must carefully look to the continuing operational performance of the company when evaluating the risk of a loan.

> In reading a balance sheet, it is important to remember that property and equipment are recorded at cost and not the price at which an asset may be sold or the cost to replace the asset.

Current Assets. Current assets are cash or other balance sheet accounts that can be changed into cash within a year. They are usually presented in the order of their liquidity:

- *Cash* is the most basic current asset.
- *Cash equivalents* are not cash but can be converted into cash so easily that they are considered equal to cash. Cash equivalents are generally highly liquid, short-term investments such as U.S. government securities, commercial paper, and money market funds. Short term is generally defined as converting or maturing into cash within 90 days if it is to be considered cash.
- *Accounts receivable* represents money that clients (project owners) owe to the firm for services rendered or for goods sold (e.g., producers of asphalt and concrete) and includes retentions receivable.
- *Inventory*, in the case of a construction firm, is the material kept on hand for use in construction of projects.

Long-Term Assets. Many assets are economic resources that cannot be readily converted into cash. Common examples might be real estate holdings, facilities, and equipment. A sale of these assets may take a year or longer. These long-term, or "other," assets are expected to provide benefit for future operations.

- *Fixed assets* are tangible assets. Generally, fixed assets refer to items such as equipment, vehicles, plant, buildings, and property. On the balance sheet, these are valued at their cost. Depreciation is subtracted from all classes except land.

Liabilities. Liabilities are obligations a company owes to outside parties. They represent the rights of others to money or services of the company. Examples include bank loans, debts to suppliers, debts to employees, debts to subcontractors, and both deferred taxes and taxes currently payable to government entities. On the balance sheet, liabilities are generally broken down into current liabilities and long-term liabilities.

Overbilling
Placing a higher value on work performed early in the project and lower values on work to be completed near the end of the project, so the contractor does not experience an out-of-pocket cash flow to finance the work.

- *Current liabilities* are those obligations that are to be paid within the year, such as accounts payable, interest on long-term debts, and taxes payable. The most pervasive item in the current liability section of the balance sheet is accounts payable. Also note that "**overbillings**" are recognized as a current liability (see "Billings in excess of costs and earnings on uncompleted contracts" in Fig. 2.3).
- *Accounts payable* are debts owed to subcontractors and to suppliers for the purchase of goods and services on an open account.
- *Long-term debt* is a liability of a period greater than 1 year. It usually refers to loans made to the company. These debts are often paid in installments. These could include loans to purchase vehicles and equipment and mortgage payments for buildings and land. If this is the case, the portion of the principal to be paid in the current year is considered a current liability. Pending litigation from construction claims against the company also represents a long-term debt. These potential claim losses require special evaluation by tax accountants.

Because current liabilities are usually paid with current assets, it is important to examine the degree to which current assets exceed current liabilities. This difference is called *working capital*. Working capital is the difference between short-term assets and liabilities. It is one of the most important measures of a company's financial health and the ability to make payments on time, whether it be the weekly payroll or supplier invoices.

Owner's Equity. Owner's equity, or the net worth of the company, is the excess of assets over liabilities. It represents what the owners have invested in the business plus accumulated profits, which are retained in the company, from ongoing operations. Losses will reduce owner's equity. The formulation of the report's balance is shown by Eq. (2.1):

$$\text{Equity (net worth)} = \text{Total assets} - \text{Total liabilities} \tag{2.1}$$

Considering the balance sheet asset and liability information contained in Figs. 2.2 and 2.3, it is possible to calculate the net worth of the company without knowing exactly the types of equity that the company owners possess (Table 2.1).

At the bottom of the balance sheet in the "stockholders' equity" section (Fig. 2.3), the composition of the private investor equity is presented. Retained earnings is that part of equity resulting from earnings in excess of losses.

Liquidity and Working Capital. Management must study the balance sheet to measure a firm's liquidity, financial flexibility, profit generation ability, and debt payment ability. Financial flexibility is the ability to take effective actions to alter the amounts and timing of cash flow in response to unexpected needs and opportunities.

TABLE 2.1 Net Worth Calculations

	Dec. 31, 2016	Dec. 31, 2015
Total assets	$893,273	$787,053
Total current liabilities	408,271	386,047
Total noncurrent liabilities	190,763	135,465
Total liabilities	599,034	521,512
Net Worth	294,239	265,541

Bonding companies (sureties) will also carefully study the company's balance sheet before issuing a bond. The picture that the balance sheet reveals concerning the company's stability will therefore affect the surety's decision concerning the bonding limit (volume of work) of the construction company. Without the ability to secure a bond, the amount of work a company can undertake is limited.

Bonding companies typically use ratios to break down the balance sheet and ratios provide more pronounced indicators of financial health. Ratios make it possible to make comparisons of key balance sheet line items. This would not be possible when only viewing exclusive line items. For example, the working capital ratio (current assets divided by current liabilities) is an important and common measure of a company's financial health. Surety bonding companies like to see a working capital ratio of at least 1.2 to buffer cash flow during standard operations. The goal is to maintain a margin between assets and liabilities throughout the project and among multiple ongoing projects.

The analysis of a balance sheet can identify potential debt payment problems. These may signify the company's inability to meet financial obligations. Liquidity is the quality that an asset has of being readily convertible into cash—and cash is necessary for paying the bills. Working capital [Eq. (2.2)] is simply the amount that current assets exceed current liabilities. The relative amount of working capital is an indication of short-term financial strength.

$$\text{Working capital} = \text{Current assets} - \text{Current liabilities} \tag{2.2}$$

Working capital is the cash that flows into, through, and out of a construction company. It flows through the company (a circulatory system) as operations are conducted. This flow of working capital through the company is a lifeblood cycle, and without the "blood" the company dies. The flow of cash includes (Fig. 2.4):

1. Clients (project owners) input *contract revenues.*
2. A major portion of this revenue stream flows out of the company as direct costs to pay *direct project costs:* subcontractors, material suppliers, the contractor's labor force, and other expenses.
3. Part of the remainder is used for job *overhead costs* (this cost is allocated to and absorbed by all jobs to keep the direct cost items working).
4. The remainder is classified as gross profit (operating profit).
5. Nonjob overhead—general and administrative (G&A)—**expenses** are then deducted.

Expenses
Goods or services consumed in operating a business.

FIGURE 2.4 Flow of funds through a construction company.

6. Finally, a small portion makes it through as *income* to the company. But this is not the portion that the company gets to keep, because taxes must be paid on income.
7. However, if there is even one bad project (total cost is greater than the contract amount), profits from profitable projects must be used to complete the work (pay direct cost).

If the company operates profitably, it should generate a cash surplus. If it does not generate surpluses, it will eventually run out of cash and pass away. From the data presented in the balance sheet for Low Bid Construction Corporation (Fig. 2.5), it is clear the contractor developed a problem with liquidity and working capital in 2017 (Table 2.2). The Low Bid Corporation was forced to use up much of the past profits (retained earnings).

Recognizing its problem, Low Bid has made a provision in the assets part of the 2017 balance sheet to generate needed working capital. See the "Assets held for sale (Note 1)" row in the Assets section of Low Bid's balance sheet (Fig. 2.5). A note that accompanies the balance sheet states "in order to improve working capital, the Corporation executed a definite agreement to sell certain assets. Accordingly, $3,213,484 of assets consisting of inventories and equipment has been classified as assets held for sale on the balance sheet."

LOW BID CONSTRUCTION CORPORATION
CONSOLIDATED BALANCE SHEETS

	December 31, 2017	December 31, 2016
Assets (Note 9):		
Current Assets:		
Cash and cash equivalents (Notes 1 and 2)	$ 2,228,506	$ 1,822,598
Restricted cash (Notes 1, 2, and 16)	2,401,548	1,783,005
Accounts receivable, net (Notes 1, 3, 10, and 16)	21,377,904	14,297,564
Prepaid expenses and other	404,780	749,708
Inventory (Note 1)	3,365,750	4,288,235
Income tax receivable (Notes 1 and 11)	—	774,000
Costs and estimated earnings in excess of billings on uncompleted contracts (Notes 1 and 4)	5,294,054	9,828,009
Total Current Assets	35,072,542	33,543,119
Property and equipment, net (Notes 1, 5, 8, and 12)	15,267,791	18,111,506
Assets held for sale (Note 1)	3,213,484	—
Deferred tax asset (Notes 1 and 11)	1,957,923	873,441
Refundable deposits	55,110	176,565
Goodwill, net (Note 1)	—	1,500,733
Mineral rights and pit development, net	533,608	1,180,666
Claims receivable (Notes 1, 4, and 14)	5,968,026	—
Other assets	80,558	—
Total Assets	$ 62,149,042	$ 55,386,030
Liabilities and Stockholders' Equity:		
Current Liabilities:		
Accounts payable (Notes 6 and 10)	$ 27,025,984	$ 17,606,113
Accrued liabilities (Notes 7 and 10)	1,811,998	2,289,698
Notes payable (Note 8)	1,685,634	1,604,399
Obligations under capital leases (Note 12)	1,118,055	1,041,921
Billings in excess of costs and estimated earnings on uncompleted contracts (Notes 1 and 4)	4,625,657	6,054,814
Total Current Liabilities	36,267,328	28,596,945
Deferred tax liability (Notes 1 and 11)	2,718,734	2,272,700
Notes payable, less current portion (Notes 8 and 9)	9,484,479	7,674,608
Obligations under capital leases, less current portion (Note 12)	2,964,195	3,603,540
Total Liabilities	51,434,736	42,147,793
Commitments and contingencies (Notes 9, 10, 12 and 14)		
Stockholders' Equity:		
Preferred stock — $.001 par value; 1,000,000 shares authorized, none issued and outstanding (Note 13)	—	—
Common stock — $.001 par value; 15,000,000 shares authorized 3,559,438 issued and outstanding (Notes 13 and 17)	3,601	3,601
Additional paid-in capital	10,943,569	10,943,569
Capital adjustments	(799,147)	(799,147)
Retained earnings	566,283	3,090,214
Total Stockholders' Equity	10,714,306	13,238,237
Total Liabilities and Stockholders' Equity	$ 62,149,042	$ 55,386,030

FIGURE 2.5 Balance sheet, Low Bid Construction Corporation (the numerous notes that accompanied the balance sheet are not shown here).

TABLE 2.2 Liquidity Analysis for Low Bid Construction Corporation

	Dec. 31, 2017	Dec. 31, 2016
Current assets	$35,072,542	$33,543,119
Current liabilities	36,267,328	28,596,945
Working capital	−1,194,786	4,946,174

Current Ratio. While the magnitude of a firm's working capital, as calculated using Eq. (2.3), provides information about the ability to prosecute the firm's workload, it is also an indication of its ability to pay short-term obligations. There are other financial ratios designed to measure a company's ability to meet these obligations. Ratios provide a means to identify changes in position that are not as apparent when only reviewing the magnitude of values.

Current ratio The relation of a company's current assets to its current liabilities.

The **current ratio**, Eq. (2.3), is a tool to identify changes in a company's ability to meet short-term obligations.

$$\text{Current ratio} = \frac{\text{Current assets}}{\text{Current liabilities}} \tag{2.3}$$

Financial analysis is the process of tracking changes in a company's financial health. Ratios are method of analysis used to identify where one should look more deeply into a company's financial statements. As a rule of thumb, the current ratio for a construction company should be above 1.3:1.

When the adequacy of working capital is examined, the composition of the company's current assets should be considered. A company with a high proportion of cash to accounts receivable is in a better position to meet its current obligations. The company with a high proportion of accounts receivable must turn those into cash before it can pay the bills. Again using the Low Bid balance sheet (Fig. 2.5), the liquidity deterioration is also reflected by the cash to accounts receivable proportional change (Table 2.3).

TABLE 2.3 Cash to Accounts Receivable Analysis for Low Bid Construction Corporation

	Dec. 31, 2017	Dec. 31, 2016
Current assets—cash	$2,228,506	$1,822,598
Current assets—accounts receivable	21,377,904	14,297,564
Working capital	1:9.6	1:7.8

Working capital *increases* when a company *makes a profit on a job,* sells equipment or assets, or borrows money on a long-term note. Contractors typically resist selling equipment or engaging in long-term borrowing, both of which directly affect working capital.

Working capital *decreases* when the company *loses money on a job,* buys equipment, or pays off long-term debt. Contractors are typically eager to buy equipment and pay off debt.

CONSTRUCTION CONTRACT REVENUE RECOGNITION

In construction accounting, everything follows accounting practices used in other industries except revenue recognition. In most industries, revenue is recognized at the point of sale, because the uncertainties related to the earning process are removed and the exchange price is known. The exception for construction to this general rule of recognition at point of sale is caused by the long-term nature of construction projects. The accounting measurements associated with long-term construction projects are a bit more difficult because events and amounts must be estimated for a period of years, and even good estimates are modified because of owner-desired change orders or unexpected conditions.

There are several different methods a construction company can use to identify revenue, including the cash, straight accrual, completed-contract, and percentage-of-completion methods, but for financial statements, the only recognized method is percentage of completion.

Cash Method of Revenue Recognition

With the cash method, a contractor records revenue when payments for services are received (actual receipts to date) and records costs as bills are paid (costs paid to date). There is never an accurate picture of the company's financial condition or how individual projects are progressing (a profit or loss). Considering the project data in Table 2.4 and using the cash method of revenue recognition, the revenue to date for the project would be

Payments received to date	$630,000
Costs paid to date	$400,000
Revenue to date	**$230,000**

TABLE 2.4 OK Project Financial Data

Project Financial Data	Amount ($)
Contract amount	1,000,000
Original estimated cost	900,000
Billed to date	700,000
Payments received to date	630,000
Costs incurred to date	450,000
Forecasted cost to complete	400,000
Costs paid to date	400,000

Straight Accrual Method of Revenue Recognition

The straight accrual method considers cost as a costs-to-date (even if not yet paid) and revenue as what has been billed to date. Considering the project data in Table 2.4 and using the straight accrual method of revenue recognition, the revenue to date for the project would be

Billed to date	$700,000
Costs incurred to date	$450,000
Revenue to date	**$250,000**

Completed-Contract Method of Revenue Recognition

The completed-contract method should almost never be used and then only when

- The contractor has predominantly very short-term contracts
- The conditions for using the percentage-of-completion method cannot be met
- There are inherent hazards in the contract beyond normal, recurring business risks

Considering the project data in Table 2.4 and using the completed-contract method of revenue recognition, the revenue to date for the project would be zero at this time, because the project has not been completed.

Percentage-of-Completion Method of Revenue Recognition

The percentage-of-completion method should be used when estimates of progress toward completion, revenues, and costs are reasonably dependable, and all these conditions exist:

- The contract clearly specifies the enforceable rights regarding services to be provided and received by the parties, the consideration to be exchanged, and the manner and terms of settlement.
- The owner (buyer) can be expected to satisfy all obligations under the contract.
- The contractor can be expected to perform contractual obligations.

When using the percentage-of-completion method to calculate revenue, it is first necessary to calculate the percentage of work that has been completed:

$$\text{Percentage of completion} = \frac{\text{Cost to date}}{\text{Cost to date} + \text{Cost to complete}} \times 100\% \tag{2.4}$$

Considering the project data in Table 2.4, the percentage of completion for the project would be

$$\frac{\$450{,}000}{\$450{,}00 + \$400{,}000} \times 100\% = 52.9\%$$

The revenue to date for a project would be

$$\text{Revenue to date} = \text{Percentage of completion} \times \text{Contract price} \tag{2.5}$$

Therefore, revenue is $52.9\% \times \$1{,}000{,}000 = \$529{,}000$

$$\text{Gross profit} = \text{Revenue} - \text{Cost} \tag{2.6}$$

Revenue to date	$529,000
Cost incurred to date	$450,000
Gross profit to date	**$ 79,000**

Over/Underbilling

When the OK Project was originally bid, the anticipated profit was $100,000 (Table 2.4, contract amount minus original estimated cost). As the work has progressed, the cost to complete the work has been continually revised. The data in Table 2.4 represents a revised estimated final cost of $850,000 (Table 2.4, cost incurred to date plus forecasted cost to complete: $450,000 + $400,000). The construction company has billed the owner for $700,000, and the actual experienced cost to date is $450,000; this leaves the contractor with $250,000 in cash. For this discussion, the fact that the contractor has not received all $700,000, because the owner is holding 10% retainage, will be ignored. Part of the $250,000 is true gross profit (what the contractor has really earned, $79,000), and part is the overbilling, $171,000. The way to calculate overbilling is by subtracting the revenue to date (percentage-of-completion method) from the billed-to-date amount. This reveals the excess of the billing over the revenue. Such an undertaking demonstrates the amounts of money belonging to the contractor and to the owner.

Billed to date	$700,000
Revenue to date (percentage-of-completion method)	$529,000
Overbilling or job borrow	**$171,000**

Most contractors are not in the banking business, in the sense that they do not have the funds to finance construction of the project for the owner. Contractors, therefore, allocate more profit to those items of work that will be completed early in the project. In the case of a lump-sum project, this is done when the schedule of values is prepared. The total amount of revenue for the total work is the same; it is just that more of the revenue is received on the early items of work. The formula for calculating overbilling is given by Eq. (2.7):

$$\text{Overbilling} = \text{Billed to date} - \text{Revenue} \tag{2.7}$$

Early in a project a contractor can be in an underbilling situation where actual earned revenue is greater than the amount billed to the owner:

$$\text{Underbilling} = \text{Revenue} - \text{Billed to date} \tag{2.8}$$

Assuming that overbillings have been paid, they indicate that the contractor is "borrowing" money from the owner by billing ahead of earned revenue. While overbillings are preferable to significant underbillings, it can pose some difficulties. For example, if the borrowed cash is used for other purposes, it may not be available for payment of job costs when they are incurred later. Overbilling represents a liability from an accounting perspective since the reported value of work has not been fully earned. The contractor has a remaining liability to install work. This excess will translate into deficit spending later in the project as underbilling occurs to balance out the contract price with final payments. Underbilling represents an asset since earned work has not yet been fully paid.

Excessive overbilling for the first half of the project can cause a substantial reduction in payments in the latter half of the project. It is not uncommon for contractors to overbill at a rate of 15% or higher early in the project to establish a positive cash flow for the duration of the project. This effectively eliminates the chance for any borrowing by exercising a bank line of credit for short-term working capital. Although this has a clear cash flow advantage, this positive offset must be met later with an equal reduction in final payments.

Many unit price contracts for heavy/civil construction projects have a *mobilization* line item to offset the early expenses of starting a job. In essence, this provides an overbilling of actual work on the project since no physical work has been accomplished. Mobilization can be classified as an indirect cost item to support direct cost work items since it provides the machines and temporary facilities to build the project, such as the job trailer, security fencing, and traffic control. However, very few unit price contracts have a *demobilization* item to move equipment and temporary facilities from the site, a real cost to the contractor. These demobilization costs can be passed on to the next project with a mobilization line item but only if there is an immediate next project requiring the items.

Managers should look for underbilling late in a project. Significant underbillings *always* indicate problems and the possibility of unapproved change orders that may not get approved or paid. Underbillings indicate that the contractor is allowing the project owner to "borrow" money, in the sense that the contractor has performed work (and incurred expenses) without yet billing for it. Underbillings reduce cash flow for the individual contract as well as other contracts and the general operations of the company. An underbilling can be characterized as an unbilled receivable.

Contract Status Report

A contract status report (work in progress schedule) summarizes the financial position of each project not completely finished, accepted by the owner. It presents a comparison of work accomplished (revenue earned) against costs and a projection of the final financial outcome of each project. Table 2.5 presents the base data required to develop a contract status report. In preparing a contract status report, it is the project manager who usually is responsible for evaluating the accuracy of the estimated cost to complete a project. The contract price is the sum of the original contract amount and all approved change orders.

TABLE 2.5 It Will Cost You a Million Construction Co. Contract Financial Data

Contract	Contract Price ($)	Cost to Date ($)	Total Amount Billed to Date ($)	Estimated Cost to Complete ($)
School 1	1,000,000	450,000	700,000	400,000
School 2	1,000,000	450,000	400,000	400,000
Store 1	1,000,000	450,000	600,000	500,000
Store 2	1,000,000	450,000	450,000	460,000
Hospital	1,000,000	475,000	525,000	575,000
	5,000,000	2,275,000	2,700,000	2,335,000

The first step in developing a complete contract status report is to calculate a new or current (at the present point in time) estimate of the total project cost. Summing the cost to date of each project with its re-estimated cost to complete does this (col. 2, Table 2.6). Once that is done, it is possible to use Eq. (2.4) (cost to date as a ratio of projected total cost) to calculate the percentage complete of each project (col. 5, Table 2.6). Additionally once the current (new) estimate of the total project cost is calculated, it can be subtracted from the project contract price to obtain a current estimated profit/loss at completion (col. 3, Table 2.6). By multiplying the estimated profit at completion by the percentage complete, profit to date is obtained (col. 6, Table 2.6). Note in Table 2.6 that in the case of the Hospital project,

TABLE 2.6 "It Will Cost You a Million Construction Co." Contract Status

Contract	Contract Price ($)	Current Estimated Total Cost ($)	Estimated Profit at Completion	Cost to Date ($)	Percentage Complete	Total Profit (Loss) Recognized to Date ($)	Estimated Cost to Complete ($)
	(1)	(2)	(3)	(4)	(5)	(6)	(11)
School 1	1,000,000	850,000	150,000	450,000	53	79,500	400,000
School 2	1,000,000	850,000	150,000	450,000	53	79,500	400,000
Store 1	1,000,000	950,000	50,000	450,000	47	23,500	500,000
Store 2	1,000,000	910,000	90,000	450,000	49	44,100	460,000
Hospital	1,000,000	1,050,000	–50,000	475,000	45	–50,000	575,000
	5,000,000	4,610,000		2,275,000			2,335,000

a loss is projected at project completion. An expected loss must be recognized in the period when the loss is identified—the total expected loss.

Table 2.7 presents the complete status report with revenue and over/underbillings recognized. Project managers are expected to analyze and explain the progress of their projects. If the project manager fails to forecast a cost overrun, the report will show an underbilling. This can be particularly troublesome if the cause of the cost overrun cannot be identified.

TABLE 2.7 Contract Status Report for "It Will Cost You a Million Construction Co."

Contract Description	Contract Price	Current Re-est. Total Cost	Estimated Gross Profit at Completion	Cost to Date	Percentage Complete	Total Gross Profit (Loss) Recog. to Date
Column	(1)	(2)	(3)	(4)	(5)	(6)
Formula		4 + 11	1 – 2		4 ÷ 2	7 – 4 or 5 × 3
School 1	$1,000,000	$ 850,000	$150,000	$ 450,000	53%	$79,412
School 2	$1,000,000	$ 850,000	$150,000	$ 450,000	53%	$79,412
Store 1	$1,000,000	$ 950,000	$ 50,000	$ 450,000	47%	$23,684
Store 2	$1,000,000	$ 910,000	$ 90,000	$ 450,000	49%	$44,505
Hospital	$1,000,000	$1,050,000	–$ 50,000	$ 475,000	45%	–$22,619
	$5,000,000			$2,275,000		

	Amount Earned to Date (Revenue)	Total Amount Billed to Date	Overbillings	Underbillings	Re-est. Cost to Complete	Contract Balance
Column	(7)	(8)	(9)	(10)	(11)	(12)
Formula	5 × 1 or 4 + 6		8 – 7	7 – 8		1 – 8
School 1	$529,412	$ 700,000	$170,588		$ 400,000	$300,000
School 2	$529,412	$ 400,000		$129,412	$ 400,000	$600,000
Store 1	$473,684	$ 600,000	$126,316		$ 500,000	$400,000
Store 2	$494,505	$ 475,000		$ 19,505	$ 460,000	$525,000
Hospital	$452,381	$ 525,000	$ 72,619		$ 575,000	$475,000
		$ 2,700,000			$2,335,000	

Financial Statement Analysis

Comparative statement
A financial statement with data for two or more successive accounting periods placed in columns side by side to illustrate changes.

In and of themselves, very few of the individual figures in a financial statement are exceptionally significant. It is their relationship to other quantities, or the magnitude and direction of change with time (since a previous statement), that is important as an indicator of the company's condition. Analysis of financial statements is largely a matter of studying relationships, changes, and trends. The three commonly used analysis techniques are (1) dollar or percentage change, (2) component percentages, and (3) ratios.

Financial statement/report analysis requires that relationships and changes between items and groups of items be described. A **comparative statement** is an analysis report that shows pairwise changes in statement items for two or more successive accounting periods in a column format on a single statement (Fig. 2.3). As an aid to controlling operations, a comparative income statement is usually more valuable than a comparative balance sheet.

Dollar and Percentage Change

The dollar magnitude (amount) of change from one reporting period to another is significant, but expressing the change in percentage terms adds perspective. The dollar magnitude of change is the difference between the amount of a comparison statement (year) and a base statement (year). The percentage change is computed by dividing the amount of the change between the statements by the magnitude for the base statement. Table 2.8 presents the dollar magnitude and percentage change of Low Bid Construction Corporation's revenue and cost of revenue for the 3-yr period 2014 through 2016.

TABLE 2.8 Analysis of Low Bid Construction Corporation's Financial Situation

	2016	2015	2014
Revenue	$174,063,148	$163,573,258	$210,002,272
Cost of revenue	169,404,787	158,935,103	200,070,826
Dollar change in revenue	$10,489,890	–$46,429,014	
Percentage change in revenue	6.4%	–22.1%	
Dollar change cost of revenue	$10,469,684	–$41,135,723	
Percentage change cost of revenue	6.6%	–20.6%	

Though the company cannot be happy about the decrease in revenue shown by the comparison, one possible cause can be eliminated. Because the percentage changes are approximately the same (e.g., revenue decreases 22.1% and likewise the cost of revenue decreases 20.6%), it seems reasonable to assume that an increase in cost was not the cause of the revenue decrease experienced in 2016.

Component Percentages

Component percentages indicate the relative size of each item included within a total. Each item on a balance sheet can be expressed as a percentage of total current assets, total assets, total current liabilities, or total liabilities. Such a display of total assets illustrates the relative importance of current and noncurrent assets, as well as the relative amount of financing obtained from current creditors, by long-term borrowing, and from the owners.

Table 2.9 presents a component percentage of current assets analysis. The analysis shows that the company is increasing its cash position and decreasing accounts receivable and unbilled receivables (underbillings).

TABLE 2.9 Component Percentage Analysis of Current Assets (Thousands)

Current Assets	2016		2015		2014	
Cash	$67,054	15.7%	$53,215	13.1%	$48,310	10.5%
Accounts receivable	$175,513	41.0%	$187,311	46.3%	$222,341	48.1%
Unbilled receivables	$74,552	17.4%	$74,514	18.4%	$101,564	22.1%
Total current assets	$427,722		$405,014		$459,249	

Ratios

A ratio is the relationship expression of two quantities (items) expressed as the quotient of one divided by the other. To compute a meaningful ratio, there must be a significant relationship between the two items considered. The current ratio [discussed before in Eq. (2.3)] summarizes the relationship between current assets and current liabilities. Besides the current ratio, some other important financial ratios for construction companies are:

$$\text{Quick ratio} = \frac{\text{Cash} + \text{Receivables}}{\text{Current liabilities}} \tag{2.9}$$

The quick ratio measures the short-term liquidity of a company:

$$\text{Debt to worth} = \frac{\text{Total liabilities}}{\text{Total shareholders' equity}} \tag{2.10}$$

The debt ratio measures the firm's leverage. The margin of protection to creditors is higher when the ratio is lower because the owners are contributing the bulk of funds for the business. Properly leveraged contractors have debt-to-worth ratios of 2:1 or less.

$$\text{Receivables to payables} = \frac{\text{Receivables}}{\text{Payables}} \tag{2.11}$$

The suggested range for the receivables-to-payables ratio varies with the type of construction activity undertaken by the firm.

Based on the type of industry being studied, there are rules of thumb for evaluating financial ratios. In the case of construction companies

- Current ratio should be above 1.3:1
- Quick ratio should be above 1.1:1
- Debt to worth should be below 2.0:1
- Receivables to payables (commercial general contractors) should be around 1.5:1

- Receivables to payables (trade/heavy contractors) should be around 2.0:1
- Receivables to payables (labor-intensive trade contractors) should be around 3.0:1

One further rule of thumb concerns a construction company's capacity to do work. A firm's backlog of work should not be more than approximately ten times the company's working capital amount. This is something a surety will consider before granting a bond for new work.

EQUIPMENT COST

Equipment cost is often one of a contractor's largest expense categories, and it is a cost fraught with variables and questions. To be successful, equipment owners must carefully analyze and answer two separate cost questions about their machines:

1. How much does it cost to operate the machine on a project?
2. What is the optimum economic life of and optimum manner to secure a machine?

The first question is critical to bidding and operations planning—or, simply, estimating. The only reason for purchasing equipment is to perform profitable work. This first question seeks to identify the expense associated with productive machine work, and is commonly referred to as ownership and operating costs (**O&O**). It is usually expressed in dollars per equipment operating hour.

O&O
The ownership and operating costs of a machine.

The second question seeks to identify the optimum point in time to replace a machine and the optimum way to secure a machine. This is important because it will reduce O&O cost and thereby lower production expense, enabling a contractor to achieve a better pricing position. The process of answering this question is known as replacement analysis. A complete replacement analysis must also investigate the cost of renting or leasing a machine.

The economic analyses, which answer these cost questions, require the input of many expense and operational factors. These input costs will be discussed first and a development of the analysis procedures will follow.

Equipment Records

Data on both machine utilization and costs are the keys to making rational equipment decisions, but collection of individual pieces of data is only the first step. The data must be assembled and presented in usable formats. Many contractors recognize this need and strive to collect and maintain accurate equipment records for evaluating machine performance, establishing operating cost, analyzing replacement questions, and managing projects. Surveys of industry-wide practices, however, indicate that such efforts are not universal.

Realizing the advantages to be gained, many owners are directing more attention to accurate record keeping. Advances in computer technology have reduced the effort required to implement record systems. Several software companies offer record-keeping packages specifically designed for contractors. In many cases, the task is simply the retrieval of equipment cost data from existing accounting files.

Automation provides the ability to handle more data economically and in shorter time frames, but the basic information required to make rational decisions is still the critical item. A commonly used technique in equipment costing and record keeping is the standard rate approach. Under such a system, jobs are charged a standard machine utilization rate for every hour the equipment is employed. Machine expenses are charged either directly to the individual piece of equipment or to separate cost accounts by equipment type. This method is sometimes referred to as an internal or company rental system. Such a system usually presents a fairly accurate representation of *investment consumption*, and it properly assigns machines expenses. In the case of a company replacing machines at regular intervals and continuing in the construction business, this system allows a check at the end of each year on estimate rental rates, as the internally generated rent should equal the expenses absorbed.

The first piece of information necessary for rational equipment analysis is not an expense but a record of the machine's usage. By the act of undertaking a replacement analysis, it is presumed there is a continuing need for a machine's production capability. Therefore, before beginning a replacement analysis, the disposal–replacement question must be resolved. Is this machine really necessary? A projection of the ratio between total equipment capacity and utilized capacity provides a quick guide for the dispose–replace question.

The level of detail for reporting equipment use varies. At a minimum, data should be collected on a daily basis to record whether a machine worked or was idle. Having the operator or foreman manually record engine hours or miles on the operator's time sheet is a low-cost solution for tracking machine usage and is probably the most common method for accumulating utilization data.

Both independent service vendors and equipment companies offer real-time automatic data collection devices. These GPS/Telematics devices installed on the machine are in many cases the most effective way to collect data. A more sophisticated system will seek to identify use on an hourly basis, accounting for actual production time and categorizing idle time by classifications such as standby, down weather, and down repair.

Most of the information required for ownership and operating, or replacement analyses, is available from accounting records. There should be accounts for machine purchase expense and final realized salvage value as part of the accounting data required for tax filings. Maintenance expenses can be tracked from mechanics' time sheets, purchase orders for parts, or shop work orders. Service logs provide information concerning consumption of consumables. Fuel amounts can be recorded at fuel points or with automated systems. Fuel amounts should be cross-checked against the total amount purchased. When detailed and correct reporting procedures are maintained, the accuracy of equipment costs analysis is greatly enhanced.

Many discussions of equipment economics include *interest* as a cost of ownership. Sometimes the fleet managers make comparisons with the interest rates a bank would charge for borrowed funds or with the rate earned on invested funds. Such comparisons assume these rates are appropriate for an equipment cost analysis. It is not logical to assign different interest costs to machines purchased wholly with retained earnings (cash) as opposed to those purchased with borrowed funds. A single corporate interest rate should be determined by examination of the combined costs associated with all sources of capital funds: debt, equity, and internal.

ELEMENTS OF MACHINE OWNERSHIP COST

Ownership cost is the cumulative result of those cash flows an owner experiences whether or not the machine is productively employed on a job. It is a cost related to finance and accounting exclusively, and it does not include the wrenches, nuts and bolts, and consumables necessary to keep the machine operating.

Most of these cash flows are expenses (outflows), but a few are cash inflows. The most significant cash flows affecting *ownership cost* are machine

1. Purchase expense
2. Salvage value
3. Major repairs and overhauls
4. Property taxes
5. Insurance
6. Storage and miscellaneous

Purchase Expense

The cash outflow the company experiences in acquiring a machine is the purchase expense. It is the total delivered cost (drive-away cost), including amounts for all options, shipping, and taxes, less the cost of pneumatic tires for wheeled machines. The machine will show as an asset on the books of the company. The company has exchanged money (cash or borrowed funds), liquid assets, for a machine, a fixed asset with which the company hopes to generate profit. As the machine is used on projects, wear takes its toll, and the machine can be thought of as being used up or consumed. This consumption reduces the machine's value because the revenue stream it can generate is likewise reduced. An owner tries to account for the decrease in value by prorating the consumption of the investment over the *service life* of the machine. This prorating is known as depreciation.

The difference between the initial acquisition expense and the expected future salvage value is the expense being measured—but to simply prorate this amount will not produce a valid analysis. Such an approach neglects the timing of the cash flows. Therefore, it is recommended that each cash flow be treated separately to allow for a time value analysis and to allow for ease in changing assumptions during sensitivity analyses.

Salvage Value

Salvage value is the cash inflow a firm receives if a machine still has value at the time of its disposal. This revenue occurs at a future date.

Used equipment prices are difficult to predict. Machine condition (Fig. 2.6), the movement of new machine prices (Fig. 2.7), and the machine's possible secondary service applications affect the amount an owner can expect to receive for a used machine. A machine with a diverse and layered service potential commands a higher resale value. Medium-size dozers often exhibit rising salvage values in later years because these can have as many as seven different levels of useful life. Such a machine may first be used as a high-production machine on a dirt spread and later be used infrequently by a farmer.

Historical resale data can provide salvage value prediction guidance and can be fairly easily accessed from auction price books. By studying such historical data and recognizing the effects of the economic environment, the magnitude of salvage value prediction errors can be minimized and the accuracy of an ownership cost analysis improved.

Depreciation
An accounting method used to describe the loss in value of an asset.

Tax Saving from Depreciation. The tax savings from **depreciation** are a phenomenon of the tax system in the United States. (This may not be an ownership

FIGURE 2.6 Salvage value is dependent on machine condition.

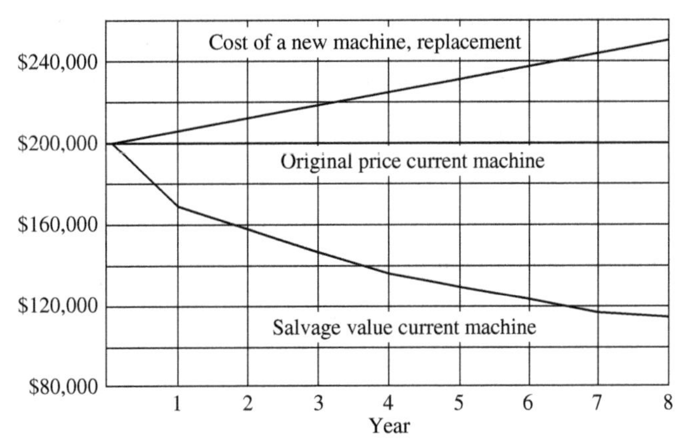

FIGURE 2.7 The movement of new machine prices; cost is one factor affecting salvage value.

cost factor under the tax laws in other countries.) Under the tax laws of the United States, depreciating a machine's loss in value with age will lessen the net cost of machine ownership. The cost saving, the prevention of a cash outflow afforded by tax depreciation, is a result of shielding the company from taxes. This is an applicable cash flow factor only if a company is operating at a profit. There are carry-back features in the tax law, so the saving can be preserved even though there is a loss in any one particular year, but the long-term operating position of the company must be at a profit for tax saving from depreciation to come into effect.

The rates at which a company can depreciate a machine are set by the revenue code. These rates usually have no relation to actual consumption of the asset (machine). Therefore, many companies keep several sets of depreciation numbers: one for depreciation tax purposes, one for corporate earnings tax accounting purposes, and one for internal and/or financial statement purposes. The first two are required by the revenue code. The third tries to accurately match the consumption of the asset based on work application and company maintenance policies.

Under the current tax laws, tax depreciation accounting no longer requires the assumption of a machine's future salvage value and useful life. The only piece of information necessary is basis. Basis refers to the cost of the machine for purposes of computing gain or loss. Basis is essential. To compute tax depreciation amounts, fixed percentages are applied to an unadjusted basis. The terminology *adjusted/unadjusted* refers to changing the book value of a machine by depreciation.

The tax law allows the postponement of taxation on gains derived from the exchange of like-kind depreciable property. If a gain is realized from a like-kind exchange, the depreciation basis of the new machine is reduced by the amount of the gain. However, if the exchange involves a disposal sale to a third party and a separate acquisition of the replacement, the gain from the sale is taxed as ordinary income.

The current tax depreciation law establishes depreciation percentages based on years of machine life. These are usually the optimum depreciation rates in terms of tax advantages. However, an owner can utilize the straight-line method of depreciation or methods that are not expressed in terms of time duration (years). Unit-of-production depreciation would be an example of a non-time–based depreciation system.

Straight-Line Depreciation. Straight-line depreciation is easy to calculate. The annual amount of depreciation D_n, for any year n, is a constant value, and thus the book value (BV_n) decreases at a uniform rate over the useful life of the machine. The calculation equations are

$$\text{Depreciation rate, } R_n = \frac{1}{N} \tag{2.12}$$

where N = number of years

$$\text{Annual depreciation amount, } D_n = \text{Unadjusted basis} \times R_n$$

Substituting Eq. (2.12) yields

$$D_n = \frac{\text{Unadjusted basis}}{N} \tag{2.13}$$

$$\text{Book value year } n,\ BV_n = \text{Unadjusted basis} - (n \times D_n) \tag{2.14}$$

Declining-Balance Depreciation. Declining-balance (DB) depreciation methods are accelerated depreciation methods. A DB method writes off larger percentages of machine cost in the early years of its life. In many cases, these methods more closely approximate the actual loss in market value with time. Declining methods range from 1.25 times the current book value divided by the life to 2.00 times the current book value divided by the life (the latter is termed *double-declining balance*). Note that although the estimated salvage value (S) is not included in the calculation, the book value cannot go below the salvage value. The following equations are necessary for using the declining-balance methods.

The symbol R is used for the declining-balance depreciation rate:

1. For 1.25 declining-balance (1.25 DB) method, $R = 1.25/N$
   For 1.50 declining-balance (1.50 DB) method, $R = 1.50/N$
   For 1.75 declining-balance (1.75 DB) method, $R = 1.75/N$
   For double-declining-balance (DDB) method, $R = 2.00/N$
2. The allowable depreciation D_n, for any year n and any depreciation rate R is

$$D_n = (\mathrm{BV}_{n-1}) \times R \tag{2.15}$$

3. The book value for any year n is

$$\mathrm{BV}_n = \mathrm{BV}_{n-1} - D_n, \text{ provided that } \mathrm{BV}_n \geq S \tag{2.16}$$

Since the book value can never go below the estimated salvage value, the DB method must be forced to intersect the value S at time N.

Tax Code Depreciation. The Modified Accelerated Cost Recovery System (MACRS) is a property depreciation system adopted by the Internal Revenue Service (IRS). Under the tax code, machines are classified as 3-, 5-, 7-, 10-, 15-, or 20-yr real property. Most pieces of construction equipment are 5-yr property, including automobiles and light-duty trucks. The appropriate depreciation rates are given in Table 2.10. Note that 1 yr is added to the property life, where a half-year convention is applied to both the first and last years of the equipment life. This implies that the equipment is put into service during the last half of the first year (July through December) and is still in service during the first half of the last year (January through June). MACRS provides for a slightly larger write-off in the earlier years of the cost recovery period since it uses a hybrid of the declining balance at the beginning life and straight line at the ending life. Like the DB method, the MACRS method does

TABLE 2.10 Tax Code–Specified Depreciation Rates for Recovery Period (Half-Year Convention*)

Year	3-yr Property	5-yr Property	7-yr Property
1	33.33%	20.00%	14.29%
2	44.45%	32.00%	24.49%
3	14.81%	19.20%	17.49%
4	7.41%	11.52%	12.49%
5		11.52%	8.93%
6		5.76%	8.92%
7			8.93%
8			4.46%

*Percentages for machines purchased earlier or later in the tax year are different.

not account for salvage value. The full set of depreciation tables showing the MACRS percentages are available in the IRS's free publication 946, *How to Depreciate Property,* available on the Internet at www.irs.gov/publications/p946.

Evaluating Depreciation Methods. Each depreciation method has a unique approach for costing machine life. Tax advantage is usually the key factor when evaluating these methods. If a machine is disposed of before the depreciation process is completed, no depreciation can be recovered in the year of disposal. Any gain, as measured against the depreciated value or adjusted basis, is treated as ordinary income.

The tax savings from depreciation are influenced by

1. Disposal method for the old machine
2. Value received for the old machine
3. Initial value of the replacement
4. Class life
5. Tax depreciation method

Based on the relationships between these elements, three distinct situations are possible:

1. No gain on the disposal—no income tax on zero gain.
2. A gain on the disposal:
   a. Like-kind exchange—no added income tax, but basis for the new machine is adjusted.
   b. Third-party sale—the gain is taxed as income; the basis of the new machine is fair market value paid.
3. A disposal in which a loss results—the basis of the new machine is the same as the basis of the old machine, decreased by any money received.

Assuming a corporate profit situation, the applicable tax depreciation shield formulas are

1. For a situation where there is no gain on the exchange:

$$\text{Total tax shield} = \sum_{n=1}^{N} t_c D_n \tag{2.17}$$

where n = individual yearly time periods within a life assumption of N years
t_C = corporate tax rate
D_n = annual depreciation amount in the nth time period

2. For a situation where a gain results from the exchange
   a. Like-kind exchange—Eq. (2.17) is applicable. It must be realized that the basis of the new machine will be affected.
   b. Third-party sale.

$$\text{Total tax shield} = \left(\sum_{n=1}^{N} t_c D_n \right) - \text{gain} \times t_c \tag{2.18}$$

Gain is the actual salvage amount received at the time of disposal minus the book value.

Basis is a determining factor in making analysis calculations, and the actual salvage derived from the machine directly affects the depreciation saving. To perform a valid analysis, the depreciation accounting practices for tax purposes and the methods of machine disposal and acquisition the company chooses to use must be carefully examined. These dictate the appropriate calculations for the tax effects of depreciation.

Major Repairs and Overhauls

Major repairs and overhauls are included under ownership cost because they result in an extension of a machine's service life. Overhauls are very common when a contractor decides to retain a piece of equipment. They can be considered an investment in a new machine. Because a machine commonly works on many different projects, considering major repairs as an ownership cost prorates these expenses to all jobs. These costs should be added to the book value of the machine and depreciated.

Taxes

In this context, taxes refer to those equipment ownership taxes charged by any government subdivision. They are commonly assessed at a percentage rate applied against the book value of the machine. Depending on location, ownership taxes can range up to about 4.5% of book value. In many locations, there will be no tax on equipment, just user fees. Over the service life of the machine, taxes will decrease in magnitude as the book value decreases.

Insurance

Insurance, as considered here, includes the cost of policies to cover fire, theft, and damage to the equipment either by vandalism or construction accident. Annual rates can range from 1% to 3% of book value. This cost can be actual premium payments to insurance companies, or it can represent allocations to a self-insurance fund maintained by the equipment owner.

Storage and Miscellaneous

Between jobs or during bad weather, a company will require storage facilities for its equipment. The cost of maintaining storage yards and facilities should be prorated to the machines requiring such shelter. Typical expenses include space rental, utilities, and the wages for laborers or watchmen. These expenses are all combined in an overhead account and then allocated on a proportional basis to the individual machines. The rate may range from nothing to perhaps 5% of book value. These facility storage costs may be difficult to assign to specific pieces of equipment and often find themselves in general and administrative accounts. Either way the machine overhead is costed; it may be best accounted in an equipment ledger with dollars, and then prorated to individual machines based on value. A common approach to avoid storage costs is to keep the machine on the project as long as possible, then mobilize it to the next project. The machine is exposed to the elements, but storage cost is eliminated.

ELEMENTS OF MACHINE OPERATING COST

Operating cost is the sum of those expenses an owner experiences by working a machine on a project. If the engine is turned on or the machine moves, the cost is classified as operating. Typical expenses include:

1. Fuel
2. Lubricants, filters, and grease
3. Repairs
4. Tires
5. Replacement of high-wear items

Operator wages are rarely included under operating costs; this is because of wage variance between jobs. The general practice is to keep operator wages as a separate cost category. Such a practice aids in estimation of machine cost for bidding purposes, as the differing project wage rates can readily be added to the total machine O&O cost. In applying operator cost, all benefits paid by the company must be included—direct wages, fringe benefits, payroll taxes, and insurance. This is another reason wages are separated. Some benefits are based on an hourly basis, some on a percentage of income, some on a percentage of income to a maximum amount, and some are paid as a fixed amount. The assumptions about project work schedule will, therefore, affect wage expense.

Fuel

Fuel expense is best determined by measurement on the job. Good service records contain fuel data by individual machines. This is a consumption record over a specific period and under specific job conditions. Hourly fuel consumption can then be calculated directly. With newer instrumentation technology and telematics, the fuel data can now be captured in real time.

When company records are not available, the manufacturer's consumption data can be used to construct fuel use estimates. The amount of fuel required to power a piece of equipment for a specific period depends on the brake horsepower of the machine and the work application. Therefore, most tables of hourly fuel consumption rates are divided according to the machine type and the working conditions. To calculate hourly fuel cost, a consumption rate is found in the tables (Table 2.11) and then multiplied by the unit price of fuel. The cost of fuel for vehicles used on public highways will include applicable taxes. However, in the case of off-road machines used exclusively on project sites, there is usually no fuel tax. In some locations, a color dye is added to the zero-tax fuel for verification by law enforcement officials. Therefore, because of the tax laws, the price of gas or diesel will

TABLE 2.11 Average Fuel Consumption—Wheel Loaders

	Type of Utilization		
Horsepower (fwhp)	Low (gal/hr)	Medium (gal/hr)	High (gal/hr)
90	1.5	2.4	3.3
140	2.5	4.0	5.3
220	5.0	6.8	9.4
300	6.5	8.8	11.8

vary with machine usage. A common practice to reduce fuel purchase costs is buying in bulk quantity and having portable bulk tanks on the project.

Fuel consumption formulas have been published for both gasoline and diesel engines. The resulting values from such formulas must be adjusted by *time and load factors* to account for working conditions. This is because the formulas are derived assuming that the engine is operating at maximum output. Two working conditions must be considered when estimating fuel consumption: (1) the percentage of time (usually per hour) the machine is actually working (*time factor*) and (2) at what percentage of rated horsepower (*throttle load factor*). When operating under standard conditions, a *gasoline engine* will consume approximately 0.06 gal of fuel per flywheel horsepower-hour (fwhp-hr). A *diesel engine* will consume approximately 0.04 gal per fwhp-hr. Advanced electronic fuel injection systems have a direct effect on fuel consumption rates, and individual manufacturers should be consulted.

Lubricants, Filters, and Grease

The cost of lubricants, filters, and grease will depend on the maintenance practices of the company and the conditions at sites where the machine is employed. Some companies follow the machine manufacturer's guidance concerning periods between lubricant and filter changes. Other companies have established their own preventive maintenance change period guidelines. In either case, the hourly cost is arrived at by (1) considering the operating hour duration between changes and the quantity required for a complete change, plus (2) a small consumption amount representing what is added between changes.

Many manufacturers provide quick cost-estimating tables (Table 2.12) for determining the cost of these items. Whether using manufacturer's data or past experience, notice should be taken about whether the data matches expected field conditions. If the machine is to be operated under adverse conditions, such as deep mud, water, severe dust, or extreme cold, the data values will have to be adjusted.

TABLE 2.12 Lubricating Oils, Filters, and Grease Cost for Crawler Tractors

Horsepower (fwhp)	Approximate Cost per Hour	
	Materials ($)	Labor ($)
<100	0.22	0.15
100 to <200	0.49	0.24
200 to <300	0.65	0.24

A formula that can be used to estimate the quantity of oil required is

$$\text{Quantity consumed gph (gal per hour)} = \frac{\text{hp} \times f \times 0.006 \text{ lb/hp-hr}}{7.4 \text{ lb/gal}} + \frac{c}{t} \qquad (2.19)$$

where hp = rated horsepower of the engine
c = capacity of the crankcase in gallons
f = operating factor
t = number of hours between oil changes

With this formula, the quantity of oil consumed per rated horsepower-hour between changes is assumed to be 0.006 lb.

Repairs

Repairs, as referred to here, mean normal maintenance-type repairs. These are the repair expenses incurred on the jobsite where the machine is operated and should include the costs of parts and labor. Major repairs and overhauls are accounted for as ownership cost.

Repair expenses increase with machine age. The U.S. Army has found that 35% of its equipment maintenance cost is directly attributable to the oldest 10% of its equipment. Instead of applying a variable rate, an average is usually calculated by dividing the total expected repair cost for the planned service life of the machine by the planned operating hours. Such a policy builds up a repair reserve during a machine's early life. The reserve will then be drawn down to cover the higher costs experienced later. As with all costs, company records are the best source of expense information. When such records are not available, manufacturers' published guidelines can be used.

Downtime. While downtime is not a repair cost chargeable to a machine, it is often the most expensive part of equipment repair. The production time lost while mechanics and parts are being located and the repair is being accomplished is costly to a project. This cost is usually at its highest rate during the first few minutes or hours of the breakdown, while operating costs are still at a maximum and before the work program has been changed. If the shutdown is a short one and only one machine is affected—for example, if a hose bursts in a self-loading scraper working alone—it causes a loss of production to one unit. The same broken hose on a loader will stop the loader, as well as the string of trucks, dozer, and compacting equipment of the spread. Good maintenance programs with scheduled downtime is the most effective practice for avoiding very costly in-service downtime.

Tires

Tires for wheel-type equipment are a major operating cost because they have a short life in relation to the "iron" of a machine. They require replacement one or more times during a machine's life. In addition, they will need repairs. Repair and replacement costs are greatly affected by care and conditions of use. Proper tire selection and good maintenance and work practices often add substantially to profits.

Good tire management includes:

1. Buying tires of the proper size, strength, tread, and speed rating for the job.
2. Using tires only for purposes, loads, and speeds for which they are designed.
3. Maintaining tires at proper pressure and in good condition.
4. Keeping excavation areas, haul roads, and fills smooth and free of spillage.

Tire cost will include repair and replacement charges. These costs are very difficult to estimate because of the variability in tire wear with project site conditions and operator skill. Tire and equipment manufacturers both publish tire life guidelines based on tire type and job application. Manufacturers' suggested life periods can be used with local tire prices to obtain an hourly tire cost. Always remember, guidelines are based on good operating practices and do not account for abuses such as overloading haul units.

Tires are about a third of a truck's operating cost. Poor management practices can abuse the tires.

Replacement of High-Wear Items

Certain items on machines and trucks such as cutting edges, bucket teeth, body liners, and cables have very short service lives in relation to machine service life. The cost of repeatedly replacing these items can be a critical operating cost. By using either past experience or manufacturer life estimates, the cost can be calculated and converted to an hourly basis.

All machine operating costs should be calculated per working hour. When the working hour is the standard, it is easy to sum the applicable costs for a particular class of machines and obtain a total operating hour cost.

MACHINE COST FOR BIDDING

The process of selecting a particular type of machine for use on a construction project requires knowledge of the cost associated with operating the machine in the field. In selecting the proper machine, a contractor seeks to achieve unit production at the least cost. This cost for bidding is the sum of the O&O expenses. O&O costs are stated on an hourly basis (e.g., $135/hr for a dozer) because it is used in calculating the cost per unit of machine production. If a dozer can push 300 cubic yards (cy) per hour and it has a $135/hr O&O cost, the cost for bidding is $0.45/cy ($135/hr ÷ 300 cy/hr). The estimator/planner can use the cost per cubic yard figure directly in unit-price work. On a lump-sum job, it will be necessary to multiply the cost/unit price by the estimated quantity to obtain the total amount to be charged. Indirect costs, such as job overhead and general and administrative expenses, are then added to the direct cost at the bid item level to produce the final bid item amount.

Ownership Cost

The outflow of cash when a machine is purchased and the inflow of money in the future when the machine is retired from service are the two most significant components of ownership cost. The net result of these two cash flows, which defines the machine's decline in value across time, is termed *depreciation*. As used in this section, depreciation is the measuring system used to account for purchase expense at time zero and salvage value after a defined period. Depreciation is expressed on an hourly basis over the service life of a machine. Do not confuse the depreciation discussed here with tax depreciation. Tax depreciation has nothing to do with consumption of the asset; it is simply an artificial calculation for tax code purposes.

Tires will be replaced many times over a machine's service life. Therefore, their cost is not included in ownership cost calculations but is considered as an operating cost.

The depreciation portion of ownership cost can be calculated by either of two methods: (1) time value or (2) average annual investment.

Depreciation—Time Value Method. The time value method will recognize the timing of the cash flows (i.e., the purchase at time zero and the salvage at a future date). The cost of the tires is deducted from the total purchase price, which includes amounts for all options, shipping, and taxes (total cash outflow minus the cost of tires). A judgment about the expected service life and a corporate cost-of-capital rate are both necessary input parameters for the analysis. These input parameters are entered into the uniform series capital recovery factor (USCRF) formula, Eq. (2.20), to determine the machine's purchase price equivalent annual cost.

$$A = P\left[\frac{i(1+i)^n}{(1+i)^n - 1}\right] \tag{2.20}$$

To account for the salvage cash inflow, the uniform series sinking fund factor (USSFF) formula, Eq. (2.21), is used. The input parameters are the estimated future salvage amount, the expected service life, and the corporate cost-of-capital rate.

$$A = F\left[\frac{i}{(1+i)^n - 1}\right] \tag{2.21}$$

where P = a present single amount of money
F = a future single amount of money, after n periods of time
A = uniform *end-of-period* payments or receipts continuing for a duration of n periods
i = the rate of interest per period of time (usually 1 yr)
n = the number of time periods

Sample Calculation: Consider a company with a cost-of-capital rate of 6% that purchases a $300,000 tractor. This machine has an expected service life of 4 yr and will be utilized 2,500 hr per year. The tires cost $45,000. The estimated salvage value at the end of 4 yr is $50,000. The depreciation portion of the ownership cost for this machine using the time value method would be

Initial cost	$300,000
Cost of tires	– 45,000
Purchase price less tires	$255,000

Now it is necessary to calculate the uniform series required to replace a present value of $255,000. Using the uniform series capital recovery factor formula

$$A = \$255{,}000\left[\frac{0.06(1+0.06)^4}{(1+0.06)^4 - 1}\right]$$

$$A = \$255{,}000 \times 0.28859 = \$73{,}591 \text{ per year}$$

Next calculate the uniform series sinking fund factor to use with the salvage value.

$$A = \$50{,}000\left[\frac{0.06}{(1+0.06)^4 - 1}\right]$$

$$A = \$50{,}000 \times 0.\ 22859 = \$11{,}430 \text{ per year}$$

Therefore, using the time value method, the depreciation portion of the ownership cost is

$$\frac{\$73{,}591/\text{yr} - \$11{,}430/\text{yr}}{2{,}500\ \text{hr/yr}} = \$24.865/\text{hr}$$

Depreciation—Average Annual Investment Method. A second approach to calculating the depreciation portion of ownership cost is the average annual investment (AAI) method.

$$\text{AAI} = \frac{P(n+1) + S(n-1)}{2n} \tag{2.22}$$

where P = purchase price less the cost of the tires
S = the estimated salvage value
n = expected service life in years

The AAI is multiplied by the corporate cost-of-capital rate to determine the cost-of-money portion of the ownership cost. The straight-line depreciation of the cost of the machine minus the salvage and minus the cost of tires, if a pneumatic-tired machine, is then added to the cost-of-money portion (interest) to arrive at the total ownership depreciation.

Sample Calculation: Using the information from the previous example, the ownership depreciation using the AAI method would be

$$\text{AAI} = \frac{\$255{,}000(4+1) + \$50{,}000(4-1)}{2 \times 4}$$

$$= \$178{,}125/\text{yr}$$

$$\text{Cost-of-money portion} = \frac{\$178{,}125\ /\text{yr} \times 6\%}{2{,}500\ \text{hr/yr}} = \$4.275/\text{hr}$$

Straight-line depreciation

Initial cost	$300,000
Cost of tires	–45,000
Salvage	–50,000
	$205,000

$$\frac{\$205{,}000}{4\ \text{yr} \times 2{,}500\ \text{hr/yr}} = \$20.50/\text{hr}$$

Total depreciation portion of the ownership cost using the AAI method

$4.275/hr + $20.500 = $24.78/hr

The difference in the calculated total depreciation portion of the ownership cost is $0.09/hr ($24.87/hr – $24.78/hr). The choice of which method to use is strictly a company preference. Basically, either method is satisfactory, especially considering the impact of the unknowns concerning service life, operating hours per year, and expected future salvage. There is no single solution to calculating ownership cost. The best approach is to perform several analyses using different assumptions and be guided by the range of solutions.

Tax Saving from Tax Code Depreciation. To calculate the tax saving from depreciation, the government tax code depreciation schedules (Table 2.10) must be used. The resulting depreciation amounts are then multiplied by the company's tax rate to calculate specific savings, using Eq. (2.17) or Eq. (2.18). The sum of the yearly saving must be divided by the total anticipated operating hours to obtain an hourly cost saving.

Sample Calculation: Assume the machine is a 5-yr property under the tax code and purchased midway through the year. There was no gain on the procurement exchange. The company's tax rate is 37%. Using the same machine and company information, the hourly tax saving resulting from tax code depreciation is calculated. First, calculate the annual depreciation amounts for each of the years.

Year	5-yr Property Rates	BV_{n-1}	D_n	BV_n
0		$ 0	$ 0	$300,000
1	20.00%	300,000	60,000	240,000
2	32.00%	240,000	96,000	144,000
3	19.20%	144,000	57,600	86,400
4	11.52%	86,400	34,560	51,840
5	11.52%	51,840	34,560	17,280
6	5.76%	17,280	17,280	
			$300,000	

Using Eq. (2.17), the tax shielding effect for the machine's service life would be

Year	D_n	Shielded Amount*
1	$60,000	$22,200
2	96,000	$35,520
3	57,600	$21,312
4	34,560	$12,787
	34,560	$12,787
	17,280	$6,394
	Total	$111,000

$^*D_n \times 37\%$

$$\text{Tax saving from depreciation} = \frac{\$111{,}000}{4\ \text{yr} \times 2{,}500\ \text{hr/yr}} = \$11.10/\text{hr}$$

Major Repairs and Overhauls. When a major repair and overhaul takes place, the machine's ownership cost will have to be recalculated. This is done by adding the cost of the overhaul to the book value at the time of the overhaul. The resulting adjusted book value is then used in the depreciation calculation, as already described. If there are separate calculations for true depreciation and for tax depreciation, both will have to be adjusted.

Taxes, Insurance, and Storage. To calculate the taxes, insurance, and storage costs, common practice is to simply apply a percentage value to either the machine's book value or its AAI amount. The expenses incurred for these items are usually accumulated in a corporate overhead account. The accumulated overhead account amount divided by the value of the equipment fleet and multiplied by 100 will provide the percentage rate to be used. A overhead rate of 10% to 15% multiplied by the AAI is a simple means to account for these equipment overhead costs.

$$\text{Taxes, insurance, and storage portion of ownership cost} = \text{rate (\%)} \times BV_n \text{ (or AAI)} \qquad (2.23)$$

Operating Cost

Figures based on actual company experience should be used to develop operating expenses. Many companies, however, do not keep good equipment operating and maintenance records; therefore, many operating costs are estimated as a percentage of a machine's book value. Even companies that keep records often accumulate expenses in an overhead account and then prorate the total back to individual machines using book value.

Fuel. The amount expended on fuel is a product of how a machine is used in the field and the local cost of fuel. Today fuel is usually offered with a *time of delivery price.* A supplier will agree to supply the fuel needs of a project, but the price will not be guaranteed for the duration of the work. Therefore, when bidding a long-duration project, the contractor must make an assessment of future fuel prices. Many highway agencies have incorporated fuel price adjustment factors in their contracts to account for uncertainty in fuel prices and dampen the volatility in unit bid prices.

To calculate hourly fuel expense, a consumption rate is multiplied by the unit price of fuel. Service records are important for estimating fuel consumption.

Sample Calculation: A 220-fwhp dozer will be used to push an aggregate stockpile. This dozer is diesel powered. It is estimated that the work will be steady at an efficiency equal to a 50-min hour. The engine will work at full throttle while loading the blade (30% of the time) and at three-quarter throttle to travel and position. Calculate the fuel consumption using the engine consumption averages. If diesel cost is \$2.67/gal, what is the expected fuel expense?

Fuel consumption diesel engine 0.04 gal per fwhp-hr

Throttle load factor (operating power):

Push load	1.00 (power) × 0.30 (% of the time) = 0.30
Travel and position	0.75 (power) × 0.70 (% of the time) = 0.53
	0.83

Time factor (operating efficiency): 50-min hour: 50/60 = 0.83

Combined factor: 0.83 × 0.83 = 0.69

Fuel consumption = 0.69 × 0.04 gal/fwhp-hr × 220 fwhp = 6.1 gal/hr

Fuel Expense: 6.1 gal/hr × \$2.67/gal = \$16.29/hr

Lubricants. The quantity of lubricants used by an engine will vary with the size of the engine, the capacity of the crankcase, the condition of the piston rings, and the number of hours between oil changes. For extremely dusty conditions, it may be desirable to change oil every 50 hr, but this is an unusual condition. It is common practice to change oil every 100 to 250 hr. However, Tier 4 Final engines with exhaust gas recirculation (EGR) systems are putting more stress on engine oil. The new CJ-4 oil standard of the American Petroleum Institute is specifically aimed at supporting EGR engines. As a result, machine manufacturers are recommending 250 to 500 hr between oil changes. But some experts recommend sampling oil at 250 hr before continuing on to 500 hr. Some engine manufacturers have approached Tier 4 Final using selective catalytic reduction (SCR) technology. But this approach requires a separate tank for the urea solution. The question of oil change interval is still in flux, with the Tier 4 Final engines likely driving further changes.

Scheduling machine service based on the gallons of fuel consumed has been shown to be more effective than scheduling based on hours or miles (kilometers) run. The quantity of the oil consumed by an engine per change will include the amount added during the change plus the makeup oil between changes.

The cost of hydraulic oil, filters, and grease will be added to the expense of engine oil. The hourly cost of filters is simply the actual expense to purchase the filters divided by the hours between changes. If a company does not keep detailed machine-servicing data, it is difficult to accurately estimate the cost of hydraulic oil and grease. The usual solution is to refer to manufacturers' published tables of average usage or expense.

Repairs. There is nothing more important to the contractor's success than careful maintenance and prompt repair of equipment. The cost of repairs is normally the largest single component of machine cost. Historical averages indicate 37% of a construction machine's cost is attributed to repairs. Some general guidelines published in the past by the Power Crane and Shovel Association (PCSA) estimated repair and maintenance expenses at 80% to 95% of depreciation for crawler-mounted excavators, 80% to 85% for wheel-mounted excavators, 55% for crawler cranes, and 50% for wheel-mounted cranes. The lower figures for cranes reflect the work they perform and the intermittent nature of their use. For hydraulic machines, two-thirds of the cost is for materials and parts and one-third for labor.

A contractor must have a fairly accurate idea of the future cost of repairing a machine—before a price can be put on its use. If good records have been kept, they provide the basis on which to estimate future expenses. If there are no records, or if new equipment and/or new jobs are so different from past experience the records do not apply, estimating must be done on the basis of industry data. Equipment manufacturers supply tables of average repair costs based on machine type and work application. These must be modified to suit particular work conditions.

Possession of a number of machines will usually cause good and bad features of individual machines to average out, and a succession of jobs is likely to smooth out the ups and downs of work conditions.

Repair expenses will increase with machine usage (age). The repair cost to establish a machine rate for bidding should be an average rate.

Tires. Tire expenses include both repair and replacement. Tire maintenance is commonly handled as a percentage of straight-line tire depreciation. Tire hourly cost can be derived simply by dividing the cost of a set of tires by their expected life, and this is how many companies prorate this expense. Tire repairs are commonly accounted as a percentage of depreciation, with rates of 10% to 15% frequently used in industry. Tire retreads are treated as an overhaul item and depreciated using the straight-line method. A more sophisticated approach is to use a time-value calculation, recognizing that tire replacement expenses are single point-in-time outlays that take place over the life of a wheel-type machine.

Sample Calculation: What is the hourly tire cost part of a machine's operating cost if a set of tires can be expected to last 5,000 hr? Tires cost $38,580 per set of four. Tire repair cost is estimated to average 16% of the straight-line tire depreciation. The machine has a service life of 4 yr and operates 2,500 hr/yr. The company's cost-of-capital rate is 8%.

Not considering the time value of money:

$$\text{Tire repair cost} = \frac{\$38{,}580}{5{,}000\text{ hr}} \times 16\% = \$1.24\text{/hr}$$

$$\text{Tire use cost} = \frac{\$38{,}580}{5{,}000\text{ hr}} = \$7.72\text{/hr}$$

Therefore, tire operating cost is $8.951/hr ($1.235/hr + $7.72/hr).

Considering the time value of money:

Tire repair cost is the same: $1.235/hr.

Calculate the number of times the tires will have to be replaced.

$$\left(\frac{4\text{ yr} \times 2{,}500\text{ hr/yr}}{5{,}000\text{ hr per set of tires}}\right) = 2\text{ sets}$$

Purchase a second set at the end of the second year.

First set: Calculate the uniform series required to replace a present value of $38,580. Using the uniform series capital recovery factor formula:

$$A = \$38{,}580\left[\frac{0.08(1+0.08)^4}{(1+0.08)^4 - 1}\right]$$

$$\frac{\$38{,}580 \times 0.301921}{2{,}500\text{ hr/yr}} = \$4.66\text{/hr}$$

Second set: The second set will be purchased 2 yr in the future. Therefore, what amount at time zero is equivalent to $38,580 at 2 yr in the future? Using the present worth compound amount factor, the equivalent time zero amount is calculated.

$$P = \frac{\$38{,}580}{(1+0.08)^2} = \$33{,}076$$

Calculate the uniform series required to replace a present value of \$33,076.

$$A = \$33,760\left[\frac{0.08(1+0.08)^4}{(1+0.08)^4 - 1}\right]$$

$$\frac{\$33,076 \times 0.301921}{2,500 \text{ hr/yr}} = \$3.99/\text{hr}$$

Therefore, considering the time value of money, tire operating cost is \$9.89/hr (\$1.24/hr + \$4.66/hr + \$3.99/hr).

High-Wear Items. Because the cost of high-wear items is dependent on job conditions and machine application, the cost of these items is usually accounted for separately from general repairs.

Sample Calculation: A dozer equipped with a three-shank ripper will be used in a loading and ripping application. Actual ripping will take place only about 20% of total dozer operating time. A ripper shank consists of the shank itself, a ripper tip, and a shank protector. The estimated life for the ripper tip is 30 hr. The estimated life of the ripper shank protector is three times tip life. The local price of a tip is \$205 and that of a shank protector is \$420. What hourly high-wear item charge should be added to the operating cost of a dozer in this application?

$$\text{Tips: } \frac{30 \text{ hr}}{0.2} = 150 \text{ hr of dozer operating time}$$

$$\frac{3 \times \$205}{150 \text{ hr}} = \$4.10/\text{hr for tips}$$

Shank protectors: Three times tip life × 150 hr = 450 hr of dozer operating time

$$\frac{3 \times \$420}{450 \text{ hr}} = \$2.80/\text{hr for shank protectors}$$

Therefore, the cost of high-wear items is \$6.90/hr (\$4.10/hr tips + \$2.80/hr shank protectors).

BUY, RENT, OR LEASE

There are three basic methods for securing a particular machine to use on a project: (1) *buying* (direct ownership, purchase), (2) *renting,* or (3) *leasing*. Each method has inherent advantages and disadvantages. Ownership guarantees control of machine availability and mechanical condition, but it requires a continuing sequence of projects to pay for the machine. Ownership may at times force a company into using obsolete equipment. Advantages of owning are marketing strength, building assets, and having the equipment on hand when needed. The calculations applicable for determining the cost of direct ownership have been developed.

Rental

The rental of a machine is a short-term alternative to direct equipment ownership. With a rental, a company can pick a machine exactly suited for the job at hand. This is particularly advantageous if the job is of short duration or if the company does not foresee a continuing need for the particular type of machine in question. Rental equipment is generally newer than owned equipment. This minimizes the effects of obsolete equipment by taking advantage of recent changes in technology. Rentals are very beneficial to a company in such situations, even though the rental charges are higher than *normal* direct ownership expense. The advantage lies in the fact that direct ownership costing assumes a continuing need for and utilization of the machine. If the utilization assumption is not valid, a rental should be considered. Another important point to consider is the fact that with a rental, the company loses the tax depreciation shield of machine ownership but gains a tax deduction because rental payments are treated as an expense.

However, rental companies have only a limited number of machines and, during the peak work season, all types are not always available. Additionally, many specialized or custom machines cannot be rented.

Firms oftentimes use rentals as a way to test a machine prior to a purchase decision. A rental provides the opportunity for a company to operate a specific make or model machine under actual project conditions. Profitability of the machine, based on the company's normal operating procedures, can then be evaluated before a major capital expenditure is approved to purchase the machine.

The general practice of the industry is to price rental rates for equipment on a daily (8 hr), weekly (40 hr), or monthly (176 hr) basis. In the case of larger pieces of equipment, rentals may be available only on a monthly basis. Rentals using a measure of horsepower-hours are commonly used in the railroad industry and have been considered for the construction industry. Cost per hour usually is less for a longer-term rental (i.e., the monthly rate figured on a per-hour basis would be less than the daily rate on an hourly basis).

Responsibility for repair cost is stated in the rental contract. Normally, in the case of tractor-type equipment, the renter is responsible for all repairs. If it is a pneumatic-tired machine (on rubber), the renting company will measure tread wear and charge the renter for tire wear. In the case of cranes and shovels, the renting company usually bears the cost of normal wear and tear. The user must provide servicing of the machine while it is being used. The renter is almost always responsible for fuel and lubrication expenses. Industry practice is that rentals are payable in advance. The renting company will require from the user certificates of insurance before the machine is shipped to the job site.

Equipment cost is very sensitive to changes in use hours. Fluctuations in maintenance expenses or purchase price barely affect cost per hour. But a decrease in use hours per year can make the difference between a cost-effective machine ownership versus renting.

The basic cost considerations to examine when considering a rental can be illustrated by a simple set of circumstances. Consider a small wheel loader with an ownership cost of \$10.96 per hour. Assume the cost is based on the machine working 2,400 hr each year of its service life. If \$10.96/hr is multiplied by 2,400 hr/yr, the yearly ownership cost is found to be \$26,304.

Checking with the local rental company, the construction firm receives rental quotes of \$3,558 per month, \$1,182 per week, and \$369 per day for this size loader. Dividing by the appropriate number of hours, these rates can be expressed as hourly costs. Likewise, by dividing the calculated hourly rental rates into the construction firm's yearly ownership cost figure (\$26,304), the operating hour break-even points can be determined (Table 2.13).

TABLE 2.13 Rental versus Ownership, Operating Hour Break-Even Points

Rental Duration	Rate	Hours	Rental Rate	Operating Hour Break-Even Point ($26,304/Rental Rate)	Operating Point Break-Even Point ($3,558/Rental Rate)
Monthly	$3,558	176	$20.22/hr	1,300 hr	—
Weekly	$1,182	40	$29.55/hr	890 hr	120 hr
Daily	$ 369	8	$46.13/hr	570 hr	77 hr

If the loader will be used for fewer than 1,300 hr but more than 890 hr, the construction company should consider a monthly rental instead of ownership. When the projected usage is fewer than 1,300 hr but greater than 120 hr, a weekly rental would be appropriate. In the case of very limited usage, that is, fewer than 26 hr, the daily rate is optimal.

When a company rents, it pays for the equipment only when project requirements dictate a need. A company that owns a machine must continue to make the equipment payments even when the machine sits idle. When investigating a rental, the critical question is usually expected *hours of usage*.

Lease

Lease A long-term agreement for the use of an asset.

A **lease** is a long-term agreement for the use of an asset. It provides an alternative to direct ownership. During the lease term, the leasing company (lessor) always owns the equipment and the user (lessee) pays the owner to use the equipment. The lessor must retain ownership rights in order for the contract to be considered a true lease by the Internal Revenue Service. The lessor will receive lease payments in return for providing the machine. The lease payments do not have to be uniform across the lease period. The payments can be structured in the agreement to best fit the situation of the lessee or the lessor. In the lessee's case, cash flow at the beginning may be low, so the lessee wants payments that are initially less. Because of tax considerations, the lessor may agree to such a payment schedule. Lease contracts are binding legal documents, and most equipment leases are noncancelable by either party.

A lease pays for the use of a machine during the most reliable years of a machine's service life. Sometimes the advantage of a lease is that the lessor provides the maintenance and servicing. This frees the contractor from hiring mechanics and service personnel and allows the company to concentrate on the task of building.

Long term, when used in reference to lease agreements, is a period of time that is long relative to the life of the machine in question. An agreement that is for a very short period, as measured against the expected machine life, is a rental. A conventional—that is, true—lease will have one of three different end-of-lease options: (1) buy the machine at fair market value, (2) renew the lease, or (3) return the equipment to the leasing company.

As in the case of a rental, a lessee loses the tax depreciation shield of machine ownership but gains a tax deduction because lease payments are treated as an expense. The most important factor contributing to a decision to lease is reduced cost. Under specific conditions, the actual cost of a leased machine can be less than the ownership cost of a purchased machine. This is caused by the different tax treatments for owning and leasing an asset. An equipment user must make a careful examination of the cash flows associated with each option to determine which results in the lowest total cost.

Working capital is the cash a firm has available to support its day-to-day operations. This *cash asset* is necessary to meet the payroll on Friday, to pay the electric bill, and to purchase fuel to keep the machines running. To be a viable business, working capital assets must be greater than the inflow of bills. A machine is an asset to the company, but it is not what the diesel fuel vendor for the machines will accept as payment.

A commonly cited advantage of leasing is that working capital is not tied up in equipment. This statement is only partly true. It is true in the case when a company borrows funds to purchase a machine, the lender normally requires that the company establish an equity position in the machine: a *down payment*. Additionally, the costs of delivery and initial servicing are not included in the loan and must be paid by the new owner.

A very important consideration in the lease decision is that corporate funds are not tied up in the up-front costs of a purchase. Leasing does not require these high early cash outflows and is often considered as 100% financing. However, most leases require an advance lease payment. Some even require a security deposit and charge other up-front costs. Many leases will allow the contractor to apply their company logo to the equipment, an important marketing tool that is typically not permitted with rentals.

Still another argument for a lease is because borrowed funds are not used, the credit capacity of the company is preserved. Leasing is often referred to as off-balance-sheet financing. A lease is considered an operating expense, not a liability, as is the case with a bank loan. With an operating lease (used when the lessee does not ultimately want to purchase the equipment), leased assets are expensed. Therefore, such assets do not appear on the balance sheet. Standards of accounting, however, require disclosure of lease obligations. It is hard to believe a *lender* would be so naive as to not consider all of a company's fixed obligations, including both loans and leases. But the off-balance-sheet lease typically will not hurt *bonding capacity,* which is important to a company's ability to bid work.

Before entering into a contract with a construction company, most owners require the contractor to post a bond guaranteeing completion of the project. A third-party surety company secures this bond. The surety closely examines the construction company's financial position before issuing the bond. Based on the financial strength of the construction company, the surety typically restricts the total volume of work a company can have under contract at any one time. This restriction is known as bonding capacity. It is the total dollar value of work under contract that a surety company will guarantee for a construction company. Prime contracting directly accrues the contract value to the bonding capacity, while joint venturing or working as a subcontractor can lessen this accrual.

Owners should make a careful examination of the advantages of a lease situation. The cash flows to be considered when evaluating the cost of a lease include:

1. Inflow, initially, of the equivalent value of the machine.
2. Outflow of the periodic lease payments.
3. Tax shielding provided by the lease payments. (This is allowed only if the agreement is a true lease. Some "lease" agreements are essentially installment sale arrangements.)
4. Loss of salvage value when the machine is returned to the lessor.

These costs all occur at different points in time, so present-value computations must be made before the costs can be summed. The total present value of the lease option should be compared to the minimum ownership costs, as determined by a time-value replacement analysis. In most lease agreements, the lessee is responsible for maintenance. If, for the lease in question, maintenance expense is the same as for the case of direct ownership, then the maintenance expense factor can be dropped from the analysis. A leased machine would exhibit the same aging and resulting reduced availability as a purchased machine.

CHAPTER 3
DEVELOPING A PLAN

When a constructor prepares a plan and cost estimate for earthwork operations, the critical attributes are (1) the ***quantities*** *involved, basically volume or weight; (2) the* ***haul distances****; and (3) the* ***grades*** *for all segments of the hauls. An earthwork volume sheet allows for the systematic recording of information and making the necessary earthwork calculations. On linear projects such as roadways, a mass diagram is an analysis tool for selecting the appropriate equipment for excavating and hauling material. On nonlinear projects, such as building foundations, landfills, and some dams, bulk excavation is necessary with calculation of earthwork quantities using a grid method and trapezoidal volumes. The final objective is to predict the production rate for a group of machines (linked-system production rate) and the cost per unit of production. Estimating is determining how to construct the specified work in the most economical manner and within the time allowed by the contract. The format of all estimates should be consistent. The critical path method (CPM) is a planning and control technique that provides an accurate, timely, and easily understood graphic depiction of the project. Its purpose is to allocate resources over time in an optimal manner and in a way that allows effective reallocation and schedule control after the project starts.*

JOB STUDY

Earthwork construction may involve clearing vegetation, stripping and storing of topsoil, excavating soil and rock, hauling the spoil to waste dumps, excavating for culverts, building abutments for bridges, raising low areas to grade by fill obtained from cuts or borrow pits, and placing topsoil on finished slopes. Usually, this work must be accomplished within a time limit set by the contract. It is desirable to get the maximum number of machines and workers on the job as soon as possible after the start, but it is more important to keep them efficiently employed.

Basic factors to be considered when planning earthwork operations may include

1. Clearing and debris disposal
2. Topsoil stripping, storage, reclamation, spreading, and planting
3. Amount, type, and moisture content of excavated material
4. Amount and type of rock excavation
5. Availability of suitable borrow material and cost of purchase

6. Haul road construction and maintenance and length of hauls
7. Quality of fill required and processing of excavated material
8. Fill compaction, shrinkage, and disposal of surplus
9. Slope finishing and protection
10. Groundwater conditions and drainage requirements
11. Earthwork to support structure construction; bridges, culverts, and retaining walls
12. Availability of proper machinery, with necessary parts and supplies
13. Labor supply and skill level
14. Weather—rain, snow, ice, dust, frozen ground, frozen equipment, mud
15. Time to complete earthwork relative to the construction schedule for the structures

Excavation may be described as unclassified or it may be divided into rock yards and soil yards. In less formal jobs, these factors may be indicated only approximately. When the quantities are large, yardages should be carefully calculated. Even if quantities are provided in the contract documents, these must be verified by the contractor.

Graphical Presentation of Earthwork

The planning process begins with the decision to submit a bid to perform the work. After reviewing the project drawings and specifications, the estimator/planner visits the site of the proposed project and gathers information about location-specific conditions such as surface topography, drainage, and access. A careful study of the site vegetation can provide clues to soil types and moisture. It is also important to draw conclusions to questions such as prior use of the site for dumping or disposal of material or wetland limits. The information developed during the site visit is recorded in a report, and the report is made part of the final project estimate.

After the site visit, the quantity of materials to be furnished or moved is determined. This process is called the take-off or quantity survey. During the take-off process, the planner must make choices concerning equipment needs, sequence of operations, and crew size. The type of earthwork to be performed will influence judgments concerning equipment and methods.

Mass excavation The requirement to excavate substantial volumes of material, usually at considerable depth or over a large area.

Structural excavation Excavation undertaken in support of structural element construction; it usually involves removing materials from a limited area.

Types of Earthwork Projects. **Mass excavation** involves moving a substantial volume of material, and the excavation work is a primary part of the project. On the Eastside project in California, the contractor building the West Dam moved over 68 million cy of material. Mass excavations are typically operations of considerable excavation depth and horizontal extent and may include requirements to drill and blast rock (e.g., the movement of consolidated materials).

Structural excavation is a different type of undertaking. The excavation work is performed to support the construction of structural elements. This work is usually performed in a confined area, it is typically vertical in extent, and the banks of the work may require support systems. The volume of excavated material is not a decisive factor as much as dealing with a limited workspace and usually the vertical movement of the material. With either type of work, allowances for the waste of materials, inclement weather, delays, and other factors affecting costs must be incorporated into the estimate. The project plans typically provide graphical information used to calculate work quantities.

The critical attributes—cost drivers—of an earthwork project are (1) the quantities involved, measured by volume or weight; (2) the haul distances; and (3) the grades for all segments of the haul routes.

Station A horizontal distance of 100 ft.

Horizontal distances along a project are referenced in **stations.** The term *station* refers to locations on a 100 base numbering system. Therefore, the distance between two adjacent stations is 100 ft. Station 1 is written 1 + 00. The plus sign is used in this system of referencing points. The term, stationing, refers to the surveyor notation for laying out a project in the field and is used on the plans to denote locations along the length of the project.

Plan view A construction drawing representing the horizontal alignment of the work.

Project Documents. Three kinds of views are presented in the project documents to show earthwork construction features.

Profile view A construction drawing depicting a vertical plane cut through the centerline of the work. It shows the vertical relationship of the ground surface and the finished work.

1. *Plan view.* The **plan view** is looking down on the proposed work and presents the horizontal alignment of features. Figure 3.1 is a plan view of a highway project; it shows the project centerline with stationing noted. The project limits are the two dark exterior lines.
2. *Profile view.* The **profile view** is a cut view (side view), typically along the centerline of the work. It presents the vertical alignment of features. Figure 3.2 is a profile view; the bottom horizontal scale shows the centerline stationing, the vertical scale gives elevation, the dashed line is the existing ground line, and the solid line is the proposed final grade of the work.

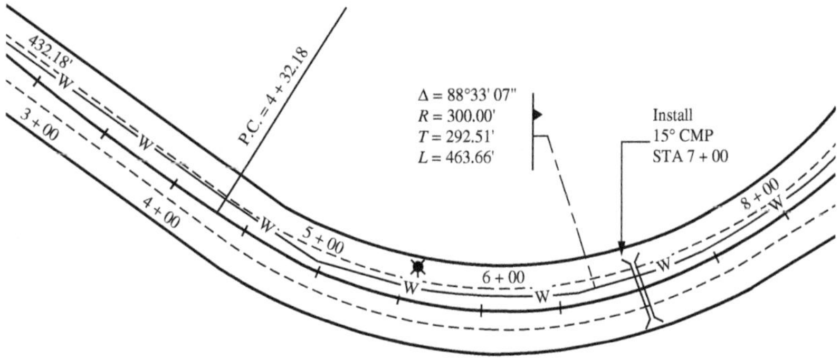

FIGURE 3.1 Plan view of a highway project.

Cross-section view A construction drawing depicting a vertical section of earthwork at right angles to the centerline of the work. Used together with centerline distances to calculate earthwork quantities.

3. *Cross-section view.* The **cross-section view** is formed by a plane cutting the work vertically and at right angles to the project's long axis. Figures 3.3 and 3.4 present fill and cut cross-sections, respectively. The straight heavy lines denote the final grade of the work, and the existing ground is shown by the thicker line.

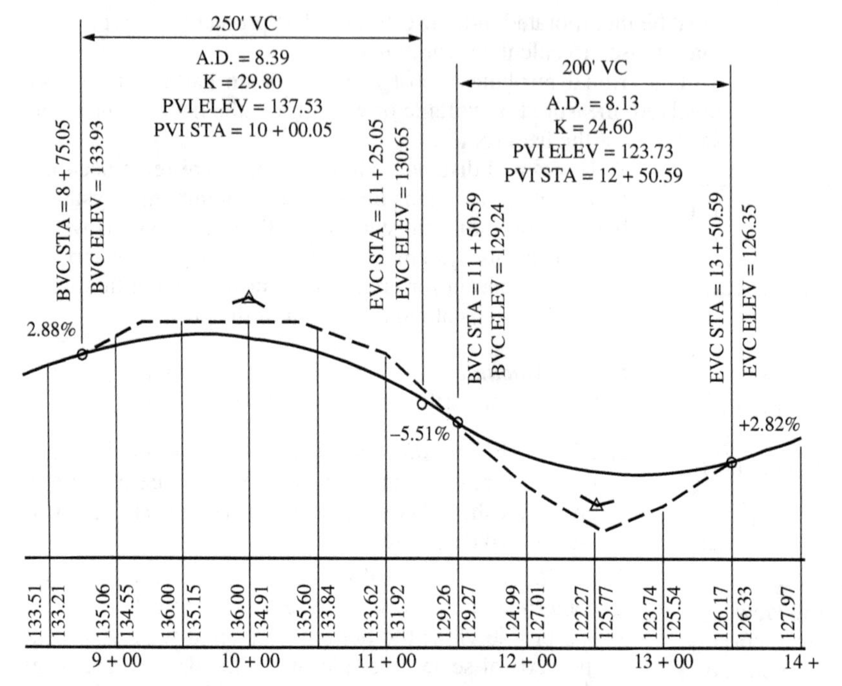

FIGURE 3.2 Profile view of a highway project.

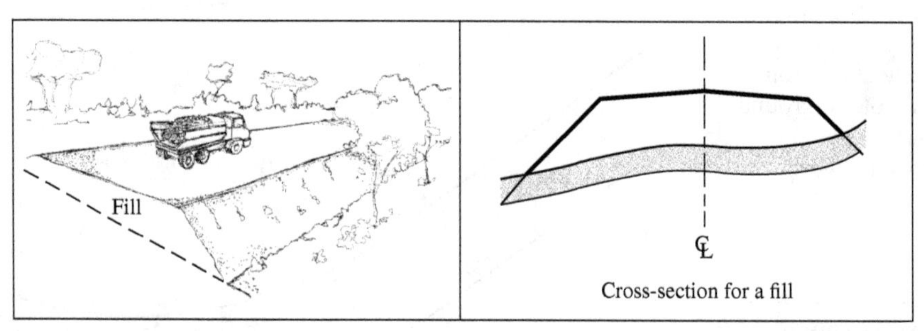

FIGURE 3.3 Earthwork cross-section of a fill situation.

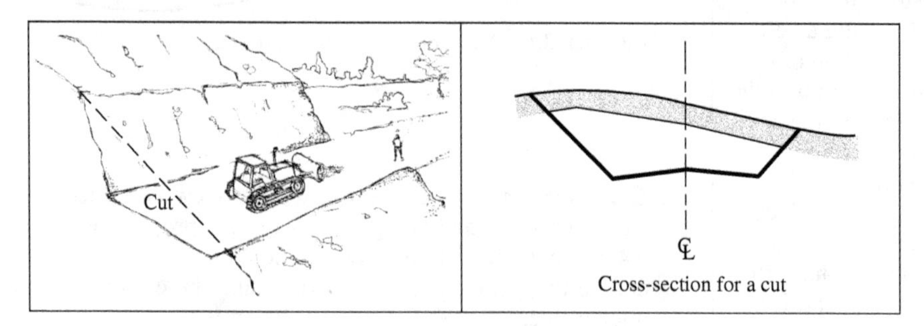

FIGURE 3.4 Earthwork cross-section of a cut situation.

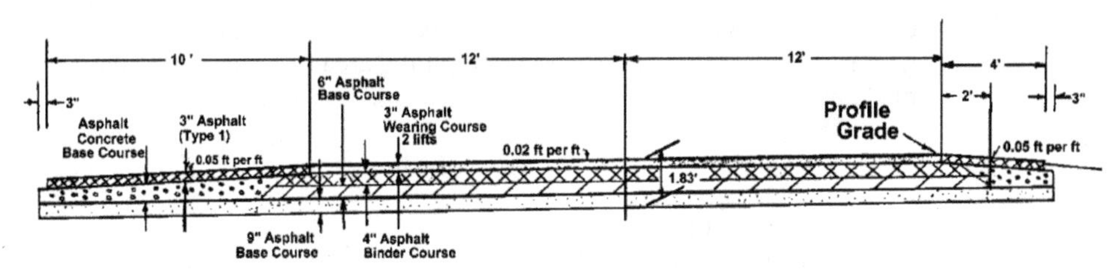

FIGURE 3.5 Cross-section view of a pavement section for an interstate highway.

> **Cross-sections**
> Earthwork drawings created by combining the project design with field measurements of existing conditions. These are typically viewed at right angles to the centerline of the project or a major project feature.

In the case of a linear project, material volumes are usually determined from the **cross-sections.** Cross-sections are pictorial drawings produced from a combination of the designed project layout and ground measurements taken in the field at right angles to the project centerline or the centerline of a project feature, such as a drainage ditch. When the ground surface is regular, field measurements are typically taken at every full station (100 ft). When the ground is irregular, measurements must be taken at closer intervals and particularly at points of change. Typical cross-sections are shown in Figs. 3.3 and 3.4.

The cross-sections usually depict the finished subgrade elevations; however, this detail should always be verified (Fig. 3.5). If the grades depicted are the top of pavement, the thickness of the pavement section must be used to adjust the earthwork computations. From the cross-section drawings, *end areas* can be computed using any of several methods.

Earthwork Quantities

Earthwork computations involve the calculation of earthwork volumes, the balancing of cuts and fills, and the planning of the most economical material hauls. The first step in estimating the cost of an earth-moving operation is calculation of the quantities. The exactness with which earthwork computations can be made depends on the extent and accuracy of field measurements represented on the drawings.

End-Area Determination. The method chosen to compute a cross-section end area will depend on the time available and the available computer or mechanical aids. Most companies use commercial computer software and digitizing tablets to determine cross-section end areas. Other methods include the use of a planimeter, subdivision of the area into geometric figures with definite formulas for areas (rectangles, triangles, parallelograms, and trapezoids), and the use of the trapezoidal formula.

Digitizing Tablet. A digitizing tablet is a board with an embedded wire mesh grid. When the cursor is traced over the board, a current is picked up by the board's grid, and the coordinates of the tracing device are passed to the computer. When a plan is taped to the board and a scale is entered, the traced plan is converted to measurements, and then a software program computes the area, length, and volume of the traced data.

Planimeter. A planimeter is a mechanical device that is used to move a tracing point around the perimeter of the plotted area. It provides a value that is then multiplied by the square scale of the figure to calculate the figure's area. A planimeter can be used on any figure no matter how irregular the figure's shape might be.

Trapezoidal Computations. The mathematics of the area computations are often based on dividing the cross-section drawing into parts. The computer can easily subdivide a cross-section drawing into a large number of strips, calculate the area of each strip, and then sum the individual areas to arrive at the area of a section. If the calculations must be made by hand, the area formulas for a triangle and a trapezoid are

$$\text{Area of a triangle} = \frac{1}{2}hw \tag{3.1}$$

where h = height of the triangle
w = base of the triangle

$$\text{Area of a trapezoid} = \frac{(h_1 + h_2)}{2} \times w \tag{3.2}$$

where (Fig. 3.6) w = distance between the two parallel sides
h_1 and h_2 = the lengths of the two parallel sides

The general trapezoidal formula for calculating area is

$$\text{Area} = \left(\frac{h_0}{2} + h_1 + h_2 + \cdots + h_{(n-1)} + \frac{h_n}{2}\right) \times w \tag{3.3}$$

where (Fig. 3.6) w = distance between the two parallel sides
$h_0 \ldots h_n$ = the lengths of the individual adjacent parallel sides

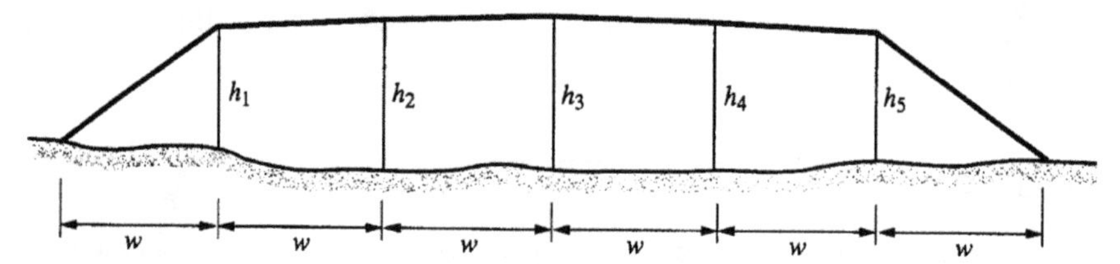

FIGURE 3.6 Division of a cross-section drawing into triangles and trapezoids.

The precision achieved using this formula depends on the number of subdivisions but is about ±0.5%.

Average end area A calculation method for determining the volume of material bounded by two cross-sections or end areas.

In the case of hillside construction, there can be both a cut area and a fill area in the same cross-section (Fig. 3.7). When making area computations, it is always necessary to calculate cut and fill areas separately.

Average End Area. The **average-end-area** method is commonly used to determine the volume bounded by two cross-sections or end areas. The volume of the solid bounded by two parallel, or nearly parallel, cross-sections is equal to the average of the two end areas times the distance between the cross-sections along their centerline (Fig. 3.8). The average-end-area formula is

$$\text{Volume [net cubic yards (cy)]} = \frac{(A_1 + A_2)}{2} \times \frac{L}{27} \tag{3.4}$$

where (see Fig. 3.8)
A_1 and A_2 = area in square feet (sf) of the respective end areas
L = the length in feet between the end areas

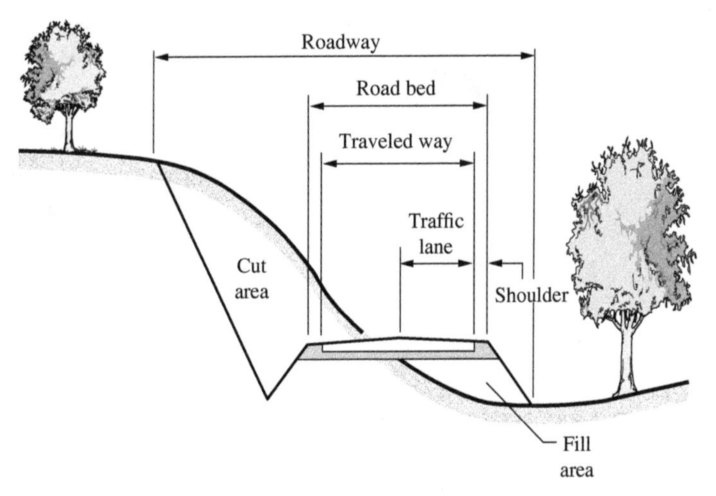

FIGURE 3.7 Cut and fill areas in the same section.

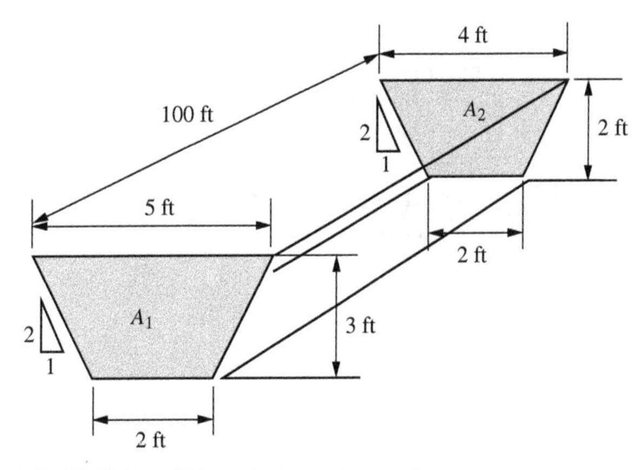

FIGURE 3.8 Volume between two end areas.

This averaging is not altogether true, because the average of the two end areas is not the arithmetic mean of many intermediate areas. The method gives volumes generally slightly in excess of the actual volumes. The precision is about ±1.0%.

Although cross-sections can be taken at any conservative interval along the centerline, judgment should be exercised, depending particularly on the irregularity of the ground and the tightness of curves. In the case of tight curves, a spacing of 25 ft is often appropriate.

Stripping The ground's upper layer of organic material. Usually, it must be removed before beginning an excavation or embankment.

Stripping. The upper layer of material encountered in an excavation is often topsoil (organic material), resulting from decomposition of vegetative matter. Such organic material is commonly referred to as **stripping.** This material is not suitable for use in an embankment, and generally it must be handled in a separate excavation operation. It can be collected and disposed of off the project or stockpiled for later use on the project to plate slopes. If the embankments are of limited

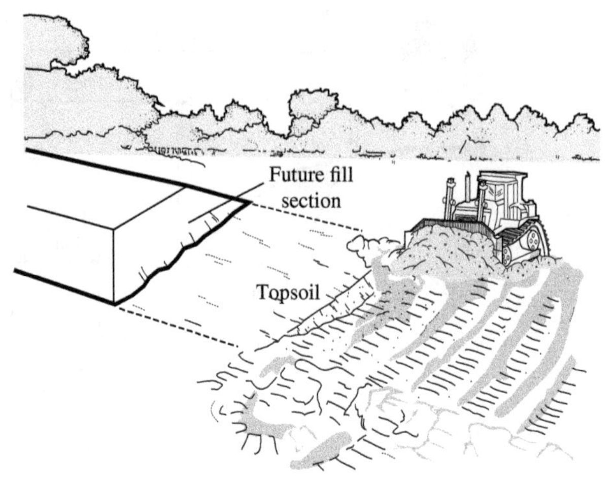

FIGURE 3.9 Stripping topsoil before building a fill.

height, the organic material below the footprint of the fill sections must be stripped before embankment placement can commence (Fig. 3.9). In the case of embankments over 5 ft in height, most specifications allow the organic material to remain if its thickness is only a few inches. When calculating the volume of cut sections, this stripping quantity must be subtracted from the net volume, as it cannot be used for embankment construction (Fig. 3.10). In the case of fill sections, the quantity must be added to the calculated fill volume (see Fig. 3.10).

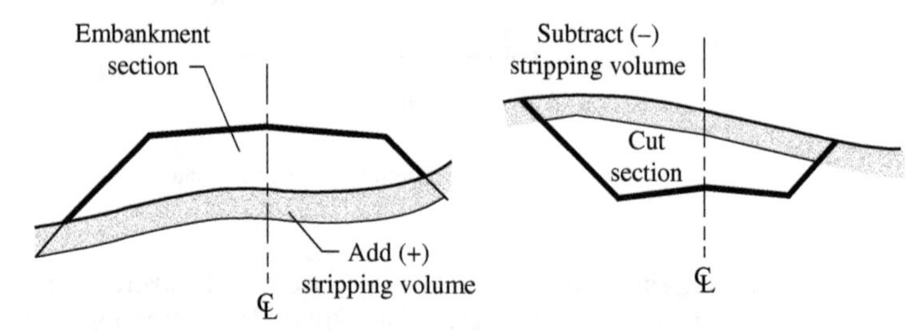

FIGURE 3.10 Effect of stripping on embankment and cut volume calculations.

Net Volume. The computed volumes from the cross-sections represent two different material states. The volumes from the fill cross-sections represent a compacted volume. If the volume is expressed in cubic yards, the notation is compacted cubic yards, ccy. In the case of cut sections, the volume is a natural, in situ volume. The term *bank volume* is used to denote this in situ volume; if the volume is expressed in cubic yards, the notation is bank cubic yards, bcy. If the cut and fill volumes are to be combined, they must be converted into compatible volumes. In Table 3.1, the conversion from compacted to bank cubic yards is made by dividing the compacted volume by 0.90. This value should be determined by soil testing of the actual material that will be handled.

TABLE 3.1 Earthwork Volume Calculation Sheet

	Station (1)	End-Area Cut (sf) (2)	End-Area Fill (sf) (3)	Volume of Cut (bcy) (4)	Volume of Fill (ccy) (5)	Stripping Cut (bcy) (6)	Stripping Fill (ccy) (7)	Total Cut (bcy) (8)	Total Fill (ccy) (9)	Adj. Fill (bcy) (10)	Algebraic Sum (bcy) (11)	Mass Ordinate (12)
(1)	0 + 00	0	0									
(2)	0 + 50	0	115	0	106	0	18	0	124	138	–138	–138
(3)	1 + 00	0	112	0	210	0	30	0	240	267	–267	–405
(4)	2 + 00	0	54	0	307	0	44	0	351	390	–390	–796
(5)	2 + 50	64	30	59	78	0	22	59	100	111	–52	–847
(6)	3 + 00	120	0	170	28	26	0	144	28	31	114	–734
(7)	4 + 00	160	0	519	0	76	0	443	0	0	443	–291
(8)	5 + 00	317	0	883	0	74	0	809	0	0	809	518
(9)	6 + 00	51	0	681	0	60	0	621	0	0	621	1,140
(10)	6 + 50	46	6	90	6	21	0	69	6	6	63	1,202
(11)	7 + 00	0	125	43	121	0	25	43	146	163	–120	1,082
(12)	8 + 00	0	186	0	576	0	81	0	657	730	–730	352
(13)	8 + 50	0	332	0	480	0	69	0	549	610	–160	–257

Structural Excavation

Before a structure can be built, there is often a need to excavate deep cuts with vertical or nearly vertical walls. These excavations require temporary earth-retaining structures to support the sides of the excavation. Although the haul distance is usually well defined—from the excavation center of mass to a specified dump location—the process of removing material from a confined pit is a challenge. The quantity of material to be excavated (bank measure) can be calculated by a grid method. The grid method requires the following:

- Existing grade elevation (a contour map or survey of the site).
- Elevation of the proposed grades.
- Difference between existing and proposed elevations.
- Grid lines applied to the surface to create individual square cells.

Grid Method. The grid method is a convenient way of computing the material to be moved from a structural excavation. The site is divided by a grid of appropriate dimensions and the elevation at the grid intersection points determined. The volume of material to be excavated or filled for a single grid (square) is determined as the average of the elevation difference at its four corners multiplied by the area of a grid square. The individual volumes of each grid square are summed to determine the volume of the excavation. This calculation work is typically done using commercial site layout and earthwork analysis software. Basically these programs use the grid process algorithm. Global positioning system (GPS) measurements can be used to efficiently collect existing site elevation data for software input. Areas of grid cells are adjusted for relative changes in site topography and the desired level of estimating accuracy, with common grid cell dimensions ranging from 15 ft squares up to 25 ft. Smaller cells provide greater resolution but in turn require more field data points.

Rougher terrain requires smaller grid dimensions to support accurate results. The squares must be of such size that no significant breaks, either in the original ground surface

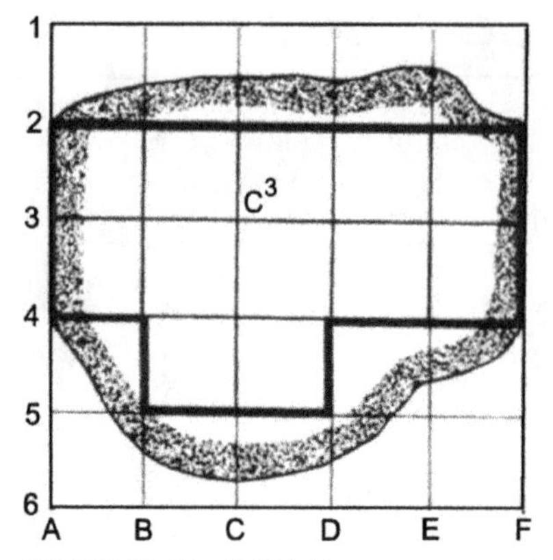

FIGURE 3.11 Grid laid over a structural excavation.

or in the bottom elevations, exist between the corners of the square or between the edges of the excavation and the nearest interior corner.

Figure 3.11 shows a structural excavation overlaid with 25-ft grid squares. To identify the various intersecting points, the lines in one direction are labeled by numbers and in the other direction by letters. Thus, the intersection of lines C and 3 would be labeled C^3.

The heavy line outlines the excavation. Within that line, determine the volume of excavation for each square and then sum the volumes.

Analyzing Earthwork Volumes

The mass diagram is an effective tool for planning linear earth-moving operations. It is easily prepared and permits the development of detailed earth-moving operational plans for road, runway, or drainage ditch type projects.

Earthwork Volume Sheet. An earthwork volume sheet can easily be constructed using a spreadsheet program. Such a sheet allows for the systematic recording of information and making the necessary earthwork calculations (see Table 3.1).

Stations. Column 1 is a listing of all stations at which cross-sectional areas have been recorded.

Area of cut. Column 2 is the cross-sectional area of the cut at each station. Usually this area must be computed from the project cross-sections.

Area of fill. Column 3 is the cross-sectional area of the fill at each station. Usually this area must be computed from the project cross-sections. Note there can be both cut and fill at a station (rows 5 and 10, see Table 3.1).

Volume of cut. Column 4 is the volume of cut between the adjacent preceding station and the station. The average-end-area formula, Eq. (3.4), is usually used to calculate this volume. This is a *bank* volume (bcy).

Volume of fill. Column 5 is the volume of fill between the adjacent preceding station and the station. The average-end-area formula, Eq. (3.4), is usually used to calculate this volume. This is a *compacted* volume (ccy).

Stripping volume in the cut. Column 6 is the stripping volume of topsoil over the cut between the adjacent preceding station and the station. This volume is often calculated by multiplying the distance between stations or fractions of stations by the width of the cut. This provides the area of the cut footprint. The footprint area is then multiplied by an average depth of topsoil to derive the stripping volume. This represents a bank volume of cut material. Usually topsoil material is not suitable for use in the embankment. The average depth of topsoil must be determined by field investigation.

Stripping volume in the fill. Column 7 is the stripping volume of topsoil under the fill between the adjacent preceding station and the station. This volume is commonly calculated by multiplying the distance between stations or fractions of stations by the width of the fill. This provides the area of the fill footprint. To derive the stripping volume, the area of the embankment footprint is multiplied by an average depth of topsoil. The stripping is a *bank* volume, but it also represents an additional requirement for fill material, compacted cubic yards of fill.

Total volume of cut. Column 8 is the volume of cut material available for use in embankment construction. It is derived by subtracting the cut stripping (col. 6) from the cut volume (col. 4) (illustrated in Fig. 3.10).

Total volume of fill. Column 9 is the total volume of fill required. It is derived by adding the fill stripping (col. 7) to the fill volume (col. 5) (illustrated in Fig. 3.10).

Adjusted fill. Column 10 is the total fill volume converted from compacted volume to bank volume.

Algebraic sum. Column 11 is the difference between column 10 and column 8. This indicates the volume of material that is available (cut is positive) or required (fill is negative) within station increments after intrastation balancing.

Mass ordinate
The cumulative material mass differential as calculated from the start through the end of the work.

Mass ordinate. Column 12 is the running total of column 11 values from some beginning point on the project profile. When the stations being summed are excavation sections, the value of this column will increase, whereas summing fill sections will result in a decrease of the column 12 values. Note: The material used within a station length is not accounted for in the mass ordinate and, therefore, is not accounted for in the mass diagram.

Mass diagram
In earthwork calculations, a graphical representation of the algebraic cumulative quantities of cut and fill along the centerline, where cut is positive and fill is negative. Used to calculate haul distance.

The **mass diagram** accounts for only material transported beyond the limits of the two cross-sections used to define the volume of material. Where there is both cut and fill between a set of stations, only the excess of one over the other is used in computing the mass ordinate. Cut material between two successive stations is first used to satisfy fill requirements between those same two successive stations before there is a contribution to the mass ordinate value. Likewise, if there is a greater fill requirement between two successive stations than there is cut available, the cut contribution is accounted for first. Only after all of the cut material is used will there be a fill contribution to the mass ordinate value. The material used between two successive stations is considered to move at right angles to the centerline of the project and, therefore, is often termed *crosshaul.* The remaining material in either case represents a longitudinal haul along the length of the project (Fig. 3.12).

Mass Diagram. Earthmoving is basically an operation where material is removed from high spots (hills) and deposited in low spots (valleys) (see Fig. 3.12). If there is a shortage, borrow is required, and if there is excess excavation (cut), it must

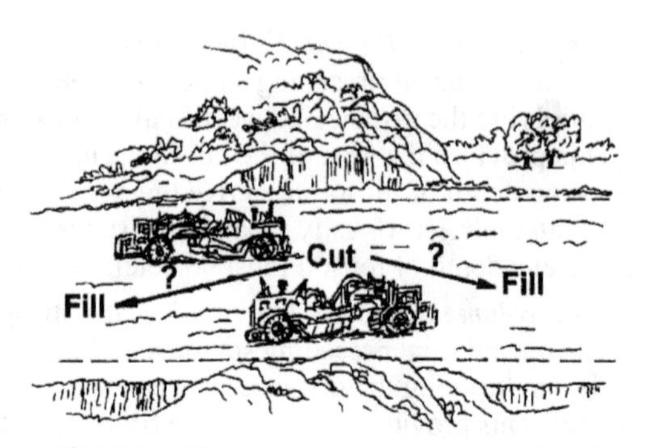

FIGURE 3.12 The mass diagram provides the information for deciding in which direction material should be hauled.

be wasted. The *mass diagram* is a method of analyzing linear earth-moving operations. It provides a graphical means for measuring haul distance (stations) in terms of earthwork volume (typically cubic yards but could be cubic meters). A station-yard is a measure of work: the movement of 1 cy through a distance of one station or 100 linear feet.

On a mass diagram graph, the horizontal dimension represents a linear dimension—stations—of a project (col. 1, see Table 3.1), and the vertical dimension (col. 12, see Table 3.1) represents the cumulative sum of excavation and embankment from some beginning point on the project profile. The diagram provides information concerning

- Quantities of materials to be moved
- Average haul distances for the movement
- The types of equipment to be considered, based on haul distance guidance

When combined with a ground profile, the average slope of haul segments can be estimated. The mass diagram is an effective tool for planning the movement of material on any project of linear extent.

Using column 1 as the horizontal scale of the graph and column 12 as the vertical scale of an earthwork volume calculation sheet, a mass diagram can be plotted. See the bottom portion of Fig. 3.13. Positive mass ordinate values are plotted above the zero datum line and negative values below. The top portion of Fig. 3.13 is the profile view of the same project.

Mass Diagram Properties. A mass diagram is a running total of the quantity of material along the project profile. It can indicate a surplus or deficit at individual points (locations). An excavation operation produces an *ascending* mass diagram curve; the excavation quantity exceeds the embankment quantity requirements. Excavation is occurring between stations A and B and between stations D and E in Fig. 3.13, and the curve is ascending. The total volume of excavation between stations A and B is obtained by projecting horizontally to the vertical axis *the mass diagram line points* at stations A and B and reading the difference of the two volumes. Conversely, if the operation is a fill situation, there is a deficiency of material and a *descending* curve is produced; the embankment requirements exceed the excavation quantity being generated. Filling is occurring between stations B and D, and the curve is descending. The volume of fill can be determined in a manner similar to the one described for the excavation.

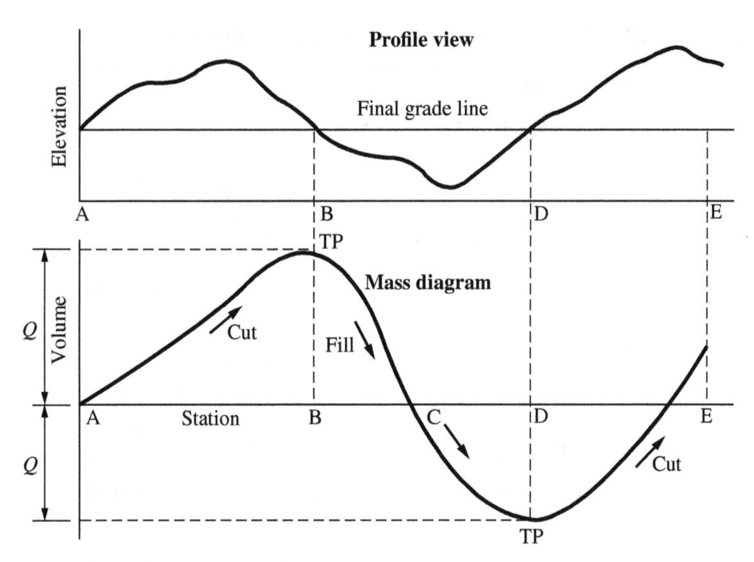

FIGURE 3.13 Properties of a mass diagram.

The *maximum* or *minimum points* on the mass diagram, where the curve transitions from rising to falling or from falling to rising, indicate a change from an excavation to fill situation or vice versa. These points are referred to as *transition points* (TPs). On the ground profile, the grade line is crossing the ground line (see Fig. 3.13 at station B and D).

When the mass diagram curve crosses the datum (or zero volume) line (as at station C), exactly as much material is being excavated (between stations A and B) as is required for fill (between stations B and C). There is no excess or deficit of material at point C in the project. The final position of the mass diagram curve above or below the datum line indicates whether the project has surplus material and a need to waste material or if there is a deficiency and a requirement to locate borrow material from outside the project limits. Figure 3.13, station E, indicates a waste situation, and excess material will have to be removed from the project.

Using the Mass Diagram. The mass diagram is a tool for selecting the appropriate equipment for excavating and hauling material. The analysis is accomplished using balance lines and calculating average haul distances.

Balance line
A horizontal line of specific length that intersects the mass diagram in two places.

Balance Lines. A **balance line** is a horizontal line and intersects the mass diagram in two places. It should have a specific length to match the maximum haul distance for different types of equipment. The maximum haul distance is the limiting economical haul distance for a particular type of equipment (Table 3.2).

Figure 3.14 shows a balance line drawn on a portion of a mass diagram. If this line were constructed for a large push-loaded scraper, the distance between stations A and C would be 5,000 ft. Between the ends of the balance line, the cut volume generated equals the fill volume required. Between stations A and C, the amount of material the scrapers will haul is dimensioned on the vertical scale, depicted by the vertical line Q. By examining either the profile view or the mass diagram, it is easy to determine the direction of haul from a cut location to a fill location. Note the arrow on the profile view in Fig. 3.14.

TABLE 3.2 Economical Haul Distances Based on Basic Machine Types

Machine Type	Economical Haul Distance
Large dozers, pushing material	Up to 300 ft*
Push-loaded scrapers	300 to 5,000 ft*
Trucks	Hauls greater than 5,000 ft

*The specific distance will depend on the size of the dozer or scraper.

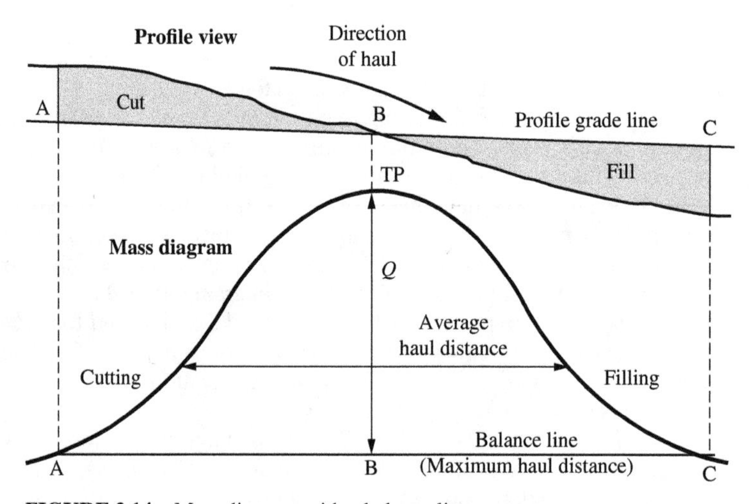

FIGURE 3.14 Mass diagram with a balance line.

In accomplishing the balanced earthwork operation between stations A and C, some of the hauls will be short, while some will approach the maximum haul distance (the distance between point A and point C). The *average haul distance* is approximately the length of a horizontal line placed one-third of the distance from the balance line in the direction of the high or low point of the curve. This is true in the case of a situation like that shown in Fig. 3.14, where the general shape of the mass diagram curve is a triangle. If the situation is similar to that shown in Fig. 3.15, where there are multiple balance lines and the area depicted is basically rectangular, the *average haul distance* is the length of a horizontal line placed midway between the balance lines. See the average haul for scrapers line shown in Fig. 3.15.

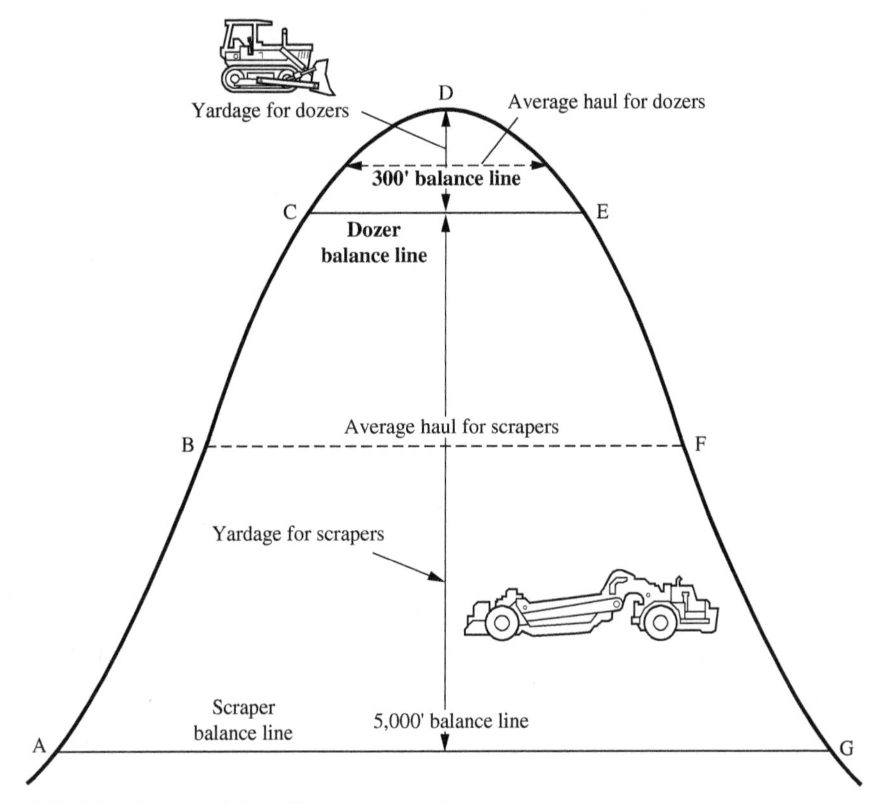

FIGURE 3.15 Two balance lines on a mass diagram.

If the curve is above the balance line, the direction of haul is from left to right (i.e., up stationing). When the curve is below the balance line, the haul is from right to left (i.e., down stationing).

Because the lengths of balance lines on a mass diagram are equal to the maximum or minimum haul distances for the balanced earth-moving operation, they should be drawn to conform to the capabilities of the equipment being proposed for use on the project. Each machine will, therefore, operate at haul distances within its range of efficiency. Figure 3.15 illustrates a portion of a mass diagram on which two balance lines have been drawn. In this situation, dozers will be used to push the short-haul material. Using dozers, the excavation between stations C and D will be placed between stations D and E. Then there will be a scraper operation to excavate the material between stations A and C and haul it to fill between stations E and G.

Average Grade. The project plans will provide a profile drawing of the work. If the mass diagram is plotted under the project profile, as shown in Figs. 3.13 and 3.16, the average haul grades of the earth-moving operations can be approximated. On the profile view, draw a horizontal line to roughly divide the cut area in half in the vertical dimension (see Fig. 3.16, line DE). Do the same for the fill area (line FG). This is a division of only the portion of the cut or fill areas defined by the balance line in question. The difference in elevation between

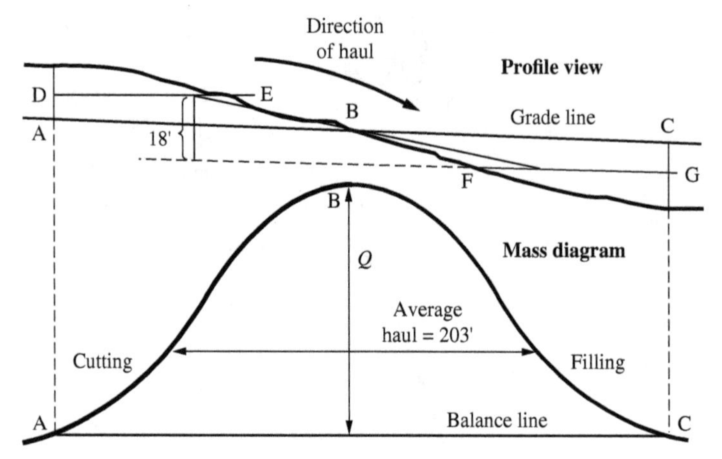

FIGURE 3.16 Using the mass diagram and profile to determine average haul grades.

these two lines provides the vertical distance to use in calculating the average grade for the haul involving the material in the balance. The average haul distance, as determined by construction of a horizontal balance line on the mass diagram, is the denominator in the grade calculation.

$$\text{Average grade (percent)} = \frac{\text{Change in elevation}}{\text{Average haul distance}} \times 100 \tag{3.5}$$

Haul Distances. The mass diagram can be used to determine average haul distances. If the values in column 8 of Table 3.1 (volume of cut) from station 0 + 00 to 8 + 50 are summed, the total is 2,188 bcy. This is the total volume of excavation within the project limits. If the positive values in column 11 (algebraic sum) are summed, the total is 2,049 bcy, which is the total volume of excavation to be moved longitudinally. The difference between these two values is the crosshaul for the project: 139 bcy. The crosshaul material is used between the two adjacent stations where it is generated. It does not move along the length of the project. Many contractors treat the crosshaul as dozer work.

Reviewing the values in column 12, Table 3.1, there is a low point, or valley, of –847 bcy at station 2 + 50 and a high point, or crest, of 1,202 bcy at station 6 + 50. The sum of the absolute values of the peaks and low points equals the total excavation that must be moved longitudinally (847 + 1,202 = 2,049 bcy). The curve can have intermediate peaks and low points (Table 3.3), and all must be accounted for when calculating the amount of material moved longitudinally.

Reviewing the values in column 12, Table 3.3, there is a low point, or valley, of –28,539 bcy at station 5 + 00, a high point, or crest, of –17,080 bcy at 8 + 00, and a second low point of –22,670 bcy at station 10 + 00. The sum of the absolute values of the peaks and low points equals the total excavation that must be moved longitudinally [(28,539 – 17,080) + (22,670 – 17,080) + (17,080 – 0) = 34,120 bcy]. Haul 1 involves 11,459 bcy, haul 2 is 5,590 bcy, and haul 3 is 17,080 bcy.

Calculating Haul Distance. The average haul distance for any of the individual hauls can also be determined by calculation. Dividing the area (stations-cubic yards) enclosed by

TABLE 3.3 Earthwork Volume Calculation Sheet for the Mass Diagram in Fig. 3.17

Station (1)	End-Area Cut (sf) (2)	End-Area Fill (sf) (3)	Volume of Cut (bcy) (4)	Volume of Fill (ccy) (5)	Stripping Cut (bcy) (6)	Stripping Fill (ccy) (7)	Total Cut (bcy) (8)	Total Fill (ccy) (9)	Adj. Fill (bcy) (10)	Algebraic Sum (bcy) (11)	Mass Ordinate (12)
										0.90	
0 + 00	0	0		0							
1 + 00	0	1,700	0	3,148	0	120	0	3,268	3,631	−3,631	−3,631
2 + 00	0	3,100	0	8,889	0	120	0	9,009	10,010	−10,010	−13,641
3 + 00	0	1,500	0	8,519	0	120	0	8,639	9,598	−9,598	−23,240
4 + 00	60	600	111	3,889	60	80	51	3,969	4,410	−4,359	−27,598
5 + 00	400	200	852	1,481	80	60	772	1,541	1,713	−941	**−28,539** ←
6 + 00	1,300	30	3,148	426	110	10	3,038	436	484	2,554	−25,985
7 + 00	2,400	400	6,852	796	120	85	6,732	881	979	5,753	−20,223
8 + 00	800	850	5,926	2,315	90	100	5,836	2,415	2,683	3,153	**−17,080** ←
9 + 00	50	1,250	1,574	3,889	5	120	1,569	4,009	4,454	−2,885	−19,965
10 + 00	95	180	269	2,648	20	10	249	2,658	2,953	−2,705	**−22,670** ←
11 + 00	200	8	546	348	60	0	486	348	387	99	−22,571
12 + 00	560	0	1,407	15	65	0	1,342	15	16	1,326	−21,245
13 + 00	1,430	0	3,685	0	100	0	3,585	0	0	3,585	−17,660
14 + 00	3,580	0	9,278	0	120	0	9,158	0	0	9,156	−8,502
15 + 00	2,600	0	11,444	0	110	0	11,334	0	0	11,334	2,833

the balance line and the mass diagram curve by the amount of material hauled (cubic yards) yields the average haul distance (stations).

The area (stations-cubic yards, usually referred to as station-yards) enclosed by the balance line, the mass diagram curve, and the zero datum line, which is haul 3 (Fig. 3.17), can be calculated using the trapezoidal formula, Eq. (3.3). The vertical value at h_0 (station 0 + 00) is 0 bcy, at h_1 is 3,631 bcy, at h_2 is 13,641 bcy, at h_3 through h_{13} is 17,080 bcy, at h_{14} is 8,502 bcy, and at h_{15} is 0 bcy. For stations 0 + 00 and 15 + 00, the beginning and ending stations in the equation (h_0 and h_n), the value is divided by 2. But 0 divided by 2 is still 0, so the area equals the sum of the verticals times the distance between the verticals, which is 1 (1 station). The sum of the verticals is 213,654 bcy, and multiplying by 1 station

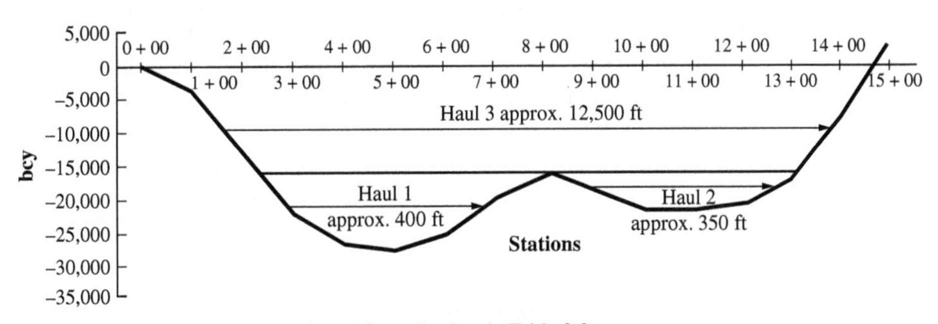

FIGURE 3.17 Mass diagram plotted from the data in Table 3.3.

produces an area of 213,645 station-bcy (sta.-yd). There is a slight error in the computation because the mass diagram curve crosses the zero balance line somewhere between station 14 + 00 and 15 + 00; therefore, the last distance is not 100 ft. But there are no data to determine exactly where the curve and line cross, other than possibly by scaling the mass diagram.

The amount of material hauled is 17,080 bcy. Therefore, the calculated average haul for haul 3 is 12.5 stations, or 12,500 ft.

$$\left(\frac{213{,}654 \text{ station-bcy}}{17{,}080 \text{ bcy}} = 12.5 \text{ stations}\right)$$

Consolidated Average Hauls. Using the individual average hauls and the quantity associated with each, a project average haul can be calculated. The calculation process is similar to the one used to calculate an averaged haul for an individual haul. Consider the three hauls depicted in Fig. 3.17 and their graphically determined average haul distances. Haul 1 is 11,459 bcy with an average haul distance of 400 ft, or four stations. Haul 2 is 5,590 bcy with an average haul of 350 ft, and haul 3 is 17,080 bcy with an average haul of 12,500 ft. By multiplying each haul quantity by its respective haul distance, a station-yard value can be determined.

Haul 1	11,459 bcy	4.0 stations	45,836 sta.-yd
Haul 2	5,590 bcy	3.5 stations	19,565 sta.-yd
Haul 3	17,080 bcy	12.5 stations	213,500 sta.-yd
	34,129 bcy		278,901 sta.-yd

If the individual station-yard values are summed and their total divided by the total quantity moved, an average haul for the project is the result. In this case, the project average haul is 8.17 stations (278,901 sta.-yd / 34,129 yd).

If all the hauls are about the same length, the estimator can consolidate the production calculations by using an averaging process. In the case of the data presented in Table 3.3 and Fig. 3.17, there are two very distinct haul situations on the project. There are two short-haul sections and one long-haul section. Each of these two situations will most likely require a different number and perhaps different type of haul units. The short-haul situation will have shorter haul times and, therefore, will require fewer haul units to achieve continuous production. The long haul will require more haul units. Assuming the haul grades are about the same, the estimator will not calculate production for every individual part of the mass diagram. This is also driven by the result of the practical situation of not mobilizing and demobilizing various numbers of machines for small differences in hauls.

Considering the data in Table 3.3, the estimator would most likely develop two production scenarios. The first scenario would be for the short hauls of hauls 1 and 2.

Haul 1	11,459 bcy	4.0 stations	45,836 sta.-yd
Haul 2	5,590 bcy	3.5 stations	19,565 sta.-yd
	17,049 bcy		65,401 sta.-yd

$$\text{Average Haul 1 and 2: } \frac{65{,}401 \text{ sta-yd}}{17{,}049 \text{ bcy}} = 3.8 \text{ stations}$$

The average haul for the two short-haul situations is 3.8 stations. The second scenario would be for the long-haul situation of haul 3.

Haul 3	17,080 bcy	12.5 stations

On some projects, it can prove economical to borrow material rather than undertake extremely long hauls. The same holds true for wasting material. The mass diagram provides the engineer the ability to analyze options.

Pricing Linear Hauls

The cost of earthwork operations will vary with the kind of soil or rock encountered and the methods used to excavate, haul, and place the material in its final deposition. It is usually not too difficult to compute the volume of earth or rock to be moved, but estimating the cost of actually performing the work depends on both a careful study of the project plans and a diligent site investigation. The site work should seek to identify the characteristics of the subsurface soils and rock that will be encountered.

The earthwork quantities and average movement distances can be determined using an earthwork quantities sheet and a mass diagram. Proper equipment and estimated production rates are determined by (1) selecting an appropriate type of machine (discussed in Chaps. 6 through 11) and (2) using machine performance data (as discussed in Chap. 5).

Spread A group of construction earthwork machines used to accomplish a specific construction task such as excavating, hauling, and compacting material.

Spread Production. To accomplish a task, machines usually work together and are supported by auxiliary machines. To accomplish a loading, hauling, and compacting task would involve an excavator (Fig. 3.18), several haul units, and auxiliary machines to distribute the material on the embankment and achieve compaction (Fig. 3.19).

Such groups of equipment are referred to as an equipment **spread**. An excavator and a fleet of trucks can be thought of as a linked system, one link of which will control the spread production. If spreading and compacting of the hauled material are required, a two-link system is created. Because the systems are linked, the capabilities of the individual components of the spread must be

FIGURE 3.18 A loader used to excavate and load haul trucks.

FIGURE 3.19 A roller used to compact the material on the fill.

FIGURE 3.20 Two-link earthwork system.

compatible in terms of overall production (Fig. 3.20). The number of machines and specific types of machines in a spread will vary with the proposed task.

The production capacity of the total system is dictated by the lesser of the production capacities of individual systems. The objective is to predict the spread production rate (*linked-system* production rate) and the cost per unit of production. In the case of the Fig. 3.17 data, the estimator would develop two production spreads. The first spread would be for the short 3.8-station haul and the second for the 12.5-station haul.

> Always ensure that a consistent set of units is used when estimating spread production. If the mass diagram quantities are expressed in bank cubic yards (bcy), excavator capacity, hauling capacity, and compaction should all be converted to bcy. The units used in the estimate are usually chosen to match those used in the owner's bid documents.

Production Example. Using the information developed in the "Consolidated Average Hauls" section, in particular, from the mass diagram in Fig. 3.17, one equipment spread will have a short 3.8-station haul and the other a long 12.5-station haul. A hoe excavator with the ability to excavate and load 280 bcy per hour will be used. Truck capacity is 12 bcy. Average truck travel speed is 10 mph for the short haul and 15 mph for the long haul. Dump time is 2 min.

$$\text{Time to load a 12-bcy truck:}\quad \frac{12\text{ bcy}}{280\text{ bcy per hr}}\times 60\text{ min/hr} = 2.57\text{ min}$$

$$\text{Time to haul 380 ft:}\quad \frac{380\text{ ft}}{88\dfrac{\text{ft}}{\text{mile}}\times\dfrac{\text{hr}}{\text{min}}\times 10\text{ mph}} = 0.43\text{ min}$$

$$\text{Time to haul 1,250 ft:}\quad \frac{1{,}250\text{ ft}}{88\dfrac{\text{ft}}{\text{mile}}\times\dfrac{\text{hr}}{\text{min}}\times 15\text{ mph}} = 0.95\text{ min}$$

where 88 is the conversion factor to feet per minute

$$\frac{5{,}280\text{ ft}}{\text{mile}}\times\frac{\text{hr}}{60\text{ min}} = 88\frac{\text{ft}}{\text{mile}}\times\frac{\text{hr}}{\text{min}}$$

Total cycle time:

	380-ft Haul	1,220-ft Haul
Load	2.57 min	2.57 min
Haul	0.43 min	0.95 min
Dump	2.00 min	2.00 min
Return	0.43 min	0.95 min
Total	5.43 min	6.47 min

$$\text{Number of trucks required for 380-ft haul:}\quad \frac{5.43\text{ min total cycle time}}{2.57\text{ min load time}} = 2.1\text{ trucks}$$

$$\text{Number of trucks required for 1,250-ft haul:}\quad \frac{6.47\text{ min total cycle time}}{2.57\text{ min load time}} = 2.5\text{ trucks}$$

It is not possible to have a fraction of a truck; therefore, for the 380-ft haul assume 2 trucks will be used and 3 for the 1,250-ft haul situation.

Production 380-ft haul (with only 2 trucks, Table 3.4, and a situation where 2.1 trucks are needed to keep the excavator fully utilized, the truck production capability will control the system production):

$$\frac{2\text{ trucks}\times 12\text{ bcy}\times 60\text{ min per hour}}{5.43\text{ min cycle time}} = 265\text{ bcy per hr}$$

TABLE 3.4 Equipment Spread 380-ft Haul (Production 265 bcy per hour)

Type of Machines	Number of Machines
Excavator—hoe	1
Trucks—12 bcy capacity	2
Dozer to spread the material on fill	1
Compactor	1
Water truck	1

Production 1,250-ft haul [loader controls production because we are using more trucks (3), Table 3.5, than the loader can support (2.5)]: 280 bcy per hour.

TABLE 3.5 Equipment Spread 1,250-ft Haul (Production 280 cy per hour)

Type of Machines	Number of Machines
Excavator—hoe	1
Trucks—12 bcy capacity	3
Dozer to spread the material on fill	1
Compactor	1
Water truck	1

Productivity
A measure of the progress of a crew or contractor against a standard such as a schedule.

Productivity: Time to Complete a Task. The most important job factor the project team must manage is **productivity**. It is the controlling factor in the equation of job cost. The cost of labor, equipment (rental or ownership), and materials is essentially fixed and cannot be changed significantly by the project management team. Productivity is different and must be managed. The effect of production on activity time is given by Eq. (3.6).

$$\text{Time to complete a task} = \frac{\text{Task quanity}}{\begin{array}{c}\text{Task production rate}\\ \text{(minimun linked system production rate)}\end{array}} \tag{3.6}$$

The total cost to complete a task is determined by the equation:

$$C_{\text{total}} = C_{\text{per hour}} \times \text{Time} \tag{3.7}$$

Therefore, if the task production rate is increased, the total cost (C_{total}) is reduced. If the task production rate is decreased, C_{total} is increased. Consider a cost of \$1,000 per hour for a crew and production is varied for producing 100 units:

$$C_{\text{total}} = \$1{,}000/\text{hr} \times \frac{100 \text{ units}}{1 \text{ unit/hr}} = \$100{,}000$$

$$C_{\text{total}} = \$1{,}000/\text{hr} \times \frac{100 \text{ units}}{10 \text{ unit/hr}} = \$10{,}000$$

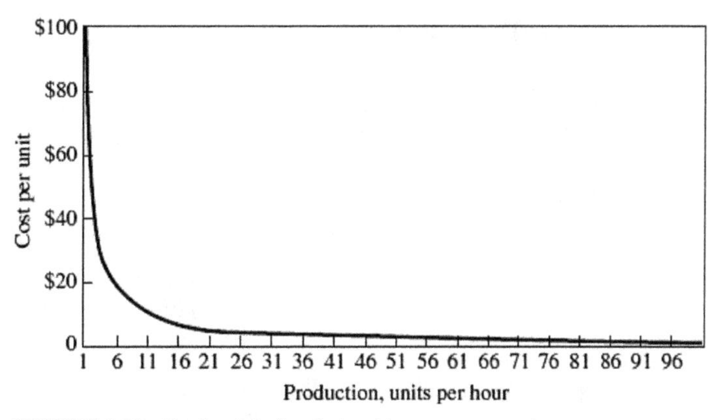

FIGURE 3.21 Productivity's relationship to cost per unit.

The relationship between C_{total} and the production rate is not a straight line. The relationship is inverse exponential, as shown in Fig. 3.21. As the curve approaches the vertical and horizontal axes, it is asymptotic (approaches zero at an infinite distance from the origin).

The next step in the estimating process is to calculate the working time required to complete a work task, as shown in Eq. (3.6).

Based on this information in the "Production Example" section, calculate the time required to complete each task. Based on the mass diagram for the two consolidated hauls, the quantity for the 380-ft haul is 17,049 bcy and for the 1,250-ft haul is 17,080 bcy.

$$\text{Short-haul time} = \frac{17{,}049 \text{ bcy}}{265 \text{ bcy per hr}} = 64.3 \text{ hr}$$

$$\text{Long-haul time} = \frac{17{,}080 \text{ bcy}}{280 \text{ bcy per hr}} = 61.0 \text{ hr}$$

Production Efficiency. In these calculations the equipment is productive for 60 mins during every hour of the workday. This is usually not the case, and each situation must be analyzed very closely. Excavators used for production work are typically kept in good operating condition, and the best operators are assigned to these machines. In the case of the trucks, a breakdown or slow truck will affect production. In the case where 3 trucks are employed and only 2.5 are required, a slow truck would have minimal effect on production. A spread's efficiency will never be 100%, and a reduced number of working minutes per hour is used to reflect production efficiency. More realistically, in this example it would be proper to assume something like 50 mins of productive time per hour. In the production calculation, this efficiency assumption accounts for possible breakdowns and other possible delays. Considering the two-truck situation, the revised production calculation for the short haul would be:

Production 380-ft haul (trucks control production):

$$\frac{2 \text{ trucks} \times 12 \text{ bcy} \times 50 \text{min/hr}}{5.43 \text{min cycle time}} = 220 \text{ bcy/hr}$$

This would also change the time duration to complete the work.

$$\text{Short-haul time} = \frac{17{,}049 \text{ bcy}}{220 \text{ cy/hr}} = 77.5 \text{ hr}$$

EARTHWORK ESTIMATE DEVELOPMENT

The success of a construction company is contingent on an ability to prepare fully detailed and accurate estimates for the cost of performing the work. The cost of earthwork operations will vary with the kind of soil or rock encountered and the methods used to excavate, haul, and place the material in its final deposition. It is usually not too difficult to compute the volume of earth or rock to be moved, but estimating the cost of actually performing the work depends on both a careful study of the project plans and a diligent site investigation. The site examination should seek to identify the characteristics of the subsurface soils and rock at the project site, and identify the depth of the groundwater table.

> Tasks are seldom spelled out in the contract documents but are necessary for evaluating the machine requirements and developing the cost to perform the work.

At the most detailed level, each task is usually related to and performed by a crew. The estimator develops the task description by defining the type of effort required to construct a work item. To complete a work item, multiple tasks may be necessary. Task descriptions should be as complete and accurate as possible to lend credibility to the estimate and aid in later review and analysis.

Components of an Earthwork Estimate

The earthwork quantities and average movement distances can be determined using the techniques described in the "Earthwork Quantities" and "Mass Diagram" sections of this chapter. Proper equipment and estimated production rates are determined by (1) selecting an appropriate type of machine and (2) using machine performance data (as discussed in Chap. 5). The costs to be matched with the selected machines are derived as described in Chap. 2.

Labor. Because it varies so widely from place to place, operator pay is often not included in machine costs when preparing an estimate. Therefore, estimators must be very careful not to leave operator pay out of their cost figures. Supervision at the superintendent level is considered an overhead expense. Foremen may also be charged to overhead, but typically they are carried as an operating expense. An operator is usually paid for more hours than the machine works. A full day's pay is earned in many cases just for reporting, whether the machine runs or not. The operator is certainly kept on the payroll during short delays for adjustment and repair and when standing by during various job delays. The operator may be paid for extra nonoperating time when the machine is serviced. If an operator is paid on an annual salary basis, the wage should be divided by the number of hours the equipment works or is expected to work during the year to obtain an hourly rate. Construction workers usually check in close to the start of work and are usually expected to have their equipment running and ready to go at the start of a shift.

The cost of labor is probably the most variable and difficult to estimate. The local labor market and conditions should be investigated to determine the available supply of all classes of labor and its competence. Local work practices must also be studied to ascertain their effect on productivity.

Direct labor costs are defined as base wages plus labor cost fringes (additives), including payroll taxes, fringe benefits, travel, and overtime allowances paid by the contractor for personnel who perform a specific construction task. The various union crafts in construction usually negotiate their own wage rates and working conditions. Therefore, if using union labor, these must be examined and understood individually. In addition to the actual workers, there are generally working crew foremen who receive an hourly wage and are considered part of the direct labor costs.

Indirect labor costs are wages and labor cost fringes paid to contractor personnel whose effort cannot be attributed to a specific construction task. Personnel such as superintendents, engineers, clerks, and site cleanup laborers are usually included as indirect labor costs (project overhead).

Crews. Direct labor cost requirements are broken into work. A labor crew, including equipment, usually performs each work task; therefore, the crew must be defined, its cost defined, and a production rate established for the task. Crews may vary in size and mix of skills. The number and size of each crew should be based on such considerations as having sufficient workers to perform a task within the construction schedule and the limitation of workspace. Once the crews have been developed, the task labor costs can be determined based on the production rate of the crew and the labor wage rates.

Wage Rates. A wage rate must be determined for each labor craft. This wage rate is the total hourly cost to the contractor. These rates will include the base wage rate plus labor overtime, payroll, taxes and insurance, fringe benefits, and travel or subsistence costs.

Wage rates on government-funded work are generally well defined. The Davis-Bacon Act, PL 74-403, requires a contractor performing construction in the United States for the federal government to pay not less than the prevailing rates established by the Department of Labor. A schedule of minimum rates is included in the project specifications. A higher wage rate should be used instead of the minimum wage when labor is in short supply for certain crafts, the work is in a remote area, or local rates are higher than the set scale. The Bureau of Labor Statistics provides online data on wage rates by areas of the country [www.wdol.gov/dba.aspx]. The wage rate should be adjusted to include travel time or night differential where these are a customary requirement.

Overtime and Shift Differential. The cost estimator should carefully consider the available working time in the construction schedule to accomplish each task.

Shift. A shift is the continuous time (except for breaks and meals) worked by one crew in one day. It is usually 8 hr, but it may be 7, 10, or even 12. The longer shifts usually include an overtime pay rate.

Multiple Shifts. Work may sometimes be speeded by working two or three shifts. Three shifts are commonly 8 hr each, with one crew taking over from another without any shutdown. The day shift is from 8:00 a.m. to 4:00 p.m., the "swing" until midnight, and the "graveyard" until the day gang takes over. Pay time is 8 hr, but a "lunch" period and time lost in the changing of the shifts reduce work time to less than 7½.

Two shifts may be of either 8 or 10 hr each. The job is usually shut down after each shift, for lubrication and repair work. Night work is less efficient than day because of the need for artificial light and the lessened accuracy and usually lower mental and physical vigor of the workers.

Multiple shifts may work at cross-purposes. This difficulty is somewhat less when the second crew arrives before the first departs. If there is no contact, the supervisors should meet in the idle period to discuss the work and coordinate their efforts. There should be a system for rotating workers among the shifts, but it should be administered intelligently. Night shifts are generally unpopular, but some individuals prefer them and they should not be rotated. Swapping of shifts among equally qualified workers should always be allowed. Rotation should be at rather long intervals to enable the workers to adjust to

Fringe benefits These are employee benefits an employer must pay in addition to an employee's base pay. Typically, these may be paid directly to the employee or they may be paid to various agencies on behalf of the employee. They would include health insurance, pension plans, and certain taxes.

changes in sleep and work hours. Two weeks is the shortest period that should be considered.

The productivity of a second or third shift should be adjusted to recognize the loss in efficiency compared to the day shift. A three-shift operation is normally avoided because of lower labor efficiency and the requirement to include equipment maintenance.

Overtime should be included in the labor cost computation when work in excess of 40 hr is required by the construction schedule or is the custom of labor in the local vicinity. Overtime is normally calculated as a percentage of the base wage rate. It is usually based on time and one-half, but may be double time, depending on the existing labor agreements. Tax and insurance costs are applied to overtime, but **fringe benefits** and travel and/or subsistence costs are not.

Taxes and Insurance. Wages paid to employees are subject to payroll taxes such as the Federal Insurance Contributions Act (FICA), Federal Unemployment Tax Act (FUTA), and state unemployment taxes. The FICA tax is for Social Security and Medicare, and FUTA is the federal employer tax used to fund state workforce agencies. Rates for all taxes and insurance should be verified prior to computation for these and any additional components.

Workman's compensation and *employer's liability insurance* costs applicable for the state in which the work is performed should be included in the composite wage rate. The project compensation rate is based on the classification of the major construction work and applies to all crafts employed.

Unemployment compensation taxes are composed of both state and federal taxes. Unemployment compensation tax will vary with each state, and the federal unemployment tax will be constant for all projects. Insurance rates can be obtained from the state unemployment office.

The calculation, however, is not simple because the percentages only apply to certain limited amounts of the employee's wages. As an example (for 2017) the FUTA tax rate is 6,0%. The tax applies only to the first $7,000 paid to each employee as wages during the year.

Additionally, employers who pay the state unemployment tax on a timely basis receive an offset credit of up to 5.4%, regardless of the rate of tax they pay the state. Therefore, the net federal tax rate is generally 0.6% (6.0% − 5.4%). This would equate to a maximum of $56.00 per employee, per year ($.006 \times \$7{,}000. = \$42.00$) in federal FUTA tax.

Fringe Benefits. Fringe benefits may include health and welfare, pension, and apprentice training depending on the craft and the location of the work. These summed costs are usually expressed as an hourly cost with the possible exception of vacation, which may be easily converted to an hourly cost. On federally funded projects, the type of fringe benefits and the amount for the various crafts can usually be found with the Department of Labor wage determination in the specifications. Nonunion contractors pay comparable fringe benefits directly to their employees.

Some fringe benefits and travel/subsistence costs are subject to payroll taxes. For example, vacation benefits are taxable and should be added to the basic wage rate.

Cost to Complete a Task. Contractors usually have their own standard formats for displaying cost estimate information in segregated and accumulated columns and rows. A representative format is presented in Tables 3.6 and 3.7 using the previously discussed short and long hauls from the mass diagram. The tables illustrates how to present important cost information.

TABLE 3.6 Estimate for the Short-Haul Task

Task	17,049-bcy Short Haul		3.8 Stations							
Gross Production Rate		265 bcy/hr								
Efficiency		50 min-hr								
Net Production Rate		220 bcy/hr		Total Hours		77.5 Hours				
		Unit cost								Unit Cost
	No.	Ownership	Operating	Labor	Other	Equipment	Labor	Other	Total	$/bcy
Foreman w/pickup	1	$2.98	$11.50	$70.20		$1,122.13	$5,440.18	$0.00	$6,562.32	$0.38
Hoe 3½	1	$103.98	$124.00	$53.83		$17,667.41	$4,171.58	$0.00	$21,838.99	$1.28
Dump Truck	2	$11.51	$27.80	$42.00		$6,092.69	$6,509.62	$0.00	$12,602.31	$0.74
Dozer	1	$87.50	$86.45	$49.50		$13,480.33	$3,836.03	$0.00	$17,316.36	$1.02
Compactor	1	$54.26	$58.10	$43.50		$8,707.39	$3,371.05	$0.00	$12,078.44	$0.71
Water Truck	1	$16.90	$9.80	$42.00	$5.00	$2,069.13	$3,254.81	$387.48	$5,711.42	$0.34
Spotter	1			$39.00		$0.00	$3,022.32	$0.00	$3,022.32	$0.18
					Total	$49,139.09	$29,605.59	$387.48	$79,132.16	$4.64
					Unit cost	$2.88	$1.74	$0.02	$4.64	
					Percentage of cost	62%	37%	0%		

TABLE 3.7 Estimate for the Long-Haul Task

Task	17,080-bcy Short Haul		12.2 Stations							
Gross Production Rate		280 bcy/hr								
Efficiency		50 min-hr								
Net Production Rate		233 bcy/hr		Total Hours		73.2 Hours				
		Unit Cost								Unit Cost
	No.	Ownership	Operating	Labor	Other	Equipment	Labor	Other	Total	$/bcy
Foreman w/pickup	1	$2.98	$11.50	$70.20		$1,059.94	$5,138.64	$0.00	$6,198.58	$0.36
Hoe 3½	1	$103.98	$124.00	$53.83		$16,688.14	$3,940.36	$0.00	$20,628.49	$1.21
Dump Truck	3	$11.51	$27.80	$42.00		$8,632.48	$9,223.20	$0.00	$17,855.68	$1.05
Dozer	1	$87.50	$86.45	$49.50		$12,733.14	$3,623.40	$0.00	$16,356.54	$0.96
Compactor	1	$54.26	$58.10	$43.50		$8,224.75	$3,184.20	$0.00	$11,408.95	$0.67
Water Truck	1	$16.90	$9.80	$42.00	$5.00	$1,954.44	$3,074.40	$366.00	$5,394.84	$0.32
Spotter	1			$39.00		$0.00	$2,854.80	$0.00	$2,854.80	$0.17
					Total	$49,292.88	$31,039.00	$366.00	$80,697.88	$4.72
					Unit cost	$2.89	$1.82	$0.02	$4.72	
					Percentage of cost	61%	38%	0%		

Estimating sheets similar to Tables 3.6 and 3.7 are usually generated by special estimating programs or by using a spreadsheet program. The task quantity and production rate are first calculated and then entered into the program. The machine unit costs are extracted from the company's equipment database and then multiplied by the estimated number of use or onsite hours to generate total machine costs. Labor costs are generated in a similar manner; however, because labor rates often vary from job to job, it is necessary to establish a labor rate database for each job. The labor rate used in the estimate should be a fully burdened rate: include payroll taxes, fringe benefits, travel, and overtime allowances. To generate the cost in Table 3.6 for the hoe, a machine ownership rate of $103.98 per hour and an operating rate of $124 per hour are used. These are often kept separate, as it is not uncommon to have standby machines on a job. In such a case only the ownership rate would be charged for time on the project. The labor hourly charge for this example includes a labor burden of 30%. The "Other" column is used for consumable materials. In this case, it is assumed water for compaction will be purchased. If permanent materials were involved, the table would have a material column.

Reviewing the Estimate. When reviewing an estimate, total cost is a necessary piece of information, but because experienced estimators usually have a feel for what certain types of work cost on a unit price basis, it is good practice to display unit costs. It is easier to notice an unreasonable unit cost number. When an item cost appears unreasonable, management should work backward through the item estimate, checking it in detail.

Comparing the task unit costs in Tables 3.6 and 3.7, there is a difference of $0.08 ($4.72 – $4.64) between the long- and short-haul cases. The unit cost is greater for the long-haul situation because of the greater haul distance and the need for a third truck. In the case of the Table 3.7 estimate, the haul is longer and, to maintain the production rate, a third truck was added to the spread, but this means production is controlled by the hoe. The variance in trucking cost per bcy is $0.31 ($1.05 – $0.74), but all other costs have decreased in magnitude because of a slight increase in production. The hoe is now fully utilized and its cost reduced to $0.02; it does not experience idle time waiting for trucks. A difference of $0.08 per cubic yard may seem small but on a project involving large yardages, even a single cent is significant in total cost terms—this is a difference experienced for every cubic yard moved. Assuming both tasks involved 17,000 bcy, the total difference would be $1,360. Some contractors carry earthwork estimates to three places past the decimal—a thousandth of a dollar.

Mobilization and Demobilization. Mobilization costs for equipment include the cost of loading at the contractor's yard; transportation to the construction site, including permits for the haul; unloading at the site; necessary assembly and testing; and standby costs during mobilization and demobilization. All labor, equipment, and supply costs required to mobilize the equipment should be included in the mobilization cost. Demobilization costs should be calculated in a similar manner for all the contractor expects to return to the storage yard. If it is expected that the machines will be moved directly to another project, the mobilization for the following project covers the demobilization expense.

Mobilization and demobilization costs for plant should be based on the delivered cost of the item, plus erection, taxes, and dismantling costs at the end of the project. Maintenance and repair are operating costs and should be distributed throughout work accomplishment.

SCHEDULING

The goal of planning is to minimize resource expenditures while satisfactorily completing a given task. Planning aims at producing an efficient use of equipment, materials, and labor, and ensuring coordinated effort. Effective planning requires continual monitoring of task accomplishment—progress. A comparison of actual progress to scheduled progress helps the manager identify problems early and permits development of revised plans to maintain the proper course toward the objective.

Activities

A common technique used to understand and organize a complex and multidimensional undertaking is to break the project into smaller pieces. In the terminology of scheduling, tasks or work items are referred to as activities. Each activity is a discrete task.

Planning involves the scheduling of activities. But management also includes monitoring and controlling the execution of the work; therefore, the activities provide the building blocks for focusing management attention.

According to the Associated General Contractors of America (AGC), activities have five characteristics:

1. *Activities consume time.* To be classified as an activity, a project task must consume time.
2. *Activities usually consume physical resources.* A project activity usually consumes labor, material, or equipment resources. However, there are a few construction activities, such as curing for concrete, that may not consume physical resources.
3. *Activities have definable start and finish points.* An activity represents a definable scope of work; that work has a starting point in time and, when all of the work is completed, an ending point in time.
4. *Activities are assignable.* Each activity is to be accomplished by a particular member of the construction team. This characteristic of an activity facilitates management of the work.
5. *Activities are measurable.* The progress toward completion of the activity's scope of work must be measurable.

An activity is therefore a task, function, or decision that *requires a time duration.*

Activity Duration. One of the most important steps in planning a project is estimating the time required to complete each activity. The duration of an activity is a function of the quantity of work to be accomplished and the work production rate. Work production rates are based on the planned composition of labor and equipment used to perform the task. Carelessly made estimates of production rates may cause uneconomical use of personnel, materials, equipment, and time.

When a contractor prepares a project estimate for bidding, the estimator calculates the quantities of material required and assumes a production rate to use in arriving at the cost for each work task. In estimating production rates, the estimator considers construction methods and techniques. The selected method establishes the task duration. This bid preparation information can serve as the base for calculating an activity duration.

The time unit used to express an activity's duration should match the objective of the scheduling effort. It is more common to use weeks, days, or hours as the activity time unit. In the case of a long-duration project, the use of a time unit expressed in weeks might be appropriate. For the majority of projects, the activity time unit is days. This use of days usually matches the terms in the construction contract, as it is common practice to express contractual time in either work or calendar days.

All activities in the schedule must have their duration expressed in the same time units.

Bar Charts

The bar or Gantt chart is a commonly used project planning and control tool. A bar chart presents planned activities as stacked horizontal bands against a background of dates (along the horizontal axis). Almost all of today's commercially available scheduling software can present information in bar chart formats. It is a simple and concise graphical picture for managing a project. It is also easy with a bar chart to compare planned production against actual production.

The bar chart is widely used as a construction-scheduling tool because of its simplicity, ease of preparation, and graphical format. Normally the activities are listed in chronological order according to their start date. Discontinuous bars are sometimes used on hand-drawn bar charts to represent interruptions to activity work. The better practice is to have individual activities with their own bars.

Field personnel can easily understand the schedule information presented on a bar chart—this is an important advantage for using the technique. Additionally, it is a very useful tool for preliminary planning and scheduling. If cost, equipment, or personnel requirements are superimposed on the activities, the total resources required at any time can be computed.

When a stand-alone bar chart is not generated from a critical path method diagram, the major disadvantage is that the user must have detailed knowledge of the particular project and of construction techniques.

Other disadvantages of planning solely with a bar chart are

- It does not clearly show the detailed sequence of the activities
- It does not show which activities are *critical* to the successful, timely completion of the project

Critical Path Method

The critical path method (CPM) focuses management's attention on the relationships between critical activities. It is an activity relationship representation of the project. The evaluation of critical tasks—those that control project duration—allows for the determination of project duration. Because of the size and complexity of construction projects, CPM scheduling is most often applied using computer software to make the calculation.

The CPM identifies which activities must be completed before other activities can begin.

The critical path method overcomes the disadvantages of a bar chart and provides an accurate, timely, and easily understood graphic of the project. One of the most important features of the CPM is the logic diagram. The logic diagram graphically portrays the relationships between project activities. With this additional information, it is easier to plan, schedule, and control the project.

> CPM calculations define a time window within which an activity can be performed without delaying the project.

Using the CPM to develop a schedule requires detailed investigation into all identifiable project tasks. This requires the manager to visualize the project from start to finish and estimate time and resource requirements for each task. It is good practice to also obtain input from superintendents and subcontractors. The advantages of using a CPM logic network include the following:

- Reduces the risk of overlooking essential tasks and provides a blueprint for long-range planning and coordination of the project work.
- Provides a graphic of the interrelationships between project activities (tasks).
- Focuses management's attention by identifying the critical tasks that control project duration.
- Generates information about the project so management can make rational and timely decisions if complications develop during the progression of the work.
- Enables the project manager to easily determine what resources are needed to accomplish the project and when the resources must be made available.
- Allows the project manager to quickly determine what additional resources will be needed if the project must be completed earlier than originally planned.
- Provides feedback about a completed project that empowers the project manager and the estimator to improve techniques and ensure the best use of resources on future projects.

A CPM is not a substitute for appropriate construction knowledge and planning. It does not make decisions for the project manager, nor can it contribute anything tangible to the actual construction. The greatest danger or disadvantage of a CPM is the ability of anyone with software proficiency to construct what appears to be a reasonable schedule. Unfortunately, there is no way to tell by looking at the plotted schedule whether this is a true picture of physical construction relationships or simply an attractive graphic.

Schedules are produced to aid in managing the work, and the information contained in the schedule must be conveyed to many project participants (e.g., subcontractors, suppliers, and foremen of work crews).

> **Predecessor activity** An activity that must be completed before a given activity can be started.

Activity Logic Network. The activity logic network benefits the manager by providing a graphical picture of the sequence of construction tasks. Before the diagram can be developed, the project must first be constructed mentally to determine activity relationships. The manager does this by asking the following questions for each activity on the activity list:

- Can this activity start at the beginning of the project? (Start activities)
- Which activities must be finished before this one begins, known as **predecessor activity**(ies)? (Precedence)

Successor activity An activity that cannot start until a given activity is completed.

- Which activities may either start or finish at the same time this one does? (Concurrence)
- Which activities cannot begin until this one is finished, known as **successor activity**(ies)? (Succession)

In some cases, the sequencing of activities is clear; they are dictated by a spatial sequence—perform stripping before beginning excavation and fill tasks. However, in other cases, the sequencing may be dictated by contractual conditions that are not always clear by a study of the project plans. In many cases, the manager must carefully study all project documents. Sometimes when estimators develop a schedule, they build the sequencing of activities based on an assumed equipment constraint, such as having only one shovel available.

Node On an activity-on-arrow schedule, the nodes mark the start and end points of an activity. With precedence diagramming, the nodes represent the activities of the schedule.

One way to determine these relationships is to list all activities in a columnar format with a second column to the right of the activities list titled "Preceded Immediately By (PIB)." For each activity, use the second column to list all other activity numbers (identifier codes) that must *immediately* precede the activity in question. If the activity can begin at the beginning of the entire project, write "None."

It is common practice and a necessity for most CPM algorithms to work that a network have only *one start* and *one finish* activity. Therefore, in the case of a network with multiple activities with no predecessor, a dummy "start" activity can be inserted at the beginning of the network. Similarly, a dummy "finish" activity is placed at the end of a network, and all activities that have no successor activity are tied to this final activity.

Dummy activities A pseudo-activity with a duration of zero. A dummy is a dotted line arrow and is used solely to indicate sequence.

There are two CPM logic-diagramming formats: (1) activity-on-the-arrow (AOA) and (2) activity-on-the-node (AON). With the AOA format (Fig 3.22), the arrows of the diagram represent the activities of the logic diagram. Unless the diagram is time scaled, the length of the arrows has no relationship to activity time durations. The **nodes** (events) on an AOA diagram serve only as connecting points for activities. When using the AOA diagramming format, it is often necessary to use **dummy activities** (arrows) to denote dependencies. Dummy activities consume no resources. AOA diagramming is sometimes referred to as I–J diagramming, with the I and J referring to the beginning and ending nodes, respectively, of an activity arrow. The logic of the diagram is that activities (arrows) leaving a node cannot begin until all of the activities heading into that node have been completed.

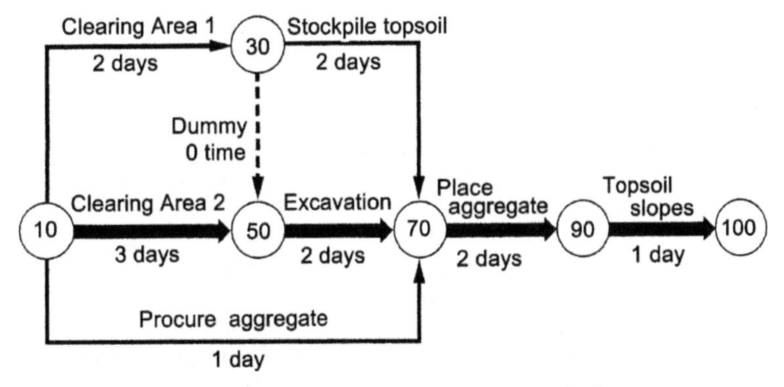

FIGURE 3.22 Example of an activity-on-the-arrow network diagram.

Activity-on-the-arrow logic diagramming was once the most popular CPM method. Today most CPM users employ the AON format, where each node represents an activity. This is sometimes referred to as "precedence diagramming." The two basic logic symbols on the precedence diagram are the *node* and precedence *arrow*, but what they represent is reversed from the AOA case.

Nodes. A node is simply a parallelogram used to represent an activity, and each activity on the activities list is represented by a node on the logic diagram. The node is of a standard shape and format and contains all the necessary activity information. Notations on each node indicate the activity's alphanumeric identifier (ID code) and duration. Sometimes additional information, such as early and late start times, early and late finish times, and required resources are also included (Fig. 3.23). Each activity in a network should have a unique identifier.

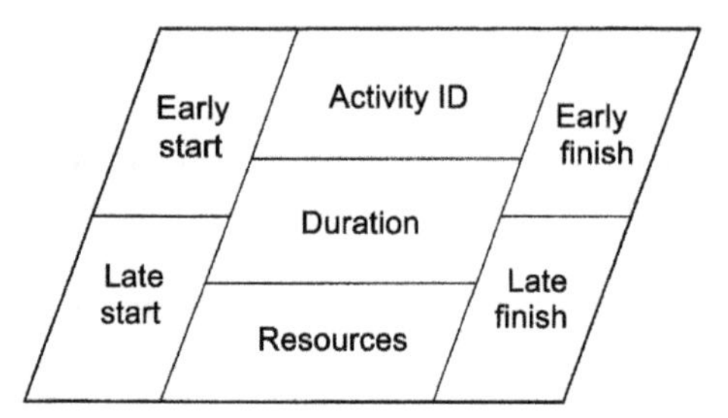

FIGURE 3.23 Example of activity node information; the specific arrangement will depend on the scheduling software used.

Activities are identified by user-defined codes. These aid in describing activities and are essential for sorting and categorizing activities when using computer-based scheduling. Each activity can have one or more activity codes. Common code categories are

- Responsibility
- Area of work
- Phase of project
- Subphase
- Type of work
- Specification section
- Bid item

When creating activity identifiers, the practice is to use a numbering system with the magnitude of the numbers increasing when moving from project start to finish. Reading of project status reports generated from the schedule is much easier when directional numbering is used. It is also the practice when first numbering activities to leave numbering gaps (e.g., 5, 10, 15, etc.). This permits easy inclusion of additional activities as planning progresses or when change orders are issued.

Precedence Arrow. The precedence arrows show the order sequence and relationship between activities (such as what activities must precede and follow another activity). The configuration of the diagram's nodes and arrows is the result of the PIB list (or the answers to the five questions previously stated concerning each activity). By the diagram logic, an

TABLE 3.8 Activity Logic Relationships

Logic Relationships	Diagram Example
SEQUENTIAL LOGIC Activity 20 cannot start until activity 10 is completed.	10 → 20
CONCURRENT LOGIC Activities 5 and 10 can proceed concurrently. **Multiple Predecessor Logic** Activity 20 cannot start until both activities 5 and 10 are completed.	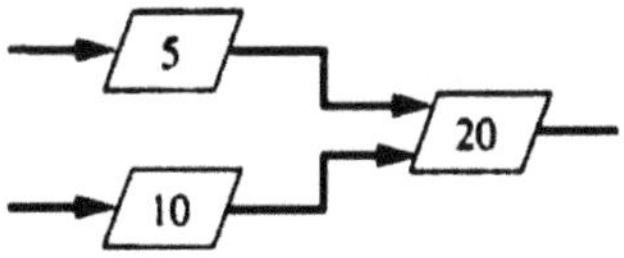 5, 10 → 20
Activities 30 and 40 cannot start until both activities 10 and 20 are completed.	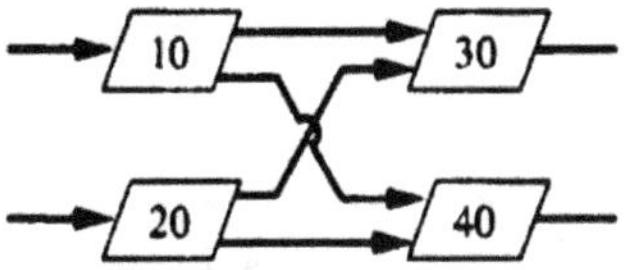 10, 20 → 30, 40
Multiple Successor Logic Activity 20 must be completed before either 30 or 40 can start, 30 can start only after 10 and 20 are completed, 40 can start immediately after 20 is completed.	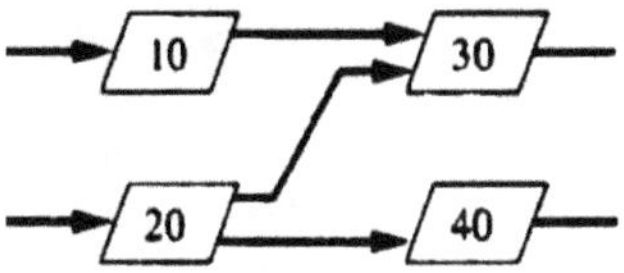 10, 20 → 30; 20 → 40

activity cannot begin until all preceding activities are complete. Table 3.8 presents several common logic relationships.

The logic network is constructed without regard to how long an activity will last or whether all necessary resources are available. It simply displays the relationships among activities, provides project understanding, and improves communications. Once the logic network has been developed, the manager together with knowledgeable superintendents and others places activity duration and resource requirements with each activity.

Schedule Calculations. The next step in the CPM process is to calculate the earliest and latest times at which the activities can occur without violating the network logic or increasing the project's overall duration. This provides the manager with a *time frame* for each activity. Within this time frame, the activity must be completed, or other activities will be delayed and then there is the possibility of creating a ripple effect and delaying the entire project. Once the activity times are calculated, it will be possible to identify the critical path activities. By carefully managing these critical activities, it is possible to reduce the time to complete the project. An activity cannot begin until all activities previous to it (arrows leading to it in the logic node) are completed.

Forward pass
A schedule calculation to determine the earliest start and finish time of the precedence diagram activities and the minimum project duration.

Forward Pass. A forward computational sequence through the logic network will yield this information:

- The earliest time each activity in the network can start and finish.
- The minimum overall duration of the project.

When performing the **forward pass** calculations, all successor activities are started as early as possible. The calculations maintain the rigor of the network logic; therefore, the earliest an activity can start is when all predecessors to the

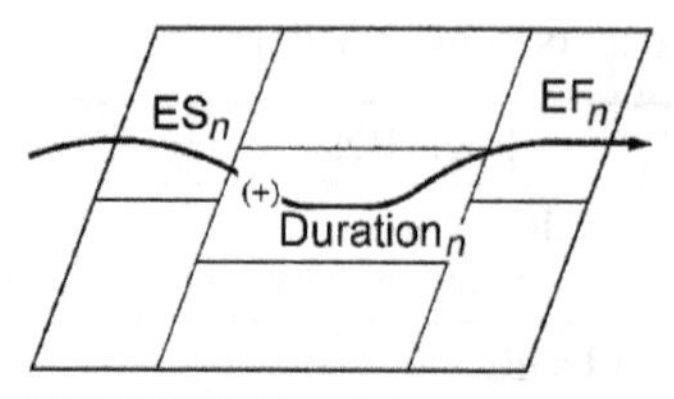

FIGURE 3.24 Calculation of an activity's early finish.

activity are completed. Furthermore, each activity is postulated to finish as soon as possible. This finish requirement yields Eq. (3.8) (Fig. 3.24).

$$\text{Early finish}_n\ (\text{EF}) = \text{Early start}_n\ (\text{ES}) + \text{Duration}_n \tag{3.8}$$

where n denotes the nth activity.

Early Start/Early Finish. The *early start time* (ES) of an activity is the earliest point in time when an activity may start, taking into account the network logic. For this text and discussion, the early start time is positioned in the upper-left corner of the activity node box. Some scheduling software may position or display this information in another location.

The starting point for performing a forward pass is the first activity in the network. In the case of the first activity of a project, the earliest time it may start is zero (the end of day 0 or the beginning of day 1).

> The event-time numbers calculated for an activity represent the end of the time period. Thus, a start or finish time of day 5 would mean the end of the fifth day (or the beginning of the sixth day).

If only one precedence arrow leads into an activity, the activity's early start time is the same as the previous activity's early finish time (at the tail of an arrow). To determine an activity's early start time when more than one arrow leads into its node, select the *largest* early finish time of all activities at the tail of the arrows [Eq. (3.9)]. Logically, an activity cannot begin until *all* preceding activities are complete (Fig. 3.25).

$$\text{Early start}_n = \text{Maximum early finish of all predecessor activities} \tag{3.9}$$

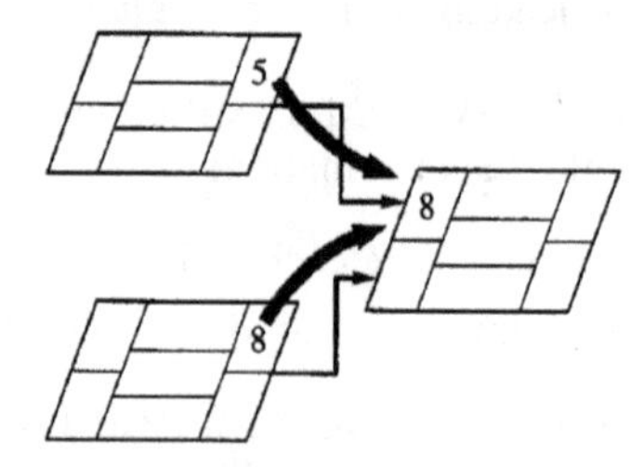

FIGURE 3.25 Calculation of early start, use largest preceding early finish.

Add the duration of each activity (center of the node in this nomenclature) to the early start time to compute the *early finish time* (EF) [Eq. (3.8)]; position the EF in the upper-right corner of the activity node. The early finish time is the earliest time the activity may finish.

Using this systematic process, work through the entire logic diagram, computing all early start and early finish times from the beginning activities to the finish of the project. This computational sequence through the logic diagram completes the forward pass. The overall duration for the project will be the EF of the last activity in the network. For the network shown in Fig. 3.26, the forward pass yields a project duration of 22 days, as determined by the sequence of construction and the time duration assigned each activity.

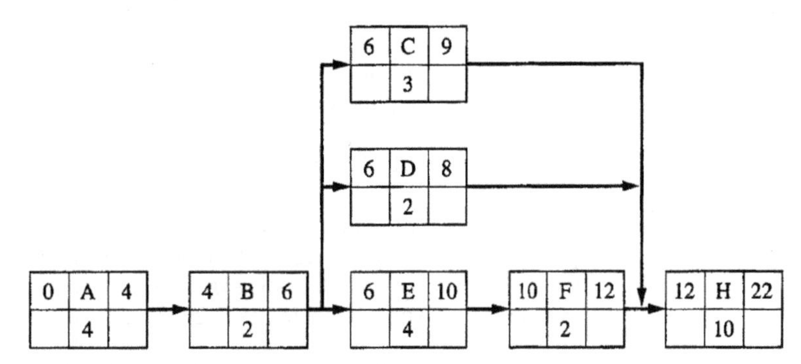

FIGURE 3.26 Forward pass calculations.

Backward pass
A schedule calculation that determines the late start and late finish times of the precedence diagram activities under the condition that the project's minimum duration be maintained.

Backward Pass. A backward computational sequence through the logic network will produce the latest point in time when each network activity can start and finish and still maintain the minimum overall project duration. The **backward pass** calculation starts with the last activity in the network. This last activity is assigned a late finish time equal to its early finish time as calculated by the forward pass.

Late Finish/Late Start. To calculate *late finish time* and *late start time* of an activity, follow the precedence arrows backward through the logic diagram (right to left). The late finish time (LF) of an activity is the latest point in time when an activity may finish without delaying the entire project, taking into account the network logic. For this text, the late finish time is positioned in the lower-right corner of the activity node box. Because the last activity in the Fig. 3.26 network (activity H) had an early finish of 22 days, the latest time when this activity can finish and not lengthen the project duration is the end of the 22nd day. The number 22 should therefore be assigned as the LF for this network and placed in the lower-right corner of the H node.

The preceding activity's late time (at the tail of the precedent arrow) is the succeeding activity's late start time (at the head of the precedent arrow). To determine an activity's late finish time when more than one arrow tail leads away from its node, choose the *least* late start time of all activities at the arrows' heads (Fig. 3.27). Logically, an activity must finish before *all* follow-on activities may begin.

To compute the late start time (LS) of an activity use Eq. (3.10), which says to subtract the activity's duration (center of the node) from its late finish time (Fig. 3.28). Position the

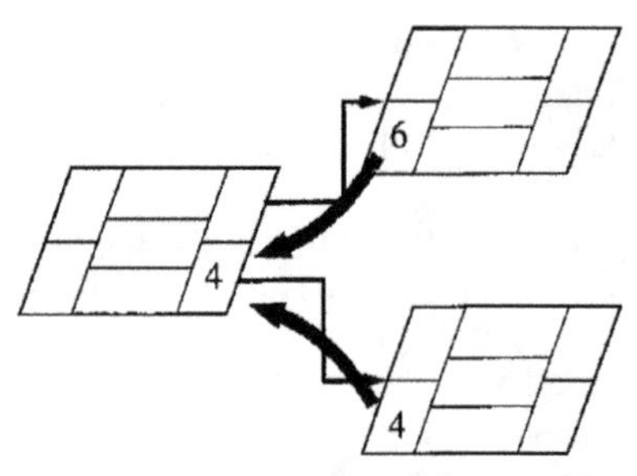

FIGURE 3.27 Calculation of an activity's late finish; use the smallest late start time of all succeeding activities.

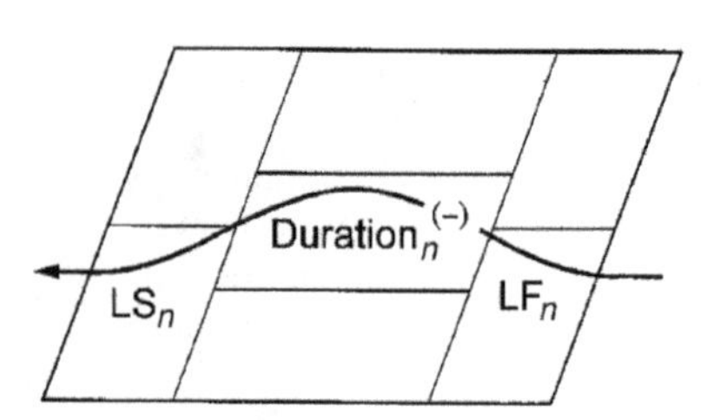

FIGURE 3.28 Calculation of an activity's late start.

LS in the lower-left corner of the activity node. The late start time is the latest time the activity may start without delaying the entire project.

$$\text{Late start}_n\text{ (LS)} = \text{Late finish}_n\text{ (LF)} - \text{Duration}_n \tag{3.10}$$

Using this backward systematic process, work through the entire logic diagram (against the arrows), computing all late finish and late start times. This computational movement back through the logic diagram is known as the *backward pass*. The late start times of the first activity must be zero. For the network shown in Fig. 3.29, the late start time of activity A is zero.

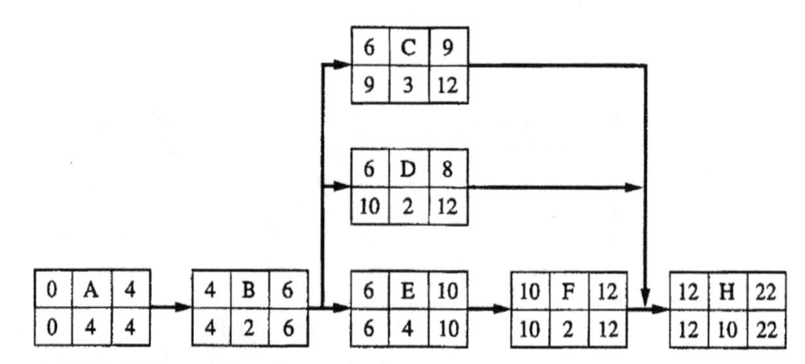

FIGURE 3.29 Backward pass calculations.

Critical Path and Critical Activities. The critical path through a schedule network is the longest time duration path through the network. It establishes the minimum overall project time duration. All activities on the critical path are, by definition, critical. A critical activity can be determined from the logic network by applying either of these rules:

- The early start and late start times for a particular activity are the same.
- The early finish and late finish times for a particular activity are the same.

For the network shown in Fig. 3.29, activities A, B, E, F, and H meet the listed rules, thus making them *critical activities*. A critical activity, if delayed by any amount of time, will delay the project's completion by the same amount of time. Critical activities are linked together, forming a path from the start activity to the finish activity called a *critical path*.

Float The time flexibility of activity performance that states the maximum allowable for not delaying a following activity or the project.

There can be more than one critical path through the network, and the critical paths may branch out or come back together at any point. All critical paths must be continuous; any critical path that does not start at the start node and end at the finish node indicates a logic error.

Critical paths are indicated on the logic diagram by methods such as double lines, bold lines, or color highlighted lines (Fig. 3.30). Any activity node not on the critical path will contain **float.** *Float* is additional time available to complete an activity beyond the activity's work duration, such as having 6 days available to do 4 days of work. Activities on the critical path have no float.

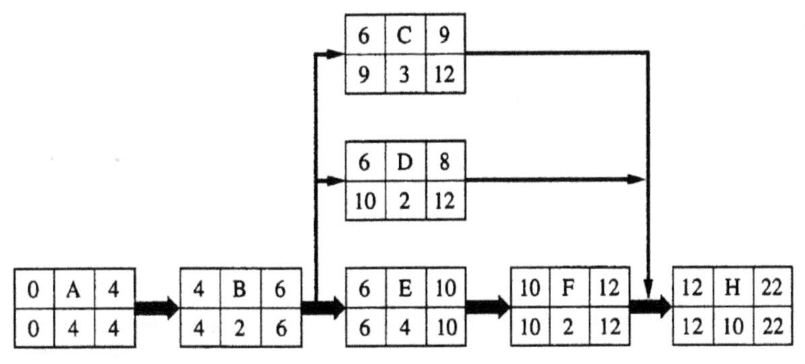

FIGURE 3.30 The critical path through a network.

Total float The amount of time an activity can be delayed without delaying project completion.

Total Float. **Total float** (TF) is the time duration an activity can be delayed without delaying project completion. Total float assumes all preceding activities are finished as early as possible and all succeeding activities are started as late as possible.

Total float for an activity can be determined by either Eq. (3.11) or Eq. (3.12).

$$\text{Total float activity}_n = \text{Late start activity}_n - \text{Early start activity}_n \tag{3.11}$$

$$\text{Total float activity}_n = \text{Late finish activity}_n - \text{Early finish activity}_n \tag{3.12}$$

where n denotes the nth activity.

Both equations yield the same result.

Total float is calculated and reported individually for each activity in a network. However, total float is a *network path attribute* and not associated with any one specific activity along the path.

Free float The time duration that an activity can be delayed without delaying the project's completion and without delaying the start of any succeeding activity.

Free Float. **Free float** is the time duration of activity delayed possible without affecting the project's completion and *without delaying the start of any succeeding activity*. Free float is the property of an activity, not a network path. Free float for an activity is determined by the equation

$$\text{Free float}_n = \text{Minimum early start of all successor activities} - \text{Early finish}_n \tag{3.13}$$

where n denotes the nth activity.

Interfering float The time available to delay an activity without delaying the project's completion, but delaying an activity into interfering float causes the delayed start of one or more following noncritical activities.

Activity C of the network shown in Fig 3.30 is not on the critical path. Therefore, the activity has float. Free float would be 3 (minimum early start of successor, 12 – EF of 9).

Interfering Float. **Interfering float** is the delay time available for an activity that will not delay the project's completion, but such activity delay will delay the start of one or more following noncritical activities. Interfering float for an activity is determined by Eq. (3.14).

$$\text{Interfering float}_n = \text{Late finish}_n - \text{Smallest early start of succeeding activity(ies)} \tag{3.14}$$

where n denotes the nth activity.

If, by the network logic, the activity in question has more than one successor, use the smallest successor ES in Eq. (3.14).

The aggregate of *free float* and *interfering float* equals the total float.

$$\text{Total float}_n = \text{Free float}_n + \text{Interfering float}_n \tag{3.15}$$

where n denotes the nth activity.

Calendar Date Schedule. If all activities are scheduled to begin on their ES dates, the schedule is known as an *early start schedule*. This logic diagram schedule provides the manager with considerable planning information. But everything to this point was done based on workday time durations. To communicate schedule information in a meaningful manner, the workday schedule needs to be converted to a calendar schedule with points in time, such as an activity start expressed as a specific date. If done manually, this is a simple counting procedure; if the schedule was created using a software package, the conversion can be done automatically.

Project start dates are usually specified in the contract documents. Other decisions or contract requirements that will affect the calendar conversion are

- *Workdays per week.* This may be specified in the contract or it can be a contractor decision.
- *Holidays that will be observed.* Again this may be specified in the contract or it can be a contractor decision.
- *Weather days.* These are a function of specific work activities, the climatic conditions at the project location, and when during the seasons the project begins.

Scheduling Activities. The calculations for the initial CPM network assume all activities will start on their early start dates, but this is not typically what happens during the daily circumstances of construction. What the CPM really does is provide the project manager with the information needed to schedule the project activities in a manner consistent with multiple project constraints. If the resource information is included with each activity, it is possible to determine the magnitude of resource requirements for each day of the project.

An activity can take place at any time within the interval defined by its early start and late finish dates and still not delay the project. This provides flexibility in when an activity is actually accomplished. Using this flexibility, a manager can exercise control over the magnitude of *resource requirements* needed on a given day. A simple scheduling process can be illustrated graphically using the activities in the Fig. 3.31 network. The critical path for this network is emphasized by heavier activity connecting arrows.

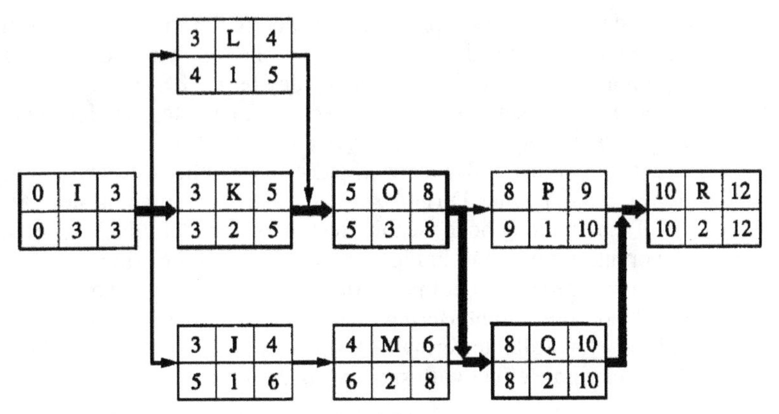

FIGURE 3.31 Scheduling network activities.

The first step is to create a bar chart of the network with all activities listed in identifier order. Ordering the activities by ID sequence makes it easier to track successor activities. After each activity, place a reminder note in parentheses of all immediately dependent activities (see the activity column of Fig. 3.32). For example, because activities P and Q cannot begin until activity O is completed, annotate activity O in the schedule like this: O (P, Q). The next step is to mark on the bar chart the time frame for each activity during which it may be performed without delaying the project or violating any of the diagram logic (sequence) relationships (see Fig. 3.32).

	Workdays											
Activity	1	2	3	4	5	6	7	8	9	10	11	12
I (J, K, L)	2	2	2									
J (M)				1	X	X						
K (O)				3	3							
L (O)				2								
M (Q)					1	1						
O (P, Q)						4	4	4				
P (R)									2			
Q (R)									3	3		
R											2	2
Sum	2	2	2	6	4	5	4	4	5	3	2	2

FIGURE 3.32 Bar chart for scheduling network activities.

Consider activity J in Fig. 3.31; the ES shows that the earliest this activity can begin is the end of day 3 (or the beginning of day 4). Thus, the beginning of day 4 to the end of day 6 (as determined from the LF) is the available time span to complete this activity (see Fig. 3.32). Because of the nature of the logic diagram, this activity cannot be scheduled earlier, because activity I must first be completed. It cannot be scheduled later, for this would delay project completion. Considering the end-of-day convention, where a start date means the end of the day or beginning of the following day, the left bracket is placed at the beginning (morning) of the following day (e.g., "ES+1").

Once the brackets are placed correctly, the next step is to formulate a trial schedule, scheduling each activity as soon as possible within the time frame, or flush with the left bracket. To schedule a particular activity, place the number of resources required inside each box along the activity line. Do not exceed the activity's duration; stop at the end of the early finish day. The remaining boxes within the brackets are left blank for now and will become either free or interfering float. In Fig. 3.32, the boxes of the critical activities have been darkened.

The crossed diagonal lines forming an × on the last two days of the activity J bracketed area denote days of interfering float. Some activities have free float (activities L, M, and P), and some have all interfering float (activity J). The placement of the ×s serves as a reminder about the ripple effect with succeeding activities.

Resources. If a loader and truck spread is identified as a controlling resource, the numbers within the activity data boxes would indicate how many spreads are required for each of the activities. It is necessary to consider only major controlling resources when developing the schedule. Each project has its own controlling resources, depending on the nature of the work. On an earthwork project, the controlling resource might be loaders. On an embankment project, the controlling resource might compaction equipment. There are usually several controlling resources, and all would have to be considered when scheduling the work. In cases where many different kinds of resources are necessary for an activity, managers may choose to use several lines contained within one set of brackets and use each line for a different type of resource. The resources required on any particular day are the sum of those required for the individual activities taking place on that day. The last row of Fig. 3.32 shows the resource requirement sum for each day of the project. Computer-scheduling algorithms are available to aid in scheduling resources on large projects.

The resource required to complete activity M is one loader and truck spread for 2 days. To show this activity scheduled as soon as possible, place the number 1 (number of loader and truck spreads) in only the first two boxes of the bracketed duration (see Fig. 3.32). Scheduling all the activities as soon as possible yields the early start schedule as shown in Fig. 3.32. All activities are scheduled to begin at their ES times.

The number of available resources is often less than the needed resources for a given day, and activities must be delayed into float whenever possible to spread resource use across the time frame of the project. When daily resource requirements exceed what is available, the manager must resource-constrain the project to have as little effect as possible on the project duration. The terminology is not completely universal, but constraining usually means that because of a resource deficiency, the project will have to be extended. However, sometimes resources can be shifted in such a way as to avoid delaying the overall project duration. The term *leveling* or *leveling project resources* has the connotation that the project duration will not be extended.

If the objective, considering the Fig. 3.31 network, is to limit the number of spreads to five or less, and to avoid a situation where there are peaks and valleys in the number of spreads required over the duration of the project, Fig. 3.33 illustrates one solution. In Fig. 3.33 activity J has been shifted 2 days into its interfering float. As a result of this action activity M had to be shifted 2 days, as it is dependent on J. Additionally, activity L was shifted 1 day, but this was into free float so no other activity was affected. If any of the critical activities are shifted to achieve the desired leveling of resources, the result will be to lengthen the project duration.

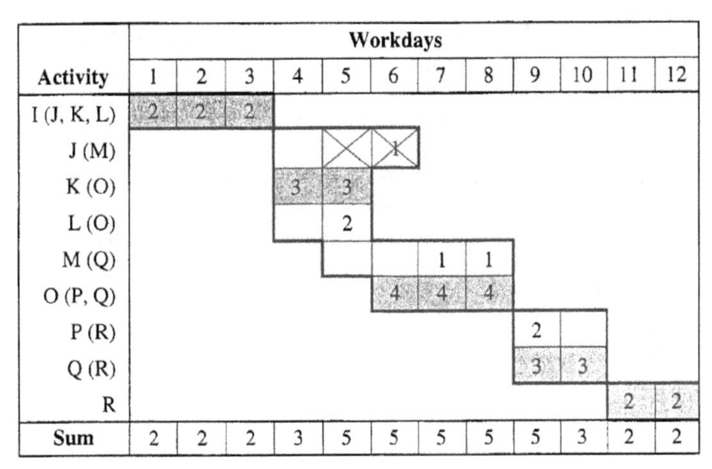

Activity	1	2	3	4	5	6	7	8	9	10	11	12
I (J, K, L)	2	2	2									
J (M)					X	X 1						
K (O)				3	3							
L (O)					2							
M (Q)							1	1				
O (P, Q)						4	4	4				
P (R)									2			
Q (R)									3	3		
R											2	2
Sum	2	2	2	3	5	5	5	5	5	3	2	2

FIGURE 3.33 A smooth resource-constrained schedule, maximum five spreads.

Planning the work according to the Fig. 3.33 schedule results in a uniform increase in the number of required spreads, a level requirement for five spreads from day 5 through day 9, and then a uniform decrease of requirements at the end of the project. This is a good situation because it is difficult to manage the work when the required number of resources is constantly changing. The problem with this resource-leveled schedule is that, with only the exception of activity P, every activity is critical. If anything should delay a critical activity, the project will be delayed.

Another possible schedule that holds the maximum number of crews to five or less is presented in Fig. 3.34. With this schedule some float has been retained for activities J, M, and P. The problem, however, is that on day 8 it is necessary to shut down one spread, and then on day 9 the spread is needed for only 1 day. It is not really possible to lay off operators for a single day, and the cost to demobilize and mobilize machines must be considered. Leveling of crew resources is always a challenge when trying to keep spreads productive.

Activity	1	2	3	4	5	6	7	8	9	10	11	12
I (J, K, L)	2	2	2									
J (M)				1	X	X						
K (O)				3	3							
L (O)					2							
M (Q)						1	1					
O (P, Q)						4	4	4				
P (R)									2			
Q (R)									3	3		
R											2	2
Sum	2	2	2	4	5	5	5	4	5	3	2	2

FIGURE 3.34 Resource-constrained schedule, maximum five spreads.

Crashing. If the CPM indicates a project duration exceeding the desired completion date, the manager should examine the logic diagram's critical path to find activity durations that may be shortened. This is known as *expediting, compressing,* or *crashing* the project. The only way to reduce the duration of the project is to shorten the duration of the critical activities. Shortening a noncritical activity will increase cost and not shorten the project duration. However, increasing the allocation of resources to critical path activities may reduce the duration of the project. Additional equipment and personnel can be committed, or the same equipment and personnel can be used for longer hours. Normally, a moderately extended workday is the most economical and productive solution. Managers may also choose to work double shifts or weekends. When expediting activities, however, consider the long-term effects on safety, morale, equipment, and consequential decreases in efficiency.

Other Management Methods

A **velocity diagram** presents a graphical picture of the relationship between time and the accomplishment of an activity (Fig. 3.35). The vertical axis of the diagram represents accomplishment of a work task: cubic yards excavated, miles of road built, kilometers of pipeline constructed. The horizontal axis presents construction time. The slope of the production line gives the activity production rate. When the vertical axis represents activity completion in terms of advancement along a linear dimension, such as the stations of a highway project, the velocity diagram delineates an activity's rate of progress in time and space. The tool focuses management attention on the rate of accomplished work.

Velocity diagram A graphic for monitoring the relationship between time and the accomplishment of an activity.

If the bid estimate production rates are based on the same time unit as those for the proposed velocity diagram, the manager can plot the diagram directly. If the time units are different, a conversion will be necessary. In the case where a CPM

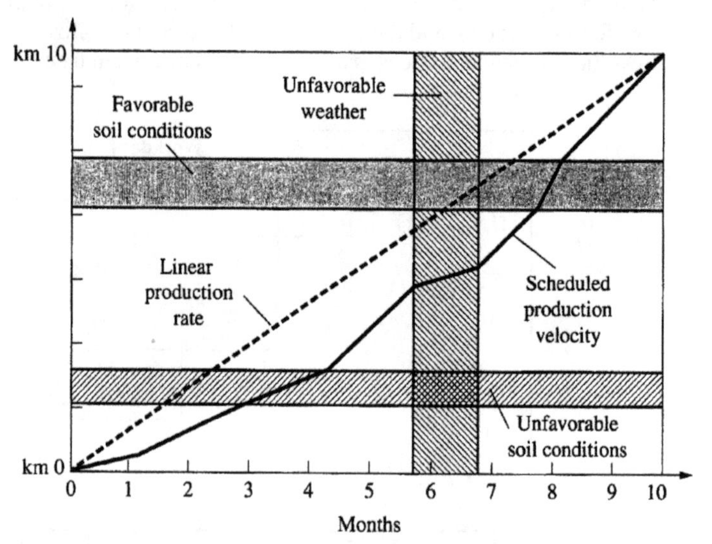

FIGURE 3.35 Velocity diagram as a scheduling tool. (*Source:* Dressler, Joachim (1980). "Construction Management in West Germany," *Journal of the Construction Division, ASCE*, Vol. 106, No. CO4. Reproduced by permission of ASCE.)

was prepared with activity duration, the production rate for a velocity diagram can easily be calculated by rearranging Eq. (3.6).

$$\text{Planned (task) production rate} = \frac{\text{Quantity of work}}{\text{Estimated duration}}$$

In the case of the Fig. 3.35 project, there are data indicating that changing ground conditions and inclement weather are considerations. Therefore, the production rates in the diagram are not linear but fully consider the impact of both physical and environmental conditions affecting the work.

By plotting actual progress on the diagram as the work progresses, the manager can measure performance against a baseline needed to achieve timely completion. The velocity diagram lends itself to efficiently managing projects that require control of time and distance between construction activities.

Linear Scheduling. A **linear schedule** is simply a series of individual velocity diagrams. They provide a graphical display of the movement of labor and equipment in time and space. Many construction projects involve groups of activities performed *consecutively* by the same crew. This is especially true in the case of linear projects such as highway or pipeline work. On such projects, it is advantageous to arrange the schedule so crews can work continuously. By scheduling the project so work can proceed continuously, costs are reduced because intervals of idle equipment and manpower are eliminated. Another important property of linear projects is that numerous activities can be carried out concurrently at different locations.

Linear schedule
A simple diagram that shows the relationships between activities, their duration, and the space where they take place at a given time.

Linear scheduling seeks to ensure that the work can proceed continuously, with no idle resources or activities interfering with one another. In the scheduling literature, this method has been referred to as repetitive-unit construction, time-space scheduling, and the vertical production method. It is an effective tool for relating activities and production rates.

As with CPM scheduling, the process of developing a linear schedule consists of four steps:

1. Identify the activities.
2. Estimate activity production rates.
3. Develop activity sequence.
4. Draft the linear schedule.

Identify Activities. The process of identifying activities has already been discussed. However, activity detail in linear schedules is usually not as extensive as in the case of a CPM schedule.

Estimate Activity Production Rates. Like velocity diagrams, linear schedules show the time and space relationship of activities. The production rate of the individual activities must be expressed in the same units. Because of how the linear schedule diagram is drafted, production of an activity (all activities) is usually expressed in terms of position.

As discussed concerning velocity diagrams, the production rate is often assumed to be linear. But the average production can be adjusted to reflect the effect of physical conditions or environmental conditions, as shown in Fig. 3.35.

Develop Activity Sequence. Even though the activities in a linear construction project can be carried out concurrently, there is usually a sequence of activities determined by the activity relationships. Due to the nature of the construction operations, activities are normally sequenced based on physical relationships. An activity starts, and after a certain amount of work is accomplished a following activity can begin and continue concurrently. Consider the bridge work shown in Fig. 3.36. Section 1 excavation must be finished before the section 1 footing activity can begin. Once the excavation for the first footing is complete, the excavation crew moves to the second footing and the carpenter crew building the footings starts work on the first footing. This process continues with the activities that follow: column, cap, and deck.

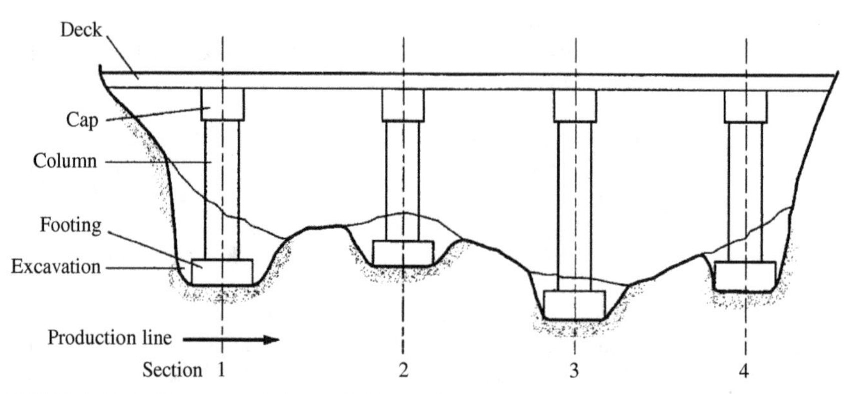

FIGURE 3.36 Linear schedule for a bridge project. (*Source:* Selinger, Shlomo (1980). "Construction Planning for Linear Projects," *Journal of the Construction Division, ASCE*, Vol. 106, No. CO2. Reproduced by permission of ASCE.)

Draft the Linear Schedule. A linear schedule is a series of individual velocity diagrams (Fig. 3.37). Time is plotted on the horizontal scale, and the common production measure is plotted on the vertical axis. The individual activity velocity diagrams are plotted in order of

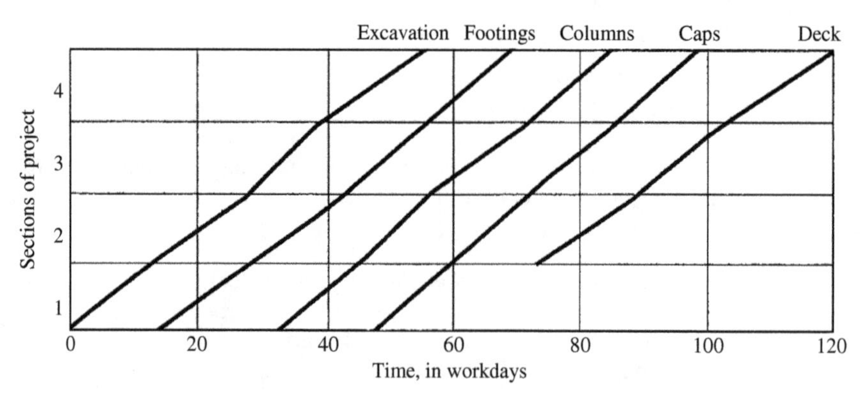

FIGURE 3.37 Linear schedule for a bridge project. (*Source:* Selinger, Shlomo (1980). "Construction Planning for Linear Projects," *Journal of the Construction Division, ASCE*, Vol. 106, No. CO2. Reproduced by permission of ASCE.)

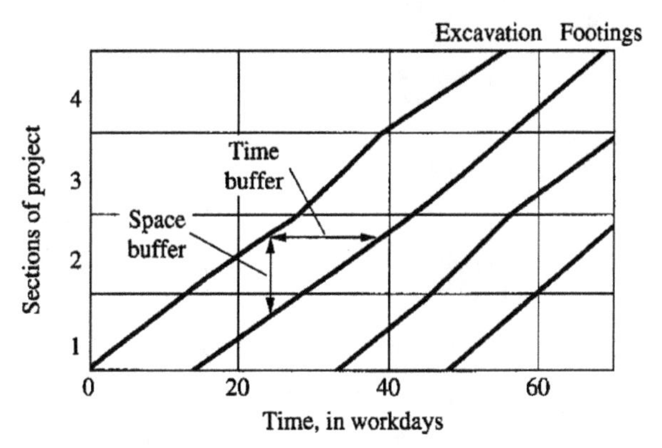

FIGURE 3.38 Linear schedule for a bridge project. (*Source:* Selinger, Shlomo (1980). "Construction Planning for Linear Projects," *Journal of the Construction Division, ASCE*, Vol. 106, No. CO2. Reproduced by permission of ASCE.)

occurrence, allowing for start and/or finish lags that ensure that the activity production lines do not overlap. An overlap would indicate a work conflict and lost productivity.

To avoid interferences, buffers are established between activities based on expected variations in production rates. The buffers can be described in terms of either time or space (Fig. 3.38).

CHAPTER 4
SOIL AND ROCK

Knowledge of the properties, characteristics, and behavior of different soil and rock types is important to both design and construction. Whereas the designer is usually more concerned about structural strength, the constructor is interested in how a material will handle during the construction process. Compacted density is the most commonly used parameter for specifying performance of earthwork construction operations because a correlation exists between a soil's physical properties and its density. The effectiveness of different compaction methods is dependent on the individual soil or rock type being manipulated.

DIGGING

Soil is loose surface material of the earth. Rock is the hard crust of the earth, which underlies and often projects through the soil cover. Geologically, all soils are considered to be rock formations. In ordinary usage, rock is something hard, firm, and stable. A contractor's definition for rock is any material that cannot be dug or loosened by available machinery, but this distinction from soil may depend more on the power, size, and digging efficiency of the machinery than on the material itself. Material to be excavated can also be roughly divided into three classes: rock, hard digging, and easy digging.

Overcoming the resistance to dig a formation will be largely affected by material hardness, coarseness, friction, adhesion, cohesion, and weight. In digging, hardness is resistance to penetration. It is increased by close packing of soil or filling of voids with finer particles or lime or other natural cements such as lime and fly ash. Clay soils are hard when dry and soft when wet. Cobbles, boulders, or hard lumps increase the power requirement for penetration. They are most troublesome when they are oversize for the machine or packed so firmly in place that they cannot slide or rotate away from the cutting edge.

As the digging edge penetrates, friction absorbs an increasing proportion of its force. It is affected by particle size and hardness, by the amount of moisture, and by the presence or absence of natural lubricants such as humus or soft clay. Adhesion is the sticking of soil to the digging parts. It may increase the friction load substantially in wet work. Cohesion is resistance to tearing apart. Firm or hard materials may split readily along bedding or cleavage planes so they can be dug rather easily from the proper direction. Relatively soft clay banks may be very difficult to dig because of strong and uniform cohesion. A tough formation lacking planes of weakness is described as "tight."

GLOSSARY OF TERMS

The following glossary defines important terms used in discussing soil and rock materials and compaction:

Aggregate, coarse Crushed rock or gravel, generally greater than ¼ in. in size.

Aggregate, fine The sand or fine-crushed stone used for filling voids in coarse aggregate. Generally less than ¼ in. in size and larger than a No. 200 sieve.

Cohesion The quality of some soil particles to be attracted to like particles, manifested in a tendency of sticking together, a property of platy clay particles.

Cohesive materials A soil with strong interparticle attractive forces.

Dust Fine particles passing through the No. 200 sieve.

Grain-size curve A graph of a material sample illustrating the percentage of grain sizes by weight.

Optimum moisture content The water content for a given compactive effort at which point the greatest dry density of a soil can be obtained.

Pavement One or more layers of rigid surfacing material used to provide high bending resistance and to distribute loads to a base layer. Pavements are usually constructed of asphalt or concrete.

Pit A rock mine where gravel deposits have been loosened by movement in the earth's crust; typically the deposits are the result of glacial or flowing water forces.

Plasticity The capability of being molded. Plastic materials do not assume their original shape after the force causing deformation is removed.

Quarry A rock mine where deposits are mechanically extracted using drilling, blasting, and crushing operations.

Rock The hard, mineral matter of the earth's crust, occurring in masses and often requiring blasting or mechanical fracture before excavation can be accomplished.

Shrinkage A volume reduction usually occurring when soils or rock are subjected to moisture or mechanical compaction.

Soil The loose surface material of the earth's crust, created naturally from the disintegration of rocks or decay of vegetation. Soil can be excavated easily using powered equipment in the field.

Swell A volume increase occurring when soils or rocks are excavated and loosened from the earth's crust.

SOIL AND ROCK PROPERTIES

Before discussing earth- and rock-handling techniques or analyzing problems involving these materials, it is necessary to first understand their physical properties. These properties have a direct effect on the ease or difficulty of handling the material, on the selection of equipment, and on achieving equipment production rates.

Steel and concrete are basically homogeneous and uniform in composition. As such, their behavior can be predicted. Soil and rock are just the opposite. By nature they are heterogeneous. In their natural state, they are rarely uniform, and their response to excavating, handling, and compacting is best understood by experience and comparing it to a similar type of material. To accomplish this, soil and rock types must be classified. Soils are classified according to particle size composition, by their physical properties, or by their behavior when the moisture content varies.

A constructor is concerned primarily with five major types of soils—gravel, sand, silt, clay, and organic matter—or with combinations of these soil types. Sieve analysis measures sizes by sifting a dried and weighed sample through a set of testing sieves (Fig. 4.1) and weighing the material retained on each screen. Screen size broadly classifies common earthwork materials as boulders, gravels, sand, and silt. If further analysis is required for particles passing through the smallest (No. 200 mesh) sieve, it is done by hydrometer. This process is based on the fact that the speed of settlement of such particles is proportional to their size.

FIGURE 4.1 Soil testing sieves.

Different agencies and specification groups denote the sizes of these types of soil differently, causing some confusion. The following size limits represent standards set forth by the American Society for Testing and Materials (ASTM):

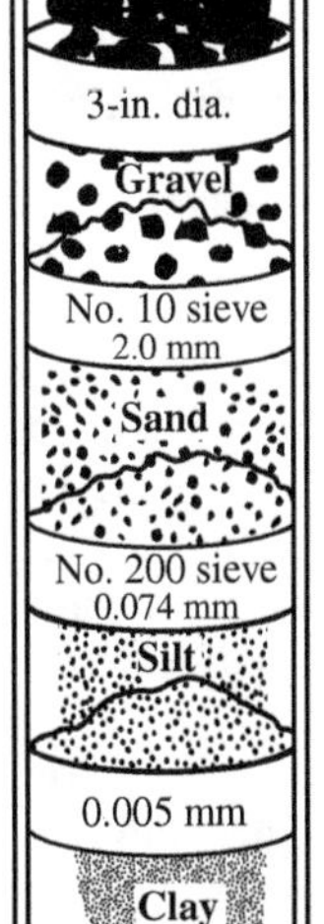

Gravel is composed of rounded or semi-round particles of rock that will pass through a 3-in. sieve and be retained on a 2.0-mm No. 10 sieve. Sizes larger than 10 in. are usually called *boulders*.

Sand is disintegrated rock with particles that vary in size from the lower limit of gravel (2.0 mm) down to 0.075 mm (No. 200 sieve). It will be classified as coarse or fine sand, depending on the grain size. Sand is a granular, non-cohesive material, and its particles have a bulky shape.

Silt or *dust* is a material finer than sand, and thus its particles are smaller than 0.075 mm but larger than 0.005 mm. It is a non-cohesive material and has little or no strength. Silt compacts very poorly by itself, but can enhance compaction of larger aggregates by filling void spaces or reorienting particle position.

Clay is a cohesive material whose particles are less than 0.005 mm. The cohesion between the particles gives clay high strength when air-dried. Clays can be subject to considerable changes in volume with variations in moisture content. They will exhibit plasticity within a range of "water contents." Clay particles are shaped like thin wafers, hence the use of the term *platy*.

Organic matter is partly decomposed vegetation. It has a spongy, unstable structure, and it will continue to decompose and is chemically reactive. If present in soils used for construction purposes, organic matter should be removed and replaced with a more suitable soil.

TABLE 4.1 Characteristics of Soils

Soil	Gravels & Sands	Silts	Clays
Grain size	Coarse grained Can see individual grains by eye alone	Fine grained Cannot see individual grains by eye alone	Fine grained Cannot see individual grains by eye alone
Characteristics	Cohesionless Non-plastic Granular	Cohesionless Non-plastic Granular	Cohesion Plastic —
Effect of water	Relatively unimportant (exception: loose saturated case with dynamic loading)	Important	Very important
Effect of grain size distribution on engineering properties	Important	Relatively unimportant	Relatively unimportant

Some broad engineering characteristics of these general soil types are presented in Table 4.1. Generally, however, the soil types are found in nature in some mixed proportions. Table 4.2 presents a classification system based on combinations of soil types.

Soils existing under natural conditions may not contain the relative amounts of desired material component types necessary to produce the properties required for construction purposes. For this reason, it may be necessary to obtain soils from several sources and then blend them for use in an engineered fill.

Borrow pit A pit from which fill material is mined.

If the material in a **borrow pit** consists of layers of different types of soils, the specifications for the project may require the use of equipment capable of excavating vertically through the layers in order to mix the soil.

Rock is formed by one of three different means (Table 4.3):

1. *Igneous* rocks solidified from molten masses.
2. *Sedimentary* rocks formed in layers, settling out of water solutions.
3. *Metamorphic* rocks are material transformed from the first two by heat and pressure.

Their respective formation processes will affect how rocks can be excavated and handled.

Categorization of Materials. In contract documents, excavation is typically categorized as common, rock, muck, or unclassified. *Common* refers to ordinary earth excavation, and the term *unclassified* reflects the lack of a clear distinction between soil and rock. The removal of common excavation will not require the use of explosives, although tractors equipped with rippers may be used to loosen consolidated formations. The specific engineering properties of the soil—plasticity, grain size distribution, and so on—will influence the selection of the appropriate equipment and construction methods.

In construction, *rock* is a material of such hardness that it cannot be removed by ordinary earth-handling equipment. This definition will be affected by technological developments in equipment. Larger and heavier machines are continually changing the limits of this definition for rock. Drilling and blasting, or some comparable method, must be used to remove rock. This normally results in considerably greater expense than for earth

TABLE 4.2 Unified Soil Classification System

Symbol	Primary	Secondary	Supplementary
GW	Coarse-grained soils	Well-graded gravels, gravel-sand mixtures, little or no fines	Wide range of grain size
GP	Coarse-grained soils	Poorly graded gravels, gravel-sand mixtures, little or no fines	Predominantly one size or a range of intermediate sizes missing
GM	Gravel mixed with fines	Silty gravels and gravel-sand-silt mixtures–may be poorly graded	Predominantly one size or a range of intermediate sizes missing
GC	Gravel mixed with fines	Clayey gravels, gravel-sand-clay mixtures, which may be poorly graded	Plastic fines
SW	Clean sands	Well-graded sands, gravelly sands, little or no fines	Wide range in grain sizes
SP	Clean sands	Poorly graded sands, gravelly sands, little or no fines	Predominantly one size or a range of sizes with some intermediate sizes missing
SM	Sands with fines	Silty sands and sand-silt mixtures, which may be poorly graded	Nonplastic fines or fines of low plasticity
SC	Sands with fines	Clayey sands, sand-clay mixtures, which may be poorly graded	Plastic fines
ML	Fine-grained soils	Inorganic silts, clayey silts, rock flour, silty very fine sands	Plastic fines
CL	Fine-grained soils	Inorganic clays of low to medium plasticity, silty sandy or gravelly clays	Plastic fines
OL	Fine-grained soils	Organic silts and organic silt-clay of low plasticity	
MH	Fine-grained soils	Inorganic silts, clayey silts, elastic silts	
CH	Fine-grained soils	Inorganic clays of high plasticity, fat clays	
OH	Fine-grained soils	Organic clays and silty clays of medium to high plasticity	
PT	Peat	Highly organic soils	

Symbol classification

COARSE-GRAINED MATERIAL

Symbol	Subdivision
G—Gravel grain size from 3 in. to No. 4 sieve size	W—Well graded, little or no fines
S—Sand grain size from No. 4 to 200 sieve size	P—Poorly graded, little or no fines
	M—Concentration of silty or nonplastic fines
	C—Concentration of clay or plastic fines

FINE-GRAINED MATERIAL

Symbol	Subdivision
M—Silt very fine grain size, floury appearance	L—Low plastic material, lean soil
C—Clay finest grain size, high dry strength—plastic	H—High plastic material, fat soil
O—Organic matter partly decomposed, appears fibrous, spongy and dark in color	

TABLE 4.3 Rock Types

Rock	Weight lb/ft^3	Weight kg/m^3	Percent of Wear	Hardness	Toughness
Granite	167	2702	4.3	18.3	11
Syenite	171	2767	3.3	18.3	15
Diorite	179	2896	3.0	18.2	17
Gabbro	185	2993	3.0	17.7	14
Peridotite	182	2944	4.0	14.2	11
Rhyolite	159	2572	3.7	18.3	19
Trachyte	170	2750	2.9	18.1	24
Andesite	166	2686	3.9	17.0	18
Basalt	177	2864	3.0	17.1	18
Diabase	186	3009	2.4	18.0	22
Limestone	165	2669	5	14.1	9
Dolomite	170	2750	5.5	14.9	9
Sandstone	164	2653	6.2	14.4	10
Chert	159	2572	9.4	18.2	12
Gneiss	172	2783	4.9	17.4	10
Schist	180	2912	4.7	16.6	13
Amphibolite	3042	3042	2.8	17.5	19
Quartizite	2734	2734	3.2	18.8	18
Eclogite	3137	3137	2.4	18.4	22
Marble	2799	2799	5.7	13.1	6

Class	Type	Family
Igneous	Intrusive (plutonic)	Granite
		Syenite
		Diorite
		Gabbro
		Peridotite
	Extrusive (volcanic)	Rhyolite
		Trachyte
		Andesite
		Basalte
		Diabase
Sedimentary	Calcareous	Limestone
		Dolomite
	Siliceous	Shale
		Sandstone
		Chert (flint)
Metamorphic	Foliated	Gneiss
		Schist
		Amphibolite
	Nonfoliated	Slate
		Quartzite
		Eclogite
		Marble

excavation. Rock excavation involves the study of the rock type, faulting, dip and strike, and explosive characteristics as the basis for selecting material removal and aggregate production equipment.

Muck or *peat* are materials still in the process of decay, and if used in an embankment, they will produce subsidence. Both are soft organic materials with high water content. Typically, they would include such things as decaying stumps, roots, logs, crops, and humus. These materials are hard to handle and can present special construction problems at their point of excavation, in transportation, and at disposal.

In many cases, the contract documents include geotechnical information. This owner-furnished information provides a starting point for an independent investigation by the constructor. Other good sources of preliminary information are topographic maps, agriculture maps, geologic maps, soil borings and surveys, agricultural maps, drilling well logs, and aerial photographs. The investigation is not complete, however, until the estimator/engineer makes an onsite visit, and in the case of large jobs, drilling or test pit exploration is conducted. In the case of a project involving rock, it is often necessary to have seismic surveys performed to identify subsurface rock strata. For one large dam project in California, 13 private seismic studies were made of the spillway rock excavation for the bidding contractors.

> You should never price an earth- or rock-handling project without first making a thorough study of the materials to be handled.

Even though many owners provide good geotechnical data compiled by qualified engineers, the designer's primary concern is how well the material will perform structurally. The constructor is interested in how the material will handle during the construction process and what volume or quantity of material is to be processed to yield the desired final structure.

Soil Weight–Volume Relationships. The primary relationships (Fig. 4.2) are expressed by Eqs. (4.1) through (4.6).

$$\text{Unit weight } (\gamma) = \frac{\text{Total weight of soil}}{\text{Total soil volume}} = \frac{W}{V} \tag{4.1}$$

$$\text{Dry unit weight } (\gamma_d) = \frac{\text{Weight of soil solids}}{\text{Total soil volume}} = \frac{W_s}{V} \tag{4.2}$$

$$\text{Water content } (\omega) = \frac{\text{Weight of water in the soil}}{\text{Weight of soil solids}} = \frac{W_w}{W_s} \tag{4.3}$$

$$\text{Void ratio } (e) = \frac{\text{Volume of voids}}{\text{Volume of soil solids}} = \frac{V_v}{V_s} \tag{4.4}$$

$$\text{Porosity } (n) = \frac{\text{Volume of voids}}{\text{Total soil volume}} = \frac{V_v}{V} \tag{4.5}$$

$$\text{Specific gravity (SG)} = \frac{\text{Weight of soil solids/volume of solids}}{\text{Unit weight of water}} = \frac{W_s/V_s}{\gamma_w} \tag{4.6}$$

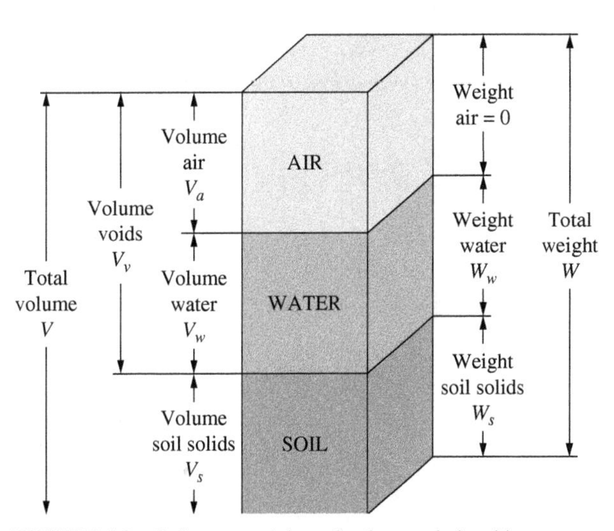

FIGURE 4.2 Soil mass weight and volume relationships.

TABLE 4.4 AASHTO Soil Classification System

General Classification	Granular Materials (35% or Less of Total Sample Passing No. 200)							Silt-Clay Materials (More Than 35% of Total Sample Passing No. 200)			
	A-1		A-3	A-2				A-4	A-5	A-6	A-7
Group Classification	A-1-a	A-1-b		A-2-4	A-2-5	A-2-6	A-2-7				A-7-5, A-7-6
Sieve analysis, percentage passing											
No. 10	50 max.										
No. 40	30 max.	50 max.	51 max.								
No. 200	15 max.	25 max.	10 max.	35 max.	35 max.	35 max.	35 max.	36 min.	36 min.	36 min.	36 min.
Characteristics of fraction passing No. 40											
Liquid limit				40 max.	41 min.	40 max.	41 min.	40 max.	41 min.	40 max.	41 min.
Plasticity index	6 max.		NP	10 max.	10 max.	11 min.	11 min.	10 max.	10 max.	11 min.	11 min.
Group index*	0		0	0		4 max.		8 max.	12 max.	16 max.	20 max.

*A group index based on a formula that considers particle size; LL and PI are given at the bottom of the table. The group index indicates the suitability of a given soil for embankment construction. A group index number of "0" indicates a good material while an index of "20" indicates a poor material.

Many more formulas can be derived from these basic relationships. Two such formulas useful in analyzing compaction specifications are

$$\text{Total soil volume } (V) = \text{Volume voids } (V_v) + \text{Volume solids } (V_s) \tag{4.7}$$

$$\text{Weight of solids } (W_s) = \frac{\text{Weight of soil } (W)}{1 + \text{water content } (\omega)} \tag{4.8}$$

If unit weights are known, which is the usual case, then Eq. (4.8) becomes

$$\gamma_d = \frac{\gamma}{1+\omega} \tag{4.9}$$

Soil Limits. Certain limits of soil consistency—liquid limit and plastic limit—were developed to serve as reference points in order to differentiate between highly plastic, slightly plastic, and non-plastic materials.

Liquid limit (*LL*). The water content at which a soil passes from the plastic to the liquid state is known as the liquid limit. High LL values are associated with soils of high compressibility. Typically, clays have high LL values; sandy soils have low LL values.

Plastic limit (*PL*). The water content at which a soil passes from the plastic to the semi-solid state is known as the plastic limit. It also refers to the lowest water content at which a soil can be rolled into a 1/8-in. (3.2-mm)-diameter thread without crumbling.

Plasticity index (*PI*). The numerical difference between a soil's liquid limit and its plastic limit is the plasticity index (PI = LL – PL). Soils with high PI values are quite compressible and have high cohesion.

On many projects, the specifications will denote a certain material gradation, a maximum LL, and a maximum PI. The American Association of State Highway and Transportation Officials (AASHTO) system of soil classification, which is the most widely used system for highway construction, illustrates this point (Table 4.4).

Volumetric Measure. For bulk materials, volumetric measure varies with the material's position in the construction process (Fig. 4.3). The same weight of a material will occupy different volumes as it is handled on the project. In general, most cohesive soils will shrink

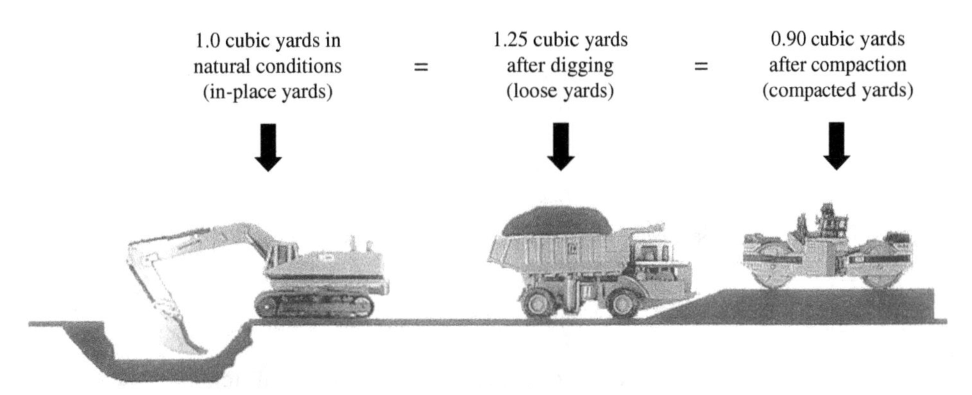

FIGURE 4.3 Material volume changes caused by processing.

10% to 30% from the bank (in situ) to compacted state. Solid rock will swell 20% to 40% from the bank to placement in an embankment. Between the bank and loose states, cohesive soils swell about 40%, and solid rock swells as much as 65%.

Soil volume is measured in one of three states:

Bank cubic yard (bcy)		1 cubic yard (cy) of material as it lies in the natural in situ state
Loose cubic yard (lcy)		1 cy of material after it has been disturbed by an excavating and loading process
Compacted cubic yard (ccy)		1 cy of material in the compacted state, also referred to as a net in-place cubic yard or embankment cubic yard

In planning or estimating a job, the engineer must use a consistent volumetric measure in any set of calculations. The necessary consistency of units is achieved by using shrinkage and swell factors. The shrinkage factor is the ratio of the compacted dry weight per unit volume to the bank dry weight per unit volume:

$$\text{Shrinkage factor} = \frac{\text{Compacted dry unit weight}}{\text{Bank dry unit weight}} \tag{4.10}$$

The volume shrinkage due to compacting a fill can be expressed as a percentage of the original bank measure weight:

$$\text{Shrinkage percentage} = \frac{(\text{Compacted dry unit weight}) - (\text{Bank unit weight})}{\text{Compacted unit weight}} \times 100 \tag{4.11}$$

The *swell* factor is the ratio of the loose dry weight per unit volume to the bank dry weight per unit volume:

$$\text{Swell factor} = \frac{\text{Loose dry unit weight}}{\text{Bank dry unit weight}} \tag{4.12}$$

The percent swell, expressed on a gravimetric basis, is

$$\text{Swell percent} = \left(\frac{\text{Bank dry unit weight}}{\text{Loose dry unit weight}} - 1\right) \times 100 \tag{4.13}$$

Table 4.5 gives representative swell values for different classes of earth. These values will vary with the extent of loosening and compaction. If more accurate values are desired

TABLE 4.5 Representative Properties of Soil and Rock

Material	Bank Weight lb/cy	Bank Weight kg/m³	Loose Weight lb/cy	Loose Weight kg/m³	Percent Swell	Swell Factor*
Clay, dry	2,700	1,600	2,000	1,185	35	0.74
Clay, wet	3,000	1,780	2,200	1,305	35	0.74
Earth, dry	2,800	1,660	2,240	1,325	25	0.80
Earth, wet	3,200	1,895	2,580	1,528	25	0.80
Earth and gravel	3,200	1,895	2,600	1,575	20	0.83
Gravel, dry	2,800	1,660	2,490	1,475	12	0.89
Gravel, wet	3,400	2,020	2,980	1.765	14	0.88
Limestone	4,400	2,610	2,750	1,630	60	0.63
Rock, well blasted	4,200	2,490	2,640	1,565	60	0.63
Sand, dry	2,600	1,542	2,260	1,340	15	0.87
Sand, wet	2,700	1,600	2,360	1,400	15	0.87
Shale	3,500	2,075	2,480	1,470	40	0.71

*The swell factor is equal to the loose dry unit weight divided by the bank dry unit weight.

for a specific project, tests should be made on several samples of the earth taken from different depths and different locations within the proposed cut. The test can be made by weighing a given volume of undisturbed, loose, and compacted earth.

Sample Calculation: An earth fill, when completed, will occupy a net volume of 187,000 cy. The borrow material that will be used to construct this fill is a stiff clay. In its "bank" condition, the borrow material has a wet unit weight (γ) of 129 lb per cubic foot (cf), a water content (ω%) of 16.5%, and an in-place void ratio (e) of 0.620. The fill will be constructed in layers of 8-in. depth, loose measure, and compacted to a dry unit weight (γ_d) of 114 lb per cf at a moisture content of 18.3%. Compute the required volume of borrow pit excavation.

Fill	Borrow
$\gamma_d = 114$ lb/cf	$\gamma_d = \dfrac{129}{1+0.165} = 111$ lb/cf

Fill		Borrow
$187{,}000\text{ cy} \times \dfrac{27\text{ cf}}{\text{cy}} \times \dfrac{114\text{ lb}}{\text{cf}}$	=	$x \times \dfrac{27\text{ cf}}{\text{cy}} \times \dfrac{111\text{ lb}}{\text{cf}}$

$$187{,}000\text{ cy} \times \frac{114}{111} = 192{,}054\text{ cy, } \textit{borrow required}$$

Note the element 114/111 is the shrinkage factor 1.03.

The key to solving this type of problem is the unit weight of the solid particles (dry weight) making up the soil mass. In the construction process, the specifications may demand that water be either expelled or added to the soil mass. In the previous example, the contractor would be required to add water to the borrow to increase the moisture content from 16.5% to 18.3%. Adjusting for the extra borrow cubic yards required to make 1 fill cy, note the water difference:

Fill		Borrow
$\gamma = 114 \times 1.183 = 135$ lb/cf		129 lb/cf
γ_d	− 114 lb/cf	− 111 lb/cf
Water	= 21 lb/cf	18 lb/cf
		× 1.03 (shrinkage factor)
		19 lb/cf

To achieve the desired fill density and water content, the contractor will have to add water. This water must be hauled in by water wagon and is not part of the in-place borrow unit weight. The quantity of water that must be added is

Fill	Borrow
Water content = 0.183	Water content = 0.165
$187{,}000 \text{ cy} \times \frac{114 \text{ lb}}{\text{cf}} \times 27 \frac{\text{cf}}{\text{cy}} \times 0.183$	$192{,}054 \text{ cy} \times \frac{111 \text{ lb}}{\text{cf}} \times 27 \frac{\text{cf}}{\text{cy}} \times 0.165$
$= 105{,}332{,}238$ lb water	$= \frac{94{,}971{,}663 \text{ lb}}{10{,}360{,}575 \text{ lb}}$ water / water

which is 1,241,941 gal, or approximately 6.5 gal per cy of borrow.

The method of soil preparation prior to compaction is not sufficiently appreciated as an important factor influencing achievement of successful results. This includes adding water or, conversely, drying the soil. The blending of the excavation material to achieve a homogeneous composition and uniform water content within a placed layer is especially important.

Constructors commonly apply what is referred to as a *swell factor* or *bulking factor* when estimating jobs. This rule-of-thumb factor should not be confused with the previously defined factors. The term *swell factor* is used in this case because of how the number is applied. The embankment yardage of the job is multiplied by this factor; it is swelled to put it in the same reference units as the borrow material. The job is then figured in *borrow yards*. This swell factor is strictly an educated *guess* of how the volume will change during handling based on past experience with similar materials. It may also reflect consideration of the project design. A case in point would be when the embankment is less than 3 ft in total height, in which case more embankment material will be required to compensate for the compaction of the natural ground below the fill. The constructor would therefore apply a higher swell factor when calculating the required borrow material for fills of minimum height.

COMPACTION SPECIFICATION AND CONTROL

Prior to preparing the specifications for a project, representative soil samples are collected and tested in the laboratory to determine material properties.

Maximum Dry Density/Optimum Moisture. A critical test for earthwork projects is the laboratory-established compaction curve. From such a curve, the maximum dry unit weight (density) and the percentage of water required to achieve maximum density are determined. This percentage of water, which corresponds to the maximum dry density for a given compactive effort, is known as the *optimum water content.* It is the amount of water required for a given soil to reach maximum density.

Figure 4.4 shows two compaction curves based on different input energy levels. The curves are plotted in dry weight (in pounds per cubic feet) against water content (percentage by dry weight). Each illustrates the effect of varying amounts of moisture on the density of a soil subjected to a given compactive effort (energy input level). The two energy levels depicted are known as *standard* and *modified Proctor tests.* The modified Proctor (higher input energy) yields a higher density at a lower water content than the standard Proctor. For the material depicted by the curves in Fig. 4.3, the optimum moisture for the standard Proctor is 16% versus 12% for the modified Proctor.

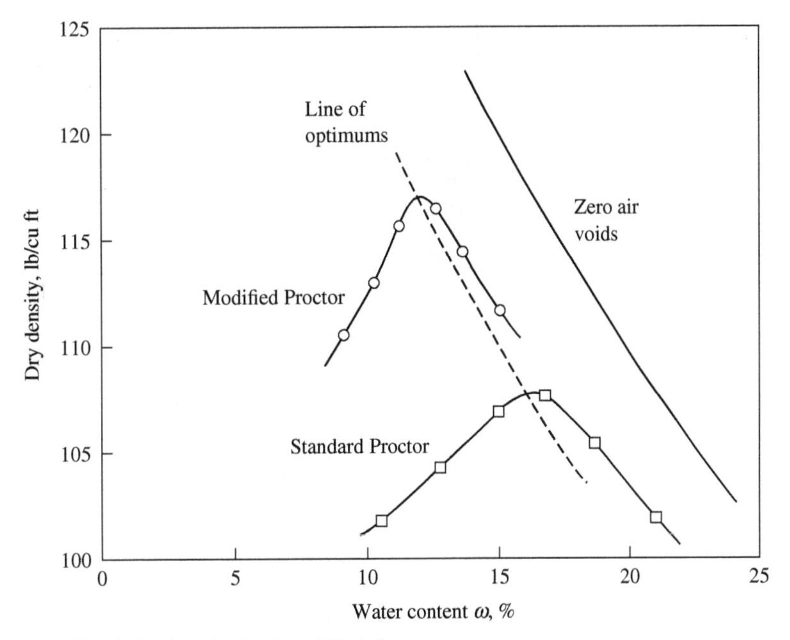

FIGURE 4.4 Standard and modified Proctor curves.

> **Proctor test**
> A method developed by R. R. Proctor for determining the moisture–density relationship in soils subjected to compaction.

The difference in optimum water content is a result of mechanical energy replacing the lubricating action of the water during the densification process. The contractor working to a modified Proctor specification (higher input energy) should plan to make more passes with the compaction equipment or to use heavier compaction equipment on the project. At the same time, there will be less of a requirement to haul water and to mix water into the material.

Compaction Tests. The most commonly used laboratory compaction test accepted by transportation agencies is the **Proctor test**. For this test, a sample of soil consisting of ¼ in. and finer material is used. The sample is placed in a steel mold in three equal layers. The cylindrical steel mold has an inside diameter of 4.0 in. and a height of 4.59 in. In the standard test, each of three equal layers is compacted by dropping a 5.5-lb rammer with a 2-in. circular base 25 times from a height of 12 in. above the specimen (Fig. 4.5). The entire specimen is removed from the mold and is immediately weighed. Then a sample of the specimen is taken and weighed. This small part of the original sample is dried to a constant weight to remove all moisture. It is then weighed again so the water content of the sample can be calculated. With the water content information, the dry weight of the specimen can be determined. The test is repeated, using varying water content specimens, until the water content that produces the maximum dry density is determined. This test is designated as ASTM D-698 or AASHTO T 99.

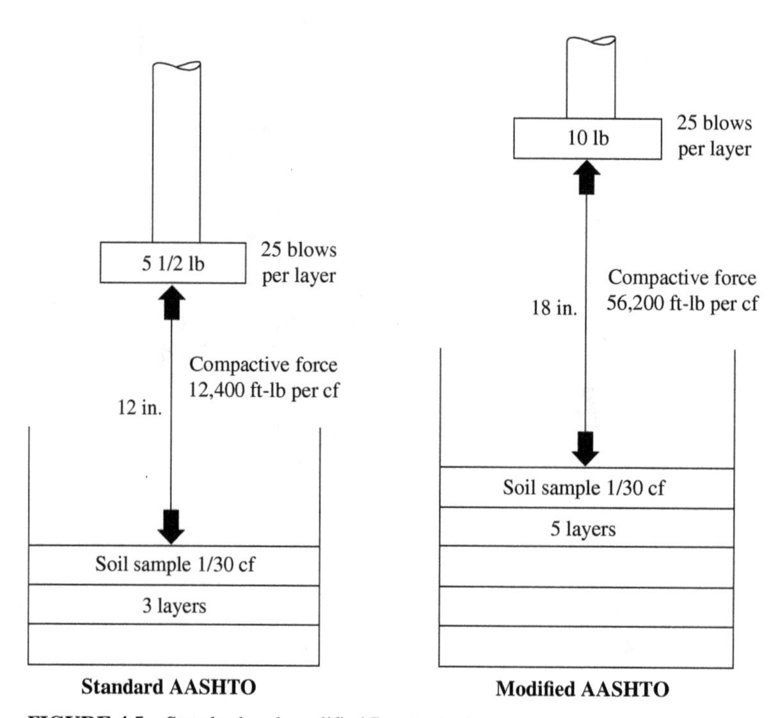

FIGURE 4.5 Standard and modified Proctor test.

The modified Proctor test, designated as ASTM D-1557 or AASHTO T 180, is performed in a similar manner, except the applied energy is approximately five times greater because a 10-lb rammer is dropped 18 in. on each of five equal layers (see Fig. 4.5).

Compaction Control. The specifications for a project may require a contractor to compact the soil to 95% to 100% of a relative density, based on the standard Proctor test or a laboratory test at some other energy level. If the maximum laboratory dry density of the soil is determined to be 120 lb per cf, and the specifications require compaction of at least 95% of maximum density, the contractor must compact the soil in the field to at least 0.95×120 lb per cf = 114 lb per cf.

Field verification tests of achieved compaction can be conducted by any of several accepted methods: sandcone, balloon, or nuclear. The first two methods are destructive tests. They involve

1. Excavating a hole in the compacted fill and weighing the excavated material
2. Determining the water content of the excavated material
3. Measuring the volume of the resulting hole by use of the sandcone or a water-filled balloon
4. Computing the density based on the obtained total weight of the excavated material and hole volume

Lift A layer of soil placed on top of previously placed embankment material. The term can be used in reference to material as spread or as compacted.

Figure 4.6 illustrates a technician performing a sandcone test on a pavement base layer consisting of moist compacted sand.

The dry density conversion can be made because the water content is known. Some difficulties are associated with such methods: (1) they are too time-consuming to conduct sufficient tests for a statistical analysis, (2) there are problems with oversized particles, and (3) there is a time delay in determining the water content. Because the tests are usually conducted on each placement **lift**, delays in testing and acceptance can delay construction operations.

FIGURE 4.6 Sandcone compaction test on a pavement base layer.

Nuclear Compaction Test. Nuclear test methods are used extensively to determine the water content and density of soils. The instrument required for this test can be easily transported to the fill and placed at the desired test location, and within a few minutes, the results can be read directly from the digital display.

The device uses the Compton effect of gamma-ray photon scattering for density determinations (Fig. 4.7) and hydrogenous thermalization of fast neutrons for moisture determinations. The emitted rays enter the ground where they are partially absorbed and reflected. Reflected rays pass through Geiger-Müller tubes in the surface gauge. Counts per minute

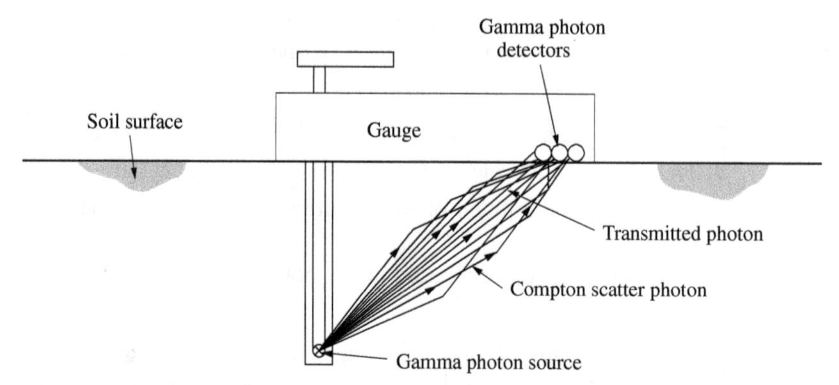

FIGURE 4.7 Nuclear gauge using the Compton effect for soil density testing.

are read directly on a reflected-ray-counter gauge and are related to moisture and density calibration curves. Lower counts returning to the detector tubes indicate higher density. Test blocks consisting of standard construction materials, such as concrete or stone, are used to calibrate the gamma-ray counts to a known reference density. Because the nuclear source will encounter "half-life" decay over time, periodic calibration is necessary to correct the count–density relationship.

Figure 4.7 illustrates the direct transmission mode for nuclear density measurement where the photons travel through the soil from the source to the detectors. This requires driving a pilot rod through the soil to enable the source to extend below the soil surface. On very compact aggregate base or asphalt pavement where driving a pilot rod is not feasible, the back-scatter mode is used with the radiation source positioned at the surface. Emitted photons pass from the source to the detector tubes just below the surface of the material layer. Figure 4.8 is a nuclear density gauge measuring granular backfill compaction in the direct transmission mode.

FIGURE 4.8 Nuclear density gauge in direct transmission mode measuring trench backfill compaction.

Advantages of the nuclear method compared with other methods include the following:

1. It decreases the time required for a test from as much as a day to a few minutes, thereby eliminating potentially excessive construction delays.
2. The engineer is better able to characterize the achieved density because more samples can be taken per unit of time and a larger sample size is possible.
3. Non-destructive testing does not require the removal of soil samples from the site of the tests.
4. It provides a means of performing density tests on soils containing large-sized or irregular-sized aggregates.
5. It reduces or eliminates, when properly calibrated and used correctly, the effect of the personal element and possible errors. Erratic results can be easily and quickly rechecked.
6. It allows a comparison of multiple gauges on a single test location to verify the accuracy of readings.

The operator of the nuclear density gauge needs to be certified because these instruments present a radiation exposure source. Additionally, the operator must exercise reasonable care to ensure the safety of bystanders. By following the instructions furnished with the instruments and wearing a nuclear dosimeter badge during operation, exposure can be kept well below the limits set by the Nuclear Regulatory Commission (NRC). In the United States and other countries, a license is required to own, possess, transport, or use nuclear-type instruments.

Laboratory versus Field. The percentage of maximum dry density is a threshold for a specific compaction effort (input energy level) and the method effort is applied. If more energy is applied in the field, a density greater than 100% of the laboratory value can be achieved. Dissimilar materials have individual curves and maximum values for the same input energy (Fig. 4.9). Well-graded sands have a higher dry density than uniform soils. As plasticity increases, the dry density of clay soils decreases.

As the curves prove, there is a point where higher moisture contents result in decreased density. This is because initially the water serves to "lubricate" the soil grains and helps the mechanical compaction operation move them into a compact physical arrangement. But the density of water is less than for soil solids, and at water contents above optimum, water is replacing soil grains in the matrix. If compaction is attempted at a water content much above optimum, no amount of effort will overcome these physical facts. The water will tend to move the grain particles away from each other and negatively affect the structural capacity development layer. Under such conditions, extra compactive effort will be wasted work. In fact, soils can be "overcompacted." Shear planes are established, and there is a large reduction in strength.

Seismic and Deflection Testing. Newer non-destructive testing technologies are being investigated for soil and rock compaction, including seismic and portable detection methods. Seismic test methods, typically used for subsurface exploration of rock and soil strata, are now being developed as alternatives to density compaction to measure modulus. Young's modulus (E), commonly referred to as a soil's elastic modulus, is an elastic soil parameter and a measure of stiffness. Mechanical waves are emitted into the compacted surface using compression (P wave), shear (S wave), and surface (R wave) waves. The stress waves, which travel and propagate outward from the contact source, are detected by accelerometer

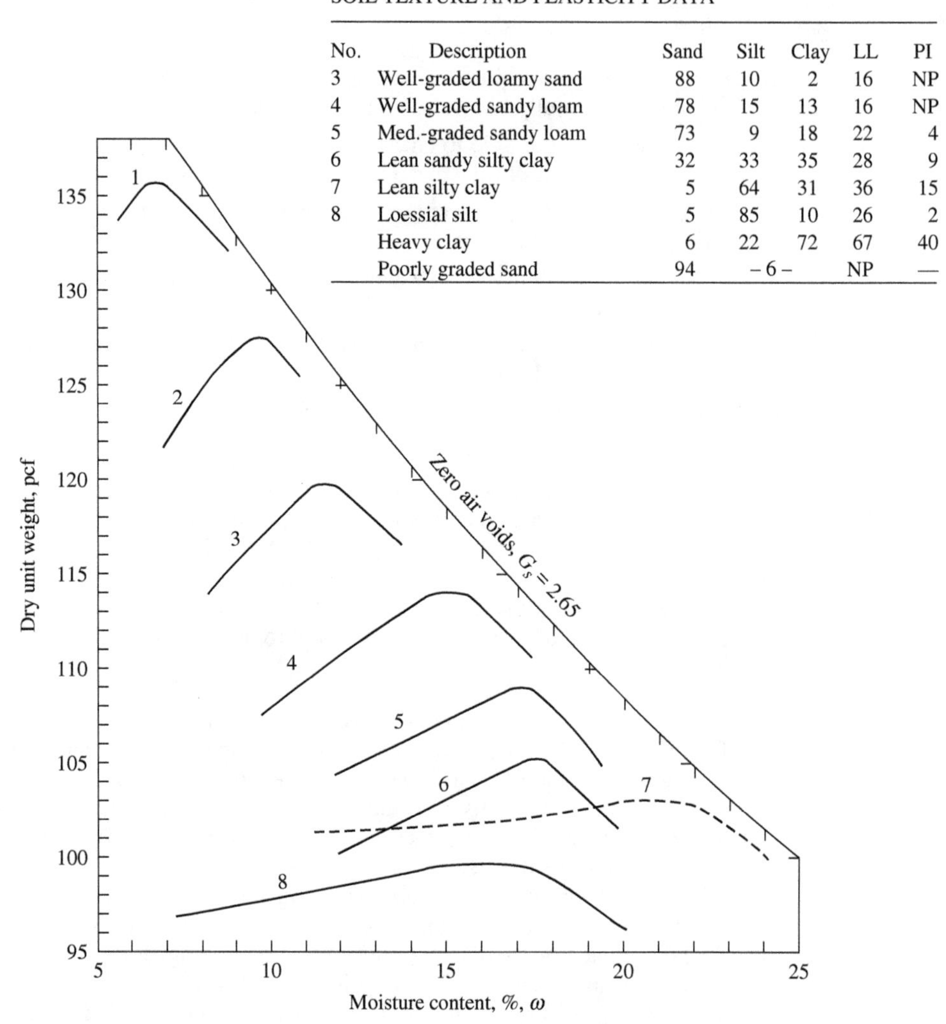

SOIL TEXTURE AND PLASTICITY DATA

No.	Description	Sand	Silt	Clay	LL	PI
3	Well-graded loamy sand	88	10	2	16	NP
4	Well-graded sandy loam	78	15	13	16	NP
5	Med.-graded sandy loam	73	9	18	22	4
6	Lean sandy silty clay	32	33	35	28	9
7	Lean silty clay	5	64	31	36	15
8	Loessial silt	5	85	10	26	2
	Heavy clay	6	22	72	67	40
	Poorly graded sand	94	– 6 –		NP	—

FIGURE 4.9 Compaction curves for eight soils compacted according to AASHTO T99. (*Source: The Highway Research Board.*)

receivers and translated into a signal. The measured difference in time between the arrival of the shear waves is used to calculate the layer modulus.

Light weight deflectometers (LWDs) are portable and lightweight test devices capable of measuring load deflection and layer modulus at point locations in the compacted layer. The LWD device consists of the weight (hammer) on a pole and the sensors (geophones) in a plate on the ground, all encompassed in one connected, portable structure. The sensors are connected to a handheld computer by wireless remote technology to estimate layer modulus using back-calculated load deflection. LWDs are rapidly gaining acceptance for use with unbound layers due to their affordability and repeatability for single-layer structural evaluation.

SOIL PROCESSING

The optimum water content for compaction varies from about 12% to 25% for fine-grained soils and from 7% to 12% for well-graded granular soils. Because it is difficult to attain and maintain the exact optimum water content, normal practice is to work within an acceptable moisture range. This range, which is usually ±2% of optimum, is based on attaining the maximum density with the minimum compactive effort.

Adding Water to Soil. If the moisture content of a soil is below the optimum moisture range, water must be added to the soil prior to compaction. When it is necessary to add water, consider the following:

- Amount of water required
- Rate of water application
- Method of application
- Sequence of construction operations
- Effects of the climate and weather

Granular materials
Soil whose particle sizes and shapes are such that they do not stick together.

Water can be added to the soil at the borrow pit or in place at the construction site. When processing **granular materials**, the best results are usually obtained by adding water in place. After water is added, it must be thoroughly and uniformly mixed with the soil.

Amount of Water Required. It is essential to determine the amount of water required to achieve a soil water content within the acceptable moisture range for compaction. The amount of water to be added or removed is normally computed in gallons per station (100 ft of length); therefore, the volume in the following formula would normally be that for one station length. The computation is based on the dry weight of the soil and compacted volume.

$$\text{Gallons} = \text{Desired dry density pounds per cf (pcf)} \times \frac{(\text{Desired water content }\%) - (\text{Water content borrow }\%)}{100} \times \frac{\text{Compacted vol. of soil (cf)}}{8.34\text{ lb per gal}} \quad (4.14)$$

The 8.34 lb per gal is the weight of a gallon of water. When adding water, it is good practice to adjust the desired moisture content 2% above optimum, but this depends on the environmental conditions (temperature and wind) and the soil type. A negative answer for Eq. (4.14) indicates that water must be removed from the borrow material before it is compacted on the fill.

Sample Calculation: Job specifications require placement of the embankment fill soil in 6-in. compacted lifts. The desired dry unit weight of the embankment is 120 pcf. From the laboratory compaction curve, the optimum water content, sometimes referred to as the optimum moisture content (OMC), of this soil is 12%. The water content of the borrow material is 5%. Again, this is from a series of tests. The roadway lift to be placed is 40 ft wide. Compute the amount of water in gallons to add on a per-station basis for each lift of material.

$$\text{Gallons per station} = 120\text{ pcf} \times \frac{12\%\text{ (OMC)} - 5\%}{100} \times \frac{40\text{ ft} \times 100\text{ ft} \times 0.5\text{ ft}}{8.34\text{ lb/gal}}$$

$$= 120\text{ pcf} \times 0.07 \times \frac{2{,}000\text{ cf}}{8.34\text{ lb/gal}}$$

$$= 2{,}015\text{ gallons per station}$$

Application Rate. Once the total amount of water has been calculated, the application rate can be calculated. The water application rate is normally calculated in gallons per square yard, using the following formula:

$$\text{Gallons per square yard} = \text{Desired dry density (pcf)} \times \frac{\%\text{ moisture added or removed}}{100} \times \text{Compacted lift thickness (ft)} \times \frac{9\text{ sf/sy}}{8.34\text{ lb per gal}} \quad (4.15)$$

Sample Calculation: Using the data from the previous sample calculation, determine the required application rate in gallons per square yard:

$$\text{Gallons per square yard} = 120\text{ pcf} \times 0.07 \times 0.5\text{ ft} \times \frac{9\text{ sf/sy}}{8.34\text{ lb per gal}}$$

$$= 4.5\text{ gal per sy}$$

Application Methods. Once the application rate has been calculated, the method of application must be determined. Regardless of which method of application is used, it is important to ensure a proper application rate and to apply the water in a uniform manner.

On most projects, the usual method of adding water to a soil is with a water distributor. Water distributors are designed to spread the correct amount of water evenly over the fill. These truck-mounted (Fig. 4.10) or towed water distributors dispense water under various pressures or by gravity feed. Many distributors are equipped with rear-mounted spray bars with machine speed and pumping rate calibrated for uniform application (Fig. 4.11). The operator can maintain the water application rate by controlling the forward speed of the vehicle.

If time is available, water can be added to soil by ponding or prewetting the area until the desired depth of water penetration is achieved. With this method, it is difficult to control the application rate. Ponding usually requires several days to achieve a uniform moisture distribution.

Reducing the Moisture Content. If a soil contains more water than desired (above the optimum moisture range), it is difficult to compact to the specified density. In such cases, steps must be taken to reduce the moisture content to within the required moisture range. Drying actions may be as simple as aerating the soil or as complicated as adding a soil stabilization agent to change the physical properties of the soil. Lime or fly ash is the typical stabilization agent for fine-grained soils. If a high water table is causing the excess

FIGURE 4.10 Water truck working on the project.

FIGURE 4.11 Water truck applying water on pavement base.

moisture, some form of subsurface drainage may be required before the soil's moisture content can be reduced.

The most common method of reducing the moisture is to scarify the soil prior to compaction. This can be accomplished with either the scarifying teeth or rippers on a motor grader or by disking the soil (Fig. 4.12). A motor grader can also use its blade to roll the soil into furrows to expose more material for drying.

Effects of Weather. Weather conditions substantially affect soil moisture content. Cold, rainy, and cloudy weather causes soil to retain water. Hot, dry, sunny, and windy weather is conducive to drying the soil. In a desert climate, evaporation claims a large amount of water intended for the soil lift. Thus, for projects in dry arid regions, the engineer might have to go as high as 6% above the optimum moisture content as a target for all water application

FIGURE 4.12 Disking to reduce moisture content.

calculations. The climate conditions will remove the extra water, and the actual moisture content during the work will fall near to the desired range.

Mixing and Blending. Whether adding water to a soil to increase the moisture content or adding a drying agent to reduce it, mixing the water or drying agent thoroughly and uniformly with the soil is essential. Even if additional water is not needed, mixing may still be necessary to ensure a uniform distribution of the existing moisture. Mixing can be accomplished using motor graders, agricultural disk harrows, or rotary cultivators. Conventional motor graders can be used to mix or blend a soil additive (water or stabilizing agent) by windrowing the material from one side of the working lane to the other.

CHAPTER 5
MACHINE FUNDAMENTALS

A constructor must select the proper equipment to relocate, place, and/or process materials economically. Analysis procedures for matching the appropriate machine to the project task require inquiry into a machine's mechanical capability. The first step in selecting a machine for a task is to calculate the power required to propel the machine and its load. This power requirement is established by two factors: (1) rolling resistance and (2) grade resistance. Equipment manufacturers publish performance data for their machines. These data in chart form enable the equipment planner to analyze a machine's ability to perform under a given set of job and load conditions.

MACHINE POWER

Earth-moving projects require the handling and processing of large quantities of bulk materials. As a consequence, the builder of such projects must select the proper equipment to relocate and/or process bulk materials economically. The decision process for matching an appropriate machine to the project task requires the constructor to take into account the properties of the material to be handled and to analyze the mechanical capabilities of the machine.

When considering a project material-handling problem, three crucial respects of the material and the task must be identified: (1) total quantity of material, (2) size of the individual pieces, and (3) the rate at which it must be moved. The quantity of material to be handled and the time constraints resulting from the project contract specifications or from expected weather conditions influence the selection of machines as to type, size, and number to be employed. Larger units generally have lower unit-production costs, but there is a tradeoff in higher mobilization and fixed costs. Moreover, machine size alternatives are subject to constraints imposed by the size of the individual material pieces to be handled. A loader used in a quarry to move shot rock must be capable of handling the largest rock sizes produced. A bottom-dump trailer may not be able to discharge material with high clay content.

Payload

The payload capacity of excavation and hauling equipment must be investigated in both volumetric and gravimetric terms. Both the struck and heaped volumetric capacity data are provided by manufacturers. Using the manufacturer's data, an analysis can be performed in terms of loose cubic yards, bank cubic yards, or compacted cubic yards.

The payload capacity of excavation buckets and hauling units is often stated by the manufacturer in terms of a loose material volume. This rated capacity is based on industry standards for the material in either a heaped condition or at a specified angle of repose. A gravimetric capacity represents the safe operational weight based on the strength of the machine's axles and structural frame.

From an economic standpoint, overloading a truck or any haul unit to improve production may appear attractive, and overloading by 20% might increase the haulage productivity by 15%, allowing for slight increases in time to load and haul. The cost per ton hauled should show a corresponding decrease, because direct labor costs will not change and fuel costs will increase only slightly. Nevertheless, this apparently favorable situation is only temporary because the advantage is being bought at the cost of premature aging of the haul unit and a corresponding increased replacement capital expense.

MACHINE PERFORMANCE

Cycle time and payload determine a machine's production rate, and machine travel speed directly affects cycle time. "Why does the machine travel at only 12 mph when its top speed is listed as 35 mph?" To answer the travel speed question, it is necessary to examine three power considerations:

- Required power
- Available power
- Usable power

Required Power

Rolling resistance The resistance of a level surface to constant-velocity motion across it.

Grade resistance The force-opposing movement of a machine up a frictionless slope.

Required power is the power needed to overcome resisting forces and cause machine motion. The magnitude of resisting forces establishes this power requirement. The forces resisting the movement of mobile equipment are (1) **rolling resistance** and (2) **grade resistance**. Therefore, required power is the power necessary to overcome the total resistance to machine movement, which is the sum of rolling and grade resistance.

$$\text{Total resistance (TR)} = \text{Rolling resistance (RR)} + \text{Grade resistance (GR)} \quad (5.1)$$

Rolling Resistance. Rolling resistance is the resistance of a level surface to constant-velocity motion across it. It is usually expressed as pounds of resistance per ton of vehicle weight or as an equivalent grade resistance. Consider a loaded truck with a gross weight of 20 tons and moving over a level road with a rolling resistance of 100 lb/ton. The force required to overcome rolling resistance and keep the truck moving at a uniform speed will be 2,000 lb (20 tons × 100 lb/ton).

This is sometimes referred to as wheel resistance or track resistance. Rolling resistance results from friction of the driving mechanism; tire flexing, in the case of pneumatic tired equipment; and the force required to shear through or ride over the supporting surface (Fig. 5.1).

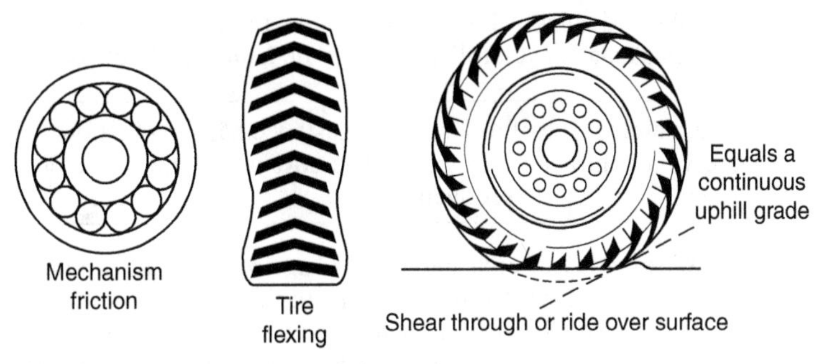

FIGURE 5.1 Mechanisms of rolling resistance.

Rolling resistance varies significantly with the type and condition of the surface over which a machine moves. Soft earth (Fig. 5.2, left) offers a higher resistance than hard-surfaced (Fig. 5.2, right) roads such as compacted earth or stone. In the case of machines with pneumatic tires, the rolling resistance varies with the size of, pressure on, and tread design of the tires. For crawler track equipment, the resistance varies primarily with the type and condition of the road surface.

FIGURE 5.2 Rolling resistance varies with the condition of the surface over which a mobile machine moves.

On a hard-surfaced road a narrow-tread, high-pressure pneumatic tire offers lower rolling resistance than a broad-tread, low-pressure tire. This is the result of the small area of contact between the tire and the road surface. If the travel surface is soft and the tire sinks into the earth, a broad-tread, low-pressure tire will offer a lower rolling resistance than a narrow-tread, high-pressure tire. A narrow tire will sink farther into the earth than the broad tire, and thus more power is needed to continually climb out of a deeper rut—a situation equivalent to climbing a steeper grade.

The rolling resistance of an earthen-haul road will not remain constant under varying climatic conditions or for the varying types of road surfaces. If the soil material is stable, highly compacted, and well maintained by a grader, and if the moisture content is kept to near optimum, it is possible to provide an earthen road surface with a rolling resistance about as low as a concrete or asphalt surface. Moisture can be added, but following an extended period of rain, it may be difficult to remove the excess moisture and the haul road will become soft and rutted; when this happens, rolling resistance will increase. Providing good surface drainage will speed the removal of the water and should permit the road to be reconditioned quickly. For a major earthwork project, it is good economy to provide graders, water trucks, and even rollers to keep the haul road in the desired condition.

> The maintenance of low-rolling resistance haul roads is one of the best financial investments a contractor can make. The cost of having a grader maintain the haul road is repaid in increased production.

A pneumatic tire sinks into the road surface until the product of bearing area and bearing capacity is sufficient to sustain the load; then the tire is always attempting to climb out of the resulting rut. The rolling resistance will increase about 30 lb/ton for each inch of tire penetration. Total rolling resistance is a function of the riding gear characteristics (independent of speed), the total weight of the vehicle, and the torque.

The estimation of off-road rolling resistance is based largely on empirical information, including experience with similar soils. Rarely are rolling resistance values based on actual field tests of haul roads. Much of the available test data come from aircraft tire research performed at the U.S. Army Waterways Experiment Station (today part of the U.S. Army Engineer Research and Development Center laboratories at Vicksburg, Mississippi). Although it is impossible to give completely accurate values for the rolling resistances for all types of haul roads and wheels, the values provided in Table 5.1 are reasonable estimates.

TABLE 5.1 Representative Rolling Resistance for Various Types of Wheels and Crawler Tracks versus Various Surfaces*

	Crawler Type Track and Wheel		Rubber Tires, Antifriction Bearings			
			High Pressure		Low Pressure	
Surface Type	lb/ton	kg/m ton	lb/ton	kg/m ton	lb/ton	kg/m ton
Smooth concrete	55	27	35	18	45	23
Good asphalt	60–70	30–35	40–65	20–33	50–60	25–30
Soil, compacted and maintained	60–80	30–40	40–70	20–35	50–70	25–35
Soil, poorly maintained	80–110	40–55	100–140	50–70	70–100	35–50
Soil, rutted, muddy, no maintenance	140–180	70–90	180–220	90–110	150–200	75–100
Loose sand and gravel	160–200	80–100	260–290	130–145	220–260	110–130
Soil, very muddy, rutted, soft	200–240	100–120	300–400	150–200	280–340	140–170

*In pounds per U.S. ton or kilograms per metric ton of gross vehicle weight.

It is possible to determine the rolling resistance of a surface by towing a truck or other vehicle whose gross weight is known along a level section at a uniform speed. The tow cable must be equipped with a dynamometer or other device capable of measuring the average tension in the cable. This tension is the total rolling resistance of the gross weight for the truck. The rolling resistance in pounds per gross ton will be

$$R = \frac{P}{W} \tag{5.2}$$

where R = rolling resistance in pounds per ton
P = total tension in tow cable in pounds
W = gross weight of mobile vehicle in tons

When tire penetration is known, an approximate rolling resistance value for a wheeled vehicle can be calculated using Eq. (5.3):

$$RR = [40 + (30 \times TP)] \times GVW \tag{5.3}$$

where RR = rolling resistance in pounds
TP = tire penetration in inches
GVW = gross vehicle weight in tons

Grade Resistance. The force-opposing movement of a machine up a frictionless slope is known as grade resistance. It acts against the total weight of the machine, whether track type or wheel type. When a machine moves up an adverse slope (Fig. 5.3), the power required to keep it moving increases approximately in proportion to the slope of the road. If a machine moves down a sloping road, the power required to keep it moving is reduced in proportion to the slope of the road. This is known as **grade assistance**.

Grade assistance The effect of gravitational force in aiding movement of a vehicle down a slope.

The most common method of expressing a slope is by gradient in percent. With a 1% slope, the surface rises or drops 1 ft vertically in a horizontal distance of 100 ft. If the slope is 5%, the surface rises or drops 5 ft per 100 ft of horizontal distance. If the surface rises, the slope is defined as plus; if it drops, the slope is defined as minus. This is a physical property not affected by the type of machine or the condition of the road, but with respect to analyzing forces, its effect is dependent upon the machine's direction of travel.

FIGURE 5.3 A scraper moving up an adverse slope.

For slopes of less than 10%, the effect of grade is to increase (for a plus slope) or decrease (for a minus slope) the required tractive effort by 20 lb per gross ton of machine weight for each 1% of grade. This can be derived from elementary mechanics by calculating the required driving force.

From Fig. 5.4, the following relationships can be developed:

$$F = W \sin \alpha \tag{5.4}$$

$$N = W \cos \alpha \tag{5.5}$$

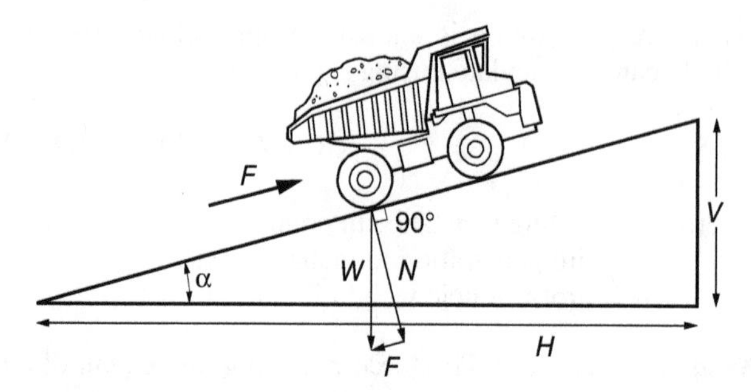

FIGURE 5.4 Frictionless slope–force relationships.

For angles less than 10°, sin $\alpha \approx$ tan α (the small-angle assumption); with that substitution:

$$F = W \tan \alpha \tag{5.6}$$

But

$$\tan \alpha = \frac{V}{H} = \frac{G\%}{100}$$

where G% is the gradient. Hence,

$$F = W \times \frac{G\%}{100} \tag{5.7}$$

If we substitute W = 2,000 lb/ton, the formula reduces to

$$F = 20 \text{ lb/ton} \times G\% \tag{5.8}$$

This formula is valid for a G up to about 10%, that is, the small-angle assumption.

Total Resistance. Total resistance equals rolling resistance plus grade resistance, or rolling resistance minus grade assistance [Eq. (5.1)]. It can also be expressed as an effective grade.

Using the relationship expressed in Eq. (5.8), a rolling resistance can be equated to an equivalent gradient:

$$\frac{\text{Rolling resistance expressed in lb/ton}}{20 \text{ lb/ton}} \tag{5.9}$$

Table 5.2 provides values for the effect of slope, expressed in pounds per gross ton or kilograms per metric ton (m ton) of vehicle weight.

By combining the rolling resistance, expressed as an equivalent grade, and the grade resistance, expressed as a gradient in percent, you can express the total resistance as an effective grade. The three terms—power required, total resistance, and effective grade—all denote the same thing. Power required is expressed in pounds. Total resistance is expressed in pounds or pounds per ton of machine weight, and effective grade is expressed in percent.

Sample Calculation: The haul road from the borrow pit to the fill has an adverse grade of 4%. Wheel-type hauling units will be used on the job, and the haul-road rolling resistance is 100 lb per ton. What will be the effective grade for the haul? Will the units experience the same effective grade for the return trip?

TABLE 5.2 The Effect of Grade on the Tractive Effort of Vehicles

Slope (%)	lb/ton*	kg/m ton*	Slope (%)	lb/ton*	kg/m ton*
1	20.0	10.0	12	238.4	119.2
2	40.0	20.0	13	257.8	128.9
3	60.0	30.0	14	277.4	138.7
4	80.0	40.0	15	296.6	148.3
5	100.0	50.0	20	392.3	196.1
6	119.8	59.9	25	485.2	242.6
7	139.8	69.9	30	574.7	287.3
8	159.2	79.6	35	660.6	330.3
9	179.2	89.6	40	742.8	371.4
10	199.0	99.5	45	820.8	410.4
11	218.0	109.0	50	894.4	447.2

*Ton or metric ton of gross vehicle weight.

Using Eq. (5.9), we obtain

$$\text{Equivalent grade (RR)} = \frac{100\text{ lb/ton rolling resistance}}{20\text{ lb/ton}} = 5\%$$

$$\text{Equivalent grade } (\text{TR}_{\text{haul}}) = 5\%\text{ RR} + 4\%\text{ GR} = 9\%$$

$$\text{Equivalent grade } (\text{TR}_{\text{return}}) = 5\%\text{ RR} - 4\%\text{ GR} = 1\%$$

where RR = rolling resistance
GR = grade resistance

Note: Effective grade is not the same for the two cases. During the haul, the unit must overcome the uphill grade; on the return, the unit is aided by the downhill grade.

Haul Routes. During the life of a project, the haul-route grades (and therefore the grade resistance experienced by a moving vehicle) may remain constant. One example of this is the trucking of aggregate from a rail yard off-load point to the concrete batch plant. In most cases, however, the haul-route grades change as the work progresses. On a liner highway or airfield project, the tops of the hills are excavated and hauled into the valleys. Early in the work, the grades are steep and reflect the existing natural ground. Over the life of the project, the grades begin to assume the design profile. Consequently, the engineer must first study the project's mass diagram to determine the direction of material movement. Then the natural ground and the design profiles specified by the contract documents must be checked to determine the grades of the proposed haul and return routes.

Site work projects are usually not linear in extent; therefore, a mass diagram is not a useful planning tool for such work. In the case of such work it is necessary to look at the cut and fill areas, lay out probable haul routes, and then check the natural and finish grade contours of the assumed haul and return routes.

This process of developing project haul routes is critical to machine productivity. Routes providing less grade resistance result in increased machine travel speed and yield a production increase. In planning a project, a constructor should always check several haul-route options before deciding on a final haul plan.

Hauling efficiency is achieved by careful haul-route planning.

Equipment selection is affected by travel distance because of the time factor distance introduces into the production cycle. All other factors being equal, increased travel distances will favor the use of high-speed, large-capacity machines. The difference between the self-loading scraper and a push-loaded scraper can be used as an illustration. The self-loading scraper will load, haul, and spread without any assisting equipment, but the extra weight of the loading mechanism reduces the scraper's maximum travel speed and load capacity. A scraper without the self-loading mechanism requires a push tractor to help it load but does not have to expend power to haul the loading mechanism with it on every cycle. Comparatively, it will be more efficient in long-haul situations, as it does not have to expend additional fuel transporting the extra machine weight.

Available Power

Internal combustion engines power most construction equipment. Because diesel engines perform better under heavy-duty applications than gasoline engines, diesel-powered machines are the workhorses of the construction industry. Additionally, diesel engines have longer service lives and lower fuel consumption, and diesel fuel presents less of a fire hazard. No matter which type of engine serves as the power source, the mechanics of energy transmission are the same.

Work and Power. Work is defined as force through a distance [Eq. (5.10)]. It is achieved when a force causes an object to move.

$$\text{Work} = \text{Force} \times \text{Distance} \tag{5.10}$$

When James Watt developed the first practical steam engine and wanted to express the work his engine could perform, he related it to a horse walking in a circle to power a pump. He used the horse analogy because the purpose of his engine was to replace the horses used to power pumping apparatuses in the mines of England. Watt defined power as the amount of work completed in a certain amount of time:

$$\text{Power} = \frac{\text{Work}}{\text{Time}} \tag{5.11}$$

Therefore, the power of a horse would be

$$\text{Power of a horse} = \frac{1{,}954{,}320 \text{ lb-ft/hour}}{60 \text{ min/hr}} = 32{,}572 \text{ lb-ft/min}$$

Watt rounded this value to 33,000 lb-ft/min (or 550 lb-ft/sec), which is the definition of one horsepower.

Torque. By burning fuel in a piston, an internal combustion engine develops a mechanical force, and the force then acts on a crankshaft with a radius, r. The crankshaft in turn drives the flywheel and gears. The gears transfer the power to the other components of the machine. The force from a rotating object, such as crankshafts (a "twisting" force), is termed *torque*. A pound-foot of torque is the twisting force necessary to support 1-lb weight on a weightless horizontal bar 1 ft from the fulcrum, or in the case of a machine it is the amount of force exerted at a distance of 1 ft from the center of the crankshaft.

The torque represented by one revolution of work is 6.2832 lb-ft (the circumference of a circle with a 1-ft radius is 6.2832 ft). The power exerted by a rotating object is the torque it exerts multiplied by the speed at which it rotates (revolutions per minute, or rpm). The relationship between horsepower and torque (T) at a specified rpm can therefore be established:

$$\text{Horsepower (hp)} = \frac{6.2832 \times \text{rpm} \times T}{33{,}000} = \frac{\text{rpm}}{5{,}252} \times T \tag{5.12}$$

Conversely, to calculate torque

$$\text{Torque (T)} = \frac{5{,}252}{\text{rpm}} \times \text{hp} \tag{5.13}$$

Horsepower Rating. Manufacturers rate machine horsepower as either gross or flywheel (sometimes listed as net horsepower). Gross horsepower is the actual power generated by the engine prior to load losses for auxiliary systems, such as the alternator, air conditioner, compressors, and a water pump. Flywheel horsepower (fwhp) can be considered usable horsepower. It is the power available to operate a machine—power the driveline—after deducting for power losses in the engine. This horsepower is sometimes listed as brake horsepower (bhp). Prior to electronic bench testing, horsepower was quantified as the amount of resistance against a flywheel brake. Although the method is no longer used, the term remains in use.

The Society of Automotive Engineers' (SAE) standardized engine-rating procedure (J1349) measures horsepower at the flywheel using an engine dynamometer. The engine is tested with all accessories installed, including a full exhaust system, all pumps, the alternator, the starter, and the emission controls. So today, equipment manufacturers measure torque on a dynamometer and then calculate horsepower by converting the radial force of torque into work units of horsepower.

Power Output and Torque

Maximum torque is not obtained at maximum rpm. In Eq. (5.13), 5,252 is divided by the rpm (5,252 ÷ rpm). Consequently, when the engine turns at an rpm less than 5,252, the effect is to produce a ratio value greater than one—to increase torque for the same hp value. Once the engine rpm increases past 5,252, the effect is to produce a ratio value less than one—to reduce the torque for the same hp value. This effect provides the engine with a power reserve. When a machine is subjected to a momentary overload, the rpm drops and the torque goes up, keeping the engine from stalling. This is commonly referred to as "lugging" the engine.

The power output from the engine, fwhp, becomes the power input to the transmission system. This system consists of the drive shaft, a transmission, planetary gears, drive axles, and drive wheels. A transmission, gear set, or gearbox is a set of gears and shafts that provide a change in the speed–power ratio. It may be a single-speed or gear reduction type in which a small gear driven by the engine meshes with a larger gear that turns the working parts; or it may be a selective or sliding gear type with several ratios.

Machines can be purchased with either a direct drive (manual or standard) or torque-converter drive. With a direct-drive machine, the operator must manually shift gears to match engine output to the resisting load. The difference in power available when considering maximum torque and torque at the governed speed is the machine's operating range for a given gear. In applications where load is constantly changing, the operation of a direct-drive machine requires a skilled operator. Operator skill is a significant factor controlling the amount of wear and tear a direct-drive machine will experience. Operators of direct-drive machines will be subjected to more operator fatigue than those on power-shift models. The fatigue factor will affect productivity.

Rimpull The tractive force between the tires of a machine's driving wheels and the surface on which they travel.

Drawbar pull The available pull that a crawler tractor can exert on a towed load.

Coefficient of traction The factor that determines the maximum possible tractive force between the powered running gear of a machine and the surface on which it travels.

A torque converter connects the machine's power source (the flywheel of the engine) with the transmission and adjusts the power output to match the load. This adjustment is accomplished hydraulically by a fluid coupling. As a machine begins to accelerate, the engine rpm will quickly reach the governed crankshaft speed, and the torque converter will automatically multiply the engine torque to provide the required acceleration force. In this process, there are losses due to hydraulic inefficiencies. If the machine is operating under constant load and at a steady whole-body speed, no torque multiplying is necessary. When no torque multiplying is needed, the transmission of engine torque can be made nearly as efficient as a direct drive transmission by locking ("lock-up") the torque converter pump and transmission together.

When analyzing a piece of equipment, the important question is the usable force developed at the point of contact between the tire and the ground (**rimpull**) for a wheel machine. In the case of a track machine, it is the force available at the drawbar (**drawbar pull**). The difference in the name is a matter of convention: both rimpull and drawbar pull are measured in the same units: pounds pull.

Rimpull. Rimpull is a term used to designate the tractive force between the tires of a machine's driving wheels and the surface on which they travel. If the **coefficient of traction** is sufficiently high, no tire slippage will occur, in which case maximum rimpull is a function of the power of the engine and the gear ratios between the engine and the driving wheels. If the driving wheels slip on the supporting surface, the maximum effective rimpull will be equal to the total pressure the tires exert on the surface multiplied by the coefficient of traction. Rimpull is expressed in pounds.

If the rimpull of a machine is not known, it can be determined from the equation

$$\text{Rimpull} = \frac{375 \times \text{hp} \times \text{eff}}{\text{speed (mph)}} \text{ (lb)} \tag{5.14}$$

This is the formulation of the available power accounting for the machine's available horsepower and operating speed. The efficiency of most tractors and trucks will range from 0.80 to 0.85. The SAE recommends using 0.85 when the efficiency is not known.

In computing the pull a tractor can exert on a towed load (speed the machine can attain), it is necessary to deduct from the rimpull of the tractor the force required to overcome the total resistance—the combination of rolling and grade. Consider a tractor that weighs 12.4 tons and whose maximum rimpull in the first gear is 13,523 lb. If it is operated on a haul road with a positive slope of 2% and a rolling resistance of 100 lb/ton, the full 13,523 lb of rimpull will not be available for towing, as a portion will be required for overcoming the total resistance arising from the haul conditions. Given the stated conditions, the pull available for towing a load will be

$$\text{Maximum rimpull} = 13{,}523 \text{ lb}$$

Force required to overcome grade resistance:

$$12.4 \text{ ton} \times (20 \text{ lb/ton} \times 2\%) = 496 \text{ lb}$$

Force required to overcome rolling resistance:

$$12.4 \text{ ton} \times 100 \text{ lb/ton} = 1{,}240 \text{ lb}$$

Total resistance: 496 lb + 1,240 lb = 1,736 lb

Power available for towing a load: 13,523 lb – 1,736 lb = 11,787 lb

Drawbar Pull. The towing force a crawler tractor can exert on a load is referred to as drawbar pull. Drawbar pull is typically expressed in pounds. To determine the drawbar pull available for towing a load, it is necessary to subtract from the total pulling force available at the engine the force required to overcome the total resistance imposed by the haul conditions. If a crawler tractor tows a load up a slope, its drawbar pull will be reduced by 20 lb for each ton of weight of the tractor for each 1% slope.

The performance of crawler tractors, as reported in the specifications supplied by the manufacturer, is usually based on the Nebraska tests. In testing a tractor to determine its maximum drawbar pull at each of the available speeds, the haul road is calculated to have a rolling resistance of 110 lb/ton. If a tractor is used on a haul road whose rolling resistance is higher or lower than 110 lb/ton, the drawbar pull will be reduced or increased, respectively, by an amount equal to the weight of the tractor in tons multiplied by the variation of the haul road from 110 lb/ton.

The drawbar pull of a crawler tractor will vary indirectly with the speed of each gear. It is highest in the first gear and lowest in the highest gear. Therefore, attainable speed is limited by the required pounds of pull for a specific resistance condition (Fig. 5.5). The specifications supplied by the manufacturer should give the maximum speed and drawbar pull for each of the gears.

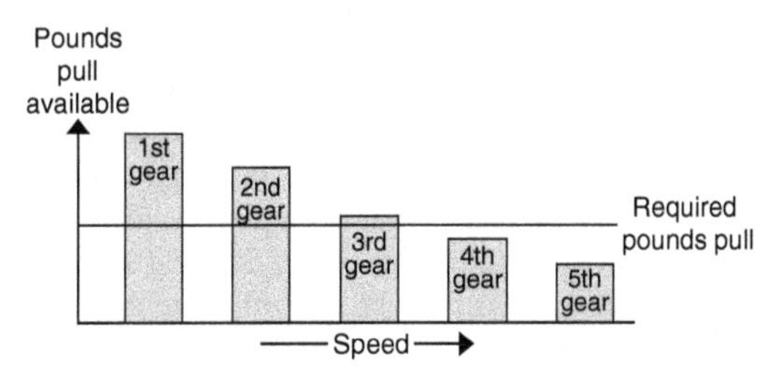

FIGURE 5.5 Total resistance limits possible speed.

Usable Power

Usable power depends on project conditions, primarily haul-road surface condition, altitude, and temperature. Underfoot conditions determine how much of the available power can be transferred to the surface to propel the machine. As altitude increases, the air becomes less dense. Above 3,000 ft, the decrease in air density may cause a reduction in horsepower output of some engines. Manufacturers provide charts detailing appropriate altitude power reductions. Temperature will also affect engine output.

Coefficient of Traction

The total energy of an engine in any machine designed primarily for pulling a load can be converted into tractive effort only if sufficient traction can be developed between the driving wheels or tracks and the haul surface. If there is insufficient traction, the full power of the engine will not be available to do work, as the wheels or tracks will slip on the surface.

Soils differ greatly in their ability to support and facilitate movement of vehicles. In terms of movement, an important characteristic is the amount of friction between the ground travel surface and the drive tires or tracks of a machine. The ability of a machine to propel itself is affected by its weight distribution. Only a portion of machine weight bearing on the drive wheels (or tracks) against the ground affects rimpull. Nonetheless, the total weight of the machine and the load, the gross vehicle weight (GVW), constitute the resistance the drive wheels or tracks must move.

Increasing the weight on slipping drive wheels increases drawbar pull in direct proportion, up to the maximum produced by the engine and gears. Shifting more of the GVW to the drive wheels increases potential traction, whereas shifting weight from drive to non-drive wheels reduces traction. But neither shift affects resistance.

Although resistance to movement is in proportion to the weight of the whole machine, ability to move the machine—if power is adequate—depends on the weight on the drive units. All-wheel-drive trucks and most crawlers keep all of their weight on the drivers. With other machines, weight distribution is critically important to performance when traction is limited.

Manufacturers usually provide detailed weight specifications for haulers, showing the total weight and its distribution, for both loaded and empty conditions.

If there is no specific information on vehicle weight distribution, a rear-drive dump truck typically carries about 50% of its empty weight and 70% of its loaded weight on the drive wheels. A front-engine scraper or wagon usually has 50% to 60% of its empty weight on the drivers and 40% to 50% of its loaded weight on the drivers.

When moving up an incline grade, vehicle weight shifts toward rear drives and away from front drives by about 1.5% for each percentage of grade. This factor increases the already considerable traction advantage of the rear-drive truck in slippery climbs.

Tire Treads. Tire treads are an important variable when assessing traction. There are many different tread designs, all of which eventually wear smooth. For most situations high lugs or cleats give best traction, but smoother tire surfaces are better on dry sand and ice. When tire chains are used, their effectiveness is reduced by high tread blocks.

Because a deep cleated tire gives the best traction on soft loam and a smooth tire on loose sand, there are likely to be intermediate surfaces on which they are equally effective, or perhaps ineffective.

The coefficient of traction can be defined as the factor by which the total weight on the drive wheels or tracks should be multiplied to determine the maximum possible tractive force between the wheels or tracks and the surface just before slipping will occur.

$$\text{Usable force} = \text{Coefficient of traction} \times \text{Weight on powered running gear} \qquad (5.15)$$

The power a machine can develop to do work is often limited by traction. The factors controlling usable horsepower are the weight on the powered running gear (drive wheels for wheel type, total weight for track type—Fig. 5.6), the characteristics of the running gear, and the characteristics of the travel surface.

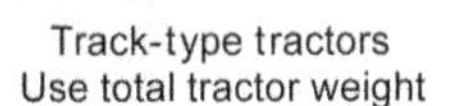

2-wheel-tractors
Use weight on drivers shown on spec sheet or 50% of GVW

Rear-drive dump truck
Use weight on drivers shown on spec sheet or 70% of GVW

FIGURE 5.6 Weight distribution on powered running gear.

The coefficient of traction between rubber tires and travel surfaces will vary with the type of tread on the tires and with the travel surface. For crawler tracks, it will vary with the design of the grosser and the travel surface. Because these exact values cannot be given, Table 5.3 gives approximate values for the coefficient of traction between rubber tires and crawler tracks, as well as surface materials and conditions. These coefficients are sufficiently accurate for most estimating purposes.

TABLE 5.3 Coefficients of Traction for Various Surfaces

Surface	Rubber Tires	Tracks
Dry concrete	0.80–1.00	0.45
Dry, clay loam	0.50–0.70	0.90
Wet, clay loam	0.40–0.50	0.70
Wet, sand and gravel	0.30–0.40	0.35
Gravel road, loose	0.35	0.50
Loose, dry sand	0.20–0.30	0.30
Dry snow	0.20	0.15–0.35
Ice	0.10	0.10–0.25

Altitude's Effect on Usable Power

The SAE standard J1349, *Engine Power Test Code—Spark Ignition and Compression Ignition—Net Power Rating Standard*, specifies a basis for a net engine power rating. The standard conditions for the SAE rating are a temperature of 60°F (15.5°C) and a sea-level barometric pressure of 29.92 in. Hg (103.3 kPa).

Engine ratings are based on a specific barometric pressure. For naturally aspirated engines, operation at altitudes above sea level will cause a decrease in available engine power as the barometric pressure decreases. A decrease in barometric pressure causes a corresponding decrease in air density, and to operate at peak efficiency the engine must have the proper amount of air. A reduction in air density affects the combustion fuel-to-air ratio in the engine's pistons.

In general, an increase in altitude will reduce the power of a four-cycle engine about 3% for each 1,000 ft about sea level. A two-cycle diesel will keep full power to 1,000 ft and then lose about 0.9% for each additional 1,000-ft rise. The better showing is due to the low-pressure blower used in the intake. A barometer drop of 1 in., as, for example, from 30.0 to 29.0 when a storm approaches, will reduce engine power about as much as an increase in altitude of 900 ft. For specific machine applications, the manufacturer's performance data should be consulted.

An increase in temperature will also reduce engine power at a rate of about 0.9% for each 10°F (5.5°C). Cooling has the reverse effect. Such power losses have a considerable effect on performance and production at high altitudes and must be considered when making production estimates.

Moreover, fuel consumption increases as power is lost with altitude, as the thinner air does not provide an efficient mixture. Special carburetor, injector, or pump adjustments may be made to save some of the fuel, but they cannot restore the power.

The effect of the loss in power due to altitude can be minimized by the installation of a turbocharger or supercharger. These mechanical forced-induction systems compress the air flowing into the combustion chamber of the engine, thus permitting sea-level performance at higher altitudes. The fundamental difference between a turbocharger and a supercharger is the unit's source of power. With a turbocharger, the exhaust stream powers a turbine, which in turn spins the compressor. The power source for a supercharger is a belt connected directly to the engine. If equipment is to be used at high altitudes for long periods, the increased performance will probably pay for installing one of these two devices.

PERFORMANCE CHARTS

Equipment manufacturers publish **performance charts** for individual machine models. The performance chart is a graphical representation of the power and corresponding speeds the engine and transmission can deliver. They enable the equipment estimator/planner to analyze a machine's ability to perform under a given set of project-imposed load conditions. The load condition is stated as either rimpull or drawbar pull. The drawbar pull/rimpull–speed relationship is inverse because as vehicle speed increases, pull decreases.

Performance charts
Graphical presentations of power and corresponding speed that an engine and transmission of a mobile machine can deliver.

Drawbar Pull Performance Chart

In the case of the track machine whose drawbar pull performance chart is shown in Fig. 5.7, the available power ranges from 0 to 160,000 lb (the vertical scale) and the speed ranges from 0 to 7.4 mph (the horizontal scale).

Assuming the power required for a certain application is 20,000 lb, this machine would travel efficiently at a speed of approximately 5 mph in first gear. This is found by first locating the 20,000-lb mark on the left vertical scale and then moving horizontally across the chart. At the intersect point of this horizontal projection and the gear curve, project a vertical line downward to the speed scale at the bottom of the chart. The horizontal projection at 20,000 lb in this case also intersects the third gear curve. If the tractor was operated in second gear, it could obtain a speed of only 3.7 mph. In the case of an application requiring a drawbar pull of 100,000 lb, the tractor would have to operate in first gear and maximum speed would be 1 mph.

Rimpull Performance Charts

Each manufacturer has a slightly different graphical layout for presenting performance chart information. However, the procedures for reading a performance chart are fundamentally the same. The steps described here are based on the typical off-road truck chart shown

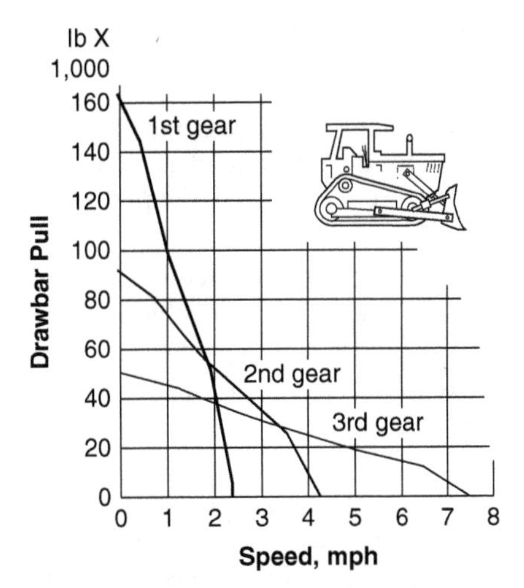

FIGURE 5.7 Drawbar pull performance chart for a crawler tractor.

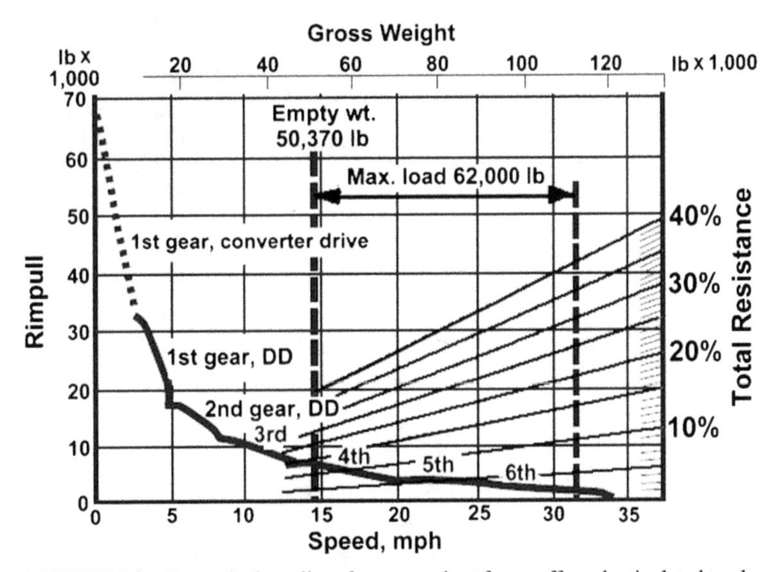

FIGURE 5.8 Example rimpull performance chart for an off-road articulated truck.

in Fig. 5.8. Some manufacturers mark the empty and gross capacity weights of the machine on the charts (see Fig. 5.8). They may also provide the maximum load—for example, 62,000 lb. If not shown on the chart, the maximum load weight will be found listed with the machine specifications.

Required Power—Total Resistance. The graphical arrangement of a rimpull chart enables a determination of machine speed using total resistance expressed either in terms of force (rimpull) or percent effective grade (grade resistance plus rolling resistance expressed as a percent grade). The following procedure is used to determine machine speed from these charts:

1. Always ensure the engine, gear ratios, and tire size of the proposed machine are the same as those for the machine identified on the chart. If the gear ratios or rolling radius of a machine's tires is changed, the performance curve will shift along both the rimpull and speed axes.
2. Estimate the rimpull (power) required—total resistance (rolling resistance plus grade resistance)—based on the reasonable job conditions.
3. Locate the power requirement value on the left vertical scale and project a line horizontally to the right intersecting a gear curve. The point of intersection of the projected horizontal line with a gear curve defines the operating relationship between horsepower and speed.
4. From the point at which the horizontal line intersects the gear curve, project a line vertically to the bottom X axis, which indicates the speed in mph (some charts provide speed in both mph and km/h). Sometimes the horizontal line from the power requirement (rimpull) will intersect the gear range curve at two different points. In such a case, the speed can be interpreted in two ways. In the case of a need for 5,600 lb of rimpull, intersecting both sixth and seventh gears (Fig. 5.9), the scraper speed could be either 15.5 or 17.3 mph.

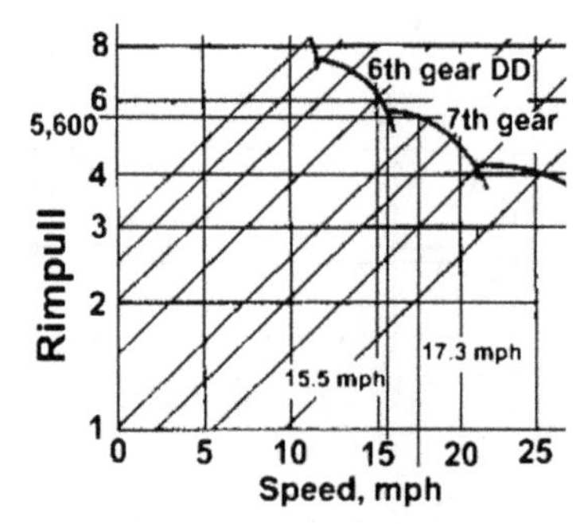

FIGURE 5.9 Rimpull performance chart—gear effect detail.

A guide in determining the appropriate speed is as follows:

- If the required rimpull is less than what was needed for the previous stretch of haul road, use the higher gear and speed.
- If the required rimpull is greater than the need on the previous stretch of haul road, use the lower gear and speed.

Effective Grade—Total Resistance. Assuming total resistance is expressed as an effective grade, the following procedure is used to determine speed:

1. Always ensure the engine, gear ratios, and tire size of the proposed machine are the same as those identified for the machine on the chart.
2. Determine the machine weight when the machine is both empty and loaded. The empty weight is the operating weight and should include coolants, lubricants, full fuel tanks,

and operator. Loaded weight depends on the density of the carried material and the load size (volume). These two weights, empty and loaded, are often referred to as the net vehicle weight (NVW) and the gross vehicle weight (GVW), respectively. The NVW (empty weight) is usually marked on the performance chart. Likewise, the GVW (gross weight), based on the gravimetric capacity of the machine, will usually be indicated on the chart. Vehicle weights are depicted on the upper horizontal scale of the chart.

3. Based on the probable job conditions, calculate a total resistance (the sum of rolling plus grade resistance, both expressed as percent grade). For the performance chart shown in Fig. 5.8, the total resistance values are those shown as the vertical scale on the right side of the chart. **Caution:** The percent grade values are tick marked as a vertical scale, but the actual total percent resistance lines run diagonally, falling from right to left. The intersection of a vertical projection from vehicle weight and a diagonal total resistance line establishes the conditions under which the machine will operate and, correspondingly, the power requirement.
4. From the intersection point of the vertical vehicle-weight projection and the appropriate total-resistance diagonal, project a line horizontally (do not proceed down the total resistance diagonal). The point of intersection of this horizontal projection with a gear curve defines the operating relationship between horsepower and speed.
5. From the point at which the horizontal line intersects the gear range curve, project a line vertically to the bottom X axis, which indicates the machine speed in mph and km/h. This is the vehicle speed for the assumed job conditions.

Performance charts are established assuming the machine is operating under standard conditions. When the machine is utilized under a different set of conditions, the rimpull force and speed must be appropriately adjusted. Operation at higher altitudes will require a percentage de-rating in rimpull approximately equal to the percentage loss in flywheel horsepower.

Not all equipment manufacturers use the same graphical format in presenting rimpull information. Figure 5.10 illustrates a two-page presentation. With this format, the resistance and gear (speed) information are separated. The arrow illustrates how to read the chart to arrive at a speed.

Retarder Performance Chart

When operating on steep downgrades, a machine's speed may have to be limited for safety reasons. A retarder is a dynamic speed-control device. Machine speed is retarded through the use of an oil-filled chamber between the torque converter and the transmission. The retarder will not stop the machine; instead, it provides speed control for long downhill hauls, reducing wear on the service brake. Figure 5.11 presents a **retarder chart** for a wheel tractor scraper.

Retarder chart A graphical presentation that identifies the controlled speed of a machine descending a slope when the magnitude of the grade assistance is greater than the rolling resistance.

When the magnitude of the grade assistance—downward slope—is greater than the rolling resistance encountered, the retarder performance chart (see Fig. 5.11) identifies the steady speed a vehicle will maintain while descending. This retarder-controlled speed is steady state, and use of the service brake will not be necessary to prevent acceleration.

A retarder performance chart is read in a manner similar to the procedure described for a performance chart; however, the total resistance (effective grade) values in this case are actually negative values. As with the rimpull chart, the

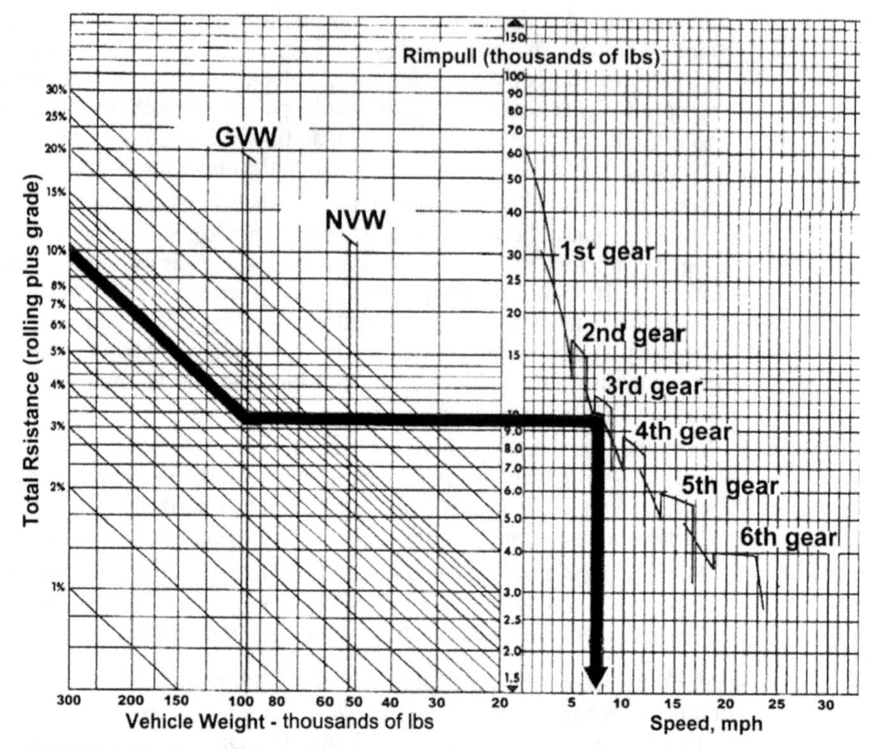

FIGURE 5.10 Example rimpull performance chart for a twin engine scraper.

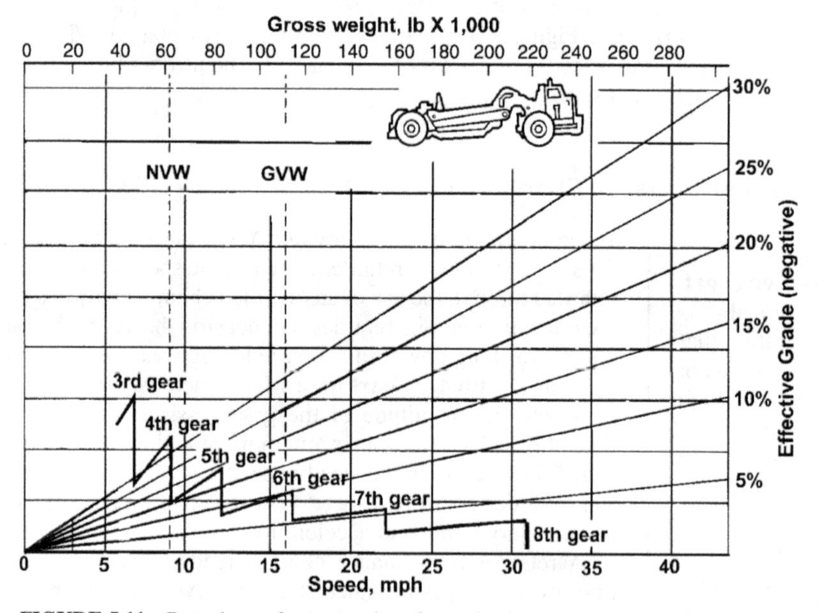

FIGURE 5.11 Retarder performance chart for a wheel tractor scraper.

horizontal line can intersect more than one gear. In a particular gear, the vertical portion of the retarder curve indicates maximum retarder effort and resulting speed. If haul conditions dictate, the operator will shift into a lower gear, and a lower speed would be applicable. Many times the decision as to which speed to select is answered by the question: How much effort will be expended in haul-route maintenance? Route smoothness is often the controlling factor affecting higher operating speeds.

Sample Calculation: A contractor proposes to use wheel tractor scrapers on an embankment job. The performance characteristics of the machines are shown in Fig. 5.12. The scrapers have a rated capacity of 14-cy struck. Empty operating weight is 69,000 lb. Based on past jobs, the average scraper load should be 15.2 bcy. The haul from the excavation area is a uniform adverse gradient of 5% with a rolling resistance of 60 lb/ton.

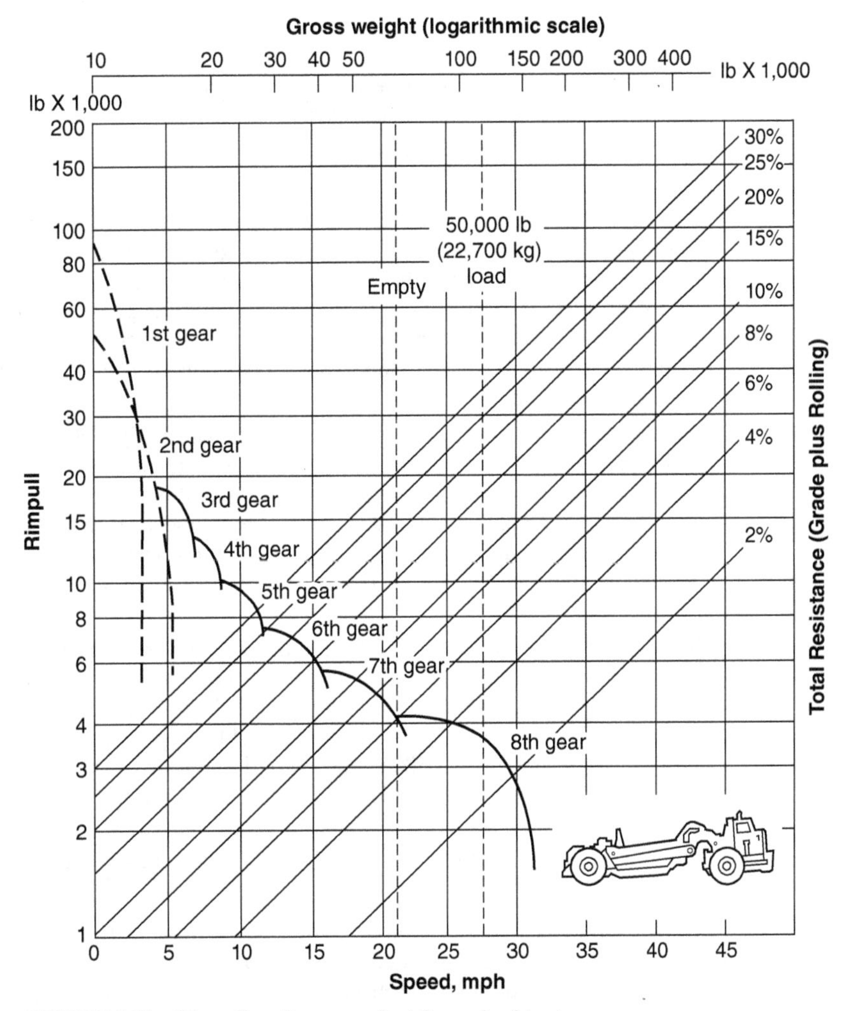

FIGURE 5.12 Rimpull performance chart for a wheel tractor scraper.

The material to be excavated and transported is a common earth with a bank unit weight of 3,200 lb/bcy.

a. Calculate the expected maximum travel speeds.

Machine weight:

Empty operating weight	69,000 lb
Payload weight, 15.2 bcy × 3,200 lb/bcy	48,640 lb
Total loaded weight =	117,640 lb

Use the loaded weight dash vertical line, 119,000 (69,000 + 50,000).

Haul conditions:

	Loaded (Haul) (%)	Empty (Return) (%)
Grade resistance	5.0	–5.0
Rolling resistance 60/20	3.0	3.0
Total resistance	8.0	–2.0

Loaded speed (haul): Using the chart in Fig. 5.12, enter the upper horizontal "gross weight" scale at 119,000 lb and the total loaded weight, and project a vertical line down to intersect with the 8.0% total resistance diagonal. From that intersection point, project a horizontal line to intersect with the gear curves. This horizontal line intersects fifth gear. By projecting a vertical line from the fifth-gear intersect point to the lower horizontal scale, the speed is determined. The loaded speed will be approximately 10 mph.

Empty speed (return): Using the retarder chart in Fig. 5.11, calculate a return speed. Enter the upper horizontal scale of Fig. 5.11 at 69,000 lb, the net vehicle weight (NVW), or empty operating weight. Because this is a commonly used weight, it is marked on the chart as a dashed line. The intersection with the 2.0% effective resistance diagonal defines the point from which to construct the horizontal line to intersect the gear curves. This horizontal line intersects eighth gear, and the corresponding speed is 31 mph.

b. If the job is at elevation 12,500 ft, what will be the operating speeds when the nonstandard barometric pressure is included in the analysis? For operating at an altitude of 12,500 ft, the manufacturer has reported that the scrapes, which have turbocharger engines, can deliver 95% of rated flywheel horsepower.

At an altitude of 12,500 ft, the rimpull that is neccssary to overcome a total resistance of 8% must be adjusted for the altitude de-rating.

$$\text{Altitude-adjusted effective resistance} = \frac{8.0}{0.95} = 8.4\%$$

Now proceed as in question **a** by locating the intersection of the total loaded weight line and the altitude-adjusted total resistance line. Projecting a horizontal line from the 8.4% altitude-adjusted effective resistance diagonal yields an intersection with the forth-gear curve and, by projecting a vertical line at that point, a speed of approximately 8 mph—a slight loss of speed. The real time effect will depend on the haul distance.

The nonstandard altitude would not affect the empty (return) speed because the machine does not require rimpull force for downslope motion. The retarder controls the machine's downhill momentum. Therefore, the empty (return) speed is still 31 mph at a 12,500-ft altitude.

MACHINE STEERING

Full hydraulic steering may be used to turn a whole section of a machine on a massive, vertical hinge. One or two double-acting cylinders are employed. This construction, called articulated steering, has been standard in self-powered overhung (four-wheel) scrapers since they first appeared. It has become the preferred construction in large four-wheel-drive tractors and some trucks.

In the scraper (Fig. 5.13), the hinge (kingpin) is near the front of the machine, and it is always the front or tractor part that swings when steering, with a result close to (although not exactly like) turning the front wheels.

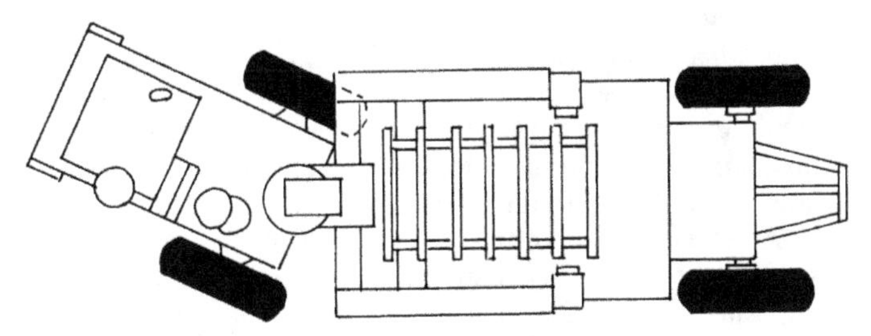

FIGURE 5.13 A scraper's articulation is forward.

Tractor loaders (Fig. 5.14) usually have the hinge midway between the axles, so both parts share equally in a turn. This action produces the effect of the four-wheel coordinated steering. Front and rear wheels run in each other's wheel path, backward and forward, and the whole weight of the tractor lines up efficiently behind any pushed load.

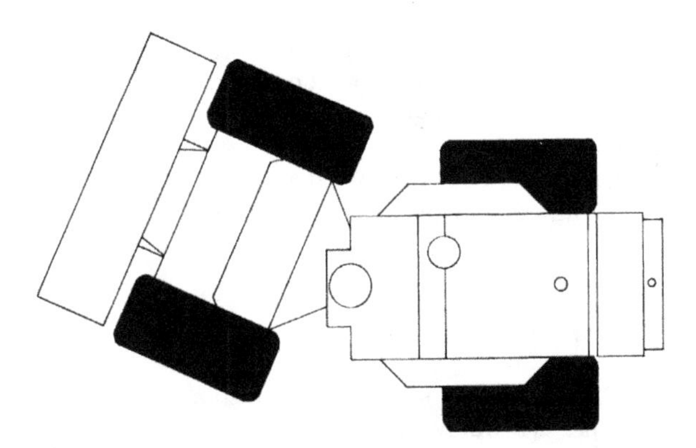

FIGURE 5.14 A tractor's articulation is centered.

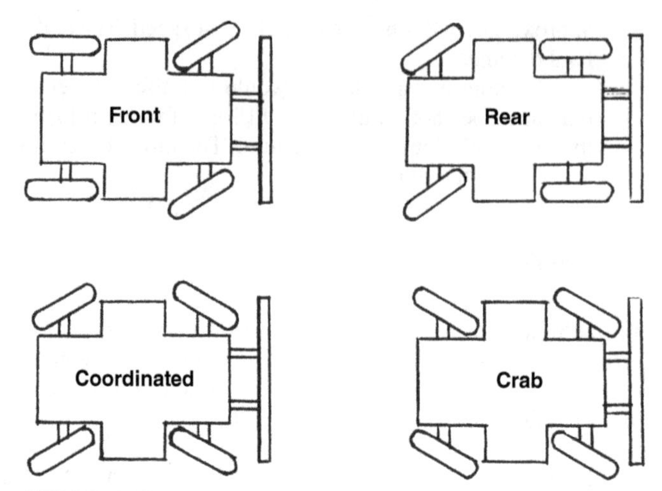

FIGURE 5.15 Front and rear steering.

Front and Rear Steering. Wheeled vehicles may be designed to steer by angling the front wheels, the rear wheels, and/or both the front and rear wheels (Fig. 5.15).

Front-wheel steer is the standard method. It is the safest and most satisfactory for most conditions, both because operators are used to it from driving cars and because the effect is the simplest. The vehicle follows the angling of the wheels. The rear wheels do not go outside the path of the front ones but trail inside it.

A vehicle with both front and rear steer can use it in two separate ways. In coordinated steering the front wheels are turned one way and the rear wheels to the same angle in the opposite direction. The wheels will then track; that is, the trailing wheels will always move in the same tracks as the leading wheels, whether the machine is moving forward or backward. This feature lessens rolling resistance on soft ground, as one pair prepares a pathway for the other.

Coordinated steering provides maximum control of direction under load, enabling the machine to keep on a straight course in spite of side forces and to force loads to change direction. Furthermore it permits accurate cutting along curved lines and short turns in proportion to the maximum angle of the wheels.

With all-wheel crab steering, both sets of wheels are turned in the same direction. If both sets are turned at the same angle, the machine will move in a straight line at an angle to its centerline. If it carries a straight dozer blade, this will meet dirt at an angle and side cast in the manner of an angle dozer. Each wheel makes its own track, giving maximum flotation.

Special results can be obtained with either coordinated or crab steering by using different angles of turn on independently controlled front and rear wheels.

SAFETY

To operate equipment safely, a person should be alert, observant, cautious, and willing to invest time and trouble to perform a task correctly. An elementary precaution is to use handholds, steps, ladders, or other helps provided for getting in and out of the seat and

FIGURE 5.16 Big machines mean restricted vision.

reaching service and inspection areas. An agile person can skip many of these but should not. Surfaces are often slippery with water, ice, or oil, and long steps and reaches make sliding and falls more likely.

Warning devices must be kept in working order. These include intermittent blowing of a horn whenever reverse gear is engaged, flashing "lollipop" lights when working along roads, and bright headlights and rear lights for night operation.

Restricted vision is a problem with increased machine size (Fig. 5.16). Before resuming work after parking, an operator must walk around the machine to make sure no one is eating or dozing in its shade. When moving, keep aware of blind spots.

Rollover protection structures are important for safety. But these can restrict vision even under good circumstances and call for increased alertness. A cab with dirty windows is very dangerous, and windows must be wiped clean as many times as needed, before and during operation. All glass must be kept clean.

Attachments should be on the ground when the machine is unattended. However, it is customary to leave crane booms up when they are supported by cables.

After shutting off the engine, move all operating levers back and forth to make sure no implements have been left in a raised position.

CHAPTER 6
TRACTORS AND DOZERS

Tractors are classified on the basis of running gear. They may be either track-laying (crawler) or wheel-type machines. A dozer is a tractor with a blade attached to push material from one location to another. It is designed to provide tractive power for drawbar work. A dozer blade has no set volumetric capacity. The amount of material the dozer moves is dependent on the quantity held by the blade during the push, and this quantity is dependent on material type and the material's moisture content. Heavy ripping of rock is accomplished by crawler dozers equipped with rear-mounted rippers. They are excellent machines for this use because of their power and the tractive force they can develop.

TRACTORS

Tractors may be either track-laying (crawler) or wheel-type machines. Dozers are tractors equipped with a front upright blade for pushing materials. These tractors are designed specifically to provide high tractive power for drawbar work while operating at lower speeds. Consistent with their purpose as a unit for drawbar work, they have a low center of gravity. This is a prerequisite for an effective dozer. The larger the difference between the line-of-force transmission from the machine and the line of resisting force, the less effective the use of the developed power. In addition to dozing, these machines are used for land clearing, ripping, assisting scrapers in loading, and towing other pieces of construction equipment. They can be equipped with either a rear-mounted winch or a ripper. For long moves between projects, or even within a project, track machines should be transported. Moving them under their own power, even at slow speeds, increases track wear and shortens the machine's operational life.

Type Characteristics

Tractors are classified on the basis of running gear:

1. Track-laying type (Fig. 6.1)
2. Wheel type (Fig. 6.2)

FIGURE 6.1 Track-type tractor.

FIGURE 6.2 Wheel-type tractor.

Crawler Tractors

Crawler tractors are track laying machines. They have a continuous track of linked shoes. This track moves in the horizontal plane across fixed rollers. At the rear of the tractor, the track passes over a vertically mounted sprocket drive wheel. As the sprocket turns, it forces the track forward or back, imparting motion to the dozer. At the tractor's front, the track passes over a vertically mounted idler wheel. The idler wheel can be adjusted to achieve the

Sprocket drive wheel

Idler wheel

proper track chain tension. This wheel is connected to a recoil device to absorb shock loads. The linked shoes of the track chain (Fig. 6.3) are made of heat-treated steel designed to resist wear and abrasion. Individual shoe plates have a protruding grouser. The grouser penetrates the ground and provides added grip and surface traction. Several companies offer tracks with rubber-covered steel shoes. These minimize damage to paved surfaces.

As discussed in Chap. 5, the usable power a machine has available to perform work is often limited by traction. This limitation is dependent on two factors:

1. Coefficient of traction for the surface being traversed
2. Weight carried by the drive wheels of the tractor

Sometimes tractor owners fill the tires of wheel-type tractors with liquid ballast to increase machine weight and overcome tractive power limitations. A mixture of calcium chloride and water is recommended as tire ballast. To add weight, steel plates can be added to the machine frame. In either case, care must be taken to ensure the weight of the machine is equally distributed between drive wheels.

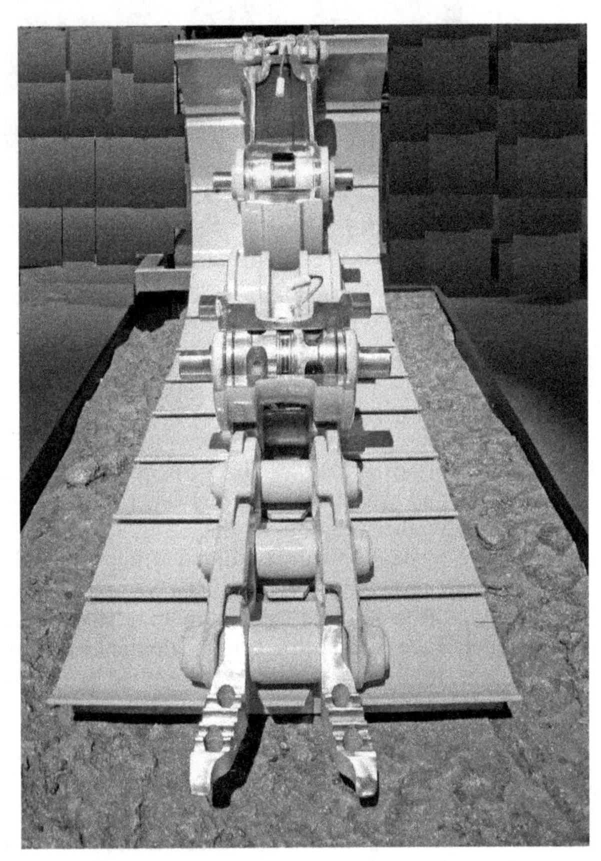

FIGURE 6.3 Track chain.

Finish work
Shaping the material to the final earthwork grade required by the specifications.

Traction or flotation requirements can be met by proper undercarriage or tire selection. A standard crawler tractor undercarriage is appropriate for general work on a range of materials, from rock to moderately soft ground. Typical ground pressure for a crawler dozer with a standard undercarriage is 6 to 9 psi. Low-ground-pressure (LGP) undercarriage configurations (Fig. 6.4) are available for dozers operating in soft ground conditions. The ground pressure exerted by a crawler dozer with an LGP undercarriage is 3 to 4 psi. LGP machines should not be used in hard ground or in rocky conditions; such a practice will reduce undercarriage life. Extra-long (XL) undercarriages increase machine stability and are available for machines dedicated to **finish work.**

FIGURE 6.4 Low-ground-pressure machine operated on soft ground.

In the case of wheel machines, wider tires provide greater contact area and increase flotation. Larger tires will reduce the developed rimpull. All rimpull charts are based on standard equipment, including tires for a machine, so the listed performance data are no longer valid.

The track-type tractor is designed for those jobs requiring high tractive effort. No other piece of equipment can provide the power, traction, and flotation needed in such a variety of working conditions. A crawler dozer can operate on slopes as steep as 45 degrees.

Both track- and wheel-type tractors are rated by flywheel horsepower (fwhp) and weight. Normally, the weight is an operating weight and includes lubricants, coolants, a full fuel tank, a blade, hydraulic fluid, the Occupational Safety and Health Administration (OSHA) rollover protective structure (ROPS), and an operator. Machine weight is important on many projects because the maximum tractive effort a tractor can provide is limited to the product of the weight times the coefficient of traction of the unit for the particular ground surface, regardless of the power supplied by the engine. Table 5.3 gives the coefficients of traction for various surfaces.

Wheel Tractors

Practically all big four-wheel-drive wheel tractors have torque converters and power shift transmissions. There may be a disconnect lever to cut out drive to one axle. It is comfortable and economical to use two-wheel drive for light loads and traveling.

Wheel tractors usually have a stay-in-position hand throttle and a foot accelerator. The throttle may be set at a low or moderate speed (low idle or fast idle), and full engine speed for heavy digging or fast travel is then obtained by pressing the accelerator.

There are several different steering systems, all power operated. Most four-wheel-drive tractors in medium and large sizes have articulated or pivot steering, with a kingpin or hinge halfway between the axles. The pivoting is powered by hydraulic rams, but most respond to the wheel in a normal automotive fashion. One set of wheels always follows the other directly, so there is no problem of watching for side swing of the wheels while turning. There is loss of stability at sharp steering angles, requiring caution on side slope operations.

An advantage of a wheel-type dozer when compared with a crawler dozer is higher travel speeds—in excess of 30 mph for some models. To attain a higher speed, however, a wheel dozer must sacrifice pulling ability. Also, because of the lower coefficient of traction between rubber tires and some ground surfaces, the wheel dozer may slip its wheels before developing its rated pulling effort. Table 6.1 provides a comparison of crawler dozer and wheel dozer utilization.

TABLE 6.1 Dozer-Type Utilization Comparison

Wheel Dozer	Crawler Dozer
Good on firm soils, concrete, and abrasive soils with no sharp-edged pieces	Can work on a variety of soils; sharp-edged pieces not as destructive to dozer, though fine sand will increase running gear wear
Best for level and downhill work	Can work over almost any terrain
Wet weather, causing soft and slick surface conditions will slow or stop operation	Can work on soft ground and over mud-slick surfaces; will exert very low ground pressures with special low ground pressure undercarriage and track configuration
The concentrated wheel load will provide compaction and kneading action to ground surface	
Good for long travel distances	Good for short work distances
Best in handling loose soils	Can handle tight soils
Fast return speeds, 8-26 mph	Slow return speeds, 5-10 mph
Can only handle moderate blade loads	Can push large blade loads

Performance Comparison

The usable rimpull force of a tractor depends on traction; therefore, the total weight of the dozer in the case of a track machine or the weight on the powered wheels of a wheel machine is a critical determinant. Even though the engine can develop a certain drawbar pull or rimpull force, all of the pull may not be available to do the work if the wheels or tracks slip. The caution is a restatement of Eq. (5.15). If the project's working surface is dry clay loam, Table 5.3 provides the following coefficient of traction factors:

Rubber tires 0.50–0.70

Track 0.90

Using the factor for tracks, 0.90, and considering a 45,560-lb track-type dozer with a power shift, the usable drawbar pull is found to be

$$45{,}560 \text{ lb} \times 0.90 = 42{,}004 \text{ lb}$$

Now consider a 45,370-lb wheel-type dozer:

$$45{,}370 \text{ lb} \times 0.60 = 27{,}222 \text{ lb}$$

The two machines have approximately the same operating weight and flywheel horsepower; yet because of the effect of traction, the track machine can supply one and a half times the usable power.

In the case of a track-type tractor and a wheel tractor operating on the same surface soil conditions, the coefficient of traction for wheel machine is much less. Therefore, a wheel-type dozer must be considerably heavier (by approximately 50%) than a crawler dozer to develop the same amount of usable force.

As the weight of a wheel-type dozer is increased, a larger engine will be required to maintain the weight-to-horsepower ratio. As ballast weight is added to the wheel-type dozer, the speed and mobility of the machine are decreased and it loses its advantage over track dozers.

Tractor Mechanics

Bull wheel The sprocket wheel used to provide positive drive to the track.

Figure 6.5 provides a cut-a-way view of the transmission, steering units, and final drive of a track-type tractor. Many larger tractors have the **bull wheel** sprocket above and slightly ahead of the rear track idler wheel, as seen in Fig. 6.6. This arrangement serves to keep the final drive above the material of the job site and provides easier servicing.

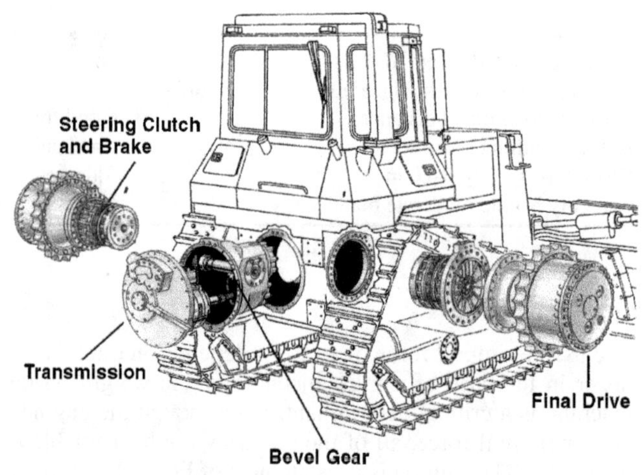

FIGURE 6.5 Detail of the powertrain in a track-type tractor. (*Courtesy of Caterpillar, Inc.*)

Engine. Internal combustion engines are used to power most tractors, with diesel engines being the most common primary power units. Some smaller machines have gasoline engines, and air-powered dozers are available for tunnel work.

Hybrid Electric Drive. With hybrid electric power machines, the electric drive sequence from a generator has 60% fewer moving parts than tractors powered through a hard drive shaft.

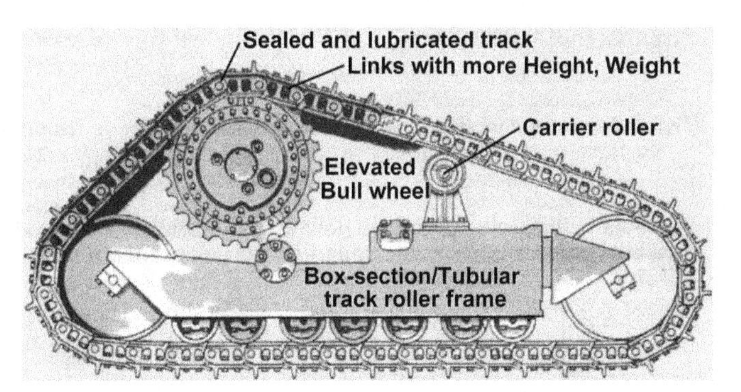

FIGURE 6.6 Elevated bull wheel. (*Courtesy of Caterpillar, Inc.*)

The greatest efficiency gains come by eliminating the torque converter and operating the engine in a much narrower range of speeds. The electric drive eliminates gear shifting and improves steering performance by 50%.

Transmission. Because the crankshaft rotation derived from the engine is usually too fast and does not have sufficient force (torque), transmissions are used to reduce the rotational speed of the crankshaft and increase the force available to do work. Transmissions provide the operator with the ability to change the machine's speed-to-power ratio so it will match the work requirements (force necessary to do work). Manufacturers provide tractors with a variety of transmissions, but the primary options are

- Direct drive
- Torque converter and power-shift transmission

Some less-than-100-hp tractors are available with hydrostatic powertrains. The small to medium, less-than-300-hp, diesel-powered machines are commonly available with either direct- or power-shift-type transmissions. Larger tractors are always equipped with power-shift transmissions.

Track Tractors with Direct Drive. When machine power is transmitted straight through the transmission as if there were a single shaft, the term "direct drive" is used to describe the mechanical arrangement. This is usually what happens when the transmission is in its highest gear. In all other gears, mechanical elements match speed and torque. Direct-drive tractors are superior when the work involves constant load conditions. A job where full blade loads must be pushed long distances would be an appropriate application of a direct-drive machine.

The specifications provided by some manufacturers list two sets of drawbar pulls for direct-drive tractors: rated and maximum. The rated value is the drawbar pull sustainable for continuous operation. The maximum drawbar pull is the pull the dozer can exert for a short period while lugging the engine, such as when passing over a soft spot in the ground when there is a temporary requirement for higher tractive effort. Thus, the rated pull should be used for continuous operation. Available drawbar pull is subject to a traction limitation imposed by slope and the nature of the surface being traversed.

Track Tractors with Torque Converter and Power-Shift Transmissions. A *power-shift* transmission can be shifted while transmitting full engine power. These transmissions are teamed with torque converters to absorb drive train shock loads caused by changes in gear ratios.

A power-shift transmission provides an efficient flow of power from the engine to the tracks and gives superior performance in applications involving variable load conditions.

Track Tractors with Hydrostatic Powertrains. Oil confined under pressure is an effective means of power transmission. A hydrostatic powertrain offers an infinitely variable speed range with constant power to both of the tractor's tracks. This type of powertrain improves machine controllability and increases operational efficiency. Hydrostatic powertrain transmissions are available in the under-130-hp-class machines.

Wheel Tractors. Most wheel tractors are equipped with torque converters and power-shift transmissions. Wheel dozers exert comparatively higher ground pressures, 25 to 35 psi.

Center Frame Track Tractors. The rigidity of the center section is obtained by making the steering clutch and transmission housings in one heavy casting or weldment with internal braces, together with heavy construction of all cases forward to the radiator. In addition, a pair of heavy side beams may run forward from the transmission case, supporting the engine, the crankcase guard, and the radiator base.

Power Take-off. The simplest power take-off (PTO) is a shaft directly connected to the transmission. The shaft extends through the rear wall of the gear case and turns when the tractor's clutch is released. It is used to power auxiliary components such as cable control units, a winch, or a hydraulic pump. If the transmission is compound, the take-off may have two gear ratios; otherwise, its speed is controlled only by the engine.

Tracks. The live axles turn large toothed wheels, called drive sprockets or bull wheels, located at the rear of the track frames or elevated, as seen in Fig. 6.7. The frames rest on the small truck or track rollers. The idlers, which are smoothed flanged wheels, are mounted on spring-cushioned yokes at the front of the frames, or at both front and back with an elevated bull wheel sprocket. Except in the case of very small machines, one or two small support rollers are mounted above each frame, to prevent excessive sagging in the upper track section. The track itself consists of a true roller chain and bolted-on shoes.

Certain types of stiff mud or wet snow may build up in the tracks and in the sprocket hollows, causing the sprocket to spin, usually with abrupt and damaging stops and starts, and will probably make the track overtight at the same time. This condition may make work difficult or impossible, as repeated hand cleaning may be required. Ice and mud shoes usually have openings in the center to permit the sprocket teeth to force the snow out,

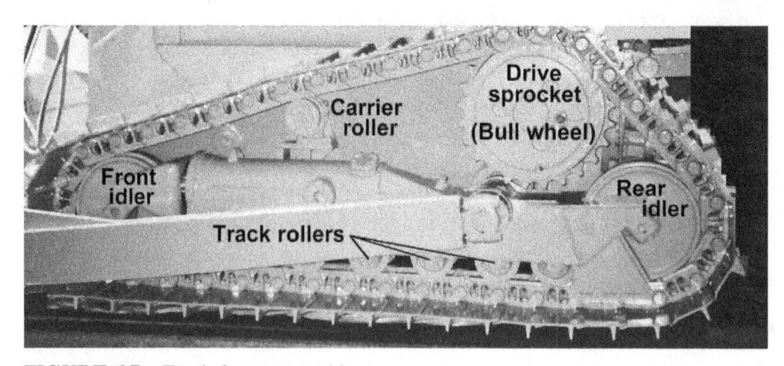

FIGURE 6.7 Track frame assembly.

leaving the inner parts comparatively free. There are also cutaway sprockets designed so mud can squeeze through openings in the tooth bottoms.

Track Wheels. The rollers and idlers are flanged in order to keep the track in line. Most machines have a mixture of single-flange and double-flange rollers on the track frames to correctly guide the track. The double-flange rollers improve track chain alignment and location, but they are more expensive. The idler customarily has a wide center flange designed to fit between the track links. The track and support rollers have outer flanges. These flanges are spaced to clasp the track rail. They may also have an inner flange. On the bottom, it is usual to alternate single- and double-flanged rollers. Rollers and idlers revolve on fixed axles. They may have tapered roller bearings or solid-sleeve types made of bronze or special metals. Good seals, to keep lubricant in and dirt out, are very important. The most successful type, the "positive" seal, consists of two finely machined rings, one attached to the axle and one to the hub, which are pressed against each other by springs. These rings fit perfectly so neither dirt nor grease can get past them. They are sufficiently hard and usually outlast other wearing parts.

Operation—Mechanical Drive

The lowest operating gear is desirable for precise work, because the slow speed gives more time to steer and make tractor and blade adjustments. The blade can, consequently, cut a grade more accurately in low gear than in second.

Gear Selection. For less exacting work, the gears should be the highest possible without lugging the engine below its governed speed and which will give the amount of control needed. Gear shifting is slow and cumbersome in some tractors because of the spur gearing used; the hand throttle setting, which makes double-clutching impractical; and track friction, which slows the machine, so operator skill is required to complete a shift rapidly before the machine stops. The gear used throughout is, therefore, usually low enough to start the load, although a higher gear might be used for movement once it is underway.

An increasing number of machines use constant-mesh gears, which shift more easily when standing and which can be shifted on the move. Speed may be increased or decreased by opening or closing the throttle or by choosing a gear. A decelerator pedal will slow the engine below the throttle setting when pushed down against a spring.

Engine electronics now permit a higher degree of powertrain integration. The technology serves to integrate the drivetrain and machine components so power and torque are adjusted to the application. Electronic machine power varies in the higher gears. The electronics also enhance operator shifting and reduce clutch loads.

Reversing. To change the travel direction of a standard crawler, first bring it to a full stop. With the clutch released, it will stop by itself in a few feet, except when going down a grade. It may be stopped in inches by its load or by applying a brake. With an ordinary transmission, move the shift lever through neutral into the desired gear, then reengage the clutch.

The reversing transmission may use sliding spur gears, which are usually somewhat clumsy or even difficult to shift, but should have helical gears with a sliding hub, which are faster and easier, or a pair of friction clutches.

Steering. Five factors are involved in steering a conventional crawler tractor: (1) the effects of the steering clutches, (2) the steering brakes, (3) the footing, (4) the grade, and

(5) the load being pulled or pushed. If one steering clutch lever is pulled back while the tractor is moving forward, its track will run idle and the other will continue to drive. The amount of turn by an unloaded tractor will depend on the resistance of the idle track to moving forward. Its resistance is made up of internal friction, which by itself will produce only very gradual change in direction and ground drag (Fig. 6.8). As a result, on a smooth, hard surface, the machine will turn very gradually. In soft ground, it will turn more sharply.

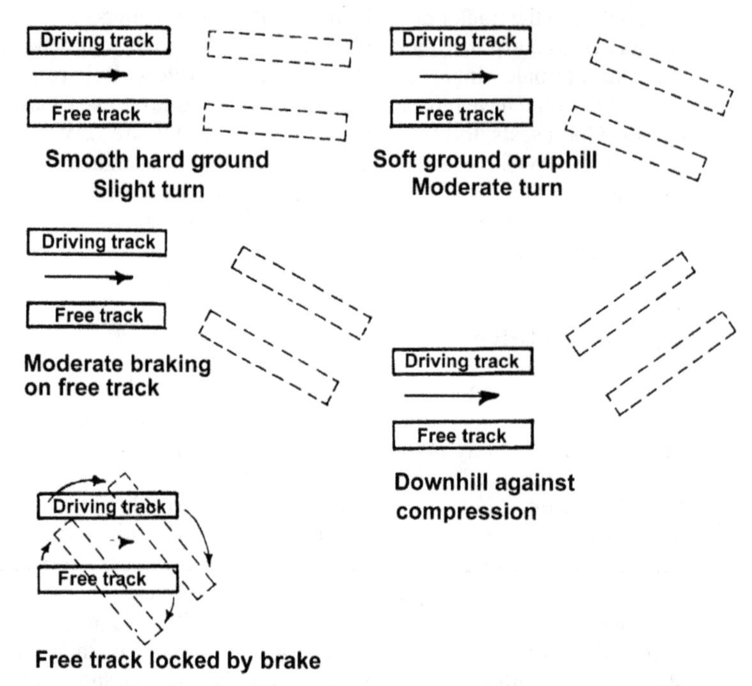

FIGURE 6.8 Steering with clutch and brake.

Rolling resistance in the free track can be increased to any desired degree by applying its steering brake. If the track is locked and traction is good, the machine will spin around. Brake application can produce any degree of turn between this and the nonbraked gradual change in direction. Braked turns are seldom smooth, as direction is usually changed in a series of steps.

When moving uphill, the free track is held back by gravity, and the weight of the machine also increases the tendency to turn. Other conditions being equal, less braking will, therefore, be required for steering when going uphill than on level ground.

If the ground is soft, both tracks will dig in. The driving track will tend to spin and dig; the braked track will push earth at the front in the direction of the turn, and oppositely at the rear. At best, turning consumes some power, and under soft conditions, the resistance built up by the displaced earth may be great enough to prevent the turn and put severe strain on the tracks. Under such conditions, turns are best made in a series of short jerks to give the tracks a chance to work away from the soil ridges before turning farther. Because the machine drives from one track while being steered, only half of its normal traction is available. This complicates maneuvering in soft ground and with a load.

A heavy load, either on the drawbar or in front of a blade, has the effect of an upgrade in supplying a turning force when drive is from one side only; the load itself resists being turned.

In general, a loaded tractor will steer more readily than a free one in response to a steering clutch only, and with greater difficulty in response to clutch and brake. Very sharp turns are usually difficult or impossible, but the location and character of the load influence this.

Even with all clutches engaged, a tractor with a heavy off-center load will tend to turn toward the loaded side. A loaded tractor using nearly its full traction when going straight will have difficulty in turning, as the driving track will be inclined to spin. It may be necessary to relieve the load by backing in order to change direction. This is usually fairly easy with a pushed load but difficult with a pulled one.

Steering by means of clutches alone works backward when going down a hill steep enough to make the tractor try to roll faster than the engine that drives it. The effect of releasing a steering clutch in this case is to allow its track to roll free and move faster than the other track, which is braked by the engine—as a result, the machine will turn away from the released clutch. Firm application of the brake on this side will cause the machine to turn toward it in the normal manner. If precise steering is required, the brake should be applied before the clutch is released.

Steering in reverse is the same process as steering forward. Use of either steering brake while its steering clutch is engaged will slow, stop, or hold the machine but has no effect on steering, as while the clutches are engaged, the axle shaft acts as a solid unit from track to track. Some crawler tractors differ from this in important respects.

Operation with Torque Converter

A torque converter causes the relative speeds of tractor and engine to vary widely. Under full throttle, a heavy load may slow the machine to a creep or even stop, and with a light load or no load, it will reach its maximum speed in the gear range being used.

The loss of speed involves an increase in pulling or pushing power, because of the torque multiplication. This saves the operator the responsibility of shifting to a lower gear, lightening the load, or even slipping the clutch, which the operator would do in a gear drive model. Under most conditions, this slowing is an efficient and desirable response to an increase in load.

Speeding up under light load is also efficient, but it is not always as welcome, as it may make it difficult or impossible to do precise work with a blade, bucket, or implement. Fortunately, it can be controlled with the decelerator pedal, which will slow both engine and tractor to the desired rate of motion.

A machine might have an output shaft governor on the torque converter. This will keep the tractor or a mounted unit such as a winch at a steady speed, regardless of load, as long as its requirement does not exceed engine power. For this, there are two throttles: one controlling a standard engine governor and the other the output shaft governor. It is usual to keep the engine throttle wide open and use the output governor to regulate the speed of operation.

DOZERS

A dozer is a tractor unit with a blade attached. The blade is used to push, shear, cut, and roll material ahead of the tractor. Dozers are effective and versatile earth-moving machines. They are used both as support and as production machines on many construction projects. These machines may be used for operations such as

- Moving earth or rock for short-haul (push) distances, up to 300 ft in the case of large dozers
- Spreading earth or rock fills
- Backfilling trenches

- Opening up pilot roads through mountains or rocky terrain
- Clearing the floors of borrow and quarry pits
- Pusher to help load tractor-pulled scrapers
- Clearing land of timber, stumps, and root mat

Blades

A dozer blade consists of a moldboard with replaceable cutting edges and end (side) bits (Fig. 6.9). Push arms and tilt cylinders or a C-frame connect the blade to the dozer (Fig. 6.10). A hydraulic system is used to raise and lower the blade. Blades vary in size and design based on specific work applications. The hardened-steel cutting edges and side bits are bolted on because they receive most of the abrasion and wear out rapidly. The bolted connection enables easy replacement. The design of some machines enables either end of the blade to be raised or lowered in the vertical plane of the blade, *tilt*. The top of the blade can be pitched forward or backward, varying the angle of attack of the cutting edge, *pitch*. Blades mounted on a C-frame can be turned from the direction of travel, *angling*. These features are not applicable to all blades, but any two of these may be incorporated in a single mount.

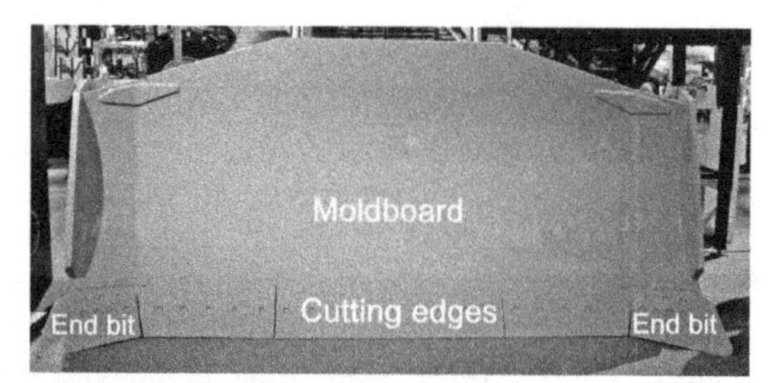

FIGURE 6.9 Parts of a dozer blade.

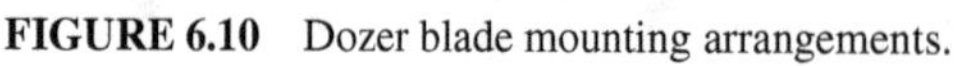

FIGURE 6.10 Dozer blade mounting arrangements.

Float. Most dozer control valves include a FLOAT position. Ordinarily FLOAT is used to smooth areas by back dragging and in pushing material on pavement or other surfaces that are hard enough to support the blade but that will be damaged if the blade is forced

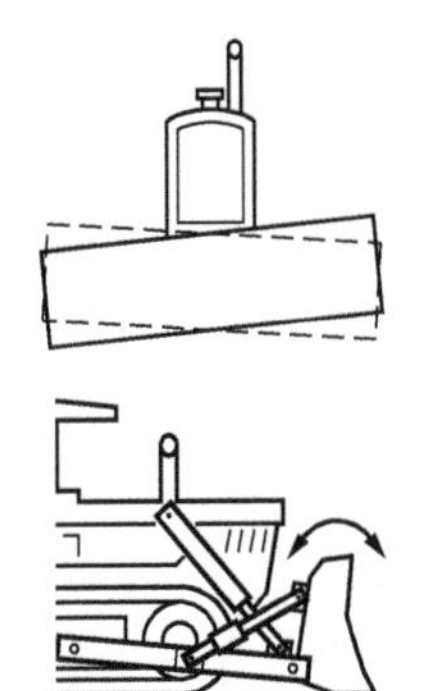

down. Because of the blade's shape, a full load tends to lift the bottom edge to the surface; transporting can be done in FLOAT.

Tilt. This movement is within the vertical plane of the blade. Tilting permits concentration of dozer driving power on a limited portion of the blade's length.

Pitch. This is a pivotal movement about the point of connection between the dozer and blade. When the top of the blade is pitched forward, the bottom edge moves back; this increases the angle of cutting edge attack.

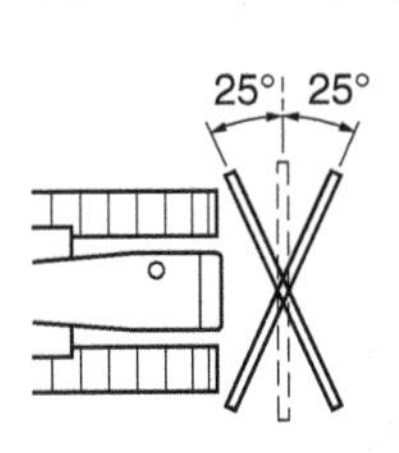

Angling. Turning the blade so it is not perpendicular to the direction of the dozer's travel is known as angling. Angling causes the pushed material to roll off the trailing end of the blade. This procedure of rolling material off one end of the blade is called side casting.

Angle blades (Fig. 6.11) can be angled to the left or right, in addition to a center, straight-across position. They may have either hydraulic or mechanical positioners. An angle blade is superior to the straight blade in light trench backfilling. Its drawbacks are its slightly greater weight, cost, and upkeep; clumsiness in restricted spaces; difficulty in turning with a load; and looseness in the joints.

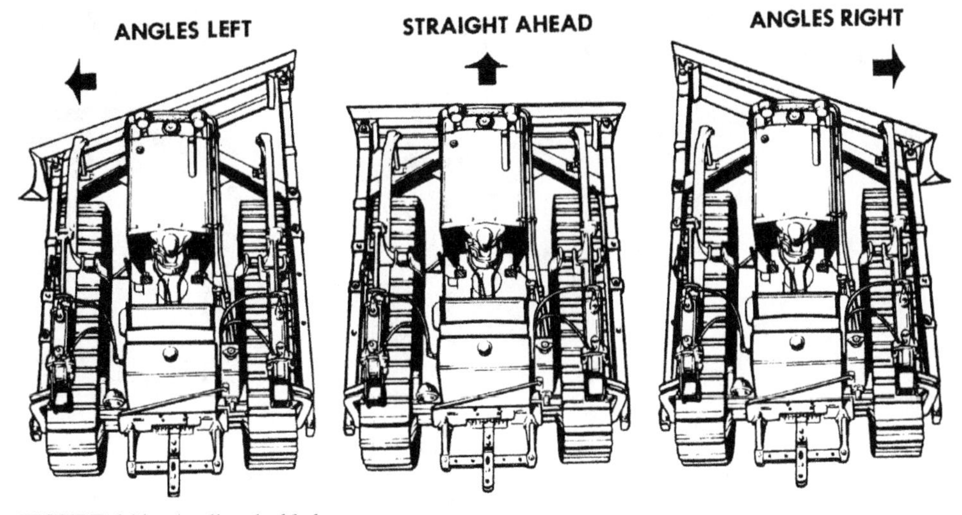

FIGURE 6.11 Angling the blade.

Blade Performance. A dozer's pushing potential is measured by two ratios:

1. Horsepower per foot of cutting edge, *cutting ratio*
2. Horsepower per loose cubic yard (lcy) of material retained in front of the blade, *load ratio*

The horsepower per foot (hp/ft) cutting ratio provides an indication of a blade's ability to penetrate and obtain a load. A higher cutting ratio indicates a more aggressive blade. The horsepower per loose cubic yard load ratio measures the blade's ability to push a load. When there is a higher ratio, the dozer can push the load at a greater speed.

The blade is raised or lowered by hydraulic rams; therefore, a positive downward force can be exerted on the blade to force it into the ground. Additionally, basic earth-moving blades are curved in the vertical plane in the shape of a flattened C. When the blade is pushed down, its cutting edge is driven into the earth. As the dozer moves forward, the cut material is pushed up the face of the blade. The upper part of the flattened C rolls this material forward. The resulting effect is to "boil" the pushed material over and over in front of the blade. The flattened C shape provides the necessary attack angle for the lower cutting edge, and at the beginning of a forward pass, the weight of the cut material on the lower half of the C helps achieve edge penetration. As the push progresses, the load in front of the blade passes the midpoint of the C and begins to exert an upward force on the blade. This "floats" the blade, reducing the penetration of the cutting edge and aiding the operator's control of the load.

Many different special-application blades can be attached to a tractor (Fig. 6.12), but five blades are common to earthwork: (1) the *straight* "S" blade, (2) the *angle* "A" blade, (3) the *universal* "U" blade, (4) the *semi-universal* "SU" blade, and (5) the *cushion* "C" blade.

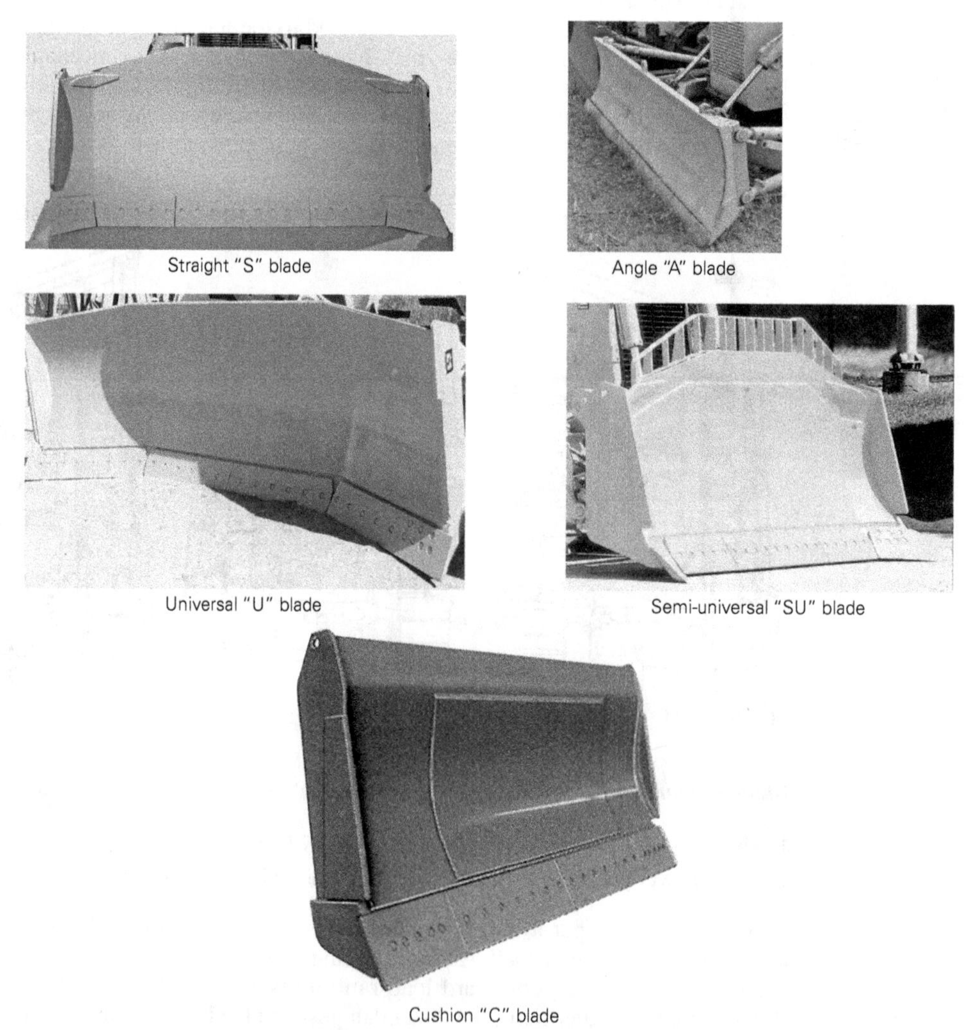

FIGURE 6.12 Common earth-moving dozer blades.

Pushing of scrapers is an important use for heavy tractors equipped with "C" blades. An ordinary dozer working as a pusher is likely to cave in the center of its blade unless it is reinforced. Reinforcement may be a plate or a cup welded outside, or it may be internal bracing. A cup will also serve to keep the scraper bumper from side-slipping.

Straight Blades "S." The straight blade is designed for short- and medium-distance passes, such as backfilling, grading, and spreading fill material. These blades have no curvature in the horizontal direction across their length and are mounted in a fixed position, perpendicular to the dozer's line of travel. Generally, a straight blade is heavy-duty, and it can normally be tilted within a 10-degree arc, increasing penetration for cutting or decreasing penetration for back-dragging material. It may be equipped with the ability to pitch, which means the operator can set one end of the cutting edge deeper into the ground to dig or pry hard materials. For easy drifting of light materials, the edges are brought to the same level—the blade is level in the horizontal plane.

Angle Blades "A." An angle blade is wider (in terms of face length) by 1 to 2 ft than an S blade. It can be angled up to a maximum of 25 degrees left or right of perpendicular to the tractor or held perpendicular to the dozer's line of travel. An A blade can be tilted, but because it is attached to the dozer by a C-frame mount, it cannot be pitched. The angle blade is very effective for side-casting material, particularly for backfilling or making sidehill cuts.

Universal Blades "U." This blade is wider than a straight blade, and the long-dimension outside edges are canted forward about 25 degrees. This canting of the edges reduces the spillage of loose material, making the U blade efficient for moving large loads over long distances. The cutting ratio is lower for the U blade than the S blade mounted on a similar size dozer. Penetration is not a prime objective of the blade's design (shape), as the lower cutting ratio relationship indicates. The U blade has a lower load ratio than would a similar S blade. This means the blade is best suited for lighter materials. Typical usages are working stockpiles and drifting loose or non-cohesive materials.

Semi-Universal Blades "SU." This blade combines the characteristics of the S- and U-blade designs. By the addition of short "wings," it has increased capacity compared to an S blade.

Cushion Blades "C." Cushion blades are mounted on large dozers when these machines are used primarily for push-loading scrapers. They have a wear-resistant center liner plate and are shorter than S blades so as to avoid pushing the blade into and cutting the rear tires of the scraper while push-loading. The shorter length facilitates maneuvering into position behind the scrapers. Rubber cushions and springs in the mounting enable the dozer to absorb the impact of contacting the scraper push block. By using a cushion blade instead of a "pusher block" to push scrapers, the dozer can clean up the cut area and increase the total fleet production. It is a blade of limited utility in pushing material and should not be used for production dozing. It cannot be tilted, pitched, or angled.

Dozer Project Employment

The first bulldozers were an adaptation of farm tractors used to plow the fields. In fact, the name "dozer" comes from a story claiming these machines replaced farm machinery powered by bulls; so now the bulls get to sleep, or doze, instead, and hence the name. These machines are the mainstay for major civil construction projects and are capable of performing a variety of tasks.

A tractor dozer of any type is worked by moving the tractor forward or, less commonly, backward and raising and lowering the blade to contact material to cut, spread, or transport it. As a dozer moves forward and digs, some of the soil cut by the blade will pile up in front of the blade and move with it, and some of it will drift off the sides, forming ridges or windrows. Resistance to the machine's movement is made up of the power absorbed in cutting and breaking up soil and in friction in the loosened soil. If the blade is lowered, more work will be done and resistance will increase, as a thick slice requires more digging power than a thin one, and the total amount of material resisting the blade is increased. If the blade is raised, the slice will thin or disappear, and the amount of earth being pushed will decrease, so both work and resistance are reduced.

In heavy digging, efficient operation involves pushing the most material without losing too much speed by engine slowing, slippage in the torque converter, or spinning the tracks. The operator starts cutting a slice that should give this result, and if the machine slows, the operator raises the blade slightly, or if the dozer is not working to capacity, the blade is lowered. The upward blade movement should be made as gradually as possible to avoid leaving a bump in the path of the tracks.

If the blade is set to cut an even depth, digging resistance will remain about the same through the pass, and the loosened and transported material will increase steadily. The increase of resistance does not slow the machine at first, as it causes the governor to open the throttle to maintain tractor speed. Once the engine is at full throttle, a further increase of material will slow the dozer, so the blade should be raised gradually to the surface of the ground where it can push the loose material without digging. Sometimes the blade is "pumped" during this lift by being dropped and raised quickly, cutting out a bit of extra material each time it is lowered.

The dozer digs and transports much more effectively downhill than on a level or uphill, and work should be arranged to work down a grade when possible.

Stripping. Dozers are excellent machines for stripping, which is the removal of a thin layer of covering material. On most projects, this term is used to describe the removal of topsoil, or organic material. Dozers are economical machines for moving material a maximum of 300 ft in the case of larger machines. The economical push distance decreases as dozer size decreases, but economical push distance also depends on the material being handled. A material with cohesion is easier to push than a granular material (like sand), which tends to run in front of the blade.

Backfilling. A dozer can efficiently accomplish backfilling by drifting material sideways with an angle blade. This enables forward motion parallel to the excavation. If a straight blade is used, the dozer will approach the excavation at a slight angle and then, at the end of the pass, turn in toward the excavation. No part of the tracks should hang over the edge of the excavation.

Caution must be exercised in making the initial pass completely across pipes and culverts. At a minimum, 12 in. of material should cover the pipe or structure before accomplishing a crossing. The diameter of the pipe, the pipe type, the distance between the sidewalls of the excavation, and the number of lines of pipe in the excavation dictate the minimum required cover. Larger-diameter pipe, larger excavation widths, and multiple lines of pipe are situations where more cover is needed before crossing the structure.

Spreading. The spreading of material dumped by trucks or scrapers is a common dozer task. Ordinarily, project specifications state a maximum loose lift thickness for compaction. Even when lift thickness limits are not stated in the contract specifications, density requirements and proposed compaction equipment will force the contractor to control the thickness of each lift. A dozer accomplishes uniform spreading by keeping the blade straight and at the desired height above the previously placed

FIGURE 6.13 Blade grade control sensors.

fill surface. The dumped material is forced directly under the blade's cutting edge. Laser blade controls are available for this type of work (Fig. 6.13).

Slot Dozing. Slot dozing is a technique whereby the blade-end spillage from the first pass or the sidewalls from previous cuts are used to hold material in front of the dozer blade on subsequent passes. When employing this method to increase production, align cuts parallel, leaving a narrow uncut section between slots. Then remove the uncut sections by normal dozing. The technique prevents spillage at each end of the blade and usually increases production by about 20%. The production increase is highly dependent on the slope of the push and the type of material being pushed.

Blade-to-Blade Dozing. Another technique used to increase dozer production is blade-to-blade dozing (Fig. 6.14), also referred to as side-by-side dozing. As the names imply, two or more machines maneuver so their blades are next to each other during the pushing phase of the production cycle. This reduces the side spillage of each machine by 50%. The extra coordination time necessary to position the machines together increases the cycle time.

FIGURE 6.14 Blade-to-blade dozing, used to increase production.

Therefore, the technique is not effective on pushes of less than 50 ft because of the excess maneuver time required. When machines operate simultaneously, delay to one machine is in effect a double delay. The combination of less spillage but increased maneuver time tends to make the total increase in production for this technique somewhere between 15% and 25%.

Breaking Piles. A pile of material may be knocked down by walking into it with the blade at the desired grade, after which it may be spread or piled elsewhere. If the heap is too large or hard for the machine to take at one pass, or if it is to be spread in more than one direction, the first pass may be made to cut away part of the pile to grade or to cut the top of it partway down. If the second method is to be used but the dozer cannot move the part it cannot reach, a ramp may be made by loosening the soil by pushing and then back-blading with down pressure, so the blade can contact the heap at a higher level.

If the pile is very large or hard in proportion to the power of the dozer, the side cut should be repeated from different angles in order to shorten the cut required for each pass. When the digging leaves a high face endangering the machine with possible collapse, the dozer should be turned toward it occasionally and driven into it with the blade held high. This should cause the bank to fall or slide without burying the side of the dozer in it.

If the sides are not accessible, a center cut may be made by first ramping up and cutting a slot down to grade, widening it to both sides. This slot, and any cut more than a few inches in depth, should be made slightly wider than the blade to avoid jamming the dozer between the walls. Jamming may be caused in very narrow cuts through rocks or material containing roots. If the blade makes grazing contact with these, they may be turned and protrude into the cut. They will then squeeze the dozer. Movement of such objects can also cause the sides of the cut to collapse. Dozers whose blades are track width or a little wider are particularly subject to getting jammed in this manner. A narrow cut also does not leave room to maneuver to get at a rock or other obstacle encountered in the ground.

Side Slopes. A dozer lacking a tilting control for the blade will cut deeper on the downhill side. On a slope, this tendency may be overcome in several ways. A shelf may be built by pushing downhill, so the dozer can start its side cut level or tipped oppositely to the slope. Or the dozer may cut and turn downhill, raising the blade, thus cutting a more or less level shelf, which can then be enlarged or graded off.

Spreading. When spreading material, the blade is held somewhat above the original surface so the material can slip under it in a smooth layer on which the tractor can move. A thin layer may be spread to the desired grade, but a thick layer should be built higher to allow for compaction. If there is not sufficient material ahead of the blade to reach the end of the area to be covered, it saves time to stop pushing as soon as the load is light and go back for more. The next bladeful will be pushed through the spot and can easily take the remnants of the first load with it. It is best to vary the path used in spreading, as it is easier to keep track of the grade if no heavy windrows are built up.

Turning. A dozer has difficulty turning while pushing a heavy load. More power is needed to swing the load than to push it straight ahead. As shown in Fig. 6.15, the turning action is similar to a lever pivoting on the braked track. Turns are easier with a wide-gauge machine, as the power arm is lengthened, and harder with an angle dozer or other dozers with the blade carried a distance ahead of the tractor, as the lengthened work arm reduces the leverage.

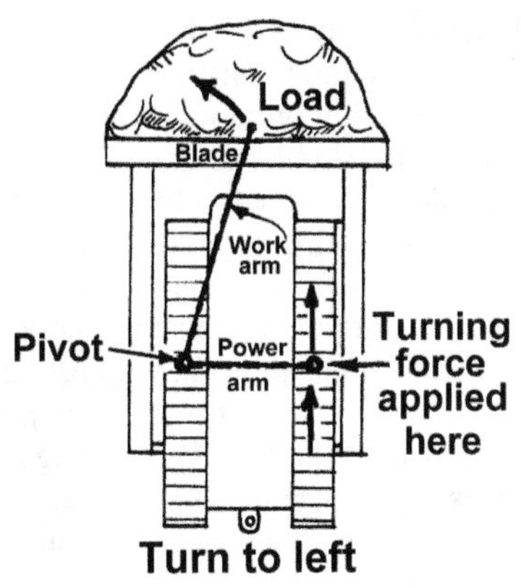

FIGURE 6.15 Leverage in swinging a load.

It is often easier to break the curve into two or more straight lines separated by angles, to pile the soil at the angle points, and to make a separate process of pushing it along the next line. The heaps at the angle points are best moved frequently instead of being allowed to pile up until it is difficult to dig.

Back-blading. After an area has been graded, it may look a bit rough because of small windrows of loose material, grouser prints, and piles left where the dozer turned. These may be smoothed down by backing over the area with the blade floating. It acts as a drag, smoothing off humps and filling hollows, but does not move enough material to change the grade.

Rocks in Cuts. Flat or irregular-shaped rocks may catch under the blade cutting edge and be pushed along with the load, increasing the resistance or rolling the rock out of the ground. The blade may sometimes be freed by shaking it up and down, but it is often necessary to stop and back up in order to get behind and under the rock.

Firmly embedded rocks, even rather small ones, may roll or slide the blade up so a bump is left on the grade. It may be necessary to back up, then move forward, forcing the blade into the ground deeply enough to firmly engage the stone and roll or push it out.

In digging out large rocks and stumps, the blade action usually is a combination of push and lift. The push action provides faster removal, but some objects may require a chiefly upward motion.

Rocks in Fills. Coarse material, such as rocks, lumps of sod, and other debris, makes grading difficult or impossible. When the blade is raised in spreading soil, the coarse material has a tendency to stay ahead of the blade to the last, although some will slip under and sometimes force the blade up, creating a situation where the slope of the grade is lost. Such pieces can be moved along a bit farther with the next pass and eventually pushed to the side.

If the stones are not to be buried but are to be left at the side for disposal, they can be worked over without making special passes. If a dozer pushes a bladeful of dirt into a loose rock, the rock will tend to drift off to the nearest side. If, during a series of pushes, each one is aimed to slide the rock in the same direction, it may be moved out of the area without any direct contact with the blade.

Pitching. Tracked vehicles pitch badly in moving over ridges, stones, or poles. After overbalancing, the machine generally falls with a crash, which can be damaging to both tractor and operator. A log or bump should be pushed out of the way or avoided if possible. Crossing, if necessary, should be done slowly and at an angle, so one side of the machine crosses the top and starts down while the other is still climbing. This slows the fall and avoids any danger of turning over backward. If the bump is a soft ridge, turning the tractor sharply while crossing the bump will cut it down.

When a dozer is digging at capacity, the tracks often spin a little, then grip and move, and then spin again. Each time they spin, they build up piles of dirt at the back. If the machine backs up in the same track, these piles will have the effect of the log, causing the back to rear up and then fall. This jolting can be avoided by keeping a wide work strip so it is not necessary to use exactly the same path backward as forward. A tractor is not much affected if only one track crosses a bump.

Gears. The pushing of material is done in low or second gear. Low gear is less demanding for the machine when doing heavy pushing, and easier for the operator when doing precise work. Second gear is considerably faster, and in clean material, good loads can be moved and smooth grades maintained. However, the higher gear makes it harder to cope with stones, hard spots, and other difficulties.

The amount of traction often determines whether a particular job can be done in second gear. A machine with narrow tracks, flat shoes, or worn grousers, or a standard machine on loose soil, spins the tracks easily. When too deep a cut is taken or an obstacle is hit, the load or shock to the engine is cushioned by slippage in the tracks, giving the operator time to raise the blade or disengage the clutch before stalling, even in a high gear.

Hill Work. Dozers may be used on moderate slopes, and wide-track models on steep slopes up to 30 degrees or more. Nevertheless, they are quite likely to overturn unless care is exercised. A machine that appears to have an ample margin of safety may be suddenly flipped on its side by running over a stone with the higher track, when at the same moment the lower track enters a hollow or soft ground. This is less apt to occur if the machine is pushing than walking, as it will then be moving slowly, will have the blade close to the ground, and will be steadied by the load. The machine also obtains some support from the windrow spilling from the downhill side of the blade.

Working on frozen slopes is hazardous, as the grousers may act as skates and allow the machine to slide uncontrollably downhill, regardless of the direction in which it is facing or trying to move. Sharp ice cleats will hold in such conditions, but soil and stones grind their points off very rapidly.

A similar danger is encountered on rock slopes, particularly shale with beds parallel to the surface. Slopes on soft fills are very treacherous, as the tip will be increased by the lower track sinking more deeply than the upper one. If a machine starts to roll over slowly, it can sometimes be saved by turning downhill and lowering the blade.

A slope that is too steep to be safely worked sideward may sometimes be graded by running the dozer along it diagonally. If it is too steep for this, soil may be pushed straight down from the high spots, moved along the bottom, then pushed up to the low places. Dozers can safely negotiate very steep upgrades and downgrades. Digging and pushing efficiency are much greater downhill and taper off to zero on steep upgrades. Steering is apt to be tricky on steep slopes, whether up or down, because of track slippage and shift in the center of gravity. Very steep grades of 25 degrees or more should be climbed forward rather than in reverse, because of better balance and traction.

Cutting should be done downhill whenever possible, and in very hard ground it may be advisable to dig it downhill, even if the spoil must then be pushed up the same hill for disposal. The engine oil-pressure gauge should be watched closely on steep work, as some engines do not lubricate properly when tilted steeply, especially at compound angles, and a low oil level that still gives adequate lubrication on a level may leave the pump dry on either upgrades or downgrades.

Where a run includes a downslope, the operator may push several loads to the top, then push most of the resulting pile down in a single pass.

Unless the ground is loose, a dozer can push much more of it than it can cut and move while cutting. Here again it may be good technique to drop one or more loads at the end of the cut, pushing the final load all the way through, along with the bulk of material piled up previously.

Cutting Hard Ground. If the blade refuses to cut down, it should usually be tipped forward and pitched. This helps it to cut when it is empty, but a load can float it out of the ground more readily. If the blade will not cut into humps or a bank, which is less usual, it should be tipped back. For general pushing, the blade should be centered or back, as this position rolls the material most effectively.

Digging in hard or stony ground will be easier if the work is arranged so cuts are made by only part of the blade. One corner of the blade should be tilted down if possible. If not, one cut can be ground down to a depth of a few inches and overlapping cuts made at the sides. If half the blade is in the air over the cut, weight per inch of edge will be doubled for the part on the ground and it will cut more effectively. The material it gets under will add weight to the blade, so it will probably do some digging in the original cut in addition to removing a substantial bit at the side.

If any considerable quantity of hard material must be dozed, it should be loosened by back-ripper teeth or by a separate ripper, or dug with a hoe or shovel.

Output. Dozer output decreases in almost direct ratio with increase in push distance. Use of scrapers, or shovel and trucks, should be considered instead of long pushes. If space is ample, the large dozer will move dirt at a lower cost per yard than a small one, and its advantage is increased in hard or rocky digging.

In restricted quarters, as in landscaping, backfilling trenches in narrow spaces, and working inside buildings, small dozers may show better productivity and lower cost than large ones.

Dozing Cycle. Most dozer digging is done in shuttle fashion with the machine facing in one direction through the dig, push, spread, and return parts of the cycle. This is because the distances covered are usually short, and turns, particularly in soft dirt, take time and spoil the grade. Consequently, it is quicker and easier to back to the cut than to make two turns in order to use a higher gear. On pushes of 100 ft or longer, the turns may be better unless the machine has a fast reverse.

Many tractors have as many reverse gears and speeds as they have forward gears and speeds. A dozer needs reverse speeds that are higher than forward ones. Backing is the unloaded part of the cycle, where lack of work should allow high speed, but some reverse gears must be powerful enough to climb steep grades and pull out of mud, and so cannot be fast. The backup part of the cycle may be put to work by using back-ripper teeth on the blade to loosen soil for the next push.

Speed in reverse may also be limited by the quality of grading done during the push. In making heavy cuts, it may not be efficient to take time and skimp loads to make a level floor with each pass, particularly if the soil is coarse. Gouges and bumps may be left by the blade and humps made by spinning tracks. The result is the need for a slow return speed to avoid pitching even when a higher gear is available.

Dozer Production Estimating

Because there is no hopper or bowl to load, a dozer has no fixed volumetric capacity. The volume of material moved is dependent on the quantity carried in front of the blade during the push. Dozer production rates are controlled by three factors:

1. Blade type
2. Type and condition of material
3. Cycle time

Blade Type. By design, straight blades roll material in front of the blades, and universal and semi-universal blades control side spillage by holding the material within the blade. Because the U and SU blades force the material to move to the center, there is a greater degree of material volumetric swell. The U or SU blade's quantity of loose material will be greater than for a similar S blade. But the ratio of this difference is not the same when considering bank cubic yards. This is because the factor to convert loose cubic yards to bank cubic yards is different for the universal-type blades and straight blades. The U or SU blade's boiling effect causes the difference.

The same type of blade comes in different sizes to fit different size dozers. Blade capacity, then, is a function of blade type and physical size. Manufacturers' specification sheets provide information concerning blade dimensions.

- Width
- Height
- Maximum digging depth
- Ground clearance at maximum lift
- Manual tilt
- Maximum hydraulic tilt
- Maximum pitch
- Angle

Type and Condition of Material. The type and condition of the material being handled affect the shape of the pushed mass in front of the dozer blade. Cohesive materials (clays) will boil and heap. Materials with a slippery quality or those with a high mica content will ride over the ground and swell out. Cohesionless materials (sands) are known as "dead" materials because they do not exhibit heap or swell properties. Figure 6.16 illustrates these material attributes.

Blade Volumetric Load

The volumetric load a blade will carry can be estimated by several methods:

1. Manufacturer's blade rating
2. Previous experience (similar material, equipment, and work conditions)
3. Field measurements

Clay material boiling in front of the blade

Cohesionless, sandy loam in front of the blade

FIGURE 6.16 Bulking attributes of materials when being pushed.

Manufacturer's Blade Ratings. Manufacturers provide blade ratings based on SAE Standard J1265. The Standard provides a uniform method for calculating blade capacity. It applies only to straight, angling, semi-U, and U-blades for crawler and wheel tractors. It applies to angling blades only in the straight (not angled) position, and it does not apply to angled blades or other tools used to side-cast materials. The Standard is used for making relative comparisons of dozer blade capacity and **not for predicting productivity in the field**.

$$Vs = 0.8WH^2 \tag{6.1}$$

$$Vu = Vs + ZH(W - Z)\tan x^{\circ} \tag{6.2}$$

where Vs = capacity of straight or angle blade in lcy
Vu = capacity of universal blade in lcy
W = the blade width in yards, exclusive of end bits
H = the effective blade height in yards
Z = the wing length measured parallel to the blade width in yards
x = the wing angle

Previous Experience. Properly documented past experience is an excellent blade load estimating method. Using this approach, to determine the total volume of material moved, it is necessary to cross-section the area both before and after the dozer excavation work is completed. It is also necessary to record the number of dozer cycles. Production studies can also be made based on the weight of the material moved. In the case of dozers, the procedures for weighing the material are normally more difficult to accomplish than surveying the volume.

Field Measurement. A procedure for measuring blade loads follows:

1. Obtain a normal blade load.
   a. The dozer pushes a normal blade load onto a level area.
   b. Stop the dozer's forward motion. While raising the blade, move forward slightly to create a symmetrical pile.
   c. Reverse and move away from the pile.

2. Measurement (Fig. 6.17).
   a. Measure the height (H) of the pile at the inside edge of each track.
   b. Measure the width (W) of the pile at the inside edge of each track.
   c. Measure the greatest length (L) of the pile. This will not necessarily be at the middle.

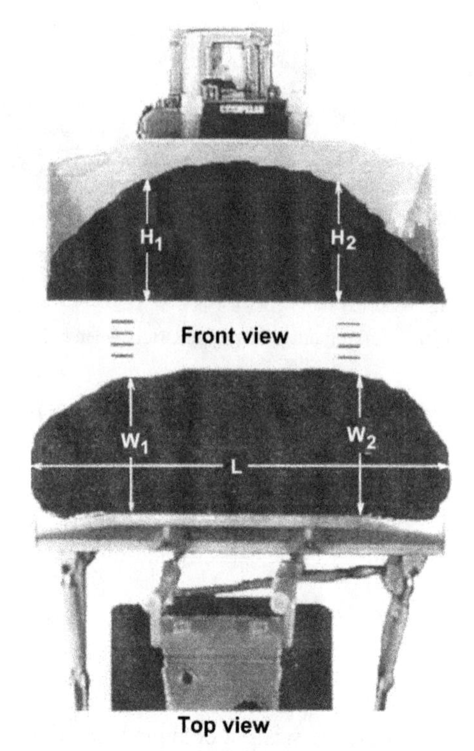

FIGURE 6.17 Measurements for calculating blade loads.

3. Computation: Average both the two height and the two width measurements. If the measurements are in feet, the blade load in lcy is calculated by the formula

$$\text{Blade load (lcy)} = 0.0139HWL \tag{6.3}$$

Production and Cost

The sum of the time required to push a load, backtrack, and maneuver into position to push again represents one dozer production cycle. The time required to push and backtrack can be calculated for each dozing situation, considering the travel distance and obtaining a speed from the machine's performance chart.

Dozing, however, is generally performed at slow speed, 1.5 to 2 mph. The lower speed is appropriate for very heavy cohesive materials. Return speed is usually the maximum attainable in the distance available. When using performance charts to determine possible speeds, remember the chart identifies instantaneous speeds. In calculating cycle duration, the estimator must use an average speed, allowing for the time required to accelerate to the attainable speed as indicated by the chart. Usually the operator cannot shift the machine past second gear in the case of distances less than 100 ft. If the distance is greater than 100 ft and the ground conditions are relatively smooth and level, maximum machine speed may be obtained. Maneuver time for power-shift dozers used in pushing material is about 0.05 min.

Production. The formula to calculate dozer pushing production in loose cubic yards per a 60-min hour is

$$\text{Production (lcy per hour)} = \frac{60 \text{ min} \times \text{blade load}}{\text{push time (min)} + \text{return time (min)} + \text{maneuver time (min)}} \tag{6.4}$$

Sample Calculation: A track-type dozer equipped with a power shift can push an average blade load of 6.15 lcy. The material being pushed is silty sand. The average push distance is 90 ft. What production, in loose cubic yards, can be expected?

Push time: 2 mph average speed (sandy material):

$$\text{Push time} = \frac{90 \text{ ft}}{5{,}280 \text{ ft/mile}} \times \frac{1}{2 \text{ mph}} \times 60 \text{ min/hr} = 0.51 \text{ min}$$

Return time: (see Fig. 5.7), second gear because less than 100 ft

Maximum speed 4 mph:

$$\text{Return time} = \frac{90 \text{ ft}}{5{,}280 \text{ ft/mile}} \times \frac{1}{4 \text{ mph}} \times 60 \text{ min/hr} = 0.26 \text{ min}$$

The Fig. 5.7 chart provides information based on a steady-state velocity. The dozer must accelerate to attain this speed. Therefore, when using such speed data, it is always necessary to make an allowance for acceleration time. Because in this example the change in speed is very small, an allowance of 0.05 min is made for acceleration time.

$$\text{Return time} = (0.26 + 0.05) = 0.31 \text{ min}$$

$$\text{Maneuver time} = 0.05 \text{ min}$$

$$\text{Production} = \frac{60 \text{ min} \times 6.15 \text{ lcy}}{0.51 \text{ min} + 0.31 \text{ min} + 0.05 \text{ min}} = 424 \text{ lcy/hr}$$

This production is based on working 60 min per hour, or an ideal condition. Job efficiency is affected by how well the work is managed in the field, the condition of the equipment, and the difficulty of the work. The efficiency of an operation is usually accounted for by reducing the number of minutes worked per hour. The efficiency factor is then expressed as working minutes per hour, for example, a 50-min hour, or a 0.83 efficiency factor. If it is necessary to calculate the production in terms of bank cubic yards (bcy), the swell factor or the percent swell for the material can be used to make the conversion.

Sample Calculation: Assume a percent swell of 0.25 for the silty sand and a job efficiency equal to a 50-min hour. What is the expected production in bank cubic yards?

$$\text{Production} = \frac{424 \text{ lcy}}{1.25} \times \frac{50 \text{ min}}{60 \text{ min}} = 283 \text{ bcy/hr} \approx 280 \text{ bcy/hr}$$

The final step is to compute the unit cost of pushing the material. A large machine should be able to push more material per hour than a small one. However, the cost to operate a large machine will be greater than the cost to operate a small one. The ratio of the total cost to operate (O&O plus operator) to the amount of material moved determines the most economical machine for the job. This ratio is the cost figure used when bidding unit price work.

Sample Calculation: The machine has an owning and operating cost of $127.10 per hour. Operators in the area where the proposed work will be performed are paid $58.50 per hour (with fringes). What is the unit cost for pushing the silty sand?

$$\text{Unit cost} = \frac{\$127.10 \text{ per hour} + \$58.50 \text{ per hour}}{280 \text{ bcy/hr}} = \$0.663 \text{ per bcy}$$

Manufacturer Production Estimation Guidance. Equipment manufacturers provide dozer production guidance through formulas and factor information. Equation (6.5) is a rule-of-thumb formula proposed by one company. This formula equates the net horsepower of a power-shift crawler dozer to an lcy production value.

$$\text{Production (lcy per 60-min hr)} = \frac{\text{net hp} \times 330}{\text{D} + 50} \tag{6.5}$$

where net hp = net horsepower at the flywheel for a power-shift crawler dozer
D = the one-way push distance in feet

Sample Calculation: A 200-net-hp power-shift dozer will be used to push material 90 ft. Use the manufacturer's formula to calculate the expected lcy production.

$$\frac{200 \times 330}{(90 + 50)} = 471 \text{ lcy per 60-min hr}$$

Again, the actual production will be less, as a 60-min hour represents an ideal situation.

Another tractor manufacturer provides a very detailed procedure for estimating dozer production. First, they provide machine-specific production curves based on push distance and "ideal" conditions. The ideal conditions are: pushing 50 ft to gain the blade load and finally dump over a vertical face. The second calculation step is to adjust the ideal curve value based on factors such as

- Transmission type
- Soil density
- Coefficient of traction
- Operator skill

Sensitivity Check. Dozer production estimating, whether using a blade volume and speed approach or manufacturer production data, should not be considered exact. The actual dozer work in the field is never under a fixed set of conditions. Therefore, when making such dozer production estimates, it is good practice to investigate the cost effects for a range of production values. With such knowledge, management can evaluate the risk resulting from estimate assumptions.

If the calculated production value was 280 bcy/hr, it is wise to investigate the cost consequences of a higher and lower production rate. Similarly, it is good practice, if the task involves moving a large quantity of material, to carry production cost value to three decimal places.

$$\text{Direct production cost} = \frac{\$185.60 \text{ per hour}}{290 \text{ bcy/hr}} = \$0.640 \text{ per bcy}$$

$$\text{Direct production cost} = \frac{\$185.60 \text{ per hour}}{280 \text{ bcy/hr}} = \$0.663 \text{ per bcy}$$

$$\text{Direct production cost} = \frac{\$185.60 \text{ per hour}}{270 \text{ bcy/hr}} = \$0.687 \text{ per bcy}$$

Therefore, the effect of achieving a production of only 270 bcy/hr would be a $0.024 ($0.687 – $0.663) increase in cost for every bcy moved, or $24,400 on a million-cubic-yard project.

Good production record keeping will greatly aid management in evaluating the reliability of production estimates.

RIPPING

The first reason for considering ripping is in aiding scrapers and dozers to dig hard soil or rock. The dozer often has the double job of ripping and pushing scrapers, or dozing.

Crawler tractors can be fitted with rear-mounted rippers (Fig. 6.18) specifically designed by the manufacturer to match tractor characteristics. Rippers are made for a range of tractor sizes, with various ripper configurations and linkage designs for depth control and adjustment of the tip's attack angle. Because of the power and tractive force available with large tractors, penetration depth of a rear ripper on such machines can be as great as 4 to 5 ft but

FIGURE 6.18 Dozer-mounted, hydraulically operated ripper.

decreases significantly with smaller, lighter tractors. A ripper is a relatively narrow-profile implement. It penetrates the earth and is pulled by the crawler tractor to loosen and split hard ground, weak rock, or old pavements and bases. Motor graders can also be equipped with rippers for light-duty applications.

Although rock has been ripped with varying degrees of success for many years, developments in methods, equipment, and knowledge have greatly increased the range of economically rippable materials. Rock types once considered to be unrippable are now ripped with relative ease and at cost reductions—including ripping and hauling with scrapers—amounting to as much as 50% when compared with the cost of drilling, blasting, loading with loaders, and hauling with trucks.

The following major developments are responsible for increased ripping capability:

- Heavier and more powerful tractors.
- Improvements in the sizes and performance of rippers to include development of impact rippers.
- Better instruments for determining the rippability of rocks.
- Improved techniques in using instruments and equipment.

Determining the Rippability of Rock

When a project involves rock excavation, the first step in the investigation is determining whether the rock can be ripped. Evaluating the rippability of a rock formation involves study of the rock type and determining the rock's density. Igneous rocks, such as granites and basaltic types, are very hard and normally impossible to rip because they lack stratification and cleavage planes. Sedimentary rocks have layered structures caused by the manner in which they were formed. This characteristic makes them easier to rip. Metamorphic rocks, such as gneiss, quartzite, schist, and slate, being a changed form from either igneous or sedimentary, vary in rippability with their degree of lamination or cleavage.

The physical characteristics can favor ripping, particularly

- Fractures, faults, and joints; these act as planes of weakness facilitating ripping.
- Weathering; the greater the degree of weathering, the more easily the rock is ripped.
- Brittleness and crystalline structure.
- High degree of stratification or lamination offers good possibilities for ripping.
- Large grain size; coarse-grained rocks rip more easily than fine-grained rocks.

Because the rippability of most rock types is related to the speed at which seismic (sound) waves travel through the rock, it is possible to use refraction seismographic methods to determine, with reasonable accuracy, whether a rock can be ripped. Refraction seismographic methods are based on Snell's Law,[1] which defines how a wave bends when it crosses the boundary between two different materials. In the earth, shear and compression waves travel with different velocities in distinct rock types. Compression waves (p-waves)

[1] Willebrord Snell (1580–1626) was a Dutch scientist who studied the behavior of light as it passed through a medium. He found a direct relationship between the sine of the angle of incidence and the sine of the angle of refraction.

travel faster than shear waves (s-waves). When a wave intersects a boundary at an angle other than 90 degrees, the wave will reflect and/or refract across the boundary, depending on the natural velocity for each material. A refracted wave then travels along the boundary between the two layers and is not transmitted into the lower layer. Measuring these waves allows the seismic velocity of the rock to be determined.

Generally, higher-seismic-velocity materials have a more crystalline dense structure that more effectively transmits the movement of seismic waves. Lower-seismic-velocity materials have a lower density, higher porosity, higher fracture content, or a higher degree of interconnected air voids—these conditions slow the transmission of waves through the material. Standardized field test equipment is available for measuring seismic velocity of the earth both near the surface and at great depths. More advanced testing technologies are available to measure both the seismic velocity and the material modulus. These provide a direct estimate of material strength.

Typically rocks exhibiting seismic waves at velocities less than 7,000 feet per second (fps) are rippable, whereas rocks with high propagate wave velocities, 10,000 fps or greater, are not rippable. Rocks with intermediate velocities are classified as marginal. Charts like Fig. 6.19 indicate rippability for a specific size dozer and for various types of soil and rock types. The information appearing in the figure is only a guide and should be supplemented with other data such as boring logs or core samples. The decision to rip or not to rip should be based on the estimated costs as compared to the costs for other excavating methods. Field tests may be necessary to determine whether a given rock can be ripped economically.

Determining the Thickness and Strength of Rock Layers. A refraction seismograph can be used to determine the top of bedrock, the thickness, and the strength of rock layers at or near the ground surface. The paths followed by seismic (sound) waves from a wave-generating source through a formation to detecting instruments (a seismograph) are illustrated in Fig. 6.20.

A geophone, which is a sound sensor, is driven into the ground at station 0. Equally spaced points 1, 2, 3, and so on are located along a line, as indicated in Fig. 6.20. Due to the geometry of refraction, it is necessary for the length of the seismic "spread" to be approximately three to five times the depth of the rock layers. A wire is connected from the

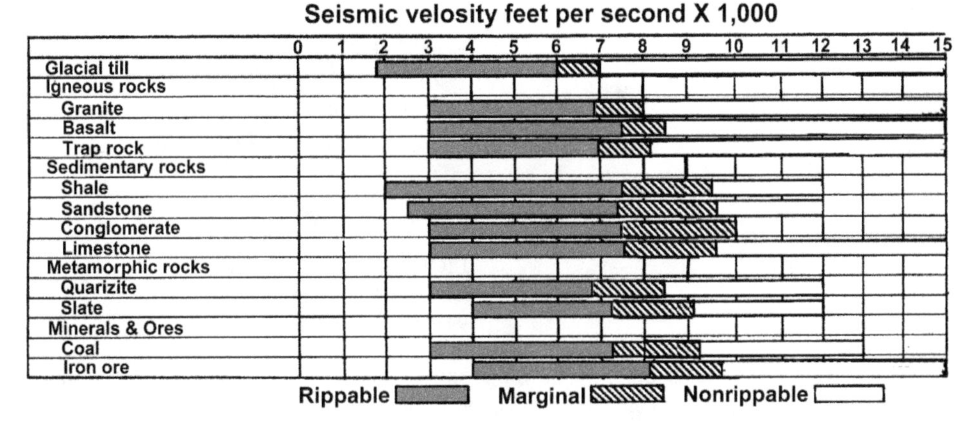

FIGURE 6.19 Rippability based on seismic wave velocities for a specific size dozer.

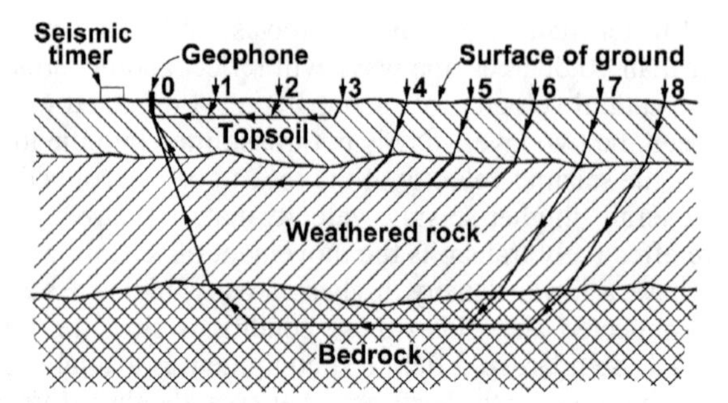

FIGURE 6.20 Path of seismic waves through earth layers with increased density with depth.

geophone to the seismic timer (Fig. 6.21), and another wire is connected from the timer to a sledgehammer or other seismic energy source. A steel plate is placed on the ground at the sensing stations in successive order. When an impulsive seismic source creates a seismic wave (the hammer strikes the steel plate), a switch closes instantly to send an electric signal, which starts the timer. At the same instant, the blow from the hammer sends seismic waves into the formation, and these waves travel to the geophone. On receipt of the first wave, the geophone signals the timer to stop recording the elapsed time. With the distance and time known, the velocity of a wave can be determined.

FIGURE 6.21 Refractive seismograph recording instrument and geophones.

As the distance from the geophone receiver to the wave source—namely, the striking hammer—is increased, waves will enter the lower and denser formation, through which they will travel at a higher speed than through the topsoil. The waves traveling through the denser formation will reach the geophone before the waves traveling through the upper soil layer arrives. The velocity through the denser formation is a higher value and will remain approximately constant as long as the waves travel through a material formation of uniform properties.

A plot such as the one shown in Fig. 6.22 yields wave velocity by measuring the distance and time associated with each layer and dividing the distance by the corresponding time:

$$\text{Velocity, } V_i = \frac{\text{Horizontal distance, } L_i}{\text{Time to travel distance } L_i} \tag{6.6}$$

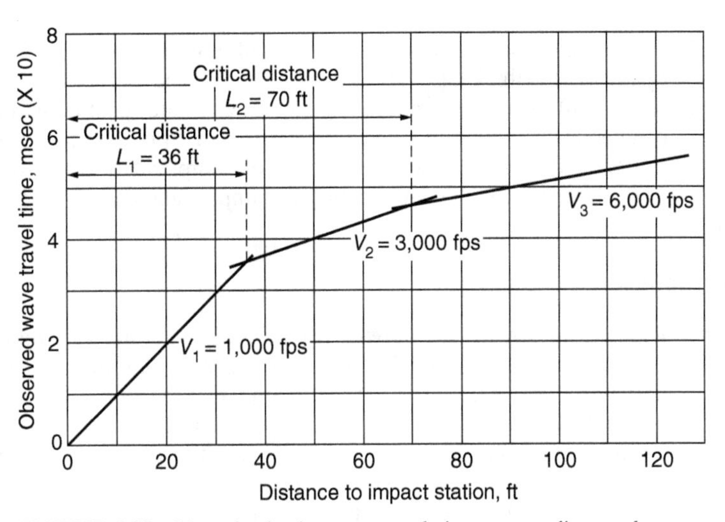

FIGURE 6.22 Plot of seismic wave travel time versus distance between source and the geophone.

The velocity remains constant as the wave passes through material of uniform density. Breakpoints in the slope of the lines indicate changes in velocity and, therefore, material density. The depth to the surface separating the two strata depends on the critical distance and the velocities in the two materials. It can be computed from the equation

$$D_1 = \frac{L_1}{2}\sqrt{\frac{V_2 - V_1}{V_2 + V_1}} \tag{6.7}$$

where D_1 = depth in feet of the first layer
L_1 = critical distance in feet
V_1 = velocity of wave in top stratum in feet per second (fps)
V_2 = velocity of wave in lower stratum in fps

Solving for D_1 from Fig. 6.22 gives

$$D_1 = \frac{36}{2}\sqrt{\frac{3{,}000 - 1{,}000}{3{,}000 + 1{,}000}} = 13 \text{ ft}$$

Thus, the topsoil has an apparent depth of 13 ft.

Equation (6.8) can be used to determine the apparent thickness of the second stratum.

$$D_2 = \frac{L_2}{2}\sqrt{\frac{V_3 - V_2}{V_3 + V_2}} + D_1\left[1 - \frac{V_2\sqrt{V_3^2 - V_1^2} - V_3\sqrt{V_2^2 - V_1^2}}{V_1\sqrt{V_3^2 - V_2^2}}\right] \tag{6.8}$$

where L_2 is 70 ft, D_1 is 13 ft, V_1 is 1,000 fps, V_2 is 3,000 fps, and V_3 is 6,000 fps.
Solving, we obtain

$$D_2 = 31 \text{ ft}$$

Ripping is normally an operation involving the top strata of a formation. Determination of the depth (thickness) of the upper two rock layers and the velocity of the upper three layers will provide the necessary data to estimate most construction jobs. The procedure can become very complicated, however, when dipping strata or changes in material properties are encountered. The investigation of complicated formations may require, in addition to seismic studies, the core drilling of samples and the excavation of test pits.

Ripper Attachments

Ripper attachments for crawler tractors are generally rear mounted. The mounting may be fixed radial, fixed parallelogram, or a parallelogram linkage with a hydraulically variable pitch (Fig. 6.23). The vertical member of the ripper forced down into the material to be ripped is known as a *shank*. A ripper tip (a tooth, point, or tap) is fixed to the lower, cutting end of the shank. The tip is detachable for easy replacement, as it constitutes the real working part of the ripper and receives all the abrasive action of the rock—it is a high-wear surface. A tip can have a service life of only 30 min to 1,000 hr, depending on the abrasive characteristics of the material being ripped.

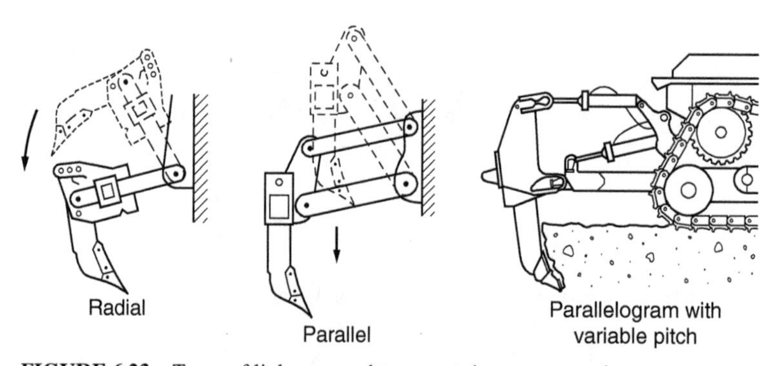

FIGURE 6.23 Types of linkages used to mount rippers on crawler tractors.

Both straight and curved shanks are available. Straight shanks are used for massive or blocky formations. Curved shanks are used for bedded or laminated rocks, or for pavements, where a lifting action will help shatter the material.

With the radial-type ripper linkage, the beam of the ripper pivots on link arms about its point of attachment to the dozer; therefore, the angle of tip attack varies with the depth the shank is depressed. This may make it difficult to achieve penetration in tough materials. The shank may also tend to "dig itself in" like an anchor.

The parallelogram-type ripper maintains the shank in a vertical position and keeps the tip at a constant angle. With adjustable parallelogram-type rippers, the operator can control the tip angle.

Heavy-duty rippers are available in either single-shank (Fig. 6.24) or multishank models (Fig. 6.18). The multishank models are available with up to five shanks in the case of smaller tractors and up to three shanks for larger tractors. The shanks are inserted into slots on the ripper frame and pin connected to the frame. This arrangement facilitates shank removal, making it possible to match the number of engaged shanks to the material properties and the project requirements. In the case of very heavy work (i.e., high-seismic-velocity material), a single center-mounted shank will maximize production. The use of multiple shanks will produce more uniform breakage if full penetration can be obtained. A single shank often rolls individual oversized pieces to the side.

FIGURE 6.24 Dozer with a single-shank ripper.

The effectiveness of a ripper depends on

- Down pressure at the ripper tip
- The tractor's usable power to advance the tip; a function of power available, tractor weight, and coefficient of traction
- Properties of the material being ripped; laminated, faulted, weathered, and crystalline structure

A rule of thumb in sizing a tractor for a ripping operation is 1 fwhp per 100 lb of down pressure on the ripper, and it must have 3 lb of machine weight per pound of down pressure to ensure adequate traction.

The number of shanks used depends on the size of the dozer, the depth of penetration desired, the resistance of the material being ripped, and the degree of breakage desired. If the material is to be excavated by scrapers, it should be broken into pieces sufficiently small to pass into the scraper, but usually the maximum size dimensions should be less than 24 to 30 in. Two shanks can be effective in softer, easily fractured materials prepared for scraper loading. Three shanks should be used only in very easy-to-rip material such as hardpan or some weak shales. Only a field test conducted at the project site will demonstrate which method, depth, and degree of breakage are appropriate and economical.

Shanks and Teeth. A tooth is made up of a long shank and the tooth proper, which may be called a point, a tip, or a cap. Teeth are detachable. They may be built up with hardface rod when worn, but this may spoil their shape and efficiency. It is better practice to weld on a forged cap. As mentioned, tooth points may have a service life of 30 min to 1,000 hr, the difference depending on the abrasive qualities of the rock. The operator should observe them every time they come out of the ground. Dull teeth are inefficient, and worn-out or broken ones allow destruction of the shank.

Operators should avoid running the tractor for extended periods "on tiptoe." Down pressure on the ripper may raise the rear of the tractor; then only the front of the tracks provides traction for forward movement. This reduces production, wears the undercarriage, and provides the operator with a rough ride.

Except when breaking hard or tough ground (pavement or frost) over soft material, teeth usually should be near full depth, with the pull beam approximately horizontal. Shallow ripping tends to be irregular in depth and causes tooth wear out of proportion to production.

Direction. On slopes it is safest to rip uphill or downhill, as large chunks might tend to lift and overturn the tractor when working perpendicular to the slope. Difficult ground might be ripped downhill only. When bedding planes or joint structure is at an angle with the surface, primary ripping is usually done against the grain. The slope of the beds will tend to pull the tooth down. However, if this results in excessive pulling up of big slabs, another direction may be used. If bedding planes are perpendicular to the surface, ripping should be across them. If parallel or nearly parallel, the tooth or teeth may tend to cut steep-sided grooves with unbroken ribs between them.

Whenever it is practical, tractors doing double duty as scraper pushers and as rippers should rip in the direction of scraper travel to avoid turns. Irregular depth and poor breakage can often be corrected by cross-ripping (ripping the same area again but at right angles to the original direction). The tractor must travel on rocks turned up the first time and may find the going very rough. It may be necessary to push some of the pieces off to the side or to smash them before cross-ripping.

The ripper has two principal uses in earthmoving. One is to make a diggable soil easier to dig, the other is to compete with explosives in loosening otherwise nondiggable material. Secondary uses include laying underground cable, cutting tree roots, and rolling out boulders.

O&O costs of a tractor will increase for machines used regularly in ripping operations. It is advisable to increase normal O&O costs by 30% to 40% if the machine is used in heavy ripping applications. Damage to the tractor and ripper is greatly affected by operator skill and attitude.

Ripping Production Estimates

Compared to moving soil, even tight clays, the cost of excavating rock by ripping and scraper loading is considerably higher. Nevertheless, ripping and scraper loading is usually much less expensive than using an alternative method to move rock, such as drilling, blasting, excavator loading, and truck hauling. Estimating ripping production is best accomplished using data from past projects and by working a test section and conducting a study of the operational methods. This enables a production determination based on weight of ripped material. Yet the opportunity to conduct such field tests is often nonexistent; therefore, most initial estimates are based on production charts supplied by equipment manufacturers.

Quick Method. By field-timing several passes of a ripper over a measured distance, an approximate production rate can be determined. The timed duration should include the turnaround time at the end of each pass. An average cycle time can be determined from the timed cycles. The quantity (volume) is determined by measuring the length, width, and depth of the ripped area. When a production rate is calculated by this method, the value is about 20% higher than an accurate cross-section study. Therefore, the quick-estimating formula is

$$\text{Ripping production (bcy/hr)} = \frac{\text{measured volume (bcy)}}{1.2 \times \text{average time (hr)}} \tag{6.9}$$

Seismic-Velocity Method. Manufacturers have developed relationships between the seismic wave velocities of different rock types and rippability. Ripping performance charts, such as those in Fig. 6.19, enable the estimator to make an initial determination of equipment suitable for ripping a particular material based on general rock-type classifications. After the initial determination of applicable machines is made, production rates for the machines are calculated from manufacturer production charts.

These charts are usually developed based on the following assumptions:

- The machine rips full time, no dozing.
- Power-shift tractor with a single-shank ripper.
- 100% operating efficiency.
- Chart is valid for all classes of rock.
- Igneous rock with seismic velocity ≥8,000 and using a tractor of 850 hp or greater, it may be necessary to reduce production by 25%.
- Igneous rock with seismic velocity ≥6,000 and using a tractor with a horsepower rating of 300 to 600, it may be necessary to reduce production by 25%.

These ripping production charts are developed from field tests conducted in a variety of materials. However, because of the extreme variations possible among materials of a specific classification, judgment is necessary when using charts provided by a manufacturer. The production rates obtained from the charts must be adjusted to reflect the actual field conditions of the project.

Operating Techniques. Ripping should be done at the maximum possible penetration depth, but it must be accomplished at a uniform depth. Tractor traction and usable force control penetration depth. For the most economical production, ripping is performed in low gear at low speed, typically 1 to 1½ mph. Speeds only slightly higher can result in a dramatic increase in operating cost from undercarriage and ripper tip wear. When ripping to load scrapers, it is best to rip in the direction of the scraper loading pattern. When removing the ripped material, always leave a "cushion" 4 to 6 in. deep of loose material. The cushion will create better underfoot conditions for the tractor doing the ripping and reduce track wear.

Take advantage of gravity and rip downhill when possible. Cross-ripping will make the pit rougher, increase scraper tire wear, and requires twice as many passes. Cross-ripping does, however, help to break up "hard spots" and large slabs of material.

DOZER SAFETY

On a project in Washington State, an operating engineer was killed when the dozer he was operating slid on ice, overturned, and came to rest at the bottom of a 150-ft embankment. The victim, who was not wearing a seat belt, was thrown from the dozer and crushed beneath it when it came to rest. The dozer's ROPS was not significantly damaged.

Supervisors are not expected to operate heavy equipment, but as part of management, they have both a moral responsibility and business reasons to ensure the equipment operators receive proper training and can recognize potential hazards associated with their machines. OSHA Construction Standards (29 CFR Part 1926) specify many safety rules. Section 1926.602(a)(2)(i) specifically states, "Seat belts shall be provided on all equipment covered by this section and shall meet the requirements of the Society of Automotive Engineers, J386-1969, Seat Belts for Construction Equipment." Yet a quick search of the Web yields many accident reports of fatalities resulting from operator failure to use seat belts.

Morally (and legally), companies have to provide employees a safe working place. From a purely business point of view, an effective safety program has the potential to drastically reduce the cost of doing business. When a company maintains an extremely low accident rate, it reaps the accompanying benefits. Practicing and demanding good safety will pay dividends throughout one's career.

CHAPTER 7
LAND CLEARING

Clearing of vegetation is usually necessary as a preliminary operation prior to moving or shaping ground. Trees must be cleared, but smaller growth also makes soil or rock difficult to handle, and the decay of contained or covered organic material will cause embankments to settle. To satisfy environmental concerns during construction, a building site must be surrounded by silting fencing. Clearing is preferably a machine job. It may be done by a wide variety of standard excavators, particularly by crawler tractors (dozers), loaders, and hoes. If a job is large and/or difficult, it will probably be economical to buy or rent one or more of the specialized clearing machines or attachments.

PROTECTION OF THE ENVIRONMENT

Excavation contractors are accustomed to moving "clean soil" and must be wary of entering contaminated sites. Work on such sites requires special permits and enhanced protection of personnel. Even if there are no contamination issues, stormwater discharges from excavation activities can significantly affect water quality. As water flows over a construction site, it can pick up and transport sediment, debris, and chemicals, dumping these into storm sewer systems or directly into rivers, lakes, or coastal waters.

Contaminated Ground

The discovery of contaminated ground presents an environmental hazard to workers and often to neighbors surrounding the project. An environmental due-diligence survey involves three stages: (1) initial assessment (any history or current evidence of contamination), then (2) investigation, and finally (3) remediation.

Assume a basement is being dug in a development and an old, unknown underground storage tank is broken and leaks heating oil into the ground. Who is to be blamed? The contractor may be in trouble if the company's only insurance is a standard, comprehensive, general liability policy. To cover this special liability, it is necessary to have an "environmental rider" on the basic liability policy.

In addition to insurance coverage, contractors should protect themselves by asking for Phase I and Phase II Environmental Site Assessments (ESAs) at prebid meetings. Standards for the Phase I and Phase II ESAs have been established by the American Society for Testing and Materials (ASTM) to address the "All Appropriate Inquiry" (AAI) aspect to the Comprehensive Environmental Response, Compensation, and Liability Act (CERCLA).

When hazardous substances have been released, CERCLA is the national policy with defined procedures for containing or removing such constituents. Additionally, CERCLA provides funding and guidance for remediation of some abandoned and contaminated hazardous waste sites. It is also necessary for a contractor to be aware of state, local, or federal regulations outside of CERCLA, as these may have additional or supplementary site assessment requirements and liability protections.

The Phase 1 report involves a review of records; a site inspection; and interviews with owners, occupants, neighbors, and local government officials. If a Phase I ESA identifies potential contamination of the site by hazardous materials, a Phase II ESA may be conducted. The Phase II ESA includes sampling and laboratory analysis to confirm the presence of hazardous materials.

Even if the reports indicate no contamination, the construction contract should have a stop-work clause—especially if excavation is involved. This allows the contractor to stop work without penalty if potential environmental hazards are found. The contractor should then contact the project owner or agent and report the finding. When an owner refuses to report a hazard to human health or the environment, the contractor may be bound to report the condition to the proper authorities.

Reporting environmental problems can be complicated because of the numerous agencies requiring notification. To avoid any oversight, report to all possible agencies—the Environmental Protection Agency (EPA), state agencies, fire departments, local planning commissions, and the National Response Commission.

Detection. The use of trenchless technologies can be helpful in locating the extent of contamination. To reach inaccessible areas under buildings or other obstructions, compact directional drilling equipment is very useful.

Horizontally drilled wells have been installed to perform remediation of subsurface groundwater and soils with pump and treat systems, such as by air spraying, soil vapor extraction, and bioremediation. Horizontal wells are similarly used to prevent contaminant migration and for characterization, that is, taking samples for evaluation, under buildings, and other areas where surface conditions prevent drilling of vertical wells. Because underground plumes often spread horizontally, fewer horizontal wells are required to treat a site than the probable number of vertical wells. The EPA believes horizontal well technology has the potential for significant cost savings on site remediation projects.

When a regulatory agency determines contaminated soil must be remediated, the owner must carefully select an experienced soil remediation contractor. This is important because the property owner is fully liable for the contractor's actions.

Treatments. The EPA encourages alternative treatment methods to clean contaminated soil or to render contaminates less harmful. These methods can be classified into five general categories: (1) biological, (2) physical, (3) immobilization, (4) chemical, and (5) thermal.

The technique chosen for a given site depends on the contaminants and the site's geology. For example, cleaning soils under an old gas station may be handled best with vapor extraction if the contamination is mostly gasoline and if the soil is not too cohesive. An important consideration when selecting a bioremediation approach to a particular situation is soil type.

Biological remediation (bioremediation) uses microorganisms such as bacteria. These eat soil contaminants and turn them into harmless—or at least less toxic—compounds. This method is preferred because it works in situ (i.e., is done in place). Nutrients and oxidizers are added to the soil to stimulate the growth of hydrocarbon-eating bacteria.

Physical methods for remediation include such processes as

1. Drawing a vacuum through wells drilled in the soil to pull out volatiles
2. Pressurizing the wells with heated air
3. In situ steam stripping, a vacuum extraction process with steam injection wells

The vapors captured in all three processes are then treated to remove the contaminants. Another physical technique uses a water-and-detergent solution to wash the contaminants from the soil.

Immobilization is a process to physically or chemically immobilize contaminants. This method may also be called solidification, whereas chemical treatment is stabilization. Dechlorination is a chemical remediation technique that reacts with chlorine in such compounds as polychlorinated biphenyls (PCBs), creating byproducts less toxic than the original contaminants.

Vitrification, incineration, and thermal desorption are in situ heat application remediation methods. Incineration is the most frequently used of any soil treatment method at federal cleanup sites, according to the EPA. A thermal desorption system puts the contaminated soil through a heated, rotating drum similar to an asphalt plant drum. The temperatures in the drum are sufficient to vaporize the contaminants for collection and treatment.

Erosion Control

The EPA cites stormwater runoff as the most common cause of surface water pollution. Consequently, the federal government has instituted the National Pollutant Discharge Elimination System (NPDES).

An NPDES stormwater permit is required when construction activities disturb 1 or more acres. When discharges are from smaller sites making up part of a larger common plan of development or sale, a permit is similarly required. The EPA promulgated the *Construction and Development Effluent Guidelines and Standards* (40 CFR Part 450) in 2009 and amended the regulations in 2014 and 2015. Construction stormwater permits include effluent limits for erosion and sediment control, pollution prevention, and site stabilization.

Construction site owners and operators are required to

- Implement erosion and sediment controls
- Stabilize soils
- Manage dewatering activities
- Implement pollution prevention measures
- Provide and maintain buffers around surface waters
- Prohibit certain discharges, such as motor fuel and concrete washout
- Utilize surface outlets for discharges from basins and impoundments

Construction projects need an erosion and sediment control plan to apply for an NPDES permit. Depending on the location of the construction site, either the state (if it has been authorized to implement the NPDES stormwater program) or EPA will administer the permit. In areas where the EPA is the permitting authority, operators of regulated construction sites are generally permitted under EPA's 2012 *Construction General Permit* (CGP).

The EPA's 2012 CGP requires compliance with effluent limits and the development of a stormwater pollution prevention plan (SWPPP). If a project owner or contractor intends to seek coverage under the 2012 CGP, it is necessary to first submit a notice of intent (NOI) using the electronic NOI (eNOI) system. Those using the system must certify that the activities proposed meet the permit's eligibility conditions and will comply with the permit's requirements. However, the use of the eNOI system is limited to construction sites where the EPA is the permitting authority.

Furthermore, it may be necessary to obtain a local permit if the community has a population of 10,000 or more. Unfortunately, the regulations vary from state to state and may be different from city to city in the same state. To determine the requirements for a specific project, it is advisable to check with the appropriate agency based on project location.

Many municipalities are now mandating the use of readily available erosion control techniques, such as exposing the smallest area of land possible for the shortest period or building a retention pond to detain runoff water far a sufficient duration so suspended sediments will settle out of the water. One state Department of Transportation seeks to limit exposed cut and fill areas based on the project's mass diagram.

The erosion and sediment control plan may include roughening the soil surface or the use of turf reinforcement mats and straw bales or other forms of check dams to control erosion and runoff flows. The plan should be prepared by or in the name of the owner-operator of the site. It is generally prepared by the project's consulting engineering or architectural firm. It is not unusual for the contractors who do earthmoving on the sites to be co-permittees, and their personnel are required to receive instruction about the NPDES regulations before any work can be performed on an NPDES-permitted site.

DISPOSAL OF CLEARING DEBRIS

Cut or uprooted vegetation must be processed or removed as part of most clearing jobs. Possible disposal options include burial, allowing time for decay; burning; shredding or chipping; removal from the area; and combinations of these methods. Disposal methods are considered first because of their impact on clearing techniques.

Nitrogen

Disposal of vegetation by any method other than burning is likely to be complicated by absorption of nitrogen during decay processes. This element is essential for life of every kind. Although abundant as a free gas in the air, its quantity in fixed or usable form in the soil is limited. The problem of nitrogen deficiency is usually not a strong objection to disposal of vegetation by any reasonable method such as surface decay, or mixing into soil, or plowing it under. Nor does it prohibit the use of wood chips for soil cover. But nitrogen deficiency must be considered in assessing both the immediate and possible long-term effects caused by the work.

The deficiency can be largely corrected by the addition of suitable amounts of nitrogen-rich fertilizer throughout the period of decay or by planting nitrogen-capturing legumes. The end result is usually substantial enrichment of the soil, as the presence of abundant nitrogen aids the conversion of vegetation to soil humus.

Surface Decay

Cut or uprooted vegetation is sometimes left on the ground to decompose. Soft material such as grass and nonwoody plants may disappear in a few weeks, but trees can make the area unusable for years.

Disadvantages of disposal by surface decay are reduced, and may even be eliminated, if woody material is reduced to small pieces as part of the clearing operation. This may be done by using a shredder or a heavy rotary mower for clearing, or a chipper to reduce the sizes of pieces after cutting. These methods are suitable for construction sites only if clearing is done long in advance of removal of topsoil. Chipped material is sometimes stockpiled for use at the end of a project as ground cover for seeded slopes.

Burial

In agricultural work, burial is often the preferred means of disposal if equipment of the proper type is available and is of sufficient size to handle the growth density and trunk sizes involved, and if the soil will permit such disposal.

Grass, weeds, brush, and small saplings may be buried intact by a brush-breaker plow or slashed, chopped, and partially or wholly buried by a heavy disk harrow. Rolling choppers (Fig. 7.1) can disintegrate and partially bury medium-size trees, including trunks.

Burial of this type is used in agricultural rather than construction clearing, and only when the ground can be left undisturbed (except for planting of a cover crop) until a large part of the material has decayed. This process may take weeks or years, depending on vegetation type and maturity, and weather.

Loose stumps are often buried, but the operation is likely to be expensive and unsatisfactory. Attached roots make stumps bulky, and cutting the roots back to the buttresses can ruin saw chains by contact with clinging soil. Even after the roots are cut back, stumps are still awkward to handle.

A fill with buried stumps will always settle badly. Spaces left in and under irregularities will gradually fill with soil, allowing the surface to sink. Decay of the wood will cause slow, long-term settlement. Rotting and settlement are minimal when the burial is in flowable mud and the organic materials are permanently wet because such conditions fill most voids and preserve the material from decay.

Burning Stumps and Debris

Where burial is not practical, burning, if permitted, is usually the most efficient method of disposal and does the least long-range damage to the environment. Nevertheless, because open burning is prohibited or severely restricted in many localities, it is worthwhile to examine the relative advantages and disadvantages of fire compared with other methods of disposal.

When debris is to be disposed of by burning, it should be piled into stacks or rows with a minimum amount of soil. Shaking the material while moving it into piles or rows will reduce the amount of soil adhering to the organic matter.

Piles of brush or trees usually contain one-fifth to one-tenth solid matter, the rest being air space. These solids are at least half water, and most of their dry weight is cellulose, lignin, and other burnables. The ash left after efficient burning is only a small percentage of the dry matter. A good fire will, therefore, reduce the vegetation to a small fraction of the

FIGURE 7.1 Rolling chopper.

original bulk. The ash residue is too fine to be good fill material, but when the quantity is insignificant, it can usually be incorporated with other soils or pushed aside without difficulty. The soil under a hot fire is rendered unfit for supporting growth for 1 to 3 years, but can be restored by plowing or ripping and introducing fertilizer.

A fire in dry material, with plenty of air, will burn with hot clean flames and produce few pollutants. However, compliance with burning regulations can be costly. Green wood and leaves, wet or dirty piles, and rather weakly burning fires will give off large amounts of smoke, containing variable quantities of methanol, methane, acetic acid, tars and oils, and carbon monoxide. Such fires may create a local nuisance.

When debris is burned while the moisture content is high, it may be necessary to provide an external source of fuel, such as diesel, to start combustion. A burner consisting of a gasoline engine–driven pump and a propeller is capable of maintaining a fire even under adverse conditions. The liquid fuel is blown as a stream into the pile of material, while the propeller furnishes a supply of air to ensure vigorous burning. Once combustion of the debris has begun, the fuel is turned off.

In general, it is best to burn machine-cleared vegetation when it is piled. A hot fire is prepared, and brush piles are pushed up on it. A new fire is made when the push becomes too long. The best results are obtained if the vegetation is uprooted and allowed to dry at least a few days before piling and burning.

Rake A frame with multiple vertical teeth (tines) mounted in the place of a solid-faced blade.

The debris dries more rapidly when scattered on the ground than in piles. Soil will tend to dry and break away from stumps and to sift out of roots and stems. The brush nearest the fire is put on first. Fires fed by a dozer tend to get choked with soil, use a **rake** blade (Fig. 7.2) to eliminate this problem. In general, matted light brush is more difficult to clear and to burn than heavy brush or small trees, as it tends to slip under the blade or to carry too much soil.

FIGURE 7.2 Rake mounted on a low-ground-pressure clearing dozer.

Dead, dry stumps can sometimes be burned without taking them out of the ground, but the process is usually slow and laborious. A standard tool for stump burning is a large kerosene torch, called a flame gun (Fig. 7.3). These produce a hot flame up to 20 in. long. When the wood starts to burn, the torch may be moved to another stump and brought back if the fire dies down. Green wood will require continuous heat for many hours.

The torch will operate most effectively if directed into a cavity with an opening in the far end so a draft can move through the opening. If the flame is aimed into a dead-end hollow, very high temperatures will be attained, but because of lack of oxygen, the wood will distill rather than burn and will be destroyed slowly. If the flame is used against the outside

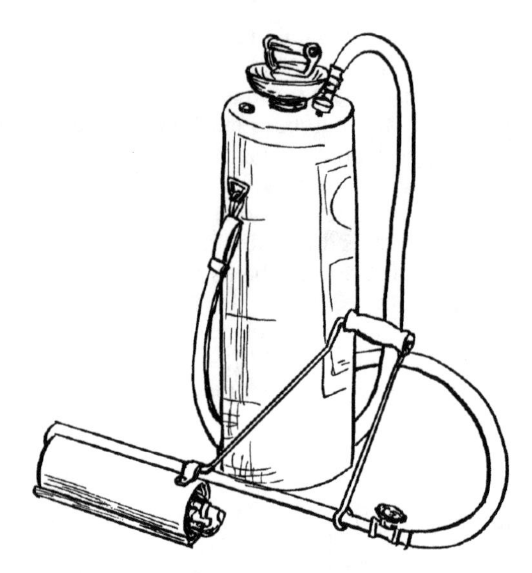

FIGURE 7.3 Flame gun.

of the stump, it should be directed upward to draw a current of fresh air between the flame and the wood.

Dry stumps may be burned by starting a wood fire alongside them and keeping it supplied with logs placed to almost touch the stump. The draft and reflection of heat will keep both surfaces burning, but the loose wood must be pressed in frequently. This method may remove only the top and outside of the stump, leaving a conical core.

Care should be taken to avoid spreading stump fires. Roots may burn underground to start surface fires at a distance. Soils rich in humus, such as swamp peat or forest loam, may burn unless saturated with water, and are very difficult to extinguish.

Stumps in a Pile. Because of difficulties in burial, the best way to dispose of a pile of stumps is to burn them. However, it is becoming increasingly difficult to obtain a permit for the work. A stump may cling to more than its own weight of soil and rock. The soil should be knocked off to make the stump burnable. This may be done by allowing the soil to dry, then rolling the stump with a dozer or other machine, or by picking it up with a grapple or clamshell and dropping.

Even if clean, green stumps are difficult to ignite. If they must be piled before burning, the base of the pile should include a layer of logs to improve air circulation. But the surest way is to burn stumps at the same time they are piled. Build a hot fire by the methods described in this text and supply it with sufficient heavy wood to keep it hot for several hours; then roll, push, or lower stumps onto it to sustain the fire.

A grapple or clamshell on a crane is the best machine to feed a stump fire, even if the stumps are pushed to it by dozers. The crane can place them properly, stack them high, and avoid pushing soil with them into the fire. Stump fires are very hot and may burn for weeks. There is usually very little smoke and few sparks.

Burning Brush. Green, wet brush and logs will burn vigorously once properly started, but considerable heat is required to boil off the sap and water and to ignite the wood. The initial heat may be obtained from a carefully built fire or by the use of flammable chemicals.

The fire should be on level ground or on a prominence. If built in a hollow or against a rock or stump, the inward flow of air will be hindered. Brush added to the top of the fire will be held up away from the heat. All flammable material should be cleared or burned away from around the site, particularly downwind. Firefighting tools should always be available.

Figure 7.4 illustrates two ways of starting the fire: andiron and tepee. The andiron consists of a pair of small logs, or rocks, or ridges of soil. These hold the logs above the ground in order to improve air circulation. Twigs and sticks, preferably dry, are laid across the andiron. These should lie close together in one direction, but should not fit together so well as to prevent air and heat going between them. No leaves or grass should be included. This pile may be ignited by burning paper, grass, or leaves under it. The material must be dry and must not be packed tightly, as this reduces the oxygen supply and the heat of the flame.

FIGURE 7.4 Andiron and tepee methods for starting fires.

A self-feeding starter may be made by tearing a section of 10 to 30 pages of newspaper into a strip and fitting it between the flat-lying base logs. Crumple the top sheet, and light it. As it burns, the heat will cause the next sheet to curl up and burn. The process repeats for every sheet, keeping a brisk fire going for long enough to ignite dry logs. No kindling is needed, but dry wood starts faster than thick, green pieces.

When the cross-sticks start to burn, more and heavier sticks are added, then partly trimmed branches, and finally, when a good bed of embers and strong flames are present, untrimmed bushes and branches. It is a good plan to put on a few logs at this time to give the fire staying power.

The tepee is similar in principle, with the sticks piled on end around the kindling. As heavier pieces are added, the tepee is crushed, but if it is burning well, this will not matter. When trying to burn roughly piled brush, the untrimmed branches may include so much airspace heat cannot cross effectively and the pile does not burn well. The fire may burn a dome-shaped hole over itself, then die down. In such a case, sticks should be poked into the fire itself to build it up, and the brush over it should be compacted by rearrangement or piling on of heavy sticks. This is tiresome work and may fail. It is better to tend the fire longer before piling on loose material to ensure it will not have to be worked over a second time.

Artificial Helps. A dying fire may be revived by the use of kerosene, fuel oil, or similar fluids. To be effective, these must be applied at the base of the pile. Putting flammable fluids on the heap itself may produce a fine flame, but it will have little kindling effect, as the evaporating fuel will absorb the downward radiating heat.

Flame guns are effective kindlers when directed into the base of a pile. A brush burner is a portable unit with a heavy fan to direct a strong air current into a pile together with the application of a mist of fuel oil for kindling the fire. Few piles can resist one for long (Fig. 7.5).

FIGURE 7.5 Brush burner.

If the fire dies down in spite of nursing, it may be best to build a new fire nearby, with greater care to avoid airspaces and coarse, green wood early in its life. Once a good fire is burning on the job, its embers may be shoveled out and used for starting other fires. This should be done rather frequently, as a long carry adds greatly to labor costs.

To start a new fire, four or five shovels of hot embers may be laid on the ground in a pile and then dry twigs and wood piled on it. Or the embers may be sifted down through piled brush. The embers give a sustained heat and consume little oxygen, so a strong new fire starts quite quickly.

Feeding. If done by hand, it usually takes at least two people cutting and dragging brush to keep one fire burning briskly. If it is allowed to burn down, it is good practice to put the unburned ends in the center hot spot before piling on more brush.

When a dozer is used, ample supplies of fuel can be brought to the fire, and it is usually well packed by the pressure of the blade and the weight of the machine if it climbs on the pile. The principal problem of dozer feeding, unless a rake is used, is soil. The soil tends to block the fire from spreading through the new material and to smother the already burning part. Every effort must be made to reduce the amount of soil by rolling and jostling piles, holding the blade high enough not to dig in.

A hot fire will burn through quite a lot of soil, but it will seldom burn clean. After it cools, the remaining stems and stumps can be sifted out by the dozer and used in building the next fire. Good results in fire feeding are obtained only if most of the new material is placed on top of the flames.

If the brush is piled a long time before being burned, dropping a match in it on a hot day may accomplish its complete removal. If it has been piled only a few hours or a few days, a fire may be built on the windward side against it but not under it. This fire may be made to spread into the heap by keeping it buried under compact brush. As the fire is fed, the heat is reflected into the pile. If the brush has leaves, it is good practice to cover any place where flames show through. A strong fire cannot be smothered with hand-piled brush.

Brush piles may be pushed on top of fires by a dozer, placed by a clamshell, or rolled on by a number of workers using long poles. If brush is being cut in an area presenting an unusual fire hazard, or the cutting is in small, scattered areas, it may be desirable to truck it to a central burning location. A continuous fire may be maintained with incoming loads dozed or hand-piled onto it, or the brush may be piled to dry and then be burned.

One excellent combination for heavy clearing and burning is a large dozer, preferably with a rake blade, and a clamshell. The dozer uproots and pushes in brush and trees, and the clamshell picks the debris up, shakes soil out, and places them on top of the fire (Fig. 7.6). A clamshell can also maintain a fire, moving in unburned ends, and can bury the fire under soil at the end of the workday. A clamshell is often the best tool for high-stacking vegetation after it has been left to rot and for burning old piles when they need rehandling.

FIGURE 7.6 Using a crane with a clam bucket to tend a fire.

Air Curtain Methods. An air curtain burner (ACB), also called a pit or trench burner, may be stationary or mobile. These methods involve fan-forced air, thereby creating a turbulent environment and accelerating incineration—up to six times faster than open-air burning. The local permit-issuing authorities are often more inclined to burning if it is done by air curtain methods.

With a trench burner, the burning is done in an earthen pit dug onsite or a berm is created to establish an above-ground trench. Ash from burning is usually left in the trench and buried. For pollution control benefits to be effective, the trench walls must be vertical.

Banking Fires. If the job is not extensive enough to justify the employment of someone at night to watch the fires, and any flammable material is nearby, fires should be buried under a few inches of clean soil at the end of the workday. Highly organic soils must not be used. The soil cover will prevent sparks from blowing, will preserve a hot bed for use in the morning, and, if the cover is not removed, may make a fair grade of charcoal.

Fire Control. Any contractors burning brush in an area subject to brush or forest fires are subject to heavy responsibility if one of their fires spreads. Also, in the presence of extensive forest fires from any cause, contractors may be required by authorities to use their workers and equipment to control them. At such a time there might not be experienced firefighting personnel available to direct the work. A brief outline of firefighting techniques is, therefore, considered appropriate.

Hand Tools. Where the material burning is largely grass and associated weeds or thin brush, a fire can be beaten out. Household brooms, occasionally dipped in water if possible, are very effective. Shovels, leafy bushes, or branches can be used with good effect. Each blow should be aimed so it directs flying sparks toward the burned area.

The fire may be starved by scraping away the vegetation just beyond the flames. This may be done with shovels, hoes, rakes, grub axes, or almost any piece of metal. A special type of firefighting tool, shaped like a heavy rake and fitted with sickle bar teeth instead of tines, is quite effective. Bushes may be cut with axes, machetes, bush hooks, or pruners.

Extinguishers. Backpack fire extinguishers, 2.5 to 6 gal, consisting of a water tank (metal, plastic, or collapsible) carried like a knapsack, a flexible hose, a hand pump, and a nozzle, are important pieces of firefighting equipment. If the grass is low or thin, spraying in the path of the fire may stop it. If the fire is strong and moving rapidly, the water may be most effectively used for putting out smoldering spots behind the beaters. Addition of a wetting agent—a small quantity of almost any detergent will do if regular compounds are not available—increases the effectiveness of the water by enabling it to soak through vegetable litter and punky wood.

Pumps. If streams or ponds are available, the contractor's pumps, particularly the light centrifugal type, are very valuable. A welder can usually make adapters to quickly attach fire hoses to the pump. The high pressures used in regular fire pumps will probably not be developed, but sufficient pressure will be available for wetting down firebreaks or making direct attacks on anything short of a crown fire.

Dozers. A dozer can put out a grass fire by starting behind the fire and straddling its line, as in Fig. 7.7. It may be able to scrape off the grass without cutting much into the ground. If this is not practical, it can skim off the sod until the load is heavy, then swing it into the burned area, or raise the blade and spread the sod over the next few feet of flames, smothering them. An angle dozer can side-cast the sod into the burned area, and a hand beater or extinguisher should follow to put out any missed spots.

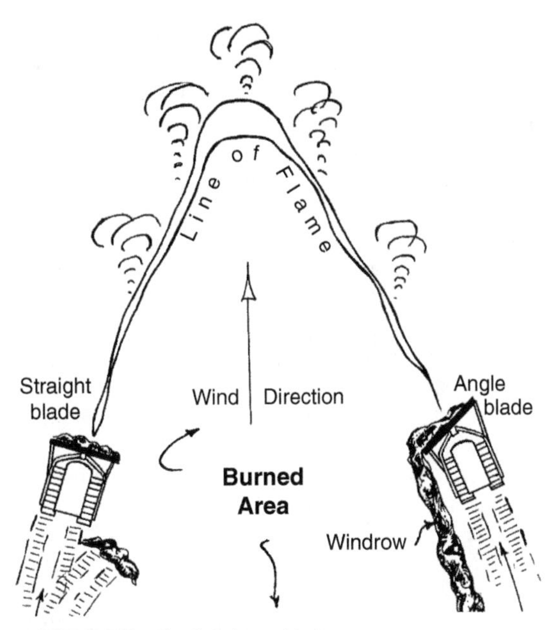

FIGURE 7.7 Firefighting with dozers.

Method of Attack. Windblown fires should not be attacked at the front, as this procedure is both dangerous and ineffective. A new fire running before a wind will assume a shape as shown in Fig. 7.7. A direct attack on the front means fighting flames several feet deep. If these should be put out, fire blowing up the sides could rekindle them in a few seconds. Pinching off the sides is both effective and reasonably safe. The fire is extinguished starting at the back, with the heat and smoke blown away from the workers. Provided a constant watch is kept behind them for rekindled hot spots, the fire cannot reclaim the extinguished area. When the front is reached, it is attacked from directly behind as well as on the sides.

If the fire is too strong for the force fighting it, the front will continue to advance, but the work on the flanks will limit its width and make the task of stopping it easier with firebreaks or backfires, or after a shift in wind direction. It can sometimes be turned by concentrating on one flank.

Firebreaks. A firebreak is any strip bare enough of flammable vegetation to delay or stop the spread of fire across it. Roads, open water, plowed fields, close-cut lawns, and even footpaths may be used. Advantage should be taken of any existing breaks when deciding where to place one to stop a burning fire. A short line is preferable, and valuable property or highly flammable areas should be protected. The break should be far enough from the fire to allow time to accomplish its completion and to start backfires; it should be in vegetation least apt to make a spark-producing or a high fire and on terrain favorable to the operation of machinery. A compromise among these features must usually be made.

A dozer may be walked along the line of the break, alternately cutting and filling, so as to mix the vegetation with soil. Hand workers with cutting or digging tools follow to cut out any spots where the fire might cross. If the brush is heavy, the dozer may turn to push heaps of it out of the path. An angle dozer or a heavy grader might be able to make a single clean cut in each direction, turning the sod and brush out from the center.

Backfires. A strong fire adds to the force of the wind moving it, somewhat as a blowtorch builds up its fuel pressure. The combined force may be enough to project a sheet of flame many feet in front of the burning line and to shower sparks for long distances forward. For this reason, the fire may cross a break of any practical width and make the area too hot for firefighters.

The principal use of the firebreak is to provide a line for starting backfires. Because the break is made on the downwind side of the fire, a new fire started along this edge burns upwind. The backfire should be made in a continuous strip along the break so it will not be able to turn and blow back. It will increase in strength as it progresses, but will be steadily farther away from the protected side. When it meets the main fire, there is liable to be a spectacular flareup and a heavy production of sparks. If the backfire was started in time, its burn area should be sufficiently wide to prevent the sparks from blowing across it. Sparks blowing across can be extinguished by personnel patrolling the break. If no shift in wind occurs, the sides of the fire can then be put out by the crew working from behind, aided by the firebreak crew.

If the break is made in a forest where the flames might crown (burn in the treetops), the trees on each side of the break should be bulldozed or cut so as to fall away from the center.

Because a change in wind direction may occur at any time, care should be taken not to start backfires prematurely and to keep workers and machines in positions where they can get away if the fire turns toward them. The burnt-over area, ponds or wet swamps, and plowed land or well-grazed pastures are suitable retreats.

Workers on the fire lines must be kept provided with food, water, and tools and relieved for rest periods. Machinery must have fuel, but other maintenance may be skimped on in dangerous situations.

Rekindling. After the spreading of a fire has been checked, it must be patrolled until all danger of it making a fresh start has passed. A grass fire in a clean field may be safe to leave within an hour, whereas wooded areas containing dead or fallen trees or rich dry soil may still be dangerous until after several soaking rains. The best time to check a burned area for hot spots is immediately after a rain or a heavy dew, as the moisture near the fires will steam.

Dead stumps can burn a long time and are difficult to extinguish unless ample supplies of water are available. Fires burning under and between logs on the ground can often be put out by moving the logs apart, or can be caused to burn out more quickly by piling additional wood on them.

The worst hazards are standing dead or hollow trees, called *snags* by lumbermen. If close to the line, snags may set fire to the unburned area by falling into it. They frequently produce sparks high above the ground, and these may drift long distances. Even thorough soaking may not extinguish them, and it may be necessary to cut them down or maintain an expensive patrol for days or weeks.

Cutting a burning tree is a dangerous job, as the cutters are in constant danger of being hit by falling pieces, and temperatures at the base may be too high for them or their tools. This job is best left to experienced firefighting crews. Snags may be pushed over by dozers, but the tops are apt to fall on the machine. A cab with maximum-strength overhead protection is needed for operator safety.

Underground Fires. Underground fires, such as occur in rich forest soils and dried-out swamps, constitute a special problem. When fire begins, it often smolders down a dead root and will burn hot and persistently. Plain water has little effect on such a fire unless applied so the area is flooded. Smaller quantities do not penetrate the deeper burning zones, where there is sufficient heat to evaporate the water in the surrounding peat and then spread through the dried material.

Special nozzles consisting of pipes long enough to reach the bottom of the fire are helpful. The lower end is plugged, and a fairly large hole is drilled in the plug to wash humus out of the way as the pipe is pushed down, and smaller holes in the side spread a soaking spray. The use of wetting agents will substantially reduce the amount of water required, and may make the difference between success and failure where the water supply is limited.

Such a fire may be confined by trenching down to inorganic or saturated soil. The digging may be quite difficult because of roots, and a backhoe or dragline might have to be used.

Peat fires spread very slowly unless they ignite surface vegetation. If equipment is not immediately available to extinguish or ditch the fire, leaves and flammable trash should be removed for 10 or more feet around it to prevent rapid spreading while arrangements are made to put it out.

Chipping

Brush, saplings, and even large trees may be fed into machines and reduced to chips of small and fairly regular size by action of a rotating toothed drum (Fig. 7.8). Larger models are equipped with grooved rollers in the throat of the machine's feed conduit. After a branch is gripped by the rollers, they pull the brush into the blades at a steady rate. These rollers are generally reversible for situations where material gets stuck. The chips may be scattered or piled in the work area or fed through a chute into trucks.

FIGURE 7.8 Brush chipper, note operator protective gear.

A small machine can be towed behind a truck and often maneuvered on the job by hand. It is hand-fed with bundles of brush and with saplings up to 4 in. in diameter. Today many chippers have mechanical feed control bars to stop the machine if someone becomes tangled with the debris during hand feeding. At the opposite end of the scale are large machines; these can handle trees with trunks up to 21 in. in diameter, without the need to even trim the branches.

The resulting chips may be an asset, may be a problem, or may have no importance. A few paper mills are able to digest wood chips, including bark and twigs, and will even pay for the material. In other areas, chips might be sold for processing into pressed wood or charcoal and distillation products. Sometimes chips can be used on the project to hold soil on slopes while vegetation becomes established or to add organic material to poor soils.

Chips from light to medium thickness vegetation may be left scattered on the ground to be incorporated with the topsoil when it is pushed off or cultivated. If the growth is heavy, the chips are likely to make the soil critically short of nitrogen and difficult to work, and may accumulate in spots as pockets of almost unmixed wood.

Chipping is usually not practical for uprooted stumps, unless they are very small in proportion to machine capacity. Their bulk and shape make it difficult or impossible for the guards to pass them or the drums to grip them, and the soil and rocks stuck to them can damage cutters and make the chips unsalable. Chips made from stumps in the ground are always contaminated with soil.

Piling chips, either by keeping a discharge chute in one direction or by dumping them from trucks off the work area, should be permitted only when they are to be reclaimed later. Such piles are likely to remain for many years before decaying sufficiently to permit growth of vegetation.

Chipping machines are expensive, consume large amounts of fuel, are extremely noisy, and may be dangerous to personnel. Their use in mass clearing is justified when there is good use for the chips, in areas of high fire danger, or where smoke cannot be tolerated. In addition, chippers are valuable in low-volume or selective clearing and trimming, where the cutting would otherwise have to be hauled away.

Removal from the Project

Removal of cleared vegetation from the project may be anything from a sound and profitable operation to a costly procedure. Where lumber or paper mills are within economical hauling distance, it may be possible to sell cut trees profitably. In some cases, the user of the wood may be glad to cut and remove the usable part of the vegetation and pay for the privilege. However, when trees are harvested for lumber, they are usually cut just above the ground line, leaving the clearing contractor the difficult job of removing the stumps without the leverage of the upper growth.

The usability and value of trees vary greatly. The quality to be found on the job influences value. Larger quantities make it worthwhile to harvest. Some users are very narrowly restricted as to species, size, straightness, and soundness. Others will take (usually at a lower price) almost any type and quantity of wood. In the case of a large job, it may be worthwhile to invest time investigating possible markets.

Firewood is another possibility. Wood cut in 2-ft lengths and split to cross-sections averaging 30 sq in. or less is often saleable.

If the contract requirement is offsite disposal at a distance, there may be several possibilities. Bulk can be greatly reduced, although possibly at considerable cost, by chipping the vegetation and hauling the chips. Otherwise, trees should be trimmed into lengths suitable for dump trucks or other hauler types, with all angles in trunks or branches cut to make them lie flat. Non–dump trucks or trailers require log-handling machines at the disposal site. Except with medium to large tree trunks or chips, these loads are likely to be mostly air.

CLEARING OPERATIONS

Dozers and loaders are basic machines for clearing, both with regular blades or buckets and with special attachments. They work best when the ground is firm and where there are no gullies, sharp ridges, or rock. Uneven surfaces make it hard to keep the blade in contact with the ground and lead to burial rather than removal of vegetation in low spots. Still, there are few places where a dozer cannot aid hand-clearing crews by clearing areas, moving logs and cut brush, cutting roads for supply trucks, or providing firebreaks.

Any dozer used for clearing work should be thoroughly protected with crankcase and radiator guards; the latter include screen with holes not over ¼ in., and accessible for removal of leaves and trash. The engine needs side guards. Minimum operator protection is a strong overhead structure.

Brush Clearing

When a dozer is working in dense undergrowth, concealment of hazards such as holes, whether natural or artificial, is always a danger. This danger can be minimized by scouting the area on foot and by moving forward in a succession of short pushes overlapping each other on the side. This enables the operator to watch from one side and to observe the nature of the ground. In addition, it avoids tangling the dozer in branches and vines.

Dozers. Dozers have a particular advantage over hand crews where briars and vines are abundant, as these are very tedious to cut but can be readily stripped off by a dozer blade, provided the operator does not take too long a pass and get the dozer caught in the tangle. Brush and small trees may be removed by a dozer moving with its blade in light contact with the ground. This will uproot or break off a number of the stems and will bend the rest over so the return trip, in the opposite direction, will remove those remaining. If the distance is short, it is best to blade the whole area in one direction and to then blade across or in the opposite direction. Individual small trees are first knocked over, then pushed out with another pass in the same direction.

Hard-baked soils will cause a high percentage of broken stems, whereas wet or sandy conditions produce uprooting. The work can be accelerated by having a laborer cut out or pick up any individual bushes requiring a second dozer pass. If the job requires removal of light stumps and roots, they may be overturned in one pass and pushed out in the next. It may be necessary to dig several inches into the soil to get good contact on a stump.

Brush heaps may be largely freed of dry, loose soil by rolling them over with the blade and shaking the blade up and down. If this is ineffective, rolling them over backward or pushing them from the side may help. A dozer with a tilt blade is very good at this work, as one corner can be used for taking out roots and pushing piles without taking much soil, and the blade can be returned to a flat position to skim off surface brush.

Crawler Tractors or Loaders with Clearing Rakes. Clearing rakes are used to grub and to pile trees after the clearing blades have worked an area (see Fig. 7.2). These add to clearing efficiency. They allow working below ground level, to take out roots as well as surface material, usually without bringing the soil with vegetation, if the soil is dry or sandy.

Like earth-moving blades, the teeth of a rake are curved in the vertical plane to form a flattened "C." With this shape, the teeth can easily be driven under roots, rocks, and boulders. As the tractor moves forward, it forces the teeth of the rake below the ground surface. The teeth will catch the roots and surface brush left from the felling operation, while allowing the soil to pass through. Rakes have an upward extension brush guard. This guard can be either a solid plate or opened spaced ribs. Some rakes have a steel center plate to protect the tractor's radiator.

The size, weight, and spacing of the rake's teeth depend on the intended application. Rakes used to grub out stumps and heavy roots must have teeth that are sufficiently strong to take the push of the tractor at full power against a single tooth. Lighter rakes with smaller and closer-spaced teeth are used for finish raking and to clear light root systems and small branches from the ground.

Rakes are often used to push, shake, and turn piles of trees and vegetation before and during burning operations. These tasks shake the soil out of the piles and improve burning. For best results, with or without rakes, the loosening and removal of brush should be done when the soil is dry if the waste is to be burned. If the soil is wet, it lumps and sticks. If the loosened material is allowed to dry, much of the soil can be shaken out while piling, by rolling and shaking.

Other Machines. A number of specialized equipment types are used for chopping or shredding brush. These may mix the minced material with the soil or leave it on the surface to rot or to be removed by other equipment or laborers. A big rotary mower mounted on the rear of a wheel tractor is highly effective up to its thickness-of-stem limit, maybe 1½ to 3 in. Vertical rotary shredders may handle larger stem sizes. Brush is mostly chopped or shattered into small pieces, but root systems are seldom disturbed. There is no suppression of regrowth.

Either a rotary or a sickle bar can be used to suppress regrowth by repeated mowing. A moldboard plow designed to slice and invert a layer of soil, thus covering the sod and leaving a rough surface, can put brush and saplings underground, cover them neatly with soil, and leave them to rot.

A big heavy disc harrow, with discs 24 in. in diameter or larger, chops brush and buries a large part of it. Big pieces may be loosened, chewed, and pushed around without burying. Both the plow and the harrow tend to create ridges and troughs in the ground surface, because they move loosened soil to the side. A rolling chopper (see Fig. 7.1) knocks down, tears, and mashes both brush and trees and cuts near-the-surface roots. A small portion of its cutting is buried.

LAND-CLEARING PRODUCTION

Crawler dozers equipped with special clearing blades are excellent machines for land clearing. Clearing land can be divided into several operations, depending on the type of vegetation, the condition of the soil and topography, the amount of clearing required, and the purpose for the clearing:

1. Removing all trees and stumps, including roots
2. Removing all vegetation above the surface of the ground only, leaving stumps and roots in the ground
3. Disposing of vegetation by stacking and burning

Several types of equipment are used, with varying degrees of success, for clearing land:

1. Crawler tractors with earth-moving blades
2. Crawler tractors with special clearing blades
3. Crawler tractors with clearing rakes

Crawler Tractors with Earth-Moving Blades

Crawler tractors with earth-moving blades were once used extensively to clear land. There are at least two valid objections to the use of tractors with earth-moving blades. Prior to felling large trees, they must excavate earth from around the trees and cut the main roots; this leaves objectionable holes in the ground and requires considerable time. Moreover, when stacking the felled trees and other vegetation, a considerable amount of earth is transported to the piles, making it difficult to maintain fires. Therefore, the use of such equipment is not recommended.

Crawler Tractors with Special Clearing Blades

There are blades specially designed for use in felling trees. The most common clearing blade is the angled blade with projecting stinger. This blade is often referred to as a "*K/G*" *blade* (Fig. 7.9). This name comes from the highly successful Rome *K/G clearing blade*, invented in the 1950s by Ernest Kissner of Lottie, Louisiana, and manufactured by the Rome Plow Company, Cedartown, Georgia.

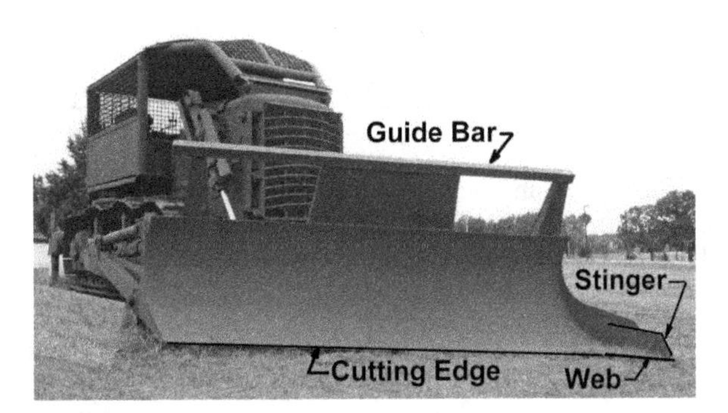

FIGURE 7.9 Rome K/G clearing blade.

The major components of a single-angle clearing blade are the stinger, web, cutting edge, and guide bar. The exterior edges of the stinger and web are sharpened like a knife blade by machining with a grinder. The stinger is in effect a protruding vertical knife. It is designed to be used as a knife to cut and split trees, stumps, and roots. The web is a horizontal knife; together the stinger and web cut the tree in both planes simultaneously (see Fig. 7.9). The blade can be tilted and is mounted at a 30-degree angle with the stinger forward. The guide bar serves to push the cut material forward and to the side of the tractor.

Clearing blades are most efficient when the tractor is operating on level ground and the cutting edges can maintain good contact with the ground surface. These blades work best when the soil holds the vegetation's root structure while the trunks are sheared. Large rocks will slow production by damaging the cutting edges of the blade.

Land-Clearing Production Estimates

Typically, clearing of timber is performed with 160- to 460-hp crawler tractors. The speed of a tractor moving through vegetation will depend on the nature of the growth and the size of the machine. It is best to estimate land clearing by using historical data from similar projects. When data from past projects are not available, the estimator can utilize the formula presented in this section as a rough guide for probable production rates. Production rates calculated strictly from the formula, however, should be used with caution.

> The estimator should always walk the project site prior to preparing a land clearing production estimate.

An observational walk through the area is necessary to obtain information needed to properly evaluate the impacting factors, such as soil condition, and to develop a complete understanding of the project requirements and possible variations from the clearing formula assumptions.

Constant Speed Clearing. When there is only small-size vegetation and it is possible to clear at a constant speed, production can be estimated by the tractor speed and the width of the pass:

$$\text{Production (acre/hr)} = \frac{\text{Width of cut (ft)} \times \text{Speed (mph)} \times 5{,}280 \text{ ft/mile} \times \text{Efficiency}}{43{,}560 \text{ sf/acre}} \tag{7.1}$$

The American Society of Agricultural Engineers' formula for estimating land-clearing production at constant speed is based on a 49.5-min hour, an efficiency of 0.825. When this efficiency is used, Eq. (7.1) reduces to

$$\text{Production (acre/hr)} = \frac{\text{width of cut (ft)} \times \text{speed (mph)}}{10} \tag{7.2}$$

The width of cut is the resulting cleared width, measured perpendicular to the direction of dozer travel. With an angled blade, this is not equal to the width (length) of the blade. Even when working with a straight blade, it may not be the same as the blade width. The width of cut should be determined by field measurement.

Sample Calculation: A 200-hp crawler dozer will be used to clear small trees and brush from a 10-acre site. By operating in first gear, the dozer should be able to maintain a continuous forward speed of 0.8 mph. An angled clearing blade will be used on the project; the width of this K/G blade is 12 ft 9 in. From past experience, the average resulting clear width will be 8 ft. Assuming normal efficiency, how long will it take to knock down the vegetation?

Using Eq. (7.2),

$$\frac{8 \text{ ft} \times 0.8 \text{ mph}}{10} = 0.64 \text{ acre/hr}$$

$$\frac{10 \text{ acres}}{0.64 \text{ acre/hr}} = 15.6 \text{ hr} \approx 16 \text{ hr}$$

Or two 8-hr workdays.

Tree Count Method—Cutting and Piling Production. The Rome Plow Company developed formulas for estimating cutting and piling production. Over the years, the production tables have been updated by other equipment manufacturers and the military. The Rome formula and tables of constants provide guidance for variable-speed clearing operations, but again use of the results should be tempered with field experience.

To develop the necessary input data for the Rome formula, the estimator must make a field survey of the area to be cleared and collect information on the following items:

1. Density of vegetation less than 12 in. in diameter:
   Dense—600 trees per acre.
   Medium—400 to 600 trees per acre.
   Light—less than 400 trees per acre.
2. Presence of hardwoods expressed in percent.
3. Presence of heavy vines.
4. Average number of trees per acre in each of the following size ranges:
   Less than 1 ft in diameter.
   1 to 2 ft in diameter.
   2 to 3 ft in diameter.
   3 to 4 ft in diameter.
   4 to 6 ft in diameter.
   The diameter of the tree is taken at chest height, or 4.5 ft above the ground. In the case of a tree with a large buttress, the measurement should be taken where the trunk begins to run straight and true.
5. Sum of diameter of all trees per acre above 6 ft in diameter at ground level.

Once the field information is collected, the estimator can enter the table of production factors for cutting (Table 7.1) to determine the time factors to be used in the Rome cutting formula, Eq. (7.3). The formula is based on the following assumptions:

- A power shift tractor.
- The ground is reasonably level (less than 10% grades).
- The machine has good footing.

$$\text{Time (min) per acre for cutting} = H[A(B) + M_1N_1 + M_2N_2 + M_3N_3 + M_4N_4 + DF] \quad (7.3)$$

where H = hardwood factor affecting total time

TABLE 7.1 Production Factors for Felling with Rome K/G Blades*

Tractor	Base Min. per acre B	1-2 ft M_1	2-3 ft M_2	3-4 ft M_3	4-6 ft M_3	Dia. > 6 ft F
		Diameter Range				
385 HP	18	0.2	0.5	1.5	4	1.2
270 HP	21	0.3	1.5	2.5	7	2.0
180 HP	28	0.5	2.0	4.0	12	4.0
120 HP	40	0.8	4.0	8.0	25	—

*Based on power-shift tractors working on reasonably level terrain (10% maximum grade) with good footing and no stones and an average mix of softwoods and hardwoods. Tractor in good operating condition, blade is sharp and properly adjusted.

Hardwoods affect overall time as follows:

75% to 100% hardwoods; add 30% to total time ($H = 1.3$)

25% to 75% hardwoods; no change ($H = 1.0$)

0% to 25% hardwoods; reduce total time 30% ($H = 0.7$)

A = tree density and the effect the presence of vines has on base time
The density of undergrowth material less than 1 ft in diameter and the presence of vines affect base time.

Dense: more than 600 trees per acre; add 100% to base time ($A = 2.0$)

Medium: 400 to 600 trees per acre; no change ($A = 1.0$)

Light: fewer than 400 trees per acre; reduce base time 30% ($A = 0.7$)

Presence of heavy vines; add 100% to base time ($A = 2.0$)

B = base time per acre for each dozer size

M = minutes per tree in each diameter range

N = number of trees per acre in each diameter range, from field survey

D = sum of diameters in foot increments of all trees per acre above 6 ft in diameter at ground level, from field survey

F = minutes per foot of diameter for trees above 6 ft in diameter

If the removal of the trees and grubbing of roots and stumps greater than 1 ft in diameter is to be accomplished in one operation, increase total time per acre by 25%. If the removal of stumps is accomplished in a separate operation, increase the time per acre by 50%. The tree count method has no correction for efficiency, as the time values in Tables 7.1 and 7.2 are based on a normal efficiency.

Sample Calculation: A 180-hp tractor equipped with a K/G blade is to be used to fell the vegetation on a highway project. Project specifications require grubbing of stumps resulting from trees greater than 12 in. in diameter. Felling and grubbing will be performed in one operation. What is the estimated rate of production for this work?

The site is reasonably level terrain with firm ground and less than 25% hardwood. The field survey gathered the following tree counts:

Average number of trees per acre, 700
1 to 2 ft in diameter, 100 trees
2 to 3 ft in diameter, 10 trees
3 to 4 ft in diameter, 2 trees
4 to 6 ft in diameter, 0 trees
Sum of diameter increments above 6 ft, none

The necessary input values for Eq. (7.3) are

$H = 0.7$, less than 25% hardwoods
$A = 2.0$, dense, >600 trees per acre

From Table 7.1 for a 180-hp dozer:

$B = 28$, $M_1 = 0.5$, $M_2 = 2.0$, $M_3 = 4.0$, $M_4 = 12$, and $F = 4.0$

$$\text{Time per acre} = 0.7[2.0(28) + 0.5(100) + 2.0(10) + 4.0(2) + 12(0) + 0(4.0)]$$
$$\text{Time per acre} = 93.8 \text{ min per acre}$$

Because the operation will include grubbing, the time must be increased by 25%.

$$\text{Time per acre} = 93.8 \text{ min per acre} \times 1.25 = 117.25 \text{ min per acre}$$

Clearing rates are often expressed in acres per hour, so for this example the rate would be 0.51 acres per hour (60 min per hr/117.25 min per acre).

Table 7.1 and Eq. (7.3) are used to calculate the time for a cutting operation. On many projects, the cut material must be piled for burning or so it can be easily picked up and hauled away. The Rome Plow Company developed a separate formula and set of constants for estimating piling-up production rates:

$$\text{Time (min) per acre for piling up} = B + M_1N_1 + M_2N_2 + M_3N_3 + M_4N_4 + DF \quad (7.4)$$

The factors have the same definitions as when used previously in Eq. (7.3), but their values must be determined from Table 7.2 for input into the piling up in Eq. (7.4). In the case of piling up grubbed vegetation, increase the total piling-up time by 25%.

TABLE 7.2 Production Factors for Piling-Up in Windrows*

Tractor	Base Min. per acre B	Diameter Range 1-2 ft M_1	2-3 ft M_2	3-4 ft M_3	4-6 ft M_3	Dia. > 6 ft F
385 HP	45	0.1	0.2	1.4	2.4	0.4
270 HP	50	0.2	0.6	2.0	4.0	1.0
180 HP	60	0.4	0.8	3.0	6.0	—
120 HP	75	0.6	1.2	5.0	—	—

*May be used with most types of raking tools. Windrows to be spaced approximately 200 ft apart.

Sample Calculation: The vegetation cut in the previous calculation must be piled. At what estimated rate can the piling up be accomplished?

$$\text{Time per acre} = 60 + 0.4(100) + 0.8(10) + 3(2) + 6.0(0) + 0(-)$$
$$= 114 \text{ min/acre}$$

Because the operation will include grubbing, the time must be increased by 25%.

$$114 \text{ min/acre} \times 1.25 = 142.5 \text{ min/acre or } 0.42 \text{ acre piled-up per hr}$$

Safety During Dozer Clearing Operations

During clearing operations with multiple dozers, it is necessary to maintain a considerable clear distance between machines. This is because falling trees can easily strike a neighboring machine if they operate close together. Operators must exercise care when pushing trees over. If the dozer follows a falling tree too closely, the stump and root mass can catch under the front of the machine. But the greatest hazard is from fire. The dozer's belly pan must be cleaned often during a clearing operation, as accumulated debris in the engine compartment can easily ignite.

TREE REMOVAL

Logging may be almost completely mechanized when large-scale operations are undertaken, with felling, trimming, bucking, and transport (or piling and burning) done by highly specialized machines. However, most clearing-for-excavation projects rely on more standard construction machines and/or hand labor. Large trees, or any trees too large to be walked down by the equipment on the job, may require special approaches.

Large trees can be pushed over with a dozer or pulled down with a cable. A dozer can generally uproot a large tree with its stump, because of leverage from a higher push or pull point and the help from the weight of the upper part of the tree. The tree tends to tear out the roots as it leans. But when very large trees are encountered, they must be dug out, a time-consuming operation (Fig. 7.10).

Pulling Small Growth

Brush and small trees often grow where they cannot be reached by machinery, because of soft or rough ground or nearness to buildings. In such cases pulling techniques will be applied to small growth. Trucks and farm tractors are usually adequate for the job. If the stems are stiff, fastening may be made high for leverage. If they are flexible, height does not matter, and the greater strength of the base may make it the best place to connect the chain or cable.

Chains tend to slide along smooth stems, and they often can be made to grip by wrapping once or twice around before fastening. Light chain with small links holds much better than coarser types. A round hook or ring should be used to make a choker. If stems are close together, it is often possible to pull several at a time by putting a single choker around the group. It will slide up until it can pull them all tight together and then should hold.

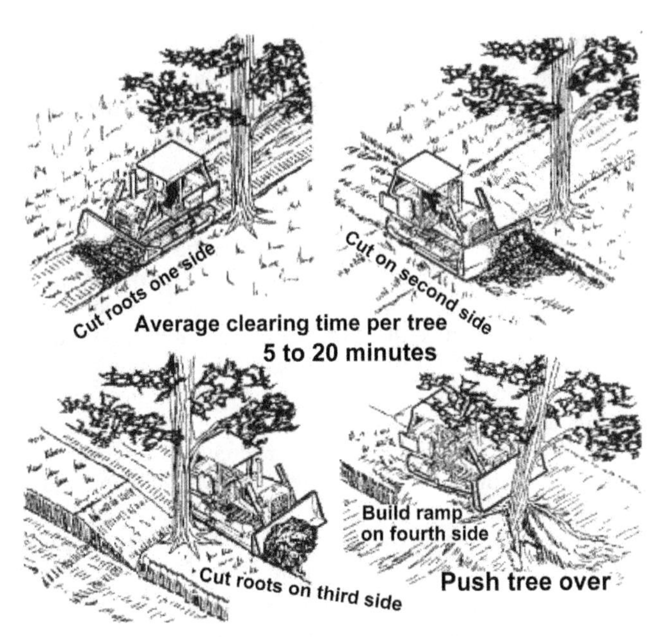

FIGURE 7.10 Tree removal by excavation and ramp.

Felling Trees

Tree trunks, even in sapling sizes, may be tricky and dangerous to push around. A stump may ride up over the top of even a large dozer blade during ordinary pushing. It may be put under tension by the pushing while another part is jammed tight—this situation can cause whipping with great force against the cab or into other machines or workers in the area when the holding force releases. An operator must be vigilant to avoid such situations.

When felled trees are pushed into tight heaps for burning, the piles may be very fire resistant. But a loose pile may also be hard to ignite. Heavy branches and crooked trunks increase this difficulty. Branches can be cut off and trunks cut into pieces to make more compact piles. Or the pieces may be placed on an existing fire by a clamshell.

Subcontracts. If trees are to be removed from a project site, an attempt should be made to sell the timber. A contractor desiring to confine operations to earthwork may be able to arrange a sale of the trees to a sawmill or wood dealer, who buys the trees on the stump and is responsible for the cutting and removal. However, a logger may fail to do the work in the time specified, and so force the contractor to clear the trees at the last moment. In making such an arrangement, the disposal of the scrap wood and brush and the height of the stumps should be specified.

A sawmill operator is interested only in large, sound trunks, whereas a pulp or firewood worker can use bulky branches. The mill will ordinarily pay the best prices but do the least work toward tract cleanup unless it has an arrangement with pulp or firewood users to take the tops and limbs.

Stump Height. From a clearing standpoint, high stumps are more easily removed than low ones. When machinery is undersized for the job, stumps are removed primarily by winching and higher heights are especially desirable. Low stumps are more difficult to cut, particularly where the trunk flares out widely at the bottom, but do not impede machines as much and can often be filled over and left. Remaining stump height from timbering may be determined by local law or lumbering custom.

Dangers. Tree cutting is dangerous work, because of the nature of the tools used and the unpredictable behavior of trees during the cutting work. In spite of best judgment, a tree may fall in an unexpected direction, even backward across wedges. A tree may fall much sooner than expected, because of concealed weakness in the base or wind.

No one should be in the cutting area except the worker(s) at the foot of the tree, and they must be alert to move quickly. The critical area should be clear of brush and litter, as tripping might have fatal consequences. Chainsaws not being used and other valuable equipment should be placed behind a tree or other protection during tree felling.

Some trees come apart while being cut. Blows of a hammer on a wedge, contact of a pusher, or even the vibration of a saw may loosen dead limbs, and they can come crashing down around the base. If the trunk is decayed, it may break as it starts down, with the lower part leaning and falling conventionally, but leaving the upper parts in the air to come straight down. The top of the tree must be watched closely during the work, and workers must not be ashamed to run instantly if anything breaks loose. The danger area is usually small and close to the trunk, although this varies with limb spread.

Leaners. A tree may be held from falling by the limbs of other trees. This may be by comparatively light contact of branches while the cut tree is nearly vertical, or because of falling into a crotch or across heavy limbs. Such a tree may be brought down by moving the lower end away from the direction of the fall. Cutting the supporting tree is too dangerous to be attempted. It is usually under extra tension and is likely to snap and fall suddenly during work. The leaning tree would then be apt to drop on the workers doing the cutting.

Uprooting Trees. With very large trees with extensive root systems, it will be difficult to uproot the stumps. In such cases it may be advisable to pull the trees over rather than to cut them down. Fasten lines as high as possible and make use of the weight of the tree. As soon as the tree is pulled toward the tractor, its center of gravity shifts to the side of the pull and aids in breaking out the roots. If a large log is chained to the tree's base on the pull side, the force of the tree's fall will be more effective at breaking roots on the pull side. The log will also serve to prevent the trunk from digging into the ground where it would be difficult to cut.

If the tree tends to break or split instead of uprooting, additional chokers may be used below the main pull point to distribute the strain and bind the trunk together. This can be done by pulling with two or more machines, or with multiple lines and blocks. A chain or cable may be restrained from sliding down by a nail or a notch, when necessary.

Trimming. When the tree is down, the branches should be cut off nearly flush with the trunk, before any other trees are dropped across it and make a tangle.

Stump Removal

The trunk part of a stump may be anywhere from a few inches to several feet in diameter. It usually flares out near ground level into a root buttresses, connected to the major roots.

Roots form a network near the ground surface, but some species of trees—pines, for example—have a taproot. This strong root extends more or less straight down from the center of the trunk and makes stump removal much more difficult. Stumps can be a major problem when clearing involves large trees. They can sometimes be cut low and left under deep fills, but usually stumps must be removed.

Stumps may be broken out by uprooting the whole tree, then disposed of as part of the tree or separately. Still it is more usual, but more work, to cut and remove the tree and then take out the stump. Stumps may be pushed out by dozers, dug out by hoes, pulled out with cables or chains, or blasted. Blasting may be combined with other methods. Occasionally it is possible to burn stumps while they are still in the ground.

Digging Out. If a stump does not yield to pushing, it must be dug out (see Fig. 7.10). This is done by trenching around it with a dozer to cut the roots. Each time the dozer cuts a big root, it may turn and push the stump to see if it is loosened. The operator will often be able to tell when it has been weakened by the way it shakes as the roots break. A ramp need be built only after an attempt to uproot from a lower level has failed.

Roots should be cut as close to the stump as the power of the dozer permits, but it is a waste of time and power to buck at a heavy root repeatedly when it could be easily broken a foot or two farther out and the stub crumpled back. If an area is to be excavated after clearing, stumps may be left until digging has undermined them and cut many of their roots, when they can be easily removed.

Excavators. A large hoe is probably the best stumper. Usually, the operator first tries direct pull to remove the stump. If it resists, it can be weakened by chopping roots on the far side and by trenching at the sides. Except in the case of very large stumps, digging and removal can be done from one position. A hoe is more effective than dozers in rocky ground and among interlocked stumps, as it can apply power in smaller spaces. It is often necessary to devote time to digging out rocks before the stump can be attacked. If the roots are strongly entrenched in bedrock or oversize boulders, the rock may have to be blasted before the stump can be pulled.

Pulling Stumps

Stumps may be pulled out of the ground. This method, although widely used, is less popular because larger machines and special attachments have made it possible to push or dig out most stumps encountered on construction jobs, without taking the extra time to rig lines. However, such equipment is not always available or it may cause too much damage, and it cannot work efficiently on soft ground.

Winches. Power to pull stumps may be supplied by almost any machine. However, the most powerful, most convenient, and best-controlled pull is obtained from winches. These are best when mounted on crawler tractors, but may be on wheel tractors or trucks, or may be portable units operated by a hand crank.

For general clearing work, the most effective tool is a dozer carrying a power winch. The winch consists of a heavy spool drum mounted on the back of the tractor and driven either hydraulically or through a power take-off (PTO) shaft from the transmission. On smaller tractors where less power is needed, most winches are driven by a hydraulic motor. Hydraulic winches offer good line pull at a precise and controllable line speed.

The winch may hold 200 or more feet of cable of a size proportionate to its power. Additional cable can be carried on a separate spool and connected to the winch cable by a choker device when needed.

Smaller winch cables are generally fastened at the working end to a short piece of chain equipped with a round hook. Larger cables may be fastened directly or through a swivel or single link to a round hook or to a wide-face cable grip hook. The cable is generally underwound on the drum (the cable runs from the work to the lower part of the drum). This gives better stability under heavy load than overwinding.

An operator should not try to operate the winch and move the tractor at the same time. Pull the load to the tractor with the winch, apply the brake, and then haul the load to the desired location.

To winch out a stump, the tractor should be placed facing directly away from it and both brakes locked. Then the cable is pulled to the stump by hand. If the brake is not used, the drum may continue to spin after being pulled and unwind and snarl the cable.

If the winch will not freewheel or the cable is very heavy, the drum is turned backward to pay it out. It is convenient to have an operator for the winch and a laborer to pull the cable. If no helper is available, the operator can stand near the winch while it turns, stripping the cable and coiling it on the ground until there is enough. The operator then stops the winch, and the helper drags the cable to the stump. The cable must then be whipped up and down and the twists worked out to avoid kinking when pulled. The winch cable may be put around the stump directly, may be hooked to a choker chain or cable, or may be run through one or more snatch blocks.

Power is applied to the winch and the cable is reeled in, with care being taken to ensure it feeds onto the drum properly. The stump may come out, or the tractor may be dragged backward. If the latter, the tractor may be anchored by a chain from the blade or front pull hook to a tree. Resistance to pull may be increased by backing the tractor against a log or bank, or by trying to pull the stump by tractor pull and allowing the tracks to spin until they have built mounds behind them. If the anchoring or blocking is effective, the stump will come out.

The drum carries a number of layers of cable, so the spool diameter is greater when full than when fully played out. It, therefore, reels in cable more slowly and powerfully on a bare drum than on a full one. On a bare drum, logging winches will give 50% to 100% more pull than the tractor itself; on a full drum, they will give the same pull as the tractor or somewhat less.

Jammed Cables. Using a nearly bare drum not only gives the greatest pull but also reduces damage to the cable. If a long cable is wound smoothly onto a drum under moderate tension and a heavy pull applied when it has built up several layers, the last wrap may squeeze between the wraps below (Fig. 7.11A). This scrapes and wears the cable and jams it so it will not spool off again. The best way to free it is to turn the drum until the catch is in the position shown in Fig. 7.11B, and jerking it, or anchoring the end and driving the tractor away. Or, in the same position on the drum, the cable may be given a couple of wraps around the drawbar and the winch turned backward, as in Fig. 7.11C.

If the cable is wound unevenly onto the drum, with the wraps crossing each other at random, it cannot be pulled down between the lower layers readily. Cross-wrapping may put severe kinks in sections of cable.

In spite of these difficulties, a long cable is desirable for general work. If reasonable effort is made to spool evenly while working, it will usually be rough enough to prevent excessive sticking, without too much bending or crushing.

Two-Part Line. Where the distance to the stump is less than one-half the cable length, a two-part line may be used by attaching a pulley to the stump and running the line from the winch around the pulley and back to the drawbar. The useful strength of the cable and the pull between the tractor and the stump are doubled. The tractor should not be anchored by

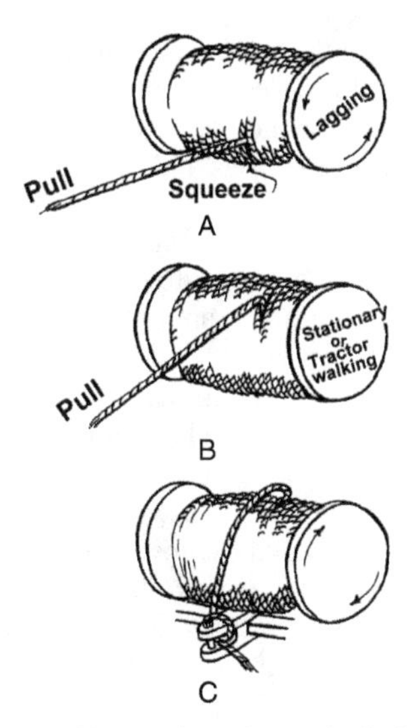

FIGURE 7.11 Freeing a jammed winch cable.

the front pull hook while using a double line anchored on the drawbar, unless the manufacturer warrants the hook is strong enough to take the strain.

Rocking. The winch and tractor each pull differently. Sometimes the pull of the tracks will do jobs the winch cannot accomplish. Use of the tractor drive helps in "rocking" stumps or trees out. The line is left slightly slack, and the tractor moved forward in low. As the line tightens, the stump may lean a few inches, then stop. When the tracks start to spin and the clutch is released, the weight of the stump, combined with the spring in the roots and in the line, will pull the tractor back. The clutch is immediately reengaged and held until the tracks spin or the engine lugs down again.

If the stump is within the tractor's power range, repeating this maneuver should gradually break it out. A long cable has more elasticity than a short one or a chain, and will be more effective at rocking. This procedure should not be allowed to degenerate into yanking, where the tractor is given a long enough slack run to be brought up with a jerk when it tightens.

Cable Breakage. Cable or chain breakage is a serious danger to both operator and helpers. A cable stretches under strain, and if it breaks suddenly, it will whip with great force. The danger to the operator is greatest if the break occurs near the tractor. The best quality cable should be used, in the largest size recommended by the winch manufacturer, if the tractor is to be anchored or used for rocking; the cable should be inspected frequently for weak spots.

Another danger inherent in the use of cable is cutting and tearing of the hands and clothing on broken wires. Preformed cable gives minimum trouble of this kind and should be used when possible. Leather-palmed gloves are good protection for the hands.

Either hemp-center or wire-center cable may be used, according to preference or the manufacturer's recommendation. Wire-center cable is about 10% stronger, size for size, is stiffer, and is not as easily deformed by crushing. It is more difficult to handle, however, and when kinked or crushed is much harder to straighten. Standard 6 × 19 (six groups of 19 wires) constructions are usually recommended.

It is good practice to work a winch at less than its maximum capacity and to avoid anchoring the tractor unless absolutely necessary. Moderate loads give long life to cables and winch parts, and avoid severe catching on the drum. If the work is heavy, strain can be reduced by the use of pulleys and multiple lines.

Multiple Lines. If pulling stumps takes the full power of the tractor or winch, it may be advisable to use **snatch blocks** to obtain greater power at slower speed. These devices, also known as blocks and pulleys, are pulleys set in frames provided with one or two round hooks or rings, usually on swivel connections. For most field work, single-pulley wheels, with a latch attachment permitting insertion of a cable from the side, are best, as cables usually carry attachments too large for threading—a tedious job even when possible.

Snatch block
A pulley block with a hinged side plate. The plate swings open to thread the winch cable through.

These blocks can be obtained in sizes to match any cable or strain. In large sizes they are very heavy, and several workers, or a loader, or light winch or other lines may be used to move them. Figure 7.12 shows several riggings using pulleys. If the lines are approximately parallel and the pulley bushings lubricated, each additional line will add about 90% to the single line pull. This puts no extra strain on the winch cable, but the chokers holding the blocks must take the combined pull of all the lines fastened to them.

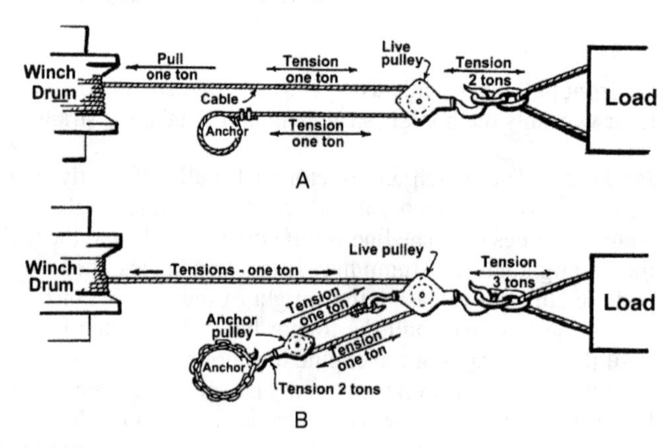

FIGURE 7.12 Use of snatch block pulleys.

The advantage obtained from the use of a block is decreased when the lines are not parallel, becoming zero when the angle between the lines is 114 degrees. Still wider angles result in loss of pull force.

Rigging. The pulling line may be a chain or cable, and the power may be direct pull by a machine, winding in of cable on a winch, or a combination of these methods with pulley blocks. The stump line is generally a choker type, as this will tighten as the pull increases. For smaller stumps a chain is preferred because it is easier to carry, safer to handle, and more resistant to abuse. However, it is much heavier than cable for the same strength, and in large sizes it is too weighty to be practical. Line pulling is preferred when the ground is too rough or soft to allow machinery to get at stumps directly and when available force needs to be increased by multiple lines.

Multiple blocks require care in rigging and pulling. Two anchors are better than one, as they spread the cables over a wider space where they are less apt to interfere with each other. Each block is best fastened to a separate choker, but one may be fastened to each end of a chain passed behind the stump if it is strong enough to take a double pull, or one to both ends of a chain given one turn around the stump. It is good practice to notch the stump for each chain used. The notches hold the chains and blocks in place so they do not slide into each other as the stump yields.

Rigging is done with the lines slack. When they are pulled tight, an inspection should be made to make sure the pulley latches have not fallen open, as a pull on an open block will bend it and cut the cable. The pulleys should be checked for jamming with debris or the chain hooks becoming disengaged. As the line is wound in, all blocks should be watched to make certain they do not collide. Lines should not be allowed to drag on each other.

Chains. A standard tow or logging chain is composed of short, straight links and carries a round hook on one end and a **grabhook** on the other. The round hook may be fastened to the chain by a ring, or a ring may be used instead of this hook. Either the round hook or the ring can be used in chokers. The hook is easier to attach and to detach, but may fall away when the chain is slack. The ring may be used by passing the grabhook through it and pulling from the grabhook end. When removing or dragging stumps, the chain near the ring is pulled through it to form a loop around the stump.

Grabhook
A hook with a nearly U-shaped inside curve so it will slip over a link of chain edgeways but will not permit the next link to slip through.

The grabhook fits over any individual chain link and will not slide along the chain. It is used to adjust the length of chain by increasing or decreasing the amount of double line, by moving it toward or away from the choker end, or by passing the chain behind a tractor drawbar pin and preventing it from being pulled out again by attaching the grabhook to the slack side, making it too large to be pulled through the space. In this case the surplus chain is slacked, and if it is long, must be hung on some part of the tractor (Fig. 7.13).

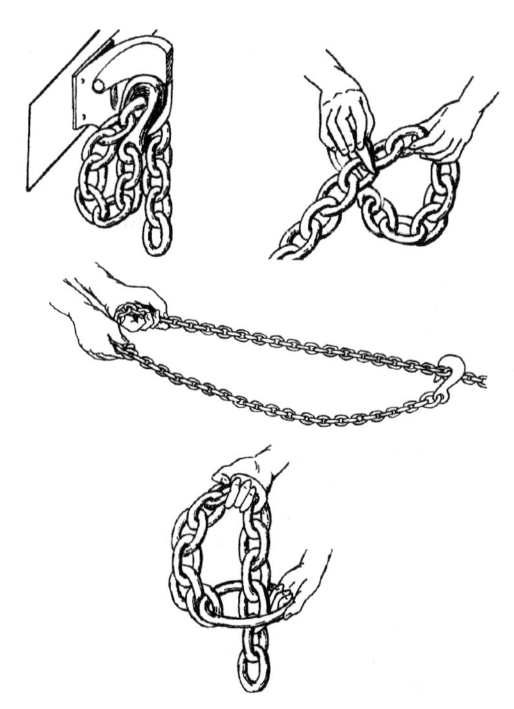

FIGURE 7.13 Grabhook uses and stump choker.

Figure 7.14 shows three ways of fastening a line to a stump. In each case, the back side of the stump is grooved by an axe. This cut will prevent the chain from squeezing off during the pull and delay its slipping off as the stump leans. The A type fastening is the easiest and usual method, pulling at the center; B is a side pull, a little harder to arrange, but it puts less of a kink in the line. It can be used with cable as well as chain and gives the advantage of a twist on the stump; C is the overhead method, which requires an inverted T notch. This gives the greatest leverage but is more likely to slip off than the others.

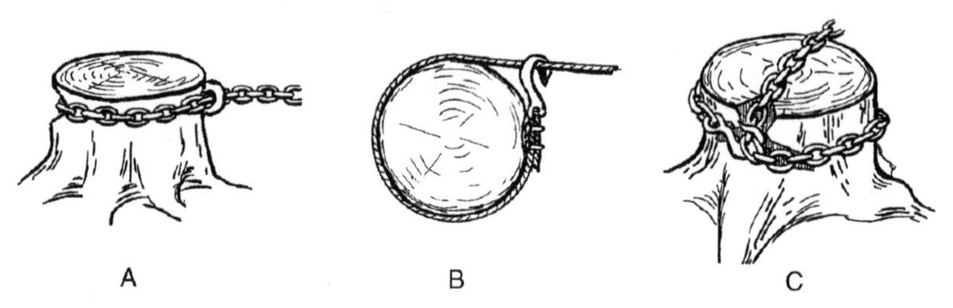

FIGURE 7.14 Methods of fastening a line to a stump.

Care should be exercised not to put loads on a twisted or kinked chain, as this will break or damage the chain. The chain can easily be checked for straightness, as the links in one plane should lie in an almost straight line.

In terms of strength, alloy steel chains weigh only about one-third as much as standard chains. Alloy chains will amply repay their much higher cost in reduced labor and fatigue and in greater efficiency. As an example, a 3/8-in. alloy chain, weighing 1.6 lb per foot is 30% stronger than the same make of 5/8-in. ordinary chain, weighing 4.1 lb per foot.

It is good practice to dip alloy chains in bright red paint so they can be easily recognized and recovered readily if mislaid.

Anchors. Anchor lines should be as low as possible. It may be best to pull the largest stumps first, using several smaller ones for anchorage if necessary. In a clearing job, there is always one last stump and no anchor. If the stump is small, it may be pulled out directly; in any case, it will respond to less elaborate artificial anchors than a large one. On the other hand, a large stump will be a dependable anchor and will prevent the need of frequently replacing rigging when anchors pull out.

The final stump may be pulled by using a living tree as an anchor. A choker should not be used under any circumstances on a tree marked for preservation. To protect the tree, padding and blocks should be used with a grabhook loop.

If no anchor is available, one can be improvised by digging a T-shaped trench 2 or more feet in depth, as shown in Fig. 7.15. A log is placed in the crossbar and the cable anchored to it and led up through the sloping trench toward the work. Load and local conditions will determine the depth of cut and size of log. In medium soil, a standard railroad tie 2 ft down should hold a horizontal pull of 5 tons. This is sometimes termed a "deadman."

Stump pulling with a winch and blocks takes more time and care, but the results are generally satisfactory.

Cables. Only the method shown in Fig. 7.14B should be used when pulling a stump with a cable choker, as the sharp bends involved in the others will cause early breakage of the cable. If a double cable line is used to reduce strain or to shorten the rope, it should not be

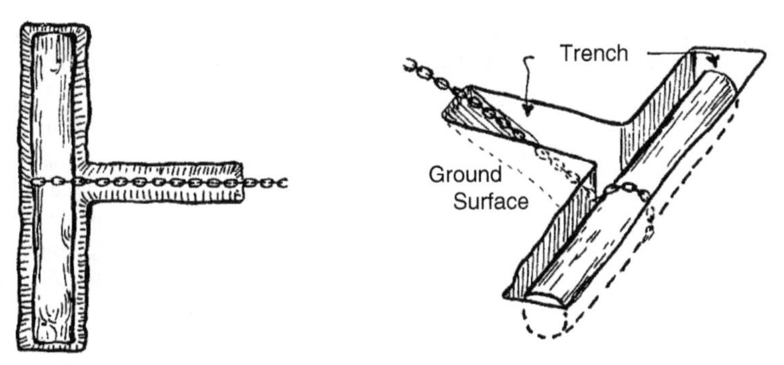

FIGURE 7.15 Constructed anchor.

bent around sharp angles. A stump is generally round and smooth enough not to cut a cable wrapped around it, and the end hooks or loops can be attached to the drawbar. If the load is angular, it is better to fasten a snatch block to it with a chain or sling choker and to run the long cable through the block pulley.

If a double cable is so wrapped around the load, care must be taken to adjust it so both ends share the strain equally, unless a single line is strong enough to take the entire pull alone.

Taproots. The presence of a taproot increases stump resistance to pushing or pulling. If the ground is hard, this root may be broken or pulled apart. If the ground is soft or the wood very tough or pliable, the pivot point may crush and the root bend so pulling effort is exerted directly against the length of the root, without benefit of leverage. In such a case, the upper roots of the stump may be torn sufficiently so the taproot can be reached and cut with an axe or a long chisel. The cut should be made while pulling, as tension makes the wood part more easily.

Pulling Clear. If the force is sufficient to uproot a stump, the roots opposite the pull break first, then those at the side, permitting the stump to be pulled onto its side. If the line does not slip off, the stump may be rolled and dragged out of the ground, but this often takes much more power than overturning the stump, and may be beyond the capacity of the machine.

If the stump will not come out all the way, the line may be slacked and a log placed or chained against the stump, as in Fig. 7.16. This log will provide a new fulcrum and aid the breaking out. Another method is to take the line off and move the tractor to pull in the opposite direction—this should free the stump without difficulty. If a number of stumps are

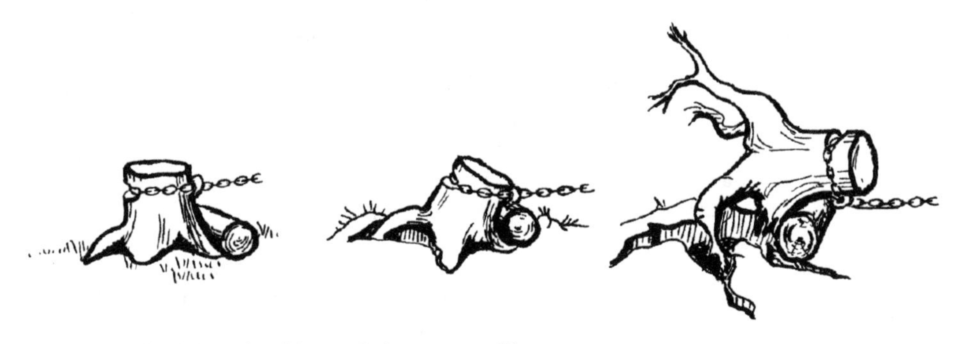

FIGURE 7.16 Using a log fulcrum during stump pulling.

being pulled, all of them may be overturned one way before pulling the tough ones in the opposite direction. Half-uprooted stumps are easily knocked out by dozers.

Resistance. A stump's resistance to uprooting varies with the direction of pull. On a slope, downhill pull is most effective. Otherwise, the stump should be pulled toward its strongest roots, as these are easier to bend than to pull apart and can be dealt with more easily when the rest of the stump is loosened. The most obvious variable in stump resistance is height. Greater height means greater leverage and easier pulling.

A buried stump is the hardest of all to pull and usually must be dug out. On filled land, two separate systems of lateral roots may be found, one under the old ground level and the other near the surface; in such cases it may be necessary to cut the trunk below the upper roots, in the same way as a taproot.

When a stump yields to pull but will not break loose, it can often be uprooted by moving it as far as possible and then slacking off to allow it to settle back. After it settles, pull again and repeat this process a number of times. This is most effective if done slowly and smoothly, whether with winch or traction.

Chopping the roots on the side opposite the pull while they are under maximum tension weakens the resistance. A moderate amount of digging will generally expose the main lateral roots. When a stump has been split by blasting, the pieces are most easily pulled away from the center, rather than across it.

Moving Timber

Crawler tractors are the standard equipment for dragging logs, but with good ground conditions they can be replaced by pneumatic-tired machines with either four-wheel or two-wheel drive.

Dragging (Skidding). If a tractor is sufficiently powerful, several logs may be pulled at a time by attaching them individually to different lines. If only one line is available, they may be fastened with a choker. The choker should be fastened well back, as logs joined in this manner often come apart while being towed.

Two-wheel-drive tractors can drag logs or bundles of logs on dry ground. Loads must be small, but these machines move quickly. If the tractor has a hydraulic lift drawbar, chain the logs to it and lift the butts off the ground; its efficiency is more than doubled, as the weight on the driving wheels is increased and log friction is greatly reduced.

Moving Short Timber. When trucks can enter the woods, cordwood and pulpwood are usually cut to size and trucked out. Because wood is much lighter than soil, a dump truck can carry several times its body rating, if the pile stays on. Placing planks, poles, or thin split logs vertically along the body sides permits high timber loads. The vertical members are held in place by the piled logs.

If the road is rough, pass a chain from the body over the top of the logs and tighten it with a binder. If the wood is cut short and rain or unforeseen mud conditions make trucking impractical, it may be dragged out by tractors.

Storage. Wood should be stored outside of the work area where it will be accessible both during and after the excavation work. Stored logs or cordwood should be stacked so as to be off the ground and well ventilated. This makes it easier to remove later and delays damage from rot and borers. Cordwood is usually stacked in easily measured units.

Personnel. If possible, experienced loggers should be employed for lumbering. They will be able to do the work much more efficiently than people with no experience.

Boulders and Buildings

A clearing area may be strewn with loose or partially buried boulders, making the work difficult. The removal of such rocks may properly be considered clearing. If sufficiently large machinery and suitable disposal points are available, the rocks may be turned or dug out and pushed away. If they are too large for easy handling and disposal, they should be broken up.

Blockhole A hole several inches deep is drilled into the rock and an explosive inserted.

Mudcap An explosive is placed in contact with a rock and fired after being covered with mud.

Breaking may be done by **blockhole** or **mudcap** blasting, backhoe-demolition hammers, or sledgehammer. Under some circumstances, a contractor may prefer to move the rocks by digging and pushing. The dozer is the standard tool for this work. Efficiency can be increased by use of a tilting blade or a heavy-duty rake blade. A dozer can move a large rock on firm ground—perhaps several times its own weight. If the stone is too large for direct pushing, it can be pushed first on one side, then on the other. If it is rounded, it can be rolled by lifting the blade while pushing. If the blade does not have enough lift to roll the rock over, it can hold the rock in a partially rolled position, with locked brakes, while the stone is blocked up. The blade may then be lowered and the push and lift repeated.

Partly buried rocks may be pushed or dug out in the same manner as stumps. The resistance they offer is usually more rigid than a stump, and if a rock will shake in the first few direct blows of the blade or bucket, it should come out. It is sometimes very difficult to get a grip on smooth sloping surfaces, so a large amount of digging may be required just to get a hold.

When a grip is obtained with a dozer blade, the rock may be raised and pushed. The blade must be kept in contact with the rock while it is lifted vertically. When it is high enough, the rock is rolled out. The rock may slip back into its hole at any time, and it is good to have a helper throw stones or logs under it so the blade can be dropped and a fresh grip obtained. If no helper is available, the operator can lock the brakes to hold pressure against the stone and do the hole filling.

If a large stone is rolled out without blocking, it may leave such a large hole that the pushing tractor may be damaged if it falls into the depression. The danger is more serious than with stumps, as rocks leave sharper-edged and harder-surfaced holes. A rock should be pushed from all angles before digging it out, as it may be susceptible to pressure from only one direction. If it is to be dug out, a bowl-shaped crater of considerable size is excavated, working on three sides, if a good grip is available at the top, or all around it if the top is smooth.

When the rock is finally loosened, it may prove so heavy the dozer cannot get it up out of the hole. One procedure in such cases is for the dozer to build a ramp into the crater. The ramp should be shaped so the machine will be pitched downward when its blade meets the rock. With gravity assisting, it should be able to push the stone a short distance up the opposite slope. The ramp is then extended and another push made until the rock is out of the hole.

Loose boulders may be pushed out of the work area and scattered, piled, or arranged in walls; or they may be buried, being either used as a fill or waste in holes. Holes may be dug to bury them.

Where many boulders are pushed into a hole, they afford dangerous footing for a dozer and may pile up above the desired grade. A moderate amount of soil, either scraped off the bank or trucked to the spot, will allow a dozer to fill in the voids between the rocks so it can safely walk across and push other boulders into place. If the area is to be finished to a grade, it pays to be liberal in supplying covering soil, for if the layer is thin, the dozer working on it may hook into freshly buried rocks and turn them into high positions.

These rocks can seldom be put back in place because of soil and other rocks getting under them, and it may be necessary to knock their tops off with hammers or explosives or dig them out and rebury them. Digging a boulder out from among others is very difficult and is likely to turn others up.

Rocks also may be trucked away from the job. They may be loaded by a loader or excavator bucket.

Or a crane may lift them by tongs, chains, or cable slings. Light chains grip well but are easily damaged. Ordinary dump truck bodies may be severely battered by oversize rocks. The floor should be protected by an extra sheet of steel, a layer of planks, or a few inches of soil.

Stone Walls. Stone walls are very common in some sections of the country and may include rocks large enough to present a problem to machinery. The big base stones are often partly or completely buried, interlocked, and bound in place by tree roots. The smaller stones may be valuable for use in masonry and may be removed by hand before or during the wrecking of the wall.

A dozer of sufficient size can walk right through the wall and scatter it around, but an undersized machine may have to start at a gateway or find a weak spot to break through and widen the hole by worrying the rocks in steps. If the wall cannot be broken from one side, try it from the other.

Foundations. Old foundations and other masonry structures usually yield readily to heavy machinery. High walls should be pulled down, as they might fall on a machine pushing them. If a foundation is too strong for available machinery, it may be weakened by blasting along the lines where it meets the floor and other walls.

SITE CLEARING SAFETY

The Occupational Safety and Health Administration regulates site clearing operations as part of Standard 1926.604, Site clearing. The rules can be summarized as follows:

- 1926.604(a)(1) Employees engaged in site clearing shall be protected from hazards of irritant and toxic plants and suitably instructed in the first aid treatment available.
- 1926.604(a)(2) All equipment used in site clearing operations shall be equipped with rollover guards meeting the requirements of this subpart. In addition, rider-operated equipment shall be equipped with an overhead and rear canopy guard meeting the following requirements:
  - 1926.604(a)(2)(i) The overhead covering on this canopy structure shall be of not less than 1/8-in. steel plate or ¼-in. woven wire mesh with openings no greater than 1 in. or equivalent.
  - 1926.604(a)(2)(ii) The opening in the rear of the canopy structure shall be covered with not less than ¼-in. woven wire mesh with openings no greater than 1 in.

CHAPTER 8
EXCAVATORS

Shovels, Hoes, Draglines, and Clamshells

Hydraulic power is the key to the versatility of modern excavators. Hydraulic front shovels are used predominantly for hard digging above track level and for loading haul units. Hydraulic hoe-type excavators are used primarily to excavate below the natural surface of the ground on which the machine rests. A dragline excavator is especially useful when there is need for extended reach in excavating or when material must be excavated from beneath the water. Clamshell excavators provide the means to excavate vertically to considerable depths.

BACKGROUND

Development of mechanical excavating machines was driven by early railroad construction. William S. Otis, a civil engineer with the Philadelphia contracting firm of Carmichael & Fairbanks, built the first practical power shovel excavating machine in 1837 (Fig. 8.1). The first "Yankee Geologist," as his machines were called, was put to work in 1838 excavating the path of a railroad between Boston and Providence, Rhode Island. The May 10, 1838, issue of the *Springfield Republican* in Massachusetts reported, "Upon the road in the eastern part of this town, is a specimen of what the Irishmen call 'digging by stame.' For cutting through a sand hill, this steam digging machine must make a great saving of labor."

Large Electric Excavators

The revolving frame for the largest (14- to 92-cy bucket capacity; 50- to 135-ton bucket payload) electric rope shovels (Fig. 8.2), used for mining operations, is more involved than for the smaller hydraulic shovels. In large, electrically controlled mining shovels, the operator may have a programmable message display (PMD) module. The newer direct current digital drive electrical control systems provide significant performance improvements over older analog drive systems. Once the module is programmed, the operator uses buttons on the PMD panel to control the operations of the shovel.

FIGURE 8.1 The Otis steam shovel; note this machine is mounted on steel wheels and ran on rails.[1]

FIGURE 8.2 Electric rope shovel in a mine; compare pick-up truck to bucket size.

HYDRAULIC EXCAVATORS

Hydraulic excavators use diesel engines to drive hydraulic pumps, motors, and cylinders, and these in turn activate the excavator's motions of digging and loading materials. The deck of the machine is supported by swing rollers, which rest on the turntable of the travel unit. It carries the engine, transmission, and operating machinery. These excavators (Fig. 8.3) may be either crawler or pneumatic tire carrier mounted, and many different specialized attachments are available for individual job applications. With the options in types, attachments, and sizes of machines, there are machines for almost any application, and each offers variations in terms of economic advantage.

[1]"Steam Excavating Machine," *London Journal of Arts and Science*, Vol. 22, 1843.

FIGURE 8.3 Hydraulic front shovel loading shot rock into a truck.

Hydraulic control of machine operation provides

- Faster cycle times
- Positive control of attachments
- Precise control of attachments
- High overall efficiency
- Smoothness and ease of operation

Crowd The forward thrust of the bucket to extract material from a bank.

Curl The circular motion (upward to excavate) of the bucket when the bucket cylinders are activated.

Boom and stick hydraulic excavators are classified by the digging motion of the bucket (Fig. 8.4). An upward motion machine is known as a "front shovel." A shovel develops breakout force by **crowd** and **curl** bucket motions away from the machine. The boom of a shovel swings in an upward arc to load; therefore, the machine requires a material bank above the running gear to work against.

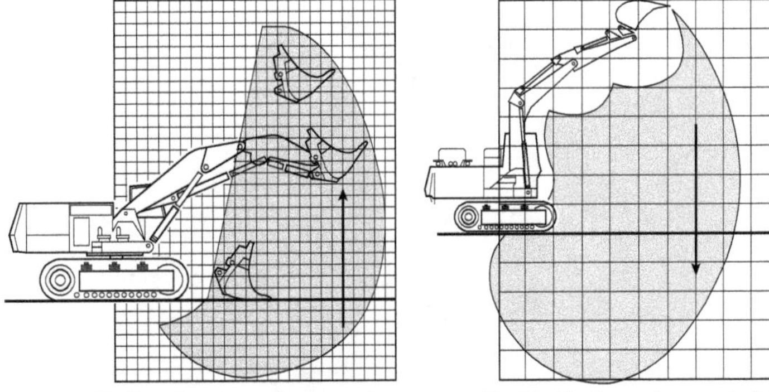

FIGURE 8.4 Digging motion of hydraulic excavators.

A "hoe" type excavator uses a downward arc motion to excavate material. It develops excavation breakout force by pulling the bucket toward the machine and curling the bucket inward. The mechanics of the downward swing dictate usage for excavating below the running gear.

CRAWLER TRAVEL UNIT

The crawler chassis for these excavators is made up of a turntable on a frame consisting of two heavy I-beams, called dead axles, used to connect two heavy track frames, which rest on the track wheels (Fig 8.5). The travel unit is considered to have the drive wheels in the rear and the idlers in the front, regardless of whether this corresponds to the front and rear position of the revolving unit (operator's cab).

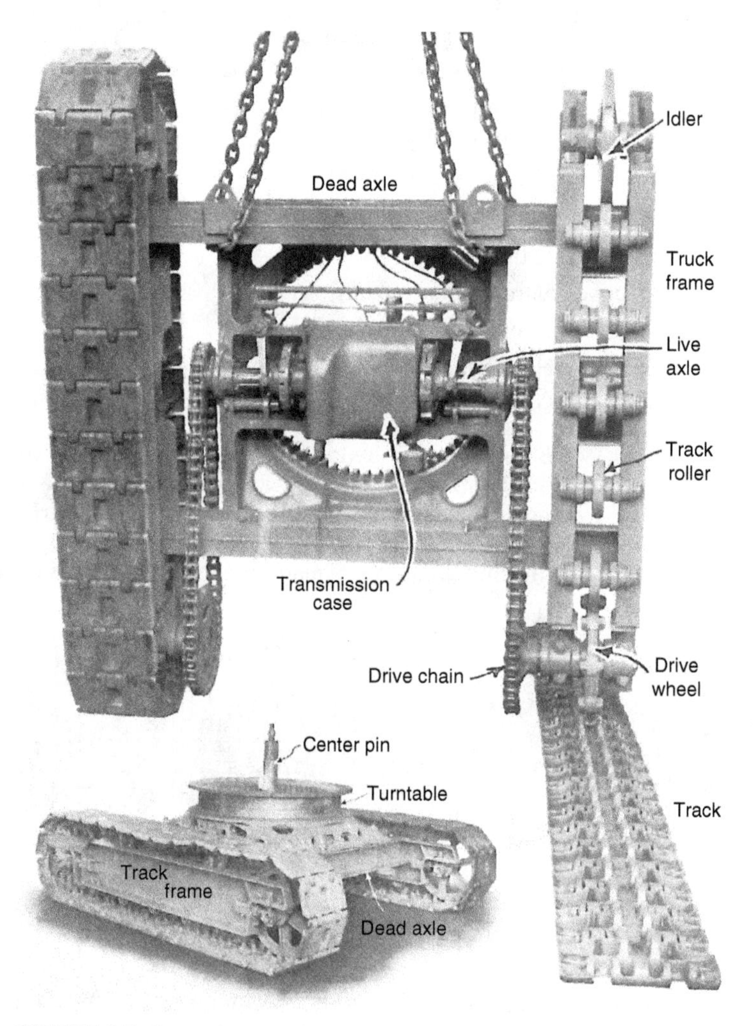

FIGURE 8.5 Lower frame and tracks.

FIGURE 8.5 (*Continued*)

Tracks

Shovel, hoe, dragline, and clamshell excavators do not depend on the tractive pull of their crawlers for their ability to load the bucket. Their tracks are made up of a number of identical shoes cut and drilled at their ends and fastened together by pins.

Track Behavior

If the machine moves forward, the drive wheel will turn so its teeth mesh with the track pads under and ahead of it, picking them up one by one and passing them around its rear and then forward over its top (Fig 8.6). As the drive wheel moves forward on the track, it pushes the truck frame and the idler ahead. If the machine is walked in reverse over good footing with the grade nearly level, the drive wheel will pull in some slack from the top of the track and pass it underneath, moving backward and pulling the truck frame and idler after it. The track it has turned underneath supports the shovel and is then carried up by friction with the idler wheel.

There are several reasons why a tracked vehicle should be walked forward when possible. With forward motion, the track is laid smoothly along the ground, and its slack hangs harmlessly in gradual curves in the top section (see Fig 8.6). The heavy strain is on the short piece from the bottom of the drive wheel to the first truck wheel supporting a good share of the shovel's weight. With reverse movement, the track may be kinked in a damaging angle, because the truck rollers must climb over or push down, and the traction stress is on the whole distance from the top of the drive wheel to the front slope of the idler. This can cause excessive wear and danger of breakage.

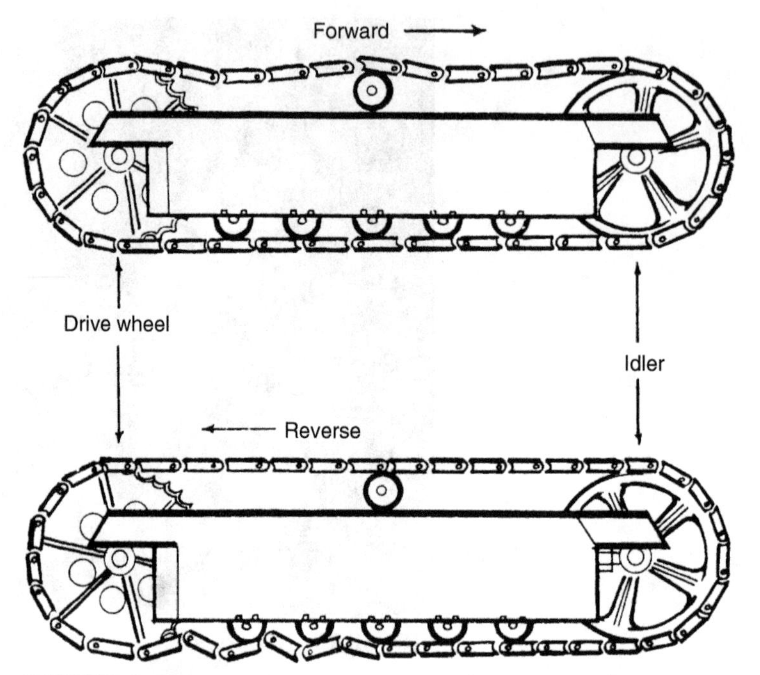

FIGURE 8.6 Excavator track behavior.

FRONT SHOVELS

Front shovels are used predominantly for hard digging above track level and for loading haul units. Loading of shot rock would be a typical application (see Fig. 8.3). Shovels are capable of developing high breakout force with their buckets. Because of their digging motion, they work best when the excavated material stands in a vertical bank (i.e., a wall of material perpendicular to the ground). Most shovels are crawler mounted and have very slow travel speeds—less than 3 mph. The parts of the shovel are designed for machine balance; each element of the front-end attachment—the shovel—is designed for the anticipated load. The front-end attachment weighs about one-third as much as the superstructure with its power parts and cab.

Size Rating of Front Shovels

The size of a shovel is indicated by the size of the bucket, expressed in cubic yards (cy). Hydraulic shovels for construction work range in capacity from 3 to 14 cy, while those for mining applications are much larger. There are three different bucket-rating standards: Power Crane and Shovel Association (PCSA) Standard No. 3, Society of Automotive Engineers (SAE) Standard J67, and the Committee on European Construction Equipment (CECE) method. All of these methods are based only on the physical dimensions of the bucket and do not address the "bucket loading motion" of a specific machine. For buckets greater than 3-cy capacity, ratings are on ¼-cy intervals and on ⅛-cy intervals for buckets less than 3 cy in size.

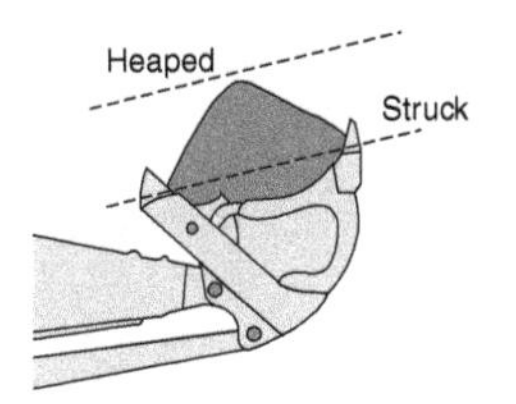

Struck Capacity. The industry definition for struck capacity is the volume actually enclosed by the bucket with no allowance for the bucket teeth.

Heaped Capacity. SAE assumes a 2:1 angle of repose without regard to material type for evaluating heaped bucket capacity.

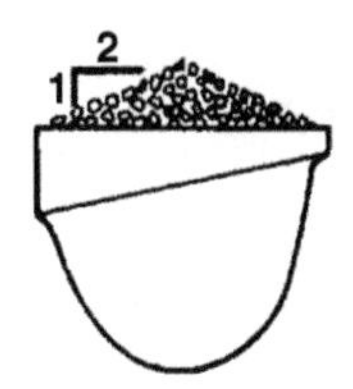

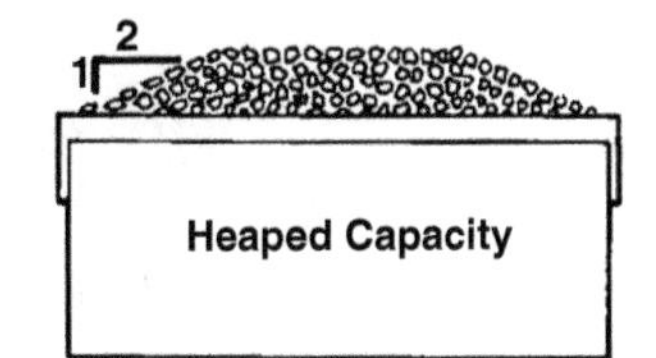

Rated-heaped capacities represent a net section bucket volume; therefore, rated-heaped capacities must be amended to "average" bucket payload based on the characteristics of the material being handled. Manufacturers usually suggest factors, termed "fill factors," for making such corrections.

Fill factor The ratio of the actual loose volume of material in the bucket compared to the bucket's rated-heaped capacity. It varies based on the type of material being handled and the type of excavator.

Fill factors account for the void spaces between individual particles of a particular type of material when it is loaded into an excavator bucket. Materials described as flowing (sand, gravel, or loose earth) should easily fill the bucket to capacity with a minimum of void space. At the other extreme are the bulky-shaped rock particles. If all the particles are of the same general size, void spaces can be significant, especially with large pieces. If the excavated material has a significant amount of oversized chunks or is extremely sticky, significant voids can also occur.

Fill factors (Table 8.1) are expressed as percentages. The adjusted volume for a bucket, considering a specific material, is calculated by multiplying the rated-heaped capacity by the fill factor. To validate fill factors, it is best to conduct field tests based on the weight of the material per bucket load.

TABLE 8.1 Fill Factors for Front Shovel Buckets

Material		Fill Factor* (%)
Bank clay, earth	(Easy digging)	95–105
Rock-earth mixture	(Easy digging)	95–105
Rock—well blasted	(Medium digging)	90–100
Rock—poorly blasted	(Hard digging)	85–95
Very hard digging		80–90

*Percent of heaped bucket capacity for bottom dump buckets.

Basic Parts and Operation

The basic parts of a front shovel are the mounting (substructure), cab, boom, stick, and bucket (Fig. 8.7). With a shovel in the correct position, near the face of the material to be excavated, the bucket is lowered to the floor of the pit with the teeth pointing into the face. A crowding force is applied by hydraulic pressure to the stick cylinder at the same time the bucket cylinder rotates the bucket through the face.

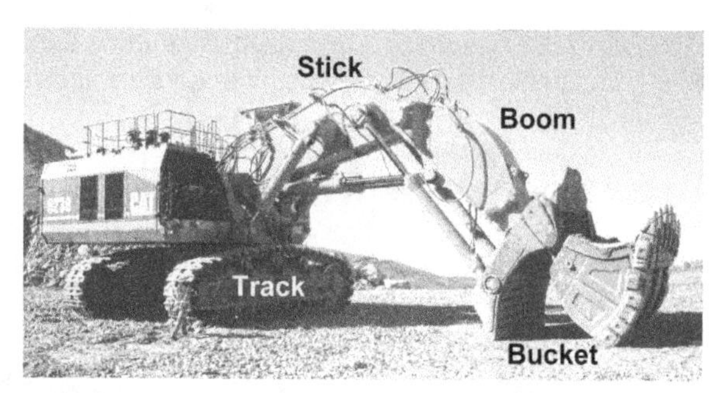

FIGURE 8.7 Basic parts of a hydraulic front shovel.

Shovel Selection

Two fundamental factors should guide the selection of a front shovel for project work: (1) cost per cubic yard of material excavated and (2) the job conditions under which the shovel will operate.

In estimating the cost per cubic yard, it is necessary to consider the

1. *Size* of the job; a job requiring the excavation of a large quantity of material may justify the higher O&O and mobilization costs of a larger shovel.
2. *Mobilization* cost of transporting the machine to the project and its assembly (a large shovel will involve more cost than a small one).
3. *Drilling and blasting* cost for a large shovel; it may be less than for a small shovel, as a large machine will handle more massive rocks than a small one. Large shovels may enable savings in drilling and blasting cost.

These job conditions should be considered in selecting a shovel:

1. If the material is hard to excavate, the bucket of the large shovel with a higher digging force will handle the material more easily.
2. If blasted rock is to be excavated, the large-size bucket will handle larger individual pieces.
3. The size of available hauling units should be considered in selecting the size of a shovel (see Fig. 8.2). If small hauling units must be used, the size of the shovel should be small, whereas if large hauling units are available, a large shovel should be used.

Haul-unit capacity should be approximately five times the excavator bucket size.

This 5:1 ratio provides for an efficient balance between excavator production capability and haul-unit production capability. Proper matching of excavator bucket size to haul-unit capacity eliminates wasted cycle time caused by mismatches between adjusted bucket volumes and haul unit volume. When there is a mismatch, it is necessary to use partial bucket loads to properly fill the haul units.

Minimum bucket dumping height is especially important when the shovel is loading haul units. It is necessary to select a shovel with physical dimensions that allow placement of a bucket in the dump position at a height above the haul unit's structure. Manufacturer specifications should always be consulted for exact values of machine dimensions and clearances.

Shovel Production

If an excavator is considered an independent machine (a one-link system), its production rate can be estimated using the following steps:

Step 1. Obtain the heaped bucket load volume from the manufacturer's data sheet. This would be a loose volume (lcy) value.

Step 2. Apply a bucket fill factor based on the type of machine and the class of material being excavated.

Step 3. Estimate a peak cycle time. This is a function of machine type and job conditions to include angle of swing, depth or height of cut, and in the case of loaders, travel distance.

Step 4. Apply an efficiency factor.

Step 5. Conform the production units to the desired volume or weight units (lcy to bcy or tons).

Step 6. Calculate the production rate.

The basic production formula is: Material carried per load × cycles per hour. In the case of excavators, this formula can be refined and written as

$$\text{Production} = \frac{3{,}600\ \text{sec} \times Q \times F \times (\text{AS:D})}{t} \times \frac{E}{60\text{-min hr}} \times \text{Volume correction} \tag{8.1}$$

Swing The rotational motion of the excavator's upper frame to the left or right.

where Q = heaped bucket capacity (lcy)
F = bucket fill factor
AS:D = angle of **swing** and depth (height) of cut correction
t = cycle time in seconds
E = efficiency (min per hour)

Volume correction for loose volume to bank volume: $\dfrac{1}{1 + \text{percent swell}}$

For loose volume to tons: $\dfrac{\text{Loose unit weight, lb}}{2{,}000\ \text{lb/ton}}$

Calculating Shovel Production. There are four elements in the production cycle of a shovel:

1. Load bucket (digging): Move bucket to the bank, fill, and raise bucket clear of the bank.
2. Swing with load: The full bucket is raised to dump height and swung over the dump target.
3. Dump load: Open bucket to dump while controlling dump height.
4. Return swing: Swing upper frame back to the bank and lower bucket to start the next cycle.

A shovel does not travel during the digging and loading cycle. Travel is limited to moving into or along the face as the excavation progresses. On average, it is necessary to move the shovel into the bank after 20 bucket loads. This movement takes, on average, 36 sec.

Typical cycle element times under average conditions for 3- to 5-cy-size shovels are

Load bucket	7–9 sec
Swing with load	4–6 sec
Dump load	2–4 sec
Return swing	4–5 sec

Larger mining shovels cycle in 25 to 45 sec, depending on their size.

The actual production of a shovel is affected by numerous factors, including the

- Class of material
- Height of cut
- Angle of swing
- Operator skill
- Condition of the shovel
- Haul-unit exchange
- Size of haul units
- Handling of oversize material
- Cleanup of loading area

Haul-unit exchange refers to the total time required for a loaded truck to clear its loading position under the excavator and for the next empty truck to be positioned for loading.

When handling shot rock, it is necessary to carefully evaluate the amount of oversize material the shovel will have to handle. A machine with a bucket whose bite width and pocket are satisfactory for the average-size pieces may spend too much time trying to handle individual oversize pieces. A larger bucket, or a larger shovel, or changing the blasting pattern should be considered when there is a large percentage of oversize material.

The use of auxiliary equipment in the loading area, such as a dozer, can reduce cleanup delays. Control of haul units and operator breaks are within the control of field management.

The capacity of a bucket is based on its heaped volume. This is a loose cubic yards (lcy) measure. To obtain the bank-measure volume of a bucket when considering a particular material, the average loose volume should be divided by 1 plus the material's swell. For example, if a 4-cy bucket is used to excavate material whose swell is 25% and the average loose volume of a bucket load is 4.25 lcy, the bank-measure volume will be $\frac{4.25}{1.25} = 3.4$ bcy. If this shovel can make 2.5 cycles/min (no allowance for lost time), the output will be $2.5 \times 3.4 = 8.5$ bcy/min, or 510 bcy/hr. This is an ideal production. Ideal production is based on digging at optimum height with a 90-degree swing and no delays.

Height of Cut Effect on Shovel Production. A loose flowing material will fill a bucket in a shorter sweep up the embankment than when chunky material is encountered. If the face against which a shovel is excavating material does not have sufficient height, it will be difficult or impossible to fill the bucket in one pass up the face. The operator will have a choice of making more than one pass to fill the bucket, which will increase cycle time, or with each cycle the partly filled bucket can be carried to the haul unit. In either case, the effect will be to reduce shovel production.

If the height of the face is considerably greater than the minimum required for filling the bucket, the operator is presented with three options. The depth of bucket penetration into the face may be reduced in order to fill the bucket in one full pass up the face. This will increase the time for a cycle. The operator may maneuver the bucket so as to begin digging above the base of the face and then remove the lower portion of the face with the next pass. Or the bucket may be run up the full height of the face and the excess material allowed to spill onto the working area of the bench. This spillage will have to be picked up later. The choice of any one of the procedures will result in lost time, compared to the time required to fill the bucket when digging at optimum height.

PCSA has published findings on the optimum height of a cut based on data from studies of small cable-operated shovels (Table 8.2). In the table, the percentage of optimum height of a cut is obtained by dividing the actual height of the cut by the optimum height for the given material and bucket and multiplying the result by 100. Thus, if the actual height of a cut is 6 ft and the optimum height is 10 ft, the percentage of optimum height of the cut is $\frac{6}{10} \times 100 = 60\%$. In most cases, other types of excavators, either track or rubber-tire loaders, have replaced the small shovels of the PCSA studies. But some general guidelines can still be gleaned from the data.

TABLE 8.2 Factors for Height of Cut and Angle of Swing Effect on Shovel Production

Percentage of Optimum Depth	Angle of Swing (Degrees)						
	45	60	75	90	120	150	180
40	0.93	0.89	0.85	0.80	0.72	0.65	0.59
60	1.10	1.03	0.96	0.91	0.81	0.73	0.66
80	1.22	1.12	1.04	0.98	0.86	0.77	0.69
100	1.26	1.16	1.07	1.00	0.88	0.79	0.71
120	1.20	1.11	1.03	0.97	0.86	0.77	0.70
140	1.12	1.04	0.97	0.91	0.81	0.73	0.66
160	1.03	0.96	0.90	0.85	0.75	0.67	0.62

Optimum height of cut ranges from 30% to 50% of the machine's maximum digging height, with the lower percentage being representative of easy-to-load materials, such as loam, sand, or gravel. Hard-to-load materials, sticky clay, or blasted rock necessitate a greater optimum height—in the range of 50% of the maximum digging height value. Common earth would require slightly less than 40% of the maximum digging height for best efficiency.

Angle of Swing Effect on Shovel Production. The angle of swing of a shovel is the horizontal angle, expressed in degrees, between the position of the bucket when it is excavating and the position where it discharges the load. The total time in a cycle includes digging, swinging to the dumping position, dumping, and returning to the digging position. If the angle of swing is increased, the time for a cycle will be increased. If the angle of swing is decreased, the time for a cycle will be decreased. Ideal production of a shovel is based on operating at a 90-degree swing and optimum height of a cut. The effect of the angle of swing on the production of a shovel is illustrated in Table 8.2. The ideal production should be multiplied by the proper correction factor to adjust the production for any given height and swing angle. Shovel production is improved by spotting the haul unit as close to the shovel as possible, but clear space must always be left for the tail swing of the shovel. This limiting distance is referred to as the *truck spotting tolerance*.

Proper excavation planning can reduce the angle of swing. For example, if a shovel is digging at optimum depth and the angle of swing can be reduced from 90 degrees to 60 degrees, the production will be increased 16%.

Sample Calculation: A shovel with a 5-cy heaped capacity bucket is loading poorly blasted rock. Figure 8.3 illustrates the situation. It is working a 12-ft-high face. The shovel has a maximum rated digging height of 34 ft. The haul units can be positioned so the swing angle is only 60 degrees. What is a conservative ideal loose cubic yard production if the ideal cycle time is 21 sec?

Step 1. Size of bucket: 5 cy

Step 2. Bucket fill factor (see Table 8.1) for poorly blasted rock: 85% to 95%; use 85%, which is a conservative estimate

Step 3. Cycle time given: 21 sec

Average height of excavation: 12 ft

Optimum height for this machine and material (poorly blasted rock):

$$0.50 \times 34 \text{ ft (max. height)} = 17 \text{ ft}$$

Percent optimum height: $\frac{12 \text{ ft}}{17 \text{ ft}} \times 100 = 71\%$

Correcting for height and swing from Table 8.2, by interpolation: 1.08

Step 4. Efficiency factor—ideal production: 60-min hour

Step 5. Production will be in lcy

Step 6. Ideal production per 60-min hour

$$\frac{3{,}600 \text{ sec/hr} \times 5 \text{ cy} \times 0.85 \text{ (fill factor)} \times 1.08 \text{ (height} - \text{swing factor)}}{21 \text{ sec/cycle}} = 787 \text{ lcy/hr}$$

Although the information given in the text and in Tables 8.1 and 8.2 is based on extensive field studies, the reader is cautioned against using the data without adjusting for the conditions on a particular project.

The estimator must always consider all job site factors before deciding on an efficiency factor to use for adjusting theoretical peak production. Experience and good judgment must be applied when selecting the appropriate efficiency factor. Based on Transportation Research Board (TRB) studies, the actual production time for shovels used in highway construction excavation operations is only 50% to 75% of available working time. Therefore, production efficiency is only a 30- to 45-min hour. The best estimating method is to develop project-specific historical data by type of machine and project factors. But the TRB study of thousands of shovel cycles provides a good benchmark for selecting an applicable efficiency factor.

Sample Calculation: A shovel with a 3-cy heaped capacity bucket is loading well-blasted rock on a highway project. The average face height is expected to be 22 ft. The shovel has a maximum rated digging height of 30 ft. Most of the cut will require a 140-degree swing of the shovel to load the haul units. What is a conservative production estimate in bank cubic yards?

Step 1. Size of bucket: 3 cy.

Step 2. Bucket fill factor (see Table 8.1) for well-blasted rock: 90% to 100%; use 90%, which is a conservative estimate.

Step 3. Cycle element times:

Load	9 sec	(because of material, rock)
Swing loaded	4 sec	(small machine, 3 cy)
Dump	4 sec	(into haul units)
Swing empty	4 sec	(small machine, 3 cy)
Total time	21 sec	

Average height of excavation: 22 ft

Optimum height: 50% of max.: 0.5×30 ft = 15 ft

Percent optimum height: $\frac{22}{15} \times 100 = 147\%$

Height and swing factor: From Table 8.2, for 147%, by interpolation approximately: 0.73

Step 4. Efficiency factor: If the TRB information is used, the efficiency would be 30 to 45 working minutes per hour. Assume 30 min for a conservative estimate.

Step 5. Class of material: well-blasted rock, swell 60% (see Chap. 4).

Step 6. Production:

$$\frac{3{,}600 \text{ sec/hr} \times 3 \text{ cy} \times 1.0 \times 0.73}{21 \text{ sec/cycle}} \times \frac{30 \text{ min}}{60 \text{ min}} \times \frac{1}{(1+0.6)} = 117 \text{ lcy/hr}$$

HYDRAULIC HOE EXCAVATORS

Hoes are used primarily to excavate below the natural surface of the ground on which the machine rests (Fig. 8.8). A hoe is sometimes referred to by other names, such as backhoe (usually applied to smaller machines) or back shovel. Hoes are excellent machines for excavating trenches and pits. Because of their positive bucket control, they are superior to draglines in operating on close-range work and loading into haul units.

Wheel-mounted hydraulic hoes (Fig. 8.9) are available with buckets up to 1 cy. Maximum digging depth for the larger machines is about 25 ft. With all four outriggers down, large wheel-mounted machines can handle 10,000-lb loads at a 20-ft radius. Wheel-mounted hoes are not production excavation machines. They are designed for mobility and general-purpose work.

The operating ranges of a hoe are illustrated in Fig. 8.10. Table 8.3 gives representative dimensions and limits of reach for hydraulic crawler-mounted hoes.

FIGURE 8.8 Crawler-mounted hydraulic hoe.

FIGURE 8.9 Small wheel-mounted hydraulic hoe.

Components of a Hydraulic Hoe

A hydraulic hoe typically has three strong structural members: the boom, the stick, and the bucket (Fig 8.11). The boom is hinged to the excavator deck at its lower end and at its upper end to the stick. The bucket or other tools are hinged to the stick. Movement at each of these hinge points is controlled by hydraulic cylinders. European manufacturers offer small hoe excavators with a boom and two sticks (Fig 8.12). These machines provide greater control for work in confined spaces.

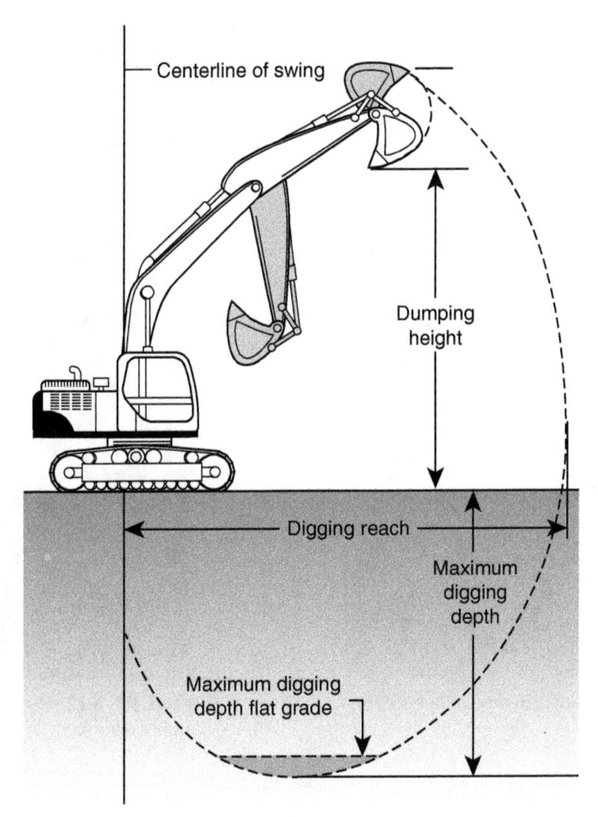

FIGURE 8.10 Operating ranges of a hydraulic hoe.

TABLE 8.3 Representative Dimensions, Limits of Reach, and Lifting Capacity of Hydraulic Crawler Hoes

Size Bucket (cy)	Stick Length (ft)	Maximum Reach @ Ground Level (ft)	Maximum Digging Depth (ft)	Maximum Loading Height (ft)	Lifting Capacity at 15 ft — Short Stick Front (lb)	Short Stick Side (lb)	Long Stick Front (lb)	Long Stick Side (lb)
3/8	5–7	19–22	12–15	14–16	2,900	2,600	2,900	2,600
¾	6–9	24–27	16–18	17–19	7,100	5,300	7,200	5,300
1	5–13	26–33	16–23	17–25	12,800	9,000	9,300	9,200
1½	6–13	27–35	17–21	18–23	17,100	10,100	17,700	11,100
2	7–14	29–38	18–27	19–24	21,400	14,500	21,600	14,200
2½	7–16	32–40	20–29	20–26	32,600	21,400	31,500	24,400
3	10–11	38–42	25–30	24–26	32,900*	24,600*	30,700*	26,200*
3½	8–12	36–39	23–27	21–22	33,200*	21,900*	32,400*	22,000*
4	11	44	29	27	47,900*	33,500*		
5	8–15	40–46	26–32	25–26	34,100†	27,500†	31,600†	27,600†

*Lifting capacity @ 20 ft
†Lifting capacity @ 25 ft

FIGURE 8.11 Arrangement of a hydraulic hoe's cylinders to develop digging forces.

FIGURE 8.12 Small hoe excavator with a boom and two sticks.

Boom. The boom is almost always of the bent or gooseneck type, concave toward the ground. It usually has one bend or angle, but may have two. This shape serves three purposes. It allows space to pull the bucket closer to the machine, permits deeper digging without interference from the tracks, and enables the operator to see past it more easily when it is raised.

The boom foot is hinged to trunnions 2 or more feet back from the deck edge. They are usually in front of the swing center, but may be behind it. If there are two points of attachment to the boom, the upper one is used for maximum digging depth and the lower for maximum dump height.

The outer end of the boom is usually prolonged into a two-piece bracket, in which the stick is held by a heavy hinge pin or pins. The stick cylinder is mounted on the boom top.

Stick. The stick, bucket stick, or arm, is hinged to the end of the boom and is connected to the stick cylinder rod at its upper (back) end and to the bucket and bucket dump arms at the bottom or front. It is usually one piece, but is extendable by telescoping, and retractable sticks are also available.

The stick's connection to the boom is not at its very end. It is slightly below the top end of the stick. The proportion between the two sections varies widely in different makes and models.

Some machines provide two places for the boom-to-stick hinge. The one closer to the bucket will supply more power for hard digging; the other will provide more speed in easier work. Extending the cylinder forces the bucket in toward the machine, crowding it into the digging.

Bucket Mounting. The bucket is connected to the stick by a hinge pin and to a triangular set of paired dump arms so it will have an arc of rotary movement around the stick hinge. When the bucket cylinder is extended, the bucket teeth move inward in a curling motion. When it is retracted, the bucket opens or extends.

When several sets of holes are provided, bucket action can be changed by moving the hinge pins. The choice is between a combination of greater speed and range of movement in bucket control, or slower motion and greater digging and breakout force. Selection depends on the work being done and the operator's preference, and is likely to be changed only under unusual conditions.

Buckets. The bucket is sometimes called a dipper or a tool (Fig. 8.13). The hoe is excellent for digging ditches and handling pipe. It is also well adapted to digging basements. The larger machines are used for general excavation and loading haul units. For efficiency in digging ditches, the bucket should cut the full required width on every pass. Therefore, buckets are usually supplied in a number of widths, ranging from 30 to 48 in. but are available down to 24 in. and up to 5 ft.

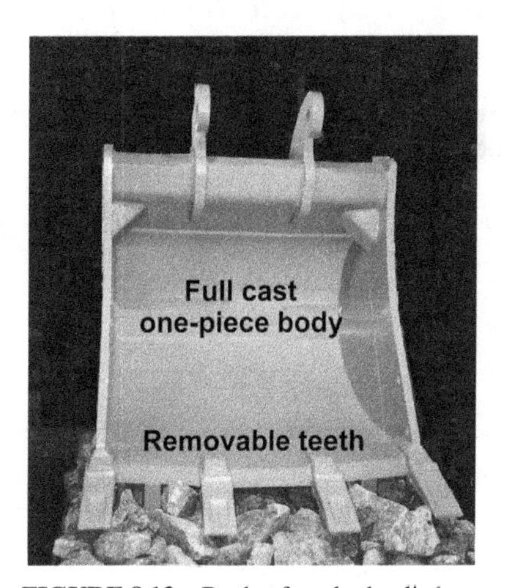

FIGURE 8.13 Bucket for a hydraulic hoe.

Narrow buckets tend to be deep in proportion to their width and may fill poorly in chunky or rocky digging. If width is the same, reducing depth from the front edge reduces capacity but may increase efficiency in loading. A standard-width bucket intended for very hard digging might be made smaller (shallower). This allows the addition of reinforcement without adding too much extra weight.

The bucket is usually slightly wider at the open or front end to reduce friction at the sides and to allow for easier dumping. Additional clearance from trench walls may be obtained and bucket cutting width increased by 2 to 8 in. by installing sidecutters. They may be fixed-width or adjustable, smooth-edged, or toothed.

Sidecutters are useful in accommodating a bucket to a wider trench, cramming more material into a narrow bucket, reducing drag in sticky soil, and reducing wear on the front edges. Wide buckets may have poor penetration. General-purpose buckets for basements and pits are usually intermediate in width and capacity. The digging edge is almost always equipped with teeth, which are removable for reversing, sharpening, or replacement.

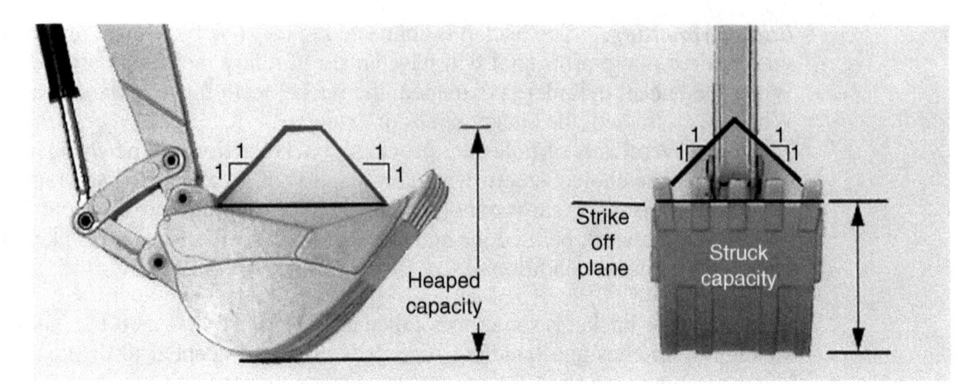

FIGURE 8.14 Hydraulic hoe bucket capacity rating dimensions.

Bucket Rating for Hydraulic Hoes. Hoe buckets are rated by PCSA and SAE standards using a 1:1 angle of repose for evaluating heaped capacity (Fig. 8.14). Buckets should be selected based on the material being excavated. By matching bucket width and bucket tip radius to the resistance of the material, it is possible to take full advantage of the hoe's potential. For easily excavated materials, wide buckets should be used. When excavating rocky material or blasted rock, a narrow bucket with a short tip radius is best. In utility work, the width of the required trench may be the critical consideration. Fill factors for hydraulic hoe buckets based on material types are presented in Table 8.4.

TABLE 8.4 Fill Factors for Hydraulic Hoe Buckets

Material	Fill Factor (% of Heaped Capacity)
Moist loam/Sandy clay	100–110
Sand and Gravel	95–110
Caliche (clay not rock)	100
Rock—poorly blasted	40–50
Rock—well blasted	60–75
Hard tough clay	80–90

Selecting a Hydraulic Hoe

When selecting a hoe for project use, consider

1. Maximum excavation depth required
2. Maximum working radius required for digging and dumping
3. Maximum dumping height required
4. Hoisting capability required, applicable when handling pipe and trench boxes

Multipurpose Tool Platform. The hydraulic boom-stick arrangement is a platform designed for multiple attachments. A quick-coupler on the end of the stick enables the machine to change attachments and perform a variety of tasks in rapid succession. There are rock drills, earth augers, grapples for land clearing (Fig. 8.15A), impact hammers (Fig. 8.15B), demolition jaws, and vibratory-plate compactors, all of which can easily be attached to the stick in place of the bucket. Additionally, there is a broad range of special-purpose buckets, such as trapezoidal buckets for digging and cleaning irrigation ditches, round-bottom buckets (Fig. 8.15C) for cast-in-place pipe operations, and clamshell buckets (Fig. 8.15D) for vertical excavation of footings. The machine's versatility is enhanced by these buckets and tools.

Rated Lifting Capacity. In storm drain and utility work, the hoe can perform both the trench excavation and handle the pipe. Manufacturers provide machine-lifting capacities (rated hoist load) based on (1) distance from the center of gravity of the load to the axis of rotation of the machine's superstructure and (2) height of the bucket lift point above the

A Hydraulic hoe with grapple

B Hydraulic ram attachment

C Special round-bottom hoe bucket

D Clamshell bucket on wheel-mounted hoe

FIGURE 8.15 The hydraulic hoe as a multipurpose tool platform.

bottom of the tracks or wheels (Fig. 8.16). Typical lifting data at only specific distances are provided in Table 8.3. To evaluate lifting ability, it is necessary to plot the data for multiple distances, as capability varies with positioning of the boom and stick (Fig. 8.17). To reach a required position, the boom and stick may have to be manipulated through various distances and depths.

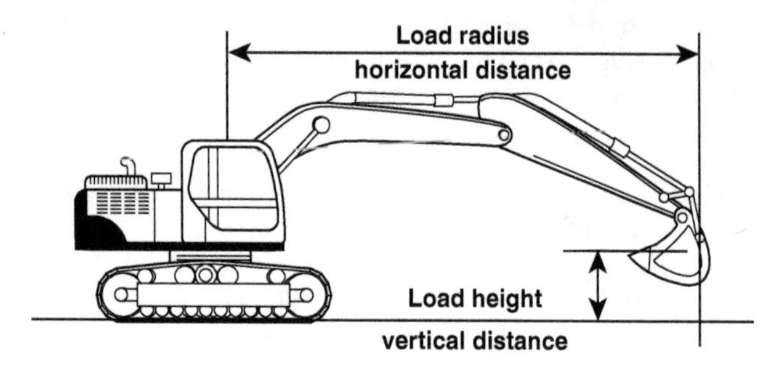

FIGURE 8.16 Hoe lifting capacity position definitions.

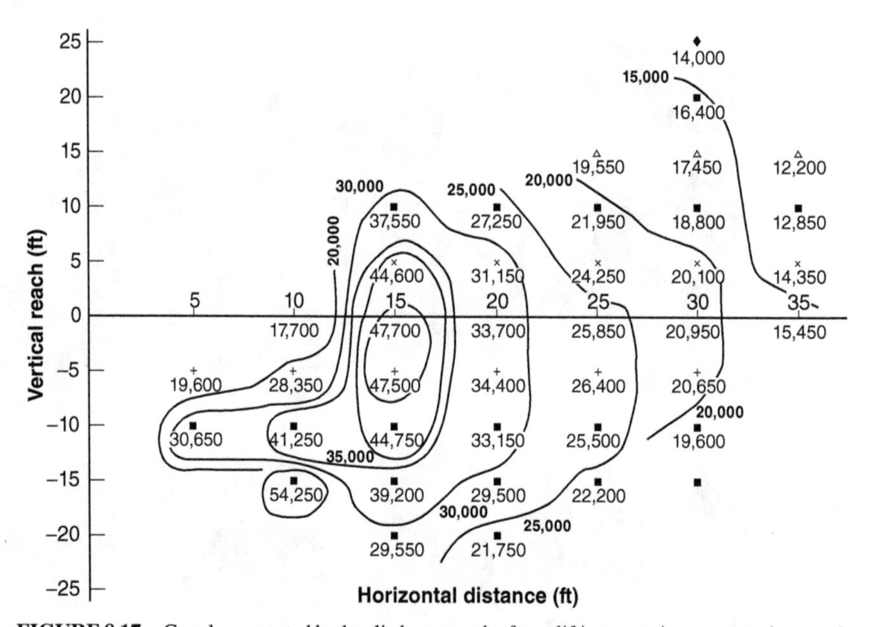

FIGURE 8.17 Crawler-mounted hydraulic hoe over the front-lifting capacity contours, in pounds.

Rated hoist load is typically established based on the following guidelines:

1. Rated hoist load shall not exceed 75% of the tipping load.
2. Rated hoist load shall not exceed 87% of the excavator's hydraulic capacity.
3. Rated hoist load shall not exceed the machine's structural capabilities.

Calculating Hydraulic Hoe Production

Hoe cycle times are approximately 20% longer in duration than those of a similar-size shovel because the hoisting distance is greater, as the boom and stick must be fully extended to dump the bucket.

Optimum depth of cut for a hoe will depend on the type of material being excavated and bucket size and type. As a rule, the optimum depth of cut for a hoe is usually in the range of 30% to 60% of the machine's maximum digging depth. Table 8.5 presents cycle times for hydraulic track hoes based on bucket size and average conditions. There are no tables relating the average hoe cycle time to variations in depth of cut and horizontal swing. Consequently, when using Table 8.5, consider those two factors when deciding on a load bucket time and the two swing times.

TABLE 8.5 Excavation Cycle Times for Hydraulic Crawler Hoes under Average Conditions*

Bucket Size (cy)	Load Bucket (sec)	Swing Loaded (sec)	Dump Bucket (sec)	Swing Empty (sec)	Total Cycle (sec)
<1	5	4	2	3	14
1–1½	6	4	2	3	15
2–2½	6	4	3	4	17
3	7	5	4	4	20
3½	7	6	4	5	22
4	7	6	4	5	22
5	7	7	4	6	24

*Depth of cut 40-60% of maximum digging depth; swing angle 30-60 degrees; loading haul units on the same level as the hoe.

The basic production formula for a hoe used as an excavator is

$$\text{Hoe (excavation) production} = \frac{3{,}600 \text{ sec} \times Q \times F}{t} \times \frac{E}{60\text{-min hr}} \times \text{Volume correction} \tag{8.2}$$

where Q = heaped bucket capacity in lcy
F = bucket fill factor for hoe buckets
t = cycle time in seconds
E = efficiency in minutes per hour

$$\text{Volume correction for loose volume to bank volume} = \frac{1}{1 + \text{percent swell}}$$

$$\text{For loose volume to tons} = \frac{\text{Loose unit weight, lb}}{2{,}000 \text{ lb/ton}}$$

Sample Calculation: A crawler hoe with a 3½-cy bucket is being considered to excavate very hard clay from a borrow pit. The clay will be loaded into trucks with a side body height of 9 ft 9 in. Soil-boring information indicates a material change at 8 ft. The material below 8 ft is an unacceptable silt and cannot be used in the embankment. What is the estimated production of the hoe in cubic yards bank measure, if the efficiency factor is equal to a 50-min hour?

Step 1. Size of bucket: 3½ cy

Step 2. Bucket fill factor (see Table 8.4): hard clay 80% to 90%; use an average of 85%

Step 3. Typical cycle element times:

Optimum depth of cut is 30% to 60% of maximum digging depth; from Table 8.3 for a 3½-cy-size hoe, maximum digging depth is 23 to 27 ft

Depth of excavation: 8 ft

$$\frac{8\text{ ft}}{23\text{ ft}}\times 100 = 34\% \geq 30\%;\text{ okay}$$

$$\frac{8\text{ ft}}{27\text{ ft}}\times 100 = 30\% \geq 30\%;\text{ okay}$$

Therefore, under average conditions and for a 3½-cy-size hoe, cycle times from Table 8.5 would be

1. Load bucket	7 sec	very hard clay
2. Swing with load	6 sec	load trucks
3. Dump load	4 sec	load trucks
4. Return swing	5 sec	
Cycle time	22 sec	

Step 4. Efficiency factor: 50-min hour

Step 5. Class of material: hard clay, swell 35% (see Chap. 4)

Step 6. Probable production:

$$\frac{3{,}600\text{ sec/hr}\times 3\tfrac{1}{2}\text{ cy}\times 0.85}{22\text{ sec/cycle}}\times\frac{50\text{ min}}{60\text{ min}}\times\frac{1}{(1+0.35)} = 300\text{ bcy/hr}$$

Can the hoe service the trucks? Check the maximum loading height from Table 8.3 for 21 to 22 ft:

$$21\text{ ft} > 9\text{ ft }9\text{ in.};\text{ okay}$$

Hoe cycle times are usually of longer duration than shovel times because, after making the cut, the hoe bucket must be raised above the ground level to load a haul unit or to position above a spoil pile. If the haul units can be spotted on the floor of the pit, the hoe bucket will be above the haul unit when the cut is completed (Fig. 8.18) and it would not be necessary to raise the bucket higher before swinging and dumping the load. Every movement of the bucket equals increased cycle time. The spotting of haul units below the level of the hoe will increase production. The cycle-time savings between loading at the same level and working the hoe from a bench above the haul units is approximately 13%.

FIGURE 8.18 Location of haul units affects hoe production.

Often, in trenching operations, the volume of material moved is not the primary concern. The critical issue is matching the hoe's ability to excavate linear feet of trench per unit of time matched to the pipe-laying production.

DRAGLINES

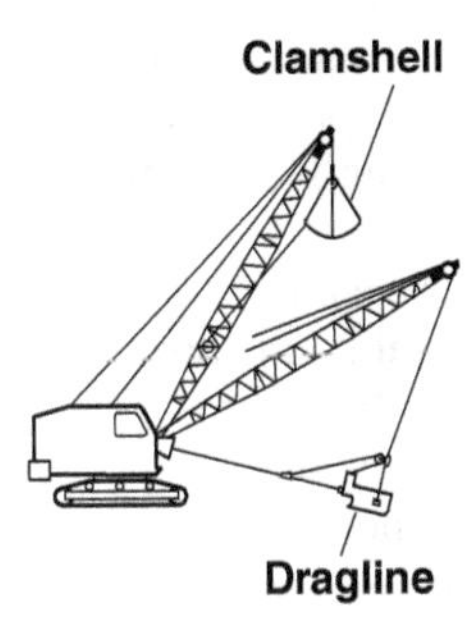

Drag buckets and clamshells are both attachments hung from a lattice-boom crane. The terms *dragline* and *clamshell* refer to the particular type of bucket used and to the digging motion of the bucket. As the name implies, a dragline works by dragging a bucket toward the machine. The clamshell bucket is designed to excavate material in a vertical direction. It works like an inverted jaw with a biting motion. With both types of excavators, the buckets are attached to the crane only by cables. Therefore, the operator has no positive control of the bucket, as with hydraulic excavators.

A dragline (Fig. 8.19) has a crane boom with a fairlead set at its foot and a bucket attached to the machine by cables. The PCSA has conducted studies concerning the performance, operating conditions, production rates, economic life, and cost of owning and operating cranes with dragline and clamshell buckets.

The dragline is a versatile machine capable of a wide range of operations. It can handle soft to medium-hard materials. The greatest advantage of a dragline over other excavators is its long reach for digging and dumping. However, a dragline does not have the positive digging force of a hydraulic shovel or hoe. Its bucket breakout force is derived strictly from the weight of the bucket and the drag pull. A bucket can bounce, tip over, or drift sideward when hard material is encountered. These weaknesses are particularly noticeable with smaller machines and lightweight buckets.

Draglines are used to excavate material and load it into hauling units, such as trucks or tractor-pulled wagons, or to deposit it in levees, dams, and spoil piles

FIGURE 8.19 A dragline excavating a large ditch.

Building levees

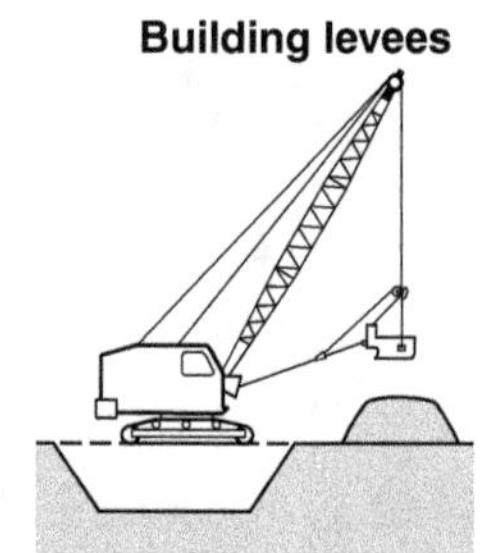

near the pits from which it is excavated. The dragline is designed to excavate below the level of the machine. A dragline usually does not have to go into the excavation or pit to remove material. It excavates material while positioned adjacent to the pit. By the process of casting its bucket into and dragging the bucket out again, it extracts material from the excavation. This is very useful when earth is removed from a ditch, canal, or pit containing water.

A dragline can load haul units positioned outside of the pit and positioned at the same level. This advantage is important when the material being excavated is wet, as the haul units do not have to go into the excavation and maneuver through the mire. When ground conditions permit, it is better, however, to position the haul units in the pit. Positioning the haul units in the excavation below the dragline will reduce hoist time and increase production.

When the excavated material can be deposited along the canal or near the pit, it is possible to use a dragline with a long boom to accomplish excavation and disposal in one operation. This eliminates the need for hauling units, reducing the cost of handling the material.

Crawler-mounted draglines can operate on soft ground conditions where wheel- or truck-mounted equipment would not be able to move. The travel speed of a crawler machine is very slow, frequently less than 1 mph, and it is necessary to use auxiliary hauling equipment to transport the unit from one job to another. Wheel- and truck-mounted lattice-boom cranes can also be rigged with dragline attachments, but this is not a common practice.

Components of a Dragline

The major components of a dragline (Fig. 8.20) are the drag bucket and a fairlead assembly. Wire ropes are used for the boom suspension, drag, bucket hoist, and dump lines. The fairlead guides the drag cable onto the drum when the bucket is being drawn in—or loaded. The hoist line, which operates over the boom-point sheave, is used to raise and lower the

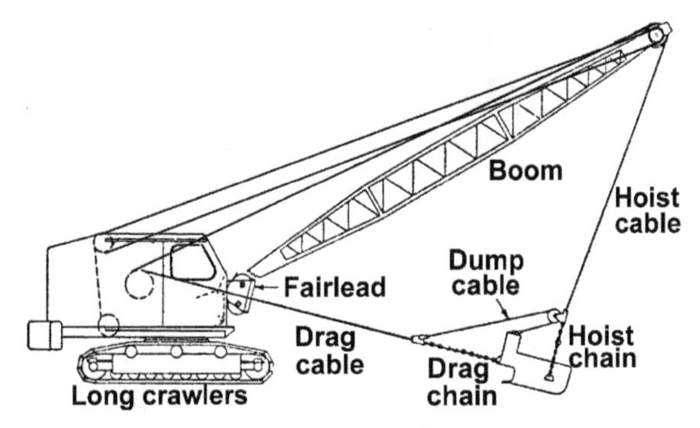

FIGURE 8.20 Basic parts of a dragline.

bucket. In the digging operation the drag cable is used to pull the bucket through the material. When the bucket is raised and moved to the dumping spot, releasing the tension on the drag cable causes the mouth (open end) of the bucket to drop vertically and gravity then draws the material out of the bucket.

Boom. The boom of a dragline or clamshell is of lattice construction. It may be of welded steel angles, angle corners, and tubular braces, or all tubular, but angular section booms are more common for dragline work because of side loading. Each boom is made up of at least two sections, tapering from their bolted center connection toward the end. The bottom section is reinforced to hinge on the boom foot pins and the top to hold the point sheaves. Additional sections, usually in lengths of 5 or 10 ft, can be placed between the upper and lower sections to obtain additional reach or dump height.

If the boom is intended for dragline work only, it carries one large sheave on the point; but if it is to also be used for a clamshell work, it has two sheaves. A smaller sheave is carried on each side for the boom support line.

Fairlead. The fairlead is a device mounted on the boom or boom foot that guides the drag cable to spool smoothly onto the drum, even when the bucket is off to the side of the boom direction. A common type (Fig. 8.21) has the front pulleys mounted in a frame on a vertical hinge and a pair of vertical sheaves in a frame behind them.

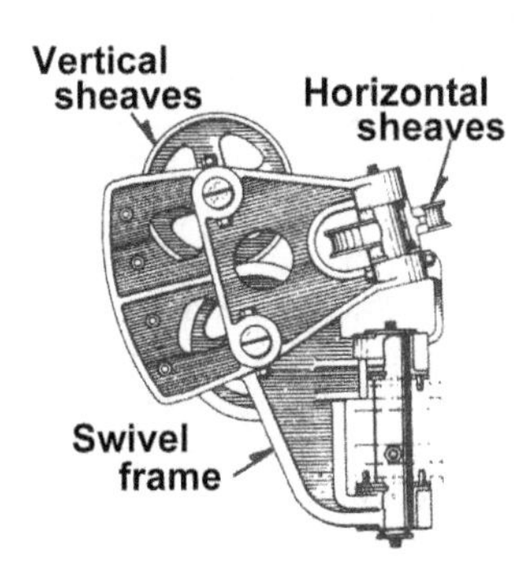

FIGURE 8.21 Four-sheave dragline fairlead.

Reeving. The dump cable runs from the top of the bucket arch over the dump sheave and forward to the drag yoke (Fig 8.22). The boom line is a standard four-part rigging. The hoist line runs from the hoist drum over a large boom point sheave and down to the dump sheave case. The drag cable runs from the drag (digging) drum through the fairlead to the drag yoke.

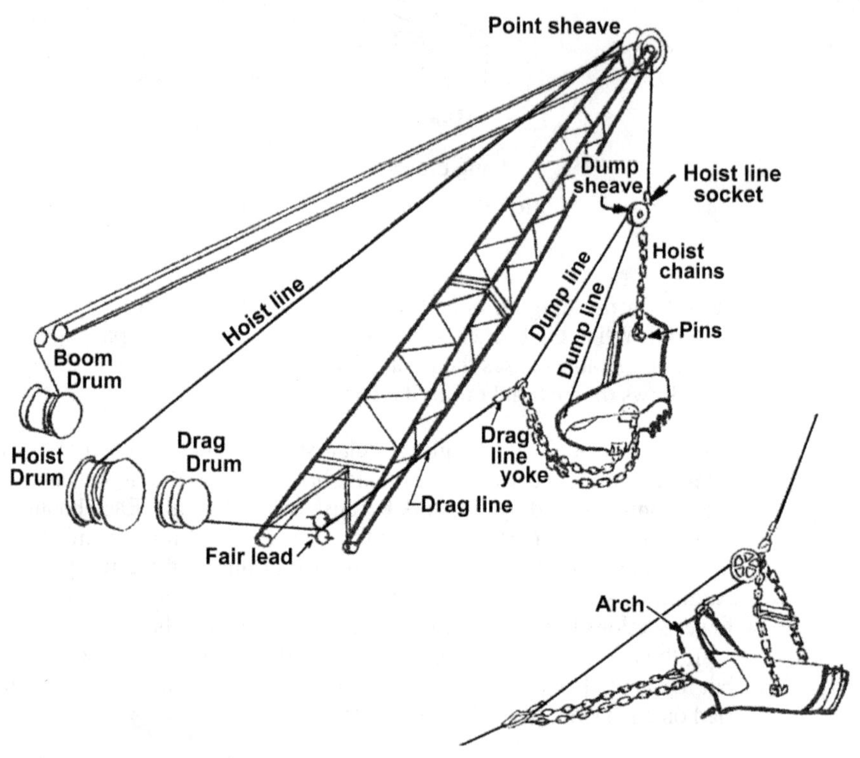

FIGURE 8.22 Dragline reeving.

Bucket Action. If the bucket is lifted with the hoist while the drag cable is slack, it will hang in fully dumped position. If tension is then put on the drag cable, it will pull on the dump cable before the slack is out of the drag chains. The dump cable will pull the front of the bucket up, toward the dump sheave (see Fig. 8.22). Releasing the drag cable will allow the dump cable to run back over the sheave, and the bucket will return to dump position, pivoting on the hoist chain pins.

If the bucket is then lowered to the ground, it either will turn to a horizontal position or will rest on its teeth and arch, depending on its balance. A pull on the drag cable will now tip the bucket forward or backward onto its teeth, and the teeth and lip of the bucket will dig in as it is dragged toward the crane. If the pull is continued with the hoist line slack, the bucket will cut to a depth determined by its weight, the angle and sharpness of its teeth, and the resistance of the soil. If a deep cut is not wanted, some tension is put on the hoist line, raising the bucket slightly. If the dump cable is long, the bucket will be raised in the rear by the hoist chains. A short dump line will cause an upward pull on the arch, raising the front of the bucket as much as or more than the rear. In either case the depth of cut is reduced. A further pull on the hoist will raise the bucket clear of the ground.

Whether the bucket will remain in the carrying position or will partially dump while being raised depends on opposing forces acting through the dump cable. The weight of the bucket front pulls down on the arch end of the cable, and the tension between the hoist and drag cables tends to stretch its drag-yoke-to-dump-sheave section and pull the bucket up. In effect, the dump cable must pinch the other two cables together in order to obtain slack to drop the front of the bucket. This pinching requires comparatively little force when the angle between the cables is small and becomes increasingly difficult as the angle flattens. Also, if the dump cable is long, it will not have to pull as strongly on the two cables to obtain slack as if it were short. The action of the bucket will therefore depend on the angle between the hoist and drag cables, the length of the dump cable, and the weight and distribution of the bucket load.

A wide angle between hoist and drag cables can be had when lifting the bucket out of the soil, either by pulling the bucket close to the crane or by keeping the boom at a low angle. Bringing the bucket all the way in usually wastes time and causes wear on the bucket and chains. This is avoided if the bucket is raised as soon as it is full. A low boom has a tendency to tip the machine when heavily loaded and often cannot be used because of obstructions or height of dump. A short dump cable makes it difficult to dump except directly under the boom point. If a live boom is used, it is possible to dig low and dump high, but this takes extra time and operator effort. The technique used will depend on the job and on the operator's preference.

Bucket. Figure 8.23 is a picture of a general-purpose bucket. A pair of drag chains is attached to the front of the bucket through brackets allowing the pull point to be moved up or down. The upper position is used for deep or hard digging as it pulls the teeth into a steeper angle. The drag chains converge in a drag yoke to which the drag cable is fastened.

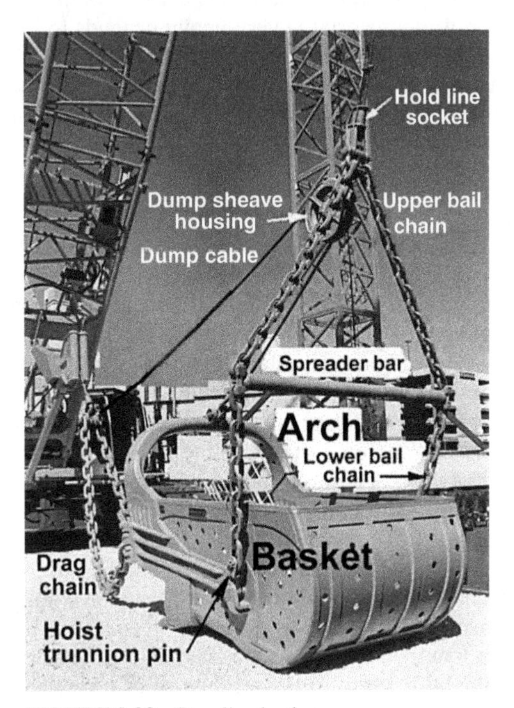

FIGURE 8.23 Dragline bucket.

The hoist (bail) chains are attached to pivot (trunnion) pins toward the rear of the bucket sides, rise vertically to a spreader bar, then converge to fasten to the dump sheave housing, which in turn is fastened to the end of the hoist cable.

Dragline buckets are made in various weights: light ones for digging soft earth and the rehandling of stockpile material, medium weight for general work, and heavy and extra heavy for deep and rocky digging. A light bucket means less weight needs to be lifted during the cycle. A heavy one has better penetration and wear resistance. Light buckets may sometimes be obtained with a toothless cutting edge, which is excellent for stripping soft topsoil, grading, and cleanup work.

Perforated or sieve buckets are standard. These have a number of holes cut in the back and sides. These are useful in wet digging, as water is pushed through the holes by incoming material and any remaining water drains out while the bucket is being lifted. The perforations are necessary to avoid profitless carrying of water and creating sloppy spoil piles. Very fluid mud or fine dry soil may be lost through the holes, but most digging, wet or dry, can be handled. Sometimes ⅜- or ½-in. chain is welded in the rear corners of solid or perforated buckets, as the slapping of the loose ends helps to dump sticky soil and to clean out thin layers remaining on the bucket sides and bottom after dumping.

The effectiveness of penetration of a dragline bucket decreases with depth below the machine, as the drag cable then pulls in a more upward direction, raising the teeth out of the soil. This can be compensated for by reversing or sharpening the teeth; by using a longer boom which, by permitting digging farther from the crane, decreases the upward angle of the drag cable; or by fastening the drag cable higher on the bucket. Larger and heavier buckets dig much better at the same depth and distance.

Choice of bucket size is determined by the materials to be handled and the length of the boom. A ¾-cy machine usually has a 40-ft boom and uses a ¾-cy general-purpose bucket. However, if the material is very heavy or tends to load in amounts greater than the bucket capacity, or if a longer boom is used without extra counterweight, or if digging is so hard or abrasive to require a heavy-duty bucket, then ⅝-cy capacity should be more satisfactory. The same machine might use a ⅞- or 1-yard light bucket on a standard boom in handling coal or dry humus.

Size of a Dragline

The size of a dragline is indicated by the size of the bucket, expressed in cubic yards. However, most machines may handle more than one size bucket, depending on the length of the boom utilized and the unit weight of the material excavated. The relationship between bucket size and boom length and angle is presented in Fig. 8.24. The crane boom can be angled relatively low when operating; however, boom angles of less than 35 degrees from the horizontal are seldom advisable because of the possibility of tipping the machine. The minimum tipping force of a dragline limits the maximum lifting capacity of the machine. Because of the machine's tipping limit, it is often necessary to reduce the size of the bucket when a long boom is used or when the excavated material has a high unit weight. When excavating wet, sticky material, and casting onto a spoil bank, the chance of tipping the machine increases because of material sticking in the bucket. The combined weight of the bucket and its load should produce a tilting force not greater than 75% of the force required to tilt the machine. A longer boom, with a smaller bucket, can be used to increase the digging reach or the dumping radius when it is not desirable to bring in a larger machine. If the material is difficult to excavate, the use of a smaller bucket will reduce the digging resistance and enable an increase in production.

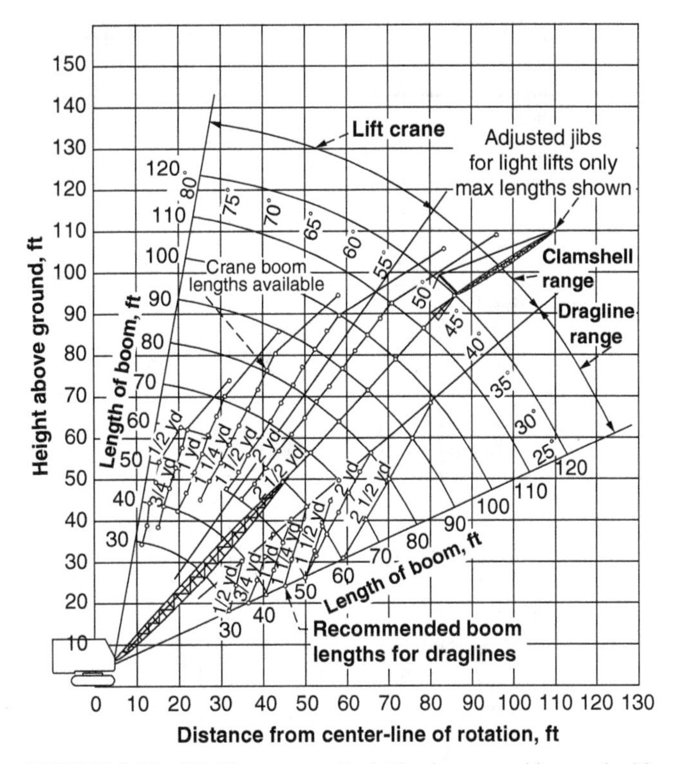

FIGURE 8.24 Working ranges of a lattice-boom machine used with different attachments.

Typical working ranges for a machine capable of handling buckets varying in size from 1¼ to 2½ cy are given in Table 8.6; see Fig. 8.25 for the dimensions given in the table.

Excavation Cycles

In the usual digging cycle, the bucket is not thrown or cast (Fig. 8.26). It is lowered into the pit with both lines taut, with the hoist brake being almost wholly released, then reapplied smoothly as the bucket is about to strike the ground, and the drag brake is released enough to allow the bucket to drop straight instead of following an arc centering on the fairlead.

When the bucket rests on the ground, the hoist cable is slackened slightly and the drag clutch engaged. The drag cable pulls the bucket, with the teeth digging in and the slice of cut piles into the bucket. If the hoist brake is locked, the bucket will move up in an arc centering on the boom point. On level ground it may pivot so the teeth dig more sharply but they no longer have the full weight of the bucket to force them into the material. When the bucket rests on the ground, the hoist cable is slackened slightly and the drag clutch engaged. The drag cable pulls the bucket with the teeth digging in and cutting a slice of material. If the hoist brake is locked, the bucket will move up in an arc centering on the boom point, and on level ground may pivot so the teeth dig more sharply but no longer have the full weight of the bucket to force them into the material.

TABLE 8.6 Typical Working Ranges for Cranes with Maximum Counterweights and Rigged for Dragline Work

J, boom length 50 ft						
Capacity (lb)*	12,000	12,000	12,000	12,000	12,000	12,000
K, boom angle (degrees)	20	25	30	35	40	45
A, dumping radius (ft)	55	50	50	45	45	40
B, dumping height (ft)	10	14	18	22	24	27
C, max. digging depth (ft)	40	36	32	28	24	20
J, boom length 60 ft						
Capacity (lb)*	10,500	11,000	11,800	12,000	12,000	12,000
K, boom angle (degrees)	20	25	30	35	40	45
A, dumping radius (ft)	65	60	55	55	52	50
B, dumping height (ft)	13	18	22	26	31	35
C, max. digging depth (ft)	40	36	32	28	24	20
J, boom length 70 ft						
Capacity (lb)*	8,000	8,500	9,200	10,000	11,000	11,800
K, boom angle (degrees)	20	25	30	35	40	45
A, dumping radius (ft)	75	73	70	65	60	55
B, dumping height (ft)	18	23	28	32	37	42
C, max. digging depth (ft)	40	36	32	28	24	20
J, boom length 80 ft						
Capacity (lb)*	6,000	6,700	7,200	7,900	8,600	9,800
K, boom angle (degrees)	20	25	30	35	40	45
A, dumping radius (ft)	86	81	79	75	70	65
B, dumping height (ft)	22	27	33	39	42	47
C, max. digging depth (ft)	40	36	32	28	24	20
D, digging reach	Depends on working conditons and operator's skill with bucket					

*Combined weight of bucket and material must not exceed capacity.

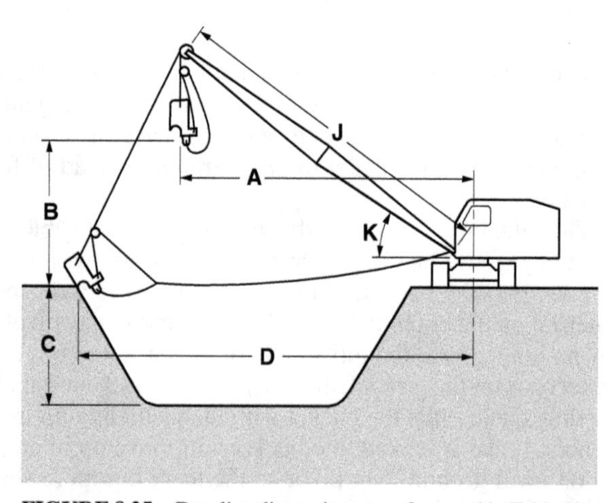

FIGURE 8.25 Dragline dimensions as referenced in Table 8.6.

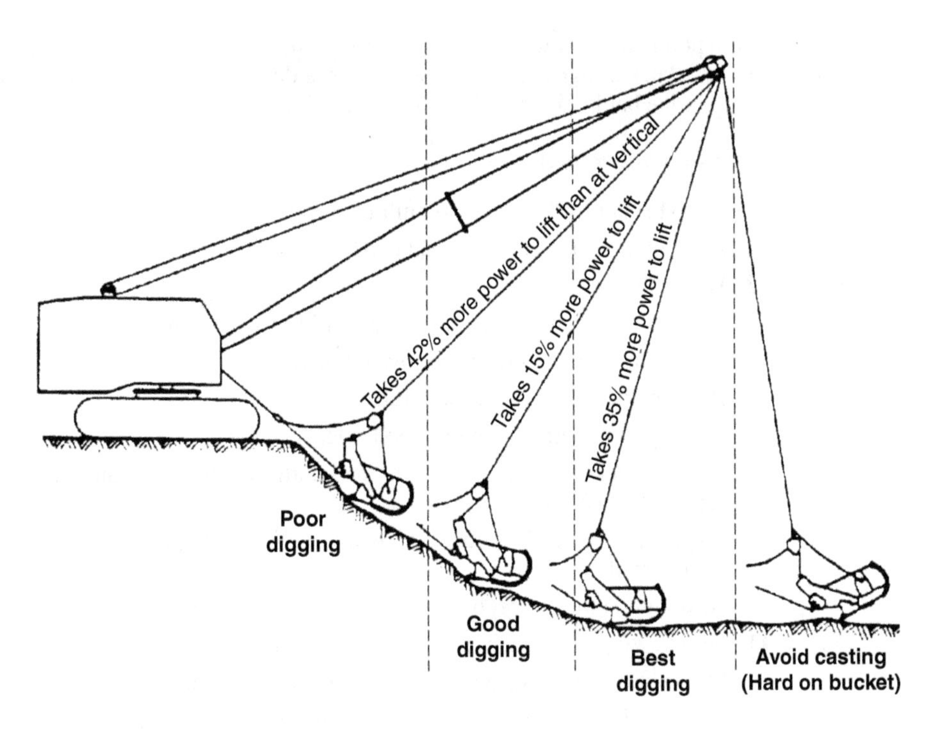

FIGURE 8.26 Dragline digging zones.

Ordinarily, the hoist brake is released enough to let the bucket cut level or follow the pit contour. If the pit slopes up toward the machine, the most favorable digging condition, it may be necessary to partially engage the hoist clutch to avoid digging in or to prevent the hoist line from becoming too slack and allowing the chains and dump sheave to slump into the bucket.

Hoisting. When the bucket is full, the hoist clutch is fully engaged and the bucket is lifted clear of the ground. If the bucket has a tendency to dump, the drag may be left engaged until the angle between drag and hoist cables is sufficiently wide to hold it. The drag clutch is then released and the hoist is continued, with the drag brake applied just enough to allow the hoist to pull the bucket forward and upward under the boom, without slackening the lines enough to dump it. If the drag brake is applied too tightly, the bucket may hit the boom.

The swing is started as soon as the bucket is clear of the ground. This is the period of heaviest load in a dragline cycle, as the machine is simultaneously lifting a load, pulling against the drag brake with the hoist clutch, and swinging. When overloads are picked up in the bucket, they may so slow the line and swing speeds as to reduce production. Production may be improved if smaller bucket loads are taken.

Dumping. Hoisting is discontinued as soon as the bucket is high enough to be dumped. When the swing is completed, the drag brake is released, partially or wholly, and the bucket swings out and dumps. A long dump cable will permit normal dumping inside the boom point; a short one will dump the bucket under the point. Raising the boom will bring the dump point closer; lowering it puts it at a greater distance. It is poor operational practice to raise the bucket higher than necessary before dumping. Clearance should be allowed so as not to strike haul units but piles can be barely cleared.

Operation Suggestions. It is best to

1. Keep bucket teeth sharp and built up to proper size.
2. Keep dump cable short so load can be picked up well away from the dragline.
3. Dig in layers.
4. Keep the digging surface sloped upward toward the machine.
5. Keep drag cable out of the excavated material.
6. Pick up bucket as soon as it is full.
7. Never pull the drag yoke into the fairlead.
8. Inspect bucket chains frequently, especially at the ends, and have them built up or replaced if worn thin.
9. Inspect the fairlead frequently—worn bushings or spacers may let sheaves wear and cut the cable.
10. Never guide the bucket by swinging the boom while digging—this can twist the boom.
11. Never swing until the bucket or load is clear of the ground—this may twist the boom.
12. Swing slowly with heavy loads and use a high boom.
13. Drop the bucket if the machine starts to tip dangerously. Do not try to dump the load as its outward swing increases the tip.
14. Never slap the bucket against the boom while hoisting.
15. Work with the machine level.
16. Never travel uphill with a high boom.

Dragline Production

The output of a dragline should be expressed in bank measure cubic yards (bcy) per hour. This quantity is best obtained from field measurements. It can be estimated by multiplying the average loose volume per bucket load by the number of cycles (bucket loads) per hour and dividing by 1 plus the swell factor for the material expressed as a fraction. For example, if a 2-cy bucket, excavating material whose swell is 25%, will handle an average loose volume of 2.4 cy, the bank-measure volume will be 1.92 cy, $\frac{2.4}{1.25}$. If the dragline can make 2 cycles/min, the output will be 3.84 bcy/min (2×1.92), or 230 bcy/hr. This is an ideal 60-min hour peak output and will not be sustainable over the duration of a project.

Dragline production will vary with these factors:

- Type of material being excavated
- Depth of cut (depth below the base of the machine's tracks)
- Angle of swing (the angle created by a set of lines running for the center point of the dragline to the point of excavation and to the point of dump)
- Size and type of bucket

- Length of boom
- Method of disposal, casting or loading haul units
- Size of the hauling units, when used
- Skill of the operator
- Physical condition of the machine
- Job conditions

Table 8.7 gives approximate dragline digging and loading cycles for various angles of swing.

TABLE 8.7 Approximate Dragline Digging and Loading Cycles for Various Angles of Swing*

Size of Dragline Bucket	Easy Digging Light Moist Clay or Loam Angle of Swing (Degrees)				Sand or Gravel Angle of Swing (Degrees)				Good Common Earth Angle of Swing (Degrees)			
(cy)	45	90	135	180	45	90	135	180	45	90	135	180
⅜	16	19	22	25	17	20	24	27	20	24	28	31
½	16	19	22	25	17	20	24	27	20	24	28	31
¾	17	20	24	27	18	22	26	29	21	26	30	33
1	19	22	26	29	20	24	28	31	23	28	33	36
1¼	19	23	27	30	20	25	29	32	23	28	33	36
1½	21	25	29	32	22	27	31	34	25	30	35	38
1¾	22	26	30	33	23	28	32	35	26	31	36	39
2	23	27	31	35	24	29	33	37	27	32	37	41
2½	25	29	34	38	26	31	36	40	29	34	40	44

*Time in seconds with no delays when digging at optimum depths of cut and loading trucks at the same level as the excavator.

Source: Power Crane and Shovel Association.

Optimum Depth of Cut. A dragline production is best when the job is planned to permit excavation at the optimum depth of the cut. Based on using short-boom draglines, Table 8.8 provides the optimum depth of cut for various sizes of buckets and classes of materials. Ideal outputs of short-boom draglines, expressed in bcy, for various classes of materials, when digging at the optimum depth, with a 90-degree swing, and no delays are presented in the table. The upper two numbers are the optimum depth of cut in feet and in meters (the value in parentheses); the lower numbers are the ideal output in cubic yards and in cubic meters (the value in parentheses).

Effect of Depth of Cut and Swing Angle on Production. Table 8.8 presents ideal dragline production capability based on digging at optimum depths with a swing angle of 90 degrees. For any other depth or swing angle, the ideal output of the machine must be adjusted by an appropriate depth-swing factor. The effect of the depth of cut and swing angle on dragline production is given in Table 8.9. In Table 8.9, the percentage of optimum depth of cut is obtained by dividing the actual depth of cut by the optimum depth for the given material and bucket. The result of the division must be multiplied by 100 to express the value as a percentage:

$$\text{Percentage of optimum depth of cut} = \frac{\text{Actual depth of cut}}{\text{Optimum depth of cut (Table 8.8)}} \times 100 \qquad (8.3)$$

TABLE 8.8 Optimum Depth of Cut and Ideal Production of Short-Boom Draglines*

Class of Material	Size of Bucket [cy (cu m)]†								
	⅜ (0.29)†	½ (0.38)†	¾ (0.57)†	1 (0.76)†	1¼ (0.95)†	1½ (1.14)†	1¾ (1.33)†	2 (1.53)†	2½ (1.91)†
Moist loam or light sandy clay	5.0	5.5	6.0	6.6	7.0	7.4	7.7	8.0	8.5
	(1.5)‡	(1.7)‡	(1.8)‡	(2.0)‡	(2.1)‡	(2.2)‡	(2.4)‡	(2.5)‡	(2.6)‡
	70	95	130	160	195	220	245	265	305
	(53)§	(72)§	(99)§	(122)§	(149)§	(168)§	(187)§	(202)§	(233)§
Sand and gravel	5.0	5.5	6.0	6.6	7.0	7.4	7.7	8.0	8.5
	(1.5)	(1.7)	(1.8)	(2.0)	(2.1)	(2.2)	(2.4)	(2.5)	(2.6)
	65	90	125	155	185	210	235	255	295
	(49)	(69)	(95)	(118)	(141)	(160)	(180)	(195)	(225)
Good common earth	6.0	6.7	7.4	8.0	8.5	9.0	9.5	9.9	10.5
	(1.8)	(2.0)	(2.4)	(2.5)	(2.6)	(2.7)	(2.8)	(3.0)	(3.2)
	55	75	105	135	165	190	210	230	265
	{42)	(57)	(81)	(104)	(127)	(147)	(162)	(177)	(204)
Hard, tough clay	7.3	80	8.7	9.3	10.0	10.7	11.3	11.8	12.3
	(2.2)	(2.5)	(2.7)	(2.8)	(3.1)	(3.3)	(3.5)	(3.6)	(3.8)
	35	55	90	110	135	160	180	195	230
	(27)	(42)	(69)	(85)	(104)	(123)	(139)	(150)	(177)
Wet, sticky clay	7.3	8.0	8.7	9.3	10.0	10.7	11.3	11.8	12.3
	(2.2)	(2.5)	(2.7)	(2.8)	(3.1)	(3.3)	(3.5)	(3.6)	(3.8)
	20	30	55	75	95	110	130	145	175
	(15)	(23)	(42)	(58)	(73)	(85)	(100)	(112)	(135)

*In cubic yards (cubic meters) bank measure (bcy) per 60-min hour.
†These values are the sizes of the buckets in cubic meters (cu m).
‡These values are the optimum depths of cut in meters (m).
§These values are the optimum ideal outputs in cubic meters (cu m).

TABLE 8.9 Effect of the Depth of Cut and Swing Angle on Dragline Production

Percentage of Optimum Depth	Angle of Swing (Degrees)							
	30	45	60	75	90	120	150	180
20	1.06	0.99	0.94	0.90	0.87	0.81	0.75	0.70
40	1.17	1.08	1.02	0.97	0.93	0.85	0.78	0.72
60	1.24	1.13	1.06	1.01	0.97	0.88	0.80	0.74
80	1.29	1.17	1.09	1.04	0.99	0.90	0.82	0.76
100	1.32	1.19	1.11	1.05	1.00	0.91	0.83	0.77
120	1.29	1.17	1.09	1.03	0.98	0.90	0.82	0.76
140	1.25	1.14	1.06	1.00	0.96	0.88	0.81	0.75
160	1.20	1.10	1.02	0.97	0.93	0.85	0.79	0.73
180	1.15	1.05	0.98	0.94	0.90	0.82	0.76	0.71
200	1.10	1.00	0.94	0.90	0.87	0.79	0.73	0.69

Calculating Dragline Production. Hourly production rates for lattice-boom cranes with dragline buckets are given in Table 8.8. These rates are based on operations at optimum cutting depth, a 90-degree swing angle, and for specific soil types. The table data are for maximum efficiency, that is, a 60-min hour. Table 8.9 gives correction factors for different depths of excavation and swing angles. Refer to Chap. 4 for soil conversion factors. Overall efficiency should be based on the expected job conditions.

Step 1. Determine an ideal production from Table 8.8, based on proposed bucket size and the type of material.

Step 2. Determine the percent of optimum depth, based on appropriate Table 8.8 data input into Eq. (8.3).

Step 3. Determine the depth of cut/swing angle correction factor from Table 8.9, using the calculated percent of optimum depth of cut and the planned angle of swing. In some cases, it may be necessary to interpolate between the values in Table 8.9.

Step 4. Decide upon an overall efficiency factor based on the expected job conditions. Draglines are seldom productively working for more than 45 min in an hour.

Step 5. Calculate a production rate. Multiply the ideal production by the depth/swing correction factor and the efficiency factor.

Step 6. Decide upon a soil conversion, if needed (see Chap. 4).

Step 7. Calculate the total hours to complete the work:

$$\text{Total hours} = \frac{\text{Cubic yards moved}}{\text{Production rate/hr}} \tag{8.4}$$

Sample Calculation: A 2-cy short-boom dragline is to be used to excavate hard, tough clay. The depth of cut will be 15.4 ft, and the swing angle will be 120 degrees. Determine the probable production for the dragline. There are 35,000 bcy of material to be excavated. How long will the project require?

Step 1. Determine the ideal production from Table 8.8, based on a 2-cy bucket size and hard, tough clay material: 195 bcy.

Step 2. Determine the percent of optimum depth of cut using Eq. (8.3). Optimum depth of cut (see Table 8.8): 11.8 ft.

$$\text{Percentage of optimum depth of cut} = \frac{15.4\text{ ft}}{11.8\text{ ft}} \times 100 = 130\%$$

Step 3. Determine the depth of cut/swing angle correction factor from Table 8.9:

Percentage of optimum depth of cut = 130%

Swing angle = 120 degrees

Depth of cut/swing angle correction factor = 0.89

Step 4. Decide upon an overall efficiency factor based on the expected job conditions. Draglines seldom work at better than a 45-min hour:

$$\text{Efficiency factor} = \frac{45\text{ min}}{60\text{ min}} = 0.75$$

Step 5. Calculate the production rate. Multiply the ideal production by the depth/swing correction factor and the efficiency factor:

$$\text{Production} = 195 \times 0.89 \times 0.75 = 130 \text{ bcy/hr}$$

Step 6. Decide upon soil conversion, if needed (see Chap. 4). It is not necessary in this example.

Step 7. Calculate the total hours using Eq. (8.4):

$$\text{Total hours} = \frac{35{,}000 \text{ bcy}}{130 \text{ bcy/hr}} = 269 \text{ hours}$$

Factors Affecting Dragline Production. To obtain the greatest operating efficiency, it is necessary to match the size of the lattice-boom crane and bucket properly. Table 8.10 gives representative capacities, weight, and dimensions for different types of dragline buckets. In selecting the most suitable size bucket for use with a given dragline, it is necessary to know the loose weight of the material being excavated. This weight should be expressed in pounds per cubic foot (lb/cf). In the interest of increasing production, it is desirable to use the largest bucket possible. The combined weight of the load and the bucket cannot exceed the capacity of the crane. This maximum will set the size of bucket for a particular job. The importance of this analysis is illustrated by referring to the information given in Table 8.10.

TABLE 8.10 Representative Capacities, Weights, and Dimensions of Dragline Buckets

	Struck	Weight of Bucket (lb)			Dimensions (in.)		
Size (cy)	Capacity (cf)	Light Duty	Medium Duty	Heavy Duty	Length	Width	Height
3/8	11	760	880		35	28	20
½	17	1,275	1,460	2,100	40	36	23
¾	24	1,640	1,850	2,875	45	41	25
1	32	2,220	2,945	3,700	48	45	27
1¼	39	2,410	3,300	4,260	49	45	31
1½	47	3,010	3,750	4,525	53	48	32
1¾	53	3,375	4,030	4,800	54	48	36
2	60	3,925	4,825	5,400	54	51	38
2¼	67	4,100	5,350	6,250	56	53	39
2½	74	4,310	5,675	6,540	61	53	40
2¾	82	4,950	6,225	7,390	63	55	41
3	90	5,560	6,660	7,920	65	55	43

Sample Calculation: The material to be handled has a loose weight of 90 lb/cf. The use of a 2-cy medium-duty bucket will be considered. A crane, rigged as a dragline, with 80 ft of boom at a 40-degree angle has a maximum safe load of 8,600 lb (Table 8.6).

The approximate weight of the bucket and its load will be

Bucket weight, from Table 8.10	= 4,825 lb
Earth, 60 cf at 90 lb/cf	= 5,400 lb
Combined weight	= 10,225 lb
Maximum safe load	= 8,600 lb

As this weight will exceed the safe load of the dragline, it will be necessary to use a smaller bucket. Try a 1½-cy bucket. The combined weight of the bucket and load will be

Bucket weight, from Table 8.10	= 3,750 lb
Earth, 47 cf at 90 lb/cf	= 4,230 lb
Combined weight	= 7,980 lb
Maximum safe load	= 8,600 lb

If a 1½-cy bucket is used, it may be filled to heaping capacity without exceeding the safe load limit of the crane.

If a 70-ft boom at a 40-degree angle, whose maximum safe load is 11,000 lb, will provide sufficient working range for excavating and disposing of the earth, a 2-cy bucket can be used and filled to heaping capacity. The ratio of the output resulting from the use of a 70-ft boom and a 2-cy bucket, compared with a -1½ cy bucket, should be approximately

Output ratio (60 cf/47 cf) × 100 = 127%

Increase in production = 27%

This does not consider the cycle time effect of the different boom lengths.

CLAMSHELL EXCAVATORS

The clamshell is a vertically operated bucket capable of working at, above, and below ground level. It is hung from a crane boom similar to a dragline bucket but for the clamshell (Fig. 8.27). There are two boom line sheaves and two operating sheaves at the point. A clamshell bucket is hung from the operating sheaves by two cables. As the name implies, it consists of two scoops hinged together. These work like the shell of a clam. Clamshells are used primarily to handle materials such as loose to medium stiff soils, sand, gravel, crushed stone, coal, and shells, and to remove materials from vertical excavations such as cofferdams, pier foundations, sewer manholes, and sheet-lined trenches.

There are hydraulic clamshell buckets mounted on the stick of hydraulic hoe–type excavators. The hydraulic excavator clamshell has limited vertical reach. Therefore, when a deep vertical excavation must be made, a crane with a cable-attached clamshell bucket is used.

Lattice-Boom Clamshells. The same wire ropes used for the crane (hook) operation can usually be used for clamshell operations. However, two additional lines must be added: (1) a secondary hoist line—the closing line—and (2) the tagline (Fig. 8.28).

Reeving. Figure 8.28 is a graphic diagram of a clamshell reeving system. The hoist (holding) and digging (closing) lines are carried over the center pair of boom point sheaves and

FIGURE 8.27 Crane using a clamshell bucket to unload a barge.

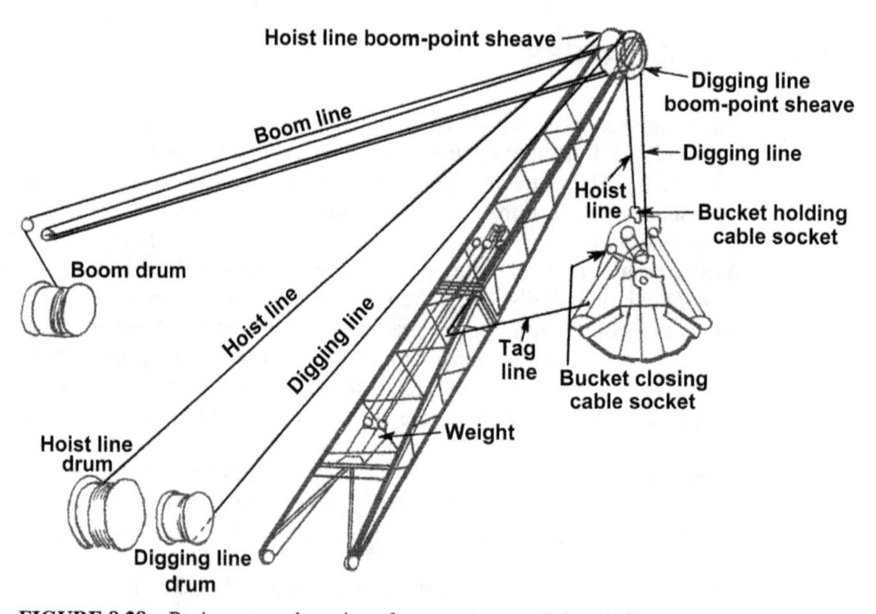

FIGURE 8.28 Basic parts and reeving of a crane-mounted clamshell.

descend directly to the bucket. The hoist line is anchored to the head beam; the closing line goes through a guide in the head beam to the sheave sets. The bucket need not be fully reeved, as the full power of all the lines is not always needed. The cable may be anchored after rounding fewer sheaves for quicker but less powerful closing.

The clamshell bucket is attached to the hoist line of the crane and to a closing line. The closing line is attached to a second cable drum on the crane. The length of the crane boom determines the height a clamshell can reach. The length of wire rope the crane's cable drums can accommodate limits the depth a clamshell can reach. A clamshell's lifting capacity varies with the length of the crane's boom, the operating radius, the size of the clam bucket, and the unit weight of material excavated.

The holding, closing, and taglines control the bucket. At the start of the digging cycle, the bucket rests on the material to be dug, with its shells open. As the closing line is reeved in, the two shells of the bucket are drawn together, causing them to dig into the material. The weight of the bucket is the only crowding action available for penetrating the material. After the bucket is closed, the holding and closing lines raise the bucket and swing it to the dumping point. The bucket is opened to dump by releasing the tension on the closing line.

Tagline. The tagline is a small-diameter cable with a spring tension winder. The tagline prevents the clamshell bucket from twisting while it is being hoisted and lowered. Being spring-loaded, the tagline does not require operator control and does not attach to the crane's operating drums.

If not spring-loaded, a winder can be mounted on the lower part of the boom. It is kept on a light tension by a weight sliding on a track inside the boom, as in Fig. 8.28. The tagline cable is ordinarily fastened to two corners of the bucket by chains. If both are on one jaw, the hinge line of the jaw will be at right angles to the cable. If one corner of each jaw is held, the opening will be in line with the cable. The bucket can also be held in a diagonal position by attaching to only one corner.

The tagline pull is usually light enough so a person can guide the bucket by hand or with a stick when it is necessary to lower it exactly into position. This is done when it is necessary to cut at a location other than where the tagline places the bucket.

Buckets. A number of different types of clamshell buckets (Fig. 8.29) are available. Each consists of two jaws hinged on a movable bar, the main shaft, and hinged on their outer ends to brackets extending to an upper bar, with the head shaft parallel to the main shaft. The head shaft is supported by the head, or head beam, and the hoist cable. If the digging clutch is engaged, the cable will pull the sheaves together, powerfully raising the center hinges of the jaws. Because the corner arms are fixed and will not allow the outer ends of the jaws to rise, the jaws pivot on their outer hinges and rotate inward and upward until they meet. The bucket is now closed and will pick up its load if raised. If the digging line is released while the bucket is held in the air by the hoist or holding line, gravity will cause the main shaft to move down, pushing the jaws downward and outward and dumping the load.

The buckets can be obtained with various types of teeth; with flat or curved lips; with toothless lips; with and without sidecutters; with ballast plates for weight; and in various weights, widths, and shapes for special conditions. Figure 8.30 shows the ways in which bucket capacity is measured. The capacity of a clamshell bucket is usually given in cubic yards. A more accurate capacity is given as water level, plate line, or heaped measure; such volumes are generally expressed in cubic feet. The water-level capacity is the capacity of the bucket if it were hung level and filled with water. The plate-line capacity indicates the capacity of the bucket following a line along the tops of the clams—this is the usual rating. The heaped capacity is the capacity of the bucket when it is filled to the maximum angle of

FIGURE 8.29 A clamshell bucket sitting on a barge.

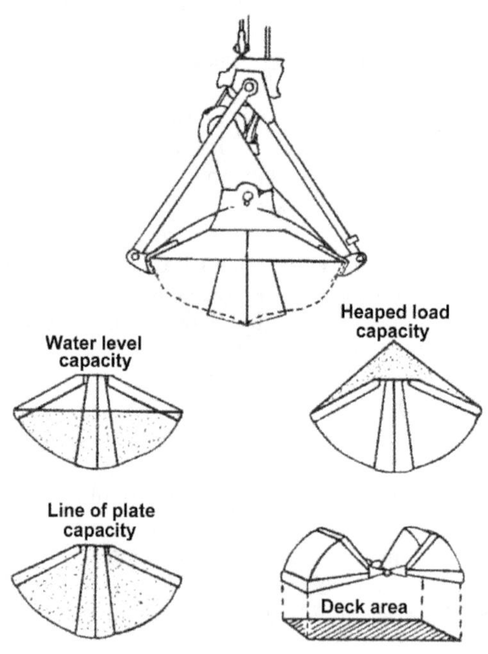

FIGURE 8.30 Clamshell bucket ratings.

repose for the given material. In specifying the heaped capacity, the angle of repose is usually assumed to be 45 degrees. The term "deck area" indicates the number of square feet covered by the bucket when it is fully open.

The more resistant the material to be excavated, the heavier the bucket must be in relation to its capacity. A good, extra heavy-duty clamshell will dig almost anything except solid rock, but its closing action when fully reeved is very slow, and its massive weight reduces the payload. A light bucket, suitable for handling soft or loose material, will have faster action and a minimum of dead weight, but it will not penetrate hard materials and will suffer damage if repeatedly banged and scraped on them. For miscellaneous work, including both hard and soft, a medium-weight, general-purpose bucket is usually selected. Some of these have provisions for adding or subtracting plates to change weight and strength.

> A small bucket must be used in mud, as suction holds it, greatly increasing the force required to lift it from the material.

The closing line should be guarded against sharp edges where it enters the bucket head, and it should be possible to reeve the bucket without using all the sheaves and without throwing the digging line off center.

Digging. In digging, the bucket is placed over the work by swinging the boom and either moving the crane or raising or lowering the boom to obtain the correct distance. The digging (drag, closing, or crowd) brake is released, causing the bucket to open, and the hoist brake is then released, allowing the bucket to contact the ground. If the ground is soft, the hoist brake will be only partially released, will be or reapplied just before the bucket lands. If the material is hard, the bucket will be allowed to fall freely, so its weight serves to drive it into the ground for a good bite. In either case, both brakes must be applied as soon as it has hit to avoid unspooling of the cable.

The digging clutch is then engaged to pull the jaws together. They first push the material inward, then curve and close under it. If the material is not too resistant and the bucket has proper weight, a full or heaping load will be gathered. The digging line will lift the bucket as soon as the jaws are tightly closed. The hoist clutch should then be engaged; it will hold the hoist cable so it does not become slack while the bucket is being raised.

Dumping. The swing is started as soon as the bucket is clear of obstructions. The distance of the bucket from the machine may have to be adjusted by raising or lowering the boom for precise dumping. This is possible only if the crane is rigged with a live boom. When the bucket is properly positioned, the digging brake is released, the jaws open by their weight and the weight of the load, and the material is dumped. The bucket is swung back to the pit in the open position and lowered or dropped for another bite.

Deep Digging. In deep ditches or shafts, it is difficult to keep the walls perpendicular, particularly if the earth is stony, as with each bite the bucket is edged a little away from the wall, thus causing the pit to grow narrower with depth. This tendency may be combated by swinging the bucket against the wall as it drops or by excavating the top of the shaft slightly larger so the minimum width will still be obtained at the bottom. It is helpful to have a bucket with sidecutters or corner teeth for this work.

Applications. Clamshell work is slow—ordinarily it moves fewer yards per hour than any other rig. This arises from the time consumed in closing the bucket and because the operator has minimum control over the position of the bucket, which is always directly under the boom point unless swung elsewhere. The bucket can be moved toward or away from the crane by raising or lowering the boom, moved around by swinging the crane, moved outwardly by centrifugal force, and moved inward as a pendulum. However, skillful operation is required to place it correctly.

A number of different buckets are required for best handle a variety of digging techniques, and there is often loss of efficiency in the use of an unsuitable or compromise type of bucket. A clamshell has the same reach as a dragline, but the bucket cannot be cast as effectively. Because most digging is done directly under the boom point, maximum depth is determined by the amount of cable the drums will carry. Dumping height is controlled by the height of the bucket under the boom point, which is variable.

Operating Suggestions. When using a clamshell

1. Keep bucket teeth sharp and built up to size.
2. Do not use more parts of line in the bucket than you need.
3. Keep the crane on solid ground.
4. Do not travel with a high boom—a bump may tip it back on the cab.
5. Do not swing uphill with a high boom.
6. Keep back from the edge of deep, wide cuts.
7. If the machine tips dangerously, release both holding and closing lines.
8. Do not hit the boom with the bucket.

Clamshell Excavator Production

It is difficult to give average production rates for a clamshell because of the variable factors affecting its operations. The critical variable factors include the difficulty of loading the bucket, the size of load obtainable, the height of lift, the angle of swing, the method of disposing of the load, and the experience of the operator. If the material must be discharged into a hopper, the time required to spot the bucket over the hopper and discharge the load will be greater than when the material is discharged onto a large spoil bank. Representative performance specifications for a clamshell-rigged crane are given in Table 8.11.

Sample Calculation: A 1½-cy rehandling-type bucket whose empty weight is 4,300 lb will be used to transfer sand from a stockpile into a hopper 25 ft above the ground. The crane's angle of swing will average 90 degrees. The average loose capacity of the bucket is 48 cf. The specifications for the crane unit given in Table 8.11 are applicable to this situation.

Time per cycle (approximate):

Loading bucket	= 6 sec
Lifting and swing load $\frac{25 \text{ ft} \times 60 \text{ sec/min}}{166 \text{ ft/min}}$	= 9 sec*
Dump load	= 6 sec
Swing back to stockpile	= 4 sec
Lost time, accelerating, etc.	= 4 sec
Total cycle time	= 29 sec or 0.48 min

*A skilled operator should lift and swing simultaneously. If this is not possible, as when coming out of a cofferdam, additional time should be allowed for swinging the load.

TABLE 8.11 Representative Performance Specifications for a Clamshell-Rigged Crane

Speeds	
Travel	0.9 mph maximum
Swing	3 rpm maximum
Rated single-line speed	
Lift clamshell	166 fpm
Dragline	157 fpm
Magnet	200 fpm
Third drum (standard travel)	185 fpm
Third drum (independent travel)	127 fpm
Rated line pulls (with standard engine)	
Lift clamshell	29,600 lb SLP
Dragline	31,400 lb SLP
Magnet	24,800 lb SLP
Third drum	25,500 lb SLP

Performance figures are based on machine equipped with standard engine.

SLP stands for Single Line Pull or 1-part line.

$$\text{Number of cycles per hour } \frac{60 \text{ min/hr}}{0.48 \text{ min/cycle}} = 125 \text{ cycles/hr}$$

$$\text{Volume per 60 minute hour } \frac{125 \text{ cycle} \times 48 \text{ cf}}{27 \text{ cf/cy}} = 222 \text{ lcy/hr}$$

If the unit operates at a 45-min hour efficiency, the probable production will be

$$222 \text{ lcy} \times \frac{45}{60} = 167 \text{ lcy/hr}$$

If the same equipment is used with a general-purpose bucket to dredge muck and sand from a sheet-piling cofferdam partly filled with water, requiring a total vertical lift of 40 ft, and the muck must be discharged into a barge, the production rate previously determined will not apply. It will be necessary to lift the bucket above the top of the cofferdam prior to starting the swing—this will increase the time cycle. Because of the nature of the material, the load will probably be limited to the water-filled capacity of the bucket, which is 33 cf. The time per cycle should be about

Loading bucket	= 8.0 sec
Lifting $\frac{40 \text{ ft} \times 60 \text{ sec/min}}{166 \text{ ft/min}}$	= 14.5 sec
Swinging 90 degrees at 3 rpm; $\frac{0.25 \text{ rev.} \times 60 \text{ sec/min}}{3 \text{ rev./min}}$	= 5.0 sec
Dump load	= 4.0 sec
Swing back	= 4.0 sec
Lowering bucket $\frac{40 \text{ ft} \times 60 \text{ sec/min}}{350 \text{ ft/min}}$	= 7.0 sec
Lost time, accelerating etc.	= 10.0 sec
Total cycle time	= 52.5 sec or 0.875 min

$$\text{Number of cycles per hour } \frac{60 \text{ min/hr}}{0.9 \text{ min/cycle}} = 67 \text{ cycles/hr}$$

$$\text{Volume per 60 minute hour } \frac{67 \text{ cycle} \times 33 \text{ cf}}{27} = 82 \text{ lcy}$$

If the unit operates at a 45-min hour efficiency, the probable production will be

$$82 \text{ lcy} \times \frac{45}{60} = 62 \text{ lcy/hr}$$

This does not consider the cycle time effect of the different boom lengths.

BACKHOE LOADERS

The success of the backhoe loader (Fig. 8.31) comes from its versatility. They are classified by horsepower. Because it is a wheel-mounted machine, job site travel speeds are as great as 25 mph. Although they are not high-production machines for any one task, their mobility and flexible use make the backhoe loader a popular tool to accomplish a variety of jobs at diverse worksite locations. It is three construction machines combined into one unit—a tractor, a loader, and a hoe. Because of its four-wheel-drive tractor capability, the backhoe loader can work in unstable ground conditions. It is an excellent excavator for digging loosely packed moist clay or sandy clay. It is not suitable for continuous high-impact diggings such as excavating hard clay or caliche.

FIGURE 8.31 Backhoe loader using the rear hoe to excavate.

Efficient operation of the backhoe depends on the type of work. Before operating the backhoe loader as a backhoe, level the machine, lower the front bucket to the ground (Fig. 8.31), and ensure the gearshift and the range shift levers are in the neutral position. Always operate with the least amount of bucket arm swing.

A large variety of booms, sticks, buckets, and attachments give the backhoe loader the versatility to excavate trenches, load trucks, clean ditches, install pipe, and break out old concrete (Fig. 8.32).

FIGURE 8.32 Backhoe loader with paving breaker attachment.

Loader Functions. A loader bucket attached to the front of the tractor enables this machine to excavate above wheel level.

Backhoe Functions. A hoe attached to the rear of the tractor enables it to excavate below wheel level. The hoe bucket can be replaced with a breaker or hammer, turning the unit into a demolition machine.

The cycle time for backhoe excavation work is controlled by the bucket swing angle and the depth of excavation. Table 8.12 provides average backhoe cycle time assuming the material moved is, at worst, moderately stiff earth.

TABLE 8.12 Average Backhoe Loader Cycle Time for Backhoe Excavation Work in Soil

Degree of Swing	Depth of Cut, feet						
	2	4	6	8	10	12	14
45°	12 sec	16 sec	22 sec	25 sec	31 sec	38 sec	46 sec
90°	22 sec	25 sec	30 sec	36 sec	42 sec	49 sec	55 sec

SPECIALTY EXCAVATORS

A variety of excavators are available for specialty applications. There are machines designed strictly for one application, such as trenching. If there is a requirement to excavate larger quantities of material from an extensive borrow area, the Holland loader is designed for such mass excavation tasks. However, if the excavation is in a busy city street, the contractor should consider a vacuum excavator.

Holland Loaders

A Holland loader is an excavating unit mounted between two crawler tractors operating in tandem. The tractors are controlled by a single operator in the lead tractor. In continuous passes through excavation areas, the loader carves material from the ground and belt-loads it into hauling units. There are both vertical- (Fig. 8.33) and horizontal-cut (Fig. 8.34) Holland loaders. Working with large-bottom dump trailers, these loaders have had production rates between 2,000 and 3,300 bcy/hr.

FIGURE 8.33 Vertical, side-cut, Holland loader.

FIGURE 8.34 Horizontal-cut Holland loader.

HYDRAULIC EXCAVATOR ACCIDENTS

A 26-year-old construction worker was killed while working in an 8-ft-deep trench, trying to remove a concrete sewer casing. Because it was impossible for the excavator operator to see the bottom of the trench where the casing was located, the victim was standing inside a trench box, giving hand signals to the operator above him. While pulling off the encasement, the bucket teeth slipped off the edge of the concrete and the excavator arm and bucket swung toward the victim, crushing him against the side of the trench box.

Bureau of Labor Statistics (BLS) data identified 346 deaths associated with excavators or backhoe loaders during the period 1992–2000. A review of the data by the National Institute for Occupational Safety and Health identified two common causes of injury:

- Being struck by the moving machine, swinging booms, or other machine components
- Being struck by quick-disconnect excavator buckets when they unexpectedly detach from the stick

Other leading causes of fatalities are rollovers, electrocutions, and machines sliding into trenches after cave-ins. Although the newer machines have rear-view cameras providing the operator in the cab with a video of what is behind the machine, caution must always be exercised when working around large machines.

Managers have a responsibility for the safety of excavator operators and those working around the machines.

- Operators should keep machine attachments at a safe distance from workers at all times.
- Workers are trained regarding safe practices when working in close proximity to heavy equipment.
- Supervisors should consider alternative working methods so workers are not placed in close proximity to heavy equipment.

CHAPTER 9
LOADERS

The loader is a versatile machine designed to excavate at or above wheel/track level. Unlike a shovel or hoe, a loader must maneuver and travel to position the bucket to load or dump. Loaders are used extensively in construction work to handle and transport bulk material, such as earth and rock, short distances and to load haul units. The hydraulic-activated lifting system exerts maximum breakout force with an upward motion of the bucket. It does not require other equipment to level or clean up the area in which it is working.

Loaders are used for digging, loading, rough grading, and limited short-distance hauling. These machines are classified on the basis of running gear:

1. Track-laying type (Fig. 9.1)
2. Wheel type (Fig. 9.2)

They may be further grouped by the capacities of their buckets or the weight the buckets can lift.

Figure 9.3 is a manufacturer's data sheet for a track loader. This loader has a hydraulic system to control a pair of push or lift arms hinged to the top of the support frame, a tractor-width bucket hinged to the front of the arms, and a pair of dump arms hinged both to the push arms and to the bucket. Track and four-wheel-drive loaders are used for heavy service, and two-wheel-drive tractors for lighter work.

LOADER MECHANICAL SYSTEMS

Track-type loaders with a bucket are usually specifically designed and differ from standard-model crawler tractors (dozers). Their tracks may be of a wider gauge and slightly longer, with an additional track roller on each side. The idler and sometimes the front roller are of extra heavy-duty design. Width is necessary for stability against side tipping when carrying raised loads. The longer tracks move the center of gravity (balance) forward so heavier loads may be excavated and carried. The heavier idler and roller construction is required by the heavy front loads.

Tractors redesigned to carry loaders do not have springs. Most have a rigid connection between the track frames and main frame at the front, thus improving stability at the expense of operator comfort and grading control.

FIGURE 9.1 Track-type loader.

FIGURE 9.2 Wheel-type loader.

If the tractor is mechanical drive, the engine clutch must be rugged, as a loader is very hard on clutches. It is likely to have ceramic discs instead of lining, or operate in a circulating and cooling oil bath. However, most loader tractors have torque converters teamed with power-shift transmissions. This construction avoids the problem of slipping a clutch and improves lugging qualities when excavating in a bank. Power shift improves flexibility and shortens cycles. A hydrostatic drive may replace the converter and transmission.

Table 9.1 provides representative specifications for track loaders. These specifications are across the ranges of commonly available track loaders.

Wheel-type loaders work on the same basic principles as track machines, but there are a number of differences in size and structure. Loader usage is strongly affected by carrier type: track or wheel. The most important units for heavy work are mounted on four-wheel-drive tractors. Smaller models are on skid-steer or conventional two-wheel-drive machines.

Dump rams and arms may be hinged on the columns above or below the lift arms, as in the track machines. They may also be on the same hinge pins or based on the lift arms

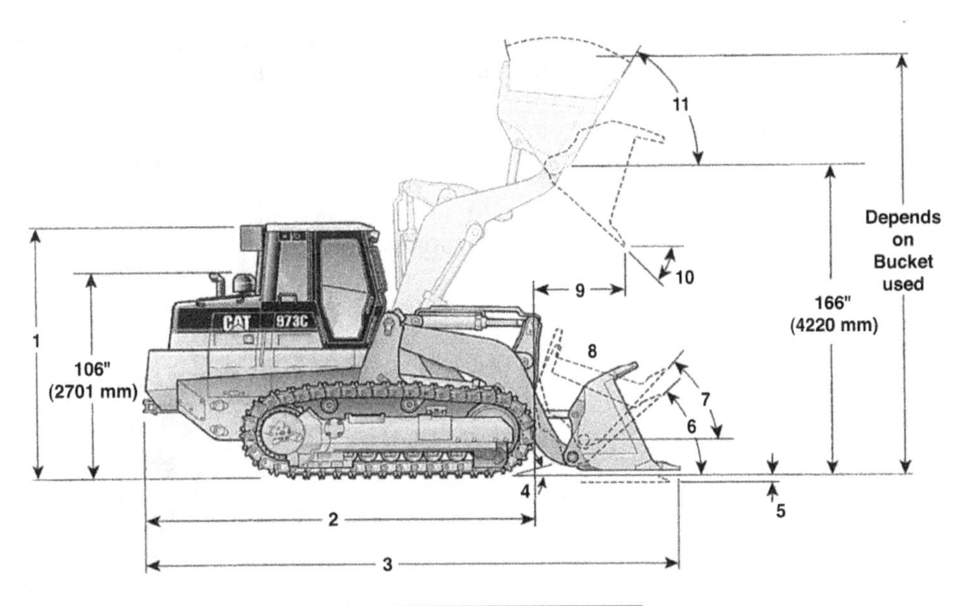

Overall machine width without bucket		
with standard track — 500 mm (19.7" shoes)	2580 mm (102")	
with wide track — 675mm 26.6"shoes)	2755 mm (108")	
Ground clearance from face to shoe	456 mm (17.9")	
Grading angle	69°	
1 Machine height to top of cab	3450 mm (136")	
2 Length to front of track	5163 mm (203")	
3 Overall machine length	depends on bucket used	
4 Carry position approach angle	15°	
5 Digging depth	depends on bucket used	
6 Maximum rollback at ground	GP 43°/MP 46°	GP = general purpose or
7 Maximum rollback at carry position	GP 50°/MP 52°	MP = multipurpose bucket
8 Bucket in carry position	—	
9 Reach at full lift position	depends on bucket used	
10 S.A.E. specified dump angle	45° (46° max.)	
11 Maximum rollback. fully raised	59°	

FIGURE 9.3 Large track loader. (*Reprinted courtesy of Caterpillar, Inc.*)

TABLE 9.1 Representative Specifications for Track Loaders

Size, Heaped Bucket Capacity (cy)	Bucket Dump Clearance (ft)	Static Tipping Load (lb)	Maximum Forward Speed (mph)	Maximum Reverse Speed (mph)	Raise/Dump/ Lower Cycle (sec)
1.00	8.5	10,500	6.5	6.9	11.8
1.30	8.5	12,700	6.5	6.9	11.8
1.50	8.6	17,000	5.9*	5.9*	11.0
2.00	9.5	19,000	6.4*	6.4*	11.9
2.60	10.2	26,000	6.0*	6.0*	9.8
3.75	10.9	36,000	6.4*	6.4*	11.4

*Hydrostatic drive.

themselves. In the two last cases, the bucket remains at a fixed angle with the lift arms as they are raised or lowered, unless changed hydraulically by automatic controls.

Four-Wheel Drive. The medium-size and large loaders on pneumatic tires are articulated, four-wheel-drive tractors (Figs. 9.2 and 9.4). Loader frame columns are forward of the swivel. Lift arms are, therefore, shorter than on the crawlers and rise to a steeper angle for the same bucket height. Most of these machines are equipped with torque converters and easy-shift or power-shift transmissions.

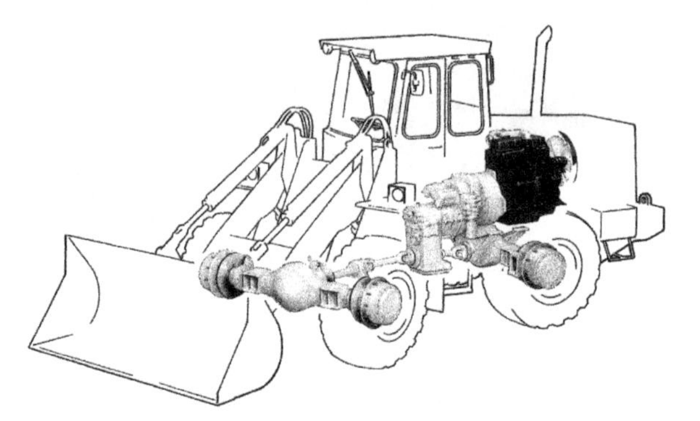

FIGURE 9.4 Power train for an articulated loader.

The operator sits high and forward, providing a good view of the bucket. The engine is in the rear. Even though the articulation turns to about 40 degrees, the turning radius is much longer than for crawlers, so more space is required to maneuver from bank to truck and back. Faster travel and usually faster shifting keep cycle time about the same as with a conventional crawler. In common with track-mounted loaders, wheel loaders can handle large boulders by carrying or pushing and can keep the pit area clean and leveled.

Wheel loaders have the advantage of quick and easy movement from one part of a job to another. They can also be driven from one job to another, but the larger loaders are so wide as to require special permits, and they are slow (30 mph or less). Therefore, it is best they be hauled on trailers when the distance is greater than a few miles.

Very large tires are generally used. They serve to provide excellent flotation, permitting work on most footings. Ground pressure is still much higher than with crawlers, but the packing effect of the tires and the more gradual turns make it possible to work easily on sandy ground, where a crawler loader would experience excessive track wear. But slippery surfaces may cause loss of both traction and steering accuracy.

Wheel Loader Operating Specifications. Representative operating specifications for wheel loaders is provided in Tables 9.2 and 9.3. These operating specification data cover the ranges of commonly available wheel loaders.

Loader Frame. The frame is composed of a massive weldment fastened to the track frames and/or the central frame. It carries the pivot or hinge pins for the push and dump arms and their hydraulic rams, and transmits the weight, thrust, and twisting strains of the loader to the tractor. The frame connections or stress points should be inspected periodically and tightened or welded as required.

TABLE 9.2 Representative Specifications for a 119-hp Wheel Loader

Speeds, forward and reverse	
Low	0– 3.9 mph
Intermediate	0–11.1 mph
High	0–29.5 mph
Operating load (SAE)	6,800 lb
Tipping load, straight ahead	17,400 lb
Tipping load, full turn	16,800 lb
Lifting capacity	18,600 lb
Breakout force, maximum	30,000 lb

Arms. The push or lift arms are hinged to the top of the columns or tower on the frame. They extend forward to hinges near the bottom of the bucket. A cross-beam or other linkage braces them near the front. The arms-and-brace assembly may be referred to as a boom.

Lift arms are raised and lowered by means of two-way cylinders in the base of each column. In principle, the dump arms are a connection between the back of the bucket, above the lift arm hinges, and the columns. This connection includes hydraulic cylinders, whose lengthening dumps the bucket and whose shortening rolls it back.

The actual mechanism is more complicated, to supply changes of leverage, stability, and/or a mechanical (parallelogram) linkage to automatically level the bucket floor as it is lifted. This last function may be taken over by hydraulic valve arrangements and controls.

Bucket. Buckets are about the same width as the outside of the tracks. The present size for standard weights ranges from 1.25 to 4.2 heaped cubic yards for crawlers. Wheel loaders have both smaller and larger sizes, from 0.78 to over 40 heaped cubic yards. The larger sizes are for handling lightweight materials.

Common loader buckets are the one-piece conventional type, general-purpose; the hinged-jaw, multipurpose; and the heavy-duty, rock bucket. There are also many special-purpose buckets and other types of attachments.

General Purpose. The general-purpose (one-piece) bucket is made of heavy-duty, all-welded steel. The replaceable cutting edges are bolted onto the bucket proper. These buckets are usually equipped with replaceable bolts on teeth, but they also come with a straight lip (edge) and no teeth.

Multipurpose. The multipurpose segmented (two-piece) hinged-jaw bucket is made of heavy-duty, all-welded steel. Multipurpose buckets are often referred to as four-in-one buckets because they can be used to dig like a normal bucket, blade, clam, and grapple. These buckets have bolted-on replaceable cutting edges. Bolt-on-type replaceable teeth are also common on these buckets.

Rock. The rock bucket is of one piece, heavy-duty construction, with a protruding V-shaped cutting edge. This protruding edge can be used for prying up and loosening shot rock. Rock buckets are heavily reinforced.

TABLE 9.3 Representative Specifications for Wheel Loaders

Size, Heaped Bucket Capacity (cy)	Bucket Dump Clearance (ft)	Static Tipping Load, @ Full Turn (lb)	Maximum Forward Speed				Maximum Reverse Speed				Raise/ Dump Lower Cycle (sec)
			First (mph)	Second (mph)	Third (mph)	Fourth (mph)	First (mph)	Second (mph)	Third (mph)	Fourth (mph)	
1.25	8.4	9,600	4.1	7.7	13.9	21	4.1	7.7	13.9	—	9.8
2.00	8.7	12,700	4.2	8.1	15.4	—	4.2	8.3	15.5	—	10.7
2.25	9.0	13,000	4.1	7.5	13.3	21	4.4	8.1	14.3	23	11.3
3.00	9.3	17,000	5.0	9.0	15.7	26	5.6	10.0	17.4	29	11.6
3.75	9.3	21,000	4.6	8.3	14.4	24	5.0	9.0	15.8	26	11.8
4.00	9.6	25,000	4.3	7.7	13.3	21	4.9	8.6	14.9	24	11.6
4.75	9.7	27,000	4.4	7.8	13.6	23	5.0	8.9	15.4	26	11.5
5.50	10.7	37,000	4.0	7.1	12.4	21	4.6	8.1	14.2	24	12.7
7.00	10.4	50,000	4.0	7.1	12.7	22	4.6	8.2	14.5	25	16.9
14.00	13.6	98,000	4.3	7.6	13.0	—	4.7	8.3	14.2	—	18.5
23.00	19.1	222,000	4.3	7.9	13.8	—	4.8	8.7	15.2	—	20.1

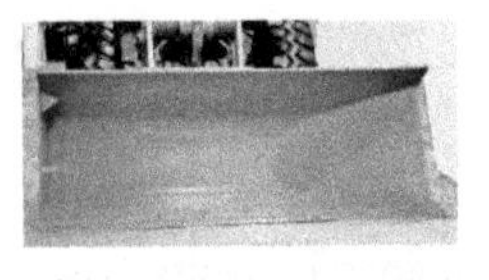

Side-Dump. A side-dump bucket allows versatility for working in confined areas, along roads in traffic, and for filling trucks. Both left- and right-hand dumping buckets are available.

Forklift. The forklift can be attached to the loader in place of a bucket.

Other Types. Many other specialty bucket types and attachments are available, including demolition buckets, plow blades for snow removal, brush rakes for clearing applications, heavy-duty sweeper brooms, and front booms designed for lifting and moving sling loads.

Light material buckets for handling coal, sawdust, or snow may be from 40% to 100% larger than standard buckets.

Teeth are standard equipment on rock buckets, and optional on standard weight. They help greatly in hard digging and in handling rock, stumps, and brush, but they interfere with grading. Their cost is partly offset by the protection they give to the cutting edge.

Design must be a compromise. The bucket should be strong enough to take punishment but light enough to raise a heavy load without overbalancing the tractor and without the bucket weight absorbing too much of the loader's lifting power.

Bucket Action. The standard bucket has three working motions. It is raised and lowered by two-way rams controlling it through the push arms, it is tilted or rolled between carrying and dumping positions by the dump rams and linkage, and it is crowded and retracted by the forward and reverse travel of the tractor.

Dumping height is the elevation above ground level of the lip of the bucket in dumped position. It may be several feet below the height of the lip in carrying position and 1½ to 2½ ft below the bucket hinges. Maximum dump height varies from 7 to 10 ft in the larger and newer machines.

The bucket will usually tilt nearly 100 degrees between dump and full-back positions (Fig. 9.3). At maximum height the dumping slope of the bottom is 45 to 50 degrees. At ground level a bucket may be kept fully dumped for float-grading while moving forward. A bucket is said to be rolled, tilted, or curled when the floor is tilted so as to retain a load.

Specific tractor design features are used to provide the needed breakout force. Obviously, the engine and hydraulic power contribute. Another is the linkage between the push arm and the dump arm of the bucket. This is usually a Z-shaped linkage capable of producing superior breakout force. This linkage is compared to the parallel linkage, which does a good job of keeping the bucket in a level position while it is being raised and improves its reach.

When a bucket is rolled back during penetration into a material, it pivots on its heel (rear of the floor) as a fulcrum; this action usually develops much more breakout force than can be provided by the hoist cylinders.

Rollback is also useful in slicing upward in hard or heavy banks. It makes it easier to pick up heavy, heaped, or sloppy loads and oversize objects. They can then be carried at a safe level, slightly above the ground.

There should be an indicator on a loader arm to show the tilt of the bucket, as this is usually difficult for the operator to observe directly. Its usefulness may be improved by a paint mark, a strip of bright tape, or a weld tack at level-on-the-ground position.

Fill Factors for Loader Buckets. The heaped capacity of a loader bucket is based on SAE Standard J742b—Front End Loader Bucket Rating. This standard specifies a 2:1 angle of repose for the material above the struck load. The rated capacity of loader buckets is expressed in cubic yards for all sizes ¾ cy or larger, and in cubic feet for all sizes under ¾ cy. Rated capacities are stated in intervals of 1 cy for buckets under ¾ cy, ⅛ cy for buckets from 1 to 3 cy, and ¼ cy for buckets over 3 cy.

The fill factor correction for a loader bucket (Table 9.4) adjusts heaped capacity based on the type of material being handled and the type of loader, wheel or track. Mainly because of the relationship between traction and developed breakout force, the bucket fill factors for the two types of loaders are different.

TABLE 9.4 Bucket Fill Factors for Wheel and Track Loaders

Material	Wheel Loader Fill Factor (%)	Track Loader Fill Factor (%)
Loose material		
Mixed moist aggregates	95–100	95–100
Uniform aggregates		
up to $\frac{1}{8}$ in.	95–100	95–110
$\frac{1}{8}-\frac{3}{8}$ in.	90–95	90–110
$\frac{1}{2}-\frac{3}{4}$ in.	85–90	90–110
1 in. and over	85–90	90–110
Blasted rock		
Well blasted	80–95	80–95
Average	75–90	75–90
Poor	60–75	60–75
Other		
Rock dirt mixtures	100–120	100–120
Moist loam	100–110	100–120
Soil	80–100	80–100
Cemented materials	85–95	85–100

Washed materials have lower fill factors (90% to 95%) in a wheel loader bucket.
Source: Reprinted courtesy of Caterpillar, Inc.

Looseness and angle of repose will affect the fill factor. Moisture content affects the angle of repose and, therefore, the amount of material carried in the loader's bucket.

Operating Loads. Once the bucket volumetric load is determined, a check must be made of payload weight. Unlike a shovel or hoe, to position the bucket to dump, a loader must maneuver and travel with the load. A shovel or hoe simply swings about its center pin and does not require travel movement when moving the bucket from loading to dump position. SAE has established operating load weight limits for loaders. The operating load, by weight, of a wheel loader is limited to less than 50% of its rated full-turn static tipping load, considering the combined weight of the bucket and the load, measured from the center of gravity of the extended bucket at its maximum reach, with standard counterweights and nonballasted tires. In the case of track loaders, the operating load is limited to less than 35% of static tipping load. The term "operating capacity" is sometimes used interchangeably for operating load. Most buckets are sized based on 3,000 lb/lcy material.

Hydraulic Systems. Loader hydraulic pumps are designed for flow capacities varying from 12 gal per minute for small loaders up to more than 100 gal per minute for the largest loaders. The relief pressure is from 2,000 to 4,500 lb per square inch. Most loaders, regardless of size, have a load-sensing system for both the hydraulic systems and transmission. For instance, a fully automatic transmission will have preset shift points, so shifting occurs at optimum torque.

The hydraulic system is equipped with filters to remove outside dirt and products of internal wear. It may be a closed system or an open one with a filtered air vent. It is important to keep oil at the proper level. Too little oil will allow the system to suck air and perform jerkily; too much may cause squirting out of the vent or building up of damaging pressure when lowering a loaded bucket.

Counterweight. On a small machine, 500 to 1,500 lb will permit lifting and carrying heavier loads, and will improve traction when carrying a load. Larger machines can carry proportionately greater rear weights.

Semi-grouser
One to three low cleats running across the track shoe.

Full grouser
A single high cleat across the track shoe.

Track Shoes. Most crawler loaders have the **semi-grouser** (three-cleat) shoes, but some have two cleat types. When shoe surfaces are flat, the tracks will spin rather readily on many footings, giving the effect of a slipping clutch. This tendency to spin cushions all parts of the tractor against shock loads, but it often interferes with steering and traction and prevents the full power of the machine from being applied to its work. Semi-grouser and other special, semi-flat shoes give better traction but will still spin rather freely under slippery conditions. **Full grousers** grip well and aid in digging but make the machine very touchy and apt to stall, building up ridges behind the tracks. Full grousers may shorten bucket life materially because of how they subject the machine and bucket to shocks and overloads.

Grousers tear up the ground when the loader turns, causing it to work itself into a hole. This is the chief objection to their use on a loader. Additionally, they can make work slow and sometimes dangerous. Flat shoes do such damage more slowly or not at all if the ground is firm. Material loosened in this manner is easily smoothed off but will dig up again on the next turn. Under such conditions truck positions may have to be changed frequently to keep the loader on good footing.

Crawler Loader Costs. In the case of crawler loaders greater than 105 hp, an owner will spend about half of the original purchase price for repair parts and labor over the life of the machine. Operating conditions can favor undercarriage life. Machines operated primarily on forgiving soil and gravel and only about 10% of their time working in demolition debris and shot rock fare best.

SKID-STEER LOADERS

Skid-steer loaders are small and very compact (Fig. 9.5). Weights range from less than 3 tons to more than 4 tons with bucket capacities from 10 cf up to 1 cy. Dumping clearance is 6 to 10 ft. Drive is usually hydrostatic, but may be by belts or chains.

The lift arms are pivoted to triangular columns behind the rear wheels and extend alongside the tractor for its full length. The cab includes steel-mesh sides to protect against operator entanglement with the arms. The entrance is at the front, across steps provided on the bucket.

The operator is tucked away in a small compartment with the machine wrapped completely around the control seat, making the whole rear of the machine a blind spot.

FIGURE 9.5 Skid-steer loaders.

They can be operated remotely. Joysticks on each side of the seat or on the remote package have multiple functions. For steering, they must be moved diagonally, straight forward/backward. When both levers are moved the same amount in the same direction, the machine will move in a straight line. If one lever is moved more than the other, the machine will turn toward the side of lesser movement. The left-hand stick usually controls bucket movement, but loading can be operated by rocker foot pedals on some machines. When there is a center pedal, it is for auxiliary hydraulically controlled equipment. Skid steer enables these units to turn in approximately their own lengths. Combined with the short base, they are capable of working in areas too restricted for effective standard loader use.

Ergonomic features are improving to help the operator of a skid steer be more comfortable. These include low-effort controls, a comfortable arm rest, a retractable seatbelt, and a suspension seat.

If the levers are moved oppositely, one side will propel forward and the other backward, and the machine will spin around. In general, a machine with a loaded bucket will tend to slide the back wheels; with an empty one it will slide the front wheels. The actual movement is rather complicated, but the result is a U-turn in little more than overall machine length.

Small size and sharp turning enable these machines to work in very restricted areas. The drive to all wheels gives them good pushing power in proportion to weight.

Attachments. One of the important attractions for a skid-steer unit of equipment is the variety of attachments it can handle. For earthwork, the attachments include

- Front-end loader bucket
- Four-in-one bucket
- A variety of other buckets
- Dozer blade
- Hydraulic breaker
- Hole drilling augur
- Trencher
- Backhoe excavator mounted on the back of the machine

Quick coupling devices for changing buckets or other front attachments are available or may be standard with some machines.

LOAD-HAUL-DUMP LOADERS

The load-and-carry method is firmly established in subsurface mining and tunnel work, where it is called load-haul-dump (LHD) transport (Fig. 9.6). These machines are specialized loading vehicles compactly designed both height-wise and width-wise, so they can work underground in tight quarters. Because of articulation and four-wheel drive, such loaders have good maneuverability.

FIGURE 9.6 Load-haul-dump (LHD) units.

Operation is usually shuttle fashion, without turns. The machine self-loads at the digging face, rolls back the bucket, backs past a chute or hoist, and usually makes a small forward turning movement to dump into it. It then straightens out and goes back to the heading.

Maximum speed for LHDs is around 17 mph, but underground haulage roads usually demand slower movement. They have four-wheel drive with articulated construction, and can climb a 40% grade with a load. The dumping height is just slightly greater than machine height: 9 to 14 ft.

FILLING THE BUCKET

Loader digging is started with the base of the bucket flat or tilted to a slight downward angle. This position gives maximum penetration into banks and high spots and cuts a smooth path on which the tracks or wheels can follow. A bucket level indicator on many machines provides the operator with angle of tilt information. The operator may adjust the bucket to the flat position by reading the indicator or by observing the bucket directly.

Excavating. For cutting down into a level surface, tip the bucket downward 10 to 30 degrees. When it has penetrated to a depth of 2 to 6 in., it should be turned up to a flat or almost flat position, while the forward motion of the tractor is continued until the bucket is filled.

This tilt sequence combines good original penetration, sturdiest position of the bucket for most of the pass, and a powerful prying effect during the change in angle. Under some soil conditions, continual minor adjustment of the angle while digging will improve cutting edge penetration.

The flat position is best for pushing a quantity of loose material, but the bucket should be turned down steeply for spreading and grading so the material will flow freely into holes. Care should be taken not to hook into solid obstructions at a steep angle, as the bucket is then in its weakest position, and leverage against the dump mechanism is at its maximum.

Penetration. The loader bucket has much less penetration ability in proportion to size than a shovel or hoe because it is larger in relation to the power and weight of the machine, has a wider digging edge, and may lack teeth. The hoist is slow in proportion to the speed of the tractor, so the bucket tends to penetrate under more material than the machine can break loose and lift.

Excavating Banks. Use a low gear to force the flat bucket into the bank toe at ground level. When resistance slows the tractor, the bucket is rolled back gradually and hoisted, while crowding with the tractor is continued (Fig. 9.7).

FIGURE 9.7 Loader bucket movement when excavating a bank.

Rolling the bucket back as it rises in the bank increases cutting efficiency by aligning the edge with its upward movement and by retracting it for a thinner slice. Crowding by the tractor tends to make the cut thicker.

The proper balance among these forces varies with the machine, the weight of the material in the bank, and the position and momentum of the bucket. It is the operator's task to balance these factors so the bucket is loaded in the minimum time.

If the bank material is hard, a thin slice may maximize production. If it is vertical, no crowding may be needed after initial penetration. If the bank slopes away from the loader, more crowding and less rollback are needed.

In general, a nearly vertical cut face in a bank not more than a foot or two higher than the lift arm hinges is most efficient for rapid loading. This is called the optimum depth. If the bank is higher than this and overhangs, the upper part should be nudged with the full bucket occasionally to bring it down. But if the overhang is substantial, excavate somewhere else if possible.

If the loader has a self-leveling device, it may be necessary to cut it off or to override it to keep proper tilt in the bank. If it does not have this device, remember not to give the bucket a high lift in full curl-back position, as this may spill part of the load over the back—and probably on the cab.

If the bank is only 2 or 3 ft high, it may be dug by keeping the bucket at final grade and running beside the bank, cutting into it as much as possible without slowing the tractor excessively. The cutting will be done by the side and floor of the bucket, which is in the bank. The soil will roll along the bucket and fill it, although the load will be heavier on the bank side.

When a tractor with a torque converter meets bank resistance, the force affects tractor speed much more than hoist speed. Even with the tracks or wheels stalled, the engine loader's hydraulic pump should maintain good speed. This has the favorable effect of slowing crowd speed relative to hoist. However, the crowding force is increased by multiplication in the converter, so the bucket is still likely to crowd in more material than it can lift.

Converters cannot be operated at full throttle under stall or near-stall conditions for more than a few seconds without excessive heat buildup and strain.

If the throttle is cut back to idle speed, the torque converter will exert almost no drive force. The loader pump will be slowed, but it will continue to exert full pressure to lift the bucket at a slower speed. This combination favors breaking out an overloaded bucket. A partially engaged throttle weakens drive without stopping it.

Some wheel loaders may be operated with the hand throttle closed. In such cases, the engine governor setting is controlled with a foot accelerator. The operator cuts engine speed by reducing foot pressure.

Some four-wheel-drive loaders have an engine (input) clutch in addition to the torque converter. This clutch is power-released when one of the brake pedals is pressed lightly or by a touch of a separate pedal. This permits disconnecting the drive (crowd) without slowing the engine or the hoist.

A crawler usually works with the hand throttle wide open. A decelerator (opposite of accelerator) pedal is depressed to slow it down. The pedal is pushed when the machine enters a bank too rapidly and is about to get stuck.

The separate responses of push and lift to load and throttle are the chief features of torque converter operation.

Payload. The bucket load volume varies with the type of bucket, power, traction, nature of the material, and operator skill. A bucket must be rolled back to carry the maximum load. When dumping close to the digging point, as in side-casting or loading a properly placed truck, maintaining a steady cycle is more important than maximum bucket load. As the distance to the dump point increases, the volume of the loads becomes more important than excavating time.

If the load must be carried over rough ground or backward up a slope, it should be limited to the weight the machine can carry easily without tipping. If the bucket does not fill sufficiently, the tractor should be backed up, the bucket lowered to floor level, and another pass made. If the load is one-sided, the second cut should be made at an angle to the bank with the empty side penetrating first.

Ramping Down. If the digging is downward, cutting a basement or a ramp, hard material may require pitching the bucket floor at a 20- to 30-degree angle to the line of the tracks and cutting in thin slices. The downward pitch of the ramp should be gradual, as the machine is nose-heavy with a loaded bucket.

Gouging. A common difficulty in digging heavy soils is excessive bucket penetration causing the bucket to be pulled down by the weight of the material. This can raise the back of the tractor off the ground. This may be controlled by keeping the bucket as flat as possible or by tipping the bucket floor into a nearly vertical position. Such a bucket position eliminates the difficulty but puts extra strain on the bucket. It may be best to let it gouge and then make an extra pass to grade off the area as often as necessary.

Transporting a Load. Backing, turning, and dumping can all be done at speed safely if the ground between the digging and dumping areas is hard and smooth. If the ground is rough, the loader must move slowly, as going over a bump or ridge with a heavy load may cause it to tip forward, with the bucket dropping to the ground and the operator's seat rising into the air. If part of the load dumps, the tractor will settle back. In crossing rough ground, the operator should be alert to lower or dump the bucket if an upset situation begins to occur. Lowering the bucket takes weight off the tractor long enough to enable it to recover balance.

Dropping a loaded bucket and stopping it abruptly in the air may burst a hoist ram hose if it does not overbalance the machine. With crawlers, it is best to cross ridges at such an angle so that one track will be partway across before the other reaches the ridge. If the ridge is soft, it may be possible to cut a more level path through it by turning sharply while on the ridge.

If the bucket is carried 2 to 4 ft above the ground, the consequences of overbalancing are unlikely to be serious. If it is high, the weight is not quite so far forward, so tipping is less likely; but if tipping occurs, it can turn the loader over. A high-held bucket can also cause side tipping.

TRUCK LOADING

Procedures for the actual digging in the bank are the same as those already described.

Maneuvering. Normal truck loading has four stages: (1) digging into the bank, (2) backing from the bank and making a partial turn to be perpendicular to the truck, (3) moving forward to the side of the truck while completing the turn and raising the bucket to clear the truck body, and (4) backing from the truck (Fig. 9.8). The bucket must be raised high enough so the downward movement of the bucket lip during dumping does not strike the truck; it is good practice to have the bucket lip high enough to clear the side of the truck box to avoid an accident while backing.

The hoist is usually completed before the truck is reached. The control is then moved to HOLD, and the loader moved so the bucket is over the truck box (Fig. 9.9).

FIGURE 9.8 Loading travel cycle for a loader.

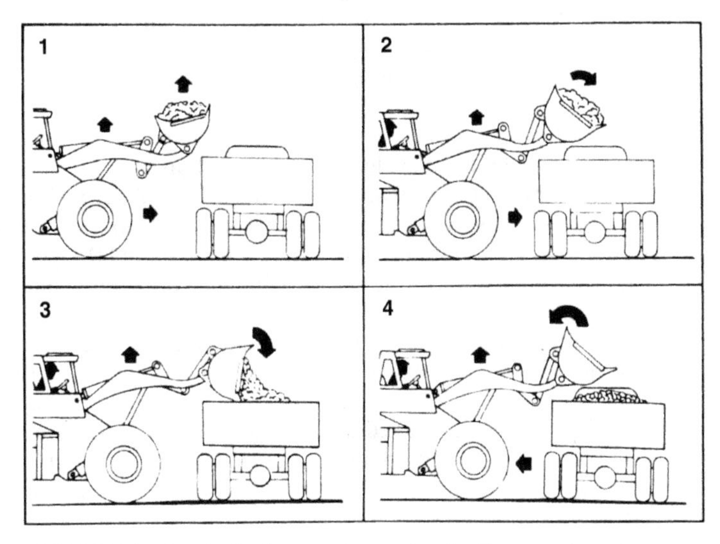

FIGURE 9.9 Loader bucket movement when loading a truck.

A good procedure is to time the lift so the bucket will just safely clear the truck as it is moved over the box and the control can be left in UP using the dump. These levels can be automatically controlled by preset electronic controls.

The loader continues forward until the bucket is as far over the body as desired or until the radiator guard touches the truck body or tire. The main clutch is released (or, with non-clutch converter drive, the throttle closed), the loader is held with a brake, and the control moved forward to dump the bucket load. The first bucket or two should be dumped slowly to reduce shock to the truck. If the material is sticky, the bucket may be shaken by banging against the dump stops by moving the dump valve lever rapidly back and forth.

The loader is shifted into reverse and backed away, with the bucket lip being raised to clear the body. When clear of the truck, the loader is stopped, put in the forward gear for digging, and headed toward the bank; the bucket is then put in the digging position and lowered during the return trip.

TABLE 9.5 Fixed Cycle Times for Loaders

Loader Size, Heaped Bucket Capacity (cy)	Wheel Loader Cycle Time* (sec)	Track Loader Cycle Time* (sec)
1.00–3.75	27–30	15–21
4.00–5.50	30–33	—
6.00–7.00	33–36	—
14.00–23.00	36–42	—

*Includes load, maneuver with four reversals of direction (minimum travel), and dump.

Table 9.5 gives fixed cycle times for both wheel and track loaders. Figure 9.8 illustrates a typical loading situation. Because wheel loaders are more maneuverable and can travel faster on smooth haul surfaces, their production rates should be higher than those of track units under favorable conditions requiring longer maneuver distances.

Spotting Trucks. When travel distance is more than minimal, it will be necessary to add travel time to the fixed cycle time. For travel distances of less than 100 ft, a wheel loader should be able to travel, with a loaded bucket, at about 80% of its maximum speed in low gear and return empty at about 60% of its maximum speed in second gear. In the case of distances over 100 ft, return travel should be at about 80% of its maximum speed in second gear. If the haul surface is not well maintained or is rough, these speeds should be reduced accordingly.

There are other possible patterns of digging and dumping. Figure 9.10, parts A and E, show the most-used methods, in which the side of the truck is at right angles to the face of the bank. This involves a quarter turn twice in each digging cycle.

In (B) the truck is parked at an angle of about 45 degrees to the bank, so the loader's turns are only half as sharp. In this position, some loads can be swung from the bank onto the truck with a very short backward movement, thus increasing loading speed. This is the best system for tracks, but because of driver resistance and indifferent supervision, it is little used.

In (C) the truck is parked parallel with the bank, and the digging is done just behind it. This involves about the same amount of turning as (B), but with greater travel distance.

In (D) the truck is parallel to the bank but at a distance from it. To place the load in the truck, the loader must make a 180-degree turn each time and a second 180-degree turn before it can return to the bank. More travel is required with this method than in others. This is the slowest and most unsatisfactory of the arrangements, but is often used in muddy or sandy pits where trucks are restricted to certain drives.

Figure 9.10(E) shows the effect of counter-rotating tracks (or wheels) to shorten loader travel.

In the case of wheel loaders, extra travel distance from bank to truck, up to 50 ft, may not result in additional loading time because of additional maneuvering space and use of higher travel speeds.

Output can usually be increased substantially by having a spotter direct the trucks or by training the drivers to be alert to the position of the loader. A driver must be ready to move the truck if the loader works away from its initial location. This is particularly important where the excavation height is shallow and truck bodies are large.

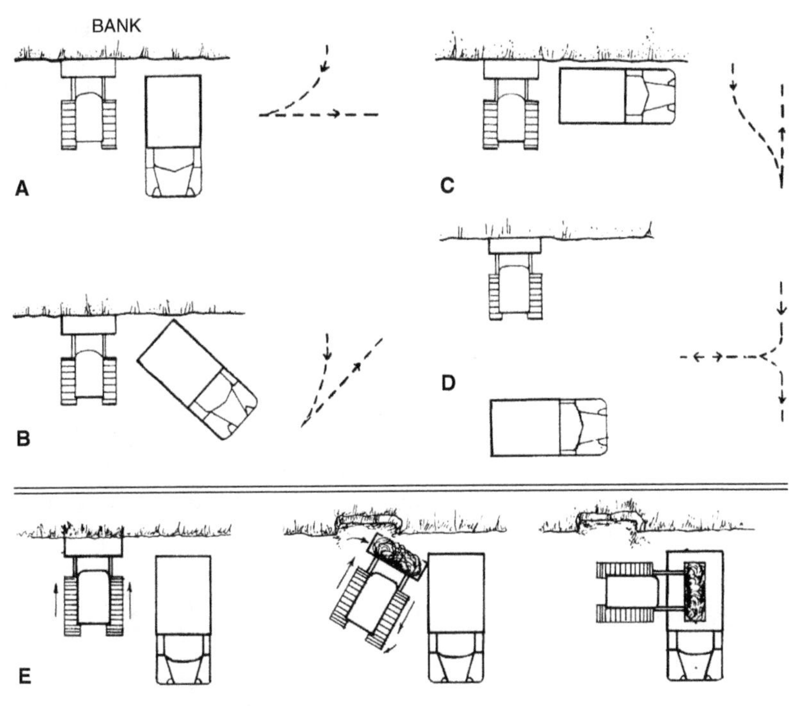

FIGURE 9.10 Truck location affects production.

Dumping into the Haul-Unit Box. The width of standard buckets varies with loader width, usually from about 6 to 13 ft. Truck boxes (including trailers and off-trucks) vary from 7 to over 25 ft in height. The dumping height of the bucket, measured from the ground to the lip of the bucket, held at maximum height with the floor inclined downward at 45 degrees, varies from 8 to 12 ft, in a few cases going as high as 20 ft.

Ordinarily, all excavators, including loaders, are matched in size to the trucks they fill. However, it is often necessary to cope with a mismatched set. If the loader is too big for the truck, care must be taken to place the load gently into the body and allow the excess to spill off the back. A well-matched body can be fully loaded by dumping in the center only. A long body is loaded by dumping alternately in the front and rear from the side, although there is a tendency to pile up too much in the center and skimp the corners. Such a body may also be loaded from the side at the front and finished by filling from the rear, over the tailgate.

For convenience and efficiency in side loading, the sideboard of the truck should be a foot or two lower than the lip of the bucket in full-lift dumped position. This makes it easy to place the load in the center or even in the far side of the body.

High Truck and Trailer Box. If a haul box is too high for convenient loading, a pile of soil is built up on one side. This heap may be moved toward the other side by pushing it with a loaded bucket held with its floor parallel with the ground and just clearing the sideboards. The bucket is then dumped and its load pushed over by the next bucketful, with this process being repeated until the body is full. Material may also be pushed over by putting the bucket in fully dumped position, dropping it just inside the body, and rotating it partway toward flat position.

If numerous large trucks are to be filled, it may save time to dig slots 2 to 3 ft deep into which they can back while the loader operates on the higher level. If the pit floor must not be torn up, a few bucketsful will build a ramp for the loader to increase working height. The ramp should be made so the loader is not tipped up steeply while dumping, as this reduces reach and increases the effort of holding position.

Distances. The loading speed is minimal if the truck is close to the digging but leaves space for the loader to turn easily. Distance is needed because the hoist is not fast enough to lift the load to the required height when travel distance is very short. When a pit is too wet or sandy for truck operation, the loader can excavate in the pit and carry the material onto firm ground where trucks can operate.

When compared to a shovel of similar rate of production, the loader has the advantages of moving around more readily, cleaning up the pit floor and moving boulders without assistance, picking up bigger rocks without chaining, and being used as a dozer while not loading. The shovel can handle harder digging, work on softer floors, and has lower repair and maintenance costs because its tracks and rollers have much less use.

LOADER PRODUCTION

Two critical factors must be considered when choosing a loader: (1) the type of material to be handled and (2) the volume of material to be handled. Loader production diminishes rapidly when used in medium-to-hard material as the digging becomes harder, or when the footing gets soft. On the other hand, an alert foreman and a good operator can often increase loader production simply by using sound procedures.

Wheel loaders work in repetitive cycles, constantly reversing direction, loading, turning, and dumping. The production rate for a wheel loader will depend on the

1. Fixed cycle time required to load the bucket, maneuver with four reversals of direction, and dump the load
2. Time required to travel from the loading to the dumping position
3. Time required to return to the loading position
4. Volume of material hauled each cycle

Loader Hauling. The large wheel loaders may carry loads in complete operation, digging, hauling, and spreading, over distances up to ¼ mile. Bucket capacity is smaller than in scrapers or trucks of comparable price, travel is slower, and possible tipping of a loaded bucket limits them to well-graded routes.

Consider a wheel loader with a 2½-cy-heaped-capacity bucket, handling well-blasted rock weighing 2,700 lb/lcy, for which the swell is 25%. The speed ranges for this loader, equipped with a torque converter and a power-shift transmission, are given in Table 9.2.

The average speeds [in feet per minute (fpm)] should be about

Hauling: all distances 0.8×3.9 mph $\times$ 88 fpm per mph = 274 fpm

Returning: 0–100 ft 0.6×11.1 mph $\times$ 88 fpm per mph = 586 fpm

Returning: over 100 ft 0.8×11.1 mph $\times$ 88 fpm per mph = 781 fpm

The effect of increased haul distance on production is shown by the tabulations in Table 9.6.

TABLE 9.6 Effect of Haul Distance on Production

Haul Distance (ft)	25	50	100	150	200
Fixed time	0.45	0.45	0.45	0.45	0.45
Haul time	0.09	0.18	0.36	0.55	0.73
Return time	0.04	0.09	0.13	0.19	0.26
Cycle time (min)	0.58	0.72	0.94	1.19	1.44
Tips per 50-min hour	86.2	69.4	53.2	42.0	34.7
Production (tons)*	262	210	161	127	105

*0.9 bucket fill factor.

Loaders have an advantage over scrapers in being able to dig into banks from the floor of a pit or a cut and in the ease with which they can build steep-sided stockpiles or dump over banks or into hoppers without the help of other equipment. A loader can supply a low or medium bin, saving either the expense of a truck ramp and hopper or rehandling by clamshell.

Job simplification may also be important. Where one unit can do the work of several machines, problems of scheduling are reduced. This factor can be particularly important with a small operation. The loader might be able to do the hauling and also take care of other work, so the need for trucks is eliminated.

Carry or Push. If material is to be moved a short distance, 200 ft or less, and the quantity is large, it may be economical to push rather than carry it. A flat bucket floating on the ground, moving between windrows of spill, will move two to three times as much material as it will carry. Loads may be even better if the push is downhill. Pushing is a standard procedure in supplying belt loaders, even over quite long distances.

Disadvantages include probable use of a lower gear, poor loads until windrows build up, rocks rolling behind the bucket into the tires, possible damage to the push route, and time spent cleaning it afterward.

Handling Oversized Material. The floor of a truck box should be protected by a layer of cushioning material before loading large rocks. This is particularly important if they must be dumped rather than lowered.

Loader buckets are not well adapted to picking up large, loose objects on the ground. They tend to push ahead of the edge unless it is dug into the ground under them, and the overhang of the back makes it hard to balance anything bulky on the bucket floor.

Rollback buckets can pick up many objects by placing the edge under them, then tilting back the bucket until the object falls into the bucket. Larger pieces, such as stumps or boulders, may be crowded against a bank, which will prevent them from falling out while the bucket is rolled back sufficiently for the object to enter. Two loaders can lift the object between them until it settles into one of the buckets.

If the object is too large to be picked up in this way or there is nothing to crowd against, it can be maneuvered onto the floor and held from falling out by a chain to the top back of the bucket. This method is especially effective with loose stumps resting upright.

For efficient odd-job use, or for crane work, it is almost essential to have a chain grab bracket, or a hook or hooks, fastened onto the top or back of the bucket. For use in the dumped position, a second bracket on the rear beam is desirable. A chain should not be anchored on a push arm and passed over the top of the bucket, as these change their relative positions during a lift and the chain is likely to be stretched and broken.

ESTIMATING TRACK LOADER PRODUCTION

The following calculation demonstrates the process for estimating track loader production.

Sample Calculation: A 2-cy track loader with the following specifications is used to load trucks from a bank of moist loam. The operation requires the loader to travel 30 ft for both the haul and return. Estimate the loader production in bank cubic yards based on a 50-min hour efficiency factor. Use a conservative fixed cycle time.

Travel speeds by gear for 2-cy track loader		
Gear	mph	fpm
Forward		
First	1.9	167
Second	2.9	255
Third	4.0	352
Reverse		
First	2.3	202
Second	3.6	317
Third	5.0	440

Assume the loader travels at an average of 80% of the specified speeds in second gear, forward and reverse. The fixed time should be based on time studies for the particular equipment and job; Table 9.5 provides fixed cycle times by loader size.

Step 1. Size of bucket: 2 cy

Step 2. Bucket fill factor (Table 9.4), moist loam, 100% to 120%; use an average of 110% and heck tipping:

Load weight:

$$2 \text{ cy} \times 1.10 = 2.2 \text{ lcy}$$

Unit weight moist loam (earth, wet) (see Chap. 4) 2,580 lb/lcy

$$2.2 \text{ lcy} \times 2{,}580 \text{ lb/lcy} = 5{,}676 \text{ lb}$$

From Table 9.1, 2-cy track machine's static tipping load is 19,000 lb

Therefore, operating load (35% static tipping) is

$$0.35 \times 19{,}000 \text{ lb} = 6{,}650 \text{ lb}$$

5,676-lb actual load < 6,650-lb operating load, therefore okay.

Step 3. Typical fixed cycle time (Table 9.5) 2-cy track loader, 15 to 21 sec; use 21 sec as a conservative estimate.

Travel loaded: 30 ft, use 80% first gear maximum speed.

$$\frac{1.9\ \text{mph} \times 80\% \times 88\ \text{fpm/mph}}{60\ \text{sec/min}} = 2.2\ \text{ft/sec}$$

Return empty: 30 ft, use 60% (less than 100 ft) of second gear maximum speed.

$$\frac{2.9\ \text{mph} \times 60\% \times 88\ \text{fpm/mph}}{60\ \text{sec/min}} = 2.6\ \text{ft/sec}$$

Fixed time	30 sec	2-cy track loader
Travel with load	13 sec	30 ft, 80% first gear
Return travel	12 sec	30 ft, 60% second gear
Cycle time	55 sec	

Step 4. Efficiency factor: 50-min hour

Step 5. Class of material: moist loam, swell 25 (see Chap. 4)

Step 6. Probable production:

$$\frac{3{,}600\ \text{sec/hr} \times 2\ \text{cy} \times 1.1}{55\ \text{sec/cycle}} \times \frac{50\,\text{min}}{60\,\text{min}} \times \frac{1}{1.25} = 96\ \text{bcy/hr}$$

ESTIMATING WHEEL LOADER PRODUCTION

The production rates for wheel loaders are determined in the same manner as for track loaders.

Sample Calculation: A 4-cy wheel loader will be used to load trucks from a processed aggregate stockpile at a quarry. The aggregate has a maximum size of 1¼ in. The haul distance will be negligible. The aggregate has a loose unit weight of 3,100 lb/cy. Estimate the loader production in tons based on a 50-min hour efficiency factor. Use a conservative fill factor.

Step 1. Size of bucket: 4 cy

Step 2. Bucket fill factor (see Table 9.4): aggregate over 1 in., 85%–90%; use 85% conservative estimate

Check tipping:

Load weight: 4 cy × 0.85 = 3.4 lcy

3.4 lcy × 3,100 lb/lcy (loose unit weight of material) = 10,540 lb

From Table 9.3: 4-cy machine static tipping load at full turn is 25,000 lb

Therefore, operating load (50% static tipping at full turn) is

$$0.5 \times 25{,}000 \text{ lb} = 12{,}500 \text{ lb}$$

10,540 lb actual load < 12,500 lb operating load; therefore okay

Step 3. Typical fixed cycle time (see Table 9.5) 4-cy wheel loader, 30 to 33 sec; use 30 sec

Step 4. Efficiency factor: 50-min hour

Step 5. Class of material: aggregate 3,100 lb per lcy

Step 6. Probable production:

$$\frac{3{,}600 \text{ sec/hr} \times 4 \text{ cy} \times 0.85}{30 \text{ sec/cycle}} \times \frac{50 \text{ min}}{60 \text{ min}} \times \frac{3{,}100 \text{ lb/cy}}{2{,}000 \text{ lb/ton}} = 527 \text{ ton/hr}$$

Consider if the 4-cy wheel loader was instead used to charge the aggregate bins at an asphalt plant.

Sample Calculation: A 4-cy wheel loader is used to charge bins at a plant. The one-way haul distance from the 1-in. aggregate stockpile to the cold bins of the plant is 220 ft. The asphalt plant uses 105 tons per hour of 1-in. aggregate. Can the loader meet this requirement?

Step 3. Typical fixed cycle time (see Table 9.5) 4-cy wheel loader, 30 to 33 sec; use 30 sec.

From Table 9.3: Travel speeds forward

First, 4.3 mph; second, 7.7 mph; third, 13.3 mph

Travel speeds reverse

First, 4.9 mph; second, 8.6 mph; third, 14.9 mph

Travel loaded: 220 ft, because of short distance and time required to accelerate and brake, use 80% maximum first gear maximum speed.

$$\frac{4.3 \text{ mph} \times 80\% \times 88 \text{ fpm/mph}}{60 \text{ sec/min}} = 5.0 \text{ ft/sec}$$

Return empty: 220 ft; because of short distance and time required to accelerate and brake, use 80% of second gear maximum speed.

$$\frac{7.7 \text{ mph} \times 80\% \times 88 \text{ fpm/mph}}{60 \text{ sec/min}} = 9.0 \text{ ft/sec}$$

Fixed time	30 sec	4-cy wheel loader
Travel with load	44 sec	220 ft, 80% first gear
Return travel	24 sec	220 ft, 80% second gear
Cycle time	98 sec	

Step 6. Probable production:

$$\frac{3{,}600 \text{ sec/hr} \times 4 \text{ cy} \times 0.85}{98 \text{ sec/cycle}} \times \frac{50 \text{ min}}{60 \text{ min}} \times \frac{3{,}100 \text{ lb/cy}}{2{,}000 \text{ lb/ton}} = 161 \text{ ton/hr}$$

161 tons/hr > 105 tons/hr requirement

The loader will meet the requirement.

LOADER SAFETY

Those working in areas where loaders are operating are exposed to the hazard of being struck or run over by the backing machine.

Incident from an OSHA Report. The operator of the loader after dumping a load into a dump truck proceeded to back up while looking over his left shoulder toward the back left side of the loader. While backing, a supervisor walked into the work area, approaching the right rear of the loader. A laborer who witnessed the incident said he saw the supervisor walk into the work area; he lost sight of him momentarily as the loader was backing up but noticed it seemed to have traveled over something. The laborer immediately motioned to the operator to stop backing up. The supervisor had been fatally struck and run over by the right front tire of the front-end loader.

OSHA standard 29 CFR 1926.602(a)(9)(ii) states:

> No employer shall permit earthmoving or compacting equipment which has an obstructed view to the rear to be used in reverse gear unless the equipment has in operation a reverse signal alarm distinguishable from the surrounding noise level or an employee signals the operator it is safe to move in reverse.

Safety Tips. Loader operators must be attentive to and follow safe operating procedures. The operator should

- Make sure the steps and handholds are clear of mud, ice, and oil/grease.
- Never walk or work under a raised loader.
- ALWAYS WEAR THE SEATBELT!
- Know the weight the loader can safely lift (see operator's manual or nameplate).
- Keep the load uphill
  - Back down when loaded and drive up when loaded.

- Raise and lower loader arms slowly and steadily.
- Allow for the extra length of the loader when making turns.
- Never move or swing a load while people are in the work area.
- Stay away from the outer edge when working along high banks and slopes.
- Watch for overhead wires and obstacles when you raise the loader.
- Travel with the load low to the ground and watch for obstructions on the ground.
- Operate the loader from the operator's seat only.
- Not lift or carry anyone on the loader, bucket, or attachments.
- Lower the loader bucket when parking or servicing.
- Ensure all parked loaders are on a firm, level surface and all safety devices are engaged.

CHAPTER 10
TRUCKS AND HAULING EQUIPMENT

Because of their high travel speeds trucks provide relatively low hauling costs. The productive capacity of a truck or truck and trailer depends on the size of the load and the number of trips it can make in a set time period. However, highway load limits and truck carrying weight may limit the volume of a load. The number of trips completed per time period is a function of cycle time. Truck cycle time has four components: (1) load time, (2) haul time, (3) dump time, and (4) return time. Tires for trucks and all other haul units should be suitably matched to the job requirements. To achieve a cost-effective operation, trucks should be balanced with the loading equipment.

TRUCKS

Because of their high travel speeds, trucks provide relatively low hauling costs when there is a need to transport excavated material, processed aggregates, or other types of construction materials. They are also used for transporting construction equipment (Fig. 10.1). The use of trucks as the primary hauling unit provides a high degree of flexibility because the number in service can usually be increased or decreased easily to permit modifications in the total hauling capacity of a fleet.

Most trucks can operate over any haul road for which the surface is sufficiently firm and smooth, and on which the grades are not excessively steep. Some units are designated "off-highway" because their size and weight are greater than the regulatory limits permitted on public highways (Fig. 10.2). Off-highway trucks are used for hauling materials in quarries and on large projects involving the movement of substantial amounts of earth and rock. On such projects, the size and costs of these large trucks are easily justified because of the increased production capability they provide.

Trucks may be classified by

- **Method of dumping the load** — Rear-dump, bottom-dump, side-dump
- **Type of frame** — Rigid-frame or articulated
- **Size and type of engine** — Gasoline, diesel, biodiesel, butane, or propane
- **Configuration of drive** — Two-wheel, four-wheel, or six-wheel
- **Transmission of power** — Direct drive or diesel-electric drive

FIGURE 10.1 Truck tractor unit towing a low-profile trailer hauling a crawler crane.

- **Wheels and axles** Number and arrangement
- **Class of material hauled** Earth, rock, coal, or ore
- **Capacity** Gravimetric (tons) or volumetric (cubic yards)

If trucks are purchased for general material hauling, units adaptable to the multipurpose use should be selected. If the trucks are to be used on a given project for a single purpose, they should be selected specifically to fit the project requirements.

FIGURE 10.2 Off-highway truck.

On-Highway Rear-Dump Trucks

Rigid-frame rear-dump trucks are suitable for hauling many types of materials (Fig. 10.3). They are probably the most common machines used for supporting earthwork and paving operations. A dump truck has four major assemblies. The chassis includes the frame, bumper, springs, dead axles, wheels, and tires. The power train is supported by the chassis and consists of the engine, clutch, transmission, driveshaft, differential, and live axles. The cab is the driver's compartment. The body assembly, which includes the load carrying box, tailgate, cab shield, and the hydraulic system and controls, is an entirely separate unit, often made by a different manufacturer and adaptable to different makes of trucks.

FIGURE 10.3 Highway rigid-frame rear-dump truck.

The shape of the load-carrying box, based on the extent of its sharp angles and corners, and the contour of the rear, through which the materials must flow during dumping, will affect the ease or difficulty of loading and dumping. The load-carrying box of trucks used to haul wet clay and similar materials must be free of sharp angles and corners. Dry sand and gravel will flow easily from a body of almost any shape (see Fig. 10.3). When hauling rock, the impact of loading on the truck body is extremely severe. Continuous use under such conditions will require a heavy-duty rock body made of high-strength steel. Even with a special body, the operator of the loading excavator must use care in placing material in the truck.

Off-highway rear-dump trucks are built to operate in mines or pits or in other types of excavations where the use of public roads is not required. These are work sites where there are no legal restrictions on body size or load weight. These trucks may be 16 to 30 ft wide, up to 50 ft long, and over 24 ft high, with loading height (body sides) between 8 and 17 ft. Their capacity range is from 40 to more than 400 tons, with body ratings of up to 300 cy struck. Gross vehicle (empty) weight may equal 1.8 times the payload capacity, down to as low as four-fifths of it. Top speeds are usually 33 to 50 mph. Road conditions and tire wear (heat buildup) limit practical speed.

The construction of an off-highway truck is heavier than in the case of highway trucks, in order to stand up under conditions of rough ground, heavy loads, and short hauls. Substantial amounts of high-strength steel may be used.

The body assembly does not have a tailgate; therefore, the body floor slopes upward at a slight angle toward the rear, typically less than 15 degrees (see Fig. 10.2). The floor shape perpendicular to the length of the body for some models is flat, whereas other models utilize a "V" bottom to reduce the shock of loading and to help center the load. Low sides and longer, wider bodies provide a better target for the excavator operator. The result of such a configuration is quicker loading cycles. Smaller off-road trucks use a traditional direct-drive transmission, whereas larger off-road trucks may employ the diesel-electric concept, where the diesel engine powers an electrical generator, which is connected to electric traction motors at the drive wheels, similar to a railroad locomotive. When moving downhill, the traction motors can retard movement by a concept known as dynamic braking, where electrical generation at the drive wheels is resisted by the generator or diesel engine.

Articulated Rear-Dump Trucks

The articulated dump truck (ADT) is specifically designed to operate through high rolling-resistance material and in rough terrain where a rigid-frame truck would have difficulty maneuvering (Fig. 10.4). An articulated joint and oscillating ring between the tractor and dump body enable all of the truck's wheels to always maintain contact with the ground. The articulation, all-wheel drive, high clearance, and low-pressure radial tires combine to produce a truck capable of moving through soft or sticky ground.

FIGURE 10.4 An articulated dump truck.

When haul-route grades are an operating factor, articulated trucks can typically climb steeper grades than rigid-frame trucks. An ADT truck can operate on grades up to 35%, whereas rigid-frame trucks can only navigate grades of 20% for short distances, and continuous grades of 15% are a more reasonable limit. A disadvantage of using an ADT is their limited volumetric capacity of 30 cy and top speeds of 35 mph, because the frame and drive train are engineered to navigate rough terrain at lower speeds.

Most ADTs are 4×4 models, but there are larger 6×6 models.

Rear-dumps, be they rigid-frame or articulated, should be considered when

1. The material to be hauled is free flowing or has bulky components
2. Hauling units must dump into restricted locations or over the edge of a bank or fill
3. There is ample maneuver space in the loading or dumping area

CAPACITIES OF TRUCKS AND HAULING EQUIPMENT

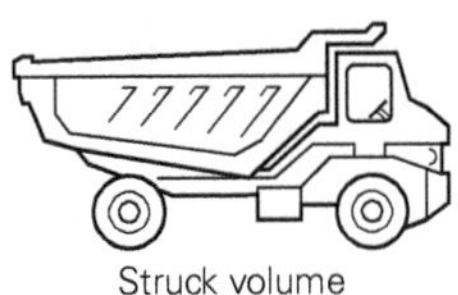

There are three methods of rating the capacities of trucks and wagons:

1. Gravimetric—the load a truck will carry, expressed as a weight
2. Struck volume—the volumetric amount a truck will carry, if the load is water level in the body assembly (dump box)
3. Heaped volume—the volumetric amount a truck will carry, if the load is heaped on a 2:1 slope above the body assembly (dump box)

The gravimetric rating is usually expressed in pounds or kilograms and the latter two ratings in cubic yards or cubic meters (Table 10.1).

TABLE 10.1 Example Specifications for a Large Off-Highway Truck

Axle	Empty Weight	Gross Weight
Front axle	72,500 lb (32,950 kg)	162,000 lb (73,600 kg)
Rear drive axle	88,500 lb (40,230 kg)	198,000 lb (90,000 kg)
Total	161,000 lb (73,180 kg)	360,000 lb (163,600 kg)
Volume		
Struck (SAE)	55 cy	42 cubic meters
Heaped (SAE 2:1)	79 cy	60 cubic meters

The struck capacity of a truck is the volume of material it will haul when it is filled level to the top of the body assembly sides (Fig. 10.5). The heaped capacity is the volume of material it will haul when the load is heaped above the body assembly sides. The standard for rating dump body capacity (SAE J1363) uses an assumed 2:1 slope. The actual heaped capacity will vary with the material being hauled. Wet earth and sandy clay can be hauled with a slope of about 1:1, whereas dry non-cohesive sand or gravel may slope to about 3:1. To determine the actual heaped capacity, it is necessary to know the struck capacity, the length and width of the body, and the slope at which the material will remain stable while the unit is moving. Smooth haul roads will permit a larger heaped capacity than rough haul roads.

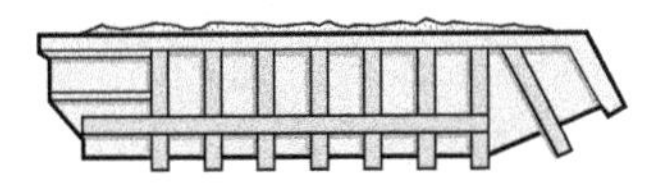

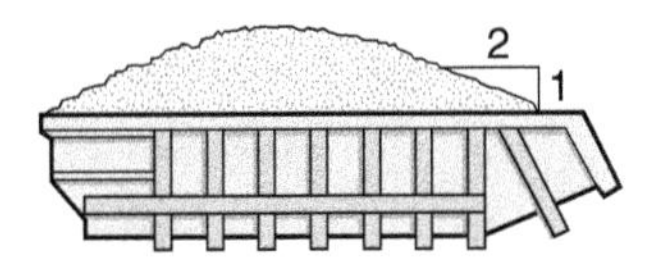

FIGURE 10.5 Measurement of volumetric capacity (SAE J1363).

The truck's weight capacity may limit the volumetric load a unit can carry. The weight capacity controls loading when hauling a material with a high unit weight, such as wet sand or metallic ore. With lighter-weight materials a truck's weight-carrying capacity will not be exceeded when the load-carrying box is filled to its heaped volumetric capacity. Always check to ensure the gravimetric weight of the volumetric load does not exceed the weight-carrying capacity of the haul unit.

Overloading will cause the unit's tires to flex too much and produce excessive internal tire temperature. Such a condition will cause permanent tire damage and increase operating costs.

In some instances, it is possible to add sideboards to increase the depth of the truck's or wagon's body assembly, thereby enabling it to haul a larger load. When this is done, the weight of the new volumes must be checked against the vehicle's gravimetric load capacity. In some cases, truck manufacturers engineer sideboards into their specifications. If the added load weight is greater than the rated gravimetric capacity, the practice will probably increase the hourly cost of operating the unit, because of higher fuel consumption, reduced tire life, and more frequent failures of parts (such as axles, springs, brakes, and transmission) and will result in higher overall maintenance costs. If the value of the extra material hauled is greater than the total increase in the cost for operating the vehicle, overloading is justified. In considering the option of sideboarding and hauling larger volumes of materials, the maximum safe loads on the tires must be checked to prevent excessive loading, which might result in considerable lost time due to tire failures.

Tires are about 35% of a truck's operating cost. Overloading abuses the tires.

The Rubber Manufacturers Association publication *Care and Service of Off-the-Highway Tires* addresses overloading and provides load and inflation tables. It can be downloaded at https://rma.org/sites/default/files/Care%20and%20Service%20of%20Off-the_Highway%20Tires.pdf.

Rigid-Frame On-Highway Rear-Dump Trucks

The ladder-type frame (Fig. 10.6) of a rear-dump truck consists of two parallel pressed-steel channels with cross braces, some of which serve as supports for the engine and transmission. The two longitudinal members are generally referred to as rails. The cross braces provide rigidity and strength, along with sufficient flexibility to withstand twisting and bending stresses encountered when operating on uneven terrain.

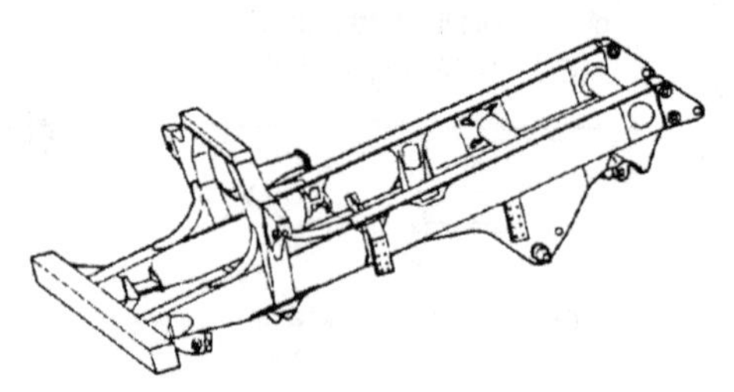

FIGURE 10.6 Truck main frame.

The front cross member is extended to the sides and serves as a bumper. For convenience in mounting bodies, the width and lengths of this section are standardized for most makes of trucks. Pull hooks should be fastened to the top of the frame members just behind the front bumper and on the rear cross member. A rear pintle or clevis hook is useful in towing. When a frame is built with heat-treated steel, it should not be heated, welded, drilled, cut, or notched. The truck frame can be damaged when the load exceeds the gross vehicle weight rating, when there is an uneven load distribution, or when there is improper positioning of the fifth wheel.

Springs. Springs are of the leaf type. Figure 10.7 shows springs for a single-drive axle. They are fastened to the frame by two shackles, one of which is a single pin hinge, and the other a U hinge to accommodate the increase in length of the spring as it is compressed.

FIGURE 10.7 Leaf type rear springs.

The rear springs carry the largest part of the load and are proportionately heavier. Each spring is fastened to the axle by a pair of U bolts and by the spring center bolt, the head of which fits into a socket in the top of the axle. The braking power on all four wheels and the driving power in the rear are transmitted to the frame through the springs, so it is important to keep their fastenings tight. If the U bolts are loose, the center bolt may shear and the axle moves out of line.

The front axle is a drop center I-beam, the rear a hollow casing that carries the differential and the axle shafts. The front axle may carry around 15,000 lb. The front wheel hubs pivot on nearly vertical kingpins held by the ends of the axles. The steering mechanism is similar to what is used in automobiles.

Brakes. The foot brakes on most small and medium trucks are hydraulic, with a vacuum. Each brake shoe has a separate wheel cylinder. Air brakes (a compressed air brake system) are a safe way of stopping large and heavy trucks. These are a friction type using compressed air to drive a piston against a brake pad. These are used when multiple trailers are towed by heavy trucks, as they provide a linked brake system. With air brakes there is "brake lag." This is the time required for the brakes to work after the brake pedal is pushed. With hydraulic brakes (used on light/medium-duty trucks), the brakes work instantly.

The parking brake is a single mechanical unit on the driveshaft behind the transmission. Its effective grip is multiplied by the rear-end gearing. It is not designed to be used for stopping the truck.

Wheels and Hubs. Brake drums are anchored to the hubs by the same studs used to hold the wheels. The wheels are steel, of either cut-out disc or spoke construction. There is a lock ring to hold the tire. The front wheels and the inner rears are mounted with the convex side out, while the outer rears have the convex side inward in order to meet the hub.

The wheel stud is pressed through the hub from the back with its head holding the drum. The inner wheel is fastened by five hollow lugs with inner and outer threads. The outer wheel is then mounted and fastened with large nuts that screw onto the outer threads of the lugs.

Both the inner and outer fastenings must be tight, as the driving and stopping power of the truck are transmitted through them. If any looseness develops, they will wear and ultimately break.

Tires. Tires for trucks and all other haul units should be suitably matched to the job requirements. Properly selecting tire sizes and the practice of maintaining correct tire air pressure will reduce rolling resistance due to the tire contact with a travel surface.

A tire supports the load by deforming where it contacts the road surface until the area in contact with the road produces a total force on the road equal to the load on the tire. Neglecting any supporting resistance furnished by the sidewalls of the tire, if the load on a tire is 5,000 lb and the air pressure is 50 psi, the area of contact will be 100 sq. in. If, for the same tire, the air pressure is permitted to drop to 40 psi, the contact area will increase to 125 sq. in. The additional area of contact will be produced by additional deformation of the tire. This will increase the rolling resistance because the tire will be continually climbing a steeper grade as it rotates. The lower pressure will also increase tire wear.

Tires are expensive and are quickly destroyed by neglect or abuse. Proper inflation can only be determined by a pressure gauge, and proper inflation is extremely important when heavy loads are carried. If the tire bulges prominently at the bottom when resting on a smooth surface, it is either soft or overloaded, and running it will develop destructive heat, weakening the fabric.

Underinflation and overloading may cause severe damage. The selected tire size and the inflated pressure should be based on the resistance, which the surface of the road offers to penetration, by the tire. For rigid road surfaces, such as concrete, small-diameter, high-pressure tires will give lower rolling resistance, whereas, for yielding road surfaces, large-diameter, low-pressure tires will give lower rolling resistance because the larger areas of contact will reduce the tire penetration depth.

Many tire failures can be traced to constant overload, excessive speed, incorrect tire selection, and poorly maintained haul roads. Underinflating tires can cause radial sidewall cracking and ply separation.

The front wheels usually carry much less weight than the rear wheels and can often safely use smaller and lighter tires. However, when all tires are of the same size, only one spare is needed, and the life of rear tires can be prolonged by a rotation program. Large front tires may rub on the frame when steered sharply.

From a safety standpoint, the front wheels should have good tires, as a front blowout on a loaded truck may put it out of control, whereas a blown-out on a rear dual will be carried temporarily by the tire next to it.

It is usually not practical to carry a spare tire on a dump truck unless the body is made to accommodate it. It cannot be mounted on the side of the body because of width restrictions. It can be placed on a reinforced cab shield, but its weight is too great for one person to handle from such a position. It is sometimes possible to make a rack for a spare under the body.

Generally, if an empty truck has a rear flat, it can return to the yard. If the trip is long, it is advisable to remove the flat tire. A front flat can be removed and one of the outer rears substituted for it. If the truck is loaded, however, it should be parked and a tire brought to it. Outer rear tires may be changed without a jack by running the inner wheel up on a block so the outer one is held clear of the pavement.

Dual Tires. When two tires are mounted on two wheels bolted to the same hub, they are referred to as dual tires. All highway dump trucks and most off-road rear dumps use dual-drive tires. Dual front tires are rare. When two tires work as a unit, they must be the same size and ply rating, have nearly the same amount of tread, and carry the same pressure. Otherwise, the larger or harder tire will carry more of the load and be damaged.

New tires of the same size but of different makes may differ in outside diameter. Tires of the same size and make but of different ply ratings are likely to differ in either the outside diameter or the loaded radius or both.

Scuffing. The two tires of a dual pair do not travel the same distance through a curve. If a truck equipped with dual 8.25-20 tires (8.25-in. tread width from sidewall to sidewall × 20-in. rim diameter) spaced 11 in. on centers makes a U-turn between curbs 60 ft apart, the outer outside tire will travel 94.2 ft and the inside one of the pair 91.4 ft, resulting in a difference in travel of 2.8 ft.

Travel distance differences between locked tires cause slippage and scuffing. Because of leverage and road crown, the outside tires slip more than the inside, so they wear faster. As the tire diameter is reduced, it carries less of the vehicle weight and is not pressed against the road surface as hard. With heavier and powerful trucks with larger size tires mounted with a wider spacing, the wear is more pronounced. It is important to rotate tires before the inside ones become overloaded and the difference prevents proper matching. The maximum permissible differences, measured in inches, are shown in Table 10.2.

TABLE 10.2 Maximum Tire Dimension Differences before Rotation

Tire Size, Inches	Diameter, Inches	Circumference, Inches
8.25 and smaller	¼	¼
9.00 and larger	½	1½

Where differences are within these limits, two practices are common. Put the two larger tires on the outside wheels to conform to the crown of the road, or put both on the right side and let the differential take care of the differences.

Measurement. Sets of duals should be checked for size differences at least every 1,000 miles, with the air pressure of the four tires being exactly the same. Replacement tires should be measured and compared before they are put on a truck.

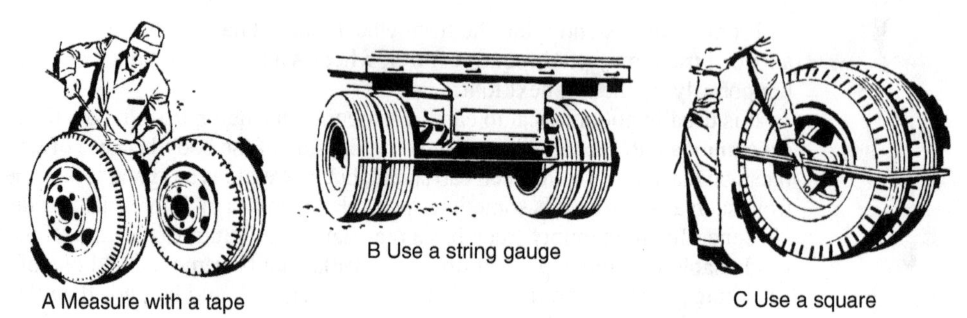

FIGURE 10.8 Methods for matching dual tires. (*Courtesy of Rubber Manufacturers Association.*)

Figure 10.8 shows three ways of checking tire sizes. In A, the circumference is measured with a tape, either before mounting or after jacking the wheel off the ground. In B, a straightedge or a taut string checks both pairs at once. In C, a "square" made of two 1-in. by 2-in. wood strips rigidly fastened to make an exact right angle can be laid along the side of the outside tire and across the treads.

Rotation. Systematic rotation of tires will prevent damaging size differences from developing. The simplest rotation is to move the right tires to the left side and the left tires to the right side, putting the inside tires on the outside and the outside tires on the inside.

Tandems. Tandem drives have two sets of axles, one in front of the other, each equipped with dual tires. With this arrangement scuffing is much greater. The outer tires wear more than the inner, and all tires are dragged sideward on turns. The side drag is hardest on the rear set.

Many tandems do not have any differential or power divider between axles; consequently, all eight tires rotate at the same speed. In such cases it is important to match the tires so the average diameter of those on one axle is within ¼ in. of the average diameter on the other axle.

Front Tires. Truck front tires for on-road haul units do not transmit driving force and carry much lighter loads; therefore, they should last much longer than those on the rear. They are, however, subject to excessive wear from being run out of line.

Front wheels do not roll exactly parallel to each other, as steering is more stable if they toe in slightly. The term "toe" denotes whether the tires turn inward or outward compared to the line parallel to the vehicle. Front wheels also have camber—they do not rest vertically on the riding surface, but both have a slight angle to the outside of the vehicle (negative camber). This is similar to how you spread your legs to steady yourself on a rocking boat. Negative camber improves vehicle handling during cornering.

The third angle from vertical for front wheels is in the front to back plane, called caster. It moves the steering pivot point from the front to back of the vehicle. It is defined as positive if the line is angled forward and negative if backward. Too much toe-in will exhibit a feathered edge at the inside of the tread design, whereas toe-out feathers the outside edge. Too much camber concentrates smooth wear on the outside of the tire; negative or reversed camber wears the insides and makes it look as if the axle is sagging. A wrong caster may result in cupping wear.

Misalignment of front wheels may result from poor adjustment, but more often it is the result of bending caused by glancing blows against curbs or banks. Severe misalignment may be felt by the driver, but if it is slight, it will show up only in tire wear. All of these alignment features support the importance of frequent tire inspection.

Wobble. Rapid tire wear, with or without cupping, may be caused by wheel wobble. The trouble may be a bent wheel, in which case only the tire is damaged. A loose or broken bearing will allow a wheel to wobble. Unless a repair is made, the wheel may come off. Defective bearings are a common cause of rear axle breakage.

Ton Miles per Hour. Tires generate heat as they roll and flex. As a tire's operating temperature increases, it will lose significant strength. The manufacturers of earthmover tires provide a ton-miles-per-hour (TMPH) limit for their tires. The TMPH is a numerical expression of the working capacity of a tire. It is good practice to calculate TMPH value every project and make a comparison with the TMPH for the tires on the proposed equipment.

$$\text{TMPH job rate} = \text{Average tire load} \times \text{Average speed during a day's operation} \tag{10.1}$$

$$\text{Average tire load (tons)} = \frac{\text{“Empty” tire load (tons)} + \text{“Loaded” tire load (tons)}}{2} \tag{10.2}$$

$$\text{Average speed (mph)} = \frac{\text{Round trip distance (miles)} \times \text{number of trips}}{\text{Total hours worked}} \tag{10.3}$$

When calculating the job's TMPH value, always select the tire carrying the highest average load. If the tires being used on the trucks have a TMPH rating of less than the project's TMPH, either the truck speed or the load, or both, must be reduced, or the trucks must be equipped with tires that have a higher rating.

Sample Calculation: An off-highway truck weighs 70,000 lb empty and 150,000 lb when loaded. The weight distribution empty is 50% front and 50% rear. The weight distribution loaded is 33% front and 67% rear. The truck has two front tires and four rear tires. The truck works an 8-hr shift hauling rock to a crusher. The one-way haul distance is 5.5 miles. The truck can make 14 trips per day. Calculate the job TMPH value for the truck.

$$\text{Total weight on two front tires (empty)} = 70{,}000 \text{ lb} \times 50\% = 35{,}000 \text{ lb}$$

$$\text{Total weight on two front tires (loaded)} = 150{,}000 \text{ lb} \times 33\% = 50{,}000 \text{ lb}$$

$$\text{Weight on individual front tire (empty)} = \frac{35{,}000 \text{ lb}}{2} = 17{,}500 \text{ lb}$$

$$\text{Weight on individual front tire (loaded)} = \frac{50{,}000 \text{ lb}}{2} = 25{,}000 \text{ lb}$$

$$\text{Average front tire load} = \frac{17{,}500 \text{ lb} + 35{,}000 \text{ lb}}{2} = 42{,}500 \text{ lb or } 10.6 \text{ ton}$$

$$\text{Total weight on four rear tires (empty)} = 70{,}000 \text{ lb} \times 50\% = 35{,}000 \text{ lb}$$

$$\text{Total weight on four rear tires (loaded)} = 150{,}000 \times 67\% = 100{,}000 \text{ lb}$$

$$\text{Weight on individual rear tire (empty)} = \frac{35{,}000 \text{ lb}}{4} = 8{,}750 \text{ lb}$$

Weight on individual rear tire (loaded) = 25,000 lb

$$\text{Average rear tire load} = \frac{8{,}750\ \text{lb} + 25{,}000\ \text{lb}}{2} = 16{,}875\ \text{lb or } 8.4\ \text{ton}$$

The front tire carries the highest average load.

$$\text{Average speed} = \frac{(2 \times 5.5\ \text{mile}) \times 14\ \text{trips}}{8\ \text{hr}}$$

Average speed = 19.25 mph

Job TMPH value = 10.6 tons × 19.25 mph

Job TMPH = 204

This means the truck needs a tire with a TMPH rating of 204 or higher for the expected job conditions.

Electronics. Trucks now have sophisticated state-of-the-art onboard computer monitoring systems. These are employed to monitor and control numerous aspects of a truck's performance. They are used to monitor truck conditions and driver behavior and can also be used to communicate with the driver. These computers are used to monitor engine conditions and communicate engine data with the driver, the back office, or both. The information provided can range from diagnostic trouble codes to routine maintenance warnings.

Dump Body. A dump body unit consists of the box or body proper, the tailgate, body hardware such as chains and pins, and optional equipment such as cab guards. The hoist, which is often sold as a separate unit, includes a subframe, pump, valve, cylinder, and the controls.

A very wide range of body and hoist configurations are available for every truck. A hoist subframe extending back from the cab is bolted onto the truck frame. It is attached at the rear by heavy hinges to the body frame.

The sides are sheet metal reinforced by flanges at the top and bottom. At the rear, heavy corner posts and the rear frame member combine to make a structure rigid enough to resist outward bending. The front and rear corners can have slots or gusset pockets for placement of sideboards. These usually consist of planks 1½ in. thick to increase the capacity of the body or to prevent spillage over the sides. Body capacity is figured level with the sides and is stamped on a plate on a front corner.

The double-acting tailgate is somewhat higher than the sides, and usually has offset hinges at the top to increase clearance for dumping bulky objects and to make closing more positive. It is made with box reinforcing. The upper hinges are equipped with removable pins. The lower hinge pin is a fixed part of the gate, but the hinge itself can be opened by means of a lever at the left front corner of the body, within reach of the driver.

If the body is flat on the subframe and the latch is open, the gate will hang in a closed position, with the lower hinge pins lying in the hollow of the latches. The body and the gate are held in this position for loading and transporting. To dump, the gate lever is pushed down and the body raised in the front, pivoting on the rear hinges. Its own weight and the pressure of the load sliding against it cause the gate to swing outward on the upper hinges. When the load is fully dumped, the body is lowered, the gate swings into closed position, and then is clamped shut.

The cab guard, which is an optional extra, is a sheet of reinforced steel, curving or angling upward and forward over the cab. When trucks are loaded by dumping material from overhead, the cab guard should be considered a necessity.

Hoist. A direct-type hoist is shown in Fig. 10.9. It consists of a hydraulic pump, a valve, and a cylinder. The pump is driven from a transmission power takeoff through shafts and universal joints, and works only when both the engine clutch and the takeoff gear are engaged.

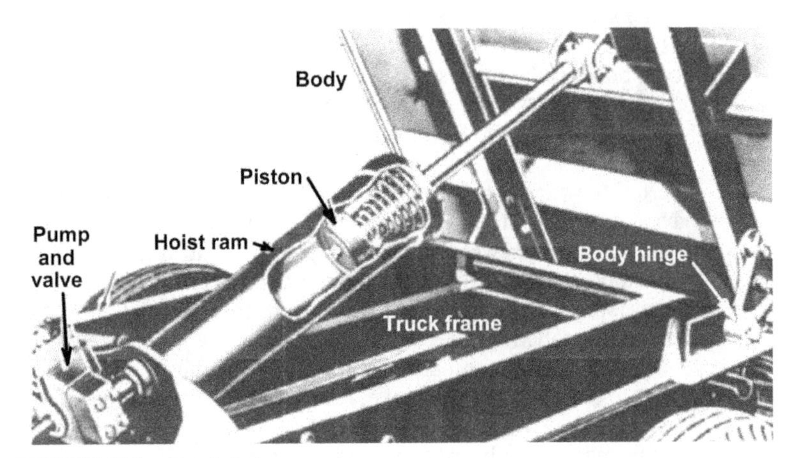

FIGURE 10.9 Simple hoist.

The single-acting ram is bolted to a cross member, which is hinged to the body subframe, and the piston rod is hinged to a crossbeam of the body. A spring may be placed between the piston and the ram head to cushion the piston when forced to the limit of its travel and to help start the body downward when pressure is released.

When the body is down, the ram slopes up slightly from the horizontal. When it is expanded, it pushes both back and up. The body hinge pins are made strong enough to resist the backward pressure so the body is forced up. Leverage is lowest and load is greatest at the start of the dump. As the body rises, a large part of the weight is transferred from the ram to the rear hinges. A number of hoist linkages are offered in which the leverage is greatest at the beginning of the dump, so the body moves slowly at first and more rapidly as it approaches the top.

Heavy Highway-Use Trucks

The Federal Motor Carrier Safety Regulations (FMCSRs) of the U.S. Department of Transportation require annual inspections of all trucks with gross vehicle weight (GVW) of 10,001 lb or more (FMCSRs Section 390.5) and trucks with GVW of 26,001 lb or more (FMCSRs Sections 382.107 and 383.5). GVW is the greater of the manufacturer's gross vehicle weight rating (GVWR) or the actual weight of the vehicle plus the load. If the transport vehicle consists of a truck and trailer, add the GVWR or actual weight of the truck to the GVWR or the actual weight of the trailer to obtain the gross vehicle weight of the combination.

Large trucks for use on highways are very similar in design to lighter ones, excepting the strength of all parts. Either gasoline or diesel engines can be obtained, but diesel is more common. Air brakes are standard. The transmission contains more speeds, and an auxiliary high-low box may be provided. Standard, rock, or quarry type bodies may be used.

Air Brakes. Heavy trucks and wheel tractors are usually equipped with brakes applied by compressed air. The four major components are the service brakes, the parking brakes, a control pedal, and an air storage tank. The air is supplied by a compressor constantly driven by the engine, which unloads, or stops pumping, when full pressure—usually 135 psi—is reached in the reservoir or receiver and resumes when it falls to 85 psi.

The service brake, used for normal driving, is applied and released by the brake pedal (also called the foot valve or treadle valve). A valve operated by a foot treadle or conventional brake pedal allows exact control of air pressure in lines leading to the brake chambers, where it acts against diaphragms to move rods and levers called slack adjusters, which apply the brakes. The front brake lines are provided with a quick-release valve. The valve drains the brakes rapidly when released and serves to prevent possible release lag from interfering with steering or vehicle balance. The rear brakes, and trailer brakes if used, may have a relay valve to feed air directly from the reservoir into the lines.

In the case of newer vehicles equipped with air brakes, the parking brakes are set using a diamond-shaped, yellow, push-pull control knob. Pulling the knob out sets the parking brakes (spring brakes), and pushing it in releases them. On older trucks, the parking brakes may be controlled by a lever.

When a truck is equipped with air brakes, it must not be operated when there is insufficient pressure to apply the brake. A low air pressure warning device is required on vehicles with air brakes. Warning of low pressure may be given by a buzzer or other audible device, but this may not be heard over engine and road noise.

Tandem Drive. The load-carrying capacity of a two-axle truck of any size can be increased by installing an extra axle in the rear. For dump use, this should be a driving axle. A truck so equipped may be called a six-wheeler, a ten-wheeler, or a tandem. The double-axle unit itself may be called a tandem or a bogie.

A tandem drive permits carrying much heavier loads in proportion to tire size and axle strength by distributing both weight and driving strains over more wheels. It improves traction but not nearly as much as a powered front axle. There are a number of different makes and types of tandem drives.

Highway Limitations. Vehicles used on public highways must not be more than 8 ft wide, and they are limited in length, gross weight, and weight on any one axle. These restrictions vary by state, but a limit of 9 tons for the load on any one axle is usual. As a result of both weight and width regulations, the highway dump truck is usually limited to a maximum carrying capacity.

A highway-use dump truck usually has a combined weight of chassis and body equal to approximately three-quarters of its rated payload, as against a usually higher ratio for off-highway haulers in comparable sizes. Highway trucks may travel at speeds over 70 mph, nearly double the speed of other haulers. As a result, the highway truck is a highly efficient hauler when operating conditions are suitable. It should not be used constantly in rough or soft pits, or on poorly maintained haul roads, or on excessively steep grades, or under large machines loading coarse rock.

Traction

Truck efficiency drops rapidly on soft ground. The wheels must constantly climb a slope to move the truck horizontally. As the front wheels sink, they have an increasing tendency to push in the manner of a sled runner instead of rolling up and out, so the force required to move the wheel is greatly increased.

If the power required to roll a driving wheel up the slope is greater than the power needed to spin the wheel in the hole, it will spin and usually dig a deeper hole. The ordinary differential will deliver its full drive power to the spinning wheel, so a standard truck will not be able to move even if only one side lacks traction.

If the front wheels are turned to climb out of a rut, they are liable to stop revolving and skid sideways so the truck cannot be steered. When the truck is put in reverse, the reaction to the twist of the driving axles tends to lift the back of the truck and to push down the front, thus reducing the weight on the driving wheels and their traction.

Similar difficulties are encountered in sand. All wheels sink somewhat and will compress under the push of a driving wheel, then shear, allowing the wheel to spin a partial turn, carrying the sand with it, then compress to hold it momentarily before shearing again. This produces a succession of shocks to the power train, while digging the wheel deeply into the sand.

Live Front Axle. The most effective way to move through such conditions is to supply driving power to both the front and rear axles of the truck. The increase in traction will vary between 50 and 200 percent, depending on the number of rear driving wheels, ground conditions, load distribution, and grade steepness.

The two principal problems connected with the all-wheel drive are (1) driving the front wheels through sharp turns and (2) the front wheels go farther than the rear wheels on curves, both backward and forward.

Figure 10.10 shows the travel path of the front wheels. The rear wheels take a smaller circle on a curve, so they turn fewer times in the same truck travel distance. On loose surfaces, this difference adjusts itself, but means should be provided to allow the front and rear drive shafts to revolve at slightly different speeds on hard pavements in order to prevent excessive wearing of tires and strain on shafts, pressure on gear teeth, and the waste of power.

A truck or other vehicle is customarily kept in two-wheel drive, and the front shaft is engaged only when extra traction is needed for either moving or stopping. Four-wheel drive greatly increases the efficiency of brakes on slippery surfaces. Front wheels do not lock as readily, resulting in better steering control.

Freewheeling Hubs. The front hubs may be fitted with freewheeling attachments so the axles will turn the wheels forward but not backward and the wheels will not turn the axles when the vehicle is moving forward. This allows the front hubs, axles, differential, and driveshaft to stop whenever drive is disconnected in the transfer case, reducing wear and noise. But in four-wheel drive, hubs that are freewheeling will not transmit power in reverse or permit the steadying effect of the front drive on braking.

When hubs are locked, a vehicle can be shifted in and out of four-wheel drive instantly at any speed, which is a safety factor in intermittently slippery conditions. With hubs freewheeling, it is necessary to stop to shift into four-wheel drive. The hubs are easily shifted into a solid connection by turning their caps with the fingers. This adjustment can be made readily if it is necessary to engage them. Control from the driver's seat is prevalent in newer hauling units. It is common practice to keep the hubs in freewheeling position except when bad footing is expected. Fingertip controls in the cab can engage the four-wheel drive system.

General Considerations. All-wheel drive is of great value wherever mud, sand, or snow is encountered. Trucks so equipped can keep a job going when conditions would make the use of two-wheel or tandem-drive trucks impossible. On highways, the extra traction is desirable for snowplowing and sometimes in towing trailers.

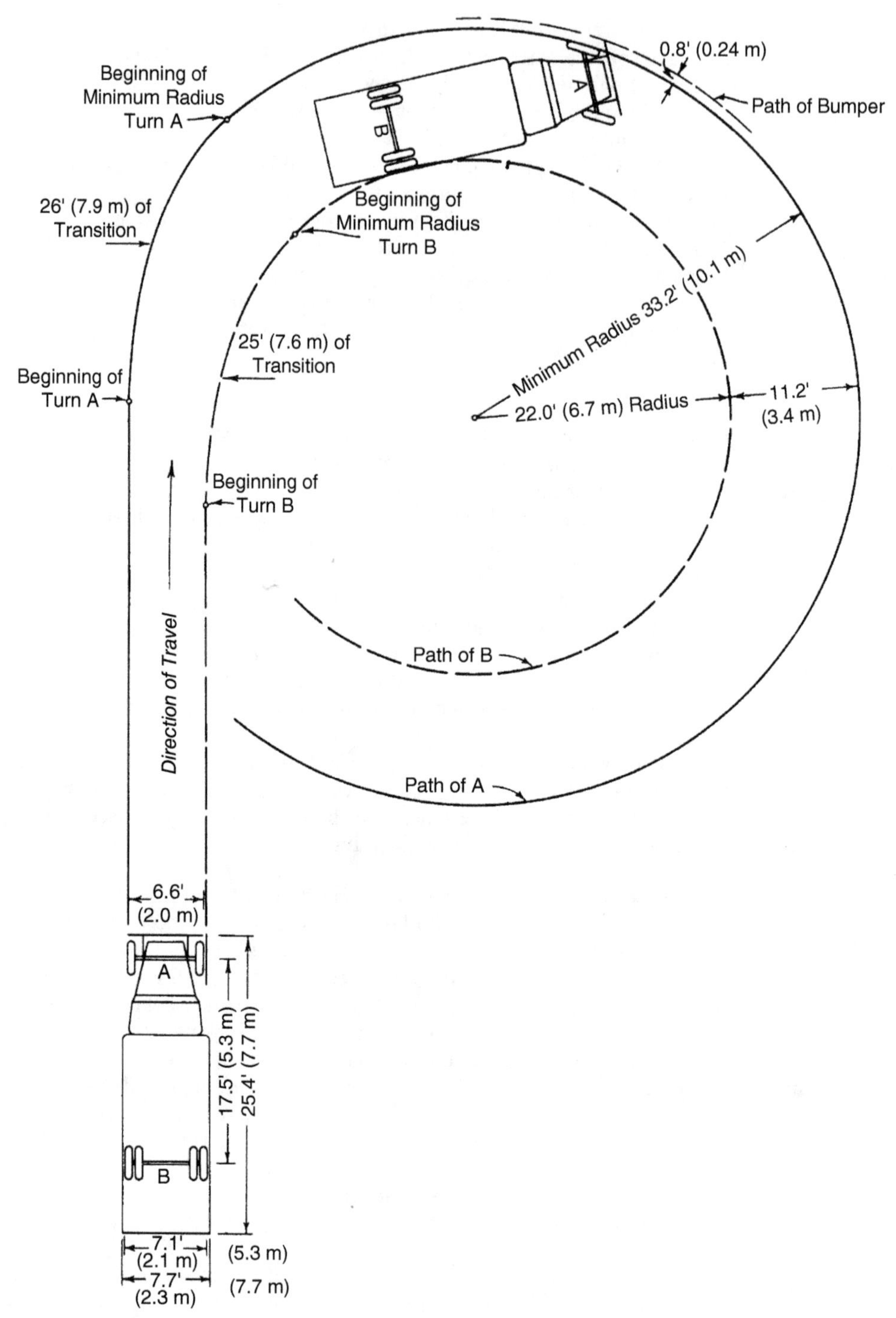

FIGURE 10.10 Right turn, two-axle truck. (*Courtesy of Department of Public Works, State of California.*)

However, the traction and flotation obtained cannot be compared with that obtained by crawler equipment. Mud must have a fairly solid bottom near the surface for any truck to operate, and very slippery surfaces require chains no matter how many wheels are driving.

All-wheel drive has several disadvantages. It is expensive, particularly when it is installed as an accessory. In return for this, the truck can carry heavier loads under much worse conditions than before conversion, so the cost is justified in many lines of work.

The most serious operating fault is steering. A standard truck can turn the front wheels at an angle of 35 degrees or more, whereas the various front-drive hubs are limited to 28 to 30 degrees. Their turning radius is therefore much longer, giving a small truck the clumsiness of a large one. This may not be particularly important on large-scale open work, but for the contractor working in woods, around residences, and in storage yards, it is a serious consideration.

Maintenance costs are somewhat higher. If a truck is to be consistently given heavy overloads, the front drive may partially pay for itself in reduced maintenance on the rear drive.

The ideal mechanical arrangement would be one that gives the driver the triple option of direct drive to the rear wheels only for easy going; a four-wheel drive with differentials for heavy loads on good footing; and a positive drive to all wheels when needed for sand, mud, or ice.

Off-Highway Trucks

Large dump trucks for off-highway use are built to operate in mines (Fig. 10.11) or in other types of excavation in which the use of public roads is not required. They are not subject to any legal restrictions in terms of size or weight.

FIGURE 10.11 Off-highway truck used in a mine.

In order to stand up under conditions of rough footing, heavy loads, and short hauls, their construction is heavier than what is used for highway trucks. Substantial amounts of high-strength steel may be used. Top speed is usually 35 to 50 mph, but haul road conditions and tire wear limit practical speed.

Construction. Components tend to be massive and comparatively simple. Road shock may be absorbed by conventional leaf springs, coil springs, nitrogen-and-oil (air-over-hydraulic) cylinders, rubber discs, oscillating bars, and various combinations of these. Horsepower range is from 125 up to 3,600. Most of these units have torque converters and power-shift transmissions, with hydraulic retarders being either standard or optional. A majority have mechanical drive, but some of the largest use a generator and electric wheel motors.

Differentials may be either standard or limited action. They are usually single reduction but may be double. Further gear reduction is obtained through planetary final drives in the wheel hubs. There are usually two axles, with drive through dual wheel sets on the rear. There are also tandems and all-wheel drives. Tires are among the largest made, up to 59/80 R63, with the same size front and rear. Steering is typically of the automotive type, with front wheels swiveling on a rigid axle, but there are also articulated models.

Body. Bodies are of the heavy-duty rock or quarry type. The sides may flare out to make a larger loading target. The floor is usually a single, heavily reinforced plate. To prevent loads from freezing down during subzero hauls, air ducts may be provided for internal heating of both floor and sides with exhaust air.

The dump body does not have a tailgate. The body floor may have a continuous upward slope from front to rear to retain the load, or it may be flat or nearly so, with an upturned chute in the rear.

Standard hoist construction is a pair of direct-acting, three-stage telescoping cylinders, with power up and partway down. Power-down permits raising the body to a very steep dumping angle—55 degrees or more—then pulling it back until gravity can lower it the rest of the way. Most materials can be dumped readily, even when backing up a grade.

Electric Drive. Some of the largest off-highway trucks are electric drive. A diesel or a turbine engine drives a generator to supply current to electric drive motors. These may be located in each wheel or in differentials between pairs of wheels. The motors use direct current on account of its superior lugging performance. The generator may produce direct current (DC) directly, or it may be an alternator whose alternating current (AC) output is converted to DC in a rectifier. The motors operate through reducing gears and are reversible.

Electric drive eliminates all gearshifts and speed ranges, automatic or otherwise. There is smooth, stepless transition from zero to maximum speed determined by the amount of reduction in the motor gearing and by grade and rolling resistance. Drive motors also supply dynamic braking to slow the hauler on downgrades and before stops.

Articulated Trucks

An articulated dump truck (ADT) is a tractor–dump body combination with the tractor part on one axle and the dump body part on another with an articulated joint between. There are articulated trucks with two or three axles, a single axle under the tractor, and either one or two axles under the dump body. Power may be delivered to all axles. In the case of three axles with at least two wheels on an axle, it is referred to as a 6 × 6 powered truck. The additional power axles give the articulated truck the ability to work in poor ground conditions.

Specifications. ADTs used in the United States are generally of sizes with 11- to 45-ton capacity. They have engines capable of delivering 100 hp to more than 400 hp. The ADTs can range from less than 35 to over 40 ft in length with outside turning radiuses up to 30 ft. Articulated dump trucks usually have high hydraulic system pressure, which means the

FIGURE 10.12 Articulated dump truck bed raised.

dumping cylinder hoists the bed faster. The bed also achieves a steeper dump angle. One model can attain a 72-degree dump angle (Fig. 10.12) in 15 sec. The combination of these two attributes, hoist speed and a steep angle, translates into quick discharge times. At least 21 ft of height clearance is required at full dump. Where there is a height limitation or if the material is sticky, there are ADTs equipped with an ejector dump body.

Suspension Systems. The tractor and dump body are two separate systems, each controlled by its own electronic control unit (ECU). The computer works to keep the truck upright by controlling pressure in the accumulator packs to which the struts are linked. The wagon system has both high- and low-pressure accumulators, and the ECU selects which to use for better performance, whether operating with the bed loaded or empty.

Trailers and Wagons

Dump bodies may be mounted on semitrailers and trailers, with either standard or special constructions. A semitrailer is a frame that has supporting axles and wheels at the rear and rests on the prime mover or tractor at the front (Fig. 10.13). Tractors for highway-type

FIGURE 10.13 Dump bodies on semitrailers.

FIGURE 10.14 Fifth wheel for truck trailer.

semitrailers are equipped with a connecting device called a fifth wheel, one variety of which is shown in Fig. 10.14.

To connect, the trailer is held at a proper height by jacks or wheeled standards, the wheels are blocked, and the tractor is backed under it. It is important to align the kingpin with the center of the fifth wheel. The upper fifth wheel or trailer hitch is contacted by the tractor fifth wheel and wedged slightly up until its knob reaches the socket, which it slips into and locks automatically. The pin and socket take the pull and thrust of towing and stopping, while the fifth wheel surface carries the weight. The knob can be released from the hitch when not under load by moving a hand lever. Off-road semis may use a kingpin arrangement.

Full trailers, often called wagons, are equipped with supporting axles and wheels at both ends, so none of their weight rests on the prime mover. The front axle may swivel or may have a steering linkage. The trailer may be fastened to a drawbar attached to the prime mover, or it carries a fifth wheel or other support for the trailer frame.

Trailers of either semi or full construction must be equipped with brakes. These are operated by air, vacuum, fluid, or electricity from the tractor, and are usually fixed to go on automatically if the connection is broken. This prevents the trailer from rolling free if it breaks away from the tractor. Brakes may be synchronized with the tractor system, controlled separately, or both.

A semitrailer can be operated in more restricted areas than a truck of the same capacity. Because the drive wheels are not at the rear, it can be backed somewhat closer to the edges of fills and farther onto soft ground than a truck. However, traction is not as good and more driving skill is required.

Bottom-Dump Trailers. Tractors towing bottom-dump trailers are economic haulers when the hauled material is free flowing, such as sand, gravel, asphalt, and reasonably dry earth (Fig. 10.15). Their tapered bodies are shaped to dump through a full-length bottom opening controlled by one or two pairs of clamshell doors. The use of bottom-dump trailers will reduce the time required to unload the material and can provide a more uniform placement over a longer distance. To take full advantage of this time saving, there must be a large, clear dumping area where the load can be spread into windrows. Bottom-dump units

FIGURE 10.15 Highway bottom-dump hauling embankment material.

are also very good for unloading into a drive-over hopper. The rapid rate and more controlled discharging of the load, as well as forward dumping movement, gives the bottom-dump wagons a time and safety advantage over rear-dump trucks.

There are both highway-sized units (see Fig. 10.15) and large off-highway units (Fig. 10.16). Relatively flat haul roads are required if maximum travel speed is to be obtained.

FIGURE 10.16 Off-highway tractor towing a bottom-dump trailer.

The doors are operated by either hydraulic or air cylinders, to slide or pivot sideward. Dumping is done with the unit moving forward, without any particular limit to travel speed. The load is deposited in a windrow. Its cross-section is determined by the width to which the doors are opened and the speed of movement. Maximum height is limited by the rear clearance of the body, but this may increase as the rear wheels ride up on the dumped material. The clamshell doors through which these units discharge their loads have a limited opening width. Difficulties may be experienced in discharging materials such as wet, sticky clay, especially if the material is in large lumps.

Tractor-towed bottom-dump trailers are economical hauling units on projects where large quantities of materials are to be transported and haul roads can be kept in reasonable condition. The bottom-dump trailer can have a single axle, tandem axles, or even triaxles. Hydraulic excavators, loaders, draglines, or belt loaders can be used to load these units.

Bottom-dumps should be considered when

1. The material to be hauled is free flowing
2. There are unrestricted loading and dump sites
3. Haul-route grades are less than 5%
4. Material is to be spread uniformly across a distance

Because of its unfavorable power–weight ratio and because there is less weight on the drive wheels of the tractor unit, thereby limiting traction, bottom-dump units have limited ability to climb steep grades.

Operating a Dump Truck

The driver of a dump truck has very poor visibility to the rear, and because of greater weight in proportion to engine power and braking capacity, the vehicle usually requires special methods of shifting gears and climbing and descending hills.

Checking. The tires are the most vulnerable and most expensive wearing parts of a truck. Tire inflation and condition are particularly important, and tires must be checked daily. If the truck has air brakes, it should not be moved until the gauge shows the safe operating pressure. Directional signals and stop lights must be in working condition for driving on a highway or on heavily traveled haul roads.

Stones stuck between dual tires should be removed immediately, as they may be thrown behind the truck. A stone can usually be pried out with a bar or board, but it is occasionally necessary to let the air out of one tire to free the stone.

Each load should be checked for objects that might fall off. These may be moved to safer positions or be pushed off in the pit.

Loading. Truck drivers may add substantially to shovel production by promptly positioning the truck for loading. If a spotter is present or the excavator operator is giving signals, directions should be followed exactly, regardless of the driver's ideas of where the truck should be. If a spot log is used, the rear wheels should be backed squarely against it. If the excavator has not moved, the truck will generally be spotted in the same place as the previous one. Following tire tracks or observing the location of spilled material makes it easy to get in the same position.

In general, a truck hauling from an excavator is positioned so the shortest practical swing is required. The shovel can reach from back to front of the body by use of the crowd mechanism when the truck faces directly away from a shovel. When positioned at a right angle (Fig. 10.17), the shovel swing controls can be used to distribute the spoil along the length of the body.

When large front shovels are used to load rock, the distance from the shovel to the center of the truck body should match the middle of the shovel bucket's crowd arc. With draglines and clamshells, it should be directly under the boom point.

FIGURE 10.17 Truck positioned at right angle to shovel.

A backhoe or a dragline loads most efficiently when it is on the bank and the trucks are on the floor of the pit. For draglines and hydraulic hoes, the difference in cycle time is the number of seconds needed to hoist the load to the higher level.

When a front-end loader is used, the truck should usually be backed against the bank at an angle of 45 degrees, so the loader makes only one-eighth of a turn to dump. Wheel loaders require more space to maneuver than crawlers.

It may be necessary to move a truck during loading. This is particularly true with belt or bucket loaders, which may themselves be moving, or if stationary, may not distribute the load throughout the body.

Curves. Curves must be taken more slowly with a load than without. The truck should be slowed before entering the curve and accelerated or held at an even speed while in it. When rounding a curve, centrifugal force pulls the truck toward the outside and is resisted by the friction of the tires on the road. If traction is good, the truck will tilt, compressing the outside springs and tires. Increase in load will both raise the center of gravity and increase the force tending to tip or slide the truck.

The force required to turn the truck is exerted by the front wheels, which slow the truck and cause it to nose down slightly. The outside wheel does more than half of the turning work because the tip of the truck adds to the weight it is carrying. This wheel is, therefore, under the greatest strain.

Skids. In normal operation the direction in which a wheel and tire are rolling controls the course of the vehicle, and considerable resistance exists against forces tending to slide the wheel sideways. This resistance decreases if the wheel is turning either faster or more slowly than required by the speed of the truck. If the wheel is locked, it moves sideways almost as readily as along its plane of rotation.

If the front wheels are locked while rounding a curve, they will no longer turn the truck, which then moves straight ahead. This type of skid most often occurs when brakes are applied too hard, but on very slippery surfaces the wheels may lock when turned sharply without brakes. When the brake is released, the wheels may revolve again, or they may remain locked until turned in the direction of the skid.

When the front wheels are again revolving, they will substantially control the direction of movement for the front of the truck until again locked by brakes or too sharp a turn.

If the rear wheels lock on a curve and the front ones do not, centrifugal force causes the rear of the truck to slide outward, pivoting on the front axle. The brakes should be immediately released and the front wheels straightened so the side pull of the turn will be stopped. If the rear wheels have skidded too far sideways to be started rolling in this manner, the tendency of the truck to spin may be checked by steering sharply in the direction of the skid. The front wheels will travel in the same direction as the back is sliding, and the truck will move diagonally sideward. The front wheels will tend to get in the line of skid and the rear wheels to start turning. If the engine stalls, it should be started immediately. Should the rear wheels slide due to spinning, traction can usually be restored by releasing the accelerator and pressing it down gradually.

Sometimes it is more important to avoid collision with a person or a particular object than to straighten out the skid. Also, once a truck has started to "pinwheel," straightening it out is a matter of skill and luck rather than rules. The safest system is to avoid skidding by driving slowly and cautiously when roads are slippery.

Backing. Most fills using rear-dump trucks require turning and backing to the edge of the fill. If possible, a fill should be arranged so the turn is made near the dump spot; the turn spot should be wide enough so the reverse gear is needed only once; turns in reverse should be toward the driver's left, and the truck should be level or facing uphill while dumping. The lowest reverse gear should be used when the load is heavy, the ground is soft or rough, or complicated steering is required.

Dumping. When dumping off the edge of a fill, the driver should back so both rear wheels will be the same distance from the edge rather than at an angle to it. If one wheel sinks in more deeply than the other, it may not be possible to either dump the truck safely or pull out with the load.

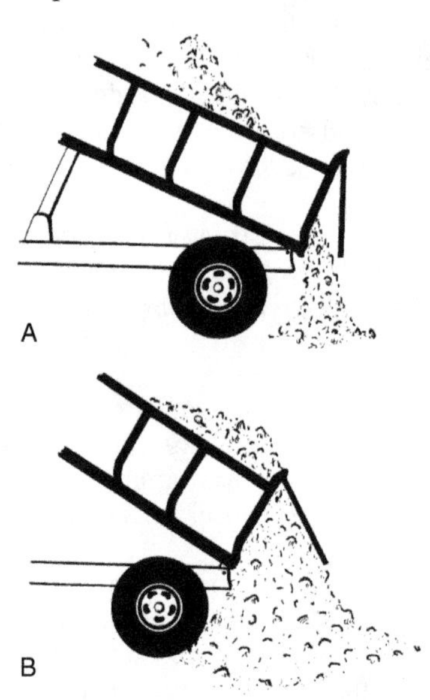

FIGURE 10.18 Dumping a load.

When the truck is in position to dump, the tailgate latch is released, the transmission power takeoff is engaged with use of the clutch and a hand control, and the hoist valve is placed in the Up position. The engine is accelerated to a moderate speed but not raced.

Most hoists are constructed so they can be left in the Up position to hold the body at its high position. However, if this appears to strain the mechanism, the valve may be moved to an intermediate Hold position as soon as the body is at its steepest angle.

As the body rises, the load slides backward along the floor and under the tailgate (Fig. 10.18A). Unless right at the edge of a bank, the load will tend to pile up until it blocks the gate, as in B. The truck is then placed in low and moved forward without disturbing the hoist controls until there is space for the remainder of the load to slide out.

If part of the load sticks in the body, the truck may be backed into the pile to shake it loose. This may have to be done several times if the load is sticky or the dumping is uphill. The clutch should be released just before the wheels hit the pile to avoid shock to the power train. If the truck rolls away from the pile on the rebound, the shaking out can be done without shifting out of reverse; otherwise, low gear is used to drive forward a few feet and reverse to hit the pile. If any considerable part of the load cannot be shaken out, it must be dislodged with a hand shovel or other tool. Most of the body can be reached from the pile and tailgate, but the upper end may require climbing up on the body running boards. Sticking can be reduced if the excavator operator will put the driest material in the bottom of the truck so the load will slide out, carrying the moist material.

Many newer haul units are equipped with truck bed vibrators to mechanically dislodge sticky materials. Vibrators may be actuated using a low-voltage electrical system, a hydraulic fluid system, or a pneumatic system. A majority of the vibrators use the pneumatic system while the truck is in a stopped position and not demanding compressed air for braking. The most effective mounting location for rear-dump trucks is toward the front of the box away from the tailgate hinge points. This allows maximum vibration of the free box end away from the hinge. When the dump box is fully raised, the vibrators are actuated to achieve maximum efficiency.

If the load includes rocks, stumps, or mats of trash large enough to have difficulty going under the tailgate, the gate should be left latched at the bottom and unfastened at the top. When the gate opens by dropping down, the load can slide over it. However, it is not safe to back against the pile to shake out the load with the gate down, as it will be bent or knocked off. If much bulky material is to be carried, it is good practice to take the gate off and use higher sideboards. In such cases the load should be piled toward the front.

> A loaded dump body should not be raised to the dumping position unless the rear wheels are level.

Any slant will cause the center of gravity of the body to shift downhill as it is raised. The twisting force exerted may break the body hinges, tear the subframe off the chassis, or overturn the truck, as shown in Fig. 10.19.

FIGURE 10.19 Dumping a load perpendicular to a slope.

When the dump is completed, the body is lowered and the latch closed. Often this is forgotten, with resulting complications in the pit when part of the next load spills out. If the gate is sprung out of shape or material sticks in its way, it may not latch on the first attempt.

Spreading. The load may be spread over a considerable length instead of being dumped in a pile. The procedure is to start the dump in the manner described, and when the body is high enough for the load to begin to spill, put the truck in gear and drive forward while continuing the hoist. If low gear is used and the hoist is fast, the spread will be short and heavy. If the hoist is slow or if it can be slowed by opening the valve only partially, the material will be spread more evenly. With any one material, the thickness of spread will be determined by the gate opening and the speed of the truck. A larger opening will increase the thickness; a higher speed will decrease it.

Surface Dumping. Dumping on the surface of a fill is usually arranged so it can be done without backing. Rear-dump trucks may be used, although their height in full-dump position makes it advisable to keep on smooth ground and move slowly. They may spread the load or dump in piles.

Such surface building is usually done by scrapers or by bottom- or side-dump trucks or wagons. The bottom-dump wagon moves rapidly while dumping. The thickness of the spread is determined by the relation between the speed and the width of the door opening. The maximum height of the windrow is determined by the clearance of the wagon.

The wagon may be followed immediately by a dozer, which spreads the material in a single pass, or a number of windrows may be built both behind and beside the front one. The latter way is economical with dozers but is liable to slow the wagons and might cause interruptions because of the necessity of traveling on the rough fill.

Trailer Operation

Semitrailer operation differs from truck backing and precautions are needed when steering sharply and when stopping. Very sharp turns may be made in proportion to the length of the whole unit, particularly when the tractor is a short coupled, off-road type, and the trailer yoke is high enough for clearance under the tractor rear tires. However, turns must be made carefully to avoid overrunning and jackknifing.

A sharp fast turn with a tractor trailer is dangerous. When the tractor is turned, the weight of the trailer pushing straight ahead tends to skid the driving wheels to the outside of the turn, increasing its sharpness—in such a case the unit may jackknife with the two parts still connected, each trying to move in different directions. This may do no more than cause a tangle of the two units, but it will be difficult to straighten without a tow. However, it can result in serious damage, overturning both units and tearing the hitch apart.

In rounding a curve, the tractor rear wheels will track toward the inside of the curve from the front, and the trailer wheels will track still farther in, as shown in Fig. 10.10. Swinging turns wide enough to keep the trailer clear and avoiding stops sudden enough to cause jackknifing are special steering requirements for forward driving.

Full trailers are much more difficult to control than semitrailers. Most drivers cannot back them more than a few feet. In general, full trailers should not be used where backing is required. Emergency measures for getting one out of a blind street include unhooking so as to tow it backward.

Rear-Dump versus Bottom-Dump Trucks. Use rear-dump trucks when

1. The material hauled is large rock, ore, shale, or a combination of free-flowing and bulky material.
2. Dumping is into restricted hoppers or over edges of a waste bank or fill.
3. The hauling unit is subject to severe loading impact when under a large shovel, dragline, or loading hopper.

4. Maximum gradeability and rapid spotting in a restricted area are required.
5. Maximum flexibility is required for hauling a variety of materials under variable job conditions.

Use bottom-dump semitrailers when

1. The material is free-flowing.
2. Maximum flotation of large single tires is required.
3. The haul route is level, allowing high-speed travel.
4. Dumping is unrestricted into a drive-over hopper, or the load is spread in windrows.
5. Safety with forward-only dumping is the primary objective.
6. Long adverse grades do not exceed 5%. This is recommended as a general rule for optimum performance but is by no means a measure of the actual gradeability of bottom-dump semitrailer trucks.

MACHINERY TRAILERS

Most excavating machines are not designed to travel under their own power on highways, particularly for long moves. Machinery trailers are used to carry such equipment from job to job. These trailers are manufactured in a wide variety of sizes and types, of which only a few can be described here.

Construction

A semitrailer has a rigid drawbar supported by the towing truck. If the wheels are in the rear, they may carry the weight of more than half the trailer and load, or only serve to stabilize against rocking if the axle is under the middle of the load. Full trailers are usually supported in the front by a swiveling axle and connected to the truck by a comparatively light-hinged draw tongue.

Any trailer should be equipped with brakes and lights operated from the truck or tractor cab. Even a small trailer can produce an unpleasant mess by jackknifing during an abrupt stop, and the side shift of its weight may put the truck out of control.

A trailer must have a deck or carrying space large enough to support or hold the machine to be carried, it must be strong enough for the job, and it must provide for loading and unloading with a minimum of difficulty and danger. It should be pulled by a truck or tractor amply powered to climb any grade it may meet and have brakes adequate for stopping when going down.

Deck Trailers. Figure 10.20 shows a medium-weight trailer—there are both level-deck and drop-deck models. Capacity range is 15 to 40 tons.

The trailer proper consists of a girder frame supporting a flat wood deck, usually hardwood planks 2 in. thick. Standard width is 8 ft. When the unit is used as a semitrailer, the gooseneck is attached to the "fifth wheel" of a tractor truck. For full trailer operation, it rests on a dolly.

The wheels may be mounted on a pair of walking beams, left and right, to allow front to rear oscillation to maintain road contact and reduce jolting. The tops of the tires are even

FIGURE 10.20 Deck trailer. (*Courtesy of Rogers Brothers Corporation.*)

FIGURE 10.21 Tandem wheel oscillation. (*Courtesy of Rogers Brothers Corporation.*)

with the top of the deck on level ground. Cutaway deck sections permit them to go higher as they oscillate. In another design, there are two axles in tandem employing leaf springs and torque arms (Fig. 10.21).

Wheels are 15 in., and tire sizes are 8:25 up to 11:00, 10 to 16 ply. Level deck height may be 28 to 36 in., with differences depending mostly on tire size. Drop decks are 6 to 9 in. lower.

Air or vacuum brakes are used on all four hubs. These are controlled from the truck cab through piping and flexible hoses with snap-on couplings. They are designed so the trailer brakes will lock on automatically if the hoses become disconnected. This serves to prevent a runaway. They can be released by bleeding the air or vacuum tank, or they may release very gradually by leakage.

The dolly (Fig. 10.22) consists of a towing platform spring-mounted on a single axle. Two perforated discs are used in the swivel connection to the gooseneck to hold grease,

FIGURE 10.22 Trailer dollies. (*Courtesy of Rogers Brothers Corporation.*)

to increase bearing surface, and to eliminate sway or whip. The draw tongue is hinged to the base of the platform. A pair of breakaway chains is attached to the frame and can be hooked to the back of the truck to hold the trailer if the tongue should become disconnected.

Heavier trailers may have three or more axles. The model shown in Fig. 10.23 is convertible between two and three. The rear and the frame and deck above it are hinged

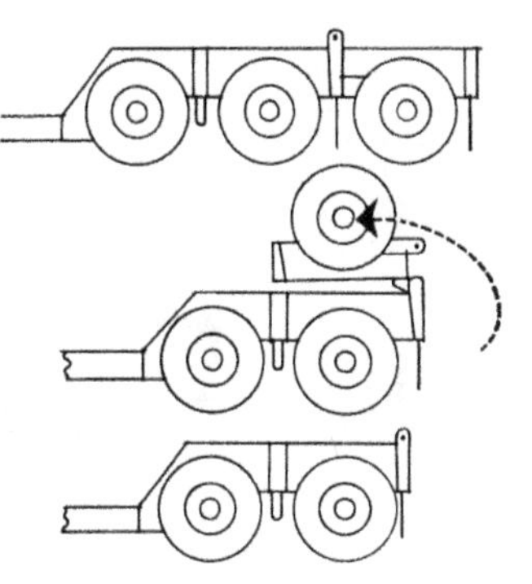

FIGURE 10.23 Flip-up third axle. (*Courtesy of Rogers Brothers Corporation.*)

to the main frame. In working position, the rear axle carries its full share of the load. When not needed, it can be unlocked, pivoted on top pins and swung up so it lies upside down on the deck. It can be disconnected and removed by pulling the pins and uncoupling the brake and electric lines. Reduction of overall length in this way reduces the need for or cost of overlength permits, makes the unit more maneuverable, and reduces tire wear.

Detachable Gooseneck. Some medium to heavy deck trailers are made so the gooseneck can be removed, leaving the front of the frame resting on the pavement or ground. Machinery can then be loaded readily using very short ramps (usually hinged to the deck) or small blocks.

The gooseneck and frame are held in alignment by removable pins or other safety locks. With these released and both brake and electric lines opened, the two units are lowered to the ground by a hydraulic jack in the gooseneck or by a line from a winch mounted on the tractor. The gooseneck is then detached from the frame and carried or dragged a short distance by the truck.

Girder or I-Beam Trailers. A trailer of this type is shown in Fig. 10.24. It is often not practical to transport shovels on deck trailers, because the combined height of trailer and shovel is too great to go under bridges along the route. This difficulty may be reduced by the use of a girder-type trailer, which supports the frame of the shovel and allows the tracks to hang a few inches off the ground.

Girder construction also eliminates the necessity of using long ramps, which may be so heavy as to require a crew to handle and are dangerous to use when slippery. Trailers built with aluminum framing can reduce dead weight but may be more susceptible to impact damage than heavier steel framing.

The tracks of a shovel on a girder trailer hang very low and may rest on the road on sharp humps or hill crests. If the tracks are not locked, they will reduce resulting drag by turning. If the machine hangs up badly, it can be walked forward while the trailer is towed at the same speed until the trailer frame will raise the tracks clear of the ground again.

Loose tracks can be prevented from hanging below the rollers by raising the upper section with a jack on the truck frame or by working pieces of wood in between the track and the bull wheel by turning it. Occasionally the tracks must be removed to obtain sufficient clearance from the ground when the machine is carried in the conventional manner.

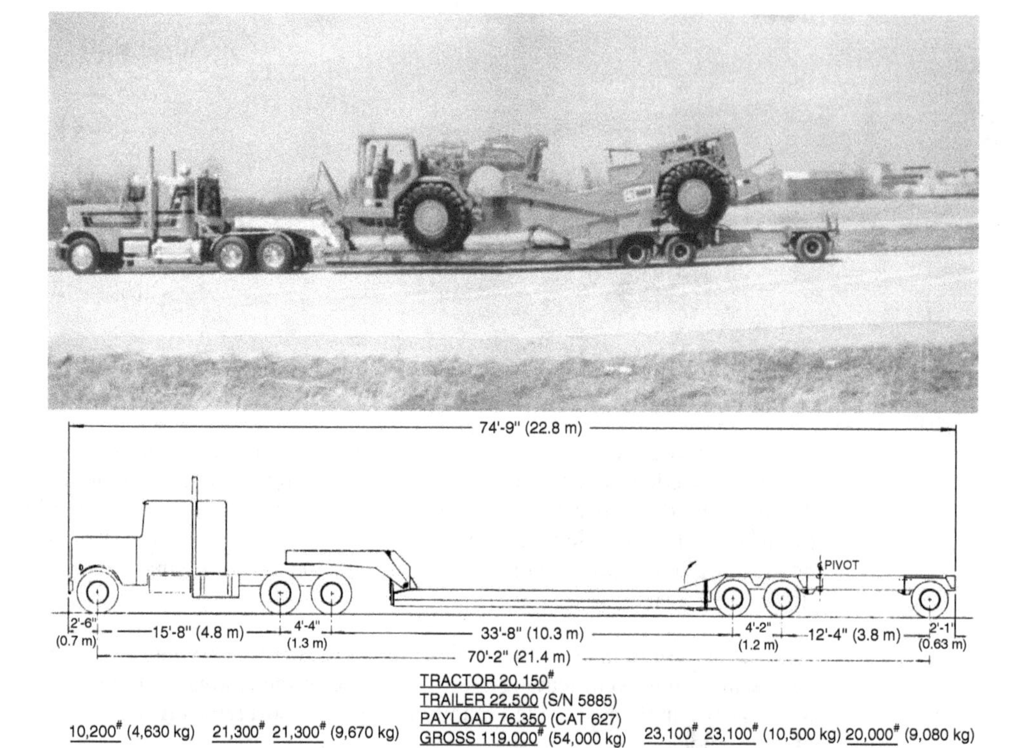

FIGURE 10.24 Loaded girder trailer, with dimensions. (*Courtesy of Talbert Manufacturing, Inc.*)

Low-Bed Trailers. Figure 10.25 shows two sizes of low-bed semitrailers used to carry machines weighing 10 tons or less. In order to reduce the overall width to 8 ft, the deck must be limited to a width of 6 ft 4 in.

The wheel stub axles and brake backplates of the lighter model are welded directly to the deck rails. The tandem wheels in the larger size are mounted on rocker arms, allowing front

FIGURE 10.25 Low-bed trailers.

and rear oscillation. The draw tongue is adjustable for height and is fastened to the truck by a vertical pin through a bracket on the rear crossmember. The connection is made by jacking the trailer and tongue to the required height and then backing the truck so the upper and lower holes in its tow bracket line up with the tongue hole. With these aligned, the pin can be inserted and locked. The trailer can tip somewhat because the pin is a loose fit in the hole.

A built-in hydraulic jack can be obtained as optional equipment. Ordinarily the trailer is blocked up when left. When this is done, it will not have to be jacked to reconnect. Deck height is 14 in. Length is 14 to 16 ft. Load is about 60% on the pintle hook of the towing truck.

Ramps are not required for loading crawler machines, and very short ramps are needed for wheeled vehicles. A pair of blocks placed behind the trailer will cause tracks to rear up so as to pass over the rear edge of the deck in loading and will ease them down in unloading.

Tilt Trailer. The deck of a tilt trailer is hinged to a subframe. The gooseneck or drawbar and the axles are parts of the subframe. When the deck is tilted, the back edge rests on the ground (Fig. 10.26), and a machine can be walked directly up the sloped deck as if it were a ramp. As its center of gravity passes the hinge, the deck automatically goes back to level position.

FIGURE 10.26 Narrow-tilt deck.

To load one of the simplest and smallest of these units, the trailer is attached to the truck, the deck lock is released, and the front of the deck is pulled up by hand until the rear rests on the ground. A machine is then driven from the ground directly up the sloping deck. When the machine has moved past the hinges, the deck will swing into carrying position and lock automatically.

The machine must be moved very slowly as it starts to overbalance the deck, which is liable to come down so rapidly as to strike a heavy blow on the lower frame and to subject the towing truck to shock. Damage to the truck can be prevented by supporting the tongue on blocks during loading.

To release blocks jammed under the tongue, move the machine to the back of the trailer with the deck locked. For transportation, the machine may be moved well forward so even if the lock fails, the deck will remain level.

Most tilt trailers are equipped with a hydraulic ram or other snubbing device, so the deck will move smoothly from loading to travel position. Heavy units may start the tilting motion by hydraulic power. Otherwise, they may be tilted by standing on the rear.

Large tilt trailers (maximum capacity about 25 tons) may have a regular gooseneck attachment to a dolly or fifth wheel. Smaller units have a rigid two-bar attachment to a special bracket on the rear of a dump or other truck. The bar usually includes a jack for raising it to the truck and for supporting it during parking and loading.

There may be a full-width solid deck over the wheels, requiring a deck height of 34 to 41 in. Slope in loading position may be 15 to 18 degrees—too steep for a weak machine or a slippery deck. This angle can be reduced by putting blocks under the edge before lowering it and putting blocks or very short ramps behind it.

Loading

Rear loading is standard practice. Self-propelled machinery is driven onto deck trailers over the back with the aid of ramps, blocks, or banks.

Ramps. Ramps may be made of planks of oak or other strong wood, 2 to 5 in. thick and 10 to 16 in. wide. A metal angle is fastened under one end of the plank and rests in an angle on a shelf on the rear of the trailer. Ramps may also be made of steel box or channel construction, or of wood reinforced with steel plates or angles.

Ramps are ordinarily so heavy two or more people are needed to handle them. It is often advantageous to place them with the machine being loaded. However, they are usually not strong enough to take the weight of a heavy machine unless supported between the ends. Short pieces of heavy timbers or railroad ties can be used to crib under the ramps.

When loading crawler machines, the point of greatest strain is near the ground where the tracks are forced upward into a climb. In unloading, bracing is particularly needed where the machine strikes the ramp as it pitches off the trailer deck. In addition to these critical points, it is good policy to have cribbing under one or two more points to reduce the length of the unsupported span. Thorough cribbing will make possible the use of lighter ramps and add materially to their life. Crawlers may be loaded either forward or backward, but are under better control when backed uphill, particularly if the tracks are loose.

Ramps supplied with a trailer are designed to load most machines on level ground. Their length should be at least 3 ft for each foot of trailer height. The incline should be considerably less than the machine is capable of climbing.

Rubber-tire machines impose a concentrated but fairly even strain on ramps. Rollers must have very slight gradients and need help from constructed material ramps or banks to maneuver onto a deck trailer. Ramps are set to line up with the tracks or wheels of the machine so it can move onto the trailer without turning. If the ramps are wider than the tracks, ramps are set so the operator can see an edge and steer accordingly. If they are narrow and the operator cannot watch them, a competent guide should direct the operator with signals.

The machine should be in its lowest gear and should move slowly, but with sufficient throttle to avoid stalling. It should not be steered sharply or stopped except in an emergency. When the top is reached, the upper end of the tracks will move upward above the tires and deck, and fall when the machine overbalances. A jarring fall can be avoided by moving very slowly. A sloping extension on the rear of the trailer will provide an intermediate grade between the ramp and deck, which reduces the abruptness of the fall. If the ramp is short, it can be extended by blocking.

Wet ramps are dangerous and should be sprinkled with sand. Steel surfaces may be too slick to climb even when sanded. Machines should never be loaded up ramps when there is snow or ice on the trailer surface.

Because the machine must walk over the tires after leaving the ramp, tracks should be inspected and any sharp pieces of metal removed. Grousers will not damage tires, but some varieties of ice cleats or chains may cause damage.

If the trailer can be backed against a bank or into a deep gutter, it may be possible to load a machine without the use of ramps and perhaps even without blocks. Lower banks or shallower gutters may be useful in making it possible to ramp up at an easier gradient than from a level.

Pieces that are wider than the trailer may be jacked and blocked up, the trailer backed under them, and the blocks knocked out. Measure all width dimensions to ensure successful trailering.

Protection of Pavements. When crawler machines are loaded or walked on paved streets, the road surface must be protected. Damage can be caused by side-scuffing during turns and by the digging in of track grousers. Crushing is particularly likely on any asphalt pavement. This is guarded against by laying planks or boards as in Fig. 10.27 to spread weight

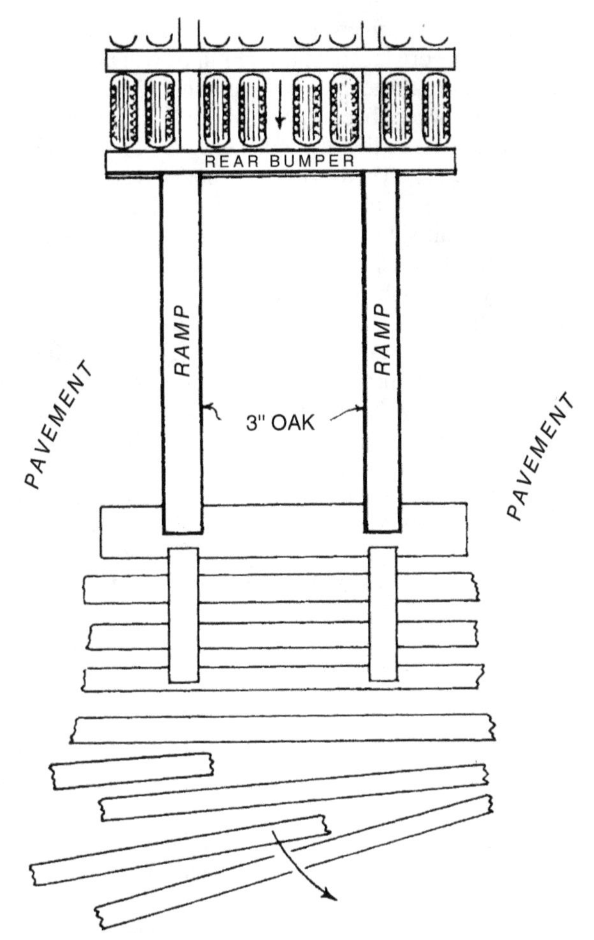

FIGURE 10.27 Protecting pavements when loading or unloading track machines.

over a wider area. Note the extra boards where the front of the tracks tends to dig in just below the ramps. For loading rubber-tired equipment from asphalt pavement, turn the wheels only when the equipment is moving. Do not turn the wheels when the equipment is stationary.

In many localities, it is unlawful to walk crawler machines on streets. Puncturing pavements and scuffing on turns can be prevented by laying thin boards under the tracks. A single thin board, narrower than the track, may protect hard pavements, whereas softer ones may need wide planks or narrow planks in pairs. If the walk path is long, the same boards can be used repeatedly. Digging in of grousers may be prevented by walking on boards in the same manner.

Side Loading. It is often more convenient to load from the side. Normal curb and sidewalk heights above the gutter permit use of shorter and lighter ramps. Drop-deck models are lower at the side than at the back. In addition, the machine may have to enter or leave the job through a driveway or other narrow access. By side loading, it can be walked directly into the drive, whereas rear ramps will put it on the street pavement and necessitate a right-angle turn, often under tight conditions. Also, it is possible to pull up alongside a roadside bank where backing would not be practical.

Trailers are not ordinarily equipped with ramp-supporting rails on the sides, but they can be welded on easily. There is usually enough recess in the girders to do this without increasing width. Unless it is blocked up, the trailer will tip sideward under the weight of the machine as it climbs on. This simplifies loading from blocks or a low bank and may make it possible for a large tractor to walk on from ground level, pulling the trailer down with its grousers. But tipping may make unblocked loading dangerous or impossible if the deck is slippery with rain.

The machine can be carried crosswise or turned on the deck. If the machine is large in proportion to the deck space, turning may be difficult or impossible. Machines narrower than the deck can turn rather readily, unless the planking is wet and slippery. The front corners of the deck should be blocked up before turning. Limit the height of these blocks, as it is important the deck not rest on them after the machine is loaded.

Transporting. When the machine is correctly positioned, it is left in gear with the clutch engaged and the brakes locked. The ramps and blocks are loaded onto the deck, and the machine is secured against sliding by blocks or chains, or both.

In most states, truck and trailer widths are limited to 8 ft. Wider vehicles must obtain special permits; the permit may limit the route and time of day of the haul.

TRUCK PRODUCTIVITY

The productivity of a truck depends on the size of its load and the number of trips it can make in a unit of time. The number of trips completed per hour is a function of cycle time. Truck cycle time has four components: (1) load, (2) haul, (3) dump, and (4) return. Load time is a function of the number of bucket cycles to fill the truck body. The haul and return times will depend on truck weight, engine horsepower, haul and return distances, and condition of the roads traversed. Dump time is a function of the type of equipment and conditions in the dump area.

When an excavator, loader, hoe, or shovel is used to load material into trucks, the size of the truck cargo body introduces several factors, which affect the production rate and the cost of handling the material. Table 10.3 compares small and large truck sizes.

Calculating Truck Productivity

The most important consideration when matching excavators and trucks is finding equipment with compatible capacities. Balancing the capacities of hauling units with the excavator bucket size and production capability is important. Matched capacities yield maximum loading efficiency and reduced project costs. When loading with hydraulic hoes, shovels, or loaders, it is desirable to use haul units whose cargo body volume is balanced with the excavator bucket volume. If this is not done, operating difficulties will develop, and the combined cost of excavating and hauling material will be higher than when a balance between trucks and excavators is achieved.

A practical rule of thumb frequently used in selecting the size of trucks is to use trucks with a capacity of four to five times the capacity of the excavator bucket. The following format can be used to balance the equipment spread and calculate truck production.

TABLE 10.3 Truck Size Comparison

Truck Size	Advantages	Disadvantages
Small (on road)	Maneuvering flexibility, especially on restricted work sites Speed, can achieve higher haul and return speeds Production, less impact if one truck breaks down Balance of fleet, easy to match number of trucks to excavator production Versatility, can haul a variety of materials: soils, rock, asphalt, snow, scrap Economics, lower initial cost	Number, more trucks increases operational dangers at load-out, along the haul road, and at the dump Labor cost, more drivers required for a given output Loading impediment, small target for excavator or loader operator Positioning time, total spotting time greater because of the number required Haul surface, generally limited to on-road or well-compacted pavements Bunching possibility with more haul units
Large (off road)	Number, fewer needed for a given output Drivers required, fewer needed for a given output, reducing cost Loading advantage, larger target for the excavator or loader operator Positioning time, frequency of spotting trucks is reduced Suited for off-road terrain Less bunching with fewer haul units	Loading time greater, especially with small excavators Loads heavier, possible damage and increased maintenance to the haul road Balance of fleet, difficult to match number of trucks to excavator production Size, may not be permitted to haul on highways Economics, higher initial cost Transportation, requires trailering between projects

Number of Bucket Loads. The first step in analyzing truck production is to determine the number of bucket loads it takes to load the truck.

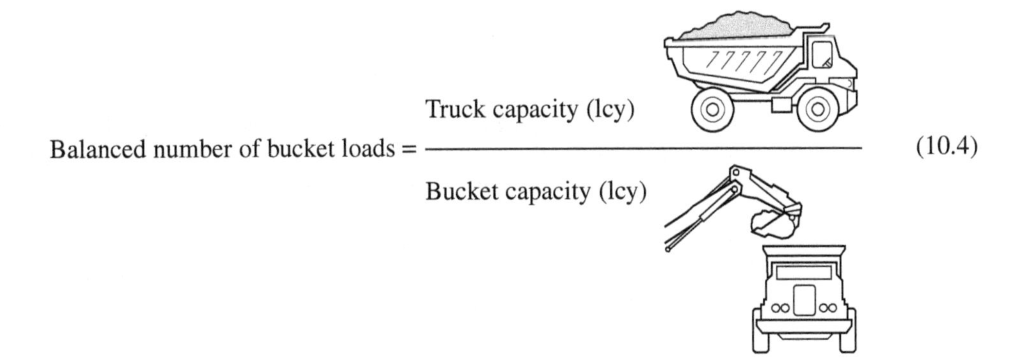

$$\text{Balanced number of bucket loads} = \frac{\text{Truck capacity (lcy)}}{\text{Bucket capacity (lcy)}} \tag{10.4}$$

Short load
Bucket or truck box not filled to capacity.

The actual number of bucket loads placed on the truck must be an integer number. It is possible to not completely fill the bucket (**short load**) to match the bucket volume to the truck volume, but such a practice is inefficient, as it results in wasted loading time.

If one less bucket load is placed on the truck, the loading time will be reduced, but the load on the truck is also reduced. Sometimes job conditions will dictate a lesser number of bucket loads be placed on the truck, such as steep grades or bunching of multiple trucks waiting to load. The truckload in such cases will equal the bucket volume multiplied by the number of bucket loads.

Next Lower Integer. For the case where the number of bucket loads is *rounded down to an integer lower* than the balance number of loads or reduced because of job conditions:

$$\text{Load time} = \text{Number of bucket loads} \times \text{Bucket cycle time} \tag{10.5}$$

$$\text{Truckload}_{\text{Short}} \text{ (volumetric)} = \text{Number of bucket loads} \times \text{Bucket volume} \tag{10.6}$$

Next Higher Integer. If the division of truck cargo body volume by the bucket volume is *rounded to the next higher integer* and the higher number of bucket loads is placed on the truck, excess material will either spill off the truck or portions remain in the bucket. In such a case, the loading duration equals the bucket cycle time multiplied by the number of bucket swings. The volume of the load on the truck now equals the truck capacity, not the number of bucket swings multiplied by the bucket volume.

$$\text{Truckload}_{\text{Full}} \text{ (volumetric)} = \text{Truck volumetric capacity} \tag{10.7}$$

Gravimetric Check. Always check the load weight against the gravimetric capacity of the truck.

$$\text{Truckload (gravimetric)} = \text{Volumetric load (lcy)} \times \text{Unit weight (loose vol. lb/lcy)} \tag{10.8}$$

$$\text{Truckload (gravimetric)} < \text{Rated gravimetric payload?} \tag{10.9}$$

Haul Time. Hauling should be at the highest safe speed and in the proper gear. To increase efficiency, use one-way traffic patterns.

$$\text{Haul Time (min)} = \frac{\text{Haul distance (ft)}}{88 \text{ fpm/mph} \times \text{haul speed (mph)}} \tag{10.10}$$

Based on the gross weight of the truck with the load, and considering the rolling and grade resistance from the loading area to the dump point, haul travel speeds can be estimated using the truck manufacturer's performance chart (Fig. 10.28).

Return Time. Based on the empty vehicle weight and the rolling and grade resistance from the dump point to the loading area, return travel speeds can be estimated using the truck manufacturer's performance chart.

$$\text{Return time (min)} = \frac{\text{Return distance (ft)}}{88 \text{ fpm/mph} \times \text{haul speed (mph)}} \tag{10.11}$$

Dump Time. Dump time will depend on the type of hauling unit and congestion in the dump area. Dumping areas are usually crowded with support equipment. Dozers or motor graders are spreading the dumped material, and multiple pieces of compaction equipment

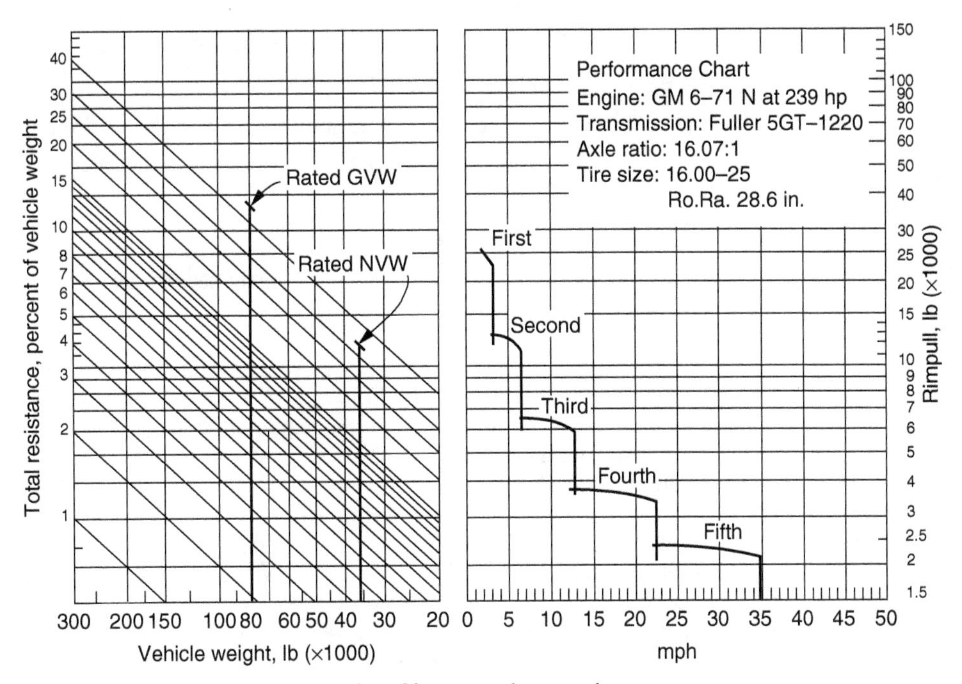

FIGURE 10.28 Performance chart for a 22-ton rear-dump truck.

may be working in the area. Rear dumps must be spotted before dumping. The truck must come to a complete stop and then back up some distance. Total dumping time in such cases can exceed 2 min. Bottom dumps will customarily dump while moving. After dumping, the truck normally turns and returns to the loading area. Under favorable conditions, a rear-dump can dump and turn in 0.7 min, but an average unfavorable time is about 1.5 min. Bottom-dumps can dump in 0.3 min under favorable conditions, but they, too, may average 1.5 min when conditions are unfavorable. Try to visualize the conditions in the dump area when estimating dump time.

Sample Calculation: Determine the speed of a truck when it is hauling a load of 22 tons up a 6% grade on a haul road with a rolling resistance of 60 lb per ton, equivalent to a 3% adverse grade. The performance chart in Fig. 10.28 applies, and the truck specifications are as follows:

Capacity

Stuck, 14.7 cy

Heaped, 2:1, 18.3 cy

Net weight empty	= 36,860 lb
Payload	= 44,000 lb
Gross vehicle weight	= 80,860 lb

It is necessary to combine the grade and rolling resistance, which gives an equivalent total resistance equal to 9% (6% + 3%) of the vehicle weight.

The procedure for using the chart in Fig. 10.28 is as follows:

1. Find the vehicle weight on the lower-left horizontal scale.
2. Read up the vehicle weight line to the intersection with the slanted total resistance line.
3. From this intersection, read horizontally to the right to the intersection with the gear curve.
4. From this intersection, read down to find the vehicle speed.

Following these four steps, the truck will operate in the second-gear range, and its maximum speed will be 6.5 mph.

The truck's performance chart should be used to determine the maximum speed for each section of haul road with a significant difference in grade or rolling resistance.

Although a performance chart indicates the maximum speed at which a vehicle can travel, the vehicle will not necessarily travel at this speed. A performance chart makes no allowance for acceleration or deceleration. Additionally, other travel route conditions and safety can control travel speed. Before using a performance chart speed in an analysis, always consider such factors as congestion, narrow haul roads, or traffic signals when hauling on public roads, because these can limit the speed to less than the value given in the chart. The anticipated effective speed is what should be used in calculating travel time. Project estimators should drive the haul route with an experienced operator to identify factors affecting haul speeds.

Truck Cycle Time. The cycle time of a truck is the sum of the load time, the haul time, the dump time, and the return time:

$$\text{Truck cycle time} = \text{Load}_{\text{time}} + \text{Haul}_{\text{time}} + \text{Dump}_{\text{time}} + \text{Return}_{\text{time}} \tag{10.12}$$

Number of Trucks Required. The number of trucks required to keep the loading equipment working at capacity is

$$\text{Balanced number of trucks} = \frac{\text{Truck cycle time (min)}}{\text{Excavator cycle time (min)}} \tag{10.13}$$

The number of trucks must be an integer number.

Integer Lower Than Balance Number. If an integer number of trucks lower than the result of Eq. (10.13) is chosen, then the trucks will control production.

$$\text{Production (lcy/hr)} = \text{Truck load (lcy)} \times \text{Number of trucks} \times \frac{60\text{ min}}{\text{Truck cycle time (min)}} \tag{10.14}$$

Integer Greater Than Balance Number. When an integer number of trucks greater than the result of Eq. (10.13) is selected, production is controlled by the loading equipment.

$$\text{Production (lcy/hr)} = \text{Truck load (lcy)} \times \frac{60\text{ min}}{\text{Excavator cycle time (min)}} \tag{10.15}$$

Balance point The situation where the loading equipment and hauling units operate at the same production level.

The **balance point** is when the loading equipment and hauling units operate at the same production level. As a rule, round down and keep the hauling units moving. This allows the loading equipment time to dress the load site, fill the bucket, and be ready to load the next hauling unit. If the single loader is constantly used, it presents a greater risk to job production if a mechanical failure were to occur. A constant 100% operation restricts routine maintenance when trucks are continually arriving to load. The breakdown of a single haul unit has a lesser impact on production than breakdown of a single loader. If there is an insufficient number of haul trucks, there will be a loss in production. If there are extra haul units, bunching can occur, driving up costs. The dump site can move closer to or farther from the load site. This feature must be considered when designing the operation. Therefore, as a starting basis, it is usually best to round down to the next integer unless conditions warrant. A cost analysis is needed to assist in this decision.

Efficiency. The production calculated with either Eq. (10.14) or Eq. (10.15) is based on a 60-min working hour. Production must, therefore, be adjusted by an efficiency factor. Longer hauling distances usually result in better driver efficiency. Driver efficiency increases as haul distances increase out to about 8,000 ft, after which efficiency remains constant. Other critical elements affecting efficiency are bunching, equipment condition, load and dump area congestion, fueling and light maintenance, operator breaks, work rules, and project planning and layout.

$$\text{Actual production} = \text{Ideal production} \times \frac{\text{Working time (min/hour)}}{60 \text{ min}} \tag{10.16}$$

Sample Calculation: A loader operating at 107 cy/hr has a 2.5-cy heaped capacity bucket and swings on a 1.2-min load cycle. The trucks have 11-cy capacity and a 30 min dump + haul + return time. A 95% fill factor applies to the loader due to material repose in the bucket. The project works 50-min hours.

$$\text{Balanced number of bucket loads} = \frac{11 \text{ cy}}{2.5 \text{ cy}} = 4.4; \text{ load a truck with 4 buckets}$$

$$\begin{aligned}\text{Truck Cycle Time} &= \text{Load} + \text{Haul} + \text{Dump} + \text{Return}\\ &= (1.2 \text{ min} \times 4 \text{ buckets}) + 30 \text{ min} = 34.8 \text{ min.}\end{aligned}$$

$$\text{Truck payload} = 2.5 \text{ cy} \times 4 \text{ buckets} \times 0.95 \text{ fill factor} = 9.5 \text{ cy}$$

$$\text{Truck production} = \frac{9.5 \text{ cy/cycle}}{34.8 \text{ min/cycle}} \times \frac{50 \text{ min}}{\text{hour}} = 13.65 \text{ cy/hr}$$

$$\text{Balanced number of trucks} = \frac{34.8 \text{ min}}{4.8 \text{ min}} = 7.25, \text{ round down to 7 trucks}$$

Each truck added to the job will increase production by 13.7 cy/hr; however, the loader will control and limit total job production at 107 cy/hr.

The next step is to determine the job production as trucks are added:

6 trucks: 13.7 cy/hr × 6 trucks = 82.2 cy/hr
7 trucks: 13.7 cy/hr × 7 trucks = 95.9 cy/hr
8 trucks: 13.7 cy/hr × 8 trucks = 109.6 cy/hr, loader controls at 107 cy/hr
9 trucks: 13.7 cy/hr × 9 trucks = 123.3 cy/hr, loader controls at 107 cy/hr

A balance curve is developed to illustrate the number of trucks with production

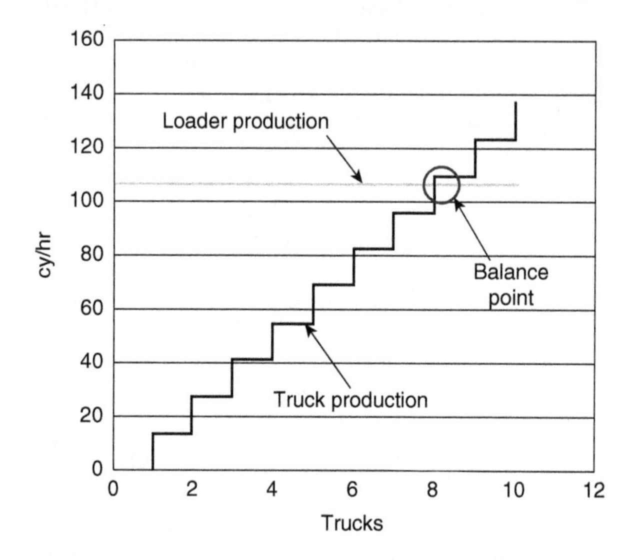

In this calculation, the production of the trucks is based on a 50-min hour. This policy should be followed when balancing an excavator or loader with hauling units because at times both will operate at maximum capacity if the number of units is properly balanced. However, the average production of a unit, excavator or truck, for a sustained period of time should be based on applying an appropriate efficiency factor to the maximum productive capacity.

Cost. A fundamental issue in optimizing production is the effect on cost. Production levels can be adjusted based on number of haul units, as described earlier, but the impact on costs must be understood. Developing unit costs provides a simple way to understand the effect of production and number of haul units.

Sample Calculation: Using production values from the previous calculation, unit costs are calculated using these hourly rates:

Loader with operator = \$90/hr
Truck with operator = \$55/hr

The unit cost in \$/cy is calculated for each truck combination ±2 trucks from the balance point of 7 trucks.

5 trucks: \$90/hr + (\$55/hr)(5 trucks) = \$365/hr ÷ 68.5 cy/hr = \$5.33/cy
6 trucks: \$90/hr + (\$55/hr)(6 trucks) = \$420/hr ÷ 82.2 cy/hr = \$5.11/cy

7 trucks: $90/hr + ($55/hr)(7 trucks) = $475/hr ÷ 95.9 cy/hr = $4.95/cy

8 trucks: $90/hr + ($55/hr)(8 trucks) = $530/hr ÷ 107 cy/hr = $4.96/cy

9 trucks: $90/hr + ($55/hr)(9 trucks) = $585/hr ÷ 107 cy/hr = $5.47/cy

The number of trucks is plotted against the unit production cost to visualize the impact.

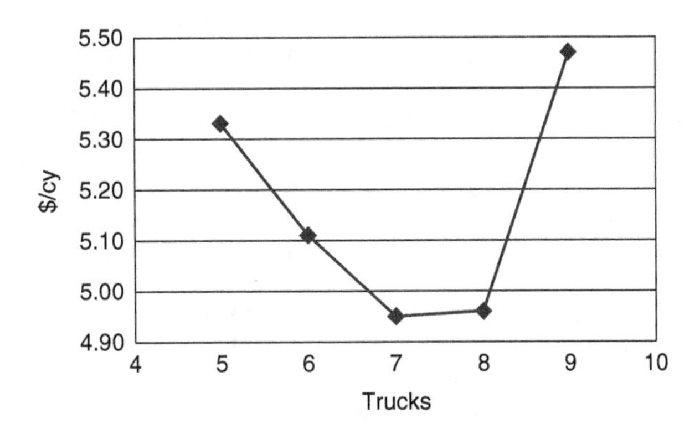

Even though the five-truck combination has the lowest hourly cost of $365/hr, it has one of the higher unit production costs: $5.33/cy. A smaller number of trucks, such as three or four, would further increase production costs. Adding more trucks beyond the 7.25-truck balance point also increases the unit production costs. The lowest production costs are the integer values on each side of the balance point: seven trucks and eight trucks. As discussed earlier, the lower integer value can be selected to avoid bunching and constant operation of the loading equipment. In this example, rounding to the higher integer value yields nearly the same cost of approximately $4.95/cy.

Production Issues. A number of other factors must be considered when matching loading and hauling units:

- Positioning the truck for loading
- Reach of the excavator
- Dumping height of the bucket
- Width of the bucket

Positioning the Truck for Loading. The truck should be positioned for loading in an efficient and safe manner. Forward-only movement of the truck in the load area is desired. It is easier and safer for the driver to position the haul unit and reduce overall truck cycle time by eliminating backing. Backing also occupies the same travel area twice—once to back and another to move forward—creating a small choke point in the operation. Backing into the load area creates a natural safety concern because the driver has limited ability to see behind the haul unit to accurately position for loading. A single loading lane should be created and adjusted as the operation continues. If backing is required, a V-pattern layout should be planned where the haul unit arrives on one leg (lane) and both backs empty and moves forward full on the second lane. As the haul unit arrives on the first lane, the driver can survey the backing lane and target the loading unit. The backing lane should be kept as straight as possible.

Reach of the Excavator. The excavator must be able to physically reach the dumping point over the cargo body of the truck. In the case of a shovel or hoe, this involves the reach of the stick and boom when extended at dump height. A loader's reach is measured from the front of its front tires to the tip of the bucket's cutting edge when the boom is fully raised and the bucket is dumped at 45 degrees.

Bucket Dump Height. Compare the excavator's bucket dump height to the height of the cargo body sides. This comparison must consider the actual configuration of the excavator and its bucket. The teeth on a rock bucket can reduce dump height by as much as a foot. Sideboarding the cargo body or using larger tires on the truck can increase its height significantly. A loader ramp can be built to achieve desired dump height.

Width of the Bucket. Compare the width of the excavator bucket to the length of the truck's cargo body. The ratio of the bucket width to cargo body length should fall within the range of 1:1.4 or 1.5. The bucket width should not make it difficult for the excavator operator to avoid striking the back of the truck cab or dumping material too close to the end of the cargo body. When bulky shaped material pieces, such as large rocks or frozen soil, land too close to the end of the dump body, they are likely to fall off the truck as it climbs a grade. Rocks spilled on the haul road can damage tires, and there will be an additional requirement for road maintenance. Even if the load does not roll off the truck, it places an undesirable weight distribution on the rear axle, which can increase axle and tire wear.

Real-Time Monitoring. Global positioning system (GPS) technology permits real-time monitoring and control of haul trucks. To determine the exact position of each haul unit, the GPS receiver must be able to lock on to three satellite signals. By measuring the difference between the time the signal was sent and the time it was received from these three satellites, the receiver can calculate latitude, longitude, and altitude of the haul unit. This technology is commonly used in highway vehicles and handheld GPS receivers.

Managing the GPS data and monitoring the haul units is a two-part system. First, the equipment must be installed into the vehicles and then a system set up at a home base to receive the information. The type of system can vary depending on the specific needs of each operation. Delivery companies, airlines, and railroads have widely adopted this technology. For example, an operation with a small fleet may only require basic vehicle tracking. This system shows location, route, stops, and speed of the truck. A company with a larger fleet or one that needs more detailed information may select a real-time GPS tracking system. These systems can provide location updates every few minutes, automatically email reports, provide digital and satellite maps, and help keep a record of vehicle maintenance. Transportation agencies have used GPS mapping approaches to monitor snowplow activities and even to identify roads requiring snow-clearing efforts. An important construction application has been with ready-mix concrete trucks hauling between a central plant and project sites spread over a large geographic area.

TRUCK SAFETY

The driver of a dump truck was killed in 2002 when he backed the truck too close to the edge of an embankment which gave way. A signal person told the driver to stop approximately 8 ft before the edge and had given the signal to dump the load. The driver, however, continued to back the truck. The ground under the rear wheels gave way, and the truck slid backward down the embankment and then rolled over, landing on its roof. The embankment berm was insufficient to stop the truck when it moved close to the edge. The truck was supplied with a lap-type safety belt, which the victim was wearing.

In 2001, a traffic control person died when a dump truck backed over him. The truck was delivering asphalt to a highway crew. The victim was directly behind the truck when it began backing.

Operating trucks and working around trucks can be hazardous. The employers of the individuals killed in these two truck accidents had well-established safety programs and written accident prevention plans. Still, two fatal accidents occurred. A 6-ft-tall person standing closer than 70 ft from the right side of a 150-ton off-highway truck cannot be seen by the driver. Employees must be mentored daily about the dangers, and management must be exceedingly proactive if accidents are to be prevented.

The investigators of these two accidents made the following recommendations:

- Conduct a hazard assessment of the work site every day, and ensure operators and drivers are aware of the hazards.
- Ensure berms are adequately built to prevent trucks from over-traveling at dump locations and on haul roads.
- Confirm heavy equipment operator conformance with all operating signals.
- Use a spotter when backing heavy equipment with blind spots.
- Develop procedures for keeping employees from entering vehicle travel paths and keeping ground personnel in clear view of those who are operating equipment.
- Wear high-visibility safety clothing.

CHAPTER 11
SCRAPERS

Both the push tractor and the scraper share in the work of obtaining the scraper load, and this sharing leads to excavation and haul economy. The dual-engine scraper has the capability to load itself without a push tractor and to climb steeper grades. The production cycle for a scraper consists of six operations: (1) loading, (2) haul travel, (3) dumping and spreading, (4) turning, (5) return travel, and (6) turning and positioning to pick up another load. A systematic analysis of the individual cycle elements is the fundamental approach for identifying the most economical employment of these machines. Scrapers are best suited for haul distances longer than 500 ft but shorter than 3,000 ft, although with very large units, the maximum distance can approach a mile.

WHEEL TRACTOR-SCRAPERS

Wheel tractor-scrapers (Fig 11.1) or bottom-dump scrapers are highly mobile excavators with a centrally located bowl used to excavate, carry, and spread the load. They are sometimes referred to as pans. These machines are available in a wide range of types and sizes. Heaped capacity is usually between 15 and 44 yards, but there are larger and smaller units. There are also models especially designed for hauling lightweight materials, like coal, and these can carry up to 73 cy.

The modern scraper is a self-powered, rubber-tire unit with hydraulic controls. Conventional single-powered axle models have power and traction sufficient for most hauling needs but require pusher-tractor assistance to load efficiently. Self-loading scrapers do not require pusher-tractor assistance—these models may have two engines with separate drive axles for extra power. There are also elevator cutter and auger self-loading models.

Scrapers are the standard tool for alternating cut and fill operations. They can operate under a wide range of conditions where the surface of the excavation is firm enough to support the weight of the scraper.

A scraper loads, hauls, and spreads in a single cycle. It loads thin layers in the cut and usually places thin layers on the fill. Therefore, its efficiency is not particularly affected by depth of cut or height of fill. The pneumatic tires of the scraper provide limited compaction of fills, but compaction equipment is still needed to achieve full densification of the lifts.

Cutting edge The ground-engaging leading edge of the scraper bowl.

A scraper is an excellent machine for bulk earthmoving and is often used as a support machine for finishing work. The **cutting edge** is carried between the front and rear wheels, eliminating pitching and providing the operator accurate cutting control. This positioning allows the operator to cut or fill precisely to grade.

FIGURE 11.1 Single-powered axle wheeled tractor-scraper.

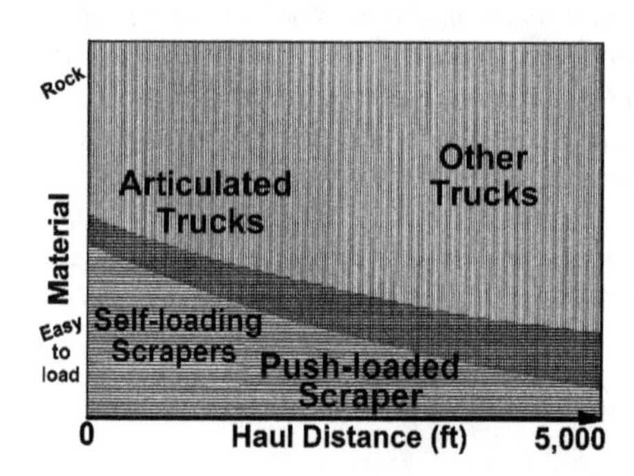

Because scrapers are a compromise between machines designed exclusively for either loading or hauling, they are not superior to function-specific equipment in either hauling or loading. Excavators, such as hydraulic hoes and front shovels or loaders, will usually surpass scrapers in terms of loading efficiency.

Trucks have faster travel speeds and, therefore, surpass scrapers in hauling, especially over long distances. However, for off-highway situations with hauls of less than a mile, a scraper's ability to both load and haul gives it an advantage. Moreover, the ability of these machines to deposit their loads in layers of uniform thickness facilitates compaction operations.

SCRAPER TYPES

There are several types of scrapers, primarily classified according to the number of powered axles or by the method of loading. Wheel-tractor machines have a hydraulically operated wheeled trailer–bowl. In the past, "pulled" two-axle scraper bowls were manufactured to be towed by crawler-tractors. These proved effective in short-haul situations, less than 600 ft one way.

crawler-tractor "pulled" two-axle scraper

Today there are single-axle "pulled" scraper bowls designed to be towed by a larger farm-type tractor with eight wheels or rubber tracks (Fig 11.2). Such light-duty towed "ag" scrapers differ from purpose-built construction scrapers primarily in weight and are not usually supported by a push tractor. They are useful machines for leveling roads, blading lots, and repairing soil erosion. The larger models can be used on construction projects and require a crawler-tractor pusher to load efficiently.

The farm-type tractor with eight wheels or equipped with rubber tracks is ideal for towing ag scrapers on softer soils or sticky clay. The tow-behind scraper arrangement frees the tractor asset for other work—disking, rental for farming planting/harvesting during peak seasons, and winter manure spreading. Most ag scrapers have a single rear axle with either two or four pneumatic tires.

FIGURE 11.2 "Ag" scraper towed by farm-type tractor on rubber tracks.

Construction-type scrapers currently available include:

Pusher-loaded (conventional)

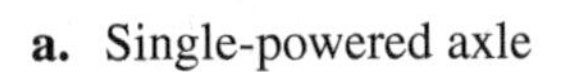

a. Single-powered axle

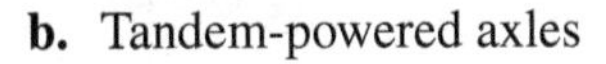

b. Tandem-powered axles

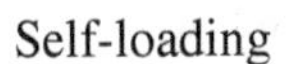

Self-loading

a. Push-pull, tandem-powered axles

b. Elevating

c. Auger

Single-Engine Scraper

On well-maintained haul roads, push-loaded wheel-type tractor scrapers have the potential for high travel speeds. Many models can achieve speeds as great as 35 mph when fully loaded. This extends the economic haul distance of the units. However, these units are at a disadvantage when it comes to individually providing the high tractive effort required for economical loading. For the single-powered axle scraper (Fig. 11.1), only a portion, on the order of 50% to 55% of the total loaded weight, bears on the drive axle.

In most materials, the coefficient of traction for rubber tires is less than for tracks. Therefore, it is necessary to supplement the loading power of these scrapers. The external source of loading power is usually a crawler-tractor pusher. Even with pusher assistance, loading costs are relatively low because both the scraper and the pusher share in providing the total power required to obtain a full load.

Single-powered axle pusher-loaded wheel-tractor scrapers are suited for jobs where haul-road rolling resistance is low and grades are minimal. They become uneconomical when

- Haul grades are greater than 5%
- Return grades are greater than 12%

Twin-Engine Scraper

Tandem-powered axle wheel-tractor scrapers have separate engines for the wheel-tractor unit and for the scraper (bowl) unit (Fig. 11.3). This doubles the scraper power, and traction is often more than doubled because the second engine is mounted at the rear, above the ejector cylinder, so all wheels are drivers. The unit may be called a two-engine, tandem-powered, or four-wheel-drive scraper.

FIGURE 11.3 Push-pull, wheel-tractor scrapers with tandem-powered axles.

The engines must be compatible with each other in terms of controls and performance, and equipped with torque converters and power-shift transmissions. They are controlled from the operator's cab in the front, and under ordinary conditions are coordinated in terms of speed and transmission ratio.

The second engine and drivetrain give better loading, as all axles are powered, and the power-to-weight ratio is much greater. The scraper has substantial advantages in terms of maneuverability and acceleration, an ability to climb steep or slippery grades, and an ability to operate independently instead of always paired. Two-engine scrapers can usually work alone in easy to medium digging. In hard digging they need, or at least benefit from, pusher help.

The twin-engine arrangement produces extra power for overcoming high rolling resistance and/or steep grades. However, even tandem-powered scrapers often require help loading. The initial cost differential between a tandem-powered scraper and a single-powered axle scraper of similar size is about 25%. Consequently, they are commonly considered a specialized machine good for opening up a job, working extremely adverse grades, or working in soft ground conditions.

Self-Loading Scrapers

Self-loading scrapers, although heavier and more costly to purchase and operate than comparable conventional scrapers, can be economical in certain applications, particularly in isolated work and for stripping materials.

Push-Pull Scrapers

These are basically tandem-powered axle scrapers with a cushioned-push block and bail mounted on the front (Figs. 11.3 and 11.4) and a hook on the rear above the usual rear push block. These features enable two scrapers to assist one another during loading by pushing and pulling one another. This arrangement is often referred as a "twin-hitch" operation.

FIGURE 11.4 Hook on the rear and cushioned-push block and bail on the front of two push-pull wheel-tractor scrapers.

To load both scrapers, the front scraper begins to load, and immediately the rear scraper comes up behind it and establishes pushing contact between the bumpers. Then the rear scraper lowers its bail onto the lead scraper's hook. It is a loose fit, so exact alignment of the two scrapers is not necessary. Loading of the front scraper is continued until the load appears adequate to the operator of the rear scraper, who then signals the front operator. The front scraper lifts its bowl, and the rear scraper lowers its bowl. The rear scraper ceases pushing, and the push block loses contact. Now the hook of the lead scraper is pulling the bail of the rear scraper. When the rear scraper has a full load, the bowl is lifted, and tension on the hook is decreased because the scraper speed will increase so the bail can be lifted automatically. The two scrapers then operate independently traveling to the fill. They will hook up again when they return to the cut. This arrangement has great advantages in terms of job efficiency with no push tractor; however, the load time is effectively doubled with one scraper assisting another. A goal is to keep loading time to under 1 min per scraper; loading that extra yard with a nearly full bowl is detrimental to hauling production.

Two push-pull scrapers can work without assistance from a push tractor, or they can function individually with a pusher when their total traction is not sufficient for achieving good loads under unfavorable conditions. When used in rock or abrasive materials, tire wear on these scrapers will increase because of more slippage from the four-wheel-drive action.

Elevating Scrapers

These are completely self-contained loading and hauling scrapers (Fig. 11.5). Pusher help is not required, and is not even useful under ordinary conditions. If a pusher is used, it is usually a crawler when the ground is too slippery for tires, and it must be operated with care to avoid damaging the elevator. Because of the scraper's mechanics of simultaneously excavating and loading, the material in the bowl is mixed and blended.

FIGURE 11.5 Elevating wheel tractor-scraper.

The bowl front of an elevating wheel tractor-scraper is an elevator made up of two roller chains carrying a number of crossbars called flights (Fig. 11.6). Therefore, these scrapers do not have an apron to retain the material in the bowl. The foot of the elevator mechanism is near the bowl cutting edge—it slopes back 40 to 45 degrees and extends above the sides of the bowl. It is driven through reduction gears, which are connected by a rigid cross shaft.

FIGURE 11.6 Chain elevator loading mechanism on the front of the scraper bowl.

Excavation of material is accomplished by lowering the bowl and rotating the elevator flights. As the flights rotate, they chop the material just ahead and above the cutting edge. The backward motion of the flights (which is considered to be "forward" rotation) carries the cut material into the bowl, along with any material loosened by the cutting edge. The flight arrangement varies among different makes and models. Loaded material cannot fall out because of the narrowness of the space between the edge and the elevator and the continued motion of the flights.

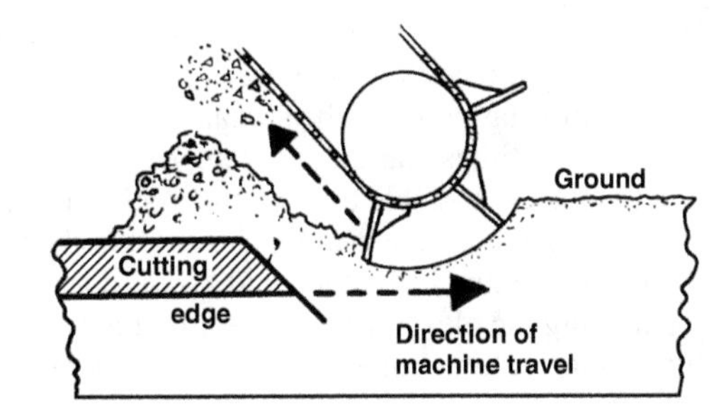

Elevator angle and height permit a full load to be acquired and held between the sloping rear side of the elevator and the ejector. The elevator is held in line at the bottom by a pair of brackets, which may be spring-loaded to permit the elevator to ride up and over boulders entering the bowl. There may be a mechanical adjustment of clearance to suit different types of digging.

The elevator may have a single or multiple forward speeds, usually between 100 and 300 ft per minute. Means to reverse may be provided as standard or optional equipment. During loading, which is done in forward gear, the elevator may use half the power of the engine. An overload valve or switch will stop the elevator if it jams.

Reverse is used for dislodging obstacles or sticky soil and occasionally to prevent soil from entering the bowl during light grading. The allocation of available power to the elevator or to the drive wheels is usually efficiently adjusted by the torque converter delivering slower motion with multiplied torque to the wheels, while a power takeoff or generator delivers full-speed power to the elevator. This difference enables the elevator to retain its rotational speed when the forward scraper motion slows in hard digging.

The general shape of the bowl is similar to a conventional scraper bowl, but its structure is different to allow for ejection. It is not practical to lift the elevator up and forward in the manner of an apron, because of a lack of space under the gooseneck. An elevating scraper has a two-part bowl floor. An opening for load discharge is created by pulling the front half of the bowl floor to the rear. This dumps part of the material. At the same time, the rear wall of the bowl is moved forward, usually in conventional ejector fashion, pushing the rest of the load out the gap left by the backsliding floor.

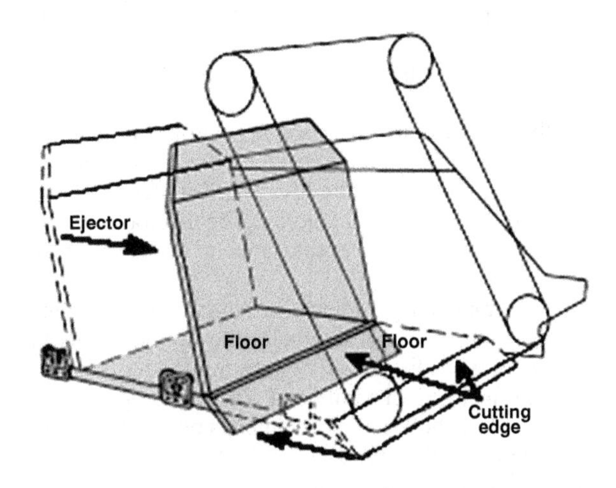

During unloading, the machine is moving forward continually, and the discharged material is struck off by the cutting edge. Because the flights have chopped and aerate the material during the loading process, the spread is generally quite uniform.

Ejection is sometimes started and stopped with an empty bowl to move the floor back far enough to make it possible to pick up a rock. The scraper is moved over the rock at full height until it is behind the elevator. The bowl is then lowered, and the object scooped up by the retracted edge, which is then slid forward to retain it.

During loading the elevator flights continuously remove the material excavated by both the flights and the cutting edge, so the material does not impose a back pressure or load to further digging.

The last yard in a load is consequently as easy to dig as the first because it does not take extra power to hoist it to the top of the heap. The problem of loading is reduced to one of cutting or chopping a slice of the ground. There is usually sufficient traction to permit this in fairly hard soil, although depth of cut may have to be reduced. Teeth can be bolted to the center section of the edge to assist digging, but they are a nuisance during spreading operations.

Self-loading makes each scraper an independent unit so it can work alone or as a member of a fleet of machines, each with equal efficiency. Coarse or hard soil is broken into small lumps or pulverized, thereby reducing voids in the load and yielding improved spreading.

The disadvantage of this scraper is that the elevator assembly is dead weight during the haul cycle and the machine is expensive compared to the cost of a conventional scraper of equal load capacity. Such scrapers are economical in short-haul situations where the ratio of haul time to load time remains low. Elevating scrapers are used for utility work, dressing behind high-production spreads, or shifting material during fine-grading operations. They are very good in small-quantity situations. No pusher is required, so there is never a mismatch between the pusher and the number of scrapers. Because of the elevator mechanism, they cannot handle large rocks or material containing rocks.

Auger Scrapers

This is another completely self-contained loading and hauling scraper. Auger wheel-tractor scrapers (Fig. 11.7) can self-load in difficult conditions, such as laminated rock, granular materials, or frozen material.

An independent hydrostatic system powers an auger located in the center of the bowl. The rotating auger (Fig. 11.8) lifts material off the scraper cutting edge and carries it to the top of the load. The lifting action is continually creating a void so new material can easily enter the bowl. This action reduces the cutting edge resistance and required rimpull. Additionally, it increases tire life. There are fewer voids, and the broken material is easily retained by the tight-closing apron. The ejector, with angled wing extensions, moves forward to completely clean out the area on each side of the auger for complete emptying of the bowl.

Auger scrapers are available in both single- and tandem-powered configurations. The two-engine auger scraper combines the two separate approaches to self-loading. Tandem power benefits loading chiefly by permitting a thicker slice of material and thereby shortening the time to attain a full bowl. The front engine may provide the majority of power for the auger, while the rear engine supplies push for the cutting edge. The principal advantage of two engines is the increased ability to work, or to work more rapidly under unfavorable

FIGURE 11.7 Auger wheel-tractor scraper.

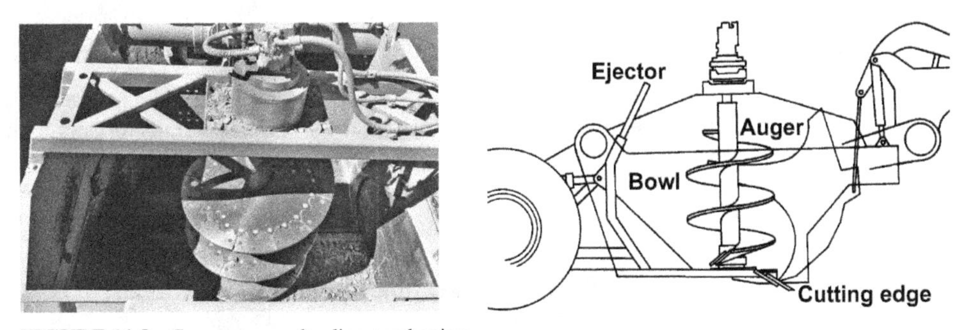

FIGURE 11.8 Scraper auger loading mechanism.

conditions, such as slippery footing in the cut or on the haul road and/or steep grades and soft ground.

As with an elevating scraper, the auger adds nonload weight to the scraper during travel. These scrapers are more costly to own and operate than conventional single- or even tandem-power machines.

SCRAPER PARTS

The wheel-tractor-scraper has two distinct sections: a tractor unit and a scraper bowl. The sections are connected by a gooseneck swivel or cushion hitch and hydraulic lines, allowing the sections to pivot. The excavating and hauling section has three basic parts: the bowl, the apron, and the ejector or tailgate (Fig 11.9).

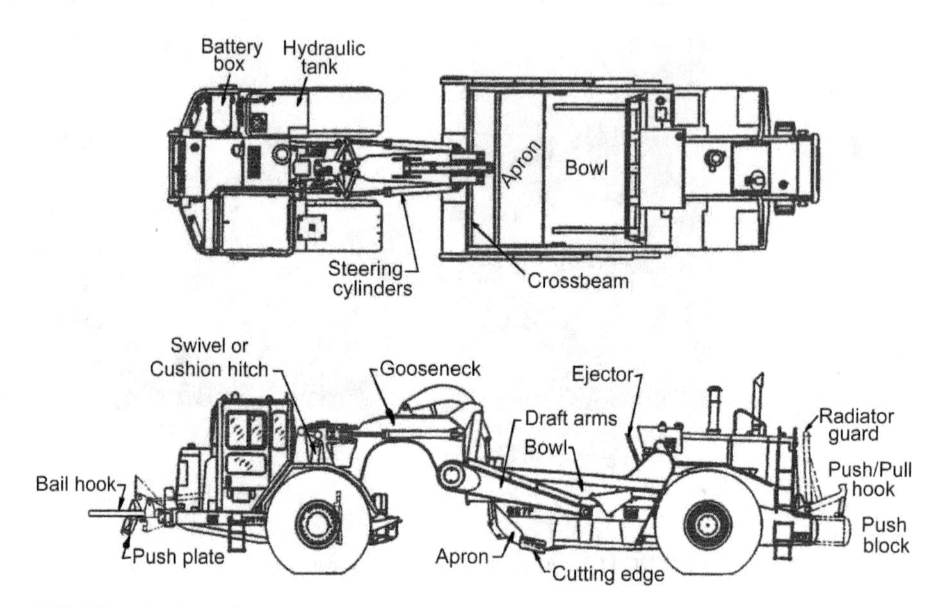

FIGURE 11.9 Identification of scraper parts.

Gooseneck

In front, the gooseneck has a vertical swivel connection with the tractor, which is usually in two parts with two pivots, upper and lower (Fig. 11.9). It permits turns of 85 to 90 degrees to each side of center. As a result, the unit is highly maneuverable and can often turn within its own length. The articulated swivel connection between the tractor and the scraper gooseneck varies in terms of construction. A pair of steering cylinders joins the tractor to the gooseneck.

Behind the swivel connection, the gooseneck arches up to allow space for the tractor wheels to pass under it on turns, then widens into a large sturdy crossbeam, and finally ends at a pair of side arms extending to the rear and somewhat downward to trunnion fastenings on the sides of the scraper bowl.

There may be a cushioning arrangement (Fig. 11.10) to reduce a riding defect called loping. Oil displacement from this cylinder is regulated by a leveling valve and resisted by a nitrogen accumulator, so a motion-dampening pressure balance is attained.

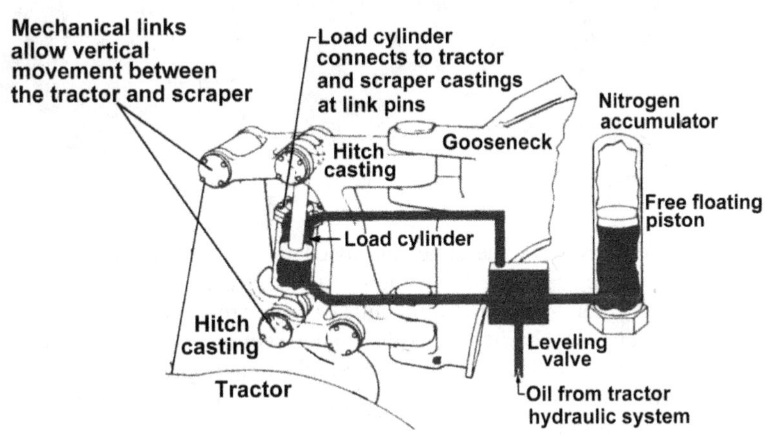

FIGURE 11.10 Cushion hitch for wheel-tractor two-axle scraper.

The gooseneck carries the lift cylinder and lever arm for the apron and a pair of hoist cylinders for the bowl. All of these may have two-way action, or be one-way with return by gravity, springs, or a counteracting cylinder.

Bowl

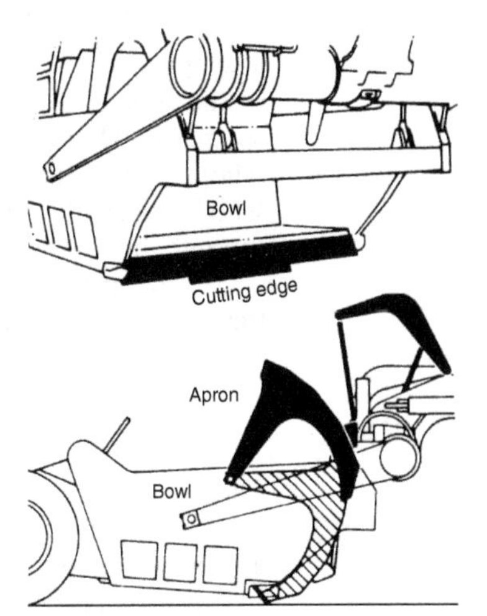

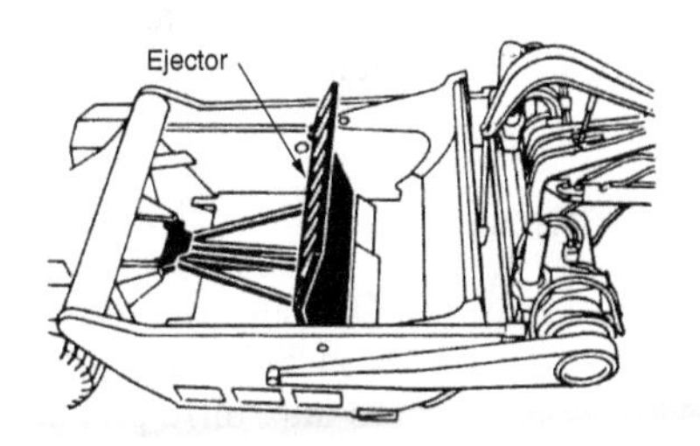

The bowl is the load-carrying portion of the scraper, and its lower front surface has a replaceable cutting edge. It is essentially a box with rigid sides, with the apron forming a vertical movable front, and a movable back ejector. Extensions of the sides behind the rear axle form a case for the ejector cylinder and support for a bumper by which the machine can be pushed. The back of the bowl is supported by the rear axle of the scraper, at the center by trunnions on the ends of the draft arms, and at the front by a pair of hydraulic cylinders suspended from the gooseneck.

The pull of the tractor is applied through the gooseneck. Most of it is transmitted by the trunnions connected at the bowl center on each side, but a variable amount comes through the lift cylinders, depending on their position.

The front edge of the floor is fitted with a cutting edge. This edge is a hard steel plate in three pieces—a wide center one and narrower ends, fastened with plow bolts with smooth side up. The sections can be removed, inverted, and reinstalled when worn on one side.

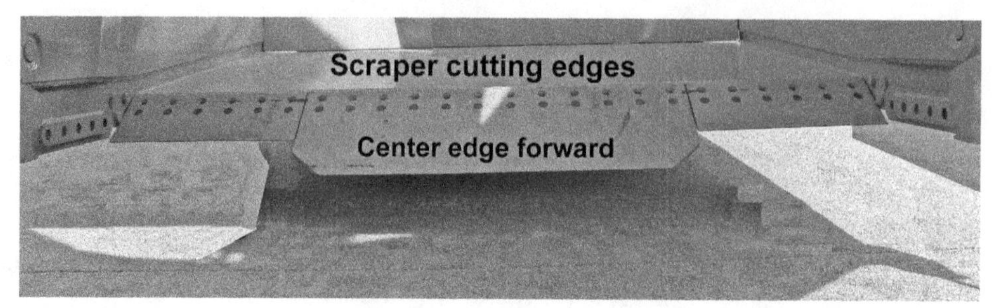

FIGURE 11.11 Scraper cutting edges, center edge forward.

For most work, the center cutting edge is set forward of the side cutting edges (Fig. 11.11). This center cutting edge is mounted back flush with the ends only when the job requires finish grading, working light cuts, or excavating sand.

Teeth may be bolted to the center section to improve penetration in tight material. Teeth can interfere with dumping and spreading and are laborious to install and remove, particularly on a worn edge. They are used more often on elevating scrapers than on conventional models.

The front vertical sides of the bowl, at their lower edge, usually have bolted-on wear plates, called sidecutters. These receive less wear than the bottom edge, but eventually need replacing.

Apron

The apron forms the front or forward side of the bowl and a variable amount of the bottom of the scraper assembly. When in a down or closed position, the apron rests against the scraper bowl at the cutting edge. When lifted, it moves upward in an arc and leaves the whole front of the bowl open. The apron is lifted and lowered by a hydraulic cylinder, which is usually linked near the base of the gooseneck. The hydraulic lines are routed through the bowl sidewalls, where they are protected from damage.

When the scraper bowl is being loaded, the apron is held at a moderate distance above the cutting edge. Excavated material slides into the bowl and rolls forward onto the apron.

Ejector

The ejector is the rear wall of the bowl. It is usually of a sliding type and can move forward horizontally, forcing the load out of the bowl. It is supported by rollers riding on the floor and on tracks welded to the sides of the bowl.

Power is provided through a two-way hydraulic cylinder inside the rear pusher block (bumper) frame of the scraper. Larger machines may have two cylinders, and they may be telescoping in design, to increase the length of push in proportion to casing length.

Controls

Operator stations are ergonomically designed for total machine control in a climate-controlled and safe environment with all controls, levers, switches, and gauges positioned to maximize productivity. A joystick to the right of the steering wheel controls the critical

scraper functions forward or reverse and the bowl's vertical motion. Left and right motions of the joystick move the ejector forward and back, respectively. The stick also has a thumb rocker switch to control the apron and two buttons, one for transmission hold and the other for a cushion hitch. The cab is forward of the bowl, offering the operator all-around visibility of the machine and the cutting edge.

THE SCRAPER-TRACTOR

The tractor includes the engine, drivetrain and drive wheels, hydraulic pumps, and operator's station. A swivel or cushion articulating joint joins the tractor to the scraper bowl.

Two-Axle (Overhung)

A typical two-axle scraper (Fig. 11.1) is powered by a tractor with a single drive axle. Stability is supplied by the attached scraper. The tractor engine projects forward from the axle, a position called overhung, with its entire weight on the drive wheels.

Drivetrain

Practically all conventional scrapers have torque converters and power-shift transmissions. The shift may be partly automatic. The transmission is typically placed behind the drive axle. Input is by a long shaft from the converter to the top of the transmission; output is a shorter shaft from the bottom forward to the differential. There may be anywhere from four to ten forward speeds and one or two reverse speeds. An eight-speed box has converter drive in the two lowest ranges and automatic shift with direct drive in the top gears. The automatic goes only up to the range set by the shift lever position. A transmission hold allows the operator to maintain converter drive for increased rimpull or to hold the current gear for enhanced control.

A retarder in the transmission may be either optional or standard. It saves wear on brakes and prevents excessive downgrade speeds. It is not for quick use, as it takes 2 to 6 sec to become effective after it is engaged.

The differential may be a nonspin design or a standard type. A standard type allows one wheel to spin. The standard unit may be fitted with a lock to force the two wheels to turn together. The throttle should be cut to stop wheel spin before engaging it, and the machine should not be turned while it is in use. It releases automatically when traction becomes equalized.

Brakes

Brakes may be of any kind suited for heavy service. Smaller units may have booster hydraulic, larger ones pressure hydraulic, air over hydraulic, or full air. A sequence valve may apply the rear scraper brakes a little before the front tractor ones to keep the rig running straight. This is a precaution against jackknifing.

With air-applied brake systems, there is usually a secondary braking system. This is a safety system in case of low air pressure—it will automatically apply the brakes if the service air pressure drops to 55 psi.

Steering

Two-axle scrapers steer by swinging around the swivel hitch kingpin. The rear wheels of the bowl do not follow the front wheels accurately on curves, but rather track inside the tractor wheels.

When moving forward, there is the need to be cautious about jackknifing. When an abrupt stop is made, particularly with the front wheels turned or when the front brakes are more effective than the rear ones, the scraper can swing forward beside the tractor, with total loss of travel direction control.

Scraper Life

Construction equipment surveys indicate 13,000 operating hours is the median life for self-propelled scrapers, but service life depends on use (primarily the type of material excavated and hauled) and can range from 8,000 to 25,000 hours.

For larger scraper models (14 cy or more capacity), there are recognized lives of primary components. Average replacement or repair lives for long-term-use machines include

- Moldboard cutting edge: 800 hours
- Tires: 3,500 hours
- Starter/alternator: 4,000 hours
- Engine and transmission: 9,000 hours
- Axles: 10,000 hours

SCRAPER PRODUCTION CYCLE

The production cycle for a scraper (Fig. 11.12) consists of six operations: (1) load, (2) haul travel, (3) dump and spread, (4) turn, (5) return travel, and (6) turn and position to gain another load:

$$T_s = \text{load}_t + \text{haul}_t + \text{dump}_t + \text{turn}_t + \text{return}_t + \text{turn}_t \tag{11.1}$$

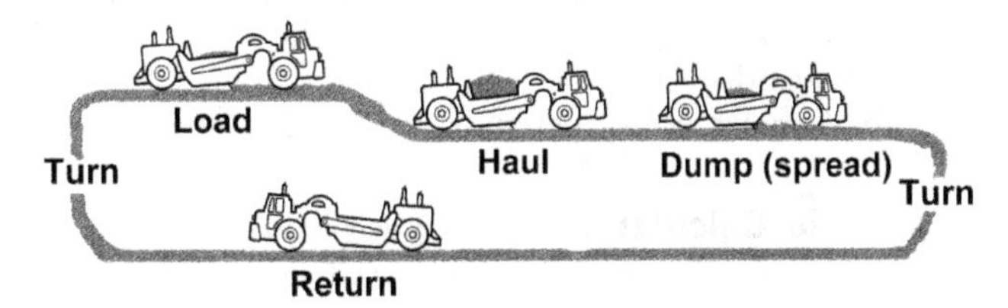

FIGURE 11.12 Scraper production cycle.

Loading time is fairly consistent, regardless of the scraper size. Even though large scrapers carry larger loads, they load just as fast as smaller machines. The larger scraper has more horsepower, and it will be matched with a larger push tractor—together these two factors yield consistent loading times. The average load time for push-loaded scrapers in common earth is 0.80 min. Tandem-powered scrapers will load slightly faster. Equipment manufacturers can supply load times for their machines based on the use of a specific push tractor. The economical load time for a self-loading elevating scraper is approximately 1 min, but this depends on the material being loaded.

When the scraper bowl is nearly full, loading an extra yard (or meter) with standard or tandem scrapers is not cost-effective. Consider the combined forces of cutting-edge ground friction and the gravimetric load already in the bowl. For shorter hauls, an almost full bowl may be more effective than a fully loaded one.

Both the haul and return time depend on the distance traveled and scraper speed, as dictated by haul road grades and rolling resistance. Hauling and returning are usually at different speeds. Therefore, it is necessary to determine the time for each separately. If the haul road has multiple grades or rolling resistance conditions, a speed should be calculated for each segment of the route. In the calculations, it is appropriate to use a short distance at a slower speed when coming out of the cut, approaching the dump area, leaving the dump area, and again when entering the cut, to account for acceleration/deceleration time. A distance on the order of 200 ft at a speed of about 5 mph in each case is usually appropriate. Extremely steep downhill (favorable) grades can result in longer travel times than those calculated. This results from an operator's inclination to downshift to keep speeds from becoming excessive. When using the machine performance charts, do not blindly accept chart speeds—always consider the human element.

Scraper Load

The volumetric load of a scraper may be specified as either the struck or heaped capacity of the bowl expressed in cubic yards. The struck capacity is the volume a scraper would hold if the top of the material was struck off even at the top of the bowl. In specifying the heaped capacity of a scraper, manufacturers usually specify the slope of the material above the sides of the bowl with the designation SAE. The Society of Automotive Engineers (SAE) specifies a repose slope of 2:1 for scrapers. The actual slope will vary with the type of material handled. Because of how a scraper loads, both of these volumes represent loose cubic volume of material (lcy).

The capacity of a scraper, expressed in bank cubic yards (bcy), can be approximated by multiplying the loose volume in the scraper bowl by an appropriate swell factor (Chap. 4). Because of the compacting effect on the material in a push-loaded scraper resulting from the pressure required to force additional material into the bowl, the swell is usually less than for material dropped into a truck by a hydraulic excavator. Therefore, the swell factor values provided in Chap. 4 should be multiplied by approximately 10% for material push-loaded into a scraper. When computing the bank measure volume for an elevating scraper, no correction is required for the Chap. 4 swell factors.

Sample Calculation: If a push-loaded scraper hauls a heaped load measuring 22.5 lcy and the appropriate material swell factor is 0.8, what is the calculated bank measure volume?

$$22.5 \text{ lcy} \times 0.8 \times 1.1 = 19.8 \text{ bcy}$$

The conventional single-engine wheel-tractor-scraper needs a pusher to obtain a full load. For normal loading, the ejector is fully retracted and the apron is held above the cutting edge so material can enter. The bowl is lowered so the cutting edge will cut a slice of ground. In hard ground, the bowl might be forced down with such force as to lift the rear scraper wheels into the air. Digging is done in low or second gear. The throttle is adjusted at the highest engine speed possible without causing the drive wheels to spin.

Depth of cut is regulated by raising or lowering the bowl. Except when working near final grade, the cut should be as deep as the scraper can handle without spinning the drive wheels. If speed is maintained, a deep cut fills the bowl fastest.

The material piles up in the bowl, with part of it falling forward on the apron (Fig. 11.13). Incoming material must be forced to rise through an increasing depth in the bowl. The loading rate slows, and a point of refusal is likely to be reached where power is not sufficient to force more material into the bowl. When this happens, the soil is pushed ahead or drifts off to the side.

If a full load has not been obtained, material may be added by pumping the bowl. The cutting edge is raised sufficiently to decrease the draft and allow the tractor to regain its full

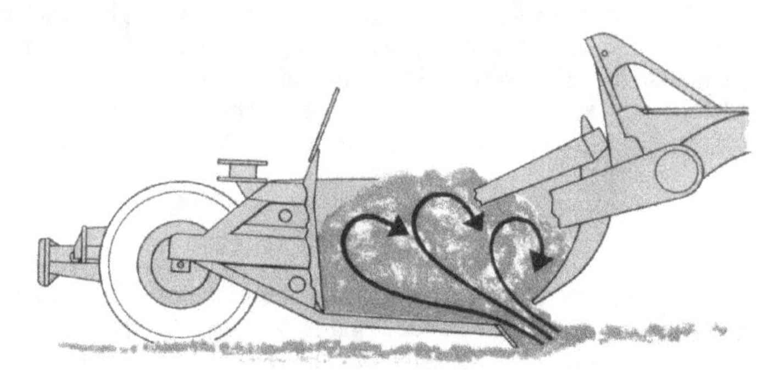

FIGURE 11.13 Material boiling into scraper bowl.

speed, and then the cutting edge is dropped several inches below the normal cutting depth. Soil will be forced into the bowl. The cutting edge is lifted as the engine starts to labor, and the process repeated if desired. However, such an action is time consuming and may reduce scraper production.

The advantage is partly the momentum of the accelerated scraper and partly the ramrod effect of the thicker layer of material punching its way up through the load. This punching effect is important. Clay and other tight soils may be loaded most effectively in thin layers, which reduce the cutting power necessary without sacrificing thrust. Sand, however, must be taken in deeper cuts and requires pumping with a much smaller load. The operational cutting depth of most scrapers ranges from 6 to 18 in.

When a satisfactory load has been obtained, the apron is lowered and the bowl edge raised slightly above the ground. This position is held for several feet before lifting to the carrying position, to spread any loose material in front of the blade and leave the cut smooth.

Push Tractor

During the loading portion of a scraper cycle, the machine needs much more power than it does at any other time. When using push tractors, the number of such tractors must be matched with the number of scrapers employed. If either the pusher or scraper must wait for the other, operating efficiency is lowered and production costs increase.

Pushers are usually tractors, either crawler or pneumatic-tire four-wheel-drive, equipped with a cushion dozer plate that has a wear-resistant center liner plate or a standard dozer blade. Because of their greater speed, pneumatic-tire pushers can service a larger number of scrapers under average conditions. However, they do not exert as strong a push in proportion to their weight as a crawler, and their usefulness declines rapidly if the cut surface becomes slippery.

It is normal for the scraper to move into position to dig, and then wait for pusher contact. But it is more efficient for the scraper to start its run, loading a few yards before the pusher makes contact. The pusher maneuvers behind the scraper, stopped or moving, and, as smoothly as possible, contact is made with the scraper's push block. A moving scraper can lessen the impact when the push dozer contacts the push block. The two machines each exert as much power as possible without excessive spinning of wheels or tracks. Most of the digging power is supplied by the pusher, and the scraper engine may be cut back to one-half or two-thirds speed.

The pusher and scraper must be in a straight line, as a pusher blade at an angle may cut scraper tires or cause jackknifing and steering problems. When a full load is obtained, the pusher drops back to engage another scraper, and the loaded scraper is shifted into a higher gear and leaves the cut area. The pusher operator has a better view of the load and may signal when this should happen.

When a scraper starts loading, the earth flows into it rapidly and easily, but as the quantity of earth in the bowl increases, the incoming earth encounters greater resistance and the rate of loading decreases quite rapidly, as illustrated in Fig. 11.14. The load-growth curve shown in this figure is for a specific scraper and pusher combination used to excavate a particular type of material; it shows the relationship between the load in the scraper and the loading time. During the first 0.5 min, the scraper loads about 85% of its maximum possible payload. During the next 0.5 min it loads only an additional 12%, and if loading is continued to 1.4 min, the gain in volume during the last 0.4 min is only 3%—an instructive example of the law of diminishing returns. In general, it does not pay to struggle to obtain a heaping scraper load—diminishing returns make the last yard a very expensive one.

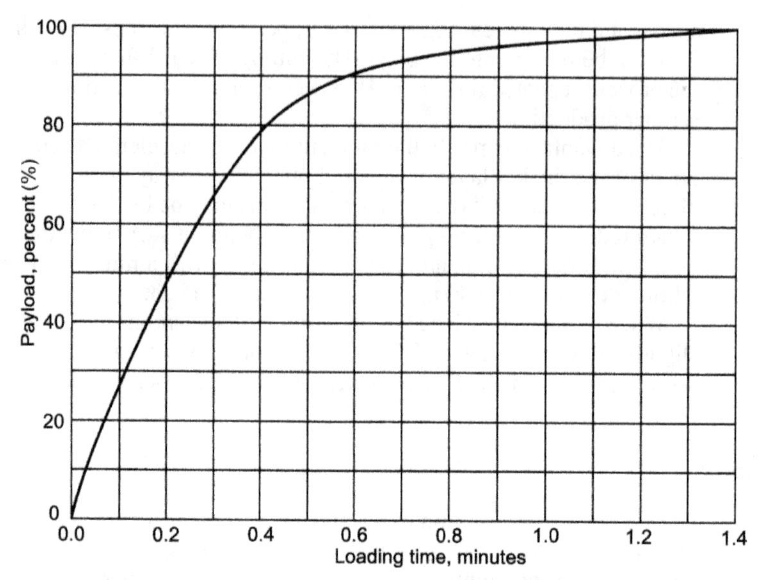

FIGURE 11.14 Typical load-growth curve for a single engine scraper and push tractor.

Pusher Cycle Time

The pusher cycle time includes the time required to push-load the scraper (the duration of pusher-scraper contact, or *contact time*) and the time required for the pusher to move into position to push-load the scraper. The cycle time for a push tractor will vary with the conditions in the loading area, the relative size of the tractor and the scraper, and the loading method. Figure 11.15 shows three loading methods. Backtrack loading is the most common method employed. It offers the advantage of always being able to load in the direction of the haul. Chain loading can be used when the excavation is conducted in a long cut. Shuttle loading is used infrequently. However, if one pusher can serve scrapers hauling in opposite directions from the cut, it is a viable method, but consider the safety of bidirectional traffic in the cut area. When the loading pattern reduces pusher and scraper moves in the pit, production is increased.

Caterpillar recommends calculating backtrack push tractor cycle time, T_p, by the formula

$$T_p = 1.4\mathrm{L}_t + 0.25 \tag{11.2}$$

where L_t = scraper load time (pusher contact time).

Pusher cycle time is a function of four components:

1. Load time of the scraper
2. Boost time (assisting the scraper out of the cut): 0.15 min
3. Maneuver time (e.g., the distance traveled): 40% of load time
4. Positioning for contact time: 0.10 min

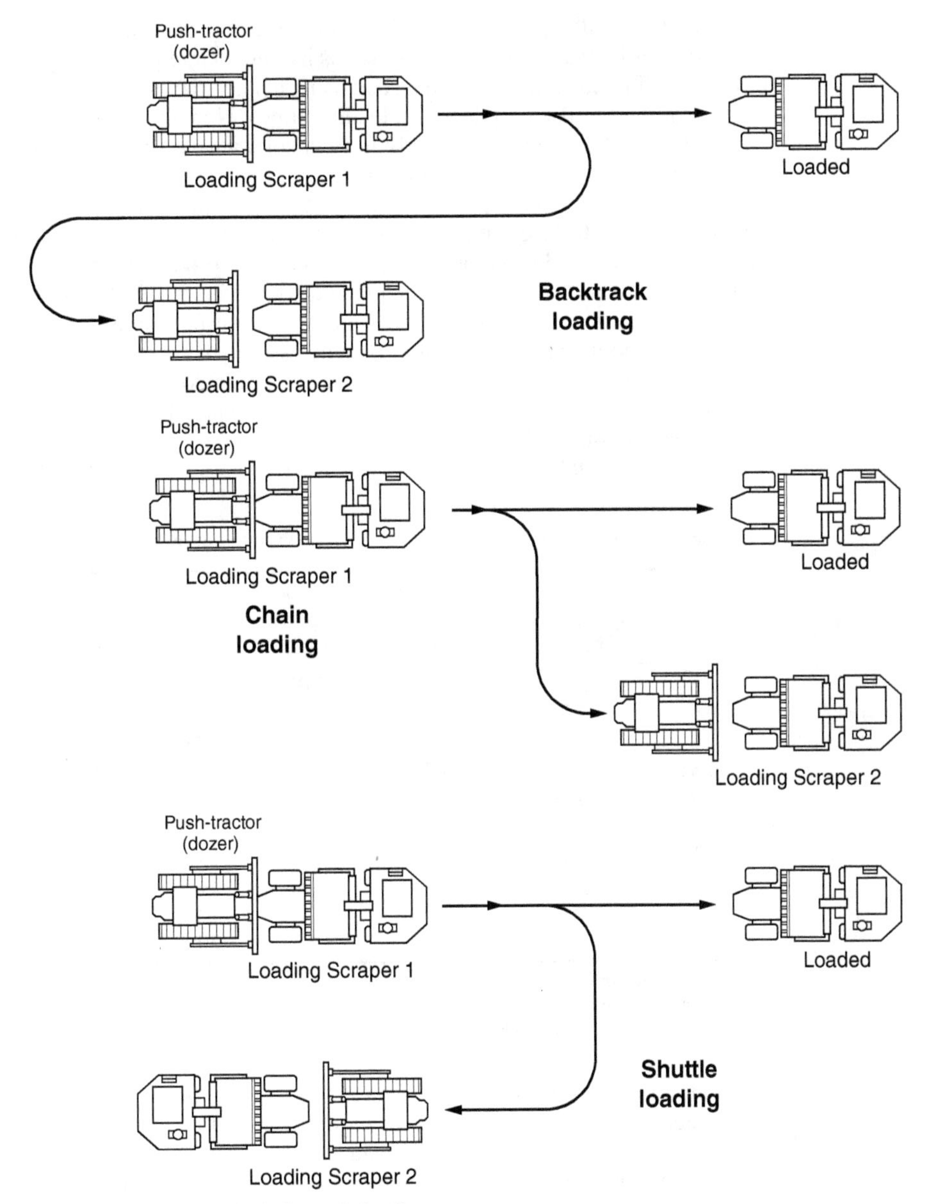

FIGURE 11.15 Methods for push-loading scrapers.

The formula links these to arrive at a total time. Pusher cycle time will be less when using the chain or shuttle method.

Assuming a scraper load time of 0.85 min (the contact time) and applying Eq. (11.2):

$$T_p = 1.4(0.85) + 0.25 = 1.44 \text{ min}$$

Favorable loading conditions will reduce loading time and increase the number of scrapers a pusher can serve. Factors such as having a large pit or cut area, ripping hard soil prior to loading, loading downgrade, and using a push tractor whose power is matched with the size of the scraper will decrease loading times. Nonetheless, multiple pushers may be required to load a scraper effectively in tight soils with no prior ripping, when very large scrapers are used, and when loading rock.

If the scraper is matched with a correctly sized push tractor, a full load should be obtained in 0.80 to 0.85 min. To spend more time forcing a small amount of additional material into the bowl is a practice resulting in a less productive cycle.

Load time is a management decision and should be made after a careful evaluation of the production and cost effects. The tendency in the field is to take too long in loading scrapers. Without a critical evaluation, many assume the practice of loading every scraper to its maximum capacity before it leaves the cut will yield the lowest cost for moving material. Numerous studies of loading practices have revealed how such a loading practice will usually reduce, rather than increase, production. Table 11.1 presents data for one loading time production study.

TABLE 11.1 Effect of Loading Time on Scraper Production (a 2,500-ft One-Way Haul Distance)

Loading Time (min)	Other Time (min)	Cycle Time (min)	Number Trips per Hour	Payload*		Production per Hour†	
				(cy)	(cu m)	(cy)	(cu m)
0.5	5.7	6.2	8.07	17.4	(13.3)	140	(107)
0.6	5.7	6.3	7.93	18.3	(14.0)	145	(111)
0.7	5.7	6.4	7.81	18.9	(14.5)	147	(112)
0.8	5.7	6.5	7.70	19.2	(14.7)	148‡	(113)
0.9	5.7	6.6	7.57	19.5	(14.9)	147	(112)
1.0	5.7	6.7	7.46	19.6	(15.0)	146	(112)
1.1	5.7	6.8	7.35	19.7	(15.1)	145	(111)
1.2	5.7	6.9	7.25	19.8	(15.2)	143	(109)
1.3	5.7	7.0	7.15	19.9	(15.2)	142	(109)
1.4	5.7	7.1	7.05	20.0	(15.3)	141	(108)

*Determined from measured performance.
†For a 50-min hour.
‡The economical time is 0.8 min.

Economical loading time is a function of haul distance. As haul distance increases, economical load time will increase. Figure 11.16 illustrates the relationship between loading time, production, and haul distance; it is presented only to demonstrate graphically the interrelationship. The economical haul distance for scrapers is usually much less than a mile. This can be recognized from the rapid rate at which production falls with increasing haul distance. For the graphed analysis, an average load time of 0.85 min was used.

Scraper Haul

The loaded scraper is driven to the fill area at the highest speed allowed by safety and reasonable comfort. Scrapers can move loads successfully over soft and rough ground, but always at the expense of speed and increased maintenance cost. Production will be increased by providing a smooth, hard surface for rubber-tire scrapers.

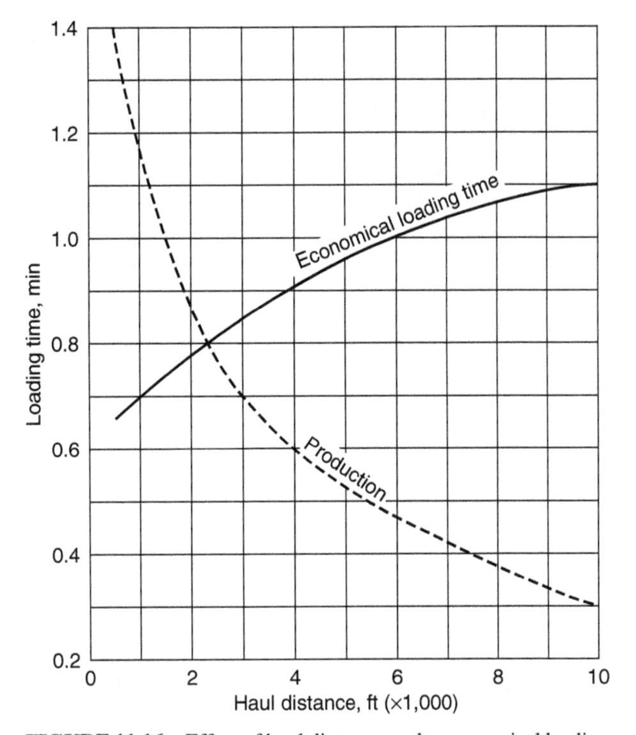

FIGURE 11.16 Effect of haul distance on the economical loading time of a scraper.

Rolling resistance (RR) is the result of a conscious supervisory decision as to how much effort (money) will be expended maintaining the haul road. As the effort (and money) expended for maintenance of haul roads increases, production will increase. The reverse is also true—if no effort is made to improve the haul road, production suffers. Graders and sometimes dozers are necessary to eliminate ruts and rough (washboard) surfaces. Water trucks (Fig. 11.17) will be required to provide moisture for compacting the haul road and for dust control. Visibility is improved by keeping roads moist. This simple action lessens the chance of accidents. Dust control also helps to alleviate mechanical wear.

Figure 11.17 Water truck on the haul road.

Rolling Resistance and Scraper Production

Representative rolling resistance values are given in Table 5.1. Rolling resistance can be expressed as either a percentage or as pounds per ton of vehicle weight. The effect of haul road conditions on rolling resistance, and thereby on scraper production and the cost of hauling earth, should always be carefully analyzed. A well-maintained haul road enables faster travel speeds and reduces the costs of scraper maintenance and repair. Figure 11.18 is from a field study of scraper haul times. The shaded area represents the range of average travel times on numerous projects. The lower boundary indicates travel times on projects with well-maintained haul roads. The upper boundary indicates poor haul roads. Considering a haul distance of 4,000 ft, the total time for the haul and return combined on a well-maintained haul road was 5.20 min. Under poor conditions, this could be as high as 7.55 min. A difference of 2.35 min in cycle time is a 4.7% production loss in a 50-min hour.

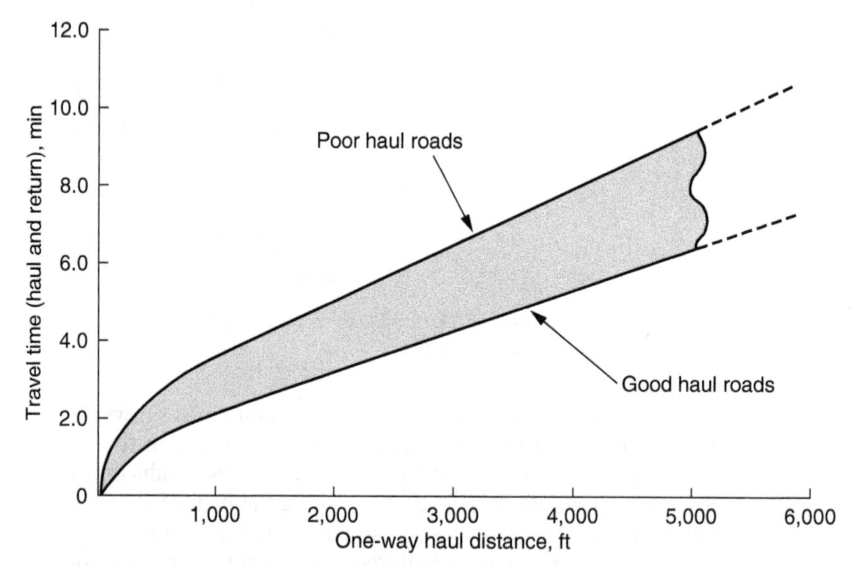

FIGURE 11.18 Average travel times for single-powered axle scrapers, capacity <25 cy, negligible grade. (*Source: U.S. Department of Transportation, FHWA.*)

Grade Resistance or Assistance

The topography of the project will dictate the haul road grades. From where the material must be excavated to where it must be hauled is a physical condition imposed by the project requirements. Still an effort should be made to plan the haul routes so as to avoid adverse grades, as they drastically reduce production. Careful layout of haul routes may achieve flatter grades, but those possibilities must usually be considered in terms of increased haul route length.

Total Resistance or Assistance

The total resistance or assistance for each segment of the haul and return routes must be determined. This is the sum of the rolling resistance and the grade resistance/assistance for each segment of a route. Total resistance or assistance together with the weight of the gross scraper weight controls haul speed. The return speed is with the empty scraper weight (EVW).

Bowl Position

A scraper bowl is carried high to avoid obstacles and low to keep the machine stable. The actual height will vary with route conditions. The bowl should never be held in the highest raised position during travel. In this position, the hoist cylinders become rigid frame members and absorb damaging stresses during travel over rough ground.

Shifting

The modern scraper shifts very easily. There may be as few as four or as many as ten forward ranges. In the case of some semiautomatics, the two lowest (digging) speeds are manual. From second gear, the operator can move the lever directly to the highest gear desired and the transmission will do the shifting. It is possible to override the automatic transmission with the lever at any time. There may be a pedal to hold it in whatever gear it is in until the pedal is released.

Shifting is guided mostly by the feel of the machine and the operator's estimate of conditions immediately ahead. But it may be controlled by dashboard gauges. It is important to keep the engine within its efficient operating speed range. If it is on the fast side, it will be necessary to shift into a higher gear. If it is near the bottom, then a lower gear must be used. If the torque converter fluid is too hot, use a lower gear.

Dumping and Spreading

Physical constraints in the dump area may dictate scraper-dumping techniques. The most common method is for the scraper to dump before turning. This uses the haul speed momentum to carry the bowl over and through the deposited material. This practice reduces the possibility of the scraper becoming stuck in the newly placed material and yields a fairly even spreading of material, thereby reducing spreading and compaction equipment effort.

In dumping a conventional scraper, the bowl is lowered until the edge will allow a layer of material the desired thickness (depth) to slide under the bowl. The apron is then raised so the opening width is sufficient for the material to exit in a continuous, smooth layer, without dropping excess in front or to the side of the bowl in windrows.

With the apron raised, the ejector is moved forward gradually, pushing material out of the bowl. Too fast a crowd will supply too much material to the cutting edge and will place unnecessary strain on the cylinders and linkage of the ejector mechanism. If the material is sticky, good results are obtained by advancing the gate slightly, then allowing it to slide back slightly and advancing it again when required. Material resting on the edge after the ejector is fully forward may often be dumped by moving the ejector back and forth. When the dump is complete, the ejector is fully retracted, the apron dropped, and the bowl raised to a travel position.

When building an embankment, spreading in thin layers supports compaction operations and reduces the need for other grading equipment. Dumping and spreading are ordinarily done heading away from the cut and at medium or high speed. In finishing a grade and under some other conditions, a low gear may be used to provide more precise spreading control.

The fill should be started at its outer edges of the embankment and built to finish the slope all the way up. This is important, as the scraper has no way of dumping over the edges, and any patch filling cannot be readily compacted.

It is also desirable to have the fill slope upward toward the edge (creating a concave surface), as this arrangement will position the weight of both tractor and scraper on the side away from the slope and reduce the danger of embankment edge failure. However, if rain is expected, the fill should be crowned to shed water.

The closeness with which a scraper can approach a high shoulder is determined by the type of soil, the degree of compaction, the degree of the side slope, the weight of the machines, and to some extent the skill of the operator. In general, fine-grained soils that are not in a muddy condition are safer than sandy soils. Working toward the edge or overhanging it slightly with a tamping roller will give good compaction.

If scrapers cannot safely go near to the edge to build a proper slope, they can spread the material a few feet away from the edge, and a grader can distribute it where needed. Embankment edges should be checked frequently for proper slope.

Elevating Scraper

The elevator stays in place during the dump. An opening for unloading is made by pulling back the cutting edge together with the front half of the floor. The ejector is then moved forward. A substantial part of the load falls by gravity through the floor opening. The ejector and floor controls are interlocked, so ejector movement does not start until the floor is completely open.

Dump Duration

Dump times vary with scraper size, but project conditions will also affect the dumping duration. Average values for dump time are presented in Table 11.2.

TABLE 11.2 Scraper Dump Times

	Scraper Type	
Scraper Size (cy) Heaped	Single Engine (min)	Tandem-Powered (min)
<25	0.30	—
25–34	0.37	0.26
35–44	0.44	0.28

Source: U.S. Department of Transportation, FHWA.

Sometimes it is necessary to negotiate the turn before dumping. This generally increases dump time and the chances of the scraper becoming stuck while dumping. The last method is to dump during the turning maneuver. This definitely increases dumping time, results in uneven spreading, and can cause scrapers to stall in the loose material. When access is limited, such as the situation at bridge abutments or when backfilling culverts, a turning/dumping maneuver may be necessary. Wet material is difficult to eject and will increase the dump time.

Return to the Cut

After completing the spread, the scraper is turned, either on the fill or after going forward to a turning point, and driven back to the cut at the highest safe speed. Safe speed might be the maximum at full governed engine speed, but usually there are limiting factors. The machine will be slowed by upgrades, soft footing, rough spots, curves, close approaches to other machines, and hazardous conditions. Accuracy of steering and efficiency of brakes must be taken into account. Two-axle machines may have a tendency to rhythmic bouncing, called loping, which is most uncomfortable and may dangerously limit operator control. Special cushion hitches may reduce this, but an operator may immobilize them during digging and spreading and then forget to reconnect these. Otherwise, a small increase or decrease in speed may steady the scraper. Loaded scrapers (or trucks) always have the right of way over empty machines.

Travel time is the sum of the times the scraper requires to traverse each segment of the haul and return routes. Based on the speeds determined from the performance or retarder charts, or the assumed speeds because of job conditions, travel time can be calculated using Eq. (11.3).

$$\text{Travel time per segment (min)} = \frac{\text{Segment distance, ft}}{88 \times \text{travel speed, mph}} \tag{11.3}$$

Turning Times

Turning time is not significantly affected by either the type or size of scraper. Based on Federal Highway Administration (FHWA) studies, the average turn time in the cut is 0.30 min, and on the fill, the average turn time is 0.21 min. The slightly slower turn in the cut is primarily caused by congestion and the necessity to spot the scraper for the pusher.

Scraper Efficiency

The term efficiency or operating efficiency is used to account for the actual productive operations in terms of an average number of minutes per hour thc machines will excavate and haul material. A 50-min hour average would yield a 0.83 (50/60) efficiency factor. The 50-min hour is a reasonable starting point if no company- and/or equipment-specific efficiency data are available. When seeking to determine scraper productivity, always visualize the work site and how the work will be performed in the field before applying an efficiency factor. If the pit will not be congested and if the dump area is wide open, a 55-min hour may be appropriate. But if the cut involves a confined area, such as excavating a ditch, or if the embankment area is a narrow bridge header, consider a reduced efficiency, maybe a 45-min hour.

OPERATIONAL CONSIDERATIONS

Scrapers are best suited for medium-haul-distance earth-moving operations. Haul distances longer than 500 ft but less than 3,000 ft are typical, although with the largest scrapers, the maximum distance can approach a mile. The selection of a particular

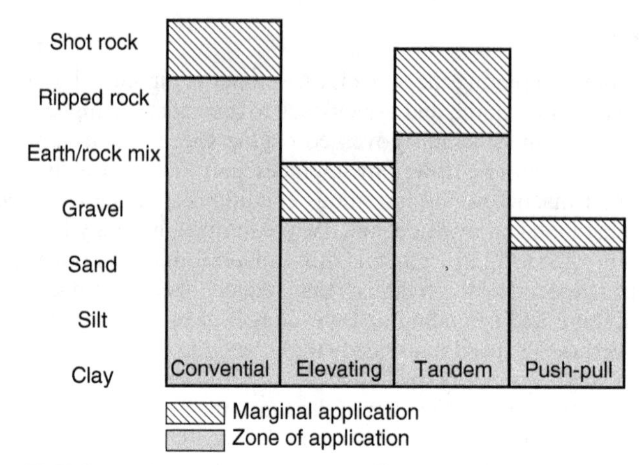

FIGURE 11.19 Zones of application for different type scrapers.

type of scraper for a project should take into consideration the type of material being loaded and transported. Figure 11.19 provides a summary of applicable scraper type based on project material type. Profitable earthwork operations are the result of careful planning and field supervision. There are several methods whereby this objective can be attained.

Ripping

Most tight soils will load faster if they are ripped ahead of scraper loading. Additionally, delays pertaining to equipment repairs will be reduced substantially, as the scraper will not be operating under as much strain. If the value of the increased production resulting from ripping exceeds the ripping cost, the material should be ripped.

When rock is ripped for scraper loading, the depth ripped should always exceed the depth to be excavated. This is done to leave a loose layer of material under the tires. This material will provide better traction and reduce the wear on the dozer tracks and scraper tires. The cost of hauling ripped or shot rock with scrapers will be higher than for moving common earth, but it can still be more economical than an excavator-truck operation. The volume of material carried will be about 70% of normal payload because of the void spaces between the bulky pieces. Repair costs will be about 150% of normal, and tire life will be about 30% to 40% of normal.

Prewetting the Soil

Some soils will load more easily if they are reasonably damp. To achieve uniform moisture conditioning of the soil, prewetting can be performed in conjunction with ripping and ahead of loading. Prewetting the soil in the cut can reduce or eliminate the use of water trucks on the fill, thereby reducing equipment congestion. Similarly, the elimination of excess moisture on the fill from water truck operations may facilitate the movement of the scrapers.

Loading Downgrade

Scrapers should be loaded downgrade and in the direction of haul when it is practical to do so. Downgrade loading results in faster loading times, whereas loading in the direction of haul both shortens the length of the haul and eliminates the need to turn in the cut with a loaded scraper. Every 1% of favorable grade is the equivalent of increasing the loading force by 20 lb/ton of the gross weight for the push tractor and scraper unit.

Dumping Operations

It is easier to compact materials dumped in thin lifts. Thick lifts require greater spreading effort and impede travel on the fill. Compaction equipment must work in patterns to be effective. An orderly pattern of dumping will make it easier to intermesh material placement and compaction operations.

Grading

Scrapers are sometimes used to smooth and patch haul roads, in grading and leveling areas where the quantity of material to be moved is not sufficient to fill the bowl, and where distances are too short for the loading-dumping cycle. For grading, the center piece of the cutting edge should be flush with the side pieces. Otherwise, finish work will be limited to the width of the center cutting edge. The location of the cutting edge between the front and rear axles provides good stability for grading, and controls are sufficiently sensitive to perform medium-fine work. But there is no means to shift material from side to side, grader fashion, so operations like crowning a road or shaping a slope are laborious.

Nonetheless, a scraper can move substantial amounts of material along a linear grade and can bring in borrow or dispose of surplus off the grade. When cuts and fills are shallow and closely spaced, the usual working position is with the apron fully lifted and the ejector most of the way forward. It is not good practice to have the ejector all the way forward, as it is then subject to twisting strains from the frame and the bowl.

The bowl is lowered or raised to give the desired depth of cut or thickness of fill, and acts much as a very steady dozer blade. If a cut is deep or long enough to provide more spoil than can be carried in front, the ejector can be retracted enough to admit some material inside the bowl. The extra material can be ejected later or used to fill low spots.

An elevating scraper must have the elevator operating in reverse or the floor partly retracted to avoid picking up loads. In either case, there is more than normal strain on the elevator flights, and the scraper should not be used in this way to cut hard surfaces.

The self-loading feature makes the elevating scraper ideal for grading work requiring only a moderate amount of offsite borrow. An elevating scraper and a grader are a good combination for this type of work. There is no definite line of separation between this light grading and heavy cuts and fills.

Utilization Efficiency

The self-powered scraper is sensitive to weather delays. The tires of a scraper's power axle quickly lose traction on rain-wet surfaces. Rain quickly makes the surface of the cut and haul roads too wet for scraper operations. On resumption of work, it may be necessary to waste the top layer, or stockpile it to dry.

Scrapers are not particularly subject to stoppage from breakdown. A pusher can usually be replaced quickly by odd-job dozers, and scrapers can sometimes load each other or they can operate for a while at partial load. They can maintain their own haul road and can continue to build a fill in the absence of grading and compacting equipment.

Maintenance

Bowl cutting edges need periodic reversal or replacement. The edges must be replaced before their lack of steel permits wear to the bowl edge. The contractor who has detailed cost records can compute the cost of these high-wear items separately. Otherwise, they are lumped together with repairs. Scrapers are often parked outdoors for long periods between jobs. They are likely to suffer serious rust damage unless they are carefully protected by paint and grease.

Repairs

Repair costs increase with weight of load, duration of load effort, roughness of the ground, project haul grades, and machine age. Severe damage may be done by boulders, ledges, and ripped rock, particularly when using powerful pushers. Some scraper models are more subject to rock damage than others.

For a rule-of-thumb calculation when lacking specific records, a scraper's nontire maintenance and repair will equal 80% of its purchase price under ordinary operation for 5,000 hr. In light service, repair costs may run only 50% of purchase; in heavy service, 150% or more.

Tires

Tire repair, recapping, and replacement may make up one-third or more of the nonlabor operating expense. If the scraper is operated at full throttle and the pusher is slow, considerable spinning of the scraper drive wheels will occur. This is wasted effort and wasted rubber, as the scraper usually supplies a modest fraction of the loading power, and a spinning wheel has less push than one whose tread is engaging the ground.

Spinning is most apt to occur with increased loading duration. The scraper operator feels the struggle of the pusher and lack of load response, and has a tendency to tramp on the throttle in an effort to help. Drive wheel spin in the cut may be reduced by driver education, running pushers in second gear, increasing pusher power, or having scrapers equipped with torque converters.

Scraper drive tires start to slip and spin on loose or muddy haul routes on grades as low as 3% to 5%. Under severe conditions of abrasive mud and steep grades, a tire can be worn smooth in 500 hr of operation.

Labor

A scraper has one operator, and no helpers or hand laborers are ordinarily needed. Pusher expenses, including labor, are divided by the number of scrapers they service. But the use of a spotter on the fill and sometimes in the excavation area will increase production.

Supervision

Full-time supervisory control should be provided in the cut. A more efficient operation will result through the elimination of confusion and traffic congestion. A spotter should always control the fill operations, being responsible for coordinating the scrapers with the spreading and compacting equipment. The spotter's job is to maintain the scraper dumping pattern. Typically, the spotter will direct each scraper operator to dump the load at the end of the preceding coverage until the end of the fill area is reached. Then the next coverage is started parallel and adjacent to the first coverage. This enables compaction equipment to work the freshly dumped material without interfering with the scrapers.

SCRAPER PERFORMANCE CHARTS

Scraper manufacturers provide specifications and performance charts for each of their machines. The chart information makes it possible to analyze and predict scraper performance under various operating conditions. Figure 11.20 presents a representative performance chart for a 20-cy, heaped-capacity, single-engine scraper.

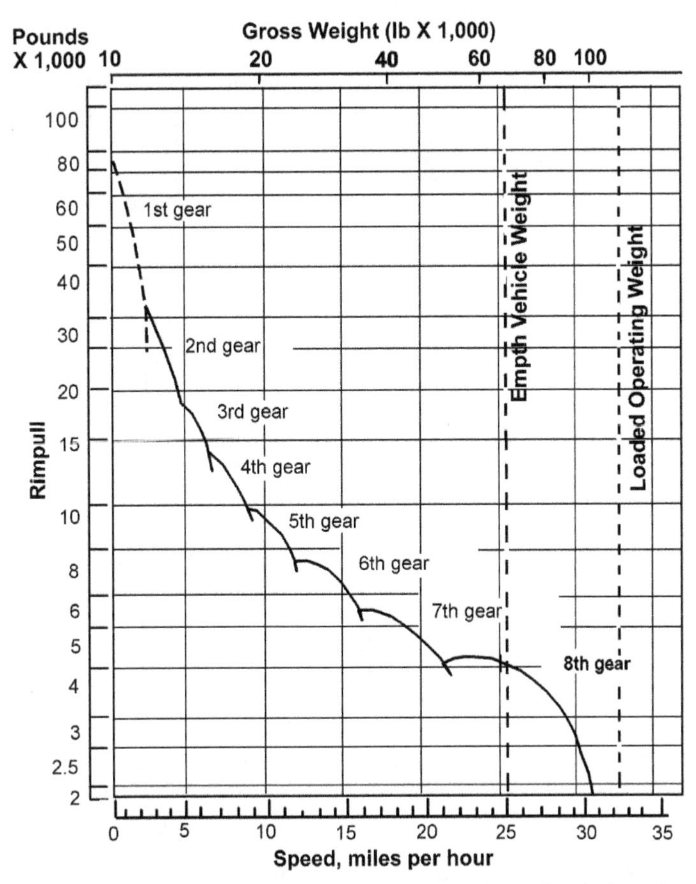

FIGURE 11.20 Performance chart for a 20-cy, heaped-capacity, single-engine scraper. (*Source: FM 5-434, Earthmoving Operations, 15 June 2000.*)

Owners sometimes add sideboards to the bowls of their scrapers. These boards allow the volumetric load the scraper carries to be increased, but an increased volume means increased load weight. If the material is lightweight, the practice will serve to increase production with little or no effect on the machine. But for normal material, the greater load can cause the scraper to carry more than the machine's rated "loaded operating weight" (Fig. 11.20). The effect will be a few more yards per load, but in time, maintenance costs will increase because of overloading. A material with a higher unit weight would aggravate this effect.

The use of rated scraper volumetric capacity data and swell values from tables can provide an estimate of the bank measure (bcy) scraper load. This is satisfactory for small jobs or if the planner has developed an accurate set of swell values for the materials commonly encountered in the work area. A better approach is to obtain actual field weights of loaded scrapers carrying a variety of materials. From such weights, the average carrying capacity of those scrapers in the fleet can be calculated directly. This can be important on large-volume excavation jobs where the significance of small differences between table values and actual measured values can have large cost effects.

SCRAPER PRODUCTION ESTIMATING

The most profitable methods and equipment to be used on any job can only be determined by a careful project investigation. The basic approach is a systematic analysis of the scraper cycle with a determination of the cost per cubic yard under the anticipated project conditions. Such an analysis is helpful when making decisions for estimating purposes, and just as importantly for scraper fleet control on the project.

Load Weight

The first step in calculating scraper production is to determine the following:

- Empty vehicle weight (EVW)
- Load weight
- Gross vehicle weight (GVW)

Empty Vehicle Weight. To determine the EVW, reference the manufacturer's data for the specific make and model of scraper being considered. When referring to the manufacturer's data, be careful about notes concerning what is included in the stated weight. Typically, empty operating weight includes a full fuel tank, coolants, lubricants, a rollover protective system (ROPS) canopy, and the operator.

Load Weight. The load weight is a function of scraper load volume and the unit weight of the material being hauled. The load volume is a loose material volume, so the unit weight needs to be a loose cubic yard unit weight, or the loose volume can be converted to match the units of the unit weight. The load weight is often referred to as the payload.

Gross Vehicle Weight. The GVW is the sum of the EVW and the load weight.

Sample Calculation: Consider a single-engine scraper with an EVW of 102,750 lb and a heaped capacity rating of 34 cy. The material to be hauled is a clay with a united weight of 3,000 lb/bcy. Push loading is required.

Based on a load time of 0.85 min, the expected load volume would be:

$$0.96 \text{ (from the Fig. 11.14 load-growth curve)} \times 34 \text{ cy} = 32.6 \text{ lcy}$$

From Chap. 4: swell factor clay = 0.74

$$\text{Load volume bank measure: } 32.6 \text{ lcy} \times 0.74 \times 1.1 = 26.5 \text{ bcy}$$

The volume is increased by 10% because the material is push-loaded into the scraper.

$$\text{Weight of load: } 26.5 \text{ bcy} \times 3{,}000 \text{ lb/bcy} = 79{,}500 \text{ lb}$$

$$\text{Gross weight (GVW)} = 182{,}250 \text{ lb } (102{,}750 \text{ lb} + 79{,}500 \text{ lb})$$

Rolling Resistance. From Table 5.1, the rolling resistance for scraper tires on a compacted and maintained soil haul road would be between 40 and 70 lb/ton of vehicle weight. Considering a worst-case scenario (a conservative analysis), the planner would use the 70 lb/ton value.

Grade Resistance. If the scraper had to climb a 4% grade from the pit to the dump location, the grade resistance would be 80 lb/ton. During the return, there would be a grade assistance situation.

Total Resistance. Therefore, for the case discussed here, the total resistance during the haul would be:

$$150 \text{ lb/ton } (70 \text{ lb/ton} + 80 \text{ lb/ton})$$

$$\frac{182{,}250 \text{ lb (GVW)}}{2{,}000 \text{ lb/ton}} \times 150 \text{ lb/ton} = 13{,}669 \text{ lb rimpull}$$

Because the rimpull charts are per 1,000 lb, on the left side of the chart in Fig. 11.20, enter at 13.7 lb and draw a horizontal line to intersect the possible gears.

The total resistance during the return would be:

$$-10 \text{ lb/ton } (70 \text{ lb/ton} - 80 \text{ lb/ton})$$

$$\frac{102{,}750 \text{ lb (GVW)}}{2{,}000 \text{ lb/ton}} \times -10 \text{ lb/ton} = -514 \text{ lb rimpull}$$

Both rimpull and retarder charts are per 1,000 lb, so it would be entered at 0.5.

Travel Speed. Steady-state travel speeds based on total vehicle weight and total resistance can be determined from the manufacturer's performance or retarder charts for the specific scrapers under consideration. If the total resistance is a positive number, the performance chart is used; if the total resistance is a negative number, the retarder chart is used.

The speeds in the charts represent a steady-state condition. They do not necessarily represent a safe operating speed for specific job site conditions. Before selecting a speed for planning purposes, visualize the project conditions and adjust chart speeds to the expected conditions. An illustration of this is hauling downhill with a full load. The chart indicates the maximum vehicle speed, but many operators are not comfortable traveling fast downhill (they realize it is an unsafe condition) and will shift into a lower gear.

Using the calculated 13.7 lb (×1,000) value and using the rimpull chart in Fig. 11.20, the scraper speed for the haul would about 5.5 mph in third gear. If the haul distance is long, the scraper might shift into fourth gear and the speed would be closer to 6 mph.

With a retarder chart, the negative value would be used, but in this case with a value close to zero, the scraper would be operated close to maximum speed—about 30 mph if the condition of the haul road is good.

Total Cycle Time. Total scraper cycle time is the sum of the times for the six operations of a production cycle—load, haul travel, dump, turn, return travel, turn to pick up another load (Fig. 11.12).

Balance Fleet. The number of scrapers a push tractor can serve is simply the ratio of the scraper cycle time to the pusher cycle time:

$$N = \frac{T_s}{T_p} \tag{11.4}$$

where N = number of scrapers per one pusher, T_s = scraper cycle time, and T_p = push tractor cycle time.

Rarely, if ever, will the value of N be an integer. Therefore, the pusher or the scrapers will be idle for some portion of time during a cycle.

Assume a total cycle time for the scrapers of 6.76 min and the pusher requiring 1.44 min to load and position to load the next scraper; then applying Eq. (11.4) the result would be:

$$N = \frac{6.76\text{ min}}{1.44\text{ min}} = 4.7$$

Consequently, the economics of using either four or five scrapers should be investigated.

Production. If the number of scrapers placed on the job is less than the number from Eq. (11.4), the scrapers will control production and the push tractor will experience idle time.

$$\begin{array}{c}\text{Production}\\\text{(scrapers controlling)}\end{array} = \frac{\text{Efficiency, min/hr}}{\text{Total cyc. time scraper, min}} \times \text{Number of scrapers} \times \text{Volume per load} \tag{11.5}$$

If the number of scrapers placed on the job is greater than the result from Eq. (11.4), the pusher will control production and the scrapers will experience idle time.

$$\text{Production (pusher controlling)} = \frac{\text{Efficiency, min/hr}}{\text{Total cyc. time pusher, min}} \times \text{Volume per load} \tag{11.6}$$

Sample Calculation: Assume only four scrapers are placed on the job and 4.7 is the balance number. Each scraper will carry a load of 26.5 bcy, and efficiency will be a 50-min hour. In such a case, Eq. (11.5) would be used to calculate spread production.

$$\text{Production (scrapers controlling)} = \frac{50 \text{ min/hr}}{6.76 \text{ min (scraper cycle time)}} \times 4 \times 26.5 \text{ bcy}$$

$$= 784 \text{ bcy/hr}$$

If five scrapers were used on the job, Eq. (11.6) would be used to calculate spread production.

$$\text{Production (pusher controlling)} = \frac{50 \text{ min/hr}}{1.44 \text{ min (pusher cycle time)}} \times 26.5 \text{ bcy}$$

$$= 920 \text{ bcy/hr}$$

Cost? If project schedule is more important than cost, then five scrapers should be used because of the increased production when compared to using four scrapers. But usually the decision concerning the number of scrapers to employ is a question of unit production cost. As an example, assume the scrapers have an O&O cost of \$132 per hour, and the O&O cost for the push tractor is \$117 per hour. Furthermore, the scraper operators for the project will be paid \$40 per hour (with fringes) and the pusher operator \$60 per hour. With this cost information, it is possible to determine the unit cost for moving the material:

4 scrapers @ \$132/hr + 1 pusher @ \$117/hr	= \$645/hr
4 operators @ \$40/hr + 1 operator @ \$60/hr	= \$220/hr
Cost per hour to employ a four-scraper spread	\$865/hr
5 scrapers @ \$132/hr + 1 pusher @ \$117/hr	= \$777/hr
5 operators @ \$40/hr + 1 operator @ \$60/hr	= \$260/hr
Cost per hour to employ a five-scraper spread	\$1,037/hr

Unit cost to move the material using a four-scraper spread:

$$\frac{\$865/\text{hr}}{784 \text{ bcy/hr}} = \$1.103/\text{bcy}$$

Unit cost to move the material using a five-scraper spread:

$$\frac{\$1{,}037/\text{hr}}{920 \text{ bcy/hr}} = \$1.127/\text{bcy}$$

The unit costs are very close. If the project has a large quantity of material to move, it would be best to use a five-scraper spread. However, if it is a project with a very limited amount of material to move, the mobilization cost of the fifth scraper might amount to more of an expense than the saving from the lower unit production cost. Many scraper projects involve moving large quantities of material; therefore, the difference of only a few pennies is important. For this same reason, when moving large quantities, many companies carry the cost figures to three places past the decimal.

The final decision concerning the size of spread (number of scrapers) to use must include consideration of the costs to mobilize the equipment and the costs of daily overhead experienced on the project. When daily overhead expenses are great, higher production is usually justified.

SCRAPER SAFETY

To achieve production, scrapers should travel in the highest gear possible based on haul road conditions. Nevertheless, they should never be operated at unsafe speeds. Operators must always wear seat belts, as uneven terrain and ruts in the haul road can cause violent pitching and bouncing. If not secured by the seat belt, such violent movement can cause the operator to be thrown from the scraper. Examples of injuries resulting from not wearing a seat belt can be found in many Occupational Safety and Health Administration (OSHA) reports posted on the Web:

> "... entered rough, rutted terrain that caused the scraper to become uncontrollable. The driver was ejected from the operating position and fell to the ground. Scraper continued on and stalled out about 50 yards further on. Scraper driver sustained multiple serious injuries from the fall."

In other cases, the operator violently turns the steering wheel, causing an accident.

Large earth-moving machines like scrapers, because of their structural mass, limit the operator's field of vision. Being "struck by" or "caught in between" are two of the leading causes of injuries and fatalities on construction projects. Personnel being "struck by" equipment or vehicles accounts for 22% of the accidents experienced on construction sites, and "caught in between" accounts for 18%. Therefore, access to the fill area and haul roads should be restricted. Nonessential personnel should not be allowed in areas where scrapers are working. All employees working on the ground are at risk and must receive basic safety indoctrination about hazardous conditions and be provided with proper reflective vests.

CHAPTER 12
STRUCTURAL EXCAVATION OPERATIONS AND EARTHEN SUPPORT

Design of construction operations in a confined structural excavation requires strategic planning. Although the haul distance is usually well defined—from excavation center of mass to a specified dump location—the process of removing material from a confined pit is a planning challenge. Dozers, backhoe excavators, and front-end loaders each have unique characteristics for excavating in restricted spaces. There is often a need to excavate deep cuts with vertical or nearly vertical walls. Temporary supports available for a deep vertical excavation include pile driving, shoring, lagging, soil nails, shotcreting, and dewatering.

PRELIMINARY WORK

Geotechnical Assessment

A necessary step before beginning a structural excavation is conducting an analysis of existing conditions. Visiting the project site is always recommended to visually perceive unique challenges associated with site topography, onsite materials, road access, and logistics for transporting materials to and from the site.

A formal means of capturing existing conditions and the anticipated elements with design and construction is preparing a written *geotechnical report*. The report is a comprehensive tool to better understand the stability and structural capacity of the soil and the anticipated behavior of existing soils or rock to excavation and handling. Typical steps in preparing a geotechnical report include:

1) Researching and reviewing previously prepared geotechnical reports or general web-based soil surveys
2) Collecting available site and vicinity information such as topographic maps, land surveys, well logs, and environmental reports
3) Sampling subsurface and groundwater information by drilling geotechnical borings
4) Conducting ground-penetrating radar scans, performing dynamic cone penetration tests, excavating test pits, and performing seismic refraction testing
5) Performing laboratory testing of collected soil and rock samples

6) Recommending foundation design(s) associated with the existing site and proposed structural loading

7) Identifying any unique geotechnical aspects of construction, especially those presenting significant challenges adversely affecting the project

Although a geotechnical report may be prepared, the actual excavation of in situ material presents a risk to the project estimator and constructor. The material properties may change and not respond as planned to laboratory tests, such as volumetric changes from swelling while excavating and shrinking while compacting (see Chap. 4). The depth and composition of soil strata are only reliable in terms of soil boring locations. Such was the case on the structural excavation for a $25 million three-story educational building with one story below grade. The soil borings indicated soil at the foundation column lines; however, between the columns and the borings there was dense limestone causing a $1 million change order at the start of the project. This example highlights the fact that geotechnical reports may not capture all aspects of the jobsite, although they do offer important insight to begin planning a deep excavation.

Site Protection

Areas adjoining a structural excavation are to be protected from construction operations for the duration of the project. Temporary fencing is one of the simpler and more effective protection methods (Fig. 12.1). Projects adjacent to busy street traffic can include standard traffic control equipment such as barricades, barrels, cones, and signage. Perforated opaque mesh may be added to the protection fence as a means to limit spectator visibility into the excavation. Perforated mesh emblazoned with the contractor's name can provide

FIGURE 12.1 Site protection fence and concrete barrier for structural excavation.

a marketing opportunity. Excavations adjacent to high-volume roadways or very deep excavations should include concrete barriers (Jersey barriers).

Tree Protection

In a structural excavation, any required clearing is likely to be of the selective type. Large trees or trees of desirable species may determine the location of the building structure or infrastructure alignment and should be guarded against damage and burial. For larger trees, protective fencing with an optional access gate and spaced signage is an effective method. For any size tree, it may be advisable to wrap the trunks of such trees in cloth or warning tape and protect them with a collar of vertical boards. In cases of the tree base being temporarily buried, the original ground line should be marked so that the fill may be removed accurately and burial or overcutting avoided.

Topsoil

Topsoil is usually present, and when reused for landscaping, it should be stored in a planned location. This involves stripping the area to be excavated and could include areas where spoil is to be piled. Topsoil stripping may be a substantial part of the cost of the excavation due to specialized removal with dozers (Chap. 7), excavators (Chap. 8), or scrapers (Chap. 11). In places where there is no well-defined topsoil or the topsoil makes up one-third or more of the spoil and mixes well with the subsoil, special stripping may not be needed.

A method of stripping topsoil that is often most economical in the long run is to remove it completely from digging areas. Figure 12.2A schematically shows this option of placing spoil in compact piles well away from the two corners of the proposed excavation. This will usually keep it clear of digging, piling, and ditching, and will leave it in a position for straight-line spreading. But it should not be put in corners where a dozer would have trouble maneuvering. Figure 12.2B shows a more usual method. The digging and piling area is stripped followed by the topsoil being pushed into two piles. If too small an allowance is made for the spoil, the topsoil may have to be moved back farther, requiring double handling and added cost, or the spoil may get buried by the fill and partly lost. In either case, it will probably interfere with backfilling operations. In Fig. 12.2C, the topsoil is pushed to

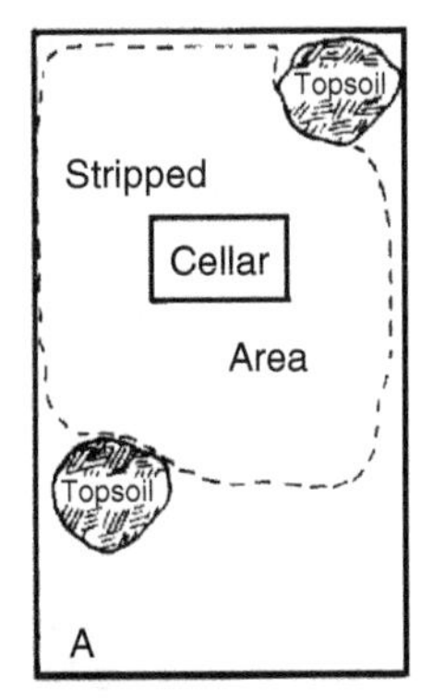

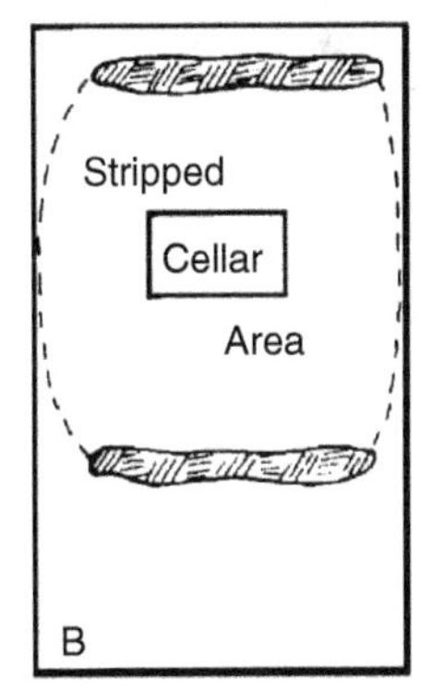

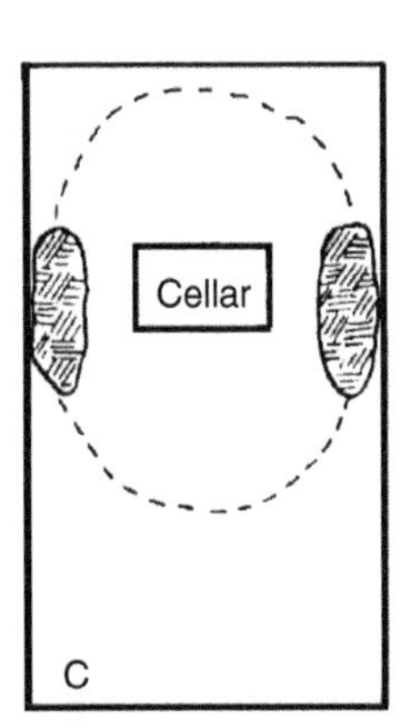

FIGURE 12.2 Topsoil piling.

FIGURE 12.3 Dozer piling topsoil in corner of a structural excavation location.

the sides and fill piled to the front and rear. A dozer stripping operation using Fig. 12.2A for a basement excavation is shown in Fig. 12.3.

Topsoil stripping operations are customarily performed by a dozer (Chap. 7) due to higher drawbar forces and relatively higher grading accuracy. Self-elevating scrapers (Chap. 11) are an excellent choice to strip topsoil to very accurate grades and are very effective for large areas and moderate-haul distances for 1,000 to 5,000 ft. Backhoe excavators will remove topsoil rapidly, and if the soil is heavy and wet, this may be the preferred choice because they do not compact the topsoil and cause it to cake. If the planned topsoil layer is thin and required for finishing, it may be deliberately mixed with some fill while stripping to increase its bulk. If the subsoil has a loose texture, there will be little or no harm to fertility, and regrading will be simplified.

Artificial Obstacles

Digging difficulties may be caused by obstructions placed by the builder. Site protection fences, jobsite trailers, toolboxes, and temporary structures are common obstructions on most projects. Batter boards used as reference for exact placement of foundations, as well as corner stakes and grade stakes, are placed in select areas on structural excavation projects. They can increase the excavation cost by restricting the movement of machinery, particularly if the structural excavation is irregularly shaped. Material laydown areas also restrict excavation production.

Depth

The depth of the excavation depends on the first floor finished elevation (FFE) relative to the original grade, as well as the depth of substructure below it, and utility trenches entering from below the foundation. The substructure may include floor slab thickness, joists, sills, crawlspace headroom, and thickness of the basement floor and gravel or crusted rock underneath. Excavation for geofoam is included in the depth estimate. Vapor barriers with plastic sheeting or bentonite slurry is not considered in the depth. For pipe laying projects, the pipe invert, pipe wall thickness, and bedding must be accounted for when calculating depth. If the full area is to be dug to the bottom of the footings, their depth must also be considered. On sloping ground, excavation depth will vary at different points.

DOZER DIGGING

Three standard machines for digging smaller structural excavations such as basements, bridge piers, and manholes are the dozer (Chap. 7), backhoe excavator (Chap. 8), and front-end loader (Chap. 9). Each has specialized features and capabilities for excavating the earth. A dozer can dig a structural excavation or basement somewhat deep if there is no heavy rock or mud, but operates best if the excavation is shallow. This is because it must dig away a considerable amount of the bank to ramp itself in and out, and the whole weight of the machine should come up out of the hole with every pass to maintain momentum and move material to the stockpile. It can push much larger loads on a level or a moderate upgrade than on the steep rises from a deep excavation, so the effect of steep grades must be factored into the operation (Chap. 7). Digging techniques vary with the operator and the locality. It is good practice to leave a ramp so trucks can back into the floor of the excavation to remove material and for convenience in delivering foundation materials to the point of use.

Several methods for dozer structural excavation are described and serve as preliminary approaches and should be followed only where they give satisfactory results to the specific project. An example considers a traditional rectangular-shaped basement excavation one story below grade. Trees have been removed and topsoil has been stripped. Stakes are set around the exterior perimeter of the digging line.

Open-Front Top Layer Method

Figure 12.4 illustrates the dozer first working along the short dimension pushing from south to north, inside the stake lines. The blade may be dropped at the front or south line for a fairly deep bite, and when filled, it is lifted to ride the load over the undisturbed ground until the north line is crossed. The soil may be dropped at the line or pushed a few feet back. The dozer then backs to the south line and takes another bite in a strip adjoining or overlapping the first. It may work the whole width of the front line, as in Fig. 12.4A, or only a section of it, before digging the area over which the soil has been pushed. The back edge of the cut is worked north by successive bites until the rear line of the excavation is reached, approximately as shown in the cross-section (B). A resulting spoil pile is shown in Fig. 12.3.

After completing the removal of the top layer, the dozer may cut out and pile a second layer in the same manner, as in Fig. 12.4C. This deeper cut will not extend to either digging line, as the slope down and the slope up are kept inside these lines. If the soil is too stiff where the blade cannot be filled in a short distance, the dozer may be worked over a short digging area only for several passes, after which the pile of loosened soil may be pushed to the spoil pile. Consider using a ripper to break up tight soils.

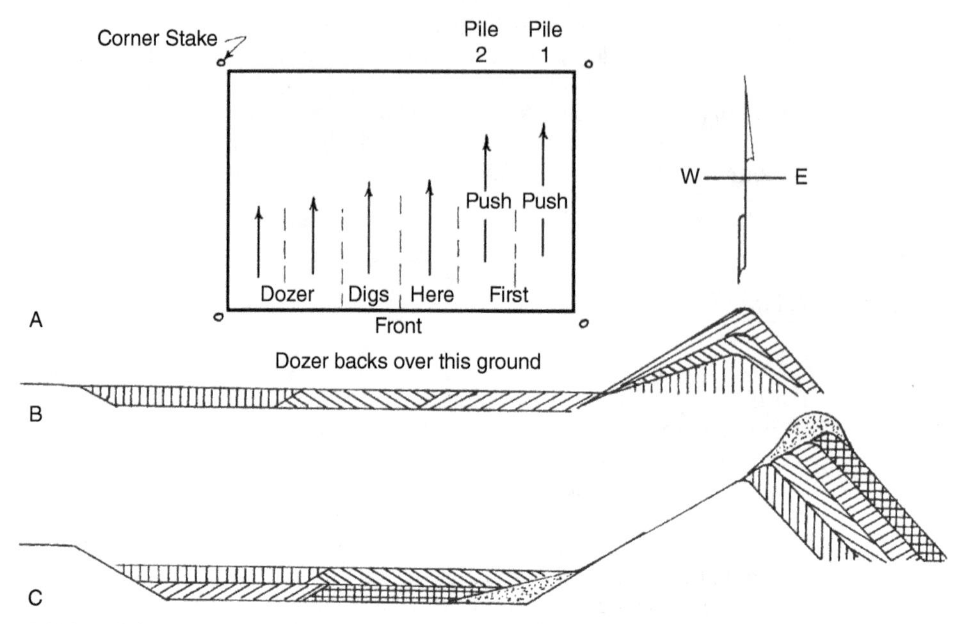

FIGURE 12.4 Basement digging sequence with a dozer.

Ramping Down

A dozer can cut hard soils but will usually not easily cut a steep downward ramp. Downward dozing has the advantage of increased drawbar pull and cutting force; however, machine stability is altered from flat horizontal dozing. Figure 12.5A shows a dozer cutting down from a line in soft soil. If digging production is good, the blade will penetrate rapidly until maximum depth is reached. A level cut is made until the machine's center of gravity moves over the cut. Then the blade will fall forward and will resume cutting until full penetration is reached, when it will level off again. This fluctuation in penetration depth may have the potential to create an unsatisfactory ramp slope.

To achieve a smoother slope, the option shown in Fig. 12.5B is to start well back from the edge with a gradual curve that is made steeper at the digging line. Cutting is regulated so the full depth of penetration will be reached as the center of gravity crosses the steepened part of the curve. As indicated, several passes are made in digging this ramp. This curve may be made steadily steeper as depth increases, because the tractor itself is at a downward slope.

If the cut cannot be made far enough back into the bank for ramping, the procedure shown in Fig. 12.5C may be followed. Soil is pushed out of the excavation into a steep pile; the dozer is backed on this and is thus pitched downward at a steep angle to cut down sharply.

Piling

The area to be occupied by the pile should be calculated to avoid double handling of material. Place the first pass at the farthest point on the base. Successive passes are pushed toward the base edge and allowed to roll over the edge. The pile is then built up in a series of wedges, with their thin ends toward the excavation.

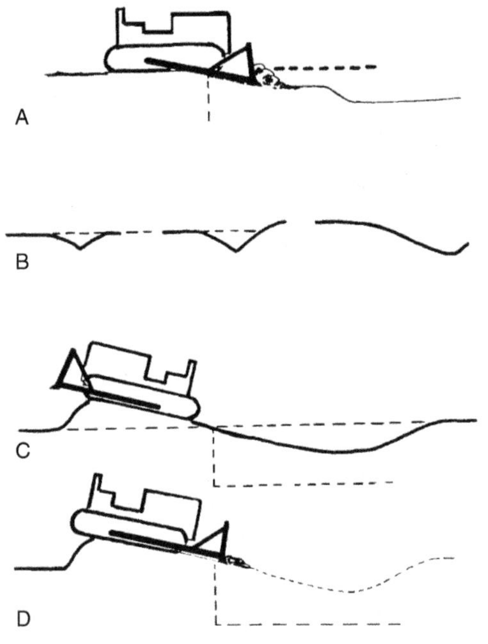

FIGURE 12.5 Cutting down from the edge.

Lower Layers

After the digging has reached the half depth, the dozer should be moved out onto the west bank and headed east along the south line of the excavation. An entrance ramp should be started several feet back of the west line, as in Fig. 12.6A, and a steep ramp cut down toward floor level. As the blade fills, the machine is swung toward the center of the hole, and the soil is left on its floor. The dozer returns to the south line, makes another cut, again swinging the spoil out into the center, pushing it somewhat farther. It thus cuts out the full-width ramp by which the dozer entered the pit from the south and occasionally shifts to pushing the soil obtained from the ramp up on the north pile. As the east edge is approached, soil may be pushed up on its bank, cutting a ramp, instead of to the north.

The north slope may then be cut away in the same manner, the spoil first being pushed up the undestroyed section of the ramp, as in Fig. 12.6B, and finally onto the west bank. These cuts result in vertical walls along the two long sides of the hole. The inside ramps may also be removed by oblique passes from near the center as in Fig. 12.6C, steering the dozer so that the blade is parallel with the bank. The slope ahead of the dozer can be gouged away in this manner, with the spoil being edged out into the open and pushed up the west bank. With the west section of ramp removed, the dozer can be turned to cut away the east part as in Fig. 12.6D.

Finishing

After removal of the north and south slopes, the floor of the cut may be deepened by pushing to the east and west. These ramps up to original grade should have their lower ends at

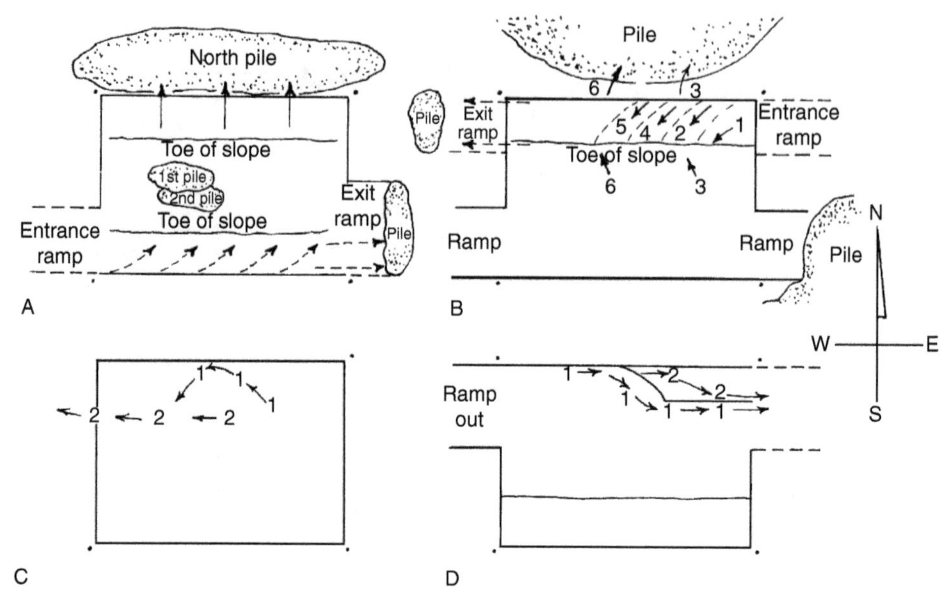

FIGURE 12.6 Cutting lower layers.

the excavation line, and will therefore be cut back deeply into the bank, as in Fig. 12.7A. Because a steep ramp means less extra excavation, the slopes should be made as steep as practical. They may be cut all the way through to begin with, for an easy gradient, and steepened as the hole deepens. The amount of soil removed for ramps can be slightly reduced by narrowing them as they go up, as in Fig. 12.7B. The ramps can be partly filled by the last loads pushed, as in Fig. 12.7C, except where space is left for supply trucks. As the bottom of the hole approaches the final grade, it should be checked with a level or a global positioning system (GPS) unit. It is not practical to dig narrow trench footing below the floor level with a dozer.

The procedure outlined should produce an excavation as shown in Fig. 12.7C with two straight walls correctly spaced. Ramping out of the short sides reduces the amount of extra digging that is one of the drawbacks of dozer work. The entire front is left free for access and storing building material.

However, the spoil may not be properly placed for backfilling and grading. In such a location, a building would usually have the fill spread around the perimeter, with particular attention to building up the front wall. Here, fill for the front grade would have to be obtained from the sides, which, in turn, might need to be partly replenished from the back. This involves extra pushing.

Irregular Basements

Figure 12.8 shows a basement with an irregular shape. This may be dug by the open-front method by cutting back the jog as indicated in Fig. 12.8A and by pushing soil for the small south room into the main basement, to be piled on the east side. Figure 12.8B shows the same basement pushed up into four piles. The jogs are handled by simply moving the ramps and piles outward the extra distance.

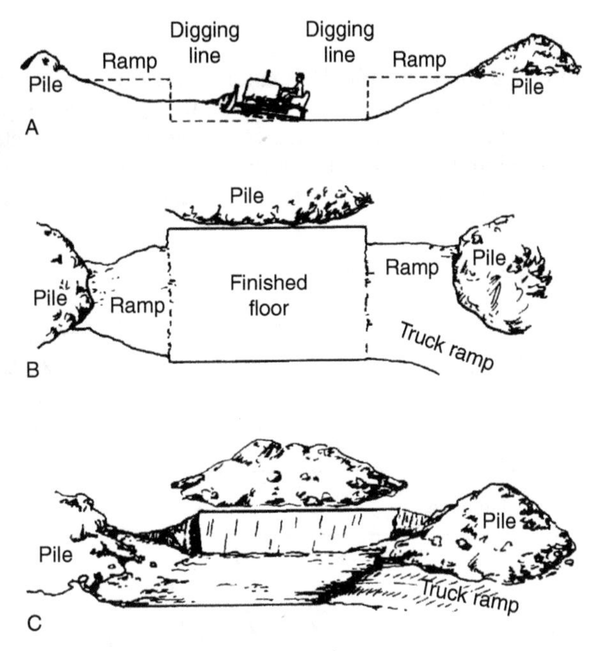

FIGURE 12.7 Finishing excavation.

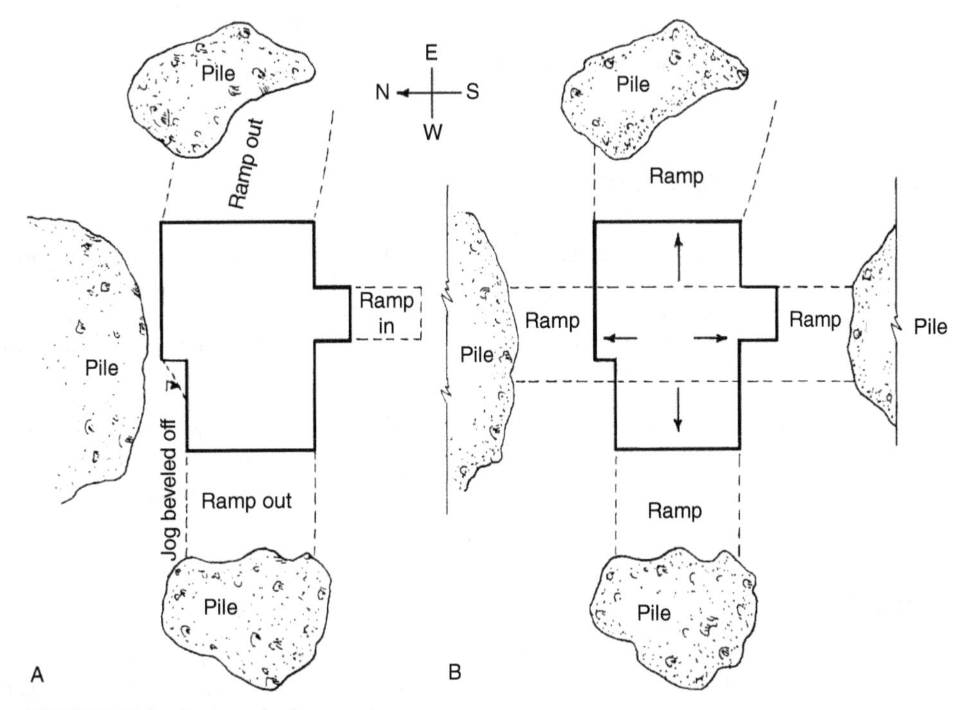

FIGURE 12.8 An irregular basement.

Limited Access

Figure 12.9 shows a difficult situation where trees or other buildings permit entrance by the dozer at only three points and where all the spoil must be pushed out at one of these points. The dozer movements indicated by the arrows are repeated at each cutting level. In soft soils, it might be possible to cut the corners with the dozer by loosening the soil and then back dragging. This type of excavation takes several times as long as open digging.

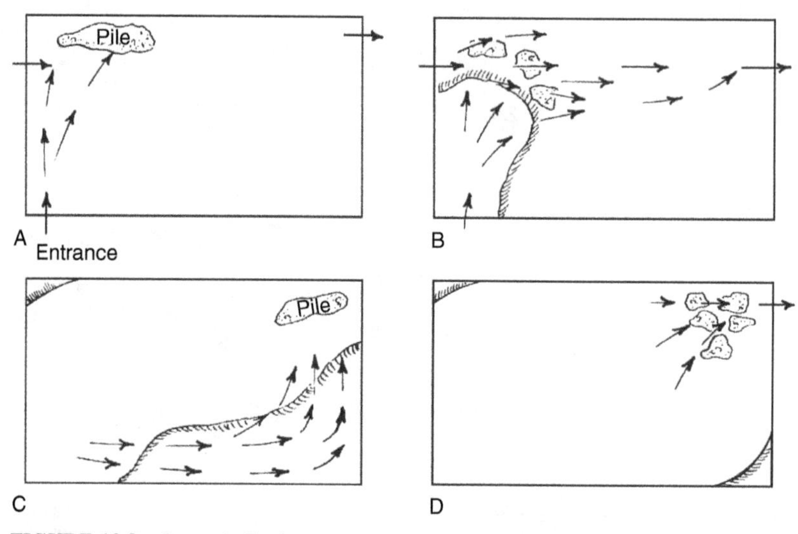

FIGURE 12.9 One-exit digging.

BACKHOE EXCAVATOR DIGGING

The backhoe (Chap. 8) is a principal machine for structural excavation work. Any size hoe is capable of shallow digging, but larger machines compare most favorably with the dozers when the hole is to be over 6 ft deep or when unfavorable bottom conditions are encountered, such as water, mud, boulders, or ledges. They are capable of ditching when necessary without changing attachments.

It is recommended that the digging lines be set outside of the required excavation, although in stiffer clays, the backhoe can do a very exact job. In addition to the corner stakes, intermediate guide stakes should be set at short intervals along the digging lines, as the operator cannot sight along these lines without assistance from a rodman, and the finished wall is established with the first cut.

Lining Up

Accurate lining up of the machine is essential for a clean job. If the cut is to begin along the south line, the excavator is positioned as in Fig. 12.10A, with the bucket about three-quarters extended and resting a few inches beyond the southwest corner. The boom and the

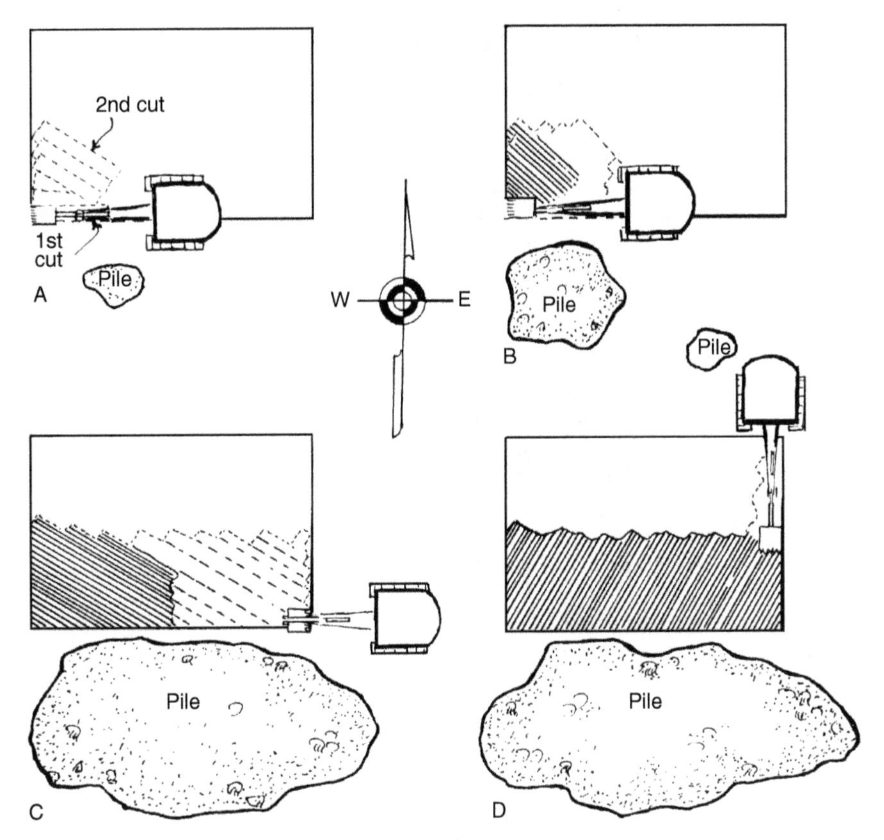

FIGURE 12.10 Basement digging with a backhoe.

tracks are parallel with the south digging line. Lining it up in this manner is simplified by marking the width of the bucket centered between the tracks.

Digging

A ditch is dug to bottom grade with its left edge on the digging line, and the spoil is dumped to the south. The far end of the ditch will curve rather sharply inward. When the backhoe has dug as much of the south wall as it can from its position, it reaches to the center of the west line and digs a trench back from there. The triangle included in these ditches is dug in layers to bottom grade and as near to straight down from the west line as possible.

The excavator is then backed up a few feet to the position in Fig. 12.10B. It can now cut the west end of the ditch almost vertical, because of the more extended position of the bucket. The south wall ditch is then extended as near the machine as possible, and material between it and the center cut down in layers to the bottom. The center line will be irregular. The spoil pile will tend to build up too sharply at the edge of the hole unless pushed back. This pushing may be done by regulating the outward swing of the bucket during the dump, so it strikes the pile at a spot where its momentum will push a considerable quantity of soil outward, without stopping its own motion. Knocking soil back should be initiated early, before the pile gets too high. The quantity of soil that can be put in a pile is greatly increased in this way.

Digging is continued in the same manner, with careful attention to a clean, level bottom until the east end is reached. The backhoe can probably cut this to a nearly vertical wall immediately in front of it, but will leave a ragged edge farther north, as shown in Fig. 12.10C. The backhoe is then turned and walked into the unexcavated north section. When its center is a half-bucket width inside the east line, as in Fig. 12.10D, it stops and shaves the end of the excavation, then trenches to the north edge. It next straddles the north line and is lined up in the same manner as before, with the bucket resting in the hole in the northeast corner.

The north section is excavated in the same manner as described for the south and completes the necessary hole. The west edge may be cleaned up, if necessary, by turning the backhoe to walk parallel to the edge so that the bucket can dig straight up. The backhoe should not be put in this position, however, unless the soil is firm and is known to have good load-bearing qualities, as a crawler machine is vulnerable to cave-ins or slumping under one track. This edge may also be trimmed from the north and south banks.

The completed excavation and spoil piles are shown in Fig. 12.11. Notice the piles are somewhat offset from the hole so the south pile can easily be used for fill on the east end and the north pile on the west end. Both ends are left open for access and material delivery.

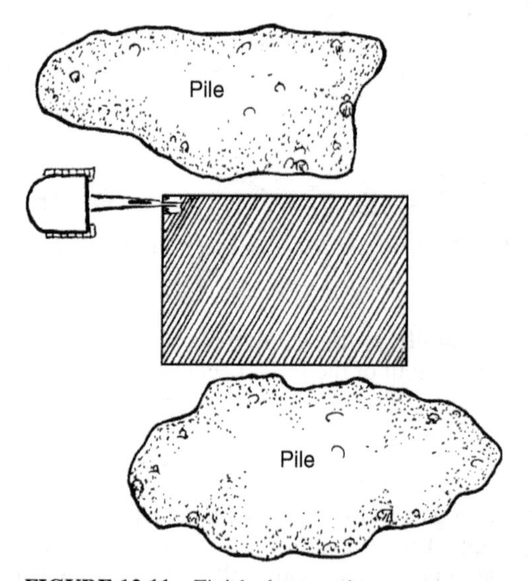

FIGURE 12.11 Finished excavation.

The north cut could have been made in the same direction as the south cut if the east ditch were shorter, but the fill would then have been concentrated toward the east end. A more finished hole could be made by starting the digging with a ditch along the west edge, dug from the south. The spoil pile would largely block the access to that side, unless the soils were piled in the basement area for rehandling. If access were not important, this ditch could be widened toward the center, reducing the amount to be piled to south and north. Existence of such a ditch would make it necessary to work the north section toward the east.

Loading Trucks

Depending on the project plan, the backhoe can load the spoil in trucks instead of dumping it on the ground. Where the piles will be so large that they will have to be dragged back, trucks may be used to take part of the material away, with the backhoe continuing to dump on the piles when no truck is in loading position. If grading plans have been prepared requiring use of the fill away from the foundation, it may be cheaper to truck it than to push it later with a dozer. However, enough of a pile should be left by the hole for backfilling between the foundation walls and the edge of the excavation.

The backhoe can dig footing trenches below the floor level where it is working parallel to the edge, as along the south, east, and north walls in Fig. 12.10. The hydraulic backhoe can dig them anywhere, but the parallel position is easiest.

Checking Grade

Cutting the bottom to proper grade is more difficult with a backhoe than with a dozer, because the

- Backhoe operator looks down at the grade rather than along it.
- It is difficult for the operator to climb down to check the grade.
- The machine cannot be moved back to correct mistakes.

It is very helpful to the operator to have someone to check the work, such as a rodman, although the operator can manage it alone if necessary, especially with newer units equipped with GPS in-cab grade measurements. The corner stakes, and preferably some other stakes, may be marked at a certain height above the bottom. In a flat site, the marks would all be 1 ft above the ground; in a sloping terrain the highest stake should be marked a little above the ground, and the others marked progressively higher.

If the operator is checking the grades alone, they may fasten a taut string between two stakes so that it will go over the spot in question and measure the distance from the floor to the string with the stick. Spots that cannot be crossed by the string or measured directly from the height of the wall may be checked from a known spot by eye. Hand, transit, or laser levels may be used.

A backhoe cannot cut a perfectly flat pit floor, because of the projection of the teeth and the tooth bases. The smoothest grade is obtained when the bottom of the bucket is used for finishing, by curling the bucket and oscillating across the floor. Interchanging buckets with smooth cutting edges is another option with quick-coupler attachments; however, the storage space for the additional bucket and impacts on production must be factored into this operation.

Irregular Shape

Figure 12.12 shows a basement of the same irregular shape as that in Fig. 12.8. The principal considerations in doing complicated excavations with a backhoe are

- Avoid digging into a trap.
- Avoid surrounding the hoe or blocking it from other work by piles of spoil.
- Work either parallel or at right angles to outside edges.

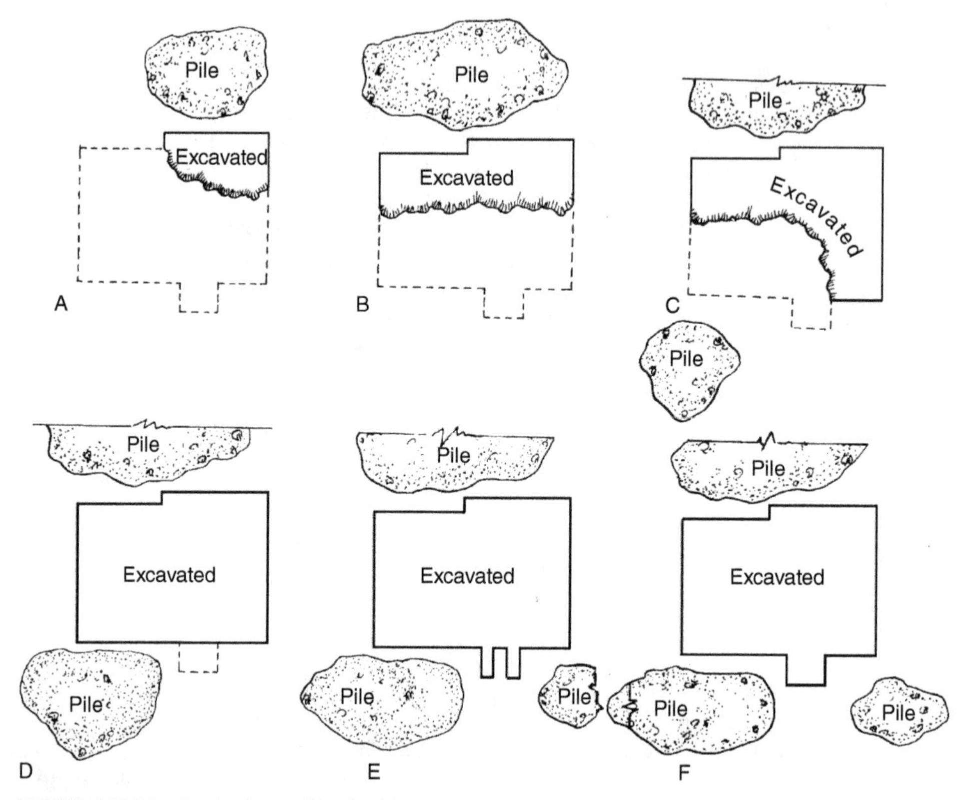

FIGURE 12.12 Cutting jogs with a backhoe.

There are several ways in which this basement can be dug. The north side can be dug from the east end, as in Figs. 12.12A and 12.12B. When the jog is reached, it is finished off with a vertical cut and the backhoe backed away and brought back in position to cut along the inner line. If the start is made at the west, the cut is brought a little beyond the jog, and the position then shifted to dig along the outer line. The machine may dig the south side by entering from the west and starting at the southeast corner.

Excavation is carried back to the west line, first ditching the edge then digging out the center. Care is taken to begin the spoil pile well west of the south room. The backhoe is then moved off to the south and up to the room. It is first lined up to the west side of this room and makes a cut from the main excavation to the south line of the room. The backhoe is then moved so as to cut the east wall of the room and then digs the rest of the room.

Another way to do this would be in the same manner as the north wall, treating the room as a double jog. However, this would involve extra digging because the bucket needs considerable width in which to cut down.

If the excavation site is a hillside, the work should be managed so the backhoe tracks will head uphill or downhill, not across. If the grade is steep, the backhoe should dig from downhill to avoid the danger of being pulled into the hole if the bucket hooks into something solid. If work must be done from the upper side, the stability of the ground should be checked, and both tracks must be securely blocked against sliding.

Other Shovel Rigs

Front shovels (refer to Figs. 8.2 and 8.3) are seldom used in structural excavations, but they have the capability. For satisfactory results, the ground should be firm at bottom grade, and the spoil should build into steep-sided piles. A ramp must be dug outside the excavation line with a slope, usually 3½ to 2½ on 1, which the shovel can climb when the job is finished. The spoil piles can be pushed back from the edge of the hole to some extent by crowding against them with a closed bucket. The walls of the hole tend to slope in at the bottom and to be somewhat jagged because of the different angles at which the bucket cuts them. Both of these features can be reduced or eliminated by careful digging, but extra time will be required, which affects production.

Clamshells (refer to Fig. 8.27) are not ordinarily used for structural excavation because they are more effective in softer materials and have a lower production rate. They are very effective in marine excavations to bridge piers and waterway structures with mucky, soft material. Clamshells have comparable grade accuracy as a backhoe, and for small, deep excavations will do the work more cheaply than a dozer. Digging is done from the top so no ramps are required. An edge may be cut with the tracks parallel with it and the tagline chains attached to one jaw, or with the tracks at right angles to the digging line and the tagline on both jaws. Either of these arrangements will permit cutting straight-sided trenches along the outer lines. The center is best cut in layers or in sections behind completed edges. A medium- or heavy-duty bucket should be used. The spoil may be placed in isolated piles, in windrows, or in trucks as desired.

FRONT-END LOADER DIGGING

A front-end loader, with a standard full-width hydraulic dump bucket, may dig an open type of excavation in the same manner as a bulldozer. It should be considered as an alternative to a dozer of the same power, because it can both penetrate stiff soil and load spoil into haul trucks. The ability of the front-end loader to cut straight ahead then back and turn with the load makes it possible to reduce the amount of excavation outside the digging lines for ramps. Front-end loaders are discussed in detail in Chap. 9, and for structural excavations, the tracked type is preferred due to higher drawbar pull at lower speeds and a lower center of gravity (Fig. 12.13).

Common methods for front-end loaders to excavate structures are the open-front method described earlier for backhoe excavators or the three-pile procedure shown in Fig. 12.14. The first layer is not cut quite to the ends of the excavation (Fig. 12.14A), and the ramps built in reaching the bottom are inside the digging lines (Fig. 12.14B). They are cut back to the steepest practical grades on the last pushes, and then one ramp, a foot or so wider than the machine, is cut into the bank. All soil left inside the digging line is then removed by being picked up and carried or pushed out the slot ramp.

The machine can carry a larger load up the ramp when moving forward than when reversing. This is because it is nose-heavy when loaded—a condition that is made worse when backing up an incline, both by the shift of the center of gravity toward the front and by the reaction from the driving torque, which pushes the front down. In ascending a grade forward, both of these forces tend to raise the front and improve stability. These effects become more pronounced as the incline becomes steeper. Turning around in a small excavation may be difficult with a loaded bucket so that it is often better to back out with a small load.

FIGURE 12.13 Tracked front-end loader beginning a structural excavation.

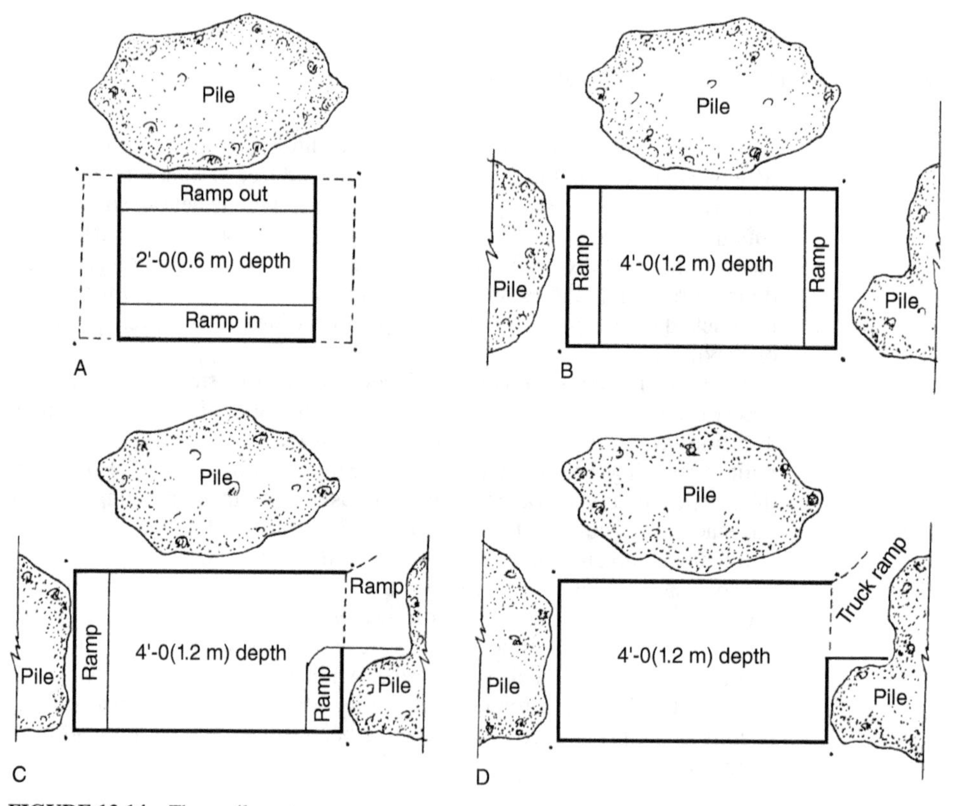

FIGURE 12.14 Three-pile system.

Cutting Walls

The excavation should be finished with vertical walls as much as possible. Long walls may be cut neatly by digging parallel to the edge, but because of limited space, much finishing must be done by working straight toward the digging line. Until the line is reached, the bank digging procedures in Chap. 9 may be followed.

A vertical face may be cut by allowing bank resistance to move the tractor back as the bucket rises. If the wall is high, the machine must be moved forward again after the bucket lip gets above the push arm hinges. This back-and-forth motion compensates for the curved path of the rising bucket. The same effect may be achieved by careful regulation of the bucket tilt, as it makes a shallower cut when rolled back than when flat.

Any jogs or irregularities can be cut by digging into the wall from the excavation without making additional ramps. It is easiest to take these out in layers as the floor is worked down, but they can also be dug after completion of the main work. Soil carried out of the hole may be spread or distributed to nearby low spots much more easily than by a dozer, and can also be readily loaded into trucks.

CONTRACTOR SETUP

A contractor engaged in structural excavation will probably have a combination of equipment, including a backhoe, a dozer, a fleet of trucks for hauling the spoil material away, miscellaneous support equipment, and drills or impact hammers for rock strata (Fig. 12.15). The backhoe may perform primary digging, while the dozer moves the material to a stockpile. If there is excess material to be hauled away, a truck hauling operation is designed (Chap. 10).

FIGURE 12.15 Multiple operations in a structural excavation.

Trucking Spoil

In large commercial excavations of the haul-away type (Fig. 12.15), one of the most important considerations is arranging for the disposal of the spoil, except when it is to be used as fill on the same project. It may be possible to sell it, or it might be necessary to pay for the privilege of wasting it on someone's property. Disposal arrangements may not only determine the price to be charged for the digging, but also the time of starting the work and the number and type of excavating and hauling units to be used (refer to Chap. 10). An inspection should be made of the dumping site, and any price quoted for fill should contain provisions for additional charges for dumping delays. Hauling restrictions during certain hours may apply in certain cities or localities.

Machinery

In a medium to large excavation, a crawler front-end loader, a backhoe, or a front shovel unit is usually preferred, unless the bottom is too wet or sandy for trucks. The backhoe needs less space, does not tear up the floor, is better in rock, and can tolerate rough or steep footing. Increasing size and decreasing depth favor the loader. It can replace the shovel on most jobs. A hydraulic backhoe of sufficient size does the neatest work.

Ramps

In most cases, the backhoe or front loader cuts a ramp down, inside the digging lines (Fig. 12.16), which must be of adequate grade and material for loaded trucks and other equipment. The grade may be between 1:5 and 1:12, depending on loaded truck capability and floor area of the excavation. The slope is made as gentle as the length and depth of the hole permit. The bottom toe of the ramp must allow enough turning radius for equipment exiting the excavation floor onto the ramp. If the plot is sloped, the ramp should be cut in from the lowest point on the edge to which trucks have access. Earth ramps are generally removed immediately upon completion of the excavation they serviced. Timber ramps

FIGURE 12.16 Shallow sloped ramp to the floor of a structural excavation.

afford less tractive resistance and better footing than earth and can be built with a steeper slope. They can be left in place during construction of the foundation for convenience in moving building material. However, timber work is expensive and is largely limited to use in excavations that are very deep in proportion to their size. Timbers are generally reserved for the matting of cranes and outriggers.

Figure 12.17 illustrates steps to consider when creating the ramp. The excavator works the west edge as shown in Fig. 12.17A using the procedures described earlier. Upon reaching the floor level of the first cut, it may continue to the back wall, or it may first make a side cut so that trucks can turn around in the pit. If the excavator works straight through to the back wall, it is then walked to the foot of the ramp again to take another slice toward the back, as in Fig. 12.17B, as double spotting of trucks is easier when working away from

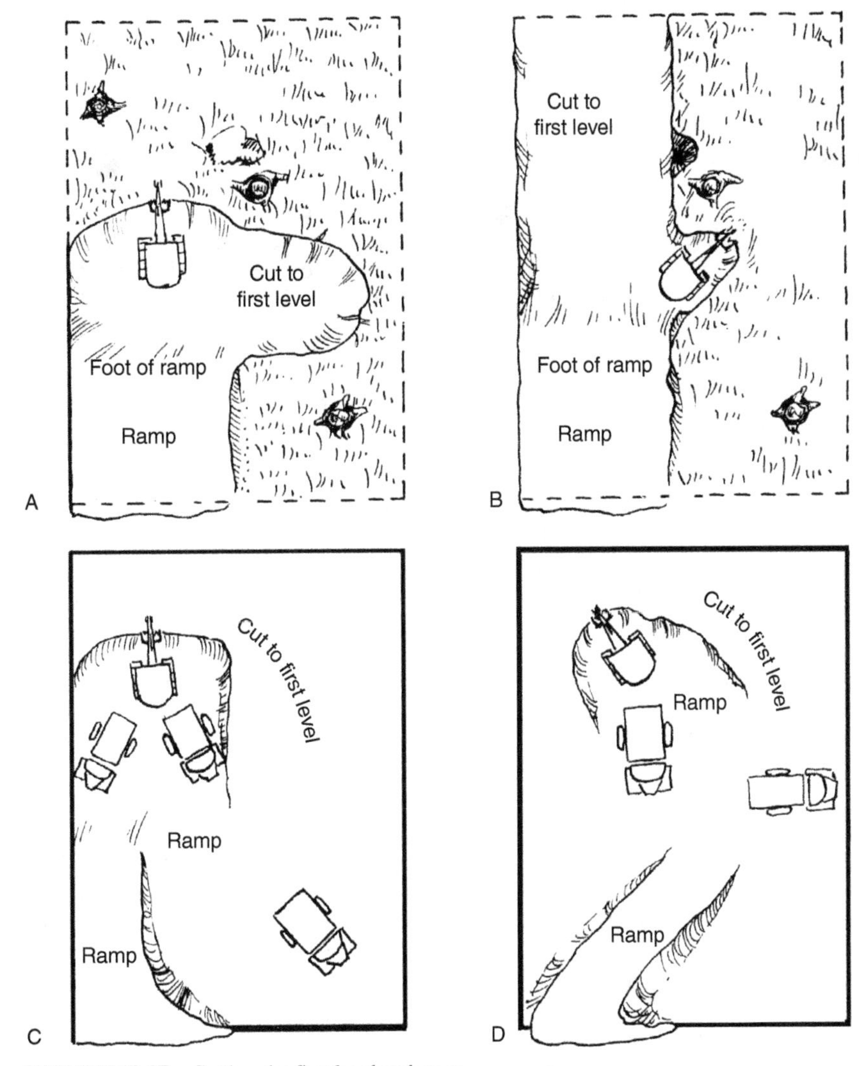

FIGURE 12.17 Cutting the first level and ramp.

the ramp than toward it. Once the pit floor is widened enough to allow trucks to turn, the digging may be extended in any direction so long as the shovel may be easily reached by trucks. Eventually the bank to the east of the ramp is dug away, and usually part of the ramp itself, leaving it wide enough for one truck only.

When the first level is complete, ramp cutting is resumed until the bottom is reached, as in Fig. 12.17C. Trucks will drive down the upper section, turn on the upper level, and go back down to the excavator. It is not necessary that the ramp continue in the same direction, but this is the most economical method where the pit is sufficiently long. Any turn must be made very wide for the convenience of the trucks. Cutting of the lower level proceeds in much the same way as for the upper, except that the floor grade must be carefully monitored. If strip footing trenches below the floor level are required, they may be dug immediately after trimming of the wall is complete and the spoil spread on the excavation floor or moved by the excavator, front-end loader, or dozer. A somewhat gentler ramp gradient could be obtained by using a diagonal or a zigzag ramp, as in Fig. 12.17D.

Ramp Removal

When the digging is complete and the ramp must be removed to create a bottom level floor, a backhoe is the best choice. A backhoe with an effective downward reach can remove the ramp from an upper level. For deep excavations with multiple levels below grade, the lower ramp is removed first and the process repeated, moving up levels. Trucks are loaded on the up-ramp side of the excavator.

Multiple Excavators

If a project is large enough to justify the use of two excavators (Fig. 12.15), there must be special consideration to sequencing operations. The excavators may operate in tandem at the same load-out location or at independent locations. Excavating at separate locations can build in separation between swinging excavators and possibly add safety to the operation. In cases of separate excavation operations, a second machine can ramp down from the opposite side of the excavation and cut through to meet the first one at the center. After this, one ramp might be used as an entrance and the other as an exit, or one might be cut away. Or the second excavator might be brought in after the first one reaches the lower level, with both excavators using the same ramp for trucking operations.

Traffic

An external factor potentially limiting the number and size of excavators in such an excavation is the traffic congestion on the street. Public traffic flow can often create a hauling production bottleneck. Project superintendents, managers, and estimators must factor in scenarios where truck velocity is impeded. Bunching of trucks from sporadic traffic flow can result in a line of empty trucks parked waiting on the street and a potential traffic hazard. Choking of truck production often results in excavator operation remaining half-idle for lack of trucks to load. In congested urban areas (Fig 12.15), traffic may be one of the principal controlling factors in digging production.

STRUCTURAL EXCAVATION SUPPORT

Excavation depths exceeding 10 or 20 ft require specialized support system planning. Lateral earth pressure is proportional to the vertical excavation depth. Therefore, the soil at the lower face tends to expand and move into the cut area. If a support is placed against the excavation surface to prevent soil movement, then the pre-excavation stress is maintained. Support is especially needed when the excavation is alongside an existing building or street. The method of support generally used is either to drive continuous sheet piling or soldier piles with timber lagging between them. This support system may be needed along the length of the adjacent building or street, depending upon excavation depth. There are other methods for supporting an excavated wall of earth such as soil nails and tiebacks.

Soldier Piles and Lagging

Soldier piles or soldier beams are H-piling set in predrilled holes around the periphery of an excavation (Fig. 12.18). Predrilling, as opposed to driving, is used to provide close control of alignment and location. The piles, typically on a 6- to 12-ft spacing, are grouted in place with weak concrete. Lagging is inserted in the steel web between the piles with wood, steel, or precast concrete panels. These are inserted behind the front pile flanges to retain the soil as the excavation proceeds in small stages downward (Fig. 12.19). If a void forms behind the lagging, it must be filled with compacted material. The walls can be strengthened by the use of anchors or bracing. The use of soldier piles and lagging is, in most cases, the least expensive temporary retaining system. If the site has a high water table, the system is not appropriate without dewatering.

FIGURE 12.18 Predrilling H-pile to support a vertical excavation along an existing building.

FIGURE 12.19 Soldier piles and lagging used to support a vertical excavation along a city street.

Soil Nailing

Soil nailing is an in situ reinforcing of the soil while it is excavated from the top down (Fig. 12.20). The array of soil nails is a passive inclusion installed in a grid pattern to create a stable mass of soil. The grouted, in-place, tension-resisting steel elements (nails) are placed in near-horizontal holes drilled into the exposed face at 3- to 6-ft centers. The face is secured by the application of reinforced shotcrete facing or wall panels (Fig. 12.20) with

FIGURE 12.20 Soil nailing with timber lagging.

FIGURE 12.21 Soil nailing with shotcrete face.

the shotcreting face method the most common (Fig. 12.21). Shotcrete is mortar mix sprayed at high velocity onto the vertical surface and reinforcing steel. A pattern of lightweight reinforcing steel or wire mesh is attached to the tensile steel at each anchoring hole. A single or series of shotcrete layers is applied, and the excavation is extended for the next lift depth until the full planned depth is reached.

A drainage system may be installed on the exposed face, as shown by the PVC pipes at the bottom of the wall in Fig. 12.21. In some cases, ungrouted soil nails are used if the nails are driven into the ground. This mass of reinforced soil functions to retain the less stable material behind it. In the right soil conditions, soil nailing is a rapid and economical means of constructing excavation support systems. It is not necessary to embed any of the structural elements below the bottom of the excavation as with soldier beams used in ground anchor walls, and this can be an advantage at some locations. They may not be appropriate at locations close to the wall perimeter.

Tieback Systems

Anchors or tiebacks eliminate obstructions in the excavation inherent in rakers (Fig. 12.22) or struts. Tieback systems are generally very effective in preventing movements of the excavation walls. Usually, the tiebacks and excavation wall are left in place after the permanent construction inside the braced excavation is complete.

In urban environments, deep vertical excavations are common for building foundations. Tieback systems are therefore used to provide an open work space in the excavation. Typically the tiebacks are spaced from 7 to 13 ft vertically and from 5 to 15 ft horizontally. Sloped rakers may be necessary on wider excavations where the horizontal assembly is not feasible (Fig. 12.23). Sloped rakers can also minimize overhead obstructions for workers and ease the flow of materials down into the excavation. Reducing the spacing is not always appropriate because the capacity of the system is reduced due to adjacent tieback zones not acting independently.

FIGURE 12.22 Rakers used in addition to soldier piles and lagging at a building foundation excavation

FIGURE 12.23 Sloped rakers for a wide building excavation used in addition to soil nails (left) and soldier piles (center).

Similar to soldier piling and soil nails, installing tieback systems is a highly technical operation, involving knowledge of soil behavior, detailed engineering calculations, and a detailed geotechnical report. The tiebacks must be solidly embedded in stable ground, preferably bedrock. Sheeting to support the tiebacks along the excavation wall must be engineered to withstand significant lateral forces. It may be necessary to drive interlocking steel sheet piling to several feet below the pit floor and, if needed, tiebacks into the ground the piling is holding back.

Horizontal timbers, called whalers, are placed along the face of the sheeting, being temporarily supported on cleats nailed to the planks. Beams or plank mats called heels are placed on firm, undisturbed soil in the pit floor, sloping down toward the wall. These are used as abutments to take the thrust of the breast timbers extending from the heels to the whalers. If the heels are firm, the spacing of breast timbers can be increased by using heavier whalers. While this bracing is being installed, an adjoining section of the wall is trimmed. This is braced in the same manner and the work continued in successive sections.

PUMPING WATER AND DEWATERING

Dewatering is an issue with structural excavations (effectively, the excavation is a bathtub). The challenge with structural excavation is devising a plan for removing the material as the machines create a pit with vertical walls, while simultaneously creating a large reservoir that may harbor water. With structural excavations, it is necessary to control the surface and subsurface hydrologic environment in such a way as to permit the structure to be constructed "in the dry." Inflow of stormwater will require the careful locating of sumps—these are usually gravel-filled—and dewatering pumps to ensure a dry excavated floor. The discharge piping must be arranged so it will not interfere with subsequent construction operations.

The depth of the groundwater table must also be considered. In cases of shallow groundwater tables, slurry walls may be constructed by excavating a deep perimeter wall using elongated excavator arms or clamshell buckets, then filling the cavity to create a heavily reinforced concrete wall. The excavation can then proceed inside the reinforced wall to the desired structural floor depth.

Pumps

Pumps are used extensively on construction projects for operations such as

1. Removing water from pits, tunnels, and other excavations
2. Dewatering cofferdams
3. Furnishing water for jetting and sluicing
4. Lowering the water table for excavations

Construction pumps (Fig. 12.24) must frequently perform under severe conditions, such as those resulting from variations in the pumping head or from handling water that is muddy or highly corrosive. The required rate of pumping may vary considerably over the duration of a project.

FIGURE 12.24 Small wheel-mounted centrifugal pump.

The factors to consider when selecting a pump for a construction application include

1. Dependability
2. Availability of repair parts
3. Simplicity to permit easy repairs
4. Economical installation and operation
5. Operating power requirements

Glossary of Pumping Terms

The following glossary defines the important terms used in describing pumps and pumping operations.

Capacity The total volume of liquid a pump can move in a given amount of time. Capacity is usually expressed in gallons per minute (gpm) or gallons per hour (gph).

Discharge head The (total) discharge head is the sum of the static discharge head plus the head losses of the discharge line.

Discharge hose The hose used to carry the liquid from the discharge side of the pump.

Self-priming The ability of a pump to separate air from a liquid and create a partial vacuum in the pump. This causes the liquid to flow to the impeller and on through the pump.

Static discharge head The vertical distance from the centerline of the pump impeller to the point of discharge (Fig. 12.25).

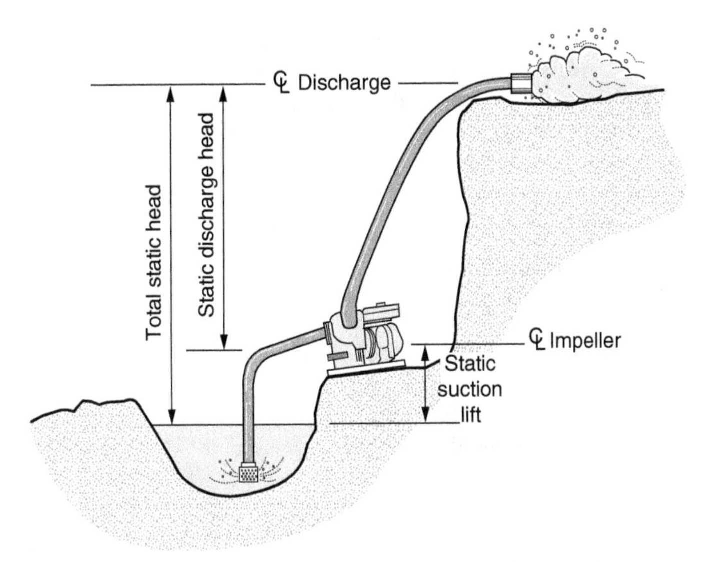

FIGURE 12.25 Dimensional terminology for pumping operations.

Static suction lift The vertical distance from the centerline of the pump impeller to the surface of the liquid to be pumped (Fig. 12.25). Suction capability is limited by atmospheric pressure. Therefore, maximum practical suction lift is 25 ft. Decreasing the suction lift will increase the volume that can be pumped.

Strainer A cover matched to the size of the pump and attached to the end of the suction hose that permits solids of only a certain size to enter the pump body.

Suction head The (total) suction head is the sum of the static suction lift plus the suction line head losses.

Suction hose The hose connected to the suction side of the pump. Suction hose is made of heavy rubber or plastic tubing with a reinforced wall to prevent it from collapsing.

Total static head The suction head plus the discharge head.

Classification of Pumps

The pumps commonly employed on construction projects may be classified as (1) displacement—reciprocating and diaphragm—or (2) centrifugal.

Reciprocating Pumps. A reciprocating pump operates as the result of the movement of a piston inside a cylinder. When the piston is moved in one direction, the water ahead of the piston is forced out of the cylinder. At the same time, additional water is drawn into the cylinder behind the piston. Regardless of the direction of movement of the piston, water is forced out of one end and drawn into the other end of the cylinder. This is classified as a double-acting pump. If water is pumped during a piston movement in one direction only, the pump is classified as single acting. If a pump contains more than one cylinder, mounted side by side, it is classified as a duplex for two cylinders, triplex for three cylinders, etc. Thus, a pump may be classified as duplex double-acting or duplex single-acting.

The capacity of a reciprocating pump depends essentially on the speed of the pump cycle and is independent of the head. The maximum head at which a reciprocating pump will deliver water depends on the strength of the component parts of the pump and the power available to operate the pump. The capacity of this type of pump may be varied considerably by varying the speed of the pump.

The advantages of reciprocating pumps are

1. They are able to pump at a uniform rate against varying heads.
2. Increasing the speed can increase their capacity.

The disadvantages of reciprocating pumps are

1. They are heavy and large pumps for a given capacity.
2. They impart a pulsating flow of water.

Diaphragm Pumps. The diaphragm pump is also a displacement type pump. The central portion of the flexible diaphragm is alternately raised and lowered by the pump rod that is connected to a walking beam. This action draws water into and discharges it from the pump. Because this type of pump will handle clear water or water containing large quantities of mud, sand, sludge, and trash, it is popular as a construction pump. The 2- and 3-in. gasoline-powered models move 50 to 85 gpm. They have the ability to handle air without losing prime and can handle water with a solid content greater than 25% by volume.

Centrifugal Pumps. A centrifugal pump contains a rotation element called an impeller. To water passing through the pump, the impeller imparts sufficient velocity to cause the water to flow from the pump even against considerable pressure. A mass of water may possess energy due to its height above a given datum or due to its velocity. The former is potential energy, whereas the latter is kinetic energy. One type of energy can be converted into the other under favorable conditions. The kinetic energy imparted to a particle of water as it passes through the impeller is sufficient to cause the particles to rise to some determinable height.

On construction projects, pumps frequently must be set above the surface of the water that is to be pumped. Consequently, self-priming centrifugal pumps are well suited for the needs of construction. Such a pump is self-priming to heights of 25 ft when in good mechanical condition. At altitudes above 3,000 ft, there is a definite effect on a pump's performance. As a general rule, a self-priming pump will lose 1 ft of priming ability for every 1,000 ft of elevation. A self-priming pump operated in Flagstaff, Arizona, at an elevation of 7,000 ft will develop only 18 ft of suction lift rather than the normal 25 ft. As the temperature of water increases above 60°F, the maximum suction lift of the pump will decrease. A pump generates heat that is passed to the water. Over a long duration of operation, as the heat increases, a pump located at a height very close to the suction maximum can lose prime.

Multistage Centrifugal Pumps. If a centrifugal pump has a single impeller, it is called a single-stage pump, whereas if there are two or more impellers and the water discharged from one impeller flows into the suction of another, it is called a multistage pump. Multistage pumps are especially suitable for pumping against a high head or pressure, as each stage imparts an additional pressure to the water. Pumps of this type are used frequently to supply water for jetting, where the pressure may run as high as several hundred pounds per square inch.

Pump manufacturers furnish curves showing the performance of their pumps under different operating conditions. A set of curves for a given pump will show the variations in capacity, efficiency, and horsepower for different pumping heads. These curves are helpful in selecting the pump that is most suitable for a given pumping condition. Figure 12.26 illustrates a set of performance curves for a 10-in. centrifugal pump. For a total head of 60 ft, the capacity will be 1,200 gpm, the efficiency 52%, and the required power 35 brake horsepower (bhp). If the total head is reduced to 50 ft and the dynamic suction lift does not exceed 23 ft, the capacity will be 1,930 gpm, the efficiency 55%, and the required power 44 bhp. This pump will not deliver any water against a total head in excess of 66 ft, which is called the "shutoff head."

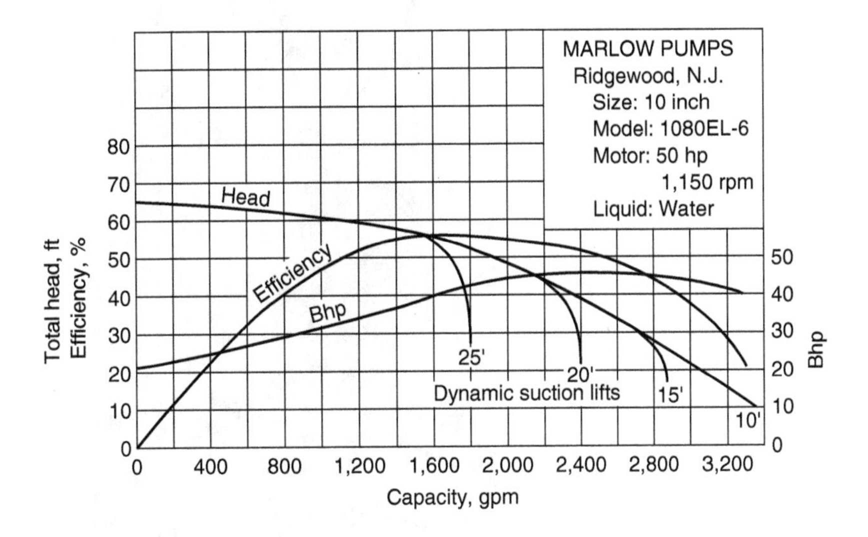

FIGURE 12.26 Performance curves for a centrifugal pump.

Because a construction pump frequently is operated under varying heads, it is desirable to select a pump with relatively flat head capacity and horsepower curves, even though efficiency must be sacrificed to obtain these conditions. A pump with a flat horsepower demand permits the use of an engine or an electric motor that will provide adequate power over a wide pumping range, without a substantial surplus or deficiency, regardless of the head.

Submersible Pumps. Figure 12.27 shows electric motor–operated submersible pumps. These are useful in dewatering tunnels, foundation pits, trenches, and similar places. With a submersible pump, there is no suction lift limitation and, of course, no need for a suction hose. Another advantage is that there are no noise problems. For construction applications, a pump made of iron or aluminum is best, as other materials are much more prone to damage when the pump is dropped. The power cord for a submersible pump should have a strain relief protector, as people tend to inadvertently lift the pump by the power cord.

There are basically two size categories for submersible pumps: small fractional horsepower size pumps and larger pumps with 1-hp or larger power units. The small pumps, typically ¼, ⅓, and ½-hp units, are for minor-nuisance dewatering applications. The 1-hp and larger pumps are for moving large volumes and/or high head conditions.

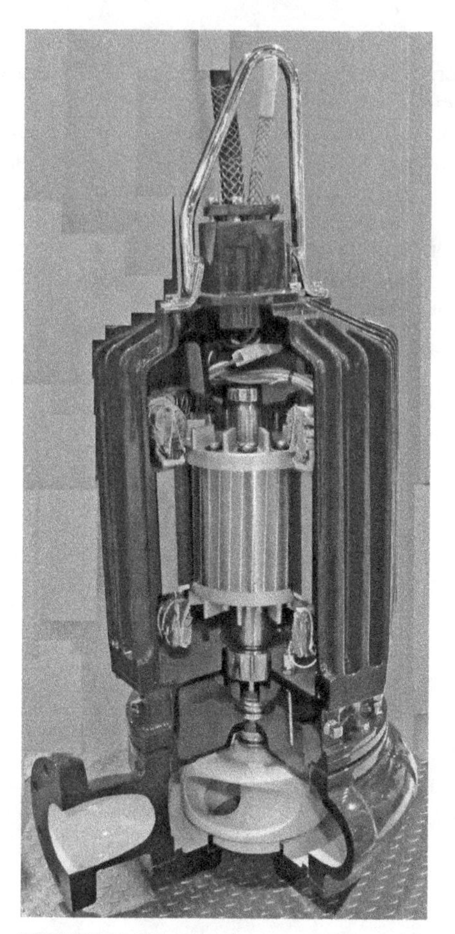

FIGURE 12.27 Electric motor–operated submersible pump.

Loss Head Due to Friction in a Pipe

Table 12.1 gives the nominal loss of head due to water flowing through new steel pipe. The actual losses may differ from the values given in the table because of variations in the diameter of the pipe and in the condition of the pipe's interior surface.

The relationship between the head of fresh water in feet and pressure in psi is given by the equation

$$h = 2.31p \tag{12.1}$$

or

$$p = 0.434h \tag{12.2}$$

where h = depth of water or head in feet
p = pressure at depth h in psi

TABLE 12.1 Friction Loss for Water, in Feet per 100 ft, of Clean Wrought-Iron or Steel Pipe*

Flow in U.S. gpm	Nominal Diameter of Pipe (in.)													
	½	¾	1	1¼	1½	2	2½	3	4	5	6	8	10	12
5	26.5	6.8	2.11	0.55										
10	95.8	24.7	7.61	1.98	0.93	0.31	0.11							
15		52.0	16.3	4.22	1.95	0.70	0.23							
20		88.0	27.3	7.21	3.38	1.18	0.40							
25			41.6	10.8	5.07	1.75	0.60	0.25						
30			57.8	15.3	7.15	2.45	0.84	0.35						
40				26.0	12.2	4.29	1.4	0.59						
50				39.0	18.5	6.43	2.2	0.9	0.22					
75					39.0	13.6	4.6	2.0	0.48	0.16				
100					66.3	23.3	7.8	3.2	0.79	0.27	0.09			
125						35.1	11.8	4.9	1.2	0.42	0.18			
150						49.4	16.6	6.8	1.7	0.57	0.21			
175						66.3	22.0	9.1	2.2	0.77	0.31			
200							28.0	11.6	2.9	0.96	0.40			
225							35.3	14.5	3.5	1.2	0.48			
250							43.0	17.7	4.4	1.5	0.60	0.15		
275								21.2	5.2	1.8	0.75	0.18		
300								24.7	6.1	2.0	0.84	0.21		
350								33.8	8.0	2.7	0.91	0.27		
400									10.4	3.5	1.4	0.35		
500									15.6	5.3	2.2	0.53	0.18	0.08
600									22.4	6.2	3.1	0.74	0.25	0.10
700									30.4	9.9	4.1	1.0	0.34	0.14
800											5.2	1.3	0.44	0.18
900											6.6	1.6	0.54	0.22
1,000											7.8	2.0	0.65	0.27
1,100											9.3	2.3	0.78	0.32
1,200											10.8	2.7	0.95	0.37
1,300											12.7	3.1	1.1	0.42
1,400											14.7	3.6	1.2	0.48
1,500											16.8	4.1	1.4	0.55
2,000												7.0	2.4	0.93
3,000													5.1	2.1
4,000														3.5
5,000														5.5

*For old or rough pipes, add 50% to friction values.

Source: *The Association of Equipment Manufacturers, Contractors Pump Bureau's* Selection Guidebook for Portable Dewatering Pumps.

TABLE 12.2 Length of Steel Pipe, in Feet, Equivalent to Fittings and Valves

	Nominal Size (in.)											
Item	1	1¼	1½	2	2½	3	4	5	6	8	10	12
90° elbow	2.8	3.7	4.3	5.5	6.4	8.2	11.0	13.5	16.0	21.0	26.0	32.0
45° elbow	1.3	1.7	2.0	2.6	3.0	3.8	5.0	6.2	7.5	10.0	13.0	15.0
Tee, side outlet	5.6	7.5	9.1	12.0	13.5	17.0	22.0	27.5	33.0	43.5	55.0	66.0
Close return bend	6.3	8.4	10.2	13.0	15.0	18.5	24.0	31.0	37.0	49.0	62.0	73.0
Gate valve	0.6	0.8	0.9	1.2	1.4	1.7	2.5	3.0	3.5	4.5	5.7	6.8
Globe valve	27.0	37.0	43.0	55.0	66.0	82.0	115.0	135.0	165.0	215.0	280.0	335.0
Check valve	10.5	13.2	15.8	21.1	26.4	31.7	42.3	52.8	63.0	81.0	105.0	125.0
Foot valve	24.0	33.0	38.0	46.0	55.0	64.0	75.0	76.0	76.0	76.0	76.0	76.0

Source: *Courtesy The Gorman-Rupp Company.*

Table 12.2 gives the equivalent length of straight steel pipe with the same loss in head due to water friction as fittings and valves.

Rubber Hose

The flexibility of rubber hose makes its use with pumps very convenient. Such hose may be used on the suction side of a pump if it is constructed with a wire insert to prevent collapse under partial vacuum. Rubber hose is available with end fittings corresponding to those for iron or steel pipe. As a "rule of thumb," total length of hose should be less than 500 ft on a centrifugal pump, and less than 50 ft for a diaphragm pump.

Hose size should match the pump size. Using a larger suction hose will increase the pumping capacity (e.g., 4-in. hose on a 3-in. pump), but it can also cause overload of the pump motor. A suction hose smaller than the pump will starve the pump and can cause cavitation. This in turn will increase the wear of the impeller and **volute**, and lead to early pump failure. A discharge hose larger than the pump will simply reduce friction loss and can increase volume if long discharge distances are involved. If a small discharge hose is used, friction loss is increased and, therefore, pumping volume reduced. Table 12.3 gives the loss in head in feet per 100 ft due to friction caused by water flowing through hose. The values in the table also apply to rubber substitutes.

Volute The housing in which the pump's impeller rotates is known as the volute. It has channels cast into the metal to direct the flow of liquid in a given direction.

Selecting a Pump

Before deciding on a pump for a given job, it is necessary to analyze all of the pump information and understand the project conditions affecting the proposed operations. Necessary information includes

1. The rate at which the water must be pumped
2. The height of lift from the existing water surface to the point of discharge
3. The pressure head at discharge, if any
4. The variations in water level at suction or discharge
5. The altitude of the project
6. The height of the pump above the surface of water to be pumped

TABLE 12.3 Water Friction Loss, in Feet per 100 ft, of Smooth-Bore Hose

Flow in U.S. gpm	Actual Inside Diameter of Hose (in.)											
	⅝	¾	1	1¼	1½	2	2½	3	4	5	6	8
5	21.4	8.9	2.2	0.74	0.3							
10	76.8	31.8	7.8	2.64	1.0	0.2						
15		68.5	16.8	5.7	2.3	0.5						
20			28.7	9.6	3.9	0.9	0.32					
25			43.2	14.7	6.0	1.4	0.51					
30			61.2	20.7	8.5	2.0	0.70	0.3				
35			80.5	27.6	11.2	2.7	0.93	0.4				
40				35.0	14.3	3.5	1.2	0.5				
50				52.7	21.8	5.2	1.8	0.7				
60				73.5	30.2	7.3	2.5	1.0				
70					40.4	9.8	3.3	1.3				
80					52.0	12.6	4.3	1.7				
90					64.2	15.7	5.3	2.1	0.5			
100					77.4	18.9	6.5	2.6	0.6			
125						28.6	9.8	4.0	0.9			
150						40.7	13.8	5.6	1.3			
175						53.4	18.1	7.4	1.8			
200						68.5	23.4	9.6	2.3	0.8	0.32	
250							35.0	14.8	3.5	1.2	0.49	
300							49.0	20.3	4.9	1.7	0.69	
350								27.0	6.6	2.3	0.90	
400									8.4	2.9	1.1	0.28
450									10.5	3.6	1.4	0.35
500									12.7	4.3	1.7	0.43
1,000										15.6	6.4	1.6

Source: *The Association of Equipment Manufacturers, Contractors Pump Bureau's* Selection Guidebook for Portable Dewatering Pumps.

7. The size of pipe to be used, if already determined
8. The number, sizes, and types of fittings and valves in the pipeline

In many cases, it is good practice to fix the pump intake with a screen (Fig. 12.28) to prevent objects of sufficient size from entering and plugging the hose or damaging the pump. Water containing leaves or other floating trash will readily clog a screen and may make necessary the placing of an outer screen of ¼- or ½-in. mesh. This outer screen should be placed far enough from the inlet so the water being drawn by the pump does not hold the rubbish against it.

Wellpoint Systems

In excavating below the surface of the ground, constructors often encounter groundwater prior to reaching the required depth of excavation. In the case of an excavation into sand and gravel, the flow of water will be large unless a method is adopted to intercept and remove the water. Dewatering, temporarily lowering the piezometric level of groundwater, is then necessary. After the construction operations are completed, the dewatering actions

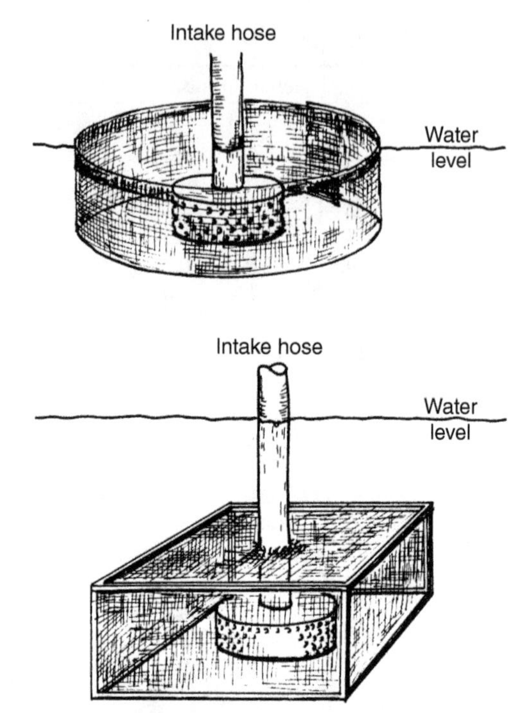

FIGURE 12.28 Trash protection for a pump intake.

can be discontinued and the groundwater will return to its normal level. When planning a dewatering activity, it should be understood that groundwater levels change from season to season as a result of many factors.

Ditches located within the limits of the excavation can be used to collect and divert the flow of groundwater into sumps from which it can be removed by pumping. However, the presence of collector ditches within the excavation usually creates a nuisance and interferes with the construction operations. A common method for controlling groundwater is the installation of a wellpoint system along or around the excavation to lower the water table below the excavation bottom, thus permitting the work to be done under relatively dry conditions.

A wellpoint is a perforated tube enclosed in a screen that is installed below the surface of the ground to collect water in order to lower the piezometric level of groundwater. The essential parts of a wellpoint are illustrated in Fig. 12.29. The top of a wellpoint is attached to a vertical riser pipe. The riser extends a short distance above the ground surface, at which point it is connected to a larger pipe called a "header." The header pipe lies on the ground surface and serves as the trunk line to which multiple risers are connected. A valve is installed between each wellpoint and the header to regulate the flow of water. Header pipes are usually 6 to 10 in. in diameter. The header pipe is connected to the suction of a centrifugal pump. A wellpoint system may include a few or several hundred wellpoints, all connected to one or more headers and pumps.

The principle by which a wellpoint system operates is illustrated in Fig. 12.30. Figure 12.31 shows how a single point will lower the surface of the water table in the soil adjacent to the point. Figure 12.31 also shows how several points, installed reasonably close together, lower the water table over an extended area.

FIGURE 12.29 Parts of a wellpoint system.

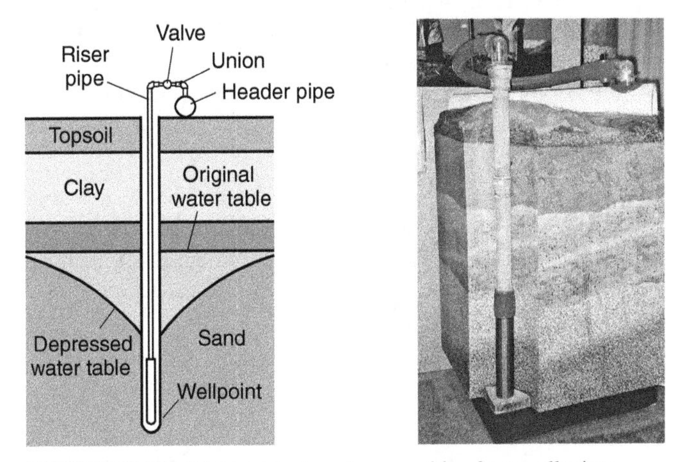

FIGURE 12.30 Water table drawdown resulting from wellpoints.

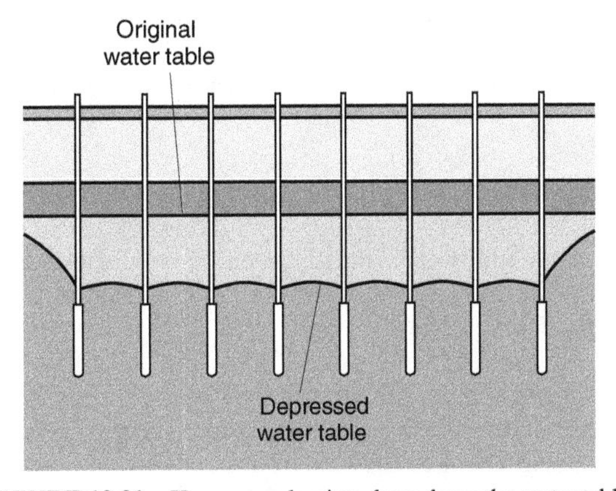

FIGURE 12.31 How several points draw down the water table.

Wellpoints will operate satisfactorily if they are installed in a permeable soil such as sand or gravel. If they are installed in a less permeable soil, such as silt, it may be necessary first to create a permeable well. A permeable well can be constructed by sinking, for each point, a 6- to 10-in.-diameter pipe, removing the soil from inside the pipe, installing the wellpoint, filling the space inside the pipe with sand or fine gravel, and then withdrawing the pipe. This leaves a volume of sand around each wellpoint to act as a water collector sump and a filter to increase the rate of flow for each point.

Wellpoints can be installed at any spacing, but usually the spacing vary from 2 to 5 ft (Fig. 12.31) along the header. The maximum height to which water can be lifted is about 20 ft. If it is necessary to lower the water table to a greater depth, one or more additional stages of wellpoints should be installed, each stage at a lower depth within the excavation.

Capacity of a Wellpoint System. The capacity of a wellpoint system depends on the number of points installed, the permeability of the soil, and the amount of water present. An engineer who is experienced in this kind of work can perform tests that will provide data to make a reasonably accurate estimate concerning the capacity necessary to lower the water to the desired depth. The flow per wellpoint may vary from 3 to 4 gpm, in the case of fine to medium sands, to as much as 30 or more gpm for coarse sand. Figure 12.32 presents approximate flow rates to wellpoints in various soil formations.

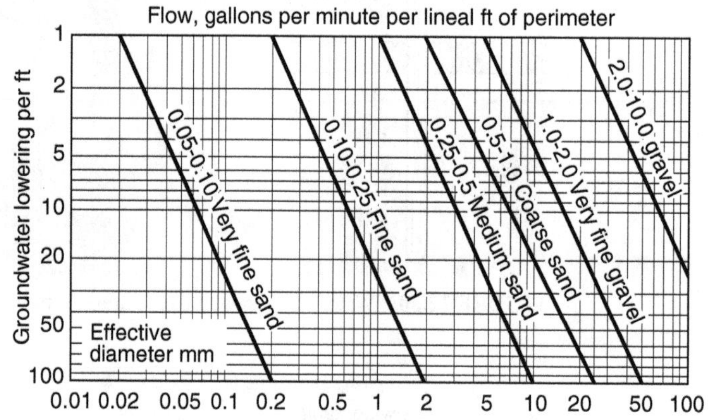

FIGURE 12.32 Approximate flow through various soil formations to a line of wellpoints. (*Source: Moretrench America Corporation.*)

The data in Fig. 12.32 will aid in the selection of the size of pumps that should be used with a wellpoint system. As an example, consider that it is necessary to dewater a pit that is 15 ft deep and that the water table is 5 ft below the surface of the ground. The soils to be encountered are fine sands. Therefore, starting with a water-lowering requirement of 10 ft (15 – 5) on the left side of the chart, proceed horizontally to the fine sand diagonal. From the intersection of the horizontal projection and the fine sand diagonal, project a vertical line down to the flow rate numbers on the bottom of the chart. Consequently, a flow of 0.5 gpm per foot of header pipe could be expected for these conditions.

Deep Wells

Another method for dewatering an excavation is the use of deep wells. Large-diameter deep wells are suitable for lowering the groundwater table at sites where

- The soil formation becomes more pervious with depth
- The excavation penetrates or is underlain by sand or coarse granular soils

In addition, there must be a sufficient depth of pervious materials below the level to which the water table is to be lowered for adequate submergence of well screens and pumps. The advantage of deep wells is that they can be installed outside the zone of construction operations.

Cofferdams

When dry excavation is carried a considerable distance below the water table without dewatering the area, the heavy walls constructed to keep out soil and water are called cofferdams. Cofferdams consisting of single rows of interlocked steel pilings that enclose areas, with interior bracing, have been used for depths exceeding 60 ft, although ordinary practice limits them to 40 ft. They are used extensively for the construction of bridge piers or shallow-water structures, such as inlets and retaining walls. Cofferdams may be installed by hammering or vibrating the piling in undisturbed ground until it reaches bedrock, or to sufficient depth below the excavation floor to be considered safe. All the sections should be placed and driven to moderate depth before any of them are driven all the way, to make sure that all joints interlock properly. The shape is generally square to rectangular. Cross-bracing supports the piling walls against lateral hydrostatic forces.

Excavation is likely to be done by clamshells in softer soils. Bracing is placed against the inside of the wall as it becomes exposed. If the soil is very porous, great difficulty may be experienced getting the water down the first few feet, as the joints between sections leak quite freely until forced together by water pressure. More or larger pumps may be used at this stage of the job than at any later time. It may be necessary to trench outside the wall to place a clay seal partway down or to partially seal the soil with cement grout. Porous soil under the bottom of the wall may permit excessive quantities of water and sand to boil up in the bottom of the excavation as the final grade is approached, so a concrete seal is poured to create an impermeable base.

Caissons

A caisson is a structure that serves to keep soil and water out of an excavation, and forms part of the permanent structure for which the excavation is made. Open-top caissons for a bridge pier are shown in Fig. 12.33. A hollow square, ring, or other shape is made of reinforced concrete, with the bottom tapered to an inside edge. In water excavations, sheet piling can be driven around the outer perimeter of the caisson to allow forming the concrete walls, installing reinforcing steel, and pouring the caisson concrete walls (Fig. 12.33). If the work starts on dry ground, it may be built in a shallow excavation where it is to be used. If the start is underwater, it is made elsewhere with walls high enough to keep out water when it is lowered into place. Transportation is usually by barge. The caisson is lowered by digging inside to undermine it and building the top to provide more weight and to keep it above ground or water as it descends. Most of the digging is done underwater.

FIGURE 12.33 Multiple caissons for a bridge pier.

When it comes to the bottom, investigation must be made to determine whether it is on bedrock or boulders. If the rock surface slopes, concrete must be pumped underneath to give it firm bearing on the low side. The pneumatic caisson has an airtight cap above the bottom, with sufficient air pressure maintained under it to keep water out. Air locks and chambers are provided for entrance and exit of personnel and material. Much of the digging is done by hand, and in deep work at high pressures workers may be limited to less than an hour of work at a stretch, with long periods spent in entering and leaving the high-pressure work chamber. Depths of over 100 ft can be achieved.

CHAPTER 13
TRENCHING AND TRENCHLESS TECHNOLOGIES

Depending on the needed width and depth of a trench excavation and the hardness of the material being excavated, there are specifically designed machines suitable for successfully and economically accomplishing trench work. If personnel must enter and work in a trench, special safety precautions must be taken, including possibly the use of temporary structures to support the sides of the open excavation. A variety of techniques are used to minimize or eliminate the need for open excavations. These methods and special equipment are commonly employed to install underground utilities. Trenchless technologies have been used successfully to place a wide variety of utilities, from small-diameter cable and conduit installations such as electric and communications lines, to large-diameter utilities such as water and sewer distribution lines and oil and gas pipelines.

TRENCHING

Most dry trenching is accomplished by conventional hydraulic hoes or small backhoes. There are, however, special wheel and ladder trenchers. In soft ground or swamps, draglines may be the best machines. Clamshells are used for deep vertical work. Shallow trenches in dry ground may be dug by hoes, graders, or dozers.

Investigation

Before an excavation of any kind is begun, the site should be inspected carefully for conditions requiring precautionary measures. This is especially important when working in a developed area with buildings and underground utilities. Always survey the adjacent properties before beginning an excavation and, if possible, before bidding the work. Document all defects such as cracking and settlement of surface features (pavements, driveways, and sidewalks), so any claim for damages after the excavation work is completed can be correctly assessed and it can be determined if the issue was preexisting or the result of the recent work.

The location of existing underground utilities (sewer pipes, electric lines, telephone, fuel, water, and gas) must be determined before excavation begins. The contractor should contact the local utility one-call system and ask them to establish the locations of all utilities. Access to the utility alignment or areas where it may be necessary to encroach on adjacent property must be identified. Land easements may exist to grant access to the project site, and these agreements usually have specific conditions. Existing permanent easements allow perpetual access to the utility alignment for its intended life. These grant the utility owner or its contractor the ability to maintain the utility. Temporary construction easements may be granted if access to the permanent easement is difficult or more working space is needed. Construction easements are typically only valid for the duration of the project as detailed in the permit. If the route of the utility line is parallel to the needed trench and within the risk area defined by a triangular wedge with a distance no less than the depth of the trench plus half of the trench width, extra precautions must be taken to ensure no damage results to the utility line.

Vacuum Excavation

In underground utility construction, there is a procedure known as *potholing* to discover where existing utility lines are located. Many government agencies require potholing so new construction operations do not damage existing utility lines. The preferred method for potholing is by vacuum excavation.

Safety is the key reason for using vacuum or hydrovac excavation techniques. "Vac" excavation units make use of powerful vacuum systems to safely excavate around buried utilities (Fig. 13.1). Vacuum excavators typically come equipped with debris tanks. The debris tanks have capacities ranging from 3.5 to 15 cy.

FIGURE 13.1 Vac-truck used to locate utilities.

The portable vac equipment is like a large vacuum cleaner using either high-pressure water or air, up to 1400 cfm, to break apart and displace the soil. They excavate a relatively small-diameter hole while collecting the spoil material in the vacuum unit's debris tank. Depending on the machine used and the soil conditions, a 12-in.-diameter, 5-ft-deep pothole can be completed in 20 min or less. Many vacuum excavation units are capable of digging much deeper, but most utility potholes are typically no more than 6 ft deep.

High-pressure water, in the case of hydrovac units, or compressed air penetrates, expands, and breaks up the soil. A high-speed fan or positive-displacement blower generates an air current producing a vacuum within the vacuum tank. A vacuum hose attached to the vacuum tank and supported by a hydraulically operated boom is guided by the operator, and a suction nozzle collects the loosened debris.

Hydrovac units simultaneously inject pressurized water into the ground and extract the liquefied soil by vacuum. Small skid-mounted or trailer mounted hydrovac units are available for use with smaller single-axle trucks. Large triple-axle, truck-mounted units are fully equipped with vacuum pump, water pump, filtration system, and debris and water tanks.

High-velocity air units work in the same manner as a hydrovac and use a suction nozzle to collect the loosened debris. A secondary air chamber in the tank reduces the airflow, causing the material to drop from the airstream, and finally a fine mesh filter captures the smaller particles. Air units eliminate the generation of slurry waste produced by hydrovac units but may not be as effective in certain types of soils. Air units permit the placement of the vacuumed soil back into the pothole after the utility location has been verified.

Vacuum units are excellent tools when it is necessary to work between existing buried electrical and communication conduits or utility pipelines. They are very good for city work where there is little or no space to perform an exploratory excavation or to stockpile the excavated material. The vacuum-excavated material can easily be transported away for disposal without the need for rehandling.

The use of vacuum excavation has been emphasized by the increased use of horizontal directional drilling (HDD). In connection with HDD, there are two main functions for vacuum excavation: (1) potholing to identify potentially conflicting cables or utilities prior to drilling and (2) handling of the drilling fluid slurry generated during HDD operations.

Trenching with a Hoe

Types. Two types of backhoes are commonly used for trenching: (1) hydraulic excavators, also known as a crawler-hoe or hoe, utilize a fully revolving superstructure mounted on a crawler-type or wheel-mounted undercarriage and (2) smaller hydraulic hoes, mounted on the back of a tractor and having a swing arc of 200 degrees or less. These machines are commonly referred to as combination backhoe loaders and are often used for shallow work or intermittent trenching work.

Backhoe Loader. The backhoe loader (Fig. 13.2) is not a high-production machine for any one task, but it provides flexibility to accomplish a variety of work tasks. It is three construction machines combined into one unit—a tractor, a loader, and a hoe. Because of its four-wheel-drive tractor capability, the backhoe loader can work when unstable ground conditions are encountered. It is an excellent excavator for digging loosely packed moist clay or sandy clay. It is not as suitable for continuous high-impact digging of hardpan soils such as with stiff clay or hard caliche.

The hydraulics provide complete control of bucket motions, with power for up and down and push and pull action. A "wrist action" bucket can be adjusted in angle during the digging pass for precise control of cutting, for tremendous breakout force on obstructions, for nonspill carrying of bucket loads to the dumping point, and for selection of place and rate of dumping.

FIGURE 13.2 Backhoe loader.

Using a Backhoe Loader. The smaller hoe, mounted on the back of a wheel tractor, is usually the most economical and efficient trenching machine for work around buildings. These machines have the advantages of being lightweight and using rubber tires, which are less likely to damage lawns and paths than the steel tracks of a crawler-type machine. Accidental damage to trees and structures caused by operator mistakes is likely to be minimal.

The small buckets used with a backhoe loader can be another advantage; these are usually narrower than the standard-size buckets of larger machines. For common pipe or wire installation trenches, they remove less soil and reduce backfill work. However, in this respect they are not nearly as productive and economical as a small trencher.

Their fast cycle and the prying power of the wrist-action bucket often enable a backhoe loader to perform trench work and produce out of proportion to their size. The tractors usually carry front-loader buckets enabling them to backfill the trenches, carry pipe, and perform other miscellaneous work on the site.

Models are available in both two-wheel and four-wheel drive, but even a two-wheel-drive backhoe loader seldom gets stuck because the weight of the hoe on the driving wheels supplies excellent machine traction. If underfoot conditions are poor, downward and outward pressure on the bucket can lift the rear wheels and push the tractor out of a rut, and if this maneuver does not move the machine, the bucket can also be used to pull it out backward.

Trench Width. Trenching with any machine is more efficient if the trench is the same width as the bucket cut. The bucket cut width is the width of the bucket plus any attached sidecutters. The bucket should be wider at the front than the back to prevent the sides of the cut from binding the bucket and to simplify dumping. Trench widths may be determined by project specifications as appropriate for the type of pipe being installed. Trench width, in combination with the placement and compaction of pipe bedding material and the trench backfill material, are three important factors affecting the loads the pipe will carry and the service life of the pipe.

Standard bucket widths, with or without sidecutters, are 18 to 42 in. Small machines may be as narrow as 12 in., and big ones 5 ft wide or more. Manufacturer specifications will often refer to the digging width of the bucket as the "bite width."

Direction of Work. A hoe should start at, and dig away from, any obstacle it cannot cross, such as a building. If there are two such obstacles, separate starts are made at each, and an extra-work connection is made between the two trenches.

If the centerline is on a grade, working in the uphill direction makes digging more difficult by reducing digging force and increasing the tendency of the bucket to pull the machine into the trench. While digging downhill is easier, the working end of the trench may fill with water if the ground is wet or the job stands unfinished during a rain. Underwater work is sloppy, inaccurate, and often unstable, leading to a poor-quality pipe installation workmanship.

Starting a Trench. The machine is positioned on the centerline of the trench, with the tracks or wheels parallel to it and the bucket extended to almost its full reach and resting on the starting point (Fig. 13.3A). This method will minimize swing angle and increase production. The tight swing also affords the operator clear visibility of the excavation and spoil pile or haul truck.

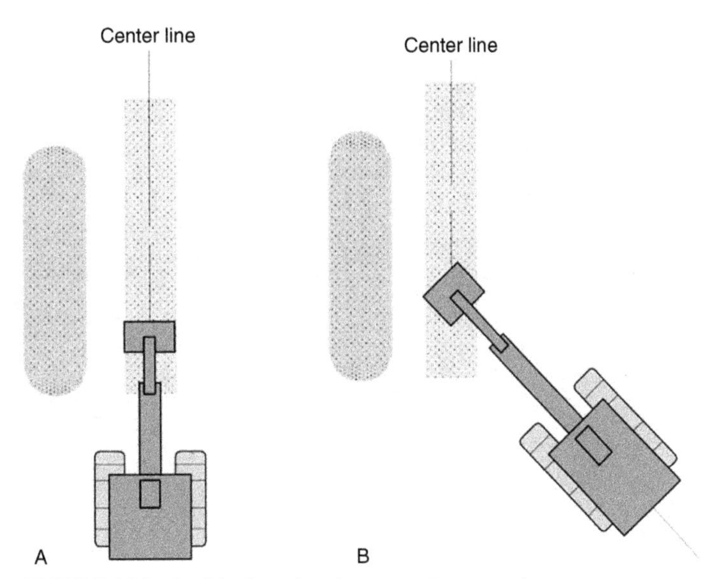

FIGURE 13.3 Positioning a hoe for excavating a trench.

The soil is taken out in layers down to the required depth. The starting point may be squared off with a vertical face from top to base. The bottom is smoothed off and checked for depth as the digging proceeds. When the desired depth has been obtained along the length the hoe can reach, the machine is walked backwards along the centerline and away from the excavated portion of trench. The distance moved may be from 2 to 12 ft before excavating the next section of trench. Short moves are made in connection with deep trenching, cutting the bottom to an exact grade, or cutting curves; longer moves are feasible for rough, shallow work.

Curves. Curves are dug as a succession of short, straight trenches, but a skilled operator can bevel the edges to produce a smooth curve. The machine stands with its center a little outside of the centerline, and digging is done in the outer half of the bucket reach. Moves are short. Marking paint on the excavation centerline can aid in maintaining this offset through the curve.

Angles. Many kinds of pipe require laying in straight lines and for the changes in alignment to be made by installing angled fittings rather than laying straight sections of pipe through a curve. The pipe trenches must match the alignment of the pipe. Angles are made by digging slightly past the angle point, then shifting the excavator to straddle the new centerline, shown in Fig. 13.3B.

Spoil Piles. Spoil from the trench is usually piled on one side, far enough from the excavation side to allow a footpath or working space between it and the trench. Common sense dictates and Occupational Safety and Health Administration (OSHA) regulations state: "Spoil piles, tools, equipment, and materials must be kept at least 2 ft. from the excavation's edge [1926.651(j)(2)]." If a large volume of soil is being moved, the pile must be pushed back by the bucket as it is built. It is not good practice to allow the spoil to come to the edge of the trench, as this can contribute to cave-in accidents. Piling on both sides is usually avoided because of the need to provide workers with access to the trench for assembling the pipe and to allow room for the backfilling operation.

When it is necessary for the excavator to push material away from the trench side, the extra pushing motion will increase cycle times and affect production. Width of a spoil pile can be estimated before the operation begins by using the loose excavated volume and the soil angle of repose. Simple sketches are useful for visualizing the spoil pile width and distance to the trench edge as illustrated in Fig. 13.4.

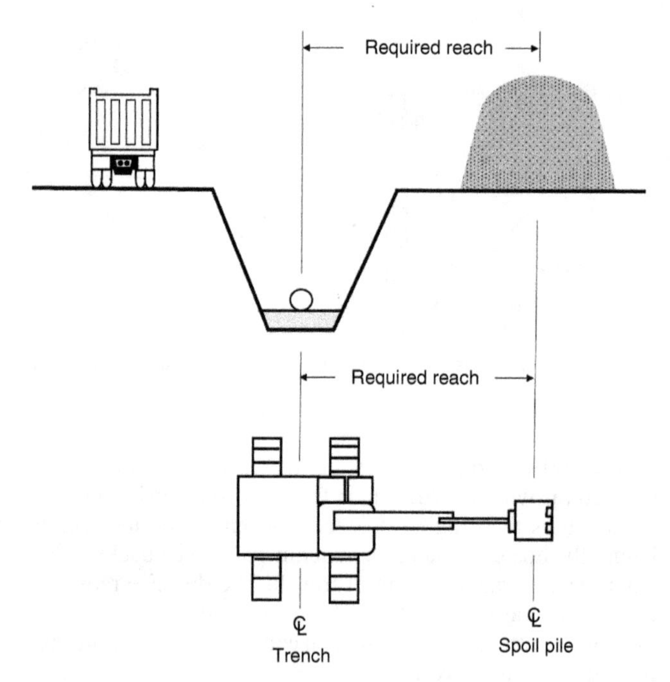

FIGURE 13.4 Illustration of working room dimensions.

Topsoil. Topsoil is salvaged during the digging by scraping it off first. If the topsoil is to be saved and placed back on top of the filled-in trench, it may be piled on the opposite side. If the volume of the spoil is not large, the topsoil may be placed on the same side as the fill but farther back, so when a dozer backfills the trench, the topsoil will be next to the blade and will reach the trench after the fill.

Sod. If sod is to be saved, it should be removed ahead of the excavator work. It may be dug by hand or cut in strips by a tractor-drawn or self-powered sod cutter. The strips left by it may be sliced in sections and piled at a safe distance by hand. The sod should be taken out at least 6 in., and preferably 1 ft, back from the trench sides to avoid damage. Removed sod cuts can be either straight slabs or rolls. Rolling the sod has the advantages of protecting the turf and creating longer slab cuts.

Production. The rate of excavating a trench depends on a number of variables, including required trench depth and width; bucket size; soil type and quantity; obstacles and hazards, both below and above ground; presence of rock; accuracy of grade required; and operational efficiency.

A shallow trench, with excavated soil piled on the edge, offers the fastest digging cycle, but the bucket will not be filled during each cycle. Deeper digging slows the cycle and means more material to move but permits better filling of the bucket. Optimum machine digging is slightly greater than half the maximum digging depth due to the high breakout force of the curled boom and the ability create a full bucket load. A narrow bucket does not fill as well as a wide one at any depth. If a trench must have a width wider than the excavator bucket, the excavation work will take more time due to the increased number of cycles required to excavate the trench wider than the bucket width.

Boulders and heavy roots slow the digging. The presence of buried pipes or conduits can cause serious delays, particularly if their exact location is unknown. A constant possibility of contacting interfering underground utilities and structures will greatly affect production. Buildings or trees interfere with maneuvering and, therefore, reduce production efficiency, as does the lack of space for the spoil pile.

It takes much longer to clean an irregular rock surface for blasting than to dig a clean trench to grade, and there is also the additional expense of drilling, blasting, and removing the blasted rock.

The need to maintain an accurate grade causes the operator to work more slowly, and occasional stops are needed to check grade or depth. The use of a pipe laser can make the process more efficient. A smooth bottom finish is produced readily by a wrist-action bucket, but with some difficulty by a rigid one. Still, experienced operators can efficiently trench to grade, and this maximizes production and minimizes the amount of pipe bedding material required. Performing additional operations such as stripping sod and topsoil separately will slow the digging from 5% to 30%.

When a trench needs to be braced during or immediately after the digging, production will be determined by the rate at which bracing is set, which is almost always much slower than the digging.

Trenching Rate Calculation. A ½-cy hoe with a 36-in.-wide bucket, including sidecutters, will be used to excavate a trench 3 ft wide and 6 ft deep in common earth. No special complications are expected.

This trench has a width of 3 ft (1 yd) and a depth of 6 ft (2 yd), so its cross-section is 18 sq ft (2 sy). Therefore, 2 cy will be removed for each lineal yard of excavation advance, or ⅔ cy per lineal foot.

This size machine can perform one cycle (load bucket, swing loaded, dump bucket, swing empty) in 13 sec (0.216 min). Assuming the trenching operation will operate at a 45-min-per-working-hour efficiency, it would then complete 208 cycles per hour.

$$\frac{45 \text{ min/hr}}{0.216 \text{ min/cycle}} = 208 \text{ cycles/hr}$$

The soil has a swell factor of 1.25 (25% swell), resulting in 0.8 (1 ÷ 1.25) bcy removed per loose cubic yard of bucket load. The bucket averages only four-fifths of a load in loose yards; its efficiency factor is 0.8. Multiplying the swell factor by the efficiency factor by the ½-cy capacity of the bucket yields

$$0.8 \times 0.8 \times 0.5 \text{ cy} = 0.32 \text{ bank yard per cycle}$$

Multiplying 208 cycles per hour by the 0.32 bucket load, the production is 66.6—use 67 bcy per hour. Because there are 2 cy to each lineal yard of trench, the trenching rate is 33.5 yd or 100.5 ft (approximately 100 ft) per hour.

$$\frac{67 \text{ bcy/hr}}{2 \text{ bcy/lineal yard}} = 33.5 \text{ lineal yards/hr}$$

A 30-in.-wide trench with a 30-in. bucket would come out about the same, as what was gained in handling smaller volume would be lost in poorer bucket efficicncy. However, if the trench were 12 ft deep, either bucket would probably be efficient.

TRENCHING MACHINES

There are both large trenching machines and small utility trenchers, including two-wheel-drive, walk-behind trenchers for trenching in soft ground conditions. Trenchers provide relatively fast digging, with positive control of trench depth and width, thereby reducing expensive finishing operations. Most large trenchers are crawler-mounted to increase their stability and to distribute the weight over a larger area. A trencher's primary advantage over an excavator is rate of production for pure trench work. The rate of trenching difference can be as high as a factor of ten.

Trenching machines are designed for excavating trenches or ditches of considerable length and having a variety of widths and depths. The term "trenching machine," as used here, applies to the wheel-and-ladder-type machines. Most general construction work requires trenches no wider than 2 ft or no deeper than 7 ft, so 95% of the machines to support such work are in the 40- to 150-hp range. Machines in this horsepower range are satisfactory for digging utility trenches for water and gas and shoulder drains on highways. Larger trenchers, having a gross horsepower rating of over 340, are used to excavate the much wider and deeper trenches necessary to lay large-diameter cross-country pipelines or water and sewage system main distribution lines.

There are fully hydrostatic trenchers with self-leveling tracks. A hydrostatic trencher delivers horsepower to the cutting chain and wheels/tracks more efficiently and tends to be smoother to operate. With self-leveling systems, a vertical trench can be maintained on uneven terrain with up to an 18.5% slope.

Wheel-Type Trenching Machines

The wheel-type trenching machine (Fig. 13.5) is operated by lowering the rotating wheel to the desired depth while the unit moves forward slowly. A wheel trencher, sometimes referred to as a digging wheel or bucket wheel, employs an endless chain wheel fitted with rim buckets to excavate. The removable buckets are equipped with cutter teeth and are available in varying widths to which sidecutters can be attached when it is necessary to increase the width of a trench. A conveyor belt below the upper part of the wheel receives the excavated material when the inverted buckets reach the top of the wheel. The conveyer

FIGURE 13.5 Wheel-type trenching machine.

belt can pile the material on either side of the trench. These machines are also known as "bucket wheel excavators."

Wheel-type trenchers are available with maximum cutting depths exceeding 8 ft and excavate trench widths from 12 in. to approximately 84 in. Many are available with 25 or more digging speeds to enable the selection of the most suitable speed for almost any job condition. Wheel-type machines are especially suited to excavating trenches for utility services placed in relatively shallow trenches. A wheel-type trencher is effective in cutting hard, tight soils. The rate of trencher advance seems to also be affected by the water content of the material being excavated. In general they can be operated for less than ladder-type trenching machines.

Ladder-Type Trenching Machines

Figure 13.6 illustrates a ladder-type trenching machine. When depth is necessary, this machine's vertical reach ability is desirable, but it is achieved by being much more costly with more moving parts. By installing extensions to the ladders or booms and by adding more buckets and chain links, it is possible to dig trenches in excess of 30 ft with large machines. Trench widths in excess of 6 ft are available. Most of these machines have booms whose lengths can be varied, thereby permitting a single machine to be used on trenches varying considerably in depth. This eliminates the need to own a different machine for varying ranges of trench depth. A machine may have 30 or more digging speeds to suit the needs of a given job.

The machine consists of two continuous chains mounted on a boom. Cutter buckets equipped with teeth are attached to the chains. In addition, shaft-mounted sidecutters can be installed on each side of the boom to increase the width of a trench. As the buckets travel up the underside of the boom, they bring out earth and deposit it on a conveyor belt for discharge to either side of the trench. As a machine moves over uneven ground, it is possible to vary the depth of the cut by adjusting the incline, but not the length of the boom.

FIGURE 13.6 Ladder-type trenching machine.

Ladder-type trenching machines have considerable flexibility with regard to trench depths and widths. However, these machines are not suitable for excavating trenches in rock or where large quantities of groundwater, combined with unstable soil, prevent the walls of a trench from remaining in place. If the soil, such as loose sand or mud, tends to flow into the trench, it may be desirable to adopt some other method of excavating.

Chain Trenchers

To excavate material beyond the capability of a bucket-type excavator, a chain trencher is employed (Fig. 13.7). These machines use a digging chain or belt mounted on a metal frame or boom, analogous to an oversized chainsaw.

FIGURE 13.7 Chain trencher.

Power-shift machines are designed for continuous work in harder soils. If the application is more flexible, then a hydrostatic machine will be a better trencher. The trencher must have the proper type of chain for the ground conditions where it will be employed. By using the correct chain for conditions, the machine's efficiency and productivity are maximized.

The cutting edge can either be reinforced sharp plates or bolt-on carbide teeth. The sharp plate is best suited for trenching softer material. When trenching rocky soil, carbide teeth would probably be the correct choice; however, in a soft material such teeth would blend the soil instead of excavating it. Carbide teeth are staggered across successive rows to provide a uniform cut. A cantilevered steel guard above the circulating chain limits rocks and material from being thrown from the return chain, it deflects material from falling directly down on the return chain, and it also shields the chain from accidental contact by workers and other equipment.

The digging parts for the smaller-size trenchers—chain, teeth, and sprocket—will typically need replacing after about 300 hours. They must be replaced as a set to facilitate even cutting and to maintain consistent wear.

Rockwheels

Rockwheel trenchers (or rock saws) are designed for narrow and deep cuts, 5 ft in depth and up to 1½ ft in width. These machines can cost up to 20% more than a chain-equipped model. They have a large circular chain with teeth, similar to a circular saw blade. When the wheel spins, the teeth cut into the ground. The circulating wheel is mounted on a powered axle. An overhead guard traps and deflects material from the trencher and protects surrounding equipment and workers. Depth of cut is generally limited to the radius of the wheel when the powered center axle contacts the ground.

Plow Trencher

A plow trencher uses a vertical blade to cut a narrow, shallow trench (Fig. 13.8). This fundamental method of cutting a linear path through the ground has been used for ages, with the more common application being in agriculture for the planting of seeds. Construction

FIGURE 13.8 Plow trencher installing fiber-optic cable.

plow trenchers are effective for shallow, narrow cuts where small-diameter utilities are installed, such as buried power cables or telecommunication lines.

A vibrating action and slow forward movement create sufficient force to separate the soil and simultaneously insert the trailing utility cable into the ground. The tractor unit can have tracks or rubber tires, with most wheeled models having an articulated frame for uneven ground and directional control. Side spoils are kept to a minimum using this method because only the narrow vibrating blade displaces the soil. The maximum depth of most plow trenchers is about 5 ft, with special models manufactured for deeper depths. Its use is generally limited to soft to stiff soils with minimal rock. In cases of thick rock, another trenching method may need to be considered.

SELECTING TRENCHING EQUIPMENT

The choice of equipment to be used in excavating a trench will depend on the

1. Job conditions
2. Required depth and width of the trench
3. Class of soil
4. Extent to which groundwater is present
5. Width of the right of way for disposal of excavated material

If a relatively shallow and narrow trench is to be excavated in firm soil, the wheel-type machine is probably the most suitable. Plow trenchers are an ideal choice for very narrow and shallow utility installations. However, if the soil is rock requiring blasting, the most suitable excavator will be a hoe. If the soil is an unstable, water-saturated material, it may be necessary to use a hoe or a clamshell and let the walls establish a stable slope. When it is necessary to install solid sheeting to hold the walls in place, either a hoe or a clamshell can excavate between the braces holding the sheeting in place.

Trenching Machine Production

Many factors influence the production rates of trenching machines. These include the class of soil; depth and width of the trench; extent of shoring required; topography; climatic conditions; extent of vegetation such as trees, stumps, and roots; physical obstructions such as buried pipes, sidewalks, paved streets, and buildings; and the speed with which the pipe can be placed in the trench. Trencher performance defined in terms of forward speed and volume of excavation is significantly affected by the resistance of the ground being excavated. Machines are design so the cutting or excavation power remains constant irrespective of machine power. An increase in ground strength (rock) reduces cutting rates exponentially.

If a trench is dug for the installation of sewer pipe under favorable conditions, a machine could dig 300 ft of trench per hour. However, an experienced pipe-laying crew may not be able to place more than 15 joints of small-diameter pipe, 10 ft long, including wyes and tees, in an hour. Thus, the speed of the trench advance will be limited to about 150 ft/hr, regardless of a machine's ability to excavate. In estimating the probable rate for digging a trench, one must apply an appropriate operating factor to the speed at which the machine could dig if there were no interferences.

GRADERS AND DOZERS

Graders can fashion shallow ditches with sloped sides rapidly and neatly. If there is no use for the excavated soil, it is usually spread and blended beside the edges. Shallow ditches are best accomplished using a grader, but dozers can accomplish the initial rough ditching.

Ditching with a Grader

Besides road grading and embankment finish work, shallow-ditch construction is a basic grader task. For better grader control and straighter ditches, a 3- to 4-in.-deep marking cut (Fig. 13.9) should be made on the first pass at the outer edge of the back slope (usually identified by slope stakes). The toe of the blade should be in line with the outside edge of the lead wheel. This marking cut provides a guide for subsequent operations.

FIGURE 13.9 Graded ditch-cut operations.

Each subsequent ditch cut is made as deep as possible without stalling or losing control of the grader. Normally, ditching cuts are made in second gear at full throttle. Start with the blade positioned so the toe is in line with the center of the lead wheel. Bring each successive cut in from the edge of the bank slope so the toe of the blade will be in line with the ditch bottom on the final cut. Figure 13.10 illustrates the steps for cutting a V-ditch. The grader's frame should be articulated when performing steps 4 and 7 of Fig. 13.10.

Marking the Cut. Ditching is normally done on the right-hand side of the grader beginning with the marking cut. The procedures for positioning the grader are as follows:

- Ensure the moldboard is high enough off the surface to allow unrestricted movement.
- Ensure the blade is pitched halfway.
- Center shift until the left lift cylinder (heel) is straight up and down.
- Rotate the moldboard so the toe is just behind the outside edge of the right front wheel at about a 45-degree angle to the frame.
- Side-shift the blade if necessary to extend the edge of the moldboard to the outside edge of the right front wheel.
- Raise the left cylinder all the way.
- Lean the front wheels to the left. The grader is now in the ditching position.

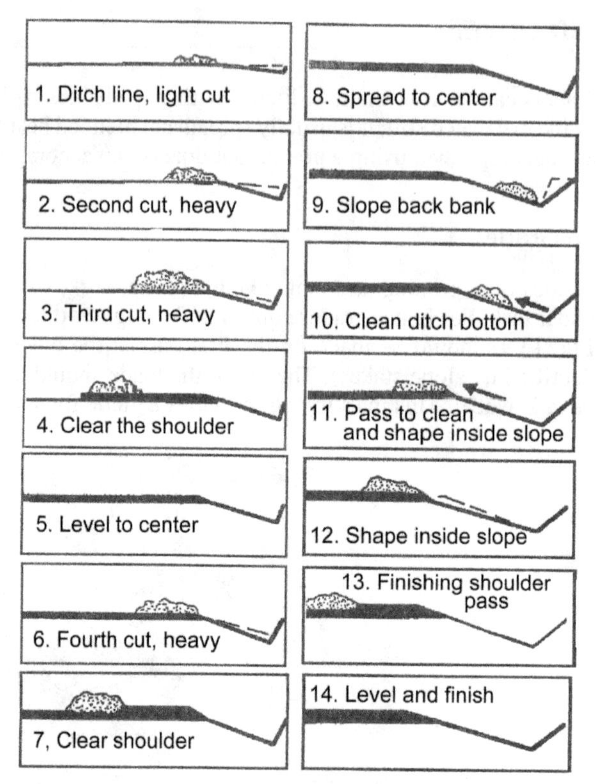

FIGURE 13.10 Steps of the V-ditching method.

Feather Raising the cutting edge ½ to 1 in. in increments while moving forward until all material passes under the moldboard.

Move the grader forward. As the right front wheel passes over the starting point of the ditch, lower the toe of the moldboard. Apply sufficient pressure to the toe to penetrate the ground surface 3 to 4 in.

Feather the material and raise the moldboard toe clear of the ground at the end of the marking cut. Continue moving forward until the rear wheels pass over and off the marking cut.

Straighten the front wheels and steer the grader to the right. Then back the grader along the outside edge of the windrow and reposition at the starting point.

Lean the front wheels to the left and begin the first cut. When making ditch cuts, windrows form between the heel of the blade and the left rear wheel. Move and level these windrows when either the ditch is at planned depth or when the windrow becomes higher than the ground clearance of the grader. This material usually forms the shoulder of a road.

Ditching with a Dozer

An angle dozer can excavate a ditch by side-casting in the same manner as a grader, but it may be harder to keep the alignment. Power-angle-tilt blades are an excellent choice for maximizing blade position and control. To cut a shallow V-ditch with a dozer, tilt the blade as shown in Fig. 13.11.

FIGURE 13.11 Angle-blade ditching.

For larger ditches, first push the material perpendicular to the centerline of the ditch as shown in Fig. 13.12A. After reaching the desired depth, push the material the length of the ditch to smooth the sides and bottom, as in Fig. 13.12B. Many times it is necessary to correct irregularities in a ditch cut this way. Attempt to remove humps and fill holes in a single pass. Use multiple passes to correct the grade.

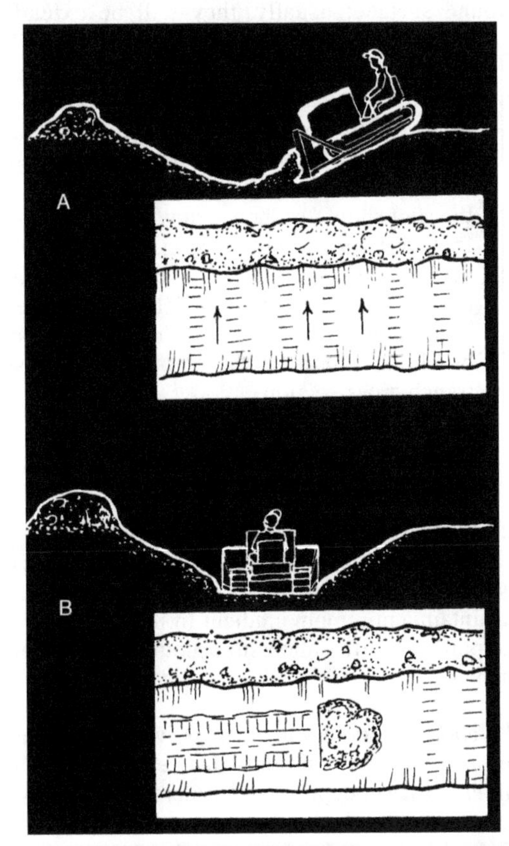

FIGURE 13.12 Bulldozer ditching.

Trenching in Rock

Trench locations will frequently contain rock too hard to be excavated by mechanical means. Occasionally the line of work may be shifted, but usually it is necessary to use an impact breaker (percussion hammer) fitted on an excavator or to blast the rock.

Soil and broken rock are removed by conventional methods and the spoil piled far back from the trench to allow space for drilling equipment. After machinery has removed the soil, the rock surface should be cleaned by hand. If the trench walls are liable to crumble and slide from drilling vibration, they should be shored or braced, even if the depth is shallow.

Open-cut blasting for trench work differs from other blasting work because of the restricted work space and because the shots are set in a line. Blasting a trench leads to the problem of leaving large boulders and causing larger-than-necessary disruption of the landscape.

When excavating a trench in rock, the diameter or width of the structural unit to be placed in the trench, whether a pipe or conduit, is the principal consideration. When considering necessary trench width, the space required for working and backfill placement requirements must be taken into account. Many specifications will specifically address backfill width. Another important consideration is the size of the excavation equipment.

The geology will have a considerable impact on the blast design. Trenches are at the ground surface; usually, they will be extending through soil **overburden** and weathered unstable rock into solid rock. This nonuniform condition must be taken into consideration. The blaster must check each individual hole to determine the actual depth of rock. Explosives are placed only in rock, not in the overburden.

Overburden
The layer of soil above the rock.

If only a narrow trench in an interbedded rock mass is required, a single row of blast holes located on the trench centerline is usually adequate. The timing of the shot should sequence down the row. The firing of the first hole provides the free face for the progression. In the situation where the trench is shallow or there is little overburden, **blasting mats** laid over the alignment of the trench may be necessary to control fly rock. In the case of a wide trench or when solid rock is encountered, a double row of blast holes is common. Small-diameter boreholes minimize the amount of explosives and limit rock fracture to the trench cross-section.

Blasting mat
A blanket, usually of woven wire rope, used to restrict or contain fly rock.

Pipe and Conduit Trenches

Most trenches are dug for the purpose of burying pipes or conduits. Conduits for electrical and telecommunication lines, and pipes for pressurized gas and water supply, run at a more or less fixed depth below the surface. Sewers, storm drains, and other gravity-flow pipes must maintain a minimum gradient from source to outlet, or else the pipeline will need to include pumps or lift stations to account for a variable depth below an irregular surface topography.

Fixed Depth. In cold-winter areas, water pipes are laid below the ground frost line. Conduit and wires are laid only deep enough to protect them against accidental excavation. In either case, depth may be increased under sharp ridges so as to provide smooth vertical curves.

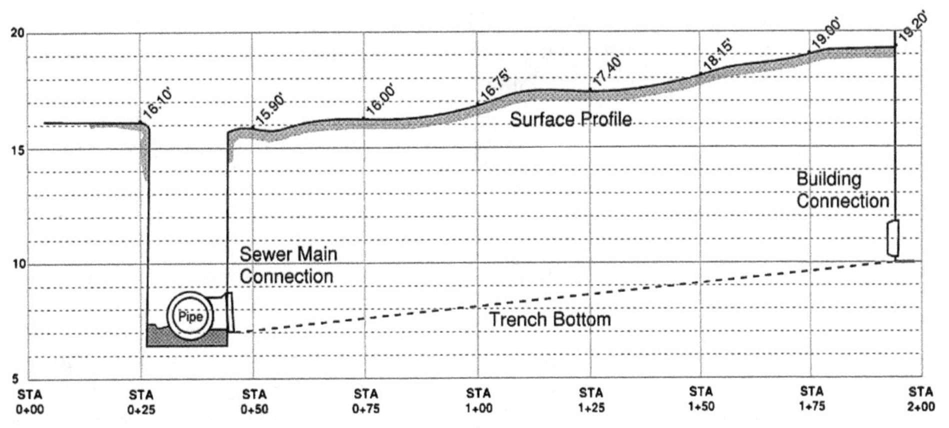

FIGURE 13.13 Setting grade for a gravity sewer line.

Gravity Flow Storm and Sanitary Sewer Lines. Close supervision is required to ensure accurate digging for a sewer or other gravity flow system. A number of methods are used to keep the trench at the proper grade.

Setting the Grade for Sewer Lines. It may be necessary to determine the grade for a sewer pipe by using either a specified gradient with at least one known elevation point or using two fixed elevation points and determining the appropriate gradient. If plans are not provided specifying the needed information, this can be done in the field by scaling the dimensions on a sheet of cross-section paper.

In Fig. 13.13, a basement has been dug and a sewer pipe needs to connect the building to a sewer main at the front of the property. An instrument is set up and the elevation of the basement floor taken. The actual elevation of the basement floor is used if it is known; otherwise, an arbitrary value can be assigned. In this example we will use an elevation of 10, which can then serve as a benchmark for the rest of the work. Another benchmark on a tree or other fixed surface spot not affected by building work should also be taken and recorded for future reference.

The figures obtained are plotted on a piece of cross-section paper, and a horizontal scale of 1 in. to 25 ft and a vertical scale of 1 in. to 5 ft are selected. If the cross-section paper contained 10 squares to the inch, for instance, the width of each printed square would then indicate 2½ ft horizontally and ½ ft vertically.

The basement is sketched in at the predetermined elevation of 10, representing the finished floor, and at a station location 1 + 95. The next point will be where the sewer service connects to the existing sewer pipe, and it is surveyed and drawn in at an elevation of 7 and a station location of 0 + 45. The surface elevations can now be surveyed and drawn on the profile at periodic intervals between the property line and the building, in this case every 25 ft. Note the surface elevations will look exaggerated because the scale is different in the horizontal and vertical orientations.

The trench bottom may now be drawn on the plot. An elevation just above the basement finished floor is where the bottom of the sewer pipe will penetrate the building, but we will simply use the finished floor elevation of 10 for the trench bottom. The invert elevation of the existing sewer main of 7 will be used as the elevation of the trench bottom at the sewer main connection point. The distance between the basement at station 1 + 95 and the existing sewer main connection point at station 0 + 45 is 150 ft of horizontal distance. The difference of the basement elevation of 10 and the sewer main connection point elevation of 7 is 3 ft of vertical distance. We can determine the slope of the sewer service by dividing the vertical distance by the horizontal distance.

$$\frac{\text{Vertical distance (rise)}}{\text{Horizontal distance (run)}} = \frac{3\ \text{ft}}{150\ \text{ft}} = 0.02 \text{ or } 2\% \text{ slope}$$

A 2% slope represents 2 ft of fall over every 100 ft, or 1 ft of fall over every 50 ft, so as the trench is excavated, it is possible to measure the elevation of our trench at any point to ensure we are installing the pipe at the proper elevation. A more precise scale can be used to provide more accurate readings.

Pipe Laser. Lasers provide means for precise control over line and grade in trenching, and in pipe placement. A pipe laser may be placed dircctly in the pipe or structure at the location where trenching will begin (Fig. 13.14). The laser is then calibrated to a specified slope, and as the trenching progresses and each new section of pipe is added, a target is placed in the end of the pipe to verify it is being laid at the proper slope and orientation. If the pipe is out of alignment with the path of the laser, workers can manually make very precise adjustments to the direction and elevation of the end of the pipe in the trench prior to backfilling.

FIGURE 13.14 Pipe laser. (*Courtesy Sitech Southwest, LLC.*)

Backfilling

Trenches dug for laying of pipe or conduit must be backfilled when the installation is complete. The removed soil or rock is pushed or pulled back into the trench. Such jobs can be handled by a number of machines, but the dozer is the standard tool for this purpose.

General Methods. If the backfill does not need to be compacted from the bottom up, it may be pushed into the trench in the ways shown in Fig. 13.15A. The dozer operates at right angles to the trench, taking as large a slice of the spoil pile as it can push. Soil that

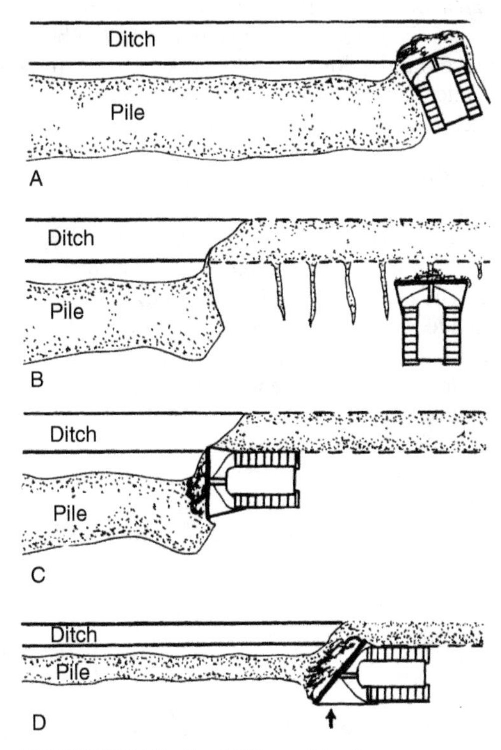

FIGURE 13.15 Backfilling with a dozer.

drifts across the blade is left in windrows and pushed in with a separate series of passes, but with poor efficiency, as in Fig. 13.15B. Or the remaining soil is pushed parallel with the ditch into the main pile, as in Fig. 13.15C, or, when the end is reached, distributed along the ditch.

Depending upon the degree of compaction and pipe diameter, there will usually be excess soil because of the space occupied by the pipe. In cases of excavations of over 6 ft in depth and pipe diameters of 8 in. or less, the excess can be negligible. The excess may be mounded over the trench and partly compacted by use of a roller, or by driving the dozer or a truck over it. Full natural settlement may take as much as a year and is liable to leave low and high spots. A complete freeze/thaw cycle from winter through spring can magnify settlement areas.

If the trench is small, it may be refilled by running a dozer with an angle blade or a grader through the pile, with the blade set to side-cast it into the ditch as in Fig. 13.15D.

A backhoe or a dragline can work from across the trench, pulling the soil toward the machine. A dragline's efficiency will be greatly increased by fastening a heavy plank or other block across the mouth of the bucket so it will not fill. By tilting the dragline bucket in the upward lift position, the bottom portion of the bucket and its weight can be used to pull material.

When pavement along the sides of the ditch is to be preserved, the best tool is a rubber-tire loader or dozer, but light or medium crawler machines, with semi-grousers or flat shoes, may be used.

Compacted Backfill. If pavement is to be immediately laid over the refilled trench, the backfill must be carefully compacted from the bottom up. This may be done by dozing or hand-shoveling fill slowly, while workers in the trench compact it with pneumatic tampers. A mechanical tamper may work from the side or straddle the trench. The top layer may be compacted by use of regular rollers.

Imported Backfill. When trenching through a roadway or under a paved area, many agencies require the trench to be backfilled with an engineered fill material, which is typically a crushed aggregate product blended to a specified gradation. This often means the entire amount of spoil produced from the trenching operation needs to be hauled to offsite disposal. The use of fill material minimizes the risk of trench settlement, which can otherwise damage the pavement above the trench.

Because the fill material is a more porous material, it promotes better draining of groundwater through the trench. The gravel may be dumped in piles in and alongside the trench and pushed in by a dozer or grader or hand-shoveled. A preferred method to avoid the mixing of clean fill stone with ground soils is a bedding box (Fig. 13.16). The width and sloped floor of the bedding box are designed for excavator buckets. As the backfill operation moves forward along the trench, the bedding box is repositioned using bucket chains or another piece of equipment, such as a loader. The loader can also charge the bedding box with clean fill dumped into stockpiles at an undisturbed location on the jobsite.

FIGURE 13.16 Bedding box for clean fill stone storage being refilled by a loader.

Trench Safety

A trench is any narrow excavation deeper than its width, with the width being 15 ft or less. Time and again noncompliance with trench safety guidelines and common sense results in the loss of life from cave-ins and entrapments (Fig. 13.17). The death rate for trench-related accidents is nearly double those for any other type of construction accident. The first line

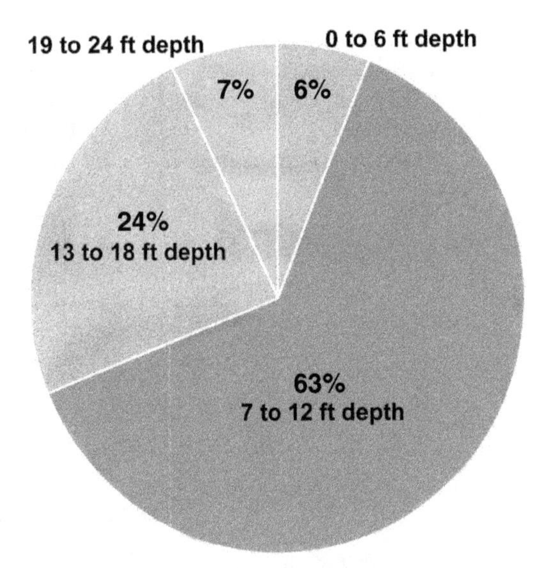

FIGURE 13.17 Deaths by depth of trench.

of defense against cave-ins is a basic knowledge of soil mechanics combined with knowledge of the material type to be excavated.

A critical trench safety question, often neglected, is previous disturbance of the material being excavated.

To protect workers from trench cave-ins, any trench measuring 5 ft or more in depth must be sloped, shored, or shielded (Fig. 13.18).

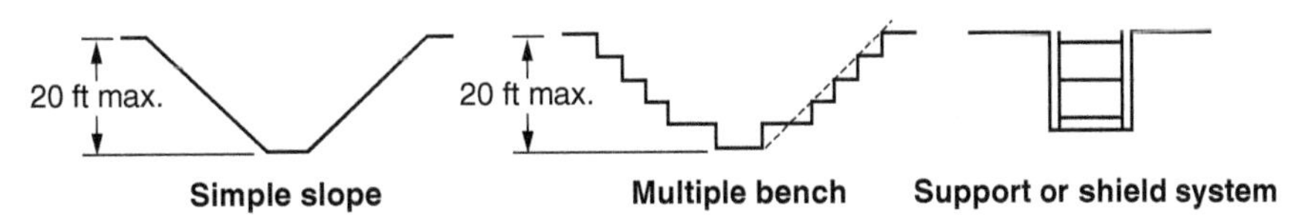

FIGURE 13.18 Methods of protecting workers in a trench.

In the case of trenches 20 ft or more in depth, OSHA regulations require design of the excavation protection by a registered professional engineer [OSHA 1926.651(i)(2)(iii) and 1926.652(b)(4)].

Sloping is a common method employed to protect workers in trenches. The trench walls are excavated in a V-shaped manner so the soil's natural angle of repose prevents cave-in. Required slope angles vary depending on the specific soil type and soil moisture. OSHA designates soils as A (clayey), B (loam), or C (sand) for minimum sloping requirements, where type A must at least be sloped at ¾:1 (53 degrees), B at 1:1 (45 degrees), and C at 1½:1 (34 degrees).

Trench shields A structural system designed to protect the workers should a trench in which they are working collapse.

Benching is a subsidiary class of sloping and involves the formation of "step-like" horizontal levels. Both sloping and benching require ample right-of-way space. When the work area is restricted or it is desired to keep the amount of necessary excavation to a minimum, shoring, sheathing, or shielding are used.

Shoring is a structural system used to apply pressure against the walls of a trench to prevent collapse of the soil. Sheathing is a barrier driven into the ground to provide support to the vertical sides of an excavation. **Trench shields** or trench boxes (Fig. 13.19) do not brace the walls of the excavation; rather, they are designed to protect the workers in case the trench walls collapse.

FIGURE 13.19 Trench box and shields protecting works in a trench.

Means of Egress. OSHA standard 1926.651(c)(2) states: "A stairway, ladder, ramp or other safe means of egress shall be located in trench excavations that are 4 feet or more in depth so as to require no more than 25 feet of lateral travel for employees." OSHA is serious about preventing trenching accidents, and every construction manager should likewise be concerned. A May 2016 trench collapse killed an employee while working on the installation of a sewer line for a house. As a result the Oregon OSHA fined the contractor $142,800 for five violations, two of which were willful violations, including failure to provide a ladder or other safe means to leave the trench.

Utility Strikes. OSHA standard 1926.651(b)(1) states: "The estimated location of utility installations, such as sewer, telephone, fuel, electric, water lines, or any other underground installations that reasonably may be expected to be encountered during excavation work, shall be determined prior to opening an excavation." Additionally, all states are required to operate a one-call system (dial 811, or www.Call811.com), which is a service for contractors when performing excavation work in an area. Contractors provide the 811 operator with the information regarding the location and limits of the planned excavation, and the one-call service notifies all utilities on record as having buried lines in the area where the excavation will take place. Those utilities then have a limited amount of time, typically 48 hours, in which to respond to the notification by marking the location of their utilities on the jobsite. This is typically done by painting a colored line or placing notification flags on the ground surface. The color of the marks corresponds to the type of utility being located, as shown in Fig. 13.20.

White	Proposed Excavation
Pink	Temporary Survey Markings
Red	Electric Power Lines, Cables, Conduit, and Lighting Cables
Yellow	Gas, Oil, Steam, Petroleum, or Gaseous Materials
Orange	Communication, Alarm or Signal Lines, Cables, or Conduit
Blue	Potable Water
Purple	Reclaimed Water, Irrigation, and Slurry Lines
Green	Sewers and Drain Lines

FIGURE 13.20 Uniform color code for marking buried utilities.

Contractors are prohibited from beginning any excavation work prior to the utility lines being exposed and protected if the proposed work will take place in the vicinity of an existing utility. Additionally, any digging adjacent to existing utilities must be done using safe excavation methods, such as hand-digging or vacuum excavation. Utility strikes can be expensive, but they can also be deadly. Gas lines and electric line strikes can cause explosions, water and sewer line strikes cause flooding and create environmentally harmful conditions, and almost every type of utility strike can result in very expensive property damage when considering a utility's loss of service. Because of these inherent risks, excavation contractors experience some of the higher general liability insurance rates in the construction industry.

HORIZONTAL DIRECTIONAL DRILLING

One of the most versatile of all the trenchless technologies is Horizontal Directional Drilling (HDD). It can be used for projects as small as a residential street or driveway crossings, or it can be used for projects as large as pipeline distribution systems installed underneath major highways and rivers.

With the HDD process, a drill rig pushes and rotates drill rods from the surface through the ground (Fig. 13.21). The drill rods exit the ground at the surface on the other side of the

FIGURE 13.21 Horizontal directional drill rig.

crossing where a product pipe or cable is attached to the drill rods and is pulled back through the borehole.

Using HDD can minimize the need for any trenching, except for where excavations may be required at locations where the utility being installed connects to an existing system. HDD installations can result in a significant reduction in the amount of surface restoration required on a project, and this can lead to significant cost savings if the feature being crossed is a road or highway. The use of HDD can also minimize traffic congestion and the increased travel times associated with work zone traffic control measures, as well as lessening the impacts to surrounding businesses and residences. Environmental impacts are similarly reduced when compared to conventional trenching operations by reducing the volume of spoil material generated and hauled away. Correspondingly, it reduces the necessary volumes of fill material required to be brought onto the project.

Small HDD installations can range from a couple hundred feet in length with a relatively small-diameter pipe, typically less than 4 in. for cables or conduits, whereas large HDD installations can range to a few thousand feet in distance with a pipeline size being anywhere from 24 to 48 in. diameter. One of the larger installations on record was a 12-in.-diameter steel propane pipeline crossing underneath a ship channel 11,653 ft with two opposing drills. This particular installation utilized what is known as the intersect method where two horizontal directional drill rigs work from each side of the crossing to connect in the middle and facilitate the installation of one continuous pipeline. While projects requiring the crossing of surface features as large as the shipping channel described here are not common, they are becoming more practical as the technology and capabilities of HDD continue to evolve.

HORIZONTAL DIRECTIONAL DRILLING MACHINES

HDD rigs come in a wide range of sizes, and very often the limiting factor in determining how long of a distance or how large of a pipe can be installed is the size of the HDD rig. The rigs are conventionally categorized into three size classes: mini rigs, midi rigs, and maxi rigs.

HDD rig size is commonly assessed based upon the amount of thrust and pullback force the machine is capable of delivering to pull a pipe through a borehole. Nevertheless, the rotational torque a machine can deliver is important. Penetration rates are determined based on the amount of torque needed to overcome the soil formation and the amount of power the drill rig can generate to turn the bit downhole. A certain amount of torque is lost transferring force down through the drill rods to the drill bit. Rotational speed also influences the available torque a drill rig can deliver downhole. An effective HDD rig will have a good balance between rotational torque and thrust/pullback capability.

Mini Rigs. Mini rigs are classified as being able to generate no more than 40,000 lb of thrust/pullback force and are capable of delivering up to 4,000 ft-lb of rotational torque. These rigs are particularly effective for projects installing small-diameter cables and conduits where installation lengths are typically less than a couple hundred feet. They are often self-contained and can be easily mobilized to a jobsite and quickly set up for maximum efficiency.

Midi Rigs. Midi rigs (Fig. 13.22) are classified as being able to generate between 40,000 lb and 100,000 lb of thrust/pullback and are capable of delivering between

FIGURE 13.22 Midi-size HDD rig.

4,000 and 20,000 ft-lb of rotational torque. These rigs are very effective for installing longer runs of small-diameter product lines, typically no more than about 2,000 ft in length, or shorter distances of larger-diameter pipelines, typically ranging anywhere from 6 to about 18 in. diameter. Midi-size rigs leverage some of the mini rig benefits, like being able to mobilize to the jobsite easily and set up quickly, but their ability to generate larger pullback forces and rotational torques enables them to meet longer distance and larger diameter needs. Midi-size rigs are sometimes used in conjunction with larger rigs to establish a pilot hole due to their quicker setup and faster cycle times before using a larger rig where more power is needed to open up the borehole and pull back larger pipes.

Maxi Rigs. Maxi rigs are classified as generating thrust and pullback forces in excess of 100,000 lb, as well as delivering more than 20,000 ft-lb of rotational torque. These large rigs are commonly employed on large-diameter pipeline projects or long-distance crossings, such as a river crossing project where the work cannot be divided into a series of shorter installations. Maxi rigs are frequently used for rock-boring applications due to their ability to generate the larger rotational torques needed to overcome the compressive strength of a rock formation. Some of the larger rigs are capable of generating well in excess of 1,000,000 lb of thrust/pullback force. Projects requiring such large forces often have many other considerations, such as how to effectively anchor the rig to the ground or how to ensure the pipe is not damaged during the installation from the forces used to pull it into the borehole. Use of a maxi rig often requires much more project engineering compared to a smaller HDD crossing project. Some examples of the engineering required for a maxi-rig HDD crossing are

- Calculations to determine the amount and types of stresses being introduced to the product pipe during pullback
- Confirming the quantity of drilling fluids required to effectively transport the cuttings out of the borehole
- Determining the minimum required and maximum allowable drilling fluid pressures to be maintained so as to effectively transport the cuttings without causing a hydraulic fracture of the soil formation

Additionally, maxi-rig operations typically require much larger staging areas to operate all the necessary equipment and, as such, incur higher mobilization costs, as it can take several truckloads to deliver the drill rig along with all the necessary ancillary support equipment.

HORIZONTAL DIRECTIONAL DRILLING OPERATIONS

Horizontal directional drilling is typically performed using a three step process: (1) pilot hole, (2) reaming, and (3) pullback.

Pilot Hole. The first step in an HDD crossing is to establish what is known as the pilot hole (Fig. 13.23). The pilot hole is used to establish the alignment for the product pipe to be installed. The HDD rig is set up and anchored to the surface on one side of the crossing, and drill rods are advanced through the ground along a planned alignment through a combination of thrusting and rotation. This is the establishment of a relatively small-diameter pilot hole. At the front of the drill rods there is a drill bit with a sloped face, referred to as a drilling assembly. When thrust is applied without rotation, the orientation of the drilling assembly is altered. Upon achieving the desired amount of directional change, the drilling assembly is then advanced, applying both thrust and rotation, which bores a hole through the soil in a relatively straight trajectory.

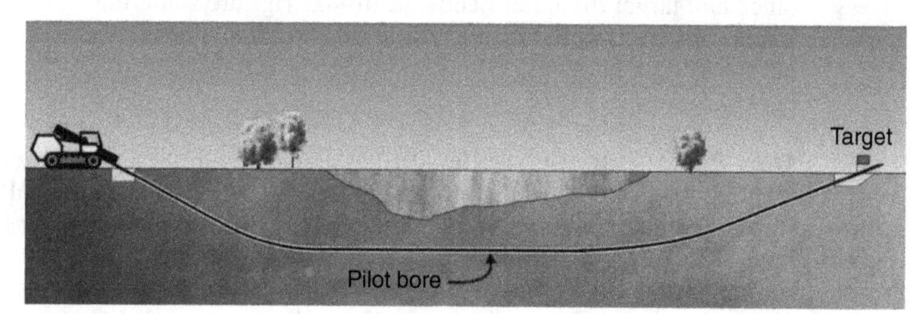

FIGURE 13.23 HDD pilot bore illustration. (*Courtesy of Samuel T. Ariaratnam.*)

Pilot hole progress is controlled with an electronic guidance system. The more commonly employed guidance systems are "walk-over" systems. These consist of a handheld receiver, a battery-powered transmitter, and a drill rig–mounted remote display allowing the rig operator to see pertinent locating information in real time (Fig. 13.24).

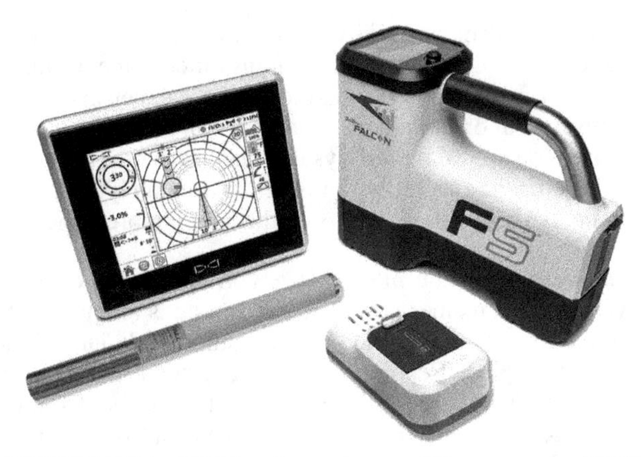

FIGURE 13.24 HDD locating equipment. (*Photo courtesy of DCI.*)

The transmitter is located just behind the drill bit in a sonde housing within the drilling assembly. When drilling the pilot hole, this transmitter emits a signal that is detectable by the receiver at the surface. The magnetic signal is used for determining direction, location, and depth. A secondary data signal includes information on the orientation of the transmitter, such as pitch (inclination) and roll (steering), as well as temperature, battery status, and in some cases annular drilling fluid pressures.

The person operating the receiver is often referred to as a "locator," and it is their responsibility to locate the in-ground position of the transmitter. Typically the drill head is advanced (often half a rod) and the drilling process is halted, the locator determines the position, depth, and direction and determines if any redirection (steering) is required. Some systems are available that provide for a more continuous process where locating takes place while drilling is ongoing.

Because the process involves a receiver detecting the signal being broadcast by the transmitter, locating equipment is subject to both passive and active types of interference, such as the magnetic field created by traffic signal loops or overhead power cables. Walk-over systems often use multiple frequencies to counter the effects of interference. An added feature of some conventional locating systems is the ability of rig operators to create a bore plan, upload it into the remote display, and track the path of the actual bore compared to the planned bore path. Many of these same systems can then be used to produce "as-built" drawings for record keeping or submission to the project owner.

A second type of guidance system is the "wire-line" method. This system eliminates the need for a locator to track the position of the drill head using the transmitter and receiver configuration. A wire-line system works by connecting a physical wire to a transmitter in the downhole assembly behind the drill bit. The wire runs through the drill rods and powers the transmitter with a DC power source at the surface. The probe then provides azimuth and inclination data indicating the subsurface position of the drill head. Additionally, a wire grid or coil can be set up on the surface, and when induced with an electric current, creates a magnetic field that the locating equipment can use to determine the position of the transmitter in the ground. This is accomplished by surveying in the corners of the surface coil and by using a computer to determine the position of the transmitter within the magnetic field; the position of the drilling assembly can then be converted to an X, Y, Z location within the coordinate system created by the survey. This technology enables "steering hands" to very precisely track the position of the transmitter as the drilling assembly advances through the ground. Wire-line systems tend to be more expensive and require more experience to use effectively, but on certain projects, they can be used to overcome some of the limitations of a conventional "walk-over" system, such as depth, battery life, and a requirement to "walk over" the drilling assembly in order to accurately find its position. Wire-line systems are commonly used with maxi rigs on the larger, more critical types of HDD bores.

Reaming. After the pilot hole is completed, the drilling assembly and electronics are removed and a reamer is attached. Reaming is the process of enlarging the borehole to the required final diameter, such that it is sufficiently large to accommodate the installation of the product pipe. Reamers (Fig. 13.25) come in many different types and sizes, each intended to work more efficiently in certain types of soil conditions. They are rotated and pulled back to the drill rig. This process may need to be done multiple times, with each pass using a slightly larger reamer than the previous pass.

FIGURE 13.25 Reamer.

The power generated by the drill rig relative to the volume and type of soil being cut is used to determine the appropriate number and size of reaming passes to use. In some instances, if the installed product pipe or cable will fit in the pilot hole, or if the HDD rig is large enough to effectively ream the hole and pull the product pipe at the same time, then the reaming step can be skipped. It is, however, good practice to ensure the borehole is adequately prepared prior to pulling in the product pipe. This will minimize the forces encountered when pulling the pipe into the bore. In order to provide the annulus diameter needed to permit the drilling fluid to flow around the product pipe, boreholes are commonly upsized to about 1.5 times the diameter of the product pipe being installed.

Pullback. The third and final step is to pull the product pipe into the borehole (Fig. 13.26). This is accomplished by attaching the pipe to the drill rods and reamer with a swivel, which permits the rods and reamer to rotate freely while keeping the pipe isolated from the rotation. Depending on the type of product being installed, it may be necessary to preassemble the pipe prior to starting the pullback. This is common on pipelines such as welded steel

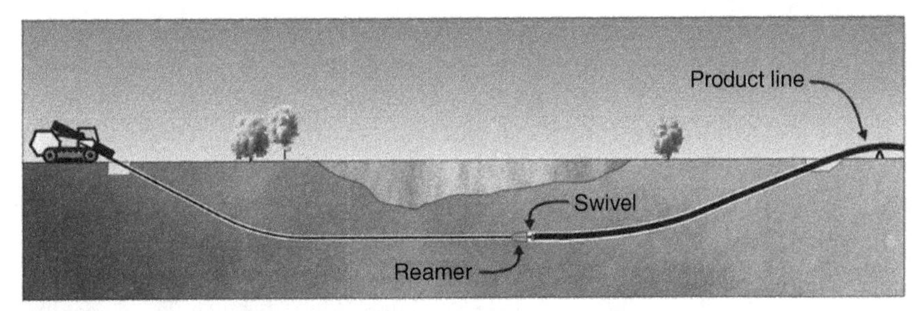

FIGURE 13.26 HDD pullback illustration. (*Courtesy of Samuel T. Ariaratnam.*)

pipe or high-density polyethylene (HDPE) pipe simply due to the time required to perform the proper welding of the pipe. Once the pullback operation begins, it should not be interrupted until the pipe has been completely installed because the amount of force necessary to initiate the pullback is generally more than the force required to keep the pipe moving through the borehole. Softer soils and clays may deform during initial drilling and then begin to compress around the drill rods and product pipe. This added friction can increase over time, especially for lengthy boreholes, requiring more time for pullback. Additionally, when the drilling fluid stops flowing through the borehole, it will begin to solidify and lose its lubricating properties. This can lead to an increase in the soil friction acting on the pipe and will require an increasing amount of power to overcome this friction and keep the pipe moving. As long as the drilling fluid is flowable, it will aid in reducing friction and facilitate a smooth pullback. On larger installations, it may be necessary for the crew to work multiple shifts to ensure a continuous pulling operation. This is because any stopping of the pullback operation introduces the risk of a stuck pipe.

Drilling Fluid. Horizontal directional drilling is a fluid-assisted process, and the drilling fluid serves a number of purposes. Drilling fluid is composed mainly of water mixed with bentonite. It is delivered into the borehole by pumping through the drill rods via a mud pump. The mud pump is often a high-pressure piston pump powered directly by the drill rig. As the drill bit rotates and cuts the soil, the drilling fluid is pumped downhole and is mixed into the surrounding soil. The bentonite aids in maintaining entrainment of the soil cuttings in the drilling mud, also known as a slurry. When the cuttings are in suspension, they can be transported out of the borehole. Because the drilling fluid is delivered under pressure, the drilling slurry is displaced and travels through the borehole, much like a conveyor belt, back to the surface where vacuum equipment can be used to handle the slurry. Depending on the size of the operation, the drilling slurry can be recycled to separate the solids from the drilling fluid, and the drilling fluid can then be recirculated back into the borehole. Recycling can greatly decrease the amount of drilling fluid onsite and can also decrease the quantity of spent drilling fluid to be disposed of upon completion of the project.

The drilling fluid also establishes a filter cake on the walls of the borehole, which helps to keep groundwater from infiltrating the borehole and diluting the drilling fluid, as well as helping to keep the drilling fluid from escaping the hole and becoming lost in the surrounding soil formation. Various drilling fluid additives are available to enhance fluid properties, such as detergents for use with sticky clays to keep the soil from sticking to the tooling and increasing rotational torques. In swelling clays, inhibitors can be used to reduce the reactivity of the clay, which reduces drilling pressures. Coarse-grained materials such as sands and gravels tend to easily fall out of suspension; as a result polymers are often used to increase the gel strength and carrying capacity of drilling fluids when working in such materials.

Finally, when drilling the pilot bore, the rotational forces being generated to cut the soil create friction, which creates a large amount of heat. This heat can easily damage the electronics used to locate the bore, so constantly supplying drilling fluid downhole will wash the electronics and keep them from overheating, ensuring their proper function.

HORIZONTAL AUGER BORING

Horizontal auger boring is sometimes referred to as "jack and bore" or simply "auger boring." It is a method used to install casing pipes under features such as roadways and rail lines. The process works by advancing a steel casing pipe through the ground. The work starts in a pit that has been excavated to below the bottom elevation of the proposed casing pipe. An auger boring machine set up on a rail system (Fig. 13.27) applies hydraulic force to "jack" the casing pipe into the soil. Simultaneously with the pushing, a rotating cutter head connected to a series of auger flights extracts the soil and pulls it away from the tunnel face. Typical installation lengths are up to 300 ft, and installed diameters can range anywhere from 8 to 84 in. or larger, although the most common installation sizes tend to be in the range of 24 to 60 in. diameter.

FIGURE 13.27 Auger boring machine. *(Photo by Barbco, Inc.)*

The spoil brought back into the pit by the auger flights can be removed with a crane-mounted clamshell bucket or by an excavator digging alongside the machine. Once the auger boring machine has driven an entire section of casing pipe, it disengages and moves backwards on the rail. Then the next section of casing pipe is placed and welded together in preparation to drive farther. The length of pipe sections capable of being installed is limited by the length of the excavation, and although productivity can be increased by eliminating welds, by using longer sections of casing pipe, this comes at the expense of a larger excavation for the push pit. Therefore, the common configuration uses pipe sections about 20 ft in length and excavations anywhere from 35 ft to 40 ft long, which accommodates both the auger boring machine and a full section of 20-ft casing pipe.

Auger boring is not a "guided" trenchless method, as are some of the other technologies, because there is no directional control at the face of the excavation. With proper planning and execution, however, auger boring can be a very precise method for installing pipelines at specified grades and elevations. Installations need to be straight, and care must be taken to properly install and align the machine prior to beginning a bore. Various techniques do exist to monitor and correct small variations in horizontal and vertical alignment, such as the use of a water level. A water level uses a small-diameter pipe attached to the top of the casing while it is being advanced; the pipe is then filled with water, and the level of the water is monitored to indicate if the vertical elevation of the casing pipe is maintained as the bore is advanced. Changes in the level of the water indicate changes to the elevation of the casing at the face of the bore.

Once the casing pipe installation has been completed, the casing pipe will usually be used to house a carrier pipe such as a sewer or water pipeline. *Casing spacers* can be used to center the pipe in the casing—this allows for greater control over the alignment of the carrier pipe. This is an important consideration for gravity flow sewer lines because they must be installed at specified elevations and at a specified slope in order to function properly. End seals can be installed on either end of the casing pipe and grout pumped into the casing to fill any voids between the casing pipe and the carrier pipe. While this process of installing a pipe inside of a pipe can generally be more expensive than utilizing other trenchless methods to install a carrier pipe directly, there are benefits. When there is a casing pipe the agency owning the pipeline can more effectively plan for future maintenance and expansion of the system because they can easily repair or replace an aging section of pipeline utilizing the same casing pipe rather than installing a new pipe underneath the surface feature or obstruction.

As with any trenchless method, proper site investigation and soil classification are key to minimizing the risks of a problematic job. Generally, auger boring is not an appropriate selection if it is necessary to install a pipeline in solid rock. Auger boring can effectively handle small amounts of cobbles and boulders, but they generally have to be smaller than about one-third the diameter of the casing pipe. This enables them to fit into the space between the inner wall of the casing pipe and between the flights of the auger to allow extraction through the casing pipe. When the machine encounters cobbles or boulders at the face of the bore too large to be "swallowed," then the bore cannot be advanced, and often the only option, other than abandoning the bore, is to retract the auger flights and have a laborer enter the casing pipe and hand-mine the obstruction. Because any bore may potentially require some hand-mining, many contractors often prefer to install casing pipes no smaller than 20 in. diameter, which is about the smallest size casing pipe needed to effectively accommodate some type of manned entry to gain access to the face of the excavation. Other conditions that can be precluded the use of auger boring are excessive groundwater conditions or running/flowing soil conditions, which can easily result in an over excavation of the material being bored and the unintentional creation of voids in the soil above the casing pipe.

Finally, because the auger boring machine needs to be located at the same elevation as the installed pipeline, consideration must be given to the depth and quantity of excavation required for the boring operation, as well as any necessary sloping and/or shoring required to protect workers from trench cave-ins.

TUNNEL BORING MACHINES

To excavate a circular cross-section through a variety of soils, tunnel boring machines (TBMs) are used in lieu of conventional hand-mining techniques. These machines mechanize the otherwise labor-intensive tunneling operations requiring workers at the face of an excavation to drill and blast the tunnel face and then excavate and muck out the spoil material. TBMs operate using a mechanically rotated cutting head advanced forward into the face of the tunnel. The excavated material is drawn into the machine and transported back through the tunnel over a series of conveyors to a train of tram cars used to transport it out of the tunnel.

The TBM advances the cutter head utilizing a series of hydraulic jacks, while the body proper is held in place by hydraulically operated grippers pressed against the tunnel wall (Fig. 13.28). There are two sets of grippers. The rear grippers hold the TBM in place while the cutter head is being advanced. Once the cutter head part of the TBM is fully extended, the grippers on the front section are engaged to hold the machine in place and the rear grippers released. The hydraulic cylinder jacks used to push the face of the TBM forward are now retracted and the rear of the TBM pulled forward.

FIGURE 13.28 TBM gripper pressed against the tunnel wall.

As the machine advances forward, tunnel liner sections are placed behind the machine. Many projects have used tunnel boring machines to construct very large-diameter tunnels for underground highways and rail lines. Some of the more noteworthy tunnel projects built using TBMs are the Channel Tunnel, more commonly referred to as the Chunnel, a 31.4-mile rail tunnel connecting France to the United Kingdom. Another notable project is the 35.5-mile-long Gotthard Tunnel under the Alps of Switzerland, one of the longest and deepest railway tunnels ever built. Longer networks of subway tunnels have been built or are being planned with the availability of TBM technology.

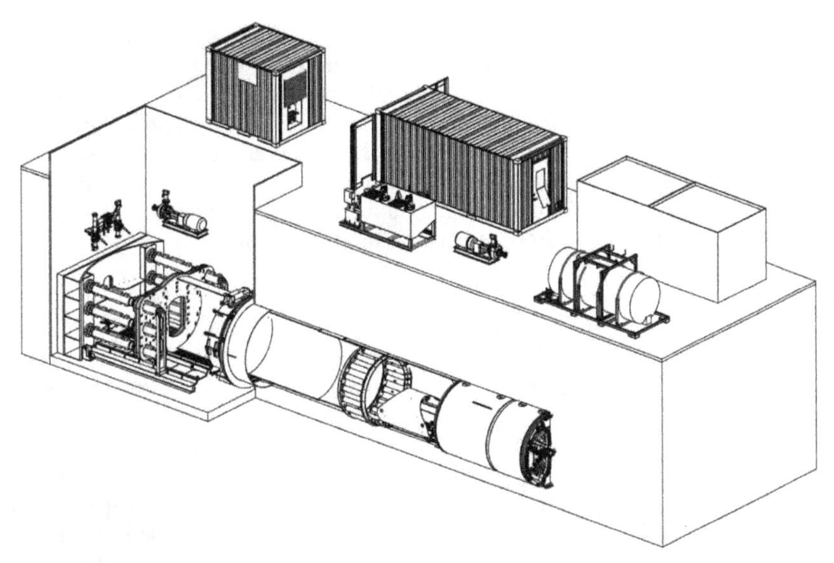

FIGURE 13.29 Microtunneling boring machine. (*Illustration courtesy of Akkerman.*)

MICROTUNNELING

Microtunneling is a term broadly applied to a number of different trenchless methods, specifically in reference to the equipment being used, but generally microtunneling is the process of constructing a relatively small-diameter utility tunnel that does not require personnel entry into the tunnel (Fig. 13.29). Slurry microtunneling is the best pipe jacking method for wet and unstable ground conditions or contaminated soils because the machine is remotely controlled to reduce risk and offers continuous support to the face. Microtunnel boring machines (MTBMs) are capable of performing installations of up to about 2,000 ft in length with pipelines in the range of 30 to 114 in. diameter. Because MTBMs are pipe jacking machines, the operation requires substantial planning and engineering. As such, microtunneling projects are typically found working on the larger end of the range of their capabilities or on more specialized projects due to the higher costs involved with the operation relative to other trenchless methods available for smaller-diameter utility installations. MTBMs are also referred to as earth pressure-balance machines, which means they do not displace more material than required to install the pipeline. By monitoring the fluid pressure at the face of the machine, they are capable of counterbalancing the hydrostatic pressure exerted on the tunnel face by the presence of groundwater. Therefore, MTBMs are appropriate for trenchless installation of deep pipelines or pipelines installed in high groundwater conditions.

A microtunneling project is typically performed by launching the MTBM from within a shaft, and the machine is advanced using a jacking frame located within the shaft. As the machine advances, sections of a special type of pipe, known as jacking pipe, are lowered into the shaft, then added behind the MTBM and advanced through the tunnel. As the length of the tunnel increases, so does the force required to advance both the pipe and the MTBM. Accordingly, the shaft is typically constructed with a thrust block to counter forces from the jacking frame. The frame provides the stability to effectively generate and transfer the force needed to advance the tunnel.

As the cutter head rotates, a series of slurry lines inject a bentonite slurry into the soil. The bentonite slurry cleans the cutters and serves to lubricate the MTBM and pipe, reducing the soil friction encountered. Additionally, the excavated soil is pulled into a slurry chamber located behind the cutter head and is mixed with the bentonite slurry, enabling the material to be pumped through a return line in the tunnel and back to the surface. On the surface, the bentonite slurry is then processed in a separation plant to remove the cuttings. This enables the recirculation of the bentonite fluid back into the tunneling operation. An operator controls the whole process from a workstation in a controlled container, which is remotely located on the surface.

Control over the alignment of the tunnel is accomplished with a laser erected on a fixed location in the shaft. This laser is pointed at a target, and an image of the target is then transmitted via a closed-circuit TV camera to the operator's cabin. Should the machine deviate from its intended alignment, the operator can make corrections by activating a series of hydraulic jacks located behind the cutter head to influence the machine's forward orientation. Because it is possible to monitor and control the alignment of the tunnel as it is being constructed, MTBMs are quite often used for sewer installations where maintaining the planned line and grade is required in order for the sewer to function properly. Several specialty tunneling navigation guidance systems are available for extended drives and tunnels with a curved alignment.

Another type of microtunneling operation is known as pilot tube guided boring, which uses a guided boring machine (GBM) system that is either a self-contained unit dedicated to the operation or is a system augmenting a conventional auger boring machine process by adding a guidance unit, which runs on the auger bore machine rails facilitating the installation of a pilot bore. Unguided bores are typically accurate to within about 1% of a planned alignment, which means a 300-ft installation can be installed to an accuracy of no less than 3 ft of deviation from a planned line and grade. Guided boring machines can increase the accuracy of the same installation to within about ¼ in. of deviation from the planned alignment.

Guided boring machines install pipe by first boring a small-diameter pilot hole through rotating and advancing of a sloped face bit (steering head) through the ground. A light-emitting diode (LED) target is placed in the steering head, and a theodolite aims for the front of the target to indicate the boring assembly's position compared to the proposed alignment. The theodolite's cross-hairs are aligned to the drive's line and grade, and the camera relays this data to a computer-controlled digital monitor, which is mounted to the jacking frame where the operator assesses the target's position (Fig. 13.30).

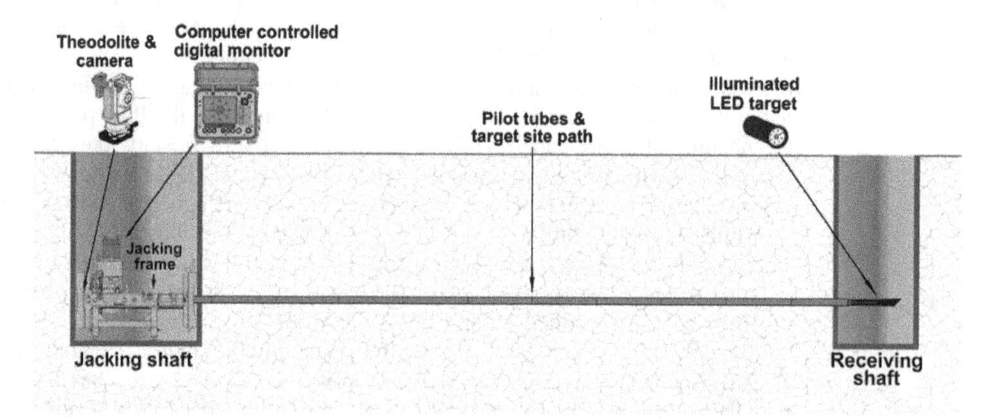

FIGURE 13.30 Guided boring machine guidance. (*Illustration courtesy of Akkerman.*)

If a line and grade adjustment is necessary, the boring assembly can be rotated to the desired orientation and pushed without rotation until the desired directional correction has been obtained. Once a pilot hole has been established, a GBM jacking frame or conventional auger boring equipment can be used to install a pipe in the pre-established pilot hole. An upsizing tool is positioned on the front of the first section of steel casing pipe to ensure the casing pipe stays centered with the pilot hole as the pipe is advanced through the ground (Fig. 13.31). This method facilitates the installation of a casing pipe while maintaining very tight control on the accuracy of the line and grade for the pipe being installed.

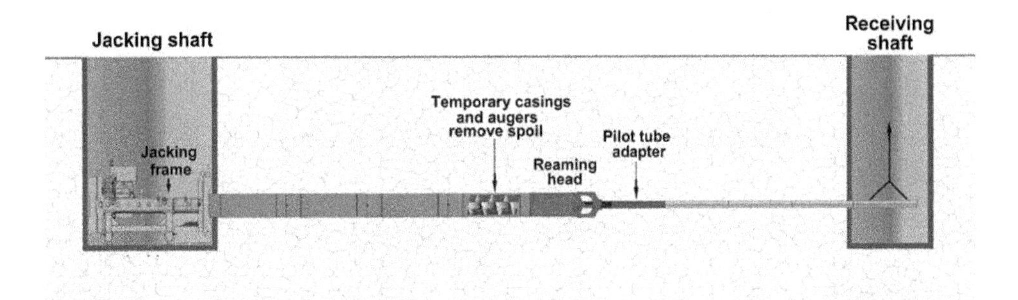

FIGURE 13.31 Guided boring machine casing installation. (*Illustration courtesy of Akkerman.*)

PIPE RAMMING

Pipe ramming (Fig. 13.32) uses a very large percussion hammer for driving an open-faced casing into the ground. This casing will be full of soil and must be cleaned out prior to installation of a carrier pipe. Pipe ramming eliminates the risk of overauguring and creating a void in the surrounding soil surface; thus, it is appropriate for granular soil conditions such as sands and gravels. Unlike other methods, pipe ramming is able to swallow any rock smaller than the open face size of the casing pipe. Pipe ramming is not a "guided" trenchless method, as there is no ability to control the direction of the pipe other than the good setup of a straight alignment at the beginning of the process.

FIGURE 13.32 Pipe ramming. (*Illustration by TT Technologies.*)

CHAPTER 14
COMPACTION, STABILIZATION, AND FINISHING

Finishing operations follow closely behind excavation or compaction of embankments. Graders are multipurpose machines used for finishing and shaping. The gradall is a utility machine that combines the operating features of a hoe, dragline, and motor grader. Gradalls are designed as versatile machines for both excavation and finishing work. Additionally, there are highly specialized trimming machines for finish grading. Compactors densify materials to increase foundation strength and stability. There are many types of compactors with each incorporating at least one method of imparting mechanical action to compact materials.

COMPACTION

With time, material will settle or compact itself naturally in a process known as consolidation. The consolidation process can be thought of as the weight of soil (the load) achieving equilibrium with its pore water pressure by seepage of the internal water. This action is governed by (1) the soil's permeability, which governs the flow; (2) the thickness of the stratum; and (3) the location of pervious boundaries, which influences the distance the water must travel. The objective of compaction is different; it seeks to quickly achieve a required density.

Obtaining a greater soil unit weight is not the direct objective of compaction. The reason for compaction is to improve soil properties by

- Reducing or preventing settlements
- Increasing strength
- Improving bearing capacity
- Developing stability
- Controlling volume changes
- Lowering permeability

Density, however, is the most commonly used parameter for specifying compaction operations because there is a direct correlation between these desired properties and a soil's density. Construction contract documents usually call for achieving a specified density, even though one of the other soil properties is the crucial objective.

Types of Compaction Equipment

Applying energy to a soil by one or more of the following methods will cause compaction:

1. Impact—sharp blow

2. Pressure—static weight

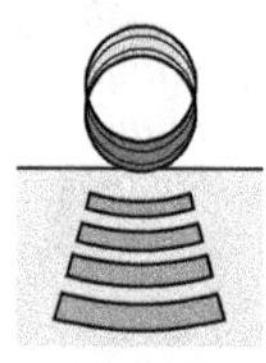

3. Vibration or oscillation—shaking

4. Kneading—manipulating or rearranging

The effectiveness of different compaction methods is dependent on the individual soil type being manipulated. Appropriate compaction methods based on soil type are identified in Table 14.1.

TABLE 14.1 Soil Types Versus the Method of Compaction

Material	Impact	Pressure	Vibration	Kneading
Gravel	Poor	No	Good	Very good
Sand	Poor	No	Excellent	Good
Silt	Good	Good	Poor	Excellent
Clay	Excellent with confinement	Very good	No	Good

Many of the available pieces of compaction equipment incorporate at least one of the densification methods, and in some cases more than one, into their performance capabilities. Many types of compacting equipment are available, including

1. Tamping rollers

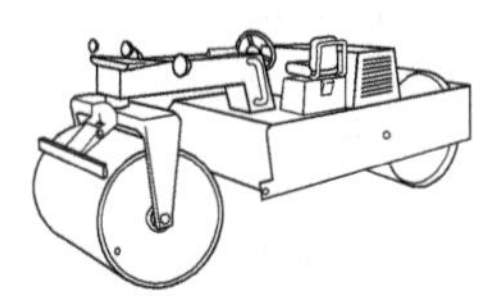

2. Smooth-drum vibratory soil compactors

3. Pad-drum vibratory soil compactors

4. Pneumatic-tired rollers

Table 14.2 summarizes the principal methods of compaction for the various types of compactors. There are also static, smooth steel-wheel rollers. These generally consist of two tandem drums, one in front and one behind. Steel three-wheel rollers are another version of this type of compactor, but they are not as common today as they were in the past. Vibratory rollers are more efficient than static steel-wheel rollers for earthwork and have largely replaced them.

TABLE 14.2 Principal Method of Compaction Used by Various Compactors

Material	Impact	Pressure	Vibration	Kneading
Sheepsfoot		X		
Tamping foot	X	X		X
Vibrating smooth	X	X	X	
Vibrating padfoot	X		X	
Pneumatic		X		X

On some projects, it may be desirable to use more than one type of compaction equipment to attain the desired results and to achieve the greatest economy.

The proper excavation and compaction equipment cannot be selected until the project soils are identified. Table 14.3 provides guidance for selecting compaction equipment based on the type of material to be compacted. As shown in the table, if required density

TABLE 14.3 Appropriate Project Compaction Equipment Based on Material Type

Material	Lift Thickness (in.)	Number of Passes	Compactor Type	Comments
Gravel	8–12	3–5	Vibrating padfoot Vibrating smooth Pneumatic	Foot psi 150–200 Diameter, length varies Tire psi 35–130
Sand	8–10	3–5	Vibrating padfoot Vibrating smooth Pneumatic Smooth static	Foot psi 150–200 Diameter, length varies Tire psi 35–65 Tandem 10–15 ton
Silt	6–8	4–8	Vibrating padfoot Tamping foot Pneumatic Sheepsfoot	Foot psi 200–400 Knob size varies Tire psi 35–50 Foot psi 200–400
Clay	4–6	4–6	Vibrating padfoot Tamping foot Sheepsfoot	Foot psi 250–500 Knob size varies Foot psi 250–500

is not achieved within four to eight coverages, a different type compactor should be considered.

Rock fills are usually spread in 18- to 48-in. lifts, with smaller cobbles in thinner lifts and larger materials in thicker lifts. Attention to spreading the material in a uniform lift is vital to achieving density during the compaction process. Consistent spreading helps to fill voids and orient the rocks so as to provide the compaction equipment with an even surface. The largest possible smooth-drum vibratory rollers are used for deep rock lifts. The combination of larger diameter and wider rolls minimizes point impact forces on rocks that may inadvertently cause fracture.

Tamping Rollers

Tamping foot compactors (Fig. 14.1) are high-speed, self-propelled, nonvibratory rollers. These rollers usually have four steel-padded wheels and can be equipped with a small blade to help level the lift. The pads are tapered with an oval or rectangular face. The pad face is smaller than the base of the pad at the drum. As a tamping roller moves over the surface, the feet penetrate the soil to produce a kneading action and sufficient pressure to mix and compact the soil from the bottom to the top of the layer. With repeated passages of the roller over the surface, the penetration of the feet decreases until the roller is said to "walk" out of the fill. Because the pads are tapered, a tamping foot roller can walk out of the lift without fluffing the soil. If it does not walk out, the roller is too heavy or the soil is too wet and the roller is shearing the soil.

Tow-behind tamping roller compactors are also available (Fig. 14.2). A trailing frame has a roller drum with attached tamping feet to provide the same kneading and pressure action to densify the material. The frame is ballasted with steel weights, or water is added to the steel drums to increase ground contact pressure. This setup has the advantage of less capital cost with self-propelled units where the tractor (see Fig. 14.2) or a dozer can be used for other tasks on the project. It has the disadvantage of no leveling blade with rubber-tired tractor units.

FIGURE 14.1 Self-propelled tamping roller equipped with leveling blade.

FIGURE 14.2 Tow-behind tamping roller compacting silty-clay material.

The working speed for these rollers is in the 8- to 12-mph range. Generally two to three passes over an 8- to 12-in. lift will achieve density, but this is dependent on the size of the roller. Four passes may be necessary in poorly-graded plastic silt or very fine clays. A tamping foot roller is effective on all soils except pure sand. To realize their true economical compaction potential, they need long uninterrupted passes so the roller can build up speed, which generates higher production. They operate at faster speeds than most other rollers.

Tamping foot compactors do not adequately compact the upper 2 to 3 in. of a lift. Therefore, if a succeeding lift is not going to be placed, it is necessary to use a pneumatic-tired or smooth-drum roller to complete the compaction or to seal the surface.

Vibrating Compactors

Vibration creates impact forces, and these forces result in greater compacting energy than an equivalent static load. This fact is the economics behind a vibratory compactor. The impact forces are higher than the static forces because the vibrating drum converts potential energy into kinetic energy. Vibratory compactors may have one or more drums. Typically on two-drum models, one drum is powered to transmit unit propulsion. Single-drum models usually have two rear rubber-tired drive wheels.

Certain types of soils such as sand, gravel, and relatively large shot rock respond quite well to compaction produced by a combination of pressure and vibration. When these materials are vibrated, the particles shift their positions and nestle more closely with adjacent particles to increase the density of the mass.

To produce the vibratory action, these rollers are actuated by an eccentric shaft internal to the drum. The eccentric shaft need only be a body that rotates about an axis other than the one through the drum's center of mass. The vibrating mass (drum) is always isolated from the main frame of the roller. Vibrations normally vary from 1,000 to 5,000 per min.

Amplitude
The vertical distance the vibrating drum or plate is displaced from the rest position by an eccentric moment.

Frequency
Number of vibrations per second or minute.

Vibration has two measurements: **amplitude**, which is the measurement of the movement, or throw, and **frequency**, which is the rate of the movement, or number of vibrations (oscillations) per second or minute (vpm). The amplitude controls the effective depth to which the vibration is transmitted into the soil, and the frequency determines the number of blows or oscillations that are transmitted in a period of time. The depth of influence for a typical 10- to 15-ton roller ranges from 2.5 ft to 4 ft. Influence depth is affected by material properties, particularly grain size. Influence depth is greater for coarse-grained soils than for fine-grained soils.

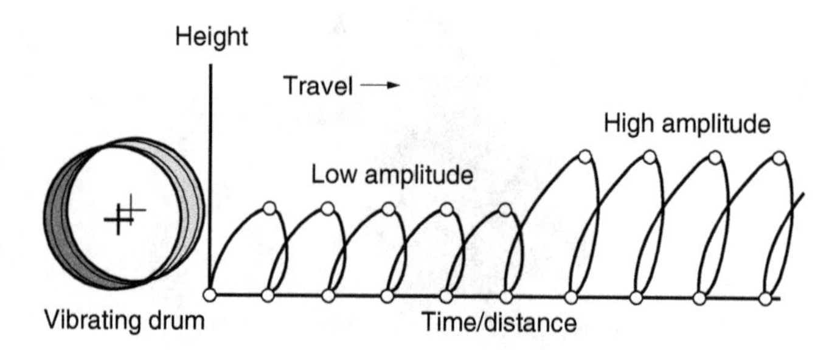

The soil particles are set in motion by the pressure waves resulting from the vibration impacts, and their motion allows the particles to rearrange into a more compact mass. In compacting granular material, frequency (the number of blows in a given period) is usually the critical parameter as opposed to amplitude. Amplitude may range from 0.01 to 0.06 in. or more, and frequency may range from 1,000 to 5,000 vpm or more. The most common settings are in a range from 2,500 to 4,000 vpm. The degree of compaction is a function of

- Force of the blows
- Frequency of the blows
- Time period over which the blows are applied

Because the characteristics of individual projects' soils are unique, appropriate force and frequency of the blows will vary by project. Every material has its own response to the impact of the blows, wherein in some cases the particles move closer together, and on other projects, the particles are displaced farther and negatively affect production. Excessive compaction in dry soils can cause hairline surface cracks to develop parallel to the drum. Simple rolling corrections are adding more moisture or adjusting the vibratory settings. A test strip is a good way to measure material response to different combinations of blows and frequency; however, the heterogeneous nature of soils and often-changing conditions across a project site lead to the requirement for multiple test strips.

The number of *impacts per foot* is a simple way to optimize the roller speed with vibratory frequency. Soils may require a minimum of 6 impacts per foot, whereas a recommended range for asphalt mixtures is 10 to 12 impacts per foot. An excessive impacts per foot can disorient the material and create surface bleeding of moisture. The frequency–time relationship accounts for the working speed requirement when using vibratory compactors. Working speed is important, as it dictates how long a particular section of the fill is compacted. A working speed of 2 to 4 mph may provide the best results when using vibratory compactors on soils or aggregate. The speed of the vibratory compactor expressed by vibratory frequency (vpm) and desired impacts per foot is

$$\text{Speed of compactor, mph} = \frac{V}{88 \times I} \tag{14.1}$$

where V = Vibratory setting in vibrations per minute
I = Desired impacts per foot
88 is the conversion factor from 88 fpm to 1 mph

For example, to achieve a desired 10 impacts per foot at a vibrator frequency of 2,800 vpm, the roller should be operated at 3.2 mph.

Intelligent compaction (IC) rollers are now manufactured with instrumentation to estimate the vibrations per foot based on measured roller speed and vibratory setting. Resistance of the roller drum with vertical deflection is measured and illustrated on a cab monitor to indicate soft areas or adequately compacted material. Digital readouts for the operator are updated in real time. If the vibrations are spaced too far apart, the soil surface may have a corrugated or "washboard" profile. The soil may be overcompacted—an indication of this condition is when the vibrating drum begins to "jump" and lose contact with the soil during one vibration cycle. In this case, the soil has achieved adequate compaction.

Smooth-Drum Vibratory Soil Compactors

The smooth-drum vibratory compactors, whether single- or dual-drum models, generate three compactive forces: (1) pressure, (2) impact, and (3) vibration. These rollers (Fig. 14.3) are most effective on granular materials, with particle sizes ranging from large rocks to fine sand. They can be used on semi-cohesive soils, with up to about 10% of the material having a plasticity index (PI) of 5 or greater. Large steel-drum vibratory rollers can be effective on rock lifts as thick as 3 ft.

> **Ballast** Weight added to the roller frame to increase ground contact pressure, usually steel weights.

Roller pressure is the weight of the roller spread across the contact area of the roller drum. Narrower, smaller-diameter drums can concentrate a higher weight per lineal foot and a higher contact pressure. Larger-diameter drums have a greater contact area but may have a lower ground contact pressure. A typical dual-drum roller weighing 10 to 15 tons may have a contact weight distributed across the drum width of 300 to 500 lb per lineal inch. One way to increase pressure for any roller size is to add **ballast**. Steel plates can be added to the frame, or water added to the drum interior.

FIGURE 14.3 Smooth-drum vibratory soil compactors.

Padded-Drum Vibratory Soil Compactors

These rollers (Fig. 14.4) are effective on soils with up to 50% of the material having a PI of 5 or greater. The edges of the pads are rolled inward, enabling them to walk out of the lift without fluffing the soil. The typical lift thickness for padded-drum units on cohesive soil is 12 to 18 in. These units are sometimes equipped with a leveling blade.

FIGURE 14.4 Padded-drum vibratory soil compactor with a leveling blade.

Small walk-behind and/or remotely controlled vibratory rollers having widths in the range of 24 to 34 in. are available (Fig. 14.5). These units are designed specifically for trench work, backfill and compaction along walls, or working in confined areas. The drums of the roller extend beyond the sides of the roller body, so the compaction can be

FIGURE 14.5 Padded-drum walk-behind vibratory roller.

accomplished adjacent to the trench or building walls. Many of these small compactors can be equipped with remote control systems, thus allowing the operator to control the roller from a safe location above the trench. Nearly all of the remote control systems use a digitized radio frequency. The combination of weight and narrow drum width can produce a typical contact weight of 250 to 300 lb per lineal inch.

Pneumatic-Tired Rollers

These surface rollers apply the principle of kneading action to effect compaction below the surface. Self-propelled models are the most common, but towed pneumatic-tired rollers are available. Pneumatics are used on small- to medium-size soil compaction jobs, primarily on bladed granular base materials. Small pneumatics are not suited for high-production, thick lift embankment compaction work. Pneumatic-tired rollers are also used in compacting asphalt, chip seals, recycled pavement, and base and sub-base materials. Because of their relatively gentle kneading action, they are well suited for intermediate and breakdown compaction of Superpave and stone mastic asphalt mixes (Fig. 14.6). The flexible tire exterior permits conformance of the tire to slightly irregular surfaces. This helps to maintain uniform density and bearing capacity, whereas a steel-drum roller would bridge over low spots while applying more pressure to high spots.

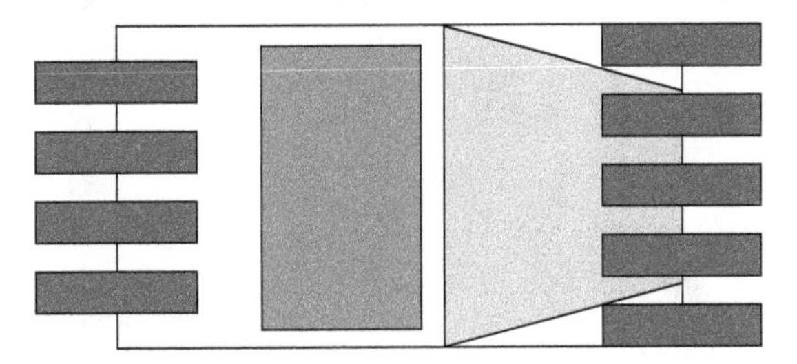

These rollers have two tandem axles, with four or five wheels on each axle. Wheels are mounted in pairs that can oscillate, or singly with spring action, allowing the tires to move

FIGURE 14.6 Self-propelled pneumatic rollers.

down into soft spots that would be bridged by a steel drum. The rear tires are spaced to track over the uncompacted surface left by the passage of the front tires. This path tracking produces complete coverage of the surface. There are also compactors in the 4- to 10-ton range with pneumatic tires on the rear axle and a smooth vibrating drum front to provide full coverage.

To increase the kneading action, pneumatic roller wheels may be mounted slightly out of line with the axle—this gives the wheels a weaving action, hence the name "wobbly wheel." The oscillating angle of the tires typically ranges from 3 to 5 degrees. Tire inflation pressures can range from 30 to over 100 psi, with a typical range of 40 to 60 psi. Lower pressures provide a flat contact area, whereas higher pressures have a curved, more-pointed contact area. Ply ratings, or tire load range, indicate the load a tire can carry at a specified pressure. By adding ballast, the weight of a unit may be varied to suit the material being compacted.

Because the area of contact between a tire and the ground surface over which it passes varies with the air pressure in the tire, specifying the total weight or the weight per wheel is not a satisfactory method of indicating the compacting ability of a pneumatic roller. Four parameters must be known to determine the compacting ability of pneumatic rollers:

1. Wheel load
2. Tire size
3. Tire ply
4. Inflation pressure

Wheel load is estimated from machine weight specifications and added ballast. Inflation pressure can be continuously measured with internal tire gauge sensors.

Variable Inflation Pressure Pneumatic Rollers. When a pneumatic roller is used to compact soil through all stages of density, the first passes over a lift should be made with relatively low tire pressures to increase flotation and ground coverage. However, as the soil is compacted, the air pressure in the tires should be increased up to the maximum specified value for the final pass. Prior to the development of newer technology with the capability of varying their tire pressure while in operation, measurement is made by stopping the roller and (1) adjusting the pressure in the tires, (2) varying the weight of the ballast on the roller, or (3) keeping rollers of different weights and tire pressures on a project to provide units to fit the particular needs of a given compaction condition.

Several manufacturers produce rollers equipped to enable the operator to vary the tire pressure without stopping the machine. The first compaction passes are made with relatively low tire pressures. As the soil is compacted, the tire pressure is increased to suit the particular conditions of the soil. The tire tread profile changes from flat to curved as the inflation pressure is increased, directly affecting the applied pressure. The use of this type of roller usually enables adequate compaction with fewer passes than are required by constant-pressure rollers.

Operation

Rolling speeds are slow. Speeds from 1½ to 3 mph (2.4 to 4.8 km/hr) are common. In rolling deep, loose material such as fill or gravel, all passes in a series except the first should be overlapped at least half the width of the drive roll. Gradual extension of the roller into the unrolled area makes possible greater concentration of weight on local ridges and high spots and keeps the rolls running at a truer grade.

In rolling a graded area with a sideslope, such as a crowned or banked road, passes should always be worked from the bottom up. The lower edges of the rolls have a tendency to push downhill, which can be best resisted by compacted material. In working uphill, the creep of soil away from the upper edge helps to preserve the slope.

A crowned road is rolled according to the pattern in Fig. 14.7, starting at one edge and working up until the center is reached by the upper roll, then moving diagonally to the opposite side and working up from there. Each rerolling is started at the bottom in the same manner.

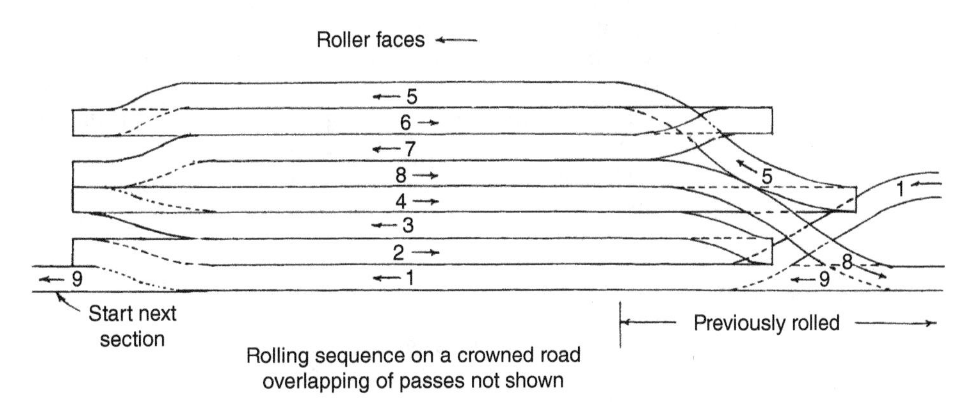

FIGURE 14.7 Rolling sequence, straight.

Banked or superelevated curves are rolled from the inner edge to the outer edge (Fig. 14.8). The transition from crown to bank is made by a diagonal from center to low side. From bank to crown, the move is from either edge straight into the adjoining low side. The meeting of these two types of grades is a convenient place to end a rolling section if the continuous-advance system is not being used.

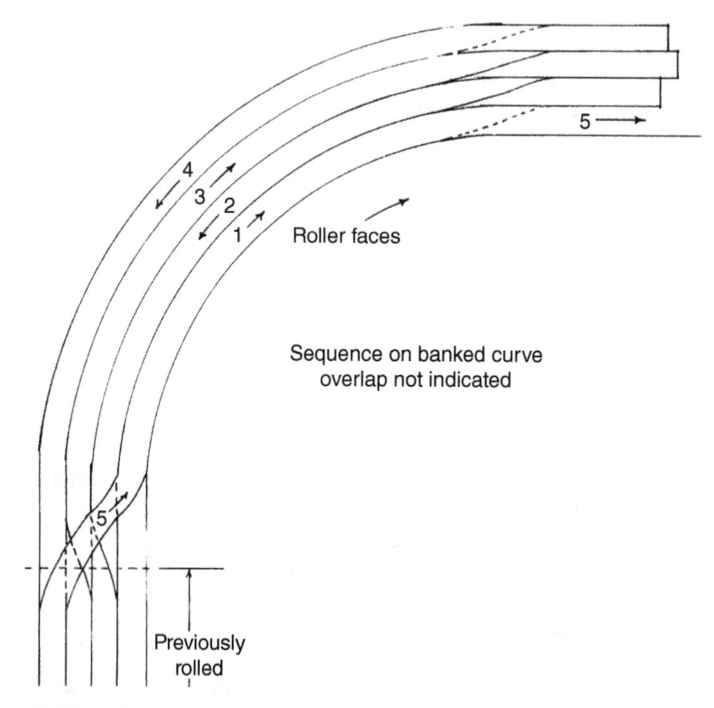

FIGURE 14.8 Rolling a banked curve.

Rolling should be continued until no advantage is noted from successive passes. The presence of too much water in the subgrade may make its compaction impossible, but long rolling will at least bring much of the water to the top where it can evaporate more readily. The waterlogged condition results in a rubbery action of the ground, in which it depresses under the rolls and springs back into nearly its original condition when they have passed. This condition may not be apparent at the start of the work, as the larger void spaces in the unconsolidated soil may be adequate to hold the water. As these spaces are reduced, however, the water is forced out and becomes a lubricant between all the particles.

Manual Compactors

It is often impractical to use rollers against walls, between manholes, in ditch bottoms, and many other places where space is restricted. Soil in such spots is compacted by tamping or the use of vibratory plate compactors.

Manually Operated Vibratory-Plate Compactors. Figure 14.9 illustrates a self-propelled vibratory-plate compactor. These are used for compacting granular soils, crushed aggregate, and asphalt concrete in locations where large compactors cannot operate. These gasoline- or diesel-powered units are rated by centrifugal force, exciter revolutions per minute, depth of vibration penetration (lift), foot-per-minute travel, and area of coverage per hour. Many of these compactors can be operated either manually as a walk-behind unit or by remote control. An eccentric cam above the contact plate delivers an impact blow to plate and the underlying soil surface at a rate of 2,000 to 3,000 vpm. Contact pressures typically range from 15 to 30 psi.

FIGURE 14.9 Self-propelled vibratory plate compactor.

Manually Operated Rammer Compactors. Electric or gasoline engine–driven rammers (Fig. 14.10) are used for compacting cohesive or mixed soils in confined areas. These units range in impact from 300 to 900 foot-pounds (ft-lb) per seconds at an impact rate up to 850 vpm, depending on the model. Performance criteria include pounds per blow, area covered per hour, and depth of compaction (lift) in inches. Rammers are self-propelled as each blow moves them ahead slightly to contact a new area.

Small compactors such as the self-propelled vibratory plate or the rammer will provide adequate compaction if

1. Lift thickness is minimal (usually 3 to 4 in.)
2. Moisture content is carefully controlled
3. Coverages are sufficient

FIGURE 14.10 Manually operated rammer.

The primary causes of density problems when backfilling utility trenches are (1) an inadequate number of coverages with the small equipment used in the confined space, (2) lifts with excessive thickness, and (3) inconsistent control of moisture. With the risk of subsidence, some projects may use a "flowable fill" of high-slump, low-strength concrete to occupy void spaces under utility pipes and raceways.

Safety Features. Hand-operated soil compactors should have

- Molded isolator rubbers to limit the amount of vibration from the exciter to the guide bar, thereby limiting exposure of the operator's hands and arms to excessive vibration.
- Hydraulic shift controls that will automatically return to a neutral position when the operator releases the shift control handle.
- Safety stop switches activated when the anticrush device (or so-called dead-man control) comes in contact with any obstruction; these switches cause the machine to stop immediately and shift into neutral.

Roller Production Estimating

The production capability of project compaction equipment must match the capability of the excavation, hauling, and spreading equipment in place on the job. Usually, excavation or hauling capability will set the expected maximum production for the work. The production formula for a compactor is

$$\text{Compacted cubic yards per hour} = \frac{16.3 \times W \times S \times L \times \text{efficiency}}{n} \tag{14.2}$$

where W = compacted width per roller pass in feet
S = average roller speed in miles per hour
L = compacted lift thickness in inches
n = number of roller passes required to achieve the required density
16.3 is the conversion factor for converting the result into cubic yards

The computed production is in compacted cubic yards (ccy), so it may be necessary to apply a shrinkage factor to convert the production to bank cubic yards (bcy), the more common way of expressing excavation and hauling production.

Sample Calculation: A padded-drum vibratory soil compactor will be used to compact a granular fill. Field tests indicate six passes of the roller operating at an average speed of 3 mph will provide the required density. The compacted lift will have a thickness of 8 in. The compacting width of this machine is 6.5 ft (66-in. drum). One bcy equals 0.85 ccy. The scraper production estimated for the project is 460 bcy/hr. How many rollers will be required to maintain this production? Assume a 45-min hour efficiency.

$$\text{Compacted cubic yards per hour} = \frac{16.3 \times 6.5 \times 3 \times 8 \times 45/60}{6}$$

$$\frac{318 \text{ ccy per hour}}{0.85} = 374 \text{ bcy/hr}$$

$$\frac{460 \text{ bcy/hr required ccy per hour}}{374 \text{ bcy/hr}} = 1.23$$

Therefore, two rollers will be required.

Roller Safety

Roller accidents are primarily the result of rollovers. A rollover can happen in many ways:

- While loading onto a trailer (drums lose traction and the roller slides off ramps/trailer).
- Operating too close to edge of an embankment, the surface under the roller gives way.

The operator can be crushed by the roller because

- The operator was not wearing a seatbelt on a rollover protection structure (ROPS)–equipped roller.
- The operator was not wearing the seatbelt correctly.
- The ROPS has been removed or lowered.

The operator's seat is for the operator only—**No Riders**.

Dynamic Compaction

The densification technique of repeatedly dropping a heavy weight onto the ground surface is known as "dynamic compaction." This process has also been described as heavy tamping, impact densification, dynamic consolidation, pounding, and dynamic precompression. For either a natural soil deposit or a placed fill, the method can produce densification to depths of greater than 35 ft. Most projects have used drop weights from 6 to 30 tons and drop heights of from 30 to 75 ft. However, on the Jackson Lake Dam project in Grand Teton National Park, a 32-ton weight from a height of 103 ft was employed.

Conventional cranes can be used for drop weights of up to 20 tons with drop heights below 100 ft. The weight is attached to the hoist line. During the drop, the hoist drum to which the line is attached is allowed to free spool, releasing the line. When heavier weights are used, specially designed dropping machines are required. With this densification technique, the in situ strata are compacted from the ground surface at their prevailing water contents. The ground vibrations produced by the method travel significant distances from the impact point. This can be a disadvantage at some locations.

The most successful projects have been those where coarse-grained pervious soils were present. The position of the water table will have a major influence on dynamic compaction employment success. It is better to use the technique only when the desired densification zone is at least 6.5 ft above the water table. Operations on saturated impervious deposits have resulted in only minor improvement at high cost and should be considered ineffective.

The depth of improvement achieved is a function of the weight of the tamper and the drop height:

$$D = n \times (W \times H)^{1/2} \tag{14.3}$$

where D = depth of improvement in meters (m)
n = an empirical coefficient, which is less than 1.0
W = weight of tamper in metric tons
H = drop height in meters

An n value of 0.5 has been suggested for many soil deposits. The use of 0.5 is a reasonable starting point; however, the coefficient is affected by the

- Type and characteristics of the material compacted
- Applied energy
- Contact pressure of the tamper
- Influence of cable drag
- Presence of energy-absorbing layers

SOIL STABILIZATION

Many soils are subject to differential expansion and shrinkage when they undergo changes in moisture content. Some soils also shift and rut when subjected to moving wheel loads. If pavements are to be constructed on such soils, it is usually necessary to stabilize them by reducing the volume changes and strengthening them to the point where they can carry the imposed load, even under adverse climatic conditions. Stabilization is a cost-effective way to use the material already on the project, rather than excavating and disposing the in-place material and hauling in clean fill. In the broadest sense, stabilization refers to any soil treatment for the purpose of increasing its natural strength. There are two kinds of stabilization: (1) mechanical and (2) chemical. In engineered construction, however, stabilization most often refers to when compaction is preceded by the addition and mixing of an inexpensive admixture, termed a "stabilization agent," which alters the chemical makeup of the soil, resulting in a more stable material.

Stabilization may be applied in place to a soil in its natural position, or mixing can take place on the fill. Also, stabilization can be applied in a plant, and then the blended material is transported to the job site for placement and compaction.

The two primary methods of stabilizing soils are

1. Incorporating lime or lime–fly ash into soils with high clay contents
2. Incorporating Portland cement (with or without fly ash) with granular or sandy soils

Fly ash is a by-product from burning coal. As such, it can be a highly variable product in terms of availability and chemical properties, thus its engineering usefulness can range from superior to extremely poor. Quality fly ash, however, can replace a portion of the lime needed to stabilize a clay-type soil. Because lime is relatively expensive and fly ash usually inexpensive, lime–fly ash stabilization of soils is often utilized.

Stabilizers

Rotary stabilizers (Fig. 14.11) are extremely versatile pieces of equipment ideally suited for mixing, blending, and aerating soil. Figure 14.11 illustrates the stabilizing agent first placed on the surface, then mixed with soil and relayed with the soil stabilizing machine. A stabilizer consists of a rear-mounted, removable-tine, rotating tiller blade, which is

FIGURE 14.11 A rotary stabilizer with lime placed on surface (left) then mixed with soil (right).

covered by a removable hood. In place, the hood creates an enclosed mixing chamber that enhances thorough blending of the soil. The tiller blade uses an uplift action to throw material against the hood. The material, deflecting off the hood, falls back onto the tiller blades for thorough blending. As the stabilizer moves forward, the material is ejected from the rear of the mixing chamber. When this happens, the ejected material is struck off by the trailing edge of the hood, resulting in a fairly level working surface. With the hood removed, the blades churn the soil, exposing it to the drying action of the sun and wind.

The soil stabilizer is often called a reclaimer because the existing soil is excavated, mixed with the stabilizing agent, and relayed or reclaimed all in one pass. Some models are equipped with a forward-mounted spray bar allowing the addition of water or stabilizing agents to the soil during the blending process. The stabilizer's use is limited to material less than 4 in. in diameter. The tines are designed to penetrate up to 22 in. below the existing surface so the unit can be used for scarifying and blending in-place (in situ) material as well as fill material; however, tine depths of 8 to 12 in. are more common to maintain forward progress.

Stabilization of Soils with Lime

In general, lime reacts readily with moderate to high plasticity clay soils, either the fine-grained clays or gravel-type clays. Such soils range in PI from 10 to 50 or more. Lime can also modify granular soils, but the greatest effectiveness occurs in clay soils. Unless stabilized, clay soils usually become very soft and unstable when water is introduced.

Soil-Lime Chemistry. In combination with compaction, soil stabilization with lime involves a chemical process. Lime, in its hydrated form [$Ca(OH)_2$], will rapidly cause cation exchange and flocculation/agglomeration, provided it is intimately mixed with the soil. A high-PI clay soil will then behave much like a material with a lower PI. This reaction begins to occur within an hour after mixing, and significant changes are realized within a few days, depending on the PI of the soil, the temperature, and the amount of lime. The observed effect in the field is a drying action.

Following this rapid soil improvement, a longer, slower soil improvement takes place, termed a "pozzolanic reaction." In this reaction, the lime chemically combines with siliceous and aluminous constituents in the soil to cement the soil particles together. Some refer to this as a "cementitious reaction," which is a term normally associated with the hydraulic action occurring between Portland cement and water, in which the two constituents chemically combine to form a hard, strong product. The confusion is increased by the fact that almost two-thirds of Portland cement is lime (CaO). But the lime in Portland cement starts out already chemically combined during manufacture with silicates and aluminates, and thus is not in an available or "free" state to combine with the clay.

The cementing reaction of the lime, as $Ca(OH)_2$, with the clay is a very slow process, quite different from the reaction of Portland cement and water, and the final form of the product is thought to be somewhat different. The slow strength with time experienced with lime stabilization of clay provides flexibility in manipulation of the soil. Lime can be added and the soil mixed and compacted, initially drying the soil and causing flocculation. Several days later, the soil can be remixed and compacted to form a dense stabilized layer that will continue to gain strength for many years. The resulting stabilized soils have been shown to be extremely durable and less susceptible to volume swell.

Lime Stabilization Construction Procedures. Lime treatments can be characterized into three classes:

1. Subgrade (or sub-base) stabilization includes stabilizing fine-grained soils in place or borrow materials, which are employed as sub-bases.
2. Base stabilization includes plastic materials, such as clay-gravels that contain at least 50% coarse material retained on a No. 40 mesh screen.
3. Lime modification includes the upgrading of fine-grained soils with small amounts of lime, 3% by weight.

There is a distinction between lime modification and stabilization. With lime-modified, no credit is accorded the layer in the structural design. It is used usually as a contractor technique to dry wet areas, to help "bridge" across underlying spongy subsoil, or to provide a working table for subsequent construction.

The basic steps in lime stabilization construction are

1. *Scarifying and pulverizing.* To accomplish complete stabilization, adequate pulverization of the clay fraction is essential. This is best accomplished with a rotary stabilizer (see Fig. 14.11). Typical cutting widths range from 6 to 8 ft and cutting depths of up to 22 in. The rotor works in an up-cut direction to lift the soil up and out of the ground. The cutting mandrel can be sloped up to 5% for adjusting the base cross-slope.
2. *Lime spreading.* Dry lime should not be spread under windy conditions. Lime slurry can be prepared in a central mixing tank and spread on the grade using standard water distributors. In adverse wind conditions when spreading dry lime, position the hopper truck immediately ahead of the stabilizer, or place and open dry lime bags immediately ahead of the stabilizer.
3. *Preliminary mixing and adding water.* During rotary mixing of the lime with the soil material, the water content should be raised to at least 5% above optimum. This may require the addition of water. A tow-behind water tank connected with hoses can provide a continuous water feed rate. Added resistance of the towed water tank on soft base material must be considered in the operation.
4. *Preliminary curing.* The lime-soil mixture should cure for 24 to 48 hr to permit the lime and water to break down (or mellow) the clay clods. In the case of extremely heavy clays, the curing period may extend to 7 days.
5. *Final mixing and pulverizing.* During final mixing, pulverization should continue until all clods are broken down to pass a 1-in. screen and at least 60% to pass a No. 4 sieve.
6. *Compacting.* The soil-lime mixture should be compacted as required by specification. Initial compaction is best accomplished with tamping or padfoot rollers, followed later by smooth-drum or pneumatic rollers.
7. *Final curing.* The compacted material should be allowed to cure for 3 to 7 days prior to placing subsequent layers. Curing can be accomplished by light sprinkling to maintain the surface in a moist condition for the desired period of time. Membrane curing is another acceptable curing method. It involves sealing the compacted layer with an asphalt emulsion. Curing is important to minimize shrinkage cracking in the hardened soil.

Stabilization of Soils with Cement

Stabilization of soils with Portland cement is an effective method of improving their structural properties. The use of Portland cement as a stabilization agent is effective as long as the soils are predominately granular and have only minor amounts of clay particles. Soils with a PI of less than 10 are likely candidates for this type of stabilization. Soils with higher amounts of clay particles are very difficult to manipulate and thoroughly mix with the cement before the cement sets; lime stabilization is best suited for clay soils. The terms "soil cement" and "cement-treated base" are often used interchangeably, and generally describe this type of stabilization. However, in some areas, the term "soil cement" refers strictly to mixing and treatment of in-place soils on the grade. The term "cement-treated base" or "lean concrete base" describes an aggregate/cement blend produced in a pugmill plant and hauled onto the grade. The amount of cement mixed with the soil is usually 3% to 10% by dry weight of the soil, with higher percentages for more fine material having a greater surface area.

As discussed in connection with lime stabilization, fly ash is plentiful in many areas, and it can be effectively used to replace a portion of the Portland cement in a soil cement treatment. Replacement percentages on an equal-weight basis can vary by soil properties, moisture content, desired strength, and overall mix design parameters. The ratio of fly ash to cement can be as low as 1:1, but more economical mixtures can range from 1:3 to 1:4. The type of coal fly ash must be considered when substituting for cement. Class C fly ash (more cementitious) and class F fly ash (more pozzolanic) have different chemical properties. Class C hydrates similar to Portland cement when water is added, whereas Class F reacts much more slowly and may require an activating agent, such as Portland cement or lime.

Soil Cement Construction Procedures. Cement stabilization involves scarifying the grade, spreading the Portland cement and fly ash uniformly over the surface of the soil (see Fig. 14.11), mixing the cement into the soil to the specified depth, compacting, fine grading, and curing. If the moisture content of the soil is low, it will be necessary to add water during the mixing operation. A continuous supply of water is necessary to ensure adequate hydration. The material sets fairly quickly and should be compacted within 30 min after it is mixed, using either tamping-foot or pneumatic-tired rollers, followed by final rolling with a smooth-wheel roller. An asphalt emulsion or another acceptable material may have to be applied to the surface to retain the moisture in the mix.

The nature of mixing in-place soil cement does not allow a normal 8-hour work schedule. When the cement is applied, the material cannot be left overnight; the processing must be completed even if overtime work is necessary.

1. ***The Project Site.*** When estimating soil cement, the configuration of the areas to be treated must be considered. Many aspects of the process can be affected by configuration. It may be necessary to travel over treated areas to obtain water for treating other areas of the project. In some instances, sites are very small and it is very difficult to load and unload equipment in a safe manner. Cement dust is another issue that should be taken into consideration. Personnel working on the grade need to have proper protective equipment, including eye protection. Cement dust can also cause damage to nearby vehicles. Always work with favorable wind conditions. Of major concern in a soil–cement operation is the amount of rock in excess of softball size. Rocks can damage the pulverizing equipment and make grading difficult. Removal of oversize rock is labor intensive, making it a critical cost consideration.
2. ***Scarifying.*** Loosening the soil, or scarifying, can help identify problem areas. If rocks are an unforeseen problem, scarifying can bring the unacceptable rocks to the surface. A good effort at rock removal will save equipment from damage. If large amounts of rock are removed, the grade must be reconfirmed after the removal process is

completed. Scarifying can also expose soft, yielding, or wet areas; roots; and hidden organic matter.

3. ***Spreading.*** During the spreading operations, the bulk application (spread rate) is checked and necessary adjustments are made. The calibration of equipment is a crucial point in federal, state, and airport work. The spread pattern of the cement is determined by site configuration. Bulker types vary somewhat, and the size of the site will determine the bulk spreader to be used. The bulker should move at a continuous rate without stopping while the product is being discharged.
4. ***Mixing.*** Mixing should be accomplished immediately after application of the cement (see Fig. 14.11). Cement hydration occurs immediately and the soil will begin to harden. The mixed soil and cement should be checked to ensure the uniformity of blend. To achieve the desired finished product, the cement must be blended thoroughly with the soil. The normal soil–cement mix procedure is in a down-cut motion using tine-type mixing teeth (stabilization). In cohesive soils, an up-cut machine with conical type bits is better.

   In the case of very dry or windy conditions, prewetting the grade can be helpful to provide an adequate material moisture content and to control dust. In extreme conditions, prewetting prior to scarifying the material may be a necessity. During the application of water, the water truck drivers should take care not to allow the discharge of water to puddle or pool. Puddles and wet spots can occur when water trucks are parked or allowed to stand idle on the grade. Such situations result in soft or yielding grade conditions. These soft spots may, in turn, become areas where compaction cannot be achieved.
5. ***Compaction.*** Initial compaction with a vibratory-pad foot roller immediately follows the mixing operation. A roller pattern should be established at the beginning of the initial compaction effort. Normally, two to three complete roller passes will be required. The depth of mix, type of soil, and required density are factors that control the number of roller passes required. The rolling must keep pace with the mixing operation.

   Following the initial compaction, the treated material should be shaped to the approximate line and grade. During grading special attention should be given to maintaining proper drainage. It is hard to make grade corrections to the cured material. The compaction effort should continue until the required density is achieved.

   The fine-grading operation should follow after acceptable density is achieved and the material has been rolled with a smooth-drum vibratory or pneumatic roller. The treated material should be kept from drying at all times during rolling, shaping, and fine grading.
6. ***Curing.*** There are several acceptable methods of curing soil cement. Some projects will specify the use of liquid asphalt curing. The bituminous materials most commonly used are emulsified asphalt SS-1, RC-250, MC-250, and RT-5. The rate of application varies from 0.15 to 0.30 gallons per square yard.

   The use of liquid asphalt has possible environmental impacts, with sticky material tracking onto adjacent roadways, requiring a cleanup operation. In many areas, the effect of runoff is of great concern, particularly for projects located near waterways. The use of liquid membrane curing compound has been proven effective. It has to be applied at a heavy rate (0.25 to 0.30 gal/sq yd) to achieve complete coverage. This can be a time-consuming operation and requires special equipment. Therefore, this has produced a preference towards the wet-cure method.

   It is necessary to keep the finished surface moist for a period of 7 days, with as little traffic as possible being allowed to travel on the newly treated material. The water truck should be careful not to make sharp turns during the initial curing. Construction equipment, particularly tracked machines, should be strictly prohibited from entering the area.

During the curing operation, the treated area must be kept from freezing for 7 days. Frost should not be allowed to form on the grade during the first 48 hr of the curing process. The use of straw (4 in. minimum) and/or poly sheeting can help reduce possible freezing, depending on the anticipated low temperature. Temperature should always be taken into consideration at the beginning of a project. Do not begin a project without a weather projection of 7 continuous days of temperatures above freezing after completion of mixing and grading.

FINISHING

Finishing, finish grading, and fine grading are all terms referring to the process of shaping materials to the required line and grade specified in the contract documents. Finishing operations follow closely behind excavation (rough grading) operations or compaction of embankments. These operations include finishing to prescribed grade those sections supporting structural members and the smoothing and shaping of slopes. On many projects, graders are used as the finishing machine. In the case of long linear projects, such as roads and airfields, there are special trimming machines to accomplish the finishing under the pavement sections.

Graders

The grader (motor grader or motor patrol) is a machine used principally in shaping and finishing, rather than in digging or transporting. It is available in sizes ranging from 45 to 500 hp and in a variety of designs.

A grader consists of a wide, controllable blade (moldboard) mounted at the center of a long-wheelbase, rubber-tire prime mover (Fig. 14.12). The blade is usually held at an angle to the direction of machine travel. It will cut, fill, and side-cast material. These functions can be performed separately or at the same time, in any combination.

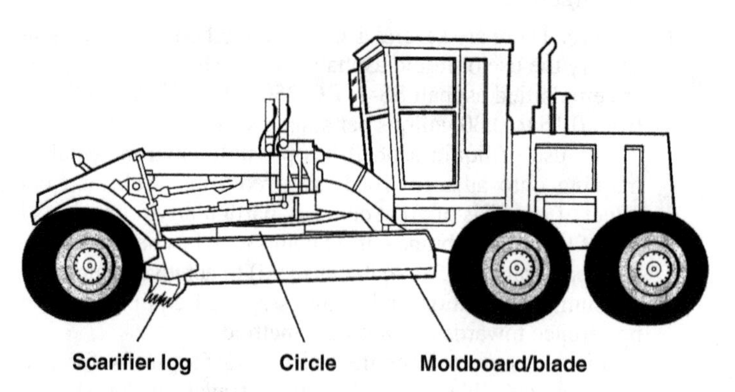

FIGURE 14.12 The components of a grader.

A grader's primary purpose is cutting and moving material with its blade. These machines are restricted to making shallow cuts in medium-hard materials; they should not be used for heavy excavation. A grader can move small amounts of material but cannot perform dozer-type work because of the structural strength, a lower drawbar pull, and location of its blade.

A grader may be used as a carrier for a number of accessories. These include snowplows and side wings, belt loaders, arm-mounted clearing tools, bank slopers, trailer drags, and disc harrows, in addition to scarifiers or rippers.

Graders are capable of progressively cutting ditches to a depth of 3 ft and for working on slopes as steep as 3:1. However, it is not advisable to run graders parallel with such steep slopes because they have a comparatively high center of gravity, and applied pressure at a critical point on the blade could cause the machine to roll over. It is more economical to use other types of equipment to cut ditches deeper than 3 ft. Graders operate at slow speeds providing a relatively high rimpull to cut and scrape the surface.

The component of the grader that actually does the work is the blade. The chassis may be articulated to provide an offset between the front and rear wheel paths. This enables the blade to reach farther into soft material while the rear wheels maintain traction on a firm surface.

Most graders are tandem drive units of the type shown in Fig. 14.13. All-wheel-drive (AWD) graders are also available and popular. They provide additional power to the ground for more efficient work in mud, gravel, sand, or snow. The diesel engine, with 125 to 500 hp, is mounted at the rear and drives four single wheels in tandem pairs through the drivetrain. The frame connection to the front axle is long and high, allowing space for carrying and manipulating the moldboards or blade under it. Some manufacturers have replaced the steering wheel and levers with joystick controls. The two joysticks, one for each hand, control steering, articulation, wheel lean, gear selection, moldboard

FIGURE 14.13 Tandem drive grader.

lift cylinder, blade float, drawbar, circle, and moldboard functions, as well as electronic throttle and manual differential lock/unlock.

The central location of the blade, and the long wheelbase of the machine, provide natural stability for the cutting edge. Smooth-acting, multidirectional control provides precision finishing ability.

Transmission. Graders need slow, powerful gearing for heavy or precise work, moderate speeds for lighter work, and travel speeds from 20 to 35 mph. There must also be a choice of reverse speeds, as some heavy work may require backing up, but most reversing is for return travel only and should be brisk. There may be six to eleven forward speeds, and two to nine reverse, in either direct drive or power shift.

Power Train. An option available on some graders is variable horsepower. When a variable-horsepower grader is shifted to higher gears, an electrical system on the transmission moves an automatic stop to a higher setting on the fuel governor. More fuel is injected, giving more power for high-speed, light-load work.

Tires. Tires are of a special heavy-duty design for grader work. They are preferably the same size all around, but the front ones may be smaller and/or lighter, unless it is an all-wheel-drive grader. To keep the blade engaged and in the proper position, graders need adequate traction.

Brakes. There are usually brakes on each of the four drive wheels to eliminate braking loads on the power train, but none on the fronts. They may be either shoe or multiple-disc design and are applied by booster-assisted hydraulic pressure. There may be a single-disc brake in the transmission, acting through the four wheels. The parking brake is usually on the lower transmission shaft, on the end opposite the drive pinion, and is operated by a hand lever.

Frame and Front Axle. The rear of the frame is a pair of beams that support the engine and the powertrain. Forward of the operator's station, these slope upward and converge into a single beam of reinforced box cross-section. This may slope down to the front axle hinge or end in a column above it.

There are also articulated frame graders. An articulated frame allows the operator to "crab steer" the grader into different operating configurations, put the front wheels over a windrow, on a slope, or down in a ditch and still keep the rear-drive wheels on solid footing.

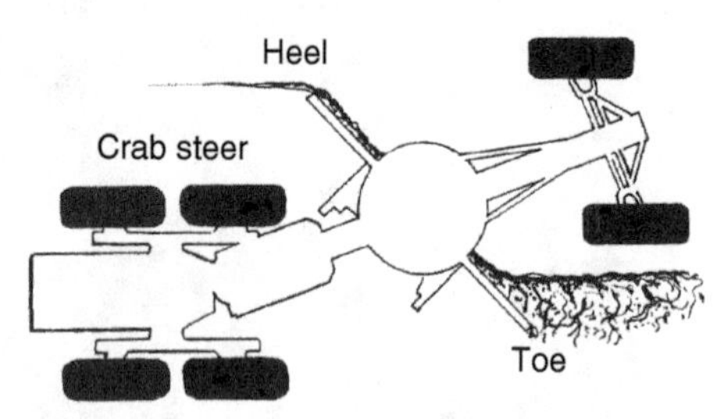

The front of the frame may be a mounting plate for front-end accessories such as a scarifier or blade with a mounted extension to serve as a narrow bumper. The front axle is compound. The lower section usually carries the weight of the machine, typically with a cambered beam high in the center for clearance over windrows, and ends below the wheel spindles. It oscillates on a central pin and is hinged to wheel spindle brackets just below the kingpins.

The upper section, of lighter construction, is substantially a straight bar (lean bar) hinged to the tops of the hub brackets. It can be moved from side to side by a hydraulic cylinder, or by mechanical means. When moved off center, it causes the front wheels to lean sideward. The angle may be as much as 18 degrees.

The front wheels can be leaned (Fig 14.14) to increase their resistance to sliding sideward on the ground because of load on the blade or steering stresses. Steering may be mechanical with a booster or by a hydraulic cylinder located behind or above the axle and operating through a more or less conventional linkage. Special arrangements may be made to avoid interference between leaning and steering.

FIGURE 14.14 Leaning front wheels.

Blade and Mounting

The blade is the primary working member of the grader. Most of the blade is a curved piece of steel called a moldboard. The term "moldboard" emphasizes the blade's function of causing dirt to roll and mix as it is moved.

Blade. Standard blade lengths in tandem graders are 12, 14, 16, or 24 ft with some 10-footers on small machines. Blade height is usually 24 or 27 in. but may be as much as 42 in. The curved vertical cross-section causes the material being pushed to have a rotary movement, rising at the bottom and falling forward at the top of the blade. This characteristic, combined with the usual sideward drift caused by working the blade at an angle, prevents spillage over the top.

The bottom is fitted with a removable and reversible wearing edge, with separate pieces (end bits) at the corners. The blade is supported and held in position by a pair of heavy curved brackets, called circle knees. They are attached to the underside of a rotatable ring, called a circle (see Fig. 14.12).

The rotating ring (circle) carries and positions the moldboard. Through elaborate hydraulics, the moldboard can be placed into many positions, either under the grader or to the side (Fig. 14.15). It can be side-shifted horizontally for increased reach outside of the tires.

FIGURE 14.15 Blade shifted to the left.

The blade is used to side-cast the material it cuts. The ends of the moldboard can be raised or lowered independent of one another. By convention, the *toe* of the blade is the foremost end of the moldboard in the direction of travel, and the *heel* is the discharge end.

Pitch (Tilt). The blade may be fastened to the knees by a hinge bar at the bottom and a curved slide at the top. This construction permits changing the pitch of the moldboard. The adjustment is controlled by hydraulic cylinders. For normal work, the blade is kept near the center of the pitch adjustment, which keeps the top of the moldboard directly over the cutting edge, but may be tipped up to 44 degrees forward and 6 degrees back.

When leaned forward, the cutting ability of the moldboard is decreased, and it has more dragging action. It will tend to ride over material rather than cut and push, and it has less chance of catching solid obstructions. A forward pitch is used to make light, rapid cuts and to blend materials. When leaned to the rear, the blade cuts readily but tends to let material spill back over the moldboard.

Blade Sideshift. Many grader operations can be made more efficient by extending the blade a greater-than-normal distance to the side (see Fig. 14.15). For extreme reach, both the circle and the blade adjustments can move the blade sideways. With power sideshift, the move is made by holding the blade clear of the ground and obstructions and sliding it sideward by means of the control valve for its hydraulic cylinder.

Drawbar and Circle. The drawbar is a V-shaped (sometimes T-shaped) connection between the front of the grader frame and the circle. It is rigid with the circle and fastened to the frame by either a ball and socket or a universal fitting that allows limited angular movement from side to side and up and down. The drawbar carries the full horizontal load on the blade, as other connections provide only vertical and side support.

The circle is a toothed ring rotated in or on a supporting frame by a circle turn mechanism. Pads and shims provide for low-friction movement and for adjustment of clearances. The circle is turned by a spur pinion gear meshing with teeth cut all around it, usually in the inside. The pinion may be turned by a shaft driven by the engine power takeoff or by a hydraulic motor, or by a direct-mounted hydraulic motor and reducing gears. In each case, movement is controlled by the operator.

Lift. The blade is lifted and lowered through the circle by a pair of arms or cylinders supported by the frame and fastened to the sides of the drawbar at the rear. Direct lift by the pair of two-way cylinders is superficially the simplest method.

Circle Sideshift. The circle sideshift, or lateral shift, swings the circle and blade to the side, usually to the right. When the move is mechanically powered, a curved track is mounted on the underside of the frame. A long, curved gear section slides on this, with action by a pinion gear attached to the frame. An arm is fastened by ball and socket joints to the right side of the gear section and the left side of the drawbar for reach out to the right.

Turning the pinion for right shift moves the gear outward and upward to the right, pulling the drawbar and circle with it. For left shift, the arm is fastened to the left side of the gear and the right side of the drawbar and the pinion rotated oppositely.

Sideshift to the right raises the blade increasingly on the right side and lowers it somewhat on the left. A hydraulic cylinder mounted on one side of the frame, with its piston connected to the opposite side of the drawbar, can be used to shift the circle.

Blade lift cylinders may be based on a hydraulic saddle or cross-piece on the grader frame. This may be designed to rotate on the frame when the locks are released. When free, it can be turned by the action of one lift cylinder and then locked in a side position.

The principal use of circle or lateral sideshift is to position the blade at a steep or vertical angle for slope work. A combination of circle and blade sideshifts will extend the blade, as seen in Fig. 14.15, to work in spots too soft to support the grader.

Blade Vertical Position. The blade can be positioned for shaping a bank or embankment slope, as seen in Fig. 14.16. This is done by moving the circle out and at an angle with a horizontal plane. The extreme angle is for the blade to be vertical or at a 90-degree angle from the horizon. This is possible either to the right or left side of the motor grader. Movements are entirely by hydraulics.

FIGURE 14.16 Blade raised.

Optional Attachments

Scarifier. The scarifier is a set of teeth used for breaking up surfaces too hard to be readily penetrated by the blade. The teeth, consisting of rather slender shanks with replaceable tips, are set in a bar with a flattened V shape, as seen in Fig. 14.17. The V is narrower than

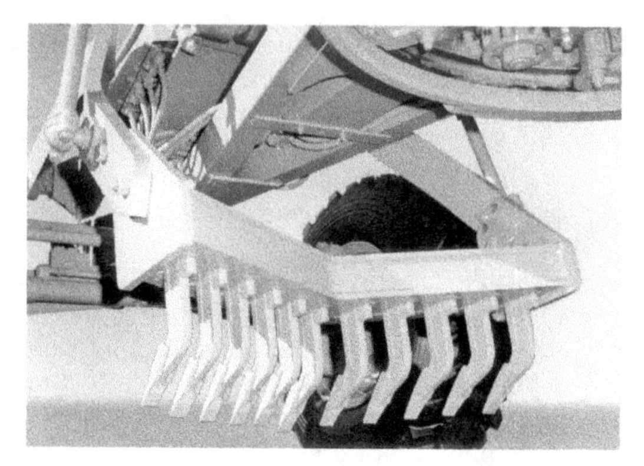

FIGURE 14.17 V-shaped scarifier mounted under the front of the grader.

the grader. It is pulled by a pair of beams hinged to the bottom front of the frame, and is usually raised and lowered by a hydraulic cylinder. Angle of penetration may be adjustable from the operator's station.

The shanks are wedged or clipped in place, and may be readily adjusted for height. A full set, often 11 in number, is used for shallow penetration and light work. For deeper penetration or slabby material, every other tooth or two out of three may be removed. Individual teeth are not designed to take the full push of the grader, and caution must be used in hard material and among rocks to avoid bending or breaking the teeth.

Scarifiers were once standard equipment but are rather less used now. They can speed up grader work and greatly reduce wear on the blade. It may be used in the same pass with the blading, or separate passes may be made for loosening a surface for shaping.

Ripper. A ripper can be mounted on the back of a grader (Fig. 14.18), may be fitted with 11 teeth for scarifying depths up to 9 in., or with 5 heavier ripper teeth for penetration to 14 in.

FIGURE 14.18 Rear-mounted ripper.

It is raised and forced down by a hydraulic cylinder. The lift frame is of the parallelogram type and keeps the teeth at the same vertical angle, whether raised or lowered to full depth.

A rear ripper will handle much heavier work than a front-mounted scarifier. The ripper can process a strip the full width of the machine, and it is not in the way of the blade. However, it adds to machine length, so is awkward when working in close quarters, usually produces coarser and more difficult-to-grade pieces, and cannot break ground while the blade is engaged.

Grader Operations

The grader is a versatile piece of equipment and plays an integral role in almost every earthwork project. Road and embankment finish work and shallow ditch cuts are basic grader operations. Grader operations are normally performed as described here.

Sidecasting. When the moldboard is set at an angle to the grader, the load being pushed spools and drifts off the trailing end of the blade (Fig. 14.19). The rolling action caused by the curve of the moldboard assists this side movement. As the blade is angled more sharply, the speed of the side-drift increases so that the dirt is not carried forward as far and a deeper cut can be made.

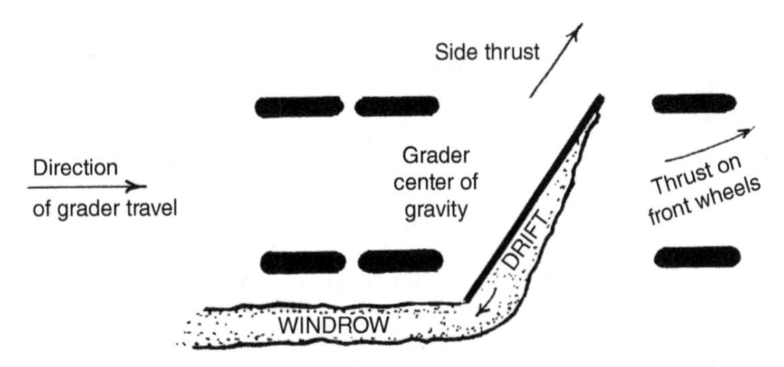

FIGURE 14.19 Grader sidecasting.

The sideward movement of the load exerts a thrust against the blade in the opposite direction, which tends to swing the front of the grader toward the leading edge. This thrust is handled by leaning the front wheels to pull against the side-drift and steering enough to compensate for any side-slipping that occurs in spite of leaning the front wheels.

Most road shaping and maintenance are done at a 25- to 30-degree angle, with straighter settings for distributing windrows and sharper ones for hard cuts and ditching.

Planing Surfaces. A grader is often used to plane or smooth off cut or fill surfaces. To do this, the moldboard is set at an angle to the surface with high material being scraped off and used to fill low spots. The operator tries to keep enough of the cut material in front of the moldboard to accomplish the necessary filling. The loosened material is moved forward and sideways to distribute it evenly.

On the next pass, the **windrow** left at the trailing edge of the moldboard is caught and moved across the blade to the heel. On the final pass, a lighter cut is made and the trailing edge (heel) of the moldboard is lifted to allow the surplus material to pass under rather than off the end. This avoids leaving a ridge. Windrows should not be piled in front of the rear wheels because they will adversely affect cutting accuracy and traction with increased rolling resistance.

Windrow
The line of loose material that spills off the trailing edge of a dozer or grader blade.

Side draft
The force on the moldboard that tends to pull the front of the grader to one side.

Moving Windrows. When a grader makes a cut, a windrow will form between the heel of the moldboard and the rear wheel. This windrow will impart a **side draft** force. The front wheels of a grader can be leaned both left and right. The grader wheels are inclined against the direction of side-draft, and the windrow can then be shifted across the grade with successive cuts. Sometimes cuts produce more material than is needed for the roadbed and shoulders. This excess material can be used as fill at other locations on the project. In that case, the excess material is drifted into a windrow and picked up by an elevating scraper. The scraper can haul the material to the appropriate location on the project.

Ditch Cuts. Normally, ditching cuts are done in second gear at full throttle. For better grader control and straighter ditches, a 3- to 4-in.-deep marking cut is made at the outer edge of the bank slope (usually identified by slope stakes or with GPS control) on the first pass. The toe of the moldboard should be in line with the outside edge of the lead tire. The marking cut provides a guide for subsequent operations. Cuts are made as deep as possible without stalling or losing control of the grader. Each successive cut is brought in from the edge of the bank slope so the toe of the moldboard will be in line with the ditch bottom on the final cut.

Haul-Road Maintenance. Haul roads should always be kept in good condition to improve haul-unit production. Graders are the best machines for maintaining haul roads. The most efficient method of road maintenance is to complete one side of a road with one pass. With this method, one side of the road is completed, while the other side is left open to traffic.

Working the material across the road from one side to the other provides leveling and maintenance of the surface. However, to maintain a satisfactory surface in dry weather, traffic-eroded material should be worked from the edges and shoulders of the road toward the center. The surface is easier to work when damp; therefore, after a rain is a good time to perform surface maintenance. If the material is too dry to be worked, a water truck may be necessary to dampen the road before grading.

Smoothing Pitted Surfaces. When fine material is present and moisture content is appropriate, rough or badly pitted surfaces can be cut smooth. The material cut from the surface is then respread over the smooth base. Again, the best time to reshape earth and gravel roads is after a rain. Dry roads should be watered using a water distributor. This ensures that the material will have sufficient moisture content to readily recompact.

Correcting Corrugated Roads. When correcting corrugated roads, care should be exercised to avoid making the situation worse. Deep cuts on a washboard surface will create moldboard *chatter*, which emphasizes rather than corrects corrugations. Scarifying may be required if the surface is too badly corrugated. With proper moisture content, the surface can be leveled by cutting across the corrugations. The operator should alternate the moldboard so the cutting edge will not follow the rough surface and cut the surface to the bottom of the corrugations. Then reshape the road surface by spreading the windrows in an even layer across the road. Compaction after shaping gives longer-lasting results.

Spreading. Graders are often used to spread and mix dumped loads. Because of their mechanical structure and operating characteristics, graders are most effective spreading and mixing free-flowing materials. Cutting or ripping well-compacted material is best accomplished with other equipment, such as dozers or excavators. Graders are best suited for material in a loose state or in a bank state with minimal resistance.

Proper Working Speeds. Always operate as fast as the skill of the operator and the condition of the grade permit. Graders should be operated at full throttle in each gear. If less speed is required, it is better to use a lower gear rather than run at less than full throttle. Correct gear ranges for various grader operations, under normal conditions, are listed in Table 14.4.

TABLE 14.4 Proper Gear Ranges for Grader Operations

Operation	Gear
Road maintenance	Second to third
Spreading	Third to fourth
Mixing	Fourth to sixth
Bank sloping	First
Ditching	First to second
Finishing	Second to fourth

Turns. When making a number of passes over a short distance (less than 1,000 ft), backing the grader to the starting point is normally more efficient than turning the machine around and continuing the work from the far end. Turns should never be made on newly laid asphalt surfaces.

Number of Passes. Grader efficiency is in direct proportion to the number of passes made. Operator skill, coupled with planning, is most important in eliminating unnecessary passes.

Tire Inflation. Overinflated tires cause less contact between the tires and the road surface, resulting in a loss of traction. Air pressure differences in the rear tires cause wheel slippage and grader bucking. To achieve good results, it is necessary to always keep grader tires properly inflated.

Grader GPS Controls

Global positioning system (GPS) technology has been widely adopted for assisting the grader operator in controlling moldboard position. GPS can achieve centimeter-level-accuracy guidance. With an integrated GPS and moldboard guidance and monitoring system, elevation control can be semi-automated. With the help of an on-board system display, the grader operator monitors the correct grade and makes necessary adjustments in the moldboard position.

Two primary types of GPS are used in construction: kinematic and differential. Kinematic GPS is widely used with handheld units that continually communicate with satellites while the operator changes location. Differential GPS, on the other hand, is more accurate because it employs a base station that communicates with the satellites and establishes a known reference point for interaction with localized GPS receivers. The localized receivers are mounted on equipment or at the job trailer and constantly communicate with the base station. A GPS receiver–equipped grader is shown in Fig. 14.20 together with a field-mounted base station to localize the control signals. One contractor sends the GPS data to the engineering project office via a radio linked for monitoring and analysis.

GPS receivers mounted on the moldboard

GPS base station

FIGURE 14.20 Grader with GPS control on a street project.

Grader Time Estimates

The following formula can be used to prepare estimates of the total time (in hours or minutes) required to complete a grader operation:

$$\text{Total Time} = \frac{P \times D}{S \times E} \tag{14.4}$$

where P = number of passes required
D = distance traveled in each pass, in miles or feet
S = speed of grader, in mph or feet per minute, fpm (use 88 fpm to change mph to fpm)
E = grader efficiency factor

Factors in the Formula. The number of passes depends on project requirements and is estimated before construction begins. For example, five passes may be needed to clean out a ditch and reshape a road surface. Travel distance per pass depends on the length of the work. An odd number of passes with the final shaping phase allows the grader to reposition to a new segment.

Speed. Grader travel speed is the most difficult factor in the formula to estimate accurately. As work progresses, conditions may require that speed estimates be increased or decreased. Work output should be computed for each rate of speed used in an operation. The speed depends largely on the skill of the operator and the type of material being handled.

Efficiency Factor. A reasonable efficiency factor for grader operations is 60%. Consider inefficiencies associated with repositioning, active haul roads, communication with personnel, and operator breaks.

Sample Calculation: Maintenance of 5 miles of haul road requires cleaning the ditches and leveling and reshaping the roadway. Use an efficiency factor of 0.60. Cleaning the ditches requires two passes in first gear (2.3 mph), leveling the road requires two passes in second gear (3.7 mph), and final shaping of the road requires three passes in fourth gear (9.7 mph). Note that distance is expressed in miles.

$$\text{Total Time} = \frac{2 \times 5 \text{ miles}}{2.3 \text{ mph} \times 0.6} + \frac{2 \times 5 \text{ miles}}{3.7 \text{ mph} \times 0.6} + \frac{3 \times 5 \text{ miles}}{9.7 \text{ mph} \times 0.6}$$
$$= 7.3 \text{ hr} + 4.5 \text{ hr} + 2.6 \text{ hr} = 14.4 \text{ hr}$$

Sample Calculation: A haul road of 1,500 ft requires leveling and reshaping. Use an efficiency factor of 0.60. The work requires two passes in second gear (3.7 mph) and three passes in third gear (5.9 mph). Note that distance is expressed in feet.

$$\text{Total Time} = \frac{2 \times 1{,}500 \text{ feet}}{88 \text{ fpm/mph} \times 3.7 \text{ mph} \times 0.6} + \frac{3 \times 1{,}500 \text{ feet}}{88 \text{ fpm/mph} \times 5.9 \text{ mph} \times 0.6}$$

$$= 15.4 \text{ min} + 14.5 \text{ min} = 29.9 \text{ min}$$

Fine-Grade Production. When used for finishing and fine-grading work such as the final shaping of a surface layer, production on a square yard per hour basis can be calculated using Eq. (14.5).

$$\text{Production (sy/hr)} = \frac{5{,}280 \times S \times W \times E}{9} \tag{14.5}$$

where S = speed of grader, in mph
W = effective width per grader pass, in feet
E = grader efficiency factor

Sample Calculation: It is required to fine-grade the subgrade of a roadway before proceeding to construct the subbase. Use an efficiency factor of 0.60. The grader will be operated in second gear (3.5 mph) for this work. The effective blade width per pass is 9 ft. Estimate the production rate for these conditions.

Using Eq. (14.5),

$$\text{Production (sy/hr)} = \frac{5{,}280 \times 3.5 \times 9 \times 0.6}{9}$$

$$\text{Production} = 11{,}088 \text{ sy/hr}$$

Note that in this example it was assumed that the fine grading could be accomplished with only one grader pass. This is not usually the case. Typically, multiple passes are required to meet the specified grade tolerance, which can be ⅛ in. in 10 ft (10 mm/m) or less.

Grader Safety

Work-zone safety efforts usually concentrate on keeping workers and construction equipment and the public separated. The National Institute of Occupational Safety and Health (NIOSH) notes that over half of work-zone fatalities are inside the work area and do not involve the public. Many work-zone fatalities are the result of workers on foot in the work zone being struck by construction vehicles moving in reverse.

Sight Lines. A 32-year-old construction surveyor died when backed over by a motor grader. Operators of large equipment, including motor graders, cannot see what is directly behind their machines. Thus, fleet managers are beginning to employ rear-vision cameras on graders and other machines. These systems enable the operator to view a small in-cab monitor via a closed-circuit wide-view camera at the rear of the vehicle. One fleet manager reported that prior to employing the cameras, they had accidents where motor graders actually backed up over the hood of vehicles. Audible warning devices are also an effective means to alert workers of backup movements.

Adjoining Activities. Many construction projects involve work along or work that crosses railroad tracks. Maybe the worst motor grader accident on record occurred in November 1961 when a grader was driven onto a rail-highway grade crossing immediately in front of an approaching train. The resulting collision and derailment of the Chicago, Rock Island, and Pacific passenger train caused injuries to 110 people, including 82 passengers and the driver of the motor grader. In 2015, a motor grader operating near railroad tracks caused derailment of a loaded coal train weighing over 14,000 tons, causing injuries to the operator and derailment of two locomotives and 19 loaded coal cars. Then later the same year, an operator of a motor grader was killed when struck by a freight train.

Safety Rules. Specific safety rules for grader operations include the following:

- All graders should be equipped with accident prevention signs and tags as per the OSHA standard 29 CFR 1926.145, "Specifications for accident prevention signs and tags," and should have a slow-moving vehicle emblem, consisting of a fluorescent yellow-orange triangle with a dark red reflective border, per 29 CFR 1910.145(d)(10).
- When operating a grader slowly on a highway or roadway, display a red flag or flashing light on a staff at least 6 ft above the left rear wheel.
- Never allow other personnel to ride on the tandem, moldboard, or rear of the grader.
- Keep the moldboard angled well under the machine when not in use.

GRADALLS

The *gradall* is a utility machine with the combined operating features of the hoe, dragline, and motor grader (Fig. 14.21). The full revolving superstructure of the unit can be mounted on either crawler tracks or wheels. The unit is designed as a versatile machine for both

FIGURE 14.21 Gradall on new pavement finishing the grade along a sidewalk.

excavating and finishing work. Being designed as a multiuse machine affects production efficiency with respect to individual applications, when compared to a unit designed specifically for a particular application.

The bucket arm of a gradall can be rotated 90 degrees or more, enabling it to be effective in reaching restricted working areas and where special shaping of slopes is required. The multipart telescoping boom can be hydraulically extended or retracted to vary digging or shaping reach. It can exert breakout force either above or below ground level.

When used in a hoe application to excavate below the running gear, its production rate will be less than a hoe equipped with an equal-size bucket. Similarly, it can perform dragline-type tasks, but it has limited reach compared to a dragline. Because the machine provides the operator with positive hydraulic control of the bucket, it can be used as a finishing tool for fine-grading slopes and confined areas, tasks normally accomplished with a grader if there were no space constraints. Wheeled units are well suited to finishing from existing or new pavement (see Fig. 14.21).

Safety. As with other excavators that rely on a rear counterweight to balance the excavating arm, there is the danger of personnel being struck or crushed between the counterweight and fixed worksite objects. Even with properly installed rearview mirrors, the operator cannot always see personnel working behind the machine. The extended boom also limits visibility around the machine. Good safety training that makes operators aware of this danger is a management responsibility.

TRIMMERS

When grading linear projects, there are a variety of large, highly specialized, but extremely versatile machines to perform fine-grade trimming. The result is better accuracy and greater production compared to fine grading with a grader. It has been reported that the production from one dual-lane trimmer is equal to that achievable with four to six graders. Another benefit is that automatic trimmers enable grade control to closer tolerances.

Operation. There are large four-track multilane trimmers and smaller three-track machines (Fig. 14.22) for bridge approach work, pavements, airfields, and parking lot applications. Both the large and small machines work using the same general control and mechanical systems. Traditional string-lines and more advanced GPS technology guide trimmer elevation.

Grade Control. Automatic trimmers use a sensor control system to establish the elevation and cross-slope of its cutting teeth, based on input from either an arm riding on a grade wire or a ski riding on an established grade. A tightly stretched wire set at a known elevation above the specified project grade and having a known offset distance from the alignment establishes a reference for the trimmer. A horizontal movable arm, mounted on the trimming machine and spring-tensioned, rides on the wire. This arm is connected to an electric switch and activates control signals for vertical movement of the trimmer. A similar vertical arm travels along the wire, steering the trimmer.

On projects where it is necessary to match an existing grade, a ski-type runner travels over the existing surface and controls the vertical elevation of the trimmer. In this case, the operator usually controls the alignment manually.

The frames of four-track multilane trimmers can provide a trimming width of 40 ft or more. These machines typically have four crawler-tracks, one at each corner. Each crawler

FIGURE 14.22 Small trimmer working off a wire grade line.

mount has a vertical member that is hydraulically adjustable either manually or by the automatic control system. With tighter urban reconstruction projects, there is a trend toward two-track trimmers.

Trimmer Assembly. The actual cutting of the grade is accomplished by a series of cutting teeth mounted on a full-width *cutter mandrel*. The cutter assembly is usually fixed at the ends, but it can break, allowing for the programming of a crown in the grade. An adjustable moldboard assembly follows the cutter mandrel and strikes off excess material. An *auger* for directing the spoil follows the moldboard. The spoil can be cast to either side of the machine or conveyed to a haul truck for removal from the project. On many large machines, the right and left augers are independently driven. There are typically *wastegates* located at each end of the auger and adjacent to the centerline of the machine. These are for depositing excess material in a windrow on the grade.

A gathering and discharge conveyor system can be added to many trimmers. These attach to the rear for removing and reclaiming material from the finished grade. The material can be directly loaded into haul units with these systems or deposited on the shoulder of the finished grade.

Production. The large, full-width trimmers have operating speeds of about 30 fpm, but this depends upon the amount (depth of cut) of material being handled. In the case of the smaller single-lane trimmers, operating speed increases significantly. Some of these machines are rated at 128 fpm, but again speed is controlled by the amount of material being cut. As operating speed is increased, there is usually a decrease in quality, with a surge of material building up against the cutting mandrel.

CHAPTER 15
COMPRESSORS AND DRILLS

Compressed air is used extensively on construction projects for powering rock drills and pile-driving hammers, and small portable tools. Air is a gas obeying the fundamental laws applicable to all gases. The loss in pressure due to friction as air flows through a pipe or a hose must be considered when deciding on the size of a pipe or hose to transport compressed air to equipment or tools. Drilling is performed for both general applications and highly specialized work. The rates of drilling rock will vary with a number of factors such as the type of drill, bit size, hardness of the rock, depth of holes, drilling pattern, terrain, and time lost in sequencing operations.

COMPRESSORS

Drilling (Fig. 15.1) and demolition tools are usually powered by compressed air. For construction, job site work, the needed air is supplied by portable compressors. These compressors are driven by standard types of industrial engines. The compressors draw in atmospheric air, compress it, and deliver it through pipes and/or hoses to the tools. Compressed air is used extensively for powering pile-driving hammers, air motors, and pumps. When dealing with air, a quantity of fluid is moved and there are resulting friction losses as it flows from the compressor through pipes and hoses. The friction losses affect the efficiency of the operation and must be considered when compressed air is used to power tools or equipment.

A compressed-air system consists of one or more compressors together with distribution lines to carry the air to the points of use. When air is compressed, it receives energy from the compressor. This energy is transmitted through a pipe or hose to the operating equipment, where a portion of the energy is converted into mechanical work. The operations of compressing, transmitting, and using air always result in a loss of energy; therefore, the overall efficiency is less than 100%.

Glossary of Gas Law Terms

Air is a gas, so it obeys the fundamental gas laws. These laws relate to the pressure, volume, temperature, and transmission of all gases. The following glossary defines the important terms used in developing and applying the gas laws to compressed air operations:

Absolute pressure The measurement of pressure relative to the zero pressure of a vacuum. It is equal to the sum of the gauge plus the surrounding atmospheric pressure, corresponding to the barometric reading. The absolute pressure is used when applying the gas laws.

FIGURE 15.1 Air compressor and rock drill.

Absolute temperature The temperature of a gas measured above absolute zero. The Kelvin scale is termed absolute and is the SI unit for temperature. Kelvin temperatures are written without a degree (°) symbol. In the English system, absolute temperature is in degrees Rankine (R).

Let T_K = Kelvin temperature, T_F° = Fahrenheit temperature, and T_R° = Celsius temperature. The relationships are then

$$T_K = \left[\frac{5}{9}(T_F^{\circ} - 32^{\circ})\right] + 273 \tag{15.1}$$

$$T_R^{\circ} = T_F^{\circ} + 459.69^{\circ} \tag{15.2}$$

It is common practice to use 460° in place of 459.69°.

Atmospheric pressure The pressure exerted at the surface of a body by a column of air in an atmosphere—the pressure in the surrounding air. It varies with temperature and altitude above sea level.

Density of air The weight of a unit volume of air, usually expressed as pounds per cubic foot (pcf). Density varies with the pressure and temperature of the air. For the standard conditions, weight of air at 59°F and 14.7 pounds per square inch (psi) absolute pressure, the density of air is 0.07658 pcf or 1.2929 kg/m^3.

Fahrenheit temperature The temperature indicated by a measuring device calibrated according to the Fahrenheit scale. For a Fahrenheit thermometer, at a pressure of 14.7 psi, pure water freezes at 32°F and boils at 212°F.

Gauge pressure The pressure exerted by the air in excess of atmospheric pressure. It is usually expressed in psi or inches of mercury and is measured by a pressure gauge or a mercury manometer.

Pressure The relationship of a force (F) acting on a unit area (A). It is usually denoted by P and can be measured in atmospheres, inches of mercury, millimeters of mercury, or pascal (Pa).

$$P = \frac{F}{A} \tag{15.3}$$

Standard conditions Because of the variations in the volume of air with pressure and temperature, it is useful to express the volume at baseline standard conditions. Standard conditions are an absolute pressure of 14.696 psi (14.7 psi is used in practice) and a temperature of 59°F (288 K).

Temperature Temperature is a measure of the amount of heat contained by a unit quantity of a material. This property governs the transfer of thermal energy, or heat between one system and another.

Vacuum Its definition is not fixed, but it is commonly taken to mean pressures below atmospheric pressure. Therefore, it is a measure of the extent to which pressure is less than atmospheric pressure. For example, a vacuum of 5 psi is equivalent to an absolute pressure of $14.7 - 5 = 9.7$ psi.

Gas Laws

Gases behave differently than solids and liquids. A gas has neither fixed volume nor shape. It assumes the shape of the container in which it is held. Three properties interact when working with a gas: pressure, volume, and temperature. Changes in one property will affect the others.

When the pressure surrounding a gas is changed, the volume of the gas responds in the opposite direction. This is Boyle's law, named after Robert Boyle (1627–1691). The relationship is expressed by the equation:

$$P_1V_1 = P_2V_2 = K \tag{15.4}$$

where P_1 = initial absolute pressure
V_1 = initial volume
P_2 = final absolute pressure
V_2 = final volume
K = a constant

When a gas undergoes a change in volume or pressure with a change in temperature, Boyle's law does not apply.

Nearly a century after Boyle, the French scientist Jacques Charles (1746–1823) made detailed measurements on how the volume of a gas was affected by temperature. The temperature–volume relationship of a gas is now known as Charles's law; mathematically it is:

$$\frac{V_1}{T_1} = \frac{V_2}{T_2} = C \tag{15.5}$$

where V_1 = initial volume
T_1 = initial absolute temperature (A)
V_2 = final volume
T_2 = final absolute temperature
C = a constant

The laws of Boyle and Charles can be combined to give the equation:

$$\frac{P_1V_1}{T_1} = \frac{P_2V_2}{T_2} = \text{a constant} \tag{15.6}$$

Equation (15.6) can be used to express the relations between pressure, volume, and temperature for any given gas, such as air.

Energy Required to Compress Air. When an air compressor increases the pressure of a given volume of air, it is necessary to furnish energy to the air. Air is drawn into the compressor at pressure P_1 and is discharged at pressure P_2. If a gas undergoes a change in volume without a change in temperature, the process is said to be *isothermal compression* or *expansion*. When a gas undergoes a change in volume without gaining or losing heat, this is referred to as *adiabatic compression* or *expansion*. For air compressors used on construction projects, the compression will not be performed under isothermal conditions. Air compressors work at conditions somewhere between pure isothermal compression (no change in temperature) and adiabatic compression (without gaining or losing heat). The actual compression condition for a given compressor can be determined experimentally.

Glossary of Air Compressor Terms

The following glossary defines the important terms pertinent to air compressors:

Aftercooler A heat exchanger used to cool the air after it is discharged from a compressor.

Compression ratio The ratio of the absolute discharge pressure to the absolute inlet pressure.

Compressor efficiency The ratio of the theoretical horsepower to the brake horsepower required by a compressor.

Discharge pressure The absolute pressure of the air at the outlet from a compressor.

Diversity factor Because not all air tools are in use simultaneously, this is the ratio of the actual quantity of air required for all uses (tools) to the sum of the individual quantities required for the average number in use.

Free air Air as it exists under atmospheric conditions at any given location.

Inlet pressure The absolute pressure of the air at the inlet to a compressor.

Intercooler A heat exchanger placed between two compression stages to remove the heat produced by air compression.

Load factor The ratio of the average load during a given period of time to the maximum rated load of a compressor.

Multistage compressor By moving the air through two or more stages of compression, the compressor produces the desired final pressure.

Reciprocating compressor By means of a piston reciprocating in a cylinder, the machine compresses the air.

Rotary (centrifugal) compressor A rotating vane or impeller type machine that, by imparting velocity to the flowing air, affects compression to the desired pressure.

Single-stage compressor The air is changed from atmospheric pressure to the desired discharge pressure in a single compression operation.

Two-stage compressor A process of compressing air in two separate operations. The first operation compresses the air to an intermediate pressure, whereas the second operation further compresses it to the desired final pressure.

Air Compressors

The capacity of an air compressor is determined by the amount of free air it can compress to a specified pressure in 1 min, under standard conditions (absolute pressure of 14.7 psi at 59°F).

Types of Compressors. The compressors used in construction work are one of three types: (1) reciprocating (or piston), (2) rotary vane, or (3) rotary screw. There are also centrifugal compressors, but these are most effective when running at full capacity and are not usually found supporting construction operations. Most portable compressors used in construction are of the rotary vane or screw type with one or two stages.

Compressor types can be further described by:

- Number of compression stages
- Cooling method (air, water, oil)
- Drive method (engine type)
- Lubrication [oil, oil-free (oil-less), where oil-free means no lubricating oil contacts the compressed air]

Normally reciprocating compressors are 1- to 50-hp units, while compressors of 100 hp and greater are typically rotary screw or centrifugal units.

Intake may be through one or several passages or manifolds. With all types of compressors, incoming air must be filtered to prevent dust from causing excessive wear, fouling moving parts, and choking passages. The diagram on the left in Fig. 15.2 portrays the airflow of a single-stage compressor. The inlet passage receives atmospheric air through a cleaner. The discharge is at full working pressure, usually 100 psi, and opens into a line to a storage tank (air receiver) from which it is piped to the tools.

The two-stage compressor (on the right in Fig. 15.2) has one or more primary or low-pressure cylinders that draw in atmospheric air, compress it to about 30 psi, and pass it through a cooling radiator (intercooler) to a secondary or high-pressure cylinder, where pressure is increased to 100 psi and discharged into a receiver.

Rating. Compressors are rated according to the number of cubic feet of atmospheric air they take in each minute when working at maximum governed speed with a specified output pressure, usually 100 psi. This cubic feet per minute rating is abbreviated to cfm or CFM. The "actual cfm" is this intake measurement, minus any losses due to leakage

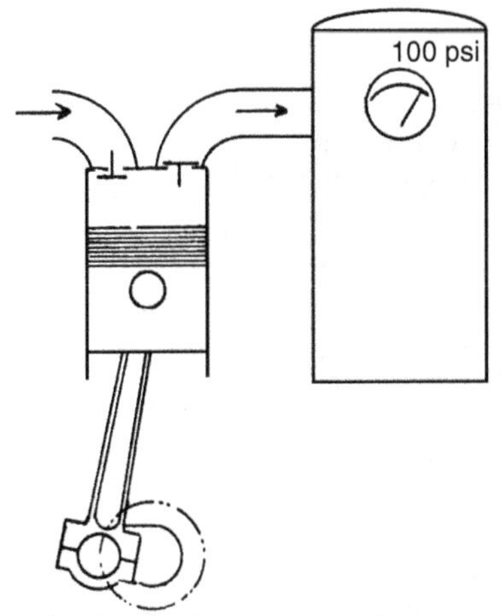

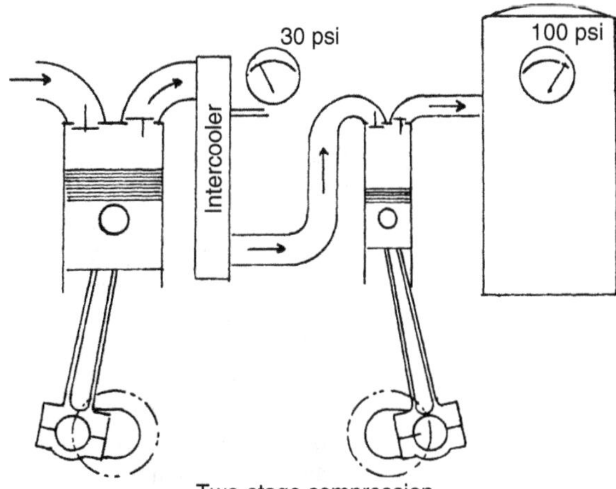

FIGURE 15.2 Airflow through a compressor.

in the pump. Actual cfm, or actual air delivery, is the volume of compressed air coming from the receiver outlet, transformed back to standard atmospheric pressure at sea level. Some garage and industrial compressors are rated according to piston displacement per minute at full speed. This figure will always be larger than the cfm in the same compressor, as air left in the space above the piston on the compression stroke expands and partly fills the cylinder on the suction stroke—as a result, the full displacement cannot be filled with new air (Fig. 15.3).

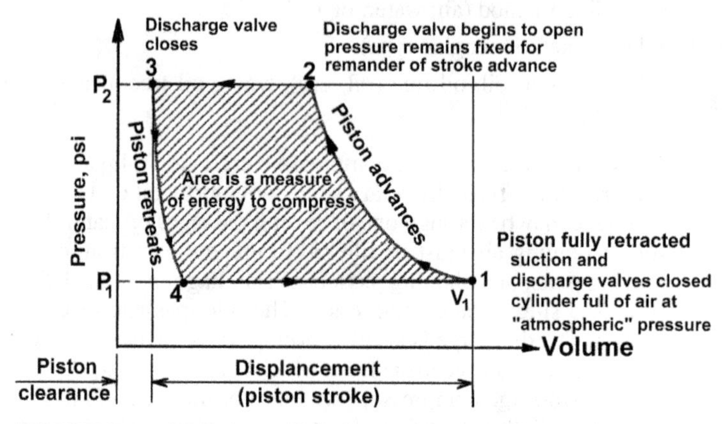

FIGURE 15.3 Air compression cycle.

The relationship between cfm and piston displacement is called volumetric efficiency. This figure may be between 65% and 80%, and depends largely on the airspace above the piston and the increase in pressure attained in the cylinder. In two-stage compressors, it is calculated on the basis of the displacement in the primary cylinders only. Do not confuse stage and acting. A reciprocating air compressor is single acting when the compression is accomplished on only one side of the piston. It is double acting where there is compression on both sides of the piston.

Pressure Control. Air-operated automatic controls are used to keep the pressure in the receiver within certain limits, usually 90 to 100 psi. In addition, the engine must have a mechanical governor to prevent it from racing when air pressure is low. During operation the engine and compressor continue turning at all times. When the receiver pressure reaches maximum, the compressor intake valves are held open so the entering suction stroke air is forced out again on the compression stroke. In this condition the compressor is said to be unloaded. The engine is throttled down at the same time the unloading takes place. In rotaries, the throttling of the engine automatically unloads the compressor.

Heat. Compression of air produces heat. Fundamental gas laws apply to the change in temperature with volume. The working parts of the compressor tend to become very hot and must be cooled by fins, water jackets, or an oil bath. The ability of air to hold moisture is decreased by pressure and increased by heat. As the hot compressed air enters the receiver, these effects are often balanced, but as the air cools in either the receiver or in the lines, extensive condensation may occur, particularly on damp days. Compressed air loses some pressure as it cools, so the compressor pumps additional air to compensate for this loss.

Stationary Compressors. Stationary compressors are generally used at fixed installations where there will be a requirement for compressed air for a long duration. One or more compressors may supply the total quantity of air. The installed cost of a single compressor will usually be less than for several compressors with the same capacity. However, several compressors provide flexibility for varying load demands and in the event of a shutdown for repairs, the entire plant does not need to be stopped.

Portable Compressors. Portable compressors are commonly used on construction sites (Fig. 15.1 and 15.4), where it is necessary to meet frequently changing job demands and where typically the need is at a number of locations. The compressors may be mounted on rubber tires or skids. They are usually powered by gasoline or diesel engines.

FIGURE 15.4 Portable air compressor.

Effect of Altitude on Compressor Capacity. The capacity of an air compressor is rated on the basis of its performance at sea level, where the normal absolute barometric pressure is about 14.7 psi. If a compressor is operated at a higher altitude, such as 5,000 ft above sea level, the absolute barometric pressure will be about 12.2 psi. Thus, at the higher altitude, air density is less and the weight of air in a cubic foot of free volume is less than at sea level.

If the air is discharged by the compressor at a given pressure, the compression ratio will be increased and the capacity of the compressor will be reduced. This may be demonstrated by applying Eq. 15.4.

Assume the 100 cf of free air at sea level is compressed to 100 psi gauge pressure with no change in temperature. Applying Eq. 15.4, we obtain

$$V_2 = \frac{P_1 V_1}{P_2}$$

where $V_1 = 100$ cf

$P_1 = 14.7$ psi absolute

$P_2 = 114.7$ psi absolute

$$V_2 = \frac{14.7 \text{ psi} \times 100 \text{ cf}}{114.7 \text{ psi}} = 12.82 \text{ cf}$$

At 5,000 ft above sea level,

$$V_1 = 100 \text{ cf}$$
$$P_1 = 12.2 \text{ psi absolute}$$
$$P_2 = 112.2 \text{ psi absolute}$$
$$V_2 = \frac{12.2 \text{ psi} \times 100 \text{ cf}}{112.2 \text{ psi}} = 10.87 \text{ cf}$$

Table 15.1 provides general guidance concerning the efficiency of reciprocating and rotary compressors operated at altitudes above sea level.

TABLE 15.1 Efficiency of Air Compressors at Various Altitudes (100 psi Gauge Output Pressure)

Altitude, ft	Single-Stage Reciprocating Compressor Percent of Efficiency	Two-Stage Reciprocating Compressor Percent of Efficiency	Rotary Compressor Percent of Efficiency
2,000	98.7	99.4	100.0
5,000	92.5	98.5	100.0
8,000	87.3	97.6	99.9
10,000	84.0	97.0	
12,000			98.6

Reciprocating Compressors

Reciprocating compressors have cylinders drawing in air during a suction stroke and discharging the air through a check valve at higher pressure during a compression stroke. These may be in the same block as the engine cylinders, but more often are in a separate block with a disk clutch connection between crankshafts.

The compressor proper consists of four or six inline cylinders in a single block. Standard industrial engine types of pistons, rods, crankshaft, bearings, and cooling system are used, and lubrication is by force feed through passages in the shaft. The compressed air is cooled by water jackets around the exhaust passages— this is the aftercooler.

Cylinders and Pistons. Compressor pistons may have either flat or convex tops. Single-stage compressors ordinarily have a number of identical in-line cylinders and pistons in a water-cooled block. A two-stage unit has a low-pressure stage requiring more cylinder capacity than the high-pressure stage. Two or three first-stage cylinders may be used for each second-stage cylinder. In such a case all cylinders can be of similar or identical size.

Such cylinders are arranged radially, as shown in Fig. 15.5, so as to operate off one throw of the crankshaft. Large models may have two or more sets, each on a separate throw. Two-stage cylinders may be either air- or water-cooled. Air cooling employs the thin cylinder walls and outside fins illustrated. The intercooler fan keeps air moving over the fins.

Heads and Valves. On two-stage reciprocating compressors, each head contains a discharge valve and passage, and usually an inlet valve and passage. The primary-stage inlet valves are fitted with an unloading device. Valves are of the automatic check type. Suction valves are similar or identical to discharge valves but are inverted to act oppositely. When

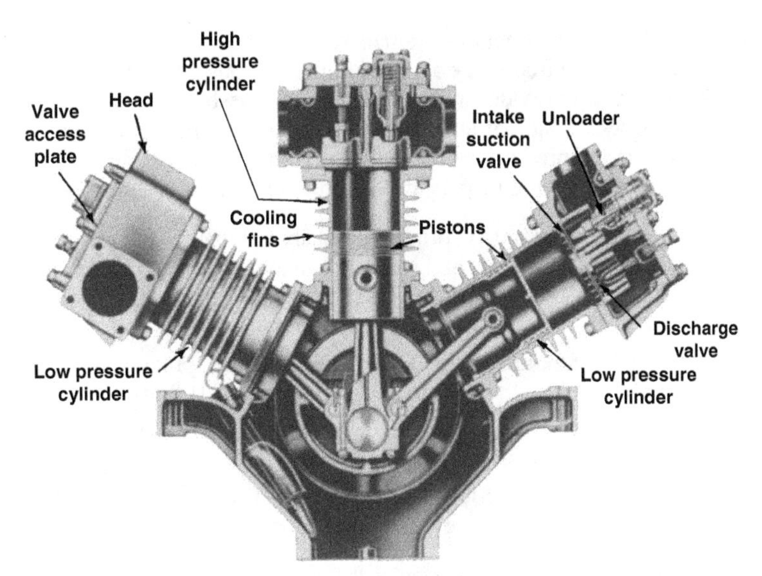

FIGURE 15.5 Two-stage reciprocating compressor cylinder arrangement.

the compressor unloads (stops compressing), air is admitted to the chamber above a diaphragm, forcing the diaphragm and plunger down until the fingers press the valve away from the seat. Air drawn in on the suction stroke is now discharged back through the inlet valve on the compression stroke, and no work is done. When air pressure above the diaphragm is released, springs push the fingers out of engagement.

This type of unloader may also be worked by air pressure against a piston attached to the plunger, rather than against a diaphragm. Valves are usually accessible for service by removal of a cover plate, without the need for disturbing other parts of the machine.

Intercooler. The intercooler, used only in two-stage machines, is a tubular cooling radiator connecting the first-stage discharge passages with the second-stage inlet. A fan mounted on the rear of the compressor provides air circulation around both of thcsc tubes and the compressor cylinders. Intercooler efficiency varies, but it reduces the temperature of the contained air to within a few degrees of atmospheric temperature. Intercooler pressure has a natural relationship to the receiver pressure, generally in the neighborhood of 30 to 100 psi. If the intercooler pressure rises disproportionately, leakage in high-pressure suction valves is often noticed.

The intercooler is fitted with a spring-loaded safety valve set for about 35 psi. This can be tripped by hand and should be blown daily to avoid the accumulation of trash, which might clog the valve. A drain valve in the bottom should also be opened daily, or more often, under pressure to blow out accumulations of water or oil. Compressors operating in high-humidity environments simply require more frequent draining. Intercoolers are sometimes fitted with an unloader or automatic drain, which consists of a blow-off valve opened by air pressure in the control tubing.

Air Receiver. The air receiver is a cylindrical tank with convex ends, usually mounted horizontally on the rear of the frame. It must conform to federal and state safety regulations, and National Board (The National Board of Boiler and Pressure Vessel Inspectors) certification is normally stamped on the receiver.

The receiver acts as a small reservoir between the compressor and the tools, thereby reducing the frequency of unloading in light use. It serves to separate moisture and oil from the air, and provides a place for draining these fluids. It is equipped with an inlet from the compressor and with one or more outgoing air lines with shutoff valves. A smaller air line goes to the pressure gauge and the automatic controls.

The following fittings are required for safe and convenient operation:

- **Safety valve** of the spring type. It should open when air pressure rises above the highest operating pressure, and it must have the capacity to let air out faster than the compressor can load the receiver tank. It should have a hand trip mechanism by which it can be opened and any sediment blown out daily.
- **Pressure gauge** for the receiver. It is standard practice to use a 200-psi face, so the needle will be vertical at a normal high pressure of 100 psi.
- **Drain valve** or cock at the low point in the receiver. This is used for draining or blowing out water, oil, and sediment.
- **Fusible plug** to reduce the chance of oil fires. It will melt if the air temperature approaches the flash point of lubricating oil vapor.

Guidance regarding the size of these fittings is provided in Table 15.2.

TABLE 15.2 Minimum Size Compressor Fittings

Compressor Size	50 cfm	105 cfm	210 cfm	315 cfm	420 cfm
Safety valve	1 in.	1¼ in.	1½ in.	2 in.	3 in.
Pressure gauge	3½ in.	3½ in.	3½ in.	6 in.	6 in.
Drain valve	½ in.	½ in.	½ in.	½ in.	½ in.
Fusible plug	⅜ in.	⅜ in.	⅜ in.	⅜ in.	⅜ in.

Air Pressure Controls. Automatic controls are necessary to keep sufficient pressure in the receiver to operate the tools and to prevent the buildup of excessive pressure. Air from the receiver is admitted to the control system by the pilot or trigger valve. Receiver air contacts a disc or a ball held on a finely machined seat by a spring. When pressure is sufficiently built up, it forces the disc off its seat. A much larger contact area of compressed air greatly increases the force against the spring, and the disc is snapped back far enough to allow air to escape quickly into a side passage to the controls.

When pressure drops sufficiently, the disc will be pushed down by the spring. As air passes the side passage, it presses against the upper surface, equalizing the receiver pressure below and allowing the spring to push the disc back on its seat. The air in the passage then leaks out through the top of the valve.

The engine throttle is either fully open or in the idling position and is liable to move frequently from one setting to the other. This wastes fuel because of the nonproductive time not under load and the extra consumption during acceleration. The change of speed and load also increases wear. Devices are available to proportion engine speed to receiver air pressure. Various designs are employed, but in principle they consist of a diaphragm or piston acted upon directly by air from the receiver pushing against the spring-held throttle lever and increasing pressure, causing the throttle to partially close. Adjustment to the desired pressure–speed relationship may be made by adjusting the tension of the spring.

These "fuel savers" may show substantial economies when the average consumption of the tools is below the capacity of the compressor, as they permit continuous operation on partial throttle. However, when the air demand is intermittent or greater than the compressor output, they are of little use.

Gauges and Controls. The instrument panel should, in addition to the usual engine gauges, carry instruments to indicate the compressor oil pressure and receiver air pressure. A two-stage compressor should also have a gauge for intercooler pressure.

The clutch control is usually an over-center lever, but in one make it is a hydraulic jack, which is pumped to disengage the clutch and released to engage it. Auxiliary levers are provided for hand operation of the throttle and the unloading mechanism.

Capacities. The air requirement of tools varies with their weight, model, and condition. In general, a 60- or 85-cfm compressor can supply two light hand drills or hammer drills, or one of medium weight. A 105- or 125-cfm can support a heavy hand drill, or two or more light ones. A wagon drill requires a 210- or 315-cfm. When tools are used intermittently, a compressor may be able to supply a larger number than is indicated by its capacity, but it will do so at the risk of reduced pressure and delays.

Mountings. Portable compressors have a variety of mountings. Rubber-mounted compressors with a capacity of 60 cfm and smaller can be moved short distances by hand; 105-cfm compressors may be towed or shifted by hand on level pavements. Larger models should be handled with a truck or tractor.

Operation. Before starting a compressor, a routine check should be made of the oil level in the engine and compressor, the fuel level, and the cooling system. The supporting surface should be nearly level. The receiver drain cock should be opened and the clutch disengaged. After starting, the engine is run for several minutes to warm, then the clutch is engaged. Soon after it turns the compressor without choking or laboring, the drain cock may be closed. If the engine is not thoroughly warm or is in poor condition, it may not be able to build up full receiver pressure immediately. If it threatens to stall, the clutch should be disengaged or the compressor unloaded by a hand control and extra warm-up time allowed.

During work the drain in the bottom of the receiver should be opened occasionally to eliminate condensation water. In hot, damp weather this can amount to several gallons per day and it is important to remove the water.

The safety valves on the receiver and the intercooler should be tripped by hand at least once per day to blow out any carbon or sludge deposits, which might prevent them from working in an emergency.

The filters in the air intake for both engine and compressor and any in the control system should be cleaned as often as necessary. Once started, the operation of the compressor is completely automatic. The oil and air pressure gauge should be checked regularly. A spare valve assembly should be kept on hand.

Carbon and Explosions. High air temperatures can cause vaporization of lubricating oil. The non-volatile residue, in combination with any dirt in the air, is liable to build up. Hard or gummy deposits will develop, interfering with valve action, and sometimes choking air passages. Because thin oils leave less residue than heavy ones, it is advisable to use the lightest oil needed for proper lubrication, which will not be pumped into the cylinders in excessive quantities.

Thorough filtering of incoming air is essential. Maintain clean filters so intake air passage is optimal. Dirty filters choke airflow and reduce operating pressure and waste fuel.

If exhaust valves leak, some of the hot compressed air which has just been forced out of the cylinder will come back in and be recompressed, resulting in very high localized pressure and excessive temperatures, which will boil off the lubricant. Under these conditions, enough oil may be present in the air to cause explosions, most often in the cylinder (dieseling) or the passages, but occasionally in the receiver or even in the tools. This danger is avoided by replacing piston rings in the compressor as soon as they allow pumping of oil, keeping valves free of carbon and dirt, and replacing broken or worn parts.

Multiple Units. Air demand too large for a single compressor may be supplied by two or more compressors discharging into a single supply system. The hookup can be arranged so one machine does most of the work or the compressors share the work equally.

The first method is most satisfactory when one machine can supply the normal load but cannot meet the peak demand. The compressor which is to carry the full burden has its pilot valve set for normal pressures, probably loading at 90 psi and unloading at 100 psi. The auxiliary has both settings about 2 psi lower. Discharge lines from both receivers should lead into a common manifold, from which air is taken by the tools.

When the compressors are started, both will run until a pressure of 98 psi is reached in the receivers and manifold. The auxiliary will then cut out and be unloaded as long as the main compressor can keep pressure above 88 psi. The compressor manufacturers need to be consulted for the piping and value arrangements to team multiple units.

Rotary Compressors

Rotary air compressors are positive displacement compressors. There are vane (or rotary vane) and screw-type rotary compressors. The early rotary compressors operated on the principle of a vane-type air motor.

Rotary Vane-Type Compressors. A cylindrical rotor is mounted off-center inside a larger cylindrical casing (Fig. 15.6). Sliding vanes are fitted into lengthwise slots in the rotor.

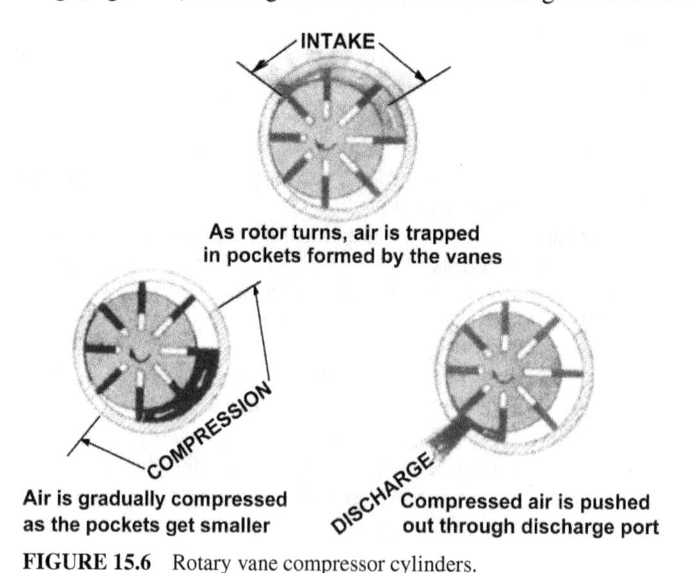

FIGURE 15.6 Rotary vane compressor cylinders.

Centrifugal force causes them to keep in contact with the casing wall whenever the rotor is turning rapidly.

The vanes divide the space between the rotor and the casing into a series of compartments. With a rotor mounted off-center, the vane is able to manipulate airflow with smaller and larger air pockets to process air through intake, compression, and discharge actions. Slots in the casing allow air to enter the compartments as they expand. As they contract, the air is compressed and is forced out an exhaust passage just before the point of closest contact between rotor and casing.

In some machines it is fairly easy to remove vanes for inspection; in others it is difficult. Compressor owners who are careful about keeping air and oil clean may get more than 6,000 hr from a set, whereas others who are less careful may have to replace them every 1,500 hr or else run serious risk of breakage.

Rotary Screw-Type Compressors. The rotary screw type of compressor uses a pair of cylinders with matching screws (Fig. 15.7) to compress the air in a single operation. There are no sliding vanes or metal-to-metal contact. Oil provides sealing and cooling. This compressor design is easy to maintain and operate. Capacity control is accomplished by variable speed and variable compressor displacement. These units are basically oil cooled (with air-cooled or water-cooled oil coolers), where the oil seals the internal clearances. Because cooling takes place inside the compressor, the working parts never experience extreme operating temperatures.

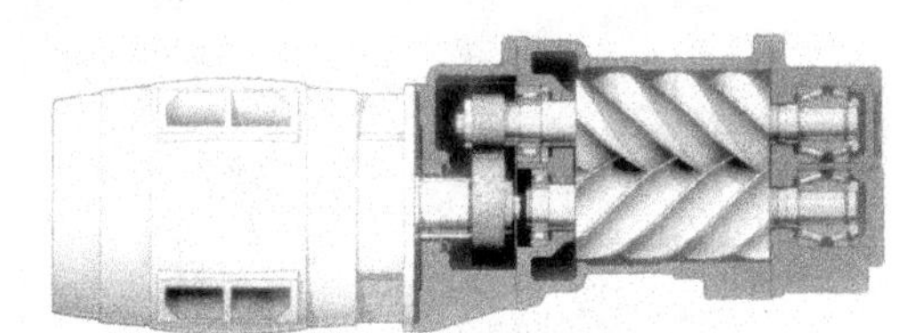

FIGURE 15.7 Rotary screw compressor. (*Courtesy of Ingersoll-Rand Company.*)

Servicing. A rotary compressor of either type requires little attention, except to make sure it has clean air and clean oil of the type and viscosity recommended by the manufacturer. The air filter must be serviced frequently to achieve efficient operation. Condensation water must be drained out of the oil filter daily, the oil filter must be in excellent condition, and oil must be changed when it starts to break down or become dirty.

The immediate result of allowing the oil to become dirty is excessive wear on the vanes or screws. They are pressed against the back or trailing edge of the rotor slots by air pressure ahead of them, and each vane may slide in and out a million times in 12 hours of operation. It is easy to understand how rapid wear is caused by lubrication failure.

Many small contractors have only occasional use for a compressor, and as a result the machine stands idle much of the time. It may not yield an adequate return on investment. However, this occasional use may be sufficiently important to justify the ownership of the machine instead of less costly use of a rental.

Compressed Air Distribution System

To provide a sufficient volume of air to multiple work locations, a compressed air distribution system is installed. The system must be designed to provide the air at pressures adequate for efficient tool operation. Any drop in pressure between the compressor and the point of use is an irretrievable loss. Therefore, the distribution system is an important element in the air supply scheme. The following are general rules for designing a compressed-air distribution system:

- Pipes should be of sufficient size so the pressure drop between the compressor and the point of use does not exceed 10% of the initial pressure.
- Each header or main line should have outlets as close as possible to the point of use. This permits shorter hose lengths and avoids large pressure drops through hose friction.
- Outlets should always be taken from the top of the pipeline to alleviate carryover of condensed moisture to tools.
- Condensation drains should be located at appropriate low points along the header or main lines.

Air Manifolds. An air manifold is a large-diameter pipe (Fig. 15.8) used to transport compressed air from one or more air compressors without a detrimental friction-line loss. The term "manifold" is also used to describe a chamber or pipe with a number of outlets used to distribute the gas to several working lines.

FIGURE 15.8 Manifold pipe with female universal crowfoot couplings for hoses.

Manifolds can be constructed of any durable pipe—black iron, stainless steel, copper, or aluminum—with proper thermal/pressure characteristics.

- Black iron or steel pipe used in compressed air systems will corrode when exposed to condensate (H_2O), and thus the pipe becomes a major source of contamination to the whole system. Compared to copper and aluminum, it is much heavier and difficult to work with but less expensive. The internal corrosion issue is much more significant with oil-free air than with lubricated compressors.

- Stainless steel is a good selection, particularly when exposed to oil-free wet air; however, threaded stainless steel often tends to leak.
- Copper pipe is a common selection for sensitive air systems.
- Aluminum has become very popular for compressed air pipe but can be damaged when used on a construction site.

Compressors are connected to the manifold with flexible hoses. A one-way check valve must be installed between the compressor and the manifold. This valve keeps manifold back pressure from possibly forcing air back into the compressor's receiver tank. The compressors grouped to supply an air manifold may be of different capacities, but the final discharge pressure of each should be coordinated at a specified pressure. In the case of construction, this is usually 100 psi. Compressors of different types should not be used on the same manifold.

Loss of Air Pressure in the Pipe. The loss in pressure due to friction as air flows through a pipe or a hose must be considered in selecting the size of a pipe or hose. Failure to use a sufficiently large pipe may cause the air pressure to drop, and the reduced pressure will not satisfactorily operate the tool to which it is providing power. Pipe size selection for an air line is a productivity (economics) issue. The efficiency of most equipment operated by compressed air drops off rapidly as the pressure of the air is reduced.

> The manufacturers of pneumatic equipment generally specify the minimum air pressure at which their equipment will operate satisfactorily. However, these values should be considered as minimum and not as desirable operating pressures. The actual pressure should be higher than the specified minimum.

Several formulas are used to determine the loss of pressure in a pipe due to friction. Equation (15.7) is a general formula:

$$f = \frac{CL}{r} \times \frac{Q^2}{d^5} \tag{15.7}$$

where f = pressure drop in psi
L = length of pipe in feet
Q = cubic feet of free air per second
r = ratio of compression
d = actual inside diameter (I.D.) of pipe in inches
C = experimental coefficient

For ordinary steel pipe, the value of C has been found to equal $0.1025/d^{0.31}$. If this value is substituted in Eq. (15.7), the result is

$$f = \frac{0.1025L}{r} \times \frac{Q^2}{d^{5.31}} \tag{15.8}$$

Table 15.3 gives the loss of air pressure in 1,000 ft of standard-weight pipe due to friction. For longer or shorter lengths of pipe, the friction loss will be in proportion to the length.

The losses given in the table are for an initial gauge pressure of 100 psi. If the initial pressure is other than 100 psi, the corresponding losses may be obtained by multiplying the

TABLE 15.3 Loss of Pressure in psi in 1,000 ft of Standard-Weight Pipe Due to Friction for an Initial Gauge Pressure of 100 psi

Free Air per Min (cf)	Nominal Diameter (in.)												
	½	¾	1	1¼	1½	2	2½	3	3½	4	4½	5	6
10	6.50	0.99	0.28										
20	25.90	3.90	1.11	0.25	0.11								
30	68.50	9.01	2.51	0.57	0.26								
40	—	16.00	4.45	1.03	0.46								
50	—	25.10	6.96	1.61	0.71	0.19							
60	—	36.20	10.00	2.32	1.02	0.28							
70	—	49.30	13.70	3.16	1.40	0.37							
80	—	64.50	17.80	4.14	1.83	0.49	0.19						
90	—	82.80	22.60	5.23	2.32	0.62	0.24						
100	—	—	27.90	6.47	2.86	0.77	0.30						
125	—	—	48.60	10.20	4.49	1.19	0.46						
150	—	—	62.80	14.60	6.43	1.72	0.66	0.21					
175	—	—	—	19.80	8.72	2.36	0.91	0.28					
200	—	—	—	25.90	11.40	3.06	1.19	0.37	0.17				
250	—	—	—	40.40	17.90	4.78	1.85	0.58	0.27				
300	—	—	—	58.20	25.80	6.85	2.67	0.84	0.39	0.20			
350	—	—	—	—	35.10	9.36	3.64	1.14	0.53	0.27			
400	—	—	—	—	45.80	12.10	4.75	1.50	0.69	0.35	0.19		
450	—	—	—	—	58.00	15.40	5.98	1.89	0.88	0.46	0.25		
500	—	—	—	—	71.60	19.20	7.42	2.34	1.09	0.55	0.30		
600	—	—	—	—	—	27.60	10.70	3.36	1.56	0.79	0.44		
700	—	—	—	—	—	37.70	14.50	4.55	2.13	1.09	0.59		
800	—	—	—	—	—	49.00	19.00	5.89	2.77	1.42	0.78		
900	—	—	—	—	—	62.30	24.10	7.60	3.51	1.80	0.99		
1,000	—	—	—	—	—	76.90	29.80	9.30	4.35	2.21	1.22		
1,500	—	—	—	—	—	—	67.00	21.00	9.80	4.90	2.73	1.51	0.57
2,000	—	—	—	—	—	—	—	37.40	17.30	8.80	4.90	2.73	0.99
2,500	—	—	—	—	—	—	—	58.40	27.20	13.80	8.30	4.20	1.57
3,000	—	—	—	—	—	—	—	84.10	39.10	20.00	10.90	6.00	2.26
3,500	—	—	—	—	—	—	—	—	58.20	27.20	14.70	8.20	3.04
4,000	—	—	—	—	—	—	—	—	69.40	35.50	19.40	10.70	4.01
4,500	—	—	—	—	—	—	—	—	—	45.00	24.50	13.50	5.10
5,000	—	—	—	—	—	—	—	—	—	55.60	30.20	16.80	6.30
6,000	—	—	—	—	—	—	—	—	—	80.00	43.70	24.10	9.10
7,000	—	—	—	—	—	—	—	—	—	—	59.50	32.80	12.20
8,000	—	—	—	—	—	—	—	—	—	—	77.50	42.90	16.10
9,000	—	—	—	—	—	—	—	—	—	—	—	54.30	20.40
10,000	—	—	—	—	—	—	—	—	—	—	—	—	25.10
11,000	—	—	—	—	—	—	—	—	—	—	—	—	30.40
12,000	—	—	—	—	—	—	—	—	—	—	—	—	36.20
13,000	—	—	—	—	—	—	—	—	—	—	—	—	42.60
14,000	—	—	—	—	—	—	—	—	—	—	—	—	49.20
15,000	—	—	—	—	—	—	—	—	—	—	—	—	56.60

values in Table 15.3 by a suitable factor. For a given rate of flow through a given size pipe, the only variable is r, the ratio of compression, based on absolute pressures, as shown in Eq. (15.8). When the gauge pressure is 100 psi, $r = 7.80$ (114.7/14.7), whereas for a gauge pressure of 80 psi, $r = 6.44$ (94.7/14.7). The ratio of these values of $r = 7.80/6.44 = 1.211$. Thus, the loss for an initial pressure of 80 psi will be 1.211 times the loss for an initial pressure of 100 psi.

Angles in the line should be kept to a minimum. If the line is reduced in size, it is preferred to use reducing couplings instead of reducing bushings to allow smoother airflow. If air may be required later between the compressor and the end of the line, tees with plugs may be used instead of couplings between lengths of pipe, so additional outlets can be added without opening and reconnecting the line.

To calculate the loss of pressure resulting from the flow of air through fittings, it is common practice to convert a fitting to equivalent lengths of pipe with the same nominal diameter. This equivalent length should be added to the actual length of the pipe in determining pressure loss. Table 15.4 gives the equivalent length of standard-weight pipe for computing pressure losses.

TABLE 15.4 Equivalent Length in Feet of Standard-Weight Pipe with the Same Pressure Losses as Screwed Fittings

Nominal Pipe Size (in.)	Gate Valve	Globe Valve	Angle Valve	Long-Radius Ell or On-Run of Standard Tee	Standard Ell or On-Run of Tee	Tee Through Side Outlet
½	0.4	17.3	8.6	0.6	1.6	3.1
¾	0.5	22.9	11.4	0.8	2.1	4.1
1	0.6	29.1	14.6	1.1	2.6	5.2
1¼	0.8	38.3	19.1	1.4	3.5	6.9
1½	0.9	44.7	22.4	1.6	4.0	8.0
2	1.2	57.4	28.7	2.1	5.2	10.3
2½	1.4	68.5	34.3	2.5	6.2	12.3
3	1.8	85.2	42.6	3.1	6.2	15.3
4	2.4	112.0	56.0	4.0	7.7	20.2
5	2.9	140.0	70.0	5.0	10.1	25.2
6	3.5	168.0	84.1	6.1	15.2	30.4
8	4.7	222.0	111.0	8.0	20.0	40.0
10	5.9	278.0	139.0	10.0	25.0	50.0
12	7.0	332.0	166.0	11.0	29.8	59.6

Sample Calculation: A 6-in. pipe with screwed fittings is used to transmit 2,000 cfm of free air at an initial 100 psi gauge pressure. The pipeline includes the following items: 4,000 ft of pipe, eight standard on-run tees, four gate valves, and eight long-radius ells. Determine the total loss of pressure in the pipeline.

Size of pipe, 6 in $\quad V = 2{,}000$ cfm

Length of pipe, 4,000 ft $\quad P_1 = 100$ psi gauge

Using the data in Table 15.4, the equivalent length of pipe will be

Pipe length	=	4,000.0 ft
Gate valves	4 × 3.5 (Table 15.4) =	14.0 ft
Standard on-run tees	8 × 15.2 (Table 15.4) =	121.6 ft
Ells, long-radius ells	8 × 6.1 (Table 15.4) =	48.8 ft
		Total 4,184.4 ft

According to Table 15.3, 0.99 psi loss per 1,000 ft of pipe. Therefore,

$$\frac{4{,}184.4\ \text{ft}}{1{,}000\ \text{ft}} \times 0.99\ \text{psi} = 4.14\ \text{psi}$$

Recommended Pipe Sizes. No book, table, or fixed data can give the correct pipe size for all installations. The correct method of determining the pipe size for a given installation is to conduct a complete engineering analysis of the project operations. Table 15.5 gives recommended sizes of pipe for transmitting compressed air for various lengths of run. This information is a useful guide for making an initial selection of pipe sizes.

TABLE 15.5 Recommended Pipe Sizes for Transmitting Compressed Air at 80 to 125 psi Gauge

Volume of Air (cfm)	Length of Pipe (ft)				
	50–200	200–500	500–1,000	1,000–2,500	2,500–5,000
	Nominal Size Pipe (in.)				
30–60	1	1	1¼	1½	1½
60–100	1	1¼	1¼	2	2
100–200	1¼	1½	2	2½	2½
200–500	2	2½	3	3½	3½
500–1,000	2½	3	3½	4	4½
1,000–2,000	2½	4	4½	5	6
2,000–4,000	3½	5	6	8	8
4,000–8,000	6	8	8	10	10

Loss of Air Pressure in Hose. The connection between a receiver or metal lines and the tools is made by flexible hose or tubing of rubber and fabric. The rubber should be neoprene or some other oil-resistant compound. Most hoses have between three and seven plies, or one to three braid layers, and may be of either wrapped or molded construction. The molded type with rayon braid fabric is lighter and more flexible for its size and strength. Heavy-duty hose with tough covers is required for mining; for quarrying rock, which breaks with sharp edges; and where the hose is subject to abuse from machines, tools, or rock falls. For less severe conditions the lighter hose is both more economical and easier to handle.

Hoses are fastened together and to other units by threaded, quarter-turn, or snap-on couplings. The threaded connections are best suited to connections that are seldom changed during the work. They take more time to assemble and disassemble than the other types but usually require less servicing. No gaskets are needed.

The quarter-turn or quick-detachable couplings are usually obtainable only for medium and small hose. The connectors in any one make are all the same size, and different-size hose can be connected without the use of bushings. Both surfaces and the gasket should be kept out of the dirt when apart and should be cleaned before coupling. Spare washers should be kept on hand, as these may be lost or damaged. For ease of operation and because of danger from the hose whipping and possible damage to eyes from blowing dirt air pressure should be turned off and the line bled before opening either type of connection.

The amount of air used by the tool is the critical factor in selecting an appropriately sized hose. When transmitting compressed air at 80 to 125 psi gauge, the guidance in Table 15.6 is valid for short hose lengths of 25 ft or less. As the required hose length increases, the nominal hose size must be increased. The loss of pressure resulting from the flow of air through hose is given in Table 15.7. These data can be used to design hoses whose length will be longer than 25 ft.

TABLE 15.6 Recommended Hose Size for Short Hose Lengths when Transmitting Compressed Air at 80 to 125 psi Gauge

Air Requirement of Tool (cfm)	Hose Nominal Size (in.)	Typical Tools
Up to 15	¼	Small drills and air hammers
Up to 40	⅜	Impact wrenches, grinders, and chipping hammers
Up to 80	½	Heavy chipping and rivet hammers
Up to 100	¾	Rock drills 35 to 55 lb, large concrete vibrators, and sump pumps
100 to 200	1	Rock drills 75 lb and drifters

Reels. Hose can be wrapped around the compressor or coiled in tool boxes when not in use. Keeping it on a reel is good policy, as it keeps it away from contact with sharp or heavy objects and avoids kinking. A dead reel, similar to the type used for garden hose, may be used for storage. If the inside end of the hose is left projecting near the axle, it may be connected to the receiver so the necessity of removing it entirely from the reel is avoided.

A more convenient device is a live reel permanently mounted on the compressor. Air from the receiver is admitted to the axle through a rotating, pressure-sealed connection, and from the axle to the hose by an ordinary coupling. The outer end of the hose should be tied to the reel when the compressor is moved to avoid unspooling.

Oilers. Most air hammers are equipped with oiling systems supplied by small reservoirs in the tool. Nevertheless, these are often neglected, need frequent attention, and may not function satisfactorily in a worn tool. A line oiler is a reservoir that feeds oil into the airstream through a needle valve. An oil mist is blown into the tool to keep it lubricated. Oil will feed only under pressure.

Oilers are built into other heavy air-operated equipment. Otherwise, they should be no more than 6 to 10 ft from the tool to be lubricated, as the air and oil may separate almost completely in more than 10 ft. Still, sufficient distance must be allowed for easy manipulation of the tools, and some lubrication will be afforded at long distances by the airstream pushing condensed oil along the inside of the hose. Some line oilers will operate in any

TABLE 15.7 Loss of Pressure in psi in 50 ft of Hose with Ending Couplings

Size of Hose (in.)	Gauge Pressure at Line (psi)	Volume of Free Air Through Hose (cfm)													
		20	30	40	50	60	70	80	90	100	110	120	130	140	150
½	50	1.8	5.0	10.1	18.1										
	60	1.3	4.0	8.4	14.8	23.5									
	70	1.0	3.4	7.0	12.4	20.0	28.4								
	80	0.9	2.8	6.0	10.8	17.4	25.2	34.6							
	90	0.8	2.4	5.4	9.5	14.8	22.0	30.5	41.0						
	100	0.7	2.3	4.8	8.4	13.3	19.3	27.2	36.6						
	110	0.6	2.0	4.3	7.6	12.0	17.6	24.6	33.3	44.5					
¾	50	0.4	0.8	1.5	2.4	3.5	4.4	6.5	8.5	11.4	14.2				
	60	0.3	0.6	1.2	1.9	2.8	3.8	5.2	6.8	8.6	11.2				
	70	0.2	0.5	0.9	1.5	2.3	3.2	4.2	5.5	7.0	8.8	11.0			
	80	0.2	0.5	0.8	1.3	1.9	2.8	3.6	4.7	5.8	7.2	8.8	10.6		
	90	0.2	0.4	0.7	1.1	1.6	2.3	3.1	4.0	5.0	6.2	7.5	9.0		
	100	0.2	0.4	0.6	1.0	1.4	2.0	2.7	3.5	4.4	5.4	6.6	7.9	9.4	11.1
	110	0.1	0.3	0.5	0.9	1.3	1.8	2.4	3.1	3.9	4.9	5.9	7.1	8.4	9.9
1	50	0.1	0.2	0.3	0.5	0.8	1.1	1.5	2.0	2.6	3.5	4.8	7.0		
	60	0.1	0.2	0.3	0.4	0.6	0.8	1.2	1.5	2.0	2.6	3.3	4.2	5.5	7.2
	70	—	0.1	0.2	0.4	0.5	0.7	1.0	1.3	1.6	2.0	2.5	3.1	3.8	4.7
	80	—	0.1	0.2	0.3	0.5	0.7	0.8	1.1	1.4	1.7	2.0	2.4	2.7	3.5
	90	—	0.1	0.2	0.3	0.4	0.6	0.7	0.9	1.2	1.4	1.7	2.0	2.4	2.8
	100	—	0.1	0.2	0.2	0.4	0.5	0.6	0.8	1.0	1.2	1.5	1.8	2.1	2.4
	110	—	0.1	0.2	0.2	0.3	0.4	0.6	0.7	0.9	1.1	1.3	1.5	1.8	2.1
1¼	50	—	—	0.2	0.2	0.2	0.3	0.4	0.5	0.7	1.1				
	60	—	—	—	0.1	0.2	0.3	0.3	0.5	0.6	0.8	1.0	1.2	1.5	
	70	—	—	—	0.1	0.2	0.2	0.3	0.4	0.4	0.5	0.7	0.8	1.0	1.3
	80	—	—	—	—	0.1	0.2	0.2	0.3	0.4	0.5	0.6	0.7	0.8	1.0
	90	—	—	—	—	0.1	0.2	0.2	0.3	0.3	0.4	0.5	0.6	0.7	0.8
	100	—	—	—	—	—	0.1	0.2	0.2	0.3	0.4	0.4	0.5	0.6	0.7
	110	—	—	—	—	—	0.1	0.2	0.2	0.3	0.3	0.4	0.5	0.5	0.6
1½	50	—	—	—	—	—	0.1	0.2	0.2	0.2	0.3	0.3	0.4	0.5	0.6
	60	—	—	—	—	—	—	0.1	0.2	0.2	0.2	0.3	0.3	0.4	0.5
	70	—	—	—	—	—	—	—	0.1	0.2	0.2	0.2	0.3	0.3	0.4
	80	—	—	—	—	—	—	—	—	0.1	0.2	0.2	0.2	0.3	0.4
	90	—	—	—	—	—	—	—	—	—	0.1	0.2	0.2	0.2	0.2
	100	—	—	—	—	—	—	—	—	—	—	0.1	0.2	0.2	0.2
	110	—	—	—	—	—	—	—	—	—	—	0.1	0.2	0.2	0.2

position, whereas others must be upright. They should always be placed in the line so the air moves in the direction of the arrow or the case. Hose between the oiler and the tool must be oil-resistant.

Diversity Factor. It is necessary to provide an adequate amount of compressed air as required by all the tools attached to the compressor. However, all tools will probably not be in operation during the same time periods. Therefore, prior to designing the compressed air system, perform an analysis of the tool use sequences to determine the probable maximum airflow needed.

As an example, if five jackhammers are on a job, normally not more than three or four will consume air (be operated) at a given time. The others will be out of use temporarily for bit changes or replacement of drill steel. Thus, the actual amount of air demanded should be based on three or four drills instead of five. The same condition will apply to other pneumatic tools.

The diversity factor is the ratio of the actual load (cfm) to the maximum calculated load (cfm) if all tools were operating at the same time. For example, if a jackhammer required 90 cfm of air, five hammers would require 450 cfm, assuming they all operate at the same time, but with only three hammers operating at a time, the demand for air would be 270 cfm. Thus, the diversity factor would be $270 \div 450 = 0.6$.

The approximate quantities of compressed air required by jackhammers and paving breakers are given in Table 15.8. The quantities are based on continuous operation at a pressure of 90 psi gauge.

TABLE 15.8 Quantities of Compressed Air Required by Pneumatic Equipment and Tools*

Equipment or Tools	Capacity or Size		Air Consumption (cfm)
	Weight (lb)	Depth of Hole (ft)	
Jackhammers	10	0–2	15–25
	15	0–2	20–35
	25	2–8	30–50
	35	8–12	55–75
	45	12–16	80–100
	55	16–24	90–110
	75	8–24	150–175
Pavingbreakers	35	—	30–35
	60	—	40–45
	80	—	50–65

*Air pressure at 90 psi gauge.

Safety. Extreme care should be exercised when working with compressed air. At close range, it is capable of putting out eyes, bursting eardrums, causing serious skin blisters, or even killing. Before using an air compressor, all pressure gauges must be in good working order.

Pneumatic power tools can be hazardous when used carelessly or improperly. Operators must perform a preoperational check of all air hoses, couplings, and connections to determine if leakage or other damage exists. Still, it is the responsibility of management to train employees in the proper use of all power tools. Employees must:

- Wear appropriate protective clothing and hearing and eye protection; additionally, gloves and respirators are often appropriate (see OSHA Standard 1926.102, Personal Protective and Life Saving Equipment).
- Install a clip or retainer to prevent attachments, such as chisels on a chipping hammer, from being unintentionally shot from the barrel.

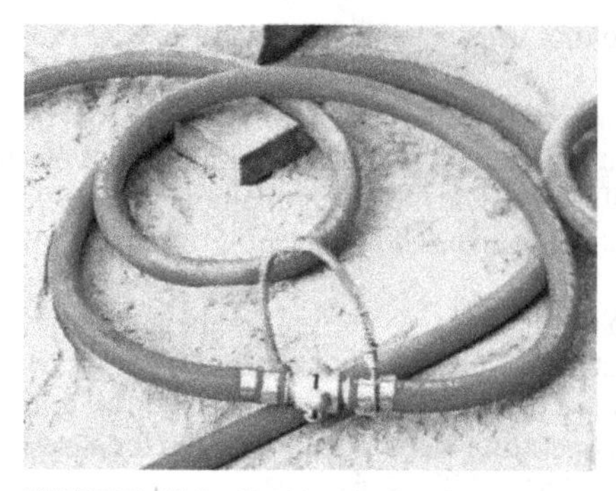

FIGURE 15.9 Positive lock wire between connected hose sections.

- Inspect all hoses to ensure they are fastened securely and cannot become accidentally disconnected. A short wire (Fig. 15.9) or positive locking device attaching the air hose to the tool will serve as an added safeguard.
- Turn off the air and disconnect the tool when repairs or adjustments are being made or the tool is not in use.
- Inspect the air hose to ensure it is in good condition and free from obstructions before connecting a pneumatic tool.
- Remove leaking or defective hoses from service. The air hose must be able to withstand the pressure required for the tool.

The surrounding work area must similarly be checked for hazards. All underground utilities should be identified prior to the start of excavating with pneumatic tools. Laborers have been electrocuted when they accidently made contact with underground power lines while digging with pneumatic tools.

DRILLS

The manner of boring a hole in hard materials did not change from ancient times until about the middle of the nineteenth century. Across that long span of time, drilling was accomplished by manpower—a man swinging a hammer against a pointed drill. The first effective pneumatic drill employed in the United States was at the eastern header of the Hoosac tunnel in western Massachusetts in June of 1866.

The purposes for which drilling is performed can vary from production work to highly specialized applications; consequently, it is necessary to select the equipment and methods best suited to the specific service. A contractor engaged in highway construction must usually drill rock under varying conditions; therefore, equipment suitable for a variety of applications would be selected. However, if equipment is needed to drill rock in a quarry where the material and conditions will not change, specialized equipment should be considered. In some instances, custom-made equipment designed for use on a single project may be justified.

Glossary of Drilling Terms

This glossary defines the important terms used in describing drilling equipment and procedures. The dimension terminology frequently used for drilling is illustrated in Fig. 15.10.

Bit The portion of a drilling tool used to cut or grind through the rock or soil by a combination of crushing and shearing actions. There are many types of bits.

Burden The horizontal distance from a rock face to the first row of drill holes or the distance between rows of drill holes.

Burden distance The distance between the rock face and a row of holes or between adjacent rows of holes (see Fig. 15.10).

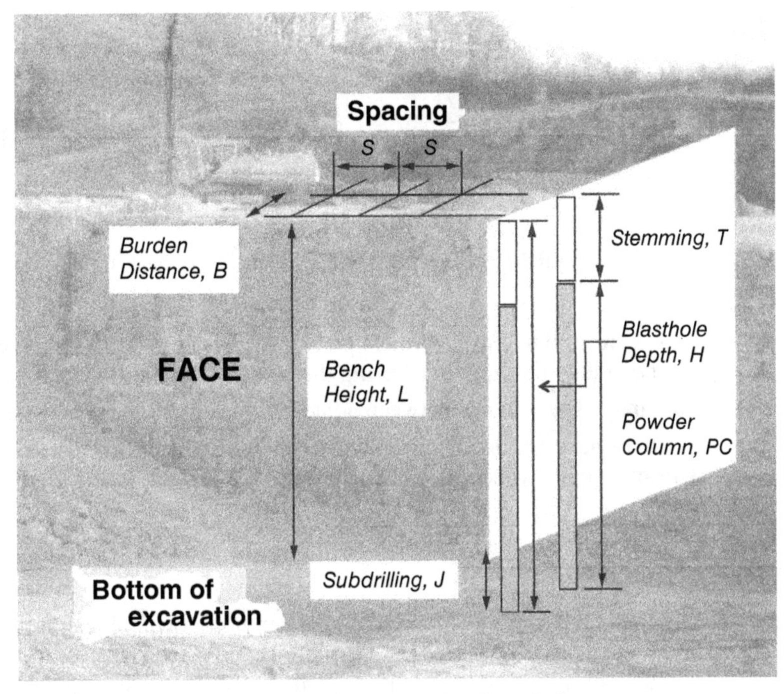

FIGURE 15.10 Drilling and blasthole dimensional terminology.

Cuttings These are the disintegrated rock particles produced by the action of the drill bit against the rock.

Drills

Abrasion The grinding effect of a bit rotation mills the rock into small particles.

Blasthole A rotary drill consisting of a steel-pipe drill stem, on the bottom of which is a roller bit. As the bit rotates, it grinds the rock.

Churn A percussion-type drill consisting of a long steel bit that is mechanically lifted and dropped to disintegrate the rock. It is used to drill deep holes, usually 6 in. in diameter or larger.

Core Drill designed for obtaining samples of rock from a hole, usually for exploratory purposes. Diamond and shot drills are used for core drilling.

Diamond A rotary abrasive-type drill whose bit consists of a metal matrix with embedded diamonds. As the drill rotates, the diamonds disintegrate the rock.

Downhole drill or downhole bit The bit and the power system providing rotation and percussion are one unit suspended at the bottom of the drill steel.

Percussion By the impact of repeated blows the rock is broken into small particles. Compressed air or hydraulic fluids can power percussion drills.

Shot A rotary abrasive-type drill whose bit consists of a section of steel pipe with a roughened surface at the bottom. As the bit is rotated under pressure, chilled-steel shot is supplied under the bit to accomplish the disintegration of the rock.

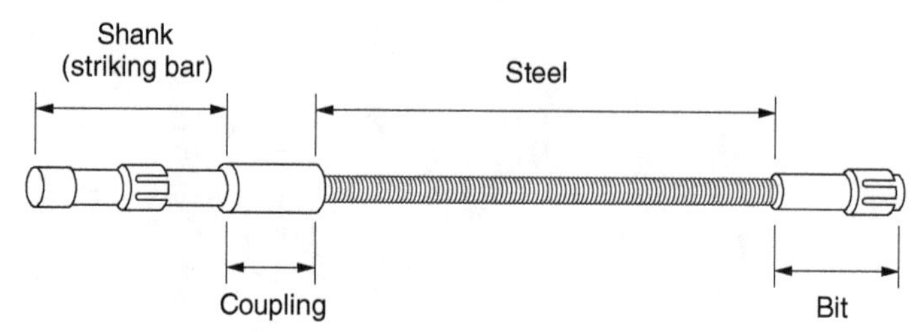

FIGURE 15.11 High-wear drilling items: shank, couplings, steel, and bit.

Shank or striker bar A short piece of steel attached to the percussion drill piston for receiving the blow and transferring the energy to the drill steel (Fig. 15.11).

Spacing The distance between adjacent holes in the same row (see Fig. 15.10).

Subdrilling The depth to which a blasthole will be drilled below the proposed final grade. This extra depth is necessary to achieve rock breakage to the required elevation.

Steels, Rods, and Pipes

Steels and rods are the connectors between non-downhole percussion drills and their bits. If only one piece is needed, it is usually called a steel; if two or more pieces are fastened together, they may be called either sectional steel or rods.

When made with tapered threads, rotary drill rods are usually called stems, pipes, or simply "rods". In moderate to deep drilling, the set of connectors and bit is called the string. In percussion work, the drill steel carries hammering and rotation with light to moderate down-pressure; with rotary drills, the steel is subjected to rotation and moderate to very heavy pressure. For both, it carries air under pressure to blow cuttings out of the hole. In larger rotaries air may be replaced by fluid.

Steel. Hollow drill steels for hand- and machine-mounted percussion drills are made in outside diameters of ⅞ to 2 in. There are standard lengths of 1 to 10, 12, 14, and 20 ft. Many odd lengths are produced by cutting down during repair work. The cross-section may be hexagonal, octagonal, or round, or the entire length of steel may have a continuous rounded thread.

One end may be forged into a shank, which fits in the chuck of a hand drill. For drifters, the shank is usually a short, separate piece, called a striking bar (see Fig. 15.11). The opposite end of the shank and both ends of most other steels carries a male thread, which may or may not be backed by a raised shoulder. Such threads are coarse, left-handed, and made in several nonmatching types.

A steel may have one tapered end to carry a tapered bit, or be forged into a cutting bit. Sections of steel (sectional steel) are fastened together by couplings (see Fig. 15.11), which are short, strong tubes with internal threads. Bits also have female threads for directly fastening to steels.

Most of the hammering impact shock is carried between rod ends, which butt against each other inside the coupling or from rod end to bit socket (bottom). If there are shoulders on the steel threads, they carry the shock through the coupling or the shoulder of the bit. If a connection is loose, or when ends or shoulders are not properly shaped, some or all of the shock will be carried by the threads, with resulting rapid deterioration.

Threads and tapers are designed to be tightened by the rotation of the drill, if jam tightly reverse twisting will not loosen them, therefore, to loosen sharp blows are applied when they are free of pressure. To prevent direct metal-to-metal contact, the threads need to be well lubricated by special heat-resistant grease. Without this, full tightening may be difficult, and impact may create tiny weld points. These prevent natural tightening while drilling and make loosening extremely difficult.

Threads and couplings deteriorate in service and require inspection after use and frequent repair. Forging machines are available to make new threaded ends with shoulders when required or to reshape old ones. Continuously threaded rod is cut to a new flat face, but must have the rim chamfered for the final 1/8 in., to about 30 degrees (Figs. 15.12 and 15.13).

Every steel must have a hollow center to carry air to the bit for hole cleaning and cooling. Care must be taken not to plug or narrow the hole with mud, trash, or repair defects.

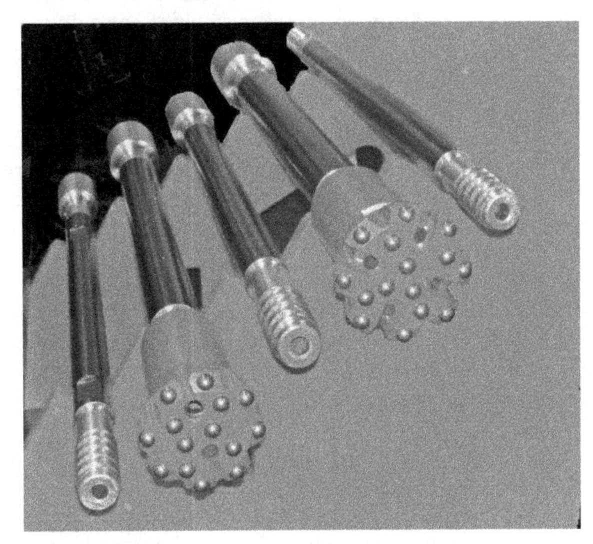

FIGURE 15.12 Drill steel with button bits.

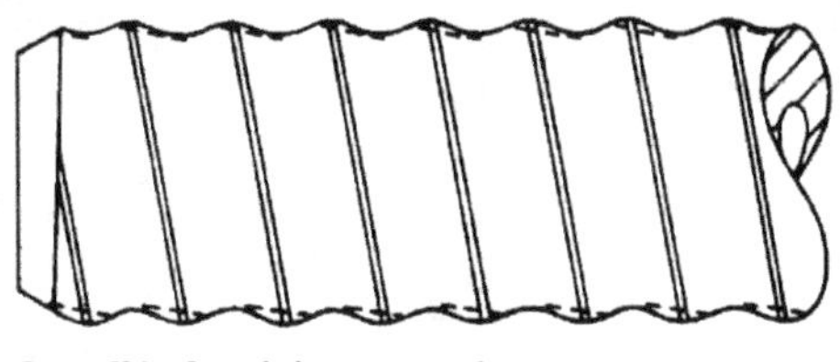

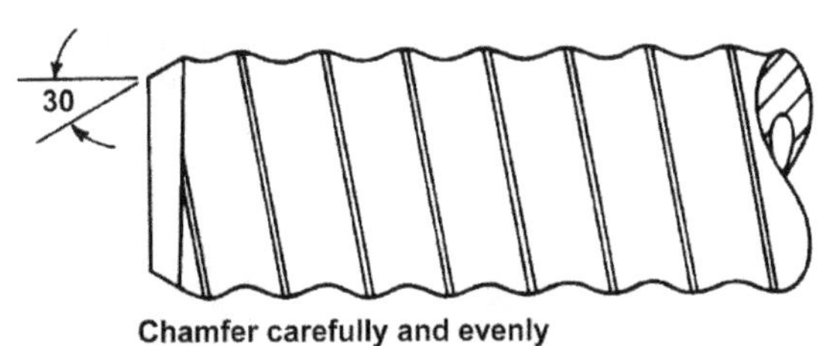

FIGURE 15.13 Trimming spiral steel. (*Courtesy of Ingersoll-Rand Company.*)

Steels are long and thin, and not nearly as strong as they look. They are easily bent by accident, by rough handling, and particularly by being blasted out of stuck holes. Their service life may be limited by metal fatigue and by cracks starting at dents and scratches on the surface. In general, long steels and thin steels are the ones most often severely damaged. A steel should never be used as a crowbar or a lever. They have almost no strength for such work.

A set of hand steels is a series in graduated lengths, from the starter to the longest piece required. For hand work, a 2-ft increase with each change is usual. The length difference between successive pieces is called the steel change.

Shanks. The three types of shanks are (1) hexagonal, (2) quarter octagonal (square, with chamfered corners), and (3) round lugged. The round shanks with two lugs are mostly used with the heavier drills and thicker steels. Shank diameter is normally the same or only slightly larger than steel diameter (see Fig. 15.11). The shank collar serves to limit the upward penetration of the steel into the chuck and enables the puller to hold the steel from sliding out of the drill.

Shank lengths, from the upper side of the collar to the end, are 3¼ in. for small steels and 4¼ in. for the larger sizes. It is important the length not vary more than 1⁄16 of an inch, and the shank end must be kept straight. In drifter drills, the shank is replaced by the striking bar, which is basically a short shank with a threaded lower end, which is kept clamped in the drill.

Drill Pipe. Down-hole and rotary drills use a heavier type of steel, called rod or pipe. It is designed to carry torque rather than impact. It uses American Petroleum Institute (API) steeply tapered threads, male on one end and female on the other. Standard lengths for deep drilling are 21, 29, and 43 ft. For mobile rigs, they are made up to suit tower height and steel change, in lengths of 20 ft, 25 ft, or longer.

Metal. There are four principal types of steel used in making drill steels (or rods): plain carbon, high carbon, alloy, and carburized. Plain carbon steel (0.80% to 0.85% carbon) is the oldest type, and is still standard for hand drills and other light hammers. It is the least expensive, usually has shoulders behind the threads, and can be rethreaded and repaired in a small shop. Outside diameters range from ⅞ to 1¼ in.

High-carbon (0.95% to 1.05% carbon) steel is used in the same sizes, but without shoulders, for somewhat heavier service with light drifters. Skill and good heat-treating equipment are needed for proper repair, unless this consists only of cutting the damaged tip off a long thread.

Various steel alloys such as manganese-chrome-molybdenum are made into 1¼- and 1½-in. diameters, which are harder, up to 45 Rockwell C scale, and highly fatigue resistant. They can be rethreaded in the field with carbide tools without heat treatment. Other repairs must be done in special shops.

Carburized steel is low carbon but surface-treated for great hardness, both inside and outside, after forming and threading. It is expensive but high performing and long lasting. It cannot be rethreaded or repaired in the field. Outside diameters range from 1 to 2 in. for use with the heaviest drills.

Percussion Bits

The bit is the essential part of a drill—it pounds against and disintegrates the rock. The success of a drilling operation depends on the ability of the bit to remain sharp under the impact of the drill against the rock. Many types and sizes of bits are available. Most bits are replaceable units fitted on the end of the drill steel (rod). Bits are available in various sizes, shapes, and hardnesses.

Design. Percussion rock bits chip or crush rock with hammer blows. Rotation of the steel changes the area of impact with each blow. Compressed air, or air and water, supplied through the center passage in the steel, and through one or more holes in the bit, blows the chipped and powdered rock away from the bit and out the drill hole. Bits are usually made of deep hardened alloy steel. They may have inserts of silicon carbide at impact areas.

Drilling rate
The total feet of hole drilled per hour by a drill.

Although bits with inserts are considerably more expensive than steel bits, the increased **drilling rate** and depth of hole obtained per bit provide an overall economy in drilling hard rock. Typical sizes for carbide-insert bits are from 1⅜ to 5 in. diameter.

Carbide-insert bits are available in four grades in order of increasing hardness (Table 15.9). Susceptibility to breakage increases with hardness. However, abrasion resistance also increases. If excessive bit breakage occurs using a specific grade, a softer grade should be tried.

TABLE 15.9 Carbide-Insert Bit Grades

Grade	Abrasion Resistance
Shock	Fair
Intermediate	Good
Wear	Excellent
Extra wear	Outstanding

The work of smashing the rock is usually done by four, but occasionally two or six, steel or insert bits, with ridge faces radial and perpendicular to the centerline of the steel. Two ridges (four wings or points) at right angles to each other are called a four-point cross, the most common construction (Fig. 15.14). If not at right angles, it is an X bit. A six-wing (six-point) design is called a rose bit.

The wings may be replaced by round buttons of carbide set in an almost continuous, slightly convex surface (Fig. 15.15). Button bits can yield faster penetration rates in a wide

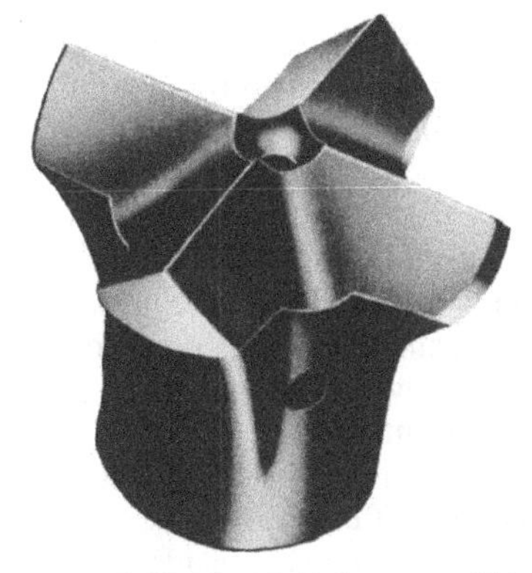

FIGURE 15.14 Multiple-use steel bit.

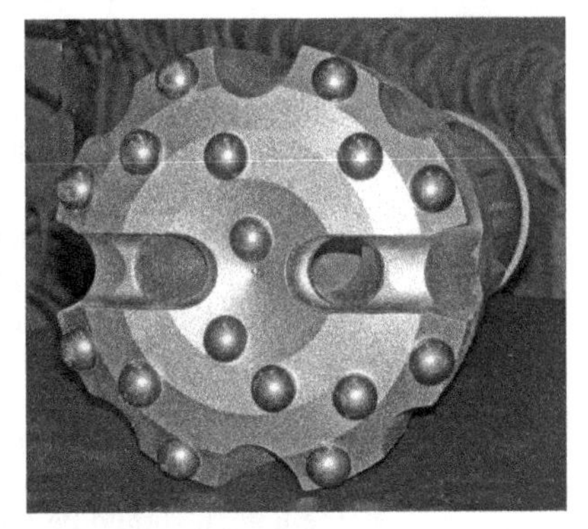

FIGURE 15.15 Button bit.

FIGURE 15.16 Worn button bit.

range of drilling applications. These bits are available in different cutting-face designs with a choice of insert grades. Most button bits are run to destruction and never reconditioned (Fig. 15.16).

The hollows between the wings are flutes. The width across the face of a pair of wings (the width of the bit) is the gauge or size. The bit is widest at the face and tapers to the back. The taper is called the gauge angle. It is usually 3 degrees in carbide bits.

Bits are selected according to hole size (the same as bit gauge), size and type of thread, type of rock, and hole footage to be expected.

Airholes. The number and placement of airholes in percussion bits varies. Small bits for light drills may have a single hole in the center of the face. This is the most effective position for removal of cuttings, but it is vulnerable to plugging. Downward movement presses it into the core of the rock, the spot the bit does not directly chip. This may inadvertently force soft or tough material into the hole.

There may be one to five holes in the volutes, or recesses, between the wings. In general, bits designed for soft and/or wet rock have more holes than those meant for hard, dry formations. A venturi bit has one or more holes to direct air back up the hole at an angle. Reverse airholes tend to break up mud collars either while they are forming or when the bit is being pulled out.

Smooth-face button bits may have a front hole somewhat off-center, and other holes in volutes are cut into the bit to provide cutting escape routes.

Carbide Inserts. The drilling speed and service life of bits can be greatly improved by carbide inserts. The material used is sintered tungsten carbide, a mixture of 9% to 12% cobalt (Co) and 88% to 91% tungsten (W), with a hardness of 87 to 91 on the Rockwell A scale. Fine powders of these materials are mixed in desired proportion and cemented into a solid by heat and pressure.

The harder inserts will give maximum footage between regrinds but may chip or break under certain conditions. The softer ones will wear more rapidly (but much more slowly than steel) but are less susceptible to breakage.

Excessive rotation speed and heavy down-pressure are likely to break or chip inserts of any grade. A carbide bit will cut faster than an all-steel bit of similar design and may drill 100 times as much footage before requiring reconditioning or replacement. Carbide bits are expensive, but under normal drilling conditions, they are more economical than steel bits.

Their use is particularly appropriate for deep holes, where their negligible taper saves the waste of drilling oversize at the top; in hard and abrasive rock, where taper, lost time, and expense of bit replacement are important; and where transportation and reconditioning of bits are difficult. In fissured rock, the large investment in carbide bits may be too risky because of the danger of their sticking in the formation.

Pieces of carbide lost in a hole as a result of chipping or falling out of a bit are extremely destructive to the rest of the bit and to other bits. If they cannot be blown or flushed out of the hole, it should be abandoned and a new hole drilled alongside.

Bit Action. With the down stroke of the drill piston, each edge of the bit is driven into the rock like a cold chisel struck with a hammer, a tiny fraction of a rotation from where it hit with the previous stroke. If the bit is sharp and rotation rapid enough, it will tend to cut off a wedge-shaped slice. If it is dull, it is more likely to crush the edge to powder to a shallower depth—a much less efficient operation.

The thickness of the slice depends on the speed of rotation, but rock properties, frequency of drill stroke, and size and sharpness of bit are also contributing factors influencing slice thickness. The depth of the cut is governed by the force of impact, plus down-pressure, through the drill string. Most drills have a fixed blow frequency at proper air pressure and full throttle. Rifle bar drills have a fixed rotation rate, so the operator can only control down-pressure. The pressure is not so strong with handheld hammers, but may be regulated up to a ton or more in drifters and downholes.

Too little pressure may reduce depth of chipping and, therefore, the rate of drilling. Too much down-pressure prevents the bit from getting back up on top of the chipped rock for a clean stroke and may reduce footage even more. In addition, it subjects the whole mechanism to damaging strain.

The depth of penetration for each blow is usually very slight. The wings follow each other closely, overlapping their work, and cut a round hole with smooth sides. In soft rock, however, the bit may penetrate so rapidly a wing will cut a spiral thread in the wall, which is not entirely removed by the following wing. This process, which is called rifling, results in a threaded, undersized hole, and as a result it may be difficult or impossible to remove the bit.

Rifling may be reduced or eliminated by supporting the drill to reduce down-pressure or by using dull bits, rose bits, or X bits. A dull bit spreads blows over too large a surface and does no chipping and very little crushing. In addition, it sets up a powerful dragging effect, which results in overtightening thread connections and greatly increases strain on the drill string and the rotation mechanism.

Efficient drilling is absolutely dependent on removing rock chips and dust as fast as they form.

Reconditioning. A few bits are designed to be used until worn beyond usefulness and then discarded. These include steel bits with surface hardening only and bits with shallow carbide inserts or buttons (see Fig. 15.16). One-use bits may have a pilot consisting of advanced cutting edges near the center. These facilitate starting a hole, cut more rapidly in some formations, and avoid the nuisance of rehandling. Most bits, however, are designed to be reconditioned several times.

As a bit is used, it wears along the cutting edges and on the sides of the wings. The sidewear is the more serious, as it soon reverses the taper so the edge becomes narrower than the trailing part of the bit. This causes the bit to drive into a hole of less diameter so the bit sticks. The degree of wear that will cause sticking varies with the type of rock.

Worn bits of the multiuse type can be reconditioned on special machines. Generally it is not economical to attempt to sharpen a bit by hand. A deep hardened bit can usually be ground several times. However, gauge size is reduced by reconditioning. The bit is cut back from the sides until the reverse taper is removed and also on the face to restore the edge.

A shallow hardened bit will be softened by deep grinding but may be hot-milled. It is heated, reshaped, and rehardened in a special forging machine. Many deep hardened bits will give better service if hot-milled rather than if ground.

Carbide bits may be reconditioned by hand or machine grinding, using a silicon carbide wheel. In soft, abrasive rock it is important to use bits with pronounced taper to avoid excessive sticking and possible loss. In fissured rock where sticking does not depend on bit design, the large investment in carbide bits is a risky proposition and is not recommended.

Hole Taper. As a bit wears, the cut hole will have a smaller diameter. In soft rock, wear may be slight. In hard or abrasive rock, the bit may wear badly with limited use. Holes taper smoothly with bit wear and in steps at bit changes. Whenever a bit is changed while drilling a hole, the new bit must be smaller to avoid sticking. If all new bits are used, the reduction must be ⅛ in. Used bits can be compared by setting their edges together.

In blasthole drilling the starting bit must be large enough to allow for the taper and still have a hole at the bottom for the quantity of explosive required.

Rotary Bits and Drive

The rotary bit cuts, chips, and/or grinds without hammering. A rotating table or head above ground supplies the rotation. This combination of rotating motion, down-pressure, and cleaning air is carried by hollow drill pipe with threaded connections. The bits are usually of the roller or tricone type with special adaptions to being cleaned by air instead of mud. Rotation of the bit against the bottom causes the cones to rotate, setting up a chipping and crushing action, with a variable amount of scraping and abrading.

In general, large teeth with wide spacing in each row are used for soft formations, as shown in Fig. 15.17, and tooth size is progressively reduced for harder drilling. Teeth are either steel or full carbide inserts. Steel teeth are surfaced with tungsten carbide or other hard facing. Full carbide inserts are used for the hardest and most abrasive formations.

FIGURE 15.17 Rotary tricone bit for soft rock.

In standard bits, the airstream comes from the bits behind the cones, tending to clean the hole bottom and blow the cuttings out of the drilled hole. This construction is best for all formations.

The jet bit directs the airstreams directly on the bottom. This avoids sandblasting the teeth and grinding with fine particles sucked into the airstream. Part of the stream of clean air may go through the bearings to cool and lubricate them.

The blasthole rotary drive is usually a rotary head fastened through a chuck to the top of the drill rod. It is turned by an air, hydraulic, or electric motor with variable speeds from 0 up to 150 rpm or more. The head is raised and lowered with the drill string by the feed mechanism.

Rigs built primarily for downhole drilling can usually be changed over for rotary bit applications. However, rotation may not be as fast or down-pressure as great as with machines built primarily for rotary drilling. Bit footage may be as little as 5 ft, but the usual expectation is from several hundred to several thousand. Service life is affected by the drillability of the formation, the design of the bit, and the manner of use and abuse.

Downhole Hammer (Down the Hole; DTH)

When drilling deep holes, it can often be more efficient to place the driving mechanism in the hole with the bit—this is termed a downhole hammer. Such a drill eliminates having to transmit the rotation and percussion forces through the drill steel. Typically these units are air-operated hammers (Fig. 15.18). They can be operated underwater by maintaining a higher air pressure in the hammer than the pressure outside. Air or water can be used to clear cuttings from the hole.

FIGURE 15.18 Cutaway of a downhole drill.

The drill is very slender in proportion to its weight and strength. It has no rotation mechanism in itself. Air is exhausted through the bit and carries chips to the surface, around the outside of the rods. Bit sizes vary from 2 to 42 in. They weigh from 13 to 175 lb. Strokes may be as short as ⅞ in. or as long as 5 in.

Blows per minute range from 500 to 2,700 in standard units and to 3,800 in high-pressure designs. Lubrication is supplied by heat-resistant drill oil added to the air from the compressor.

The drill rod above supplies air, keeps the hammer in contact with the rock at proper pressure, and causes hammer rotation. The rotation, usually at the rate of 15 or 25 rpm, is supplied by an air or hydraulic motor.

A carbide insert bit is held in the drill chuck by a split collar and a nut. In taking out the bit, the nut is loosened or "broken" by a hydraulic wrench and then spun off by the rotary mechanism.

Downhole drills are made to operate at the usual construction machinery air pressure of 100 psi. High-pressure models may run at 250 up to 500 psi. There is a tendency to increase pressures for greater efficiency. A pressure of 100 psi at the receiver usually means about 90 psi at the drill. Back pressure from the restricted openings in the bit and from the rising column of air and chips around the rods is likely to be about 20 psi, leaving an effective pressure of only 70 psi to operate the drill.

Higher pressure is usually accompanied by a decrease in stroke and an increase in frequency and drilling rate. The possible damage to the drill from higher operating pressures must be weighed against the increase in production.

The downhole drill is used for making holes in hard, medium, and soft rock (Table 15.10). Downhole drills have an advantage in softer rock where limiting the percussion action to only the bit assembly minimizes disturbance to the soft borehole walls and the potential for friction losses. Because none of its striking force is absorbed by the rod (steel for big and

TABLE 15.10 Drill Type Application

	Percussion Drills (Top Hammers)	Rotary Drills (Blasthole Drills)	Down Hole Hammer (Down the Hole, DTH)
Applications	Difficult terrain Straight holes	For weaker rocks Massive operations Many of holes	Practical for bad geology Long straight holes Constant penetration deep holes
Holes 1 to 6 in.	Very cost effective		
6 in. < Holes < 8 in.	Effective		
Large diameters		Effective to 15 in.	
Energy efficient	Good		Not very
Suitable	Very hard rock types		High flushing capability Low noise level

deep working drills is usually called rod or pipe), its working depth is limited only by the ability of the airstream to keep cuttings blown out of the hole and the capacity of the rotary bearings to carry the weight of a long string of rods.

The exhaust air rising from the bit should have a velocity of almost 3,000 fpm (feet per minute) to clean the hole. In rapid drilling with coarse chips 5,000 fpm may be needed.

When the hole is large in proportion to rod diameter, air velocity will be low because it has a wide passageway. It can be increased by using an auxiliary compressor, decreasing the size of the bit, or using larger-diameter rods.

A principal disadvantage is the danger of losing a whole drill as a result of a rock fall or formation of a mud collar. For this reason, use may not be prudent in badly fractured formations or in wet shale or other muddy strata.

The volume needed for blowing only can be worked out approximately by using the formula:

$$(\text{Hole diam.})^2 - (\text{rod diam.})^2 \times 16.4 = \text{cfm} \tag{15.9}$$

Hole Cleaning

As the bit cuts the rock into chips, sand, and dust, these particles must be removed promptly. Debris remaining at the bottom of the hole can form a packed layer and prevent the bit from striking the rock. Furthermore, slowing the circular flow of air will restrict important air pressure to drive the downhole bit and cut rock. Debris forming up-hole in a collar around the drill steel can inhibit the outflow of cuttings and cause excessive drilling friction and efficiency losses.

The basic method of chip removal is by a current of compressed air entering the hollow steel at the drill and emerging from the bit in holes at its front or bottom (one) and sides (none to four). These can be seen in Figs. 15.14 and 15.15. Some of the air is exhausted from below the piston (puff blowing) at moderate pressure and part is blown air admitted directly from the high-pressure chamber supplying the drill, by putting the throttle in BLOW position. In a properly balanced rig, the air for removing cuttings from the bottom of the hole has sufficient volume and velocity to carry the chips up and out of the hole.

> **Hole cleaning rule of thumb:** To clear the hole, a minimum of 5,000 fpm of up-hole velocity is needed. This is the speed at which air exits the hole.

Nevertheless, the actual amount of air required will vary based on the density of the material drilled and the size of the cuttings resulting from the type of bit used. The chips and coarse sand will pile up in a ring around the top of the hole, fine sand will go a little farther, and dust will drift with the wind. Chips and sand must occasionally be pushed back or removed from the edge of a deep hole, so as not to slide back into the hole.

Dust. Breathing crystalline silica dust can lead to silicosis, a deadly lung disease for which there is no known effective treatment. Consequently, it is important to prevent worker exposure to dust containing crystalline silica. Exposure to crystalline silica has also been linked to lung cancer, kidney disease, reduced lung function, and other disorders. Additionally, it contaminates exposed lubricant and lubricated surfaces, often turning them into grinding tools. OSHA has issued a Final Rule to protect workers from exposure to respirable crystalline silica.

On most construction projects, dust suppression is required by law. There are three principal methods to control dust in air exhausted from drill holes: (1) vacuum separation, (2) wet drilling, and (3) foam. Wet systems are efficient but may freeze in the winter.

Vacuum (Filter) Collectors. A hood may be placed over the drill hole to contain dust. The hood is sealed on its sides by dust curtains. They will have a gasket fitted around the rod (a rod wiper or dust seal) and be connected by flexible tubing to a container.

A vacuum pulls the air out of this through dry filters and/or cyclonic separators, leaving dust and chips behind (Fig. 15.19). The clean air is discharged to the atmosphere, and the rock dust is removed as often as necessary.

There may be considerable leakage of outside air into the hood, particularly on rough ground. This is not serious, as even a weak vacuum will remove the dust completely, leaving harmless coarse particles behind. The hood may depend principally on the drill rod for support on steep slopes.

The collector is usually built into a mobile tower drill and often into crawler drills. Skid- or wheel-mounted units are used for smaller crawlers and for hand drills. One collector may have several intake hoses to service a number of holes at the same time. Power may be gasoline or electric. There are also venturi systems for hand drills whose vacuum is created by the flow of compressed air.

A separate vacuum apparatus is suited to quarries with fairly regular work surfaces and drilling patterns, but can be a nuisance in rough and pioneer work. It permits efficient bit and hole cleaning by dry air, and this is a great advantage.

FIGURE 15.19 Vacuum and filter system for downhole drilling.

Dry systems require careful maintenance of the drill deck shroud. When changing filters, the use of respirators by workers is appropriate and may be necessary because of the exposure limits set by OSHA.

Water Injection. With a water injection system, a fine amount of water is sprayed into the airstream. These systems are more effective than vacuum collection, but the water can decrease the rate of drill penetration because of the increased density of the cutting around the bit. The material density increase will correspondingly require more air for chip removal. A water system has the added requirement of water tanks. The water turns part of the dust to a mud or slurry. This mud can produce mud collars, and these make pulling the bit difficult.

Dust Foam. Dust may be suppressed by dampening it at the bit. Water in a tank is mixed with a small amount of dust foam and fed slowly into the hole-cleaning air. The foam breaks the surface tension of the water, enabling it to coat dust particles thoroughly so the dust sticks together in little balls, and these are much too heavy to float in the air.

Unfortunately, the dampening slows the escape of the material from the face, so part of it is still there for the next blow. This regrinding absorbs power, weakens impact on the rock, and wears the bit. The loss in footage in damp drilling as compared with dry drilling may be 10% in hard rock.

The foam is necessary for efficiency. Without it, much more water is needed (wet drilling) and action is less uniform. In winter the foam–water mix needs antifreeze. Alcohol (methanol) is standard for this. Fuel oil, which may be kept in suspension by the detergent, is sometimes used.

Foam dampening is a widely used dust-suppression system. It requires a lesser volume of water than continuous liquid water spray. Any crawler drill can easily carry a tank of 30 to 50 gal capacity, enough for a full shift of drilling. Skid-, trailer-, or even wheelbarrow-mounted tanks can be used for smaller units. Aside from refilling, little maintenance is required. The principal disadvantage is reduced production with increased bit wear.

Noise

Percussion drilling is an inherently noisy operation. While it hammers metal against metal and metal against rock, the drill admits and ejects high-pressure air at a rapid frequency. Exhaust from air motors adds to the problem. There are methods of lowering the noise level noticeably. One means is to install mufflers, usually called silencers, on direct exhausts of pressure air from the drill piston, as in the tool seen in Fig. 15.20. These are usually simple devices, involving abrupt changes in direction of airflow. They are designed to cause minimum back pressure. On small drills it is a major problem to keep mufflers, small and light and sufficiently out of the way so as not to interfere seriously with the use of the tool.

Exhaust air coming out of the hole has a much lower velocity and is not an important noise source. The blow of a drill piston on a striker bar or shank is the most difficult noise to reduce. Some improvement can be obtained by reducing the resonance of the parts involved, as by a vibration-absorbing clamp. But the only way to substantially reduce it is to encase the entire drill in a sound-absorbent structure. This can create extreme problems for the operator and service technicians.

Bit-on-rock noise is diminished by depth, and is further reduced by foam dust suppression, as the bubbles are an effective absorber of noise. A downhole drill becomes much quieter as soon as it enters the hole and might be considered on noise-sensitive jobs, but a high-frequency rattling sound is common to downhole drills.

FIGURE 15.20 Muffler on hand-operated breaker. (*Courtesy of Ingersoll-Rand Company.*)

Percussion Rock Drills (Top Hammers)

Percussion drilling accomplishes disintegration of rock by hammer impacts to the bit. Most of the light, handheld breakers and rock drills (Fig. 15.21) do not rotate the drill steel like large rock drills. An anvil block may be used to transmit the blow of the piston to the steel. There are similarly air-driven pick hammers, chipping hammers, sheeting drivers, and backfill tampers.

FIGURE 15.21 Hand-operated air tools.

These drills may vary in size from handheld units such as jackhammers to large crawler-mounted rigs. Most pneumatic rock drills used in excavation are percussion tools. The bit is usually threaded onto the end of a hollow steel rod. Weights of these drills range from 10 to 500 lb. The light ones, up to 30 lb, are always hand-operated. Medium drills, up to 80 lb, are generally handheld but may be mounted on frames equipped with hand or power feeding mechanisms. Heavier models are almost always frame-mounted and power-fed.

Hand rock drills are called sinkers, hammers, and jackhammers.

Jackhammers. As the compressed air flows through a hammer, it causes a piston to reciprocate at a speed up to 2,200 blows per minute, which produces the hammer effect. The energy of this piston is transmitted to a bit through the drill steel. Air flows through a hole in the drill steel and bit to remove the cuttings from the hole and to cool the bit. The drill steel is rotated slightly following each blow so the cutting edges of the bit will not strike at the same place each time. Jackhammers are handheld, primarily air-operated, percussion-type drills; some are powered by electric motors, used primarily for drilling in a downward direction. They are classified according to their weight, such as 45 or 55 lb. A complete drilling unit consists of a hammer, drill steel, and bit.

Although jackhammers can be used to drill holes in excess of 20 ft deep, they are seldom used for holes exceeding 10 ft in depth. The heavier hammers will drill holes up to 2½ in. in diameter. Drill steel is usually supplied in 2-, 4-, 6-, and 8-ft lengths. Hammering chips, flakes, or crushes the rock; rotation gives the bit a fresh striking surfaces; and exhaust and air blown directly through the steel remove the cuttings. To minimize the vibration felt, manufacturers have patented spring-loaded designs, which allow the handle to move with the tool's movement instead of against it.

A 55-lb medium-weight drill is shown in Fig. 15.22. The three principal parts of the drill body are the upper end or back head, the cylinder, and the front head. These are machined

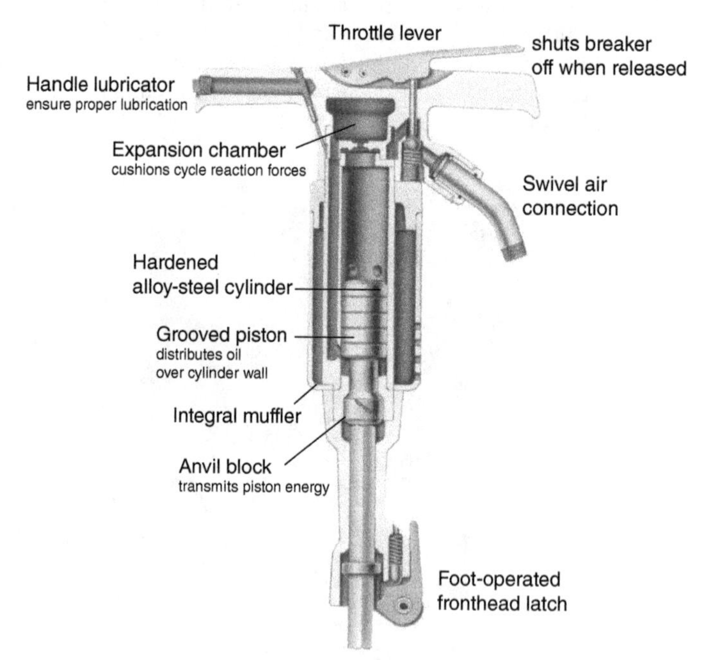

FIGURE 15.22 Hand-operated air hammer. (*Courtesy of Ingersoll-Rand Company.*)

to fit each other accurately without gaskets and are held together by a pair of alloy steel bolts, called assembly rods. The steel is held in position during work by the steel puller clamp. It is released by foot pressure on the projecting lever (Fig. 15.22).

Air reaches the drill through a flexible hose connected to a curved metal receiving tube with a swivel connection to the back head and then to the throttle valve. The throttle can be set in closed, wide-open, and several intermediate positions to regulate the speed of drill action. The piston moves rapidly up and down in the cylinder. On the downstroke its stem strikes the upper end of the drill steel. It is kept in alignment by the cylinder bore, rifle bar, stem bearing, and chuck.

As the piston completes the downstroke, it drives the steel down (or the drill up), until the collar reaches the steel puller. The piston is then forced up by air entering the bottom of the cylinder through the automatic valve located around the upper part of the rifle bar, and the weight of the drill causes it to slide down along the shank until the chuck rests on the collar again.

The shank must be of the correct length. A drill will not operate properly on a longer or shorter stroke than its design distance. Standard length is 3¼ in. from the collar to the end.

From the throttle, air goes through a passage in the back head, past the pawls on the rifle bar, into the valve, which directs it alternately to the top and bottom of the cylinder. Air exhausted from the top of the cylinder passes into the open air while the hammer is working. Exhaust from below the piston goes through a passage into the piston stem bearing and downward along the chuck driver splines into the space below the piston stem. Here it enters the hole in the center of the steel, which takes it to the bit, where it serves to blow rock chips out of the hole during drilling.

In deep holes or soft rock, the amount of air provided by this puff blowing may not remove all chips from the hole. All the air from the throttle can be turned down the steel by moving the blower or throttle valve to the BLOW position. A bypass in the back head can be connected to divert some air into the steel while leaving the bulk of it to run the hammer. This provides more cleaning in proportion to drilling.

Lubrication is provided by a reservoir around the piston stem bearing, which is filled through a plug in the side. This oil is forced by air pressure to points where it directly lubricates some surfaces or is carried by the airstream to others. However, it is safer to use a line oiler.

Power Feeds. Drills of 75 lb or more are usually supplied with supports and automatic feeds. A hand drill without handles, but having a pneumatic cylinder to support the drill and furnish thrust for light overhead drilling, is referred to as an air leg drill. These drills are used mainly for underground work. Air is admitted to the leg cylinder by a control valve, causing it to expand sufficiently to keep the bit in contact with the rock as the hole deepens. Such drills are referred to as stopers. The leg and the drill of a standard stoper make up a single straight-line unit. The operator places the bit and steel in the drill, places the pointed bottom of the leg (called the stinger), on the floor or in a wall crevice, and opens the valve to extend the feed leg until the bit is held against the overhead rock in the drilling position. During drilling, air is fed to the leg cylinder to keep an even pressure on the bit. When the steel is all in the hole or the leg is fully extended, the leg is retracted, pulling the steel out of the hole, and a longer steel is fixed to the drill.

Stoper operation is tricky. The point may slip out of position and even jump onto a foot (safety shoes are a must for stoper operators), the moderately heavy machine must be supported until the hole is collared. The hole is said to be collared when the bit has cut a sufficient depth that it will not bounce out. Collaring may prove difficult because of the strain of keeping the stinger down and handling the drill.

The stoper is intended primarily for overhead drilling, as in other positions its weight is more difficult to handle. In the case of holes inclined away from the vertical, the operator should stay on the uphill side, out from under the hole.

When the air feed leg is hinged to the drill, the machine is called a jackdrill or a jackleg drill. The leg arrangement permits convenient use in horizontal and angle drilling, as well as vertical. For an angled hole against a wall, a jackdrill may be set up with the leg and drill almost in line, so the feed supplies both support and push, or at a sharp angle where the leg supports and the operator pushes. The feed tends to push the steel against the top of the hole, but a slight down-pressure on the drill will keep it in line.

Mechanized drill jumbos have almost completely replaced the employment of jackdrills. Jumbos are operator-controlled mobile units with multiple drill arms. Rock tunneling involves drilling a number of holes in the face at the same time. The drills and their operators are supported on a moveable platform, called a jumbo. A variety of machines are available, including single or double platforms to support drill runners, and hand drills, with arrangements of columns and swivels to support either jackhammers or automatic-feed drills.

Hand Drilling

The rock drill can be carried by hand or truck to the work area and connected to a flexible air line from the compressor. A short starter steel, 1 ft to 30 in. in length, is fitted with a bit and clamped in the chuck. This bit should be the largest size that is to be used in the hole. The dirt and litter surrounding the area of the holes should be cleared away so there is no interference with starting the hole or the possibility of it sliding into the hole during drilling.

If the rock surface is horizontal, the drill is held vertically and the throttle partly opened. If the bit wanders on the surface instead of cutting it, the bit can be held in position with a foot pressed against the lower end of the steel. If the rock slopes, the steel should be held at a right angle to the slope instead of straight up and down, until a notch can be dug to hold it in position.

Once the hole is collared, the throttle can then be fully opened. The drill can be simply supported against leaning over, or if it tends to bounce, it can be pushed down with one or both hands. In soft rock it is not advisable to push down on the drill, as this increases any tendency toward rifling or binding. It is sometimes necessary to pull up slightly to avoid these difficulties. In hard rock, extra weight is helpful, particularly if the drill is light.

While the hole is shallow, the automatic puff blowing should keep the hole clean. However, it is good practice to occasionally turn the valve to BLOW. The amount of extra chips and dust raised will indicate whether puff blowing is clearing the hole.

Deepening. When the hole is cut so the drill is close to the ground, the throttle should be closed and the drill lifted until the bit and steel are out of the hole. The steel is then removed and another steel about 2 ft longer is locked by the chuck. If it is not worn much, the starting bit may be put on the second steel.

The bit is detached by holding the steel firmly and striking the bit upon any projection so as to unscrew it. The steel may be held under the foot and the bit struck with a hammer. Some operators prefer to use wrenches—either a pair of Stillsons or special wrenches—for changing bits. When installing a bit, it is hand-tightened and then tapped to seat it firmly.

As the hole deepens, it becomes increasingly necessary to blow out chips. Some hammers can be adjusted to send down an extra stream of air while drilling, which reduces the need to stop drilling to blow.

The thread is left-handed. If another bit is used, it should be slightly smaller than the starter. The easiest way to compare sizes is to put the two bits together. Drilling is resumed with the second steel. When the second steel is mostly in the hole, it is replaced with a longer steel, using the same or a smaller bit. This process is repeated until the hole is finished. Its depth is readily measured from the length of the steels.

If air and dust stop coming out of the hole, it should be blown immediately. If it will not blow, the steel should be pulled out. The trouble is usually a plugged bit. The cuttings can be removed with a thin punch, a nail, or an ice pick. An attempt to drill without a flow of air will overheat and destroy the bit and probably cause a collar of compacted cuttings to form just over it. A collar makes it very difficult to pull the bit out of the drill hole.

Whenever possible, holes should be drilled vertically. This puts the weight of the hammer on the bit instead of on the operator and simplifies keeping the hole straight.

Jamming. There are several conditions under which getting the steel and bit out of the hole is more of a problem than the drilling. Generally, the principal cause of sticking is the use of worn bits. In most types of ground, the bit gives a warning before it jams. This may be a slower penetration, producing fine rock dust instead of chips, twisting of the drill in the operator's hands, or stopping rotation of the steel. The condition of the hole may be roughly checked by raising the drill a few inches occasionally to see if it is free.

In hard or brittle rock, a worn bit may function satisfactorily, but when used in a soft formation, it will stick immediately. Worn bits are also dull bits, and penetration of the rock is slowed by their use. It is uneconomical to use a worn bit because excessive wear may make it impractical to recondition.

In soft rock a bit may rifle the hole. This trouble may be expected whenever the drilling rate is very rapid. This happens more often with new, sharp bits rather than with worn ones. Precautions that can be taken include use of special X or rose bits, using a lighter drill or partly supporting the weight of a heavy drill, grinding back the cutting edges of the bit, or reconditioning bits by grinding the sides to restore the taper without sharpening the edges.

Seams of dirt or finely disintegrated rock in a formation will cause plugging of the bit's airhole. At the same time there is rapid penetration and dirt falls on the top of the bit, where it may compact into a hard, tight collar.

When the speed of penetration increases suddenly, drilling should be stopped immediately and the hole blown. Drilling can then be resumed on partial throttle with frequent, full blowing. If air stops coming out of the drill hole, the steel should be pulled and the hole in the bit opened.

If the ground is composed of sloping layers of rock of varying hardness, or seams or cracks slope across the line of drilling, the hole will tend to drift down the slope. A curved hole will bind the steel.

Drilling on a dip at right angles to the seams should prevent this trouble. Often an intermediate direction will be satisfactory. In vertical drilling, curving can often be prevented by using light pressure and sharp bits and arranging for the drill to get maximum air pressure. This may involve cleaning the drill air filter, shortening and straightening the line from the compressor, or temporarily shutting down other units using air from the same source.

Pulling Steels. In taking a steel over 3 ft long out of a hole, the usual procedure is to unclamp the steel puller, set the hammer aside, and lift the steel by hand. However, if the steel is stuck, the hammer should be left attached and lifted, with the throttle alternately in drilling and blowing positions. The heavy vibration and rotation of the steel and the air pressure help to break the drill free.

Two people—one on each handle—can lift the drill much more effectively than one person. If the sticking is caused by spiraling (rifling), it may be possible to extract the steel by turning it against the direction of rotation while lifting. If two people are working, one person can lift the drill with the throttle closed while the other person turns the steel and drill with a wrench. If the operator is alone, remove the drill, place the wrench just under the shank collar, and pull the wrench upward while turning it. This may cause the steel to unscrew from the bit. If the hole is full depth, a steel bit can be sacrificed for economy's sake. If it is not full depth, leaving the bit in will make it necessary to drill another hole.

If the bit is carbide, every effort should be made to salvage it by direct pull. Usually the steel and bit can be raised together by jacking. No really suitable jack is generally available, but a chain jack or a regular bumper jack with a chain grab fastened to the lift hook can be used. The jack is placed on the edge of the hole, the end of a light chain given a double turn around the steel and hooked, and the chain pulled tight and caught in the jack head. The best grip is just under the collar, but a small wood wedge will generally enable it to hold anywhere on the steel.

This jack does not pull straight up, and the jack and steel will tip as the steel rises. This may cause bending of the steel or the jack if lifting is continued too long on one hold. It is a good to release the chain and fasten it again farther down before bending becomes severe.

A chain fastened to wagon drill feeds, loader buckets, bulldozer blades, cable control units, or any type of shovel or crane may be used for steel pulling. If all methods of extraction fail, a parallel hole is drilled nearby and the steel picked out of the loosened rock after the blast. It is generally bent by the explosion, and unless it lands on top, may be further twisted by a shovel.

Care of the Drill. A drill is built of special steels and is very finely machined. It, therefore, costs many times as much per pound as a shovel or dozer, and it deserves much better care than it usually receives.

The oil reservoir is built into the bottom of the cylinder and may require filling from one to four times a day, depending on the make and condition of the drill. This oil is forced or sucked into lubricating passages and into the airstream during work. The oil mist in the exhaust air gives an indication of the amount being used. A line oiler should be used instead of or in addition to the reservoir for best results, as it has a larger capacity and is more reliable.

The frequency with which either a reservoir or an oiler must be checked is best determined by experience. If exhaust air ceases to carry oil, the cause should be found immediately. Air drills have very slight clearances and experience a severe temperature drop. Air often enters them at a temperature of 100°F or more and is exhausted close to freezing. In addition, water condensing from the air or leaking from a water tube may wash oil off critical parts. Special oils designed for these conditions will give better service than standard lubricants.

Every possible precaution should be taken to keep the oil clean. This is a difficult matter when the air is full of rock dust. Dirt may clog oil or air passages or score sliding parts. Use of an inlet air filter may prevent trouble with scoring or clogging if it is cleaned frequently. Most filters are so small that only a small amount of debris can reduce the air passage enough to starve the drill.

A drill needs periodic cleaning, with the interval depending on the quality of air it receives. An oil pumping compressor, very hot air, or deteriorated hose may foul it in a day or less. With clean air, it may function well for weeks.

A quick cleanout can be done by disconnecting the air hose, pouring a cup of kerosene or fuel oil into the drill, reconnecting the air hose, and running the drill idle for about a minute.

The hose is again disconnected, a cup of rock drill oil poured in, and the drill operated again. If it is very dirty, several doses of a mixture of three parts kerosene and one part oil can be used. At longer intervals, it is a good practice to take the drill to the shop, disassemble, thoroughly clean, and check for worn parts.

Booms. Air drills and feeds of the type carried by crawler rigs may be mounted on hydraulically controlled booms (Fig. 15.23). Such booms can be mounted in any desired number on any type of base large and strong enough to carry them.

FIGURE 15.23 Drifter drill with hydraulic feed.

The boom is attached to its support by a universal or two-hinge bracket. A hoist ram raises and lowers the boom, and a side ram swings it right and left. The two controls can be used together to place it in a wide range of positions.

The drill tower or mast is fastened to the outer end of the boom by a double-hinge clamp connection called a cone. There may be a hydraulic ram to move the mast back and forth along this clamp to adjust the reach, and there may be hydraulic means to lower (dump) and tilt (roll) it. A compressor may be mounted on the same machine.

Drifters. The larger and heavier percussion drills are mounted either on a traveling carriage (tractor) (Fig. 15.24) or a frame. The combination drill and mount is known as a drifter. These tools can drill in the downward, horizontal, or upward direction. These drills are used extensively in rock excavation, mining, and tunneling. Either air or water can be used to remove the cuttings. Drifter drills are similar to jackhammers in operation, but they are larger and are used as mounted tools for downward, horizontal, or upward drilling.

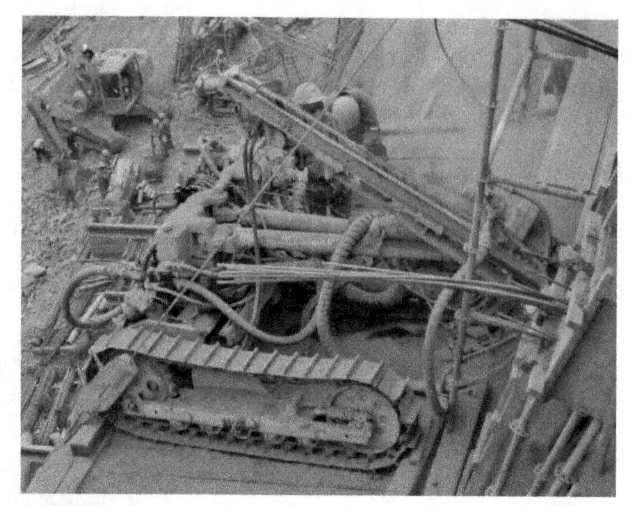

FIGURE 15.24 Drifter working from a platform—drilling is dusty work.

The drifter's weight is usually sufficient to supply the necessary feed pressure for downward drilling. But when used for horizontal or upward drilling, a hand-operated screw or a pneumatic or hydraulic piston supplies the feed pressure. They are generally limited to hole diameters of 5 in. or less and depths to 80 ft. The carriages for these drills are manufactured in many sizes and with different mast arrangements. Drills mounted on the largest carriages are capable of drilling to almost 200 ft.

The tracks are the type used on crawler tractors but they are narrower and lighter, usually 10 in. wide. Single grouser shoes are standard. These are lightweight machines with a low center of gravity; consequently, they are extremely maneuverable and can be winched up steep slopes or positioned on temporary platforms (Fig. 15.24).

Travel is termed tramming, a reference derived from mining. Maneuverability is excellent, with each track driven independently forward or backward. The brakes usually go on automatically when tramming controls are put in neutral. These brakes can (and must) be locked out of operation by a manual control if the machine is to be towed. Travel speeds are slow: 2 or 2½ mph. Speed while being towed should not exceed 5 mph.

Drill rotation may be powered by rifle bar, as described for the smaller drill; by rifle bar with double sets of pawls to permit reversing; or by a reversible air motor turning the chuck through reduction gears. Rifle bar rotation is simple and economical in terms of air, but it lacks flexibility.

Reverse rotation is used in unscrewing (breaking) steels and bits and in pulling back out of holes. The drill assembly includes a striking bar, which is a short piece of hollow steel. The upper end receives the blows of the piston; the lower end is threaded for coupling to the drill steel. It is held in line by a chuck bushing and other parts and in place by a threaded front head or chuck housing cap, which usually has lugs to allow loosening by striking with a hammer or other tool.

Blow air is carried by an air tube from the back head through the center of the piston (in the rifle bar, if there is one) and into the striking bar. The tube is a sliding fit in the bar. If it is loose, it will leak high-pressure air and cause "short stroking." If too tight, it will probably break.

The throttle controls for drilling, blowing, and rotating may be mounted on the back head, but are usually at an operator's station where other movements of the machine may also be controlled.

Travel. Crawler drills (see Fig. 15.24) are moved between jobs on trailers, or if distances are short, may be towed very slowly. For self-propelling (tramming), the operator walks behind or beside the machine within reach of the track controls. Usually there are two levers, one for each side. Moving a lever forward turns the track forward, while backward action moves the track back. An automatic brake goes on when the lever is centered.

Steering is done by varying the relative speed and/or forward-reverse direction of the two tracks. When they go in opposite directions, the machine will spin-turn in its own length. Very precise maneuvering and spotting are possible.

Tramming is preferably done with the mast folded back to a near-horizontal position for maximum stability. The rig usually pulls the compressor with it by means of a pintle hook or drawbar connection on any but the shortest moves. When moving independently, the operator must exercise care to avoid damaging the hose by stretching, kinking, or walking over it.

Setup. To set up for a single vertical hole, the operator should choose the most convenient and stable position for the machine. If ground conditions permit, drill in the area between the line of tracks and just forward.

Raise the boom to an angle of between 25 and 30 degrees, and set the mast vertical. Position the foot of the mast so the hole marker lines up with the steel centralizer either by swinging the boom and extending or retracting it (if it is extendible) or by raising or lowering it. Recheck the mast for vertical alignment, then slide it down until the foot rests solidly on the ground.

The crawlers give the machine a good grip on most types of ground and the brakes lock automatically, so the machine is sufficiently heavy to stay in place. If there is any question of its sliding even a fraction of an inch, block the tracks.

Starting the Hole. Using the feed motor move the drifter (drill) to the top of the mast. Make sure all threads on the steels, couplers, and striking bar are the same. Grease both threads on the first length of drill steel. Screw a good bit onto one end and a coupling on the other, finger tight. These are left-hand threads. Swing the centralizer arms open, then stand the steel (or rod) on its bit in the centralizer.

Move the drill down with the feed so the striking bar threads engage the rod coupling. With the rotation motor, turn the bar counterclockwise until it is firmly seated in the coupling. Be sure *not* to start hammering.

Lower the feed until the bit rests firmly on the ground. Close and latch the centralizer. Using the drifter's rotation motor, turn the striking bar counterclockwise until the rod and bit turn with it and the threads appear firmly seated.

Start the hammer action with the drill at partial throttle, continue rotation at moderate speed, and feed it down slowly. When the bit has penetrated any overburden and has collared the hole in rock, open the centralizer arms wide.

Drilling. Rotation should be fairly rapid when penetrating overburden or other loose material, but should be slow to moderate in rock. Fast rotation keeps giving the edges thicker slices of rock than they can break efficiently, thus slowing penetration while using more air. It is also likely to cause the drill string to whip and vibrate and break prematurely.

Pressure should hold the bit firmly against the bottom, but excessive pressure does not allow proper rebound from the stroke. A method to add more downforce on the drill string is to slightly lift the front tracks off the ground to transfer machine weight to the drill guide and shank. Proper alignment of drill steel must be maintained during this weight shift. Insufficient pressure on the drill string, which usually means partly supporting the weight of the string on the feed, prevents the bit from following through on the stroke, reducing penetration. Moreover, it will loosen threads and heat the bottom of the drifter.

A good stream of blow air is essential for cleaning the hole. It is the most important factor in penetrating overburden or soft rock, as these can readily plug the holes in the bit. If this happens, it will be necessary to pull the bit out for cleaning.

There should be a continuous stream of chips, sand, and dust coming out of the hole. When drilling with foam, the dust should be in little balls, but it should still come out. The volume and coarseness of this material give an indication of bit action. The positive indication of progress is how fast the rod goes down the hole. The drill cuttings are generally waste material and can be collected in plastic bags to cover and protect the open borehole from contamination after drilling.

Adding Steel. If the hole depth is greater than the length of the first rod, it will be necessary to add another rod. This is done when the striking bar coupling has almost reached the centralizer (drill steel support). Before raising the rod or rods in the hole, check the hole, then turn off the hammer, continue rotation slowly, and raise and lower the bit a foot or so to make sure it is not trapped by cuttings. Then turn off the rotation, lower the bit to the bottom, and alternately start and stop the hammer action of the drifter a few times to loosen the top-of-the-steel coupling threads. Lifting the string so it bounces may help. The ringing note of the steel changes when loosening occurs. It is important to stop rotation shutoff while hammering to loosen.

Worn threads place added wear on newer threads and can cause risk in equipment downtime. Greasing of threads is important in reducing wear and helping with the uncoupling of rods. Operate the drill rotation motor in reverse by turning the striking bar clockwise and unscrewing it from the coupling. Grease the striking bar threads, and raise it to the top of the mast. Grease the threads on another section of steel, turn a coupling onto one end by hand, and lift it to line up with the striking bar. Fit the bottom of the steel into the coupling on the piece still in the hole, and turn to catch the threads. Lower the striking bar so the threads enter the upper coupling.

Tighten the threads by slowly rotating counterclockwise. Blow air and rotate slightly, then start the hammer action after both threads are thoroughly seated. This operation is repeated for each added steel.

Withdrawing the Steel. When the hole is drilled to the required depth, stop the hammer action and turn the blow air on full while rotating the string slowly. Continue until the air from the hole is clean.

Stop the rotation and blowing, and loosen the threads as described in the previous section. Raise the string by running the last steel to the top of the mast. Next, close and latch the centralizer (support) arms, then lower the string until the coupling below the top steel section rests on the centralizer; this will support the weight of the string still in the hole.

> The centralizer must be latched with the coupling above it; otherwise, the string below can be lost in the hole.

Rotate the drill slowly in reverse (clockwise) and raise it until the striking bar thread is loosened from its coupling. Stop the rotation, and unscrew the steel by hand from its bottom coupling, which is resting on the centralizer. Remove the steel section.

Lower the striking bar until its thread engages the coupling on the centralizer. Rotate the bar to thread it loosely into the coupling to sufficiently support it. Stop the rotation and raise the string to take its weight off the centralizer. Open the centralizer arms. Run the striking bar up the mast until the next coupling is above the centralizer. Close and latch the centralizer, and lower the string until the coupling rests on the centralizer. Then repeat the steps for each section of steel in the hole. The steels should be laid out in a definite pattern. Their positions should be changed in the string on the next hole to equalize wear, and fatigue will be about equal throughout the set being used.

The bit is usually removed after the steel has been laid aside. It should be inspected for wear, chipping, and loss of inserts. If the bit is unscrewed over the hole, the hole should be covered by a board.

Problems. The drill string does not always come out easily. If the ground is wet or if water is being added to blow air to suppress dust, the drill chips and dust may combine with the water to build a mud collar above the bit (Fig. 15.25). Sometimes it is thick enough to just allow space for the rod, but a thickness of 1/16 in. may be enough to cause trouble.

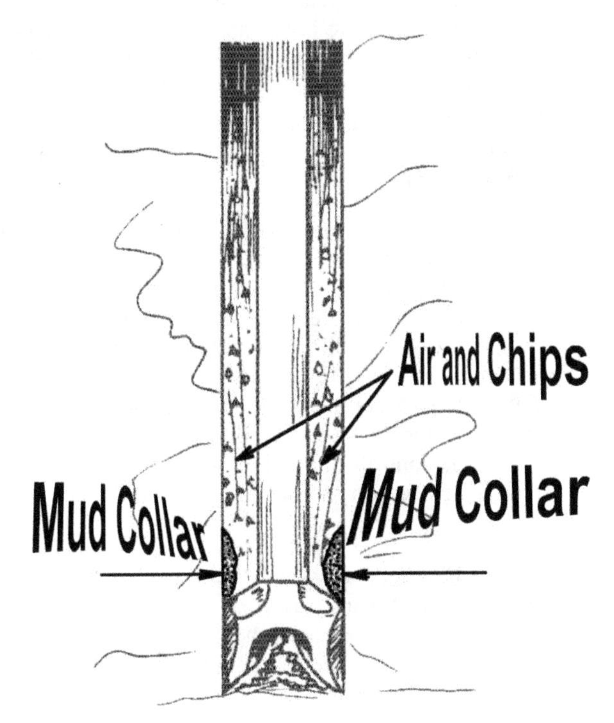

FIGURE 15.25 Mud collar.

The bit is tapered so it can follow the cutting edges without dragging and binding on the sides of the hole. However, if the hole becomes partly clogged, this taper can cause a wedging action between the bit and the hardened mud collar.

The operator can usually get through this resistance by a combination of force (the feed can usually lift a ton or more), rotation, and hammering, accompanied by maximum blowing of air. The skill of breaking a mud collar comes with experience.

The most important advice is to not get stuck. Collars usually form because of improper drilling. The most common error is not supplying sufficient blow air in proportion to the rate of penetration. A weak airstream allows an accumulation of material just above the bit. Another cause of sticking, more common in larger-diameter holes, is rock pieces falling out of the hole wall onto the top of the bit and forming a wedge. The operator will then have to struggle to overcome the resistance in much the same way as described for a mud collar.

The stream of air and cuttings out of the hole should be observed regularly, as changes in its character usually indicate the need for changes in the drilling technique. Finer grain and/or reduced output means harder rock or a dull or broken bit, requiring either another bit or increased pressure. Coarser grain and/or increased penetration rate means softer rock, with possible need for faster rotation and/or decreased pressure to reduce the danger of a mud collar. Chips of carbide detached from a bit will grind up the rest of the bit. Repeated blowing might get them out, or the hole may have to be abandoned. Visually inspecting the cuttings for metal flakes and inspecting the drill bit between boreholes is recommended.

Angle Drilling. Setup procedures for angle drilling are the same in principle as for vertical holes but are harder to visualize. The first hole may take extra time and planning. If the direction of the hole is above horizontal, the rods will have a tendency to slide out of the hole. The coupling is consequently kept on the opposite side of the centralizer from the drill while adding or withdrawing steels. If the angle is nearly horizontal, the blow air may travel along the top, leaving excessive drilling debris on the bottom. It is a good practice to pull the steel back toward the mouth of the hole occasionally, with full blow air, to drag this material toward the opening.

Alignment. Both the hole and the line of motion of the drill must be in close alignment. If the tracks of the drill move, the boom sinks, or the mast tilts during drilling, alignment will be lost, creating problems. Misalignment between the direction of the hole and the direction of the bit and steel feed puts a strain on the drill string and on the mast and the positioning mechanism. Drilling force will be lost by flexing of rods and friction against the sides of the hole. The life of both rods and couplings will be shortened by bending stresses. Failure may occur instantly. Misalignment is generally the fault of the operator.

When misalignment is detected promptly, the operator can reposition the mast by using the boom and cone controls or moving the whole machine. However, if the hole is crooked, it will probably be necessary to abandon the work and with care drill another hole closely alongside.

Lubrication. Chassis lubrication is conventional. There are a number of high-pressure grease fittings and a few gear case reservoirs. These need daily or weekly attention, according to type of use and the maintenance schedule supplied with the machine. The drifter and other air-operated units are lubricated by oil in the airstream. One recommendation for lubricator adjustment is to turn the needle valve handle to closed, then open it two turns. Then open the main air hose at the drifter, let air blow until it shows an oil mist, and then reconnect.

Visible mist is the standard and dependable indicator of the presence of oil in the air. But it does not tell how much. Correct proportion is indicated on many of these machines by a slow seepage of oil out of the bottom of the chuck housing and down the striker bar while the drill is operating. If this area is dry, there is not enough oil and the lubricator valve should be opened farther. Visually inspecting the cuttings for metal flakes and inspecting the drill bit between boreholes is recommended.

The oil tank must be full at the start of the shift, and it should be checked at least once during the shift. Many expensive components can be ruined by working briefly without oil.

Rotary Drills

With rotary drilling, the rock is ground away by applying a down pressure on the drill steel and bit and at the same time continuously rotating the bit in the hole. To remove the rock cuttings and cool the bit, compressed air is constantly forced down the drill steel and through the bit during this process. Rotary or blasthole drills are self-propelled. They can be mounted on a truck or on crawler tracks (Figs. 15.26 and 15.27). Rigs are available to drill holes of different diameters and to depths of approximately 300 ft. These drills are suitable for drilling soft to medium rock, such as hard dolomite and limestone, but are not suitable for drilling harder igneous rocks. Penetration rates can vary from 1½ ft/hr in dense,

FIGURE 15.26 Track-mounted rotary (blasthole) drill.

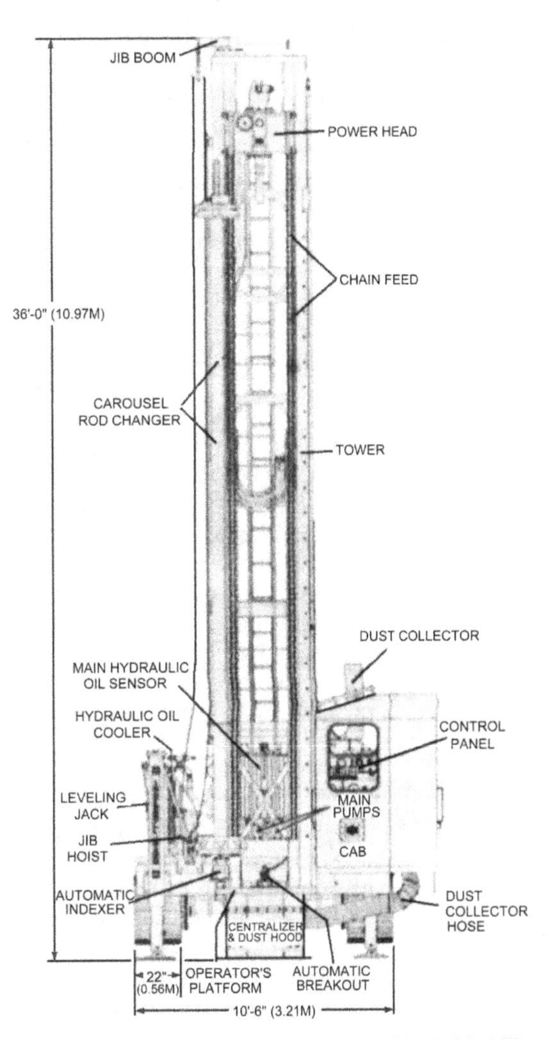

FIGURE 15.27 Carrier-mounted rotary (blasthole) drill.

hard dolomite to 50 ft/hr in limestone. The speed of drilling is regulated by pressure delivered through a hydraulic feed.

The mast on these rigs is often called a tower. It is hinged directly to the travel base, with hydraulic cylinders to raise the tower to working position and lower it for travel. Most units drill only with the tower vertically fixed to the carrier, but leveling mechanisms compensate for uneven ground or to drill at angles. Common tower heights are 30 to 60 ft, making it possible to use feed lengths of 20 to 55 ft.

The tower is usually at the rear of a truck carrier, which then becomes the "front" of the whole unit. This construction permits accurate spotting and enables the whole machine to keep farther away from dangerous edges. The mounting may be on a track (crawler) carrier with individually controlled tracks. When crawler tracks are driven separately by reversible air or hydraulic motors, they can maneuver accurately in limited space. However, these are not suited for rough or steep ground. With the tower down, maximum grade ability is about 30%. Travel speed is between 0.6 and 2.5 mph. Truck mounting is preferred where job locations are changed frequently, as in water well drilling.

The machine is lifted off the tracks and leveled before drilling by means of three or four hydraulic jacks separately controlled from the operator's station. There are usually two jacks at the tower end of the carrier and one or two at the rear. Check valves hold pressure in the jacks if the hydraulic system fails. The hydraulic pump may be driven mechanically or by an air motor. Digital bubble level gauges in the operator's cab measure tower vertical position. Older drill rigs have a bubble or pendulum level on the tower showing position from vertical.

The drill rod is ordinarily rotated from the top by an air, hydraulic, or electric motor. There may be two or three speeds, or infinitely variable rates from 0 to 80, 100, or higher rpm. The rotary head must include a swivel head through which compressed air enters the rod.

The feed mechanism may be an air motor and chain, a two-way hydraulic ram connected directly or operating a chain through a pulley, or a cable and drum crowd-retract with an electric drive for hoisting and lowering for feed control.

The feed keeps the bit in contact with the rock and provides proper drilling pressure (pulldown). This may be as much as 25 tons with these mobile machines. The feed is reversed to pull the rods and bit out of the hole. The tower has one side open for handling the rods. It has rod storage and a hydraulically controlled rod changer for adding rods to the drill string or for pulling out.

The air requirement is 250 to 1300 cfm, mostly for blow air. A compressor is usually mounted on the drill's tractor or truck, making it a completely self-contained unit. Small or special models may tow a compressor or use piped air from a stationary source. Power may be supplied by the carrier engine through a power takeoff, by a separate diesel or gasoline engine, or by electric motors.

The compressor or compressors may be the only power source, with all drill functions operated by air motors. Some machines are powered by the carrier engine through a power takeoff or by a separate diesel engine, with various combinations of air, hydraulic, and mechanical drive.

These rigs can handle rotary bits with diameters up to 15 in. and may also be used with downhole drills up to 9 in. in diameter. Augers are sometimes used, especially for soil borings. Models designed primarily for downhole equipment may have slower rotation and less down-pressure than those meant for rotary work.

Conventional drifter drills of 4½ or 5¼ in. may be used for drilling holes up to 4 in. in the conventional manner. Rotation may be provided by either the head or the drill. Depth is limited by the absorption of the hammer impact by a long drill steel. Efficiency starts to decline at 30 ft in hard rock and at somewhat greater depths in soft material.

Downhole and rotary drills are limited in depth by the ability of the rotary table bearings to carry the weight of long rods and by the ability of the airstream to keep the cuttings cleaned out of the hole. Specifications detail the compressor cfm, engine horsepower, torque, pulldown force, and many measures that size the correct drill for the job.

Drilling Soil

Various types of equipment are used to drill holes in soil. In the construction and mining industries, holes are drilled into earth for many purposes, including but not limited to

1. Obtaining soil samples for test purposes
2. Extracting oil and natural gas deposits
3. Locating and evaluating deposits of aggregate suitable for mining
4. Installing cast-in-place piles or shafts for structural support
5. Enabling the driving of load-bearing piles into hard and tough formations
6. Providing wells for supplies of water or for deep drainage purposes
7. Providing shafts for ventilating mines, tunnels, and other underground facilities
8. Providing horizontal holes through embankments, such as those for the installation of utility conduits

Sizes and Depths of Borings. Most holes drilled into soil are produced by augers attached to the lower end of a shaft called a "kelly bar." An external motor rotates this bar, which is supported by a truck, a tractor (Fig. 15.28), a skid mount, or a crane.

Augers have a helix or screw thread on drill rods. Drilling resistance and/or control of the rate of feed prevents the thread from penetrating in proportion to its turning speed, so the material cut by the bit is gripped by the threads and forced out of the hole by a screw conveyor action. The flights are made in sections proportional to the feed length of the drill unit. They are connected to each other by bolting or pinning.

The sizes of holes drilled into soil may vary from a few inches to more than 12 ft. Vertical holes are usually limited to a depth of 100 ft or less, while horizontal holes may extend over 250 ft. To increase the diameter of a hole at its bottom, an underreamer is attached to the lower end of the drill shaft. This mechanism enables

FIGURE 15.28 Auger-type drill mounter on an excavator boom.

the driller to gradually change the diameter of the cutter. The purpose of the enlargement is to increase the bearing area under a shaft-type concrete footing.

Boring Head. The boring heads on the auger are bits of the drag or fishtail type. These cut by rotary scraping. To meet different soil conditions, they are available in a variety of designs. There is usually an advanced center or pilot cutter. The teeth, called fingers (Fig. 15.29), are generally detachable. They may be set in a separate head or in the leading edge of the auger flight. Cutting edges are of steel hardened by various processes, or of tungsten carbide. Worn steel teeth may be built up with borium or carbide hardfacing. The head should be slightly larger than the auger flights so they will not bind in the hole.

FIGURE 15.29 Fingers on the auger boring head.

Most auger drill machines will handle a variety of boring head sizes. With any one machine, increasing the size of the hole decreases the speed of penetration and maximum depth obtainable, and limits the hardness of drillable material.

As a rule of thumb, maximum depth is inversely proportional to auger diameter: If a machine will drill down 50 ft with a 12-in. auger, it should not be expected to go much deeper than 25 ft with a 24-in. auger. Drilling rates vary with the power of the unit, the hole size, and the material being bored. Operator skill is also important.

Uses. Augers are primarily earth drills but can penetrate soft to medium rock such as shale, soft limestone, and sandstone. They are widely used for soil testing; prospecting for minerals; creating blastholes; placing pipes and conduits under fills, pavements, or obstructions; and for deep footings for buildings (Fig. 15.30) or bridges.

FIGURE 15.30 Auger drill for a building foundation shaft.

Casing A steel pipe used as a lining to prevent borehole collapse.

Unstable Soil. When drilling holes through unstable soils, such as mud, sand, or gravel containing water, it may be necessary to use a **casing**. Sometimes the shaft is drilled entirely through the unstable soil and then a temporary steel casing is installed in the hole to eliminate caving and groundwater entry. An alternative method is to add sections to the casing as drilling progresses until the hole is completed to the full depth of unstable soil. The remainder of the hole can in many cases then be completed without additional casing. When the hole is filled with concrete, the casing is pulled before the concrete sets.

This type of foundation has been used extensively in areas whose soils are subject to changes in moisture content to considerable depth. By placing the footings below the zone of moisture change, the effects of soil movements due to such changes are minimized.

Removal of Cuttings. Several methods are used to remove the cuttings from the holes in soils. One method for cutting removal is to attach a continuous auger to the drill head (right side of Fig. 15.30). The drill head is the actual cutting tool at the bottom of the drill stem. The auger extends from the drill head to above the ground surface. As the drill shaft and the auger rotate, the cuttings are forced to the top of the hole where they are removed and wasted. However, the depth of a hole for which this method can be used is limited by the diameter of the hole, the class of soil, and the moisture content of the soil.

Another method for removing the cuttings is to attach the drill head to only a section of the auger. When the auger section is filled with cuttings, it is raised above the surface of the ground and rotated rapidly in reverse to free it of the cuttings (see Fig. 15.29).

A third method of removing the cuttings is to use a combination drill head and cylindrical bucket (Fig. 15.31). The diameter of the bucket is the same as the diameter of the hole. As the bucket rotates, steel cutting blades attached to the bottom of the bucket force the cuttings up and into the bucket. When the bucket is filled, it is raised to the surface of the ground and emptied.

FIGURE 15.31 Auger–drill bucket being raised from a casing. The black cylinder piece above the auger is used to engage the casing and drive it into the hole.

A fourth method of removing the cuttings is to force air and water through the hollow kelly bar and drill shaft to the bottom of the hole and then upward around the drill shaft. The air or water carries the cuttings to the surface of the ground for disposal.

Drilling Methods and Production

Many factors affect the selection of drilling equipment. Among these are

1. Purpose of the holes, such as blasting, exploration, or grout injection
2. Nature of the terrain: Rough terrain may dictate track-mounted drills
3. Required depth of holes
4. Hardness of the rock or the density and cohesion of the soil
5. Extent to which the rock formation is broken or fractured

6. Size of the project (total linear amount of drilling)
7. Availability of water for drilling purposes: Lack of water favors dry drilling

For small-diameter, shallow blastholes, especially on rough terrain where larger drills cannot operate, it is usually necessary to use track-mounted drills or even jackhammers, though the production rates will be low and the costs higher. For blastholes up to about 6 in. in diameter and up to about 50 ft deep, where machines can operate, the choice may be track-mounted, rotary-percussion drills. With each addition of a drill rod necessary to reach a greater depth, penetration production is reduced. There is approximately a 20% reduction in penetration production with the addition of the second rod and another 20% loss with the addition of the third rod. With the addition of a fourth rod the reduction is roughly another 10%.

For drilling holes from 6 to 12 in. in diameter and from 50 to 300 ft deep, the rotary or blasthole drill is usually the best choice, but the type of rock affects the drilling method and bit selection.

The Drilling Pattern

In the case of holes to be loaded with explosives, the blast design (see Fig. 15 10) sets the drilling pattern. The pattern is the repeated distance between the drill holes in both directions, usually stated as "burden distance × spacing." This pattern will vary with the type of rock, the maximum permissible rock breakage size, and the depth of the blasted rock **face**. The blast design and drilling pattern in turn set the diameter of the hole, depth of the hole, and total linear footage of drilling requirements.

> **Face** The vertical height from the floor of a rock shelf to the level at which drilling is accomplished.

Drilling operations for rock excavation where the material will be used in an embankment fill must consider the project specifications concerning the maximum physical size of individual pieces placed in the fill. The blast design will be developed so as to produce rock sizes small enough to enable most of the blasted material to be handled by the excavator and/or to pass into the crusher opening without secondary blasting. Although meeting either condition is possible, the cost of excess drilling and greater amounts of explosives to produce such material can be high. The extra production cost may warrant a pattern yielding some oversized rocks. The oversized rocks will still have to be handled on an individual basis, possibly with a headache ball or dropping a fabricated steel cylinder from a loader bucket.

If small-diameter holes are spaced close together, better distribution of the explosives will result in a more uniform rock breakage. However, when the additional cost of drilling more holes (i.e., more drilling footage) exceeds the value of the benefits resulting from better breakage, the close spacing is not justified.

Large-diameter holes enable greater explosive loading per hole, making it possible to increase the spacing between holes and thereby reducing the number of holes and the cost of drilling. However, this decision can add to the cost of rock crushing operations. The goal is to seek a balance between blasting and crushing costs while meeting specifications.

Drilling Penetration Rates

The drill penetration rate will vary with a number of factors such as the type of drill and bit size, hardness of the rock, depth of holes, drilling pattern, terrain, and time lost because of sequencing other operations.

If not determined by actual field tests, the prediction of a penetration rate for estimating purposes is guided by technical data, but such estimates must be tempered by practical field experience—drilling is art as much as science.

Penetration rate is affected by specific rock properties:

- Hardness
- Texture
- Tenacity
- Formation

Hardness. A scientific definition of hardness is a measure of a material's resistance to localized plastic deformation. It is a characteristic of a material and not a fundamental physical property. Many hardness tests involve indentation, with hardness reported as resistance to scratching. The principal factors controlling rock hardness are porosity, grain size, and grain shape. In drilling practice, the term hardness is typically used in reference to the crystalline solid.

Friedrich Mohs (1773–1839)
A German mineralogist who in 1822 established a scale of hardness for classifying rocks.

The **Mohs** hardness classifications are based on the resistance of a smooth surface to abrasion—the ability of one mineral to scratch another—and rate hardness by a 10-point scale, with talc rated as 1, the softest, and diamond rated as 10, the hardest (Table 15.11). Mohs scale does not represent an actual measured hardness. If hardness is measured by instruments and a scale of 0 to 1,000 is used to classify different rocks, with diamond as 1,000, corundum, which is 9 on Mohs scale, would be rated at 250. The relationship of rock hardness using Mohs scale as an indicator and drill penetration rate is presented in Table 15.12.

TABLE 15.11 Mohs Scale for Rock Hardness

Rock	Moh Number	Scratch Test
Diamond	10	Will scratch glass
Schist	5	Knife
Granite	4	Knife
Limestone	3	Copper coin
Potash	2	Fingernail
Gypsum	2	Fingernail

TABLE 15.12 Relationship between Mohs Hardness Rating and Drill Penetration Rate

Hardness	Drilling Speed
1–2	Fast
3–4	Fast–medium
5	Medium
6–7	Slow–medium
8–9	Slow

The *Vickers* test, which yields a Vickers Hardness Number (VHN), is a more scientific method of stating hardness (Table 15.13). For the Vickers test, a pyramid-shaped diamond indenter with a 136-degree angle between opposite faces is pressed into the surface of a specimen with a force of from 2.2 to 110.2 lb. Both diagonals of the indent are measured with a microscope. The hardness value of this test is considered not to be load-dependent—an advantage in classifying materials.

Rocks are composed of mineral combinations. Thus, a Vickers Hardness Number Rock (VHNR) was developed to account for the hardness of individual minerals within the rock. The VHNR is a composite value determined by percentage, weighing each mineral's

TABLE 15.13 Comparison of Hardness Classifications

Mineral	Moh's Number	Vickers Hardness Number
Diamond	10	1,600
Corundum	9	400
Topaz	8	200
Quartz	7	100
Apatite	5	48
Fluorite	4	21
Calcite	3	9
Gypsum	2	3
Talc	1	1

TABLE 15.14 VHNR for a Sample of Gneiss

Mineral	Percentage	VHN		Contribution to Total Hardness
Quartz	30	1,060		318
Plagioclase	63	800		504
Amphibole	2	600		12
Biotite	5	110		6
			VHNR	840

hardness contribution to arrive at a single hardness value (Table 15.14). The VHNR provides a good parameter for measuring the life of drill bits.

Texture. The term texture in relation to rock refers to the grain structure—degree of crystallinity, grain size and shape, and geometric relationships between the grains. A loose-grained, structured rock (porous, cavities) can be drilled quickly. If the grains are large enough to be seen individually (granite), the rock will drill at a medium rate. Fine-grained rocks are drilled slowly.

Tenacity. The term tenacity refers to the ability of a substance to resist breakage. Terms such as brittle and malleable are used to describe rock tenacity, or breaking characteristics. The impact of rock tenacity on drilling rates is shown in Table 15.15.

TABLE 15.15 Effect of Rock Tenacity on Drilling Speed

Breaking Characteristics	Drilling Speed
Shatters	Fast
Brittle	Fast–medium
Shaving	Medium
Strong	Slow–medium
Malleable	Slow

Formation. The structure of the rock mass, its formation, affects drilling speed. Solid rock masses tend to drill quickly. If there are horizontal strata (layers), the rock should drill between a medium and fast rate. Rock with dipping planes drills at a slow to medium rate. Dipping planes also make it difficult to maintain drill hole alignment. All of these factors should be carefully considered when estimating drilling penetration rate without the benefit of actual field tests.

Historical drill penetration rates based on very broad rock-type classifications are shown in Table 15.16. These rates should be used only as an order-of-magnitude guide. Actual project estimates need to be based on the results of project site drilling tests.

TABLE 15.16 Order-of-Magnitude Drilling Production Rates

		Direct Penetration Rate		Estimated* Production Rate—Good Conditions	
Bit Size	Drill type Compressed Air	Granite (ft/hr)	Dolomite (ft/hr)	Granite (ft/hr)	Dolomite (ft/hr)
	Rotary-Percussion				
3½	750 cfm @ 100 psi	65	125	35	55
3½	900 cfm @ 100 psi	85	175	40	65
	Downhole				
4½	600 cfm @ 250 psi	70	110	45	75
6½	900 cfm @ 350 psi	100	185	65	90
	Rotary				
6¼	30,000 lb pulldown	NR	100	NR	65
6¾	40,000 lb pulldown	75	120	30	75
7-7/8	50,000 lb pulldown	95	150	45	85

NR—Not recommended.
*Estimated production rates are for ideal conditions, but they do account for all delays including blasting.

Estimating Drilling Production

The first step in estimating drilling production is to make an assumption about the type of equipment to employ. The type of rock to be drilled will guide this first assumption. Information useful in making such a decision is presented in Table 15.16. Nevertheless, the final decision on type of equipment should only be made after test drilling the specific formation. The drilling test should yield data on penetration rate based on bit size and type. Once a drill type and bit are selected, the format given in Fig. 15.32 can be used to estimate production.

(1) Depth of hole:	(a) ________	ft face,	(b) ________	ft drill
(2) Penetration rate:	________	ft/ min		
(3) Drilling time:	________	min	(1b)/(2)	
(4) Change steel:	________	min		
(5) Blow hole:	________	min		
(6) Move to next hole:	________	min		
(7) Align steel:	________	min		
(8) Change bit:	________	min		
(9) Total time:	________	min		
(10) Operating rate:	________	ft/ min	(1b)/(9)	
(11) Production efficiency:	________	min/ hr		
(12) Hourly production:	________	ft/hr	(11) × (10)	

FIGURE 15.32 Format for estimating drilling production.

Total Depth of Hole. Usually when drilling for blasting purposes, it is necessary to subdrill below the desired finish grade of the excavation. This is because when explosives are fired in blastholes, rock breakage is not normally achieved to the full bottom depth of the hole. This extra drilling depth is dependent on the blast design. Blasthole drilling is influenced by hole diameter, hole spacing, pounds of explosive per cubic yard of rock, and firing sequence. Normally 2 or 3 ft of extra depth (subdrilling) is required. For example, even though the depth to finish grade is 25 ft (see Fig. 15.32, item 1a), it may be necessary to actually drill 28 ft (see Fig. 15.32, item 1b).

Penetration Rate. Penetration rate (item 2 in Fig. 15.32) is the rate at which the drill penetrates the rock. This is usually developed by drilling test holes and is based on a specific bit size and type. In an attempt to make drilling penetration rate estimating more scientific, a drilling rate index (DRI) has been developed in Europe. DRI is an indirect method of predicting drillability. It is based on two laboratory tests: the brittleness value (S20) and a Siever J-value (SJ). The S20 is the percentage by weight of rock, from the original sample, that passes through a 11.2-mm screen after pounding with a 14-kg impactor 20 times. The original sample is composed of crushed rock passing the 16-mm screen and retained on the 11.2-mm screen. The SJ is determined by miniature drilling with a certain bit geometry, bit weight, and number of rotations to measure depth of penetration.

As general guidance, a DRI of 65 indicates good drillability and a value of 37 indicates poor drillability. For a DRI of 65, tests using standard rotary drills yield an average penetration rate of 15.3 in./min ± 15 in., and, for a DRI of 37, the average penetration rate is 9.8 in./min ± 10 in. The results between those two values are linear. Tests with percussion drills vary by manufacturer. As an example, for a DRI of 65, one manufacturer's drill might penetrate at a rate of 36 in./min, whereas a second manufacturer's drill might achieve 75 in./min. Therefore, based on the DRI there can be no general statement about penetration rates for percussion drills.

Drilling Time. With knowledge of both the depth of the hole and drilling rate, it is possible to calculate the time required for the drill to penetrate the rock (item 3 in Fig. 15.32).

Fixed Time. Fixed drilling time consists of changing steel (adding drill steel and pullback and uncoupling of steel), blowing or cleaning the hole, moving the drill, and aligning the steel over the next hole (Table 15.17).

Changing Steel. If the drilling depth is greater than the length of the drill steel, it will be necessary to add steel during the drilling process and to remove steel when coming out of the hole (item 4 in Fig. 15.32). For track-mounted percussion drills, the two

TABLE 15.17 Fixed Drilling Times

	Equipment and Site Condition			
Operation	Percussion Drilling, Clean Bench (min)	Percussion Drilling, Uneven Bench/Terrain (min)	Downhole Drilling, Even Bench (min)	Rotary Drilling, Even Bench (min)
Add one steel	0.4	0.4	2.2	2.0
Pull one steel	0.6	0.6	2.5	2.8
Pull last steel	—	—	0.6	1.0
Move	1.4	2.2–2.9	6.0	7.0
Align				2.0

TABLE 15.18 Average Weights for Drill Steel

Size (in.)	Length (ft)	Weight (lb)
1.50	10	60
1.50	12	72
1.75	10	82
1.75	12	98

standard steel lengths are 10 ft and 12 ft. Average weights for these steel lengths are given in Table 15.18. A driller needs about 0.5 min or less to add or remove a length of steel. The process is semi-automated and repetitive.

The single-pass capability (length of steel) of rotary drills varies considerably, in the range of 20 to 60 ft. The dimensions (diameter and length) of the steel increase with drill hole diameter. Nearly all large rigs have mechanized steel handling, and the time to change a piece of steel is approximately constant for all diameters but varies with length. One study of the time to add and remove drill rods on rotary drills using 20-ft lengths of steel found it took an average of 1.1 min to add and 1.5 min to remove the rods.

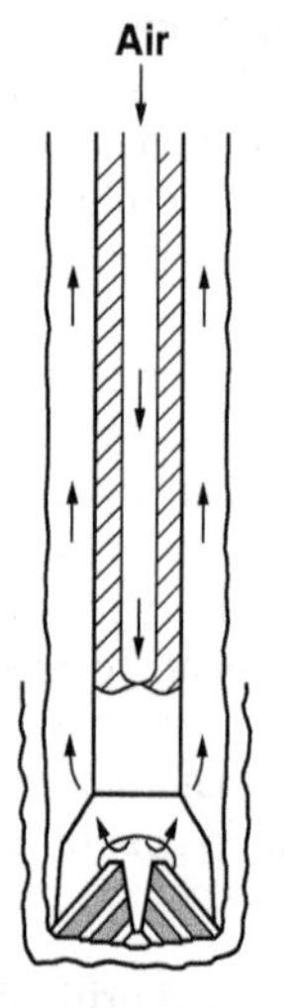

Blow Hole. After the grinding of the rock is completed, it is good practice to blow the hole (item 5 in Fig. 15.32) to ensure all cuttings are removed. However, some drillers prefer to simply drill an extra foot and pull the drill steel without blowing the hole clean.

Move the Drill. The time required to move (item 6 in Fig. 15.32) between drill hole locations is a function of the distance (blasting pattern) and terrain. Small track-mounted percussion drills can move at only 1 to 3 mph. Track-mounted rotary drills with their high masts can move at a maximum speed of about 2 mph. Remember, hole spacing is often less than 20 ft, and the operator is maneuvering to place the drill over an exact spot, so travel speed is slow. Some drill rigs are equipped with global positioning system (GPS) technology. With GPS the operator can accurately place the drill over the prescribed hole location. If a high-mast drill must traverse unlevel ground, it may be necessary to lower the mast before moving and then raise it again when the move is accomplished. This will significantly lengthen the movement time.

Align. Once over the drilling location, the mast or steel must be aligned (item 7 in Fig. 15.32). In the case of a large high-mast drill, the entire machine is leveled using hydraulic jacks. This usually takes about 1 min. A surface guide mounted at the bottom rod changer adjacent to the bit is depressed into the surface to prevent horizontal movement of the drill string.

Couplings Short steel pieces of pipe with interior threads (see Fig. 15.11). Couplings are used to hold pieces of drill steel together or to the shank. The percussion energy is transferred through the mated steel, not the coupling; therefore, the coupling must allow the two pieces of drill steel to butt together.

Change the Bit. Drill bits, steel, and **couplings** are high-wear items, and the time required to replace or change each affects drilling production. High VHNR values (VHNR ≥800) indicate high abrasiveness and, therefore, increased bit wear, whereas low values (VHNR ≤300) mean prolonged bit life. A time allowance must be made for changing bits, shanks, couplings, and steel (item 8 in Fig. 15.32). Table 15.19 gives the average life of these high-wear items based on drill footage and type on rock. The data in Table 15.19 also enable calculation of bit change frequency.

Efficiency. Finally, as with all production estimating, the effect of job and management factors must be taken into account (item 11 in Fig. 15.32). Based on drilling operation studies in Australia, the portion of drilling time when the bit is actually drilling seems to be from 70% to 75% of the total machine time. Bit activities in some cases required as much as 23% of the total time, and delays for maintenance, breakdown, and survey consumed the remaining.

Experienced drillers working on large projects with good equipment should be able to achieve a 45-min production hour. But as Australian studies found, good

TABLE 15.19a Igneous Rock: Average Life for Drill Bits and Steel in Feet

Drill Bits		Igneous Rock				
		High Silica LA <20 (Rhyolite)	High Silica 20< LA <50 (Granite)	Medium Silica LA < 50 (Granite)	Low Silica LA <20 (Basalt)	Low Silica LA >20 (Diabase)
(in.)	Type	(ft)	(ft)	(ft)	(ft)	(ft)
3	B	250	500	750	750	1,000
3	STD	NR	NR	NR	NR	750
3½	STD	NR	NR	NR	750	1,500
3½	HD	200	575	1,000	1,400	2,000
3½	B	550	1,200	2,500	2,700	3,200
4	B	750	1,500	2,800	3,000	3,500
Rotary Bits						
5	ST	NR	NR	NR	NR	NR
5-7/8	ST	NR	NR	NR	NR	NR
6¼	ST	NR	NR	NR	NR	NR
6¾	ST	NR	NR	NR	NR	800
6¾	CB	NR	NR	1,500	2,000	4,000
7-7/8	CB	NR	1,700	2,400	3,500	6,000
Downhole Bits						
6½	B	500	1,000	1,800	2,200	3,000
Drill Steel						
Shanks		2,500	4,500	5,800	5,850	6,000
Couplings		700	700	800	950	1,100
Steel	10 ft	1,450	1,500	1,600	1,650	2,200
Steel	12 ft	2,200	2,600	3,000	3,500	5,000
5 in.	20 ft	25,000	52,000	60,000	75,000	100,000

B = button, CB = carbide button, HD = heavy duty, ST = steel tooth, STD = standard, NR = not recommended, LA = The LA abrasion test, which is widely used as an indicator of the relative quality or competence of mineral aggregates.

equipment and the proper bits are critical to realizing good production. If the situation is sporadic drilling with qualified people, a 40-min or lower production hour might be more appropriate. The estimator needs to consider specific project requirements and the skill of the available labor pool before deciding on the appropriate production efficiency. It is common on many large-scale drilling operations to have two drill rigs work in tandem to meet production targets.

Sample Calculation: A project utilizing experienced drillers will require drilling and blasting of high-silica, fine-grained sandstone. Field drilling tests achieved a penetration rate of 120 ft/hr with a 3½ HD bit on a rotary-percussion drill operating at 100 psi. The drills to be used take 10-ft steel. The blasting pattern will be a 10- × 10-ft grid with 2 ft of subdrilling required. On the average, the specified finish grade is 16 ft below the existing ground surface. Determine the drilling production based on a 45-min hour.

Using the format in Fig. 15.32:

(1) Depth of hole	(a) 16 ft face	(b) 18 ft drill (16 ft + 2 ft)
(2) Penetration:	2.00 ft/min	(120 ft ÷ 60 min)

TABLE 15.19b Metamorphic Rock: Average Life for Drill Bits and Steel in Feet

Drill Bits (in.)	Type	Metamorphic Rock: High Silica LA <35 (Quartzite) (ft)	Medium Silica Low Mica (Schist) (Gneiss) (ft)	Medium Silica High Mica (Schist) (Gneiss) (ft)	Medium Silica LA <25 (Metalatite) (ft)	Low Silica LA >45 (Marble) (ft)
3	B	200	1,200	1,500	800	1,300
3	STD	NR	800	900	400	850
3½	STD	NR	1,300	1,700	850	1,600
3½	HD	NR	1,800	2,200	1,200	2,100
3½	B	450	3,000	3,500	2,000	3,300
4	B	600	3,300	3,800	2,300	3,700
Rotary Bits						
5	ST	NR	NR	NR	NR	NR
5-7/8	ST	NR	NR	NR	NR	1,200
6¼	ST	NR	NR	NR	NR	2,000
6¾	ST	NR	NR	750	NR	4,500
6¾	CB	NR	3,700	4,200	1,200	9,000
7-7/8	CB	NR	5,500	6,500	2,200	13,000
Downhole Bits						
6½	B	500	2,700	3,200	1,500	4,500
Drill Steel						
Shanks		5,000	5,700	6,200	5,550	5,800
Couplings		900	1,000	1,200	750	800
Steel	10 ft	1,700	2,100	2,300	1,500	1,600
Steel	12 ft	3,000	3,300	3,800	2,800	3,000
5 in.	20 ft	50,000	90,000	100,000	85,000	175,000

B = button, CB = carbide button, HD = heavy duty, ST = steel tooth, STD = standard, NR = not recommended, LA = The LA abrasion test, which is widely used as an indicator of the relative quality or competence of mineral aggregates.

(3)	Drilling time:	9.00 min	(18 ft ÷ 2 ft/min)
(4)	Change steel:	1.00 min	(1 add & 1 remove at 0.5 ea.)
(5)	Blow hole:	0.18 min	(about 0.1 min per 10 ft of hole)
(6)	Move 10 ft:	0.45 min	(10 ft at ¼ mph)
(7)	Align steel:	0.50 min	(not a high-mast drill)
(8)	Change bit:	0.09 min	$4 \text{ min} \times \left(\frac{18 \text{ ft per hole}}{850\text{-ft life (Table 15.19c)}}\right)$
(9)	Total time:	11.22 min	
(10)	Operating rate:	1.60 ft/min	(18 ft ÷ 11.22 min)
(11)	Production efficiency:	45 min/hr	
(12)	Hourly production:	72.0 ft/hr	[(45 min/hr) × (1.60 ft/min)]

High-Wear Items. Table 15.19 provides average values of expected life of the high-wear items—the bit, shank, couplings, and drill steel (see Fig. 15.11). A shank is the short piece of steel fixed to the drill chuck and transmits the drill's impact energy to the drill steel. Couplings are used to connect the sections of drill steel together.

TABLE 15.19c Sedimentary Rock: Average Life for Drill Bits and Steel in Feet

Drill Bits		Sedimentary Rock				
(in.)	Type	High Silica Silica Fine Grain (Sandstone) (ft)	Medium Silica Coarse Grain (Sandstone) (ft)	Low Silica Fine Grain (Dolomite) (ft)	Low Silica Fine-Med. Grain (Shale) (ft)	Low Silica Coarse Grain (Conglomerate) (ft)
3	B	800	1,200	1,300	2,000	1,800
3	STD	NR	850	900	1,500	1,200
3½	STD	NR	1,500	1,800	3,000	2,500
3½	HD	850	2,000	2,200	3,500	3,000
3½	B	2,000	3,100	3,500	4,500	4,000
4	B	2,500	3,500	2,000	5,000	4,800
Rotary Bits						
5	ST	NR	1,000	NR	8,000	6,000
5-7/8	ST	NR	2,500	NR	15,000	13,000
6¼	ST	NR	4,000	4,000	18,000	14,000
6¾	ST	500	6,000	8,000	20,000	15,000
6¾	CB	2,000	8,000	10,000	25,000	20,000
7-7/8	CB	3,000	10,000	15,000	25,000	20,000
Downhole Bits						
6½	B	2,500	3,500	5,500	7,500	6,000
Drill Steel						
Shanks		5,000	5,500	6,000	7,000	6,500
Couplings		1,000	1,200	1,500	2,000	1,750
Steel	10 ft	2,000	2,300	2,500	4,000	3,500
Steel	12 ft	4,500	5,000	6,000	7,500	7,000
5 in.	20 ft	65,000	250,000	200,000	300,000	250,000

B = button, CB = carbide button, HD = heavy duty, ST = steel tooth, STD = standard, NR = not recommended.

Sample Calculation: Assuming project conditions and drilling equipment are as described in the preceding Sample Calculation, what is the expected life, in number of holes completed, for each of the high-wear drilling items?

For an average hole depth of 18 ft, the following number of holes can be completed per each replacement.

High-Wear Item	Average Life (ft) (Table 15.19c)	Number of 18-ft Holes in High-Silica, Fine-Grained Sandstone
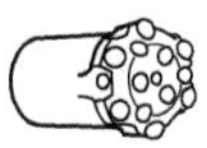 3½ HD bit	850	850/18 = 47
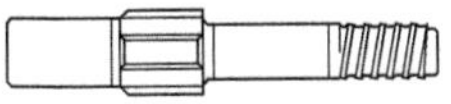 shanks	5,000	5,000/18 = 278
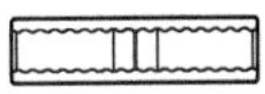 couplings	1,000	1,000/18 = 56
 steel	2,000	2,000/18 = 111

Rock Production. Drilling for blasting is only one part of the process of excavating rock; thus, when analyzing drilling production, it is good to consider the cost and output in terms of cubic yards of rock excavated. With the 10- × 10-ft pattern and 18 ft of drilling used in the previous Sample Calculation, the rock yield would be 59.3 cy.

$$\frac{10\text{ ft}\times 10\text{ ft}\times 16\text{ ft}}{27\text{ cf/cy}} = 59.3\text{ cy}$$

Though the drilling depth was 18 ft, the excavation depth is only 16 ft. Therefore, each foot of drill hole produces 3.3 cy of bank measure (bcy) rock.

$$\frac{59.3\text{ cy}}{18\text{ ft}} = 3.3\text{ cy/ft}$$

If the hourly drilling production is 72 ft, then the rock production is 238 cy (72 ft × 3.3 cy/ft). This should be matched to the blasting production and to the loading and hauling production. For example, if the loading and hauling capability is 500 cy/hr, it will be necessary to employ two drills.

In calculating cost, it is good practice to make the analysis in terms of both feet of hole drilled and cubic yards of rock produced. Considering only the high-wear items of the two Sample Calculations, if bits cost $200 each, shanks $105, couplings $50, and a 10-ft steel $210, what is the cost per cubic yard of rock produced?

Bits	$200 ÷ 850 ft = $0.235/ft
Shanks	$105 ÷ 5,000 ft = $0.021/ft
Couplings	(2 × $50) ÷ 1,000 ft = $0.100/ft
Steel	(2 × $210) ÷ 2,000 ft = $0.210/ft
	$0.566/ft

$$\text{or } \frac{\$0.566\text{ per ft}}{3.3\text{ cy per ft}} = \$0.172\text{ per cy}$$

Understanding Cost. Penetration rate, feet drilled per hour, increases with greater downward force on the bit. It will also increase with increased bit rotational speed. But penetration rate increases driven by these two factors will reduce bit life. Figure 15.33 is a graphic

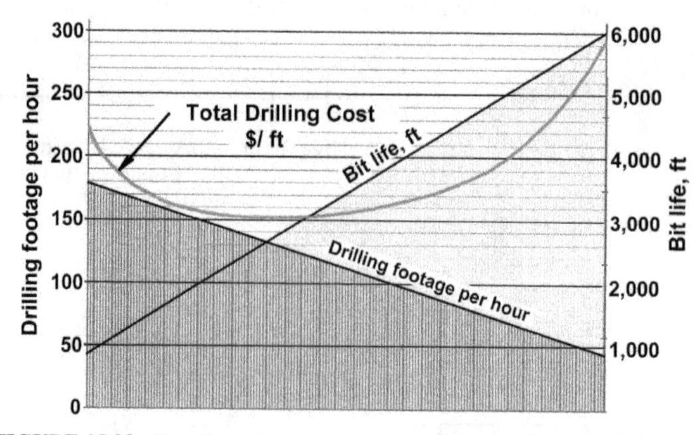

FIGURE 15.33 Relationships between penetration rate ft/hr, and drilling cost.

of the relationships between penetration rate, feet per hour, bit life, total feet drilled, and cost per foot of drilled hole. The lowest cost per unit produced ($/ft) is found at an optimum drill rate, not at the highest penetration rate.

GPS and Computer Monitoring Systems

Current technology for monitoring drilling and blasting operations is progressing from using external and somewhat subjective measurement methods of performance (i.e., time sheets and foreman- or operator-generated drilling footage reports) to onboard data acquisition and management systems. Drills with GPS systems for positioning the drilling bit on the exact borehole location are becoming more prevalent. These systems reduce the survey work required for laying out drilling patterns. The drill pattern design data, with hole attributes of depth and angle, are loaded into the drill's onboard computer system; then based on the GPS positioning information, the drilling position is displayed on a liquid crystal display (LCD) screen, enabling the operator to position the drill without the aid of stakes or markings on the ground. There are similarly electronic sensor systems to provide production-monitoring capabilities, including the recording and displaying of penetration rate, rotary torque and speed, bailing air pressure and volume, pulldown pressure, and hole depth. The bailing air pressure and volume are important to productivity and bit life, as the pressure and volume of air are critical elements for cooling the bit and clearing cuttings from the face of the bit to prevent regrinding of material. These monitoring systems can provide detailed knowledge about the relative hardness of the strata being drilled, and strata information can be imported into blast design programs to improve explosive loading calculations.

Most of these computer-based monitoring systems are being developed and used in the mining industry. But the technology has made its way into the construction industry as well. Mining industry studies of online drill-monitoring capability, which furnishes detailed information on rock quality and fracture zones, indicate such data can provide a 20% to 25% saving in blasting cost. Accurate rock property data obtained by computerized drill sensors enable a computer-based explosives truck to generate loading designs on the bench. The onboard computer in the explosives truck can formulate and control the loading of the blasthole with the exact quality and type of explosive within minutes after the drill has moved to the next blasthole.

CHAPTER 16
BLASTING

When there is a requirement to remove rock, blasting should be considered, as it is almost always more economical than mechanical excavation. In the blasting of rock, breakage results from the sustained gas pressure buildup in the borehole caused by the explosion. Every blast must be designed to meet the conditions of the rock formation and the desired final product. The blast design will affect such considerations as type of equipment and the bucket fill factor of the excavators used to move the broken rock. The resulting size of blasted rock must also be considered for hauling and crushing operations. Blasting safety depends on careful planning and faithful observation of proper blasting practices.

PURPOSES OF ROCK EXCAVATION

For the purposes of this discussion, rock is defined as material that requires loosening by explosives in order to be dug economically by machinery. Surface excavation of rock is done chiefly for the following purposes:

1. ***Stripping***—The removal and wasting of any type of rock and dirt in order to uncover valuable ore deposits.
2. ***Cutting***—Removal primarily to lower the surface. In roadway, railway, airport, and building construction, the spoil is generally used for fill elsewhere on the project. In ditching, it is often used for backfill after installation of pipes.
3. ***Quarrying or mining***—Excavation of rock that has value in itself, either before or after processing. A rough distinction can be made between these two, where quarries are ordinarily concerned with the physical characteristics of the stone, and mines with its chemical composition. However, the terms will be used interchangeably in this discussion.

One excavation can involve all three classifications, as in a heavy road cut where some material is wasted, other is used for road fill, and the best rock is crushed and used for aggregate.

Blasting may be divided into a primary operation in which rock is loosened from its original position in bulk and secondary work, which consists of reducing oversize fragments and breaking back ridges and spurs. The latter is done in the same manner as other light blasting, such as breaking boulders and chipping out ledges.

Rock work may also be classified as to the type and fineness of breakage required. Quarrying of building or dimension stone involves loosening large solid pieces from the parent rock, whereas blasting for fill or crushed rock requires pieces small enough to fit in the shovel bucket, the fill layer, or the crusher.

Stripping

In most stripping work the spoil has no value, so the cheapest way of handling is often the chosen method. It is often possible to dump spoil in excavated areas from which pay material has been removed. It is common practice to shoot and dig overburdens over 100 ft (30.5 m) deep in a single layer and the use of the largest shovels and draglines as required for such work. Overburden depths exceeding 150 ft (45.7 m) are economical in some surface mining operations.

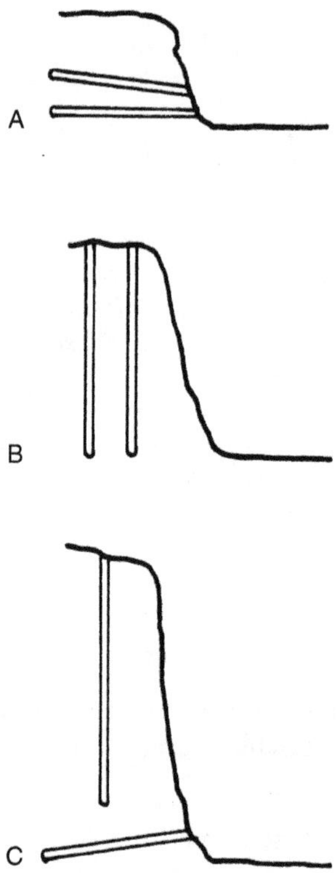

Drilling may be done horizontally from a face, as in Fig. 16.1A, vertically from the top, as in Fig. 16.1B, or in combination, as in Fig. 16.1C. Horizontal drilling has its best use when the mineral deposit is immediately under soft rock. In softer material, auger-type drills with extensions 6 to 10 ft (1.8 to 3.0 m) long and diameters of about 5 in. (12.7 cm) are used. These have a tendency to drift downward, and because distances of 30 to 75 ft (9.1 to 22.9 m) are commonly drilled, it is necessary to start them several feet above the deposit or to start them at a slight upward slant. Spacing may be 10 to 30 ft (3.0 to 9.1 m).

A

B

C

FIGURE 16.1 Horizontal and vertical drilling.

For horizontal drilling of harder bottom rock, crawler-mounted (tracked) air drills may be used, as in Fig. 16.2. The hole is usually started several feet above the bottom. Lower holes are slanted downward. Higher ones, if drilled, may be horizontal or slant upward. In cases of drilling soil nails (Chap. 12), drilling is near horizontal or slanted downward at a consistent slope among successive rows. These larger and heavier percussion drills mounted on either a tracked carriage or a frame are commonly known as a *drifter* (Fig. 16.2).

Drifters can drill in the downward, horizontal, or upward direction. One of their more effective applications is near-horizontal drilling of rock faces and stabilizing vertical walls. These drills are used extensively in rock excavation, mining, and tunneling. Either air or water can be used to remove the cuttings. Drifter drills are similar to jackhammers in operation, but they are larger and used as mounted tools for downward, horizontal, or upward drilling. The drifter's weight is usually sufficient to supply the necessary feed pressure for downward drilling. But when used for horizontal or upward drilling, a hand-operated screw or a pneumatic or hydraulic piston supplies the feed pressure.

When thick layers of the overburden are hard rock such as sandstone or limestone, holes are drilled vertically, as in Fig 16.1B or Fig. 16.1C. Rotary or downhole drills with bits from 4 to 12 in. (10.2 to 30.5 cm) in diameter are typically used (Fig 16.3). Burden distance from free face to borehole and spacing in rows commonly range from 25 to 35 ft (7.6 to 10.7 m) in high faces. Track drills are generally used to depths of 50 ft (15.2 m). Scrapers or dozers are often used to remove most of the loose soil overlying the rock before drilling. This saves considerable drilling footage, makes castings unnecessary, and, on low and medium faces, simplifies the use of track drills. If the pit area is to be regraded when work is completed, the scrapers can be used to fill the hollows between the piles left by the shovels or to place topsoil over regraded areas. Because of the tremendous size and power of the

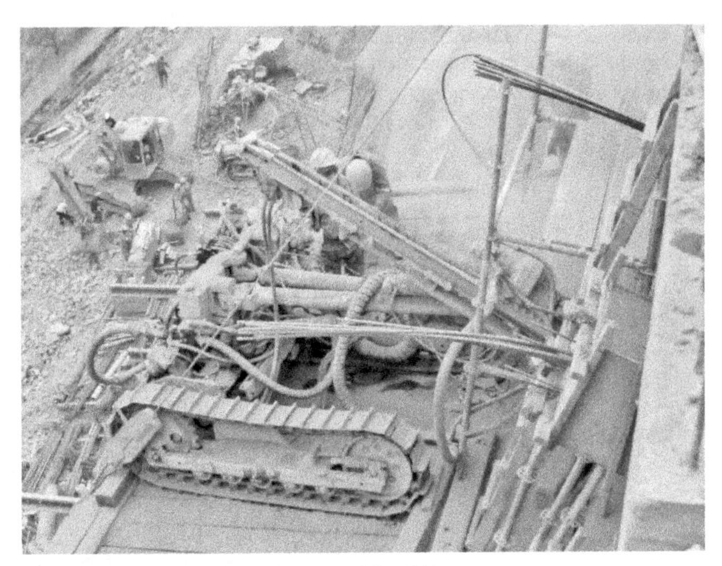

FIGURE 16.2 Horizontal drilling with a drifter.

FIGURE 16.3 Drilling overburden with a rotary drill.

excavators used in large pits, the blasting need only crack and loosen the overburden without producing fragmentation comparable to that required in a cut or quarry. Wide-spaced holes and light charges can therefore be used successfully.

Cutting

Rock cuts for highways and other construction may be of the through type as in Fig. 16.4A and the sidehill (Fig. 16.4B). Material from a sidehill cut may be thrown down to make a fill, as in Fig. 16.4C. The area to be cut should first be cleared and stripped of loose soil, and preferably of rotten rock. This may be done with dozers, scrapers, or shovels, depending on the conditions, such as the steepness of the slope. If the rock is soft, its upper surface may be loosened with rippers and removed, along with any pockets of dirt. If it is hard and irregular, extensive cleaning by hand and with small equipment may be necessary. It is desirable to remove all loose dirt, particularly if the rock is to be used as crushed aggregate for paving. Aggregate washing operations are discussed in Chap. 17.

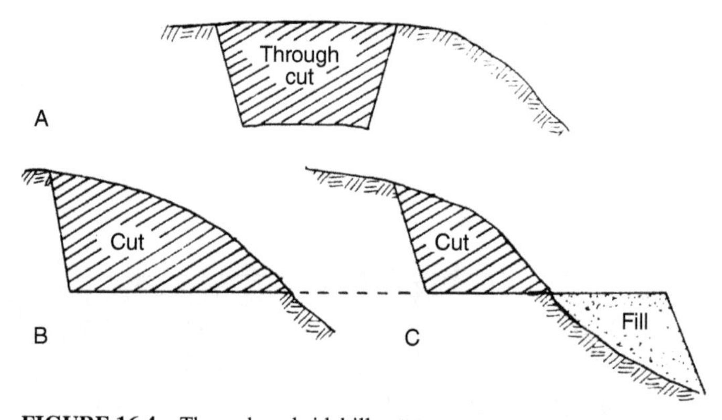

FIGURE 16.4 Through and sidehill cuts.

For deep cuts, the top layer may be drilled, shot, and removed for fill and any clean rock mined from lower levels. If the cut is 20 ft (6.1 m) or less in depth, it may be blasted in a single layer, but depths of 12 to 15 ft (3.7 to 4.6 m) are generally considered most satisfactory for track drills and digging by 1- to 2½-yd (0.76 to 1.91 cu.m) shovels or medium-size excavators or loaders. In a through cut, the full width is used as a face to provide maximum space for machinery. On a sidehill, the same technique or one or more bench faces parallel with the centerline may be planned. Degree of required fragmentation is determined largely by the depth of fill layers where the spoil is used.

Mining and Quarrying

Pit operations are largely conducted to obtain certain classes of rock or earth. The general aspects of this work are discussed in Chap. 17. Rock excavation may follow highway techniques in exploiting comparatively narrow or irregular veins, or large-scale stripping work may be necessary to make pay rock accessible to surface digging units. Pits are often distinguished by the use of high and wide faces or holes sunk below surrounding grades, with access by ramps or inclined or vertical hoists. It is advantageous to have the face wide enough so several operations can be carried out in different sections with minimum interference. The fineness of fragmentation that must be obtained by blasting is generally determined by the size of the hoppers or grizzlies on the crushers or processing machines.

BOREHOLE PATTERNS

Design and layout of borehole patterns are critical for successful blasting of rock. Important terms used in describing drilling equipment and procedures are illustrated in Fig. 16.5.

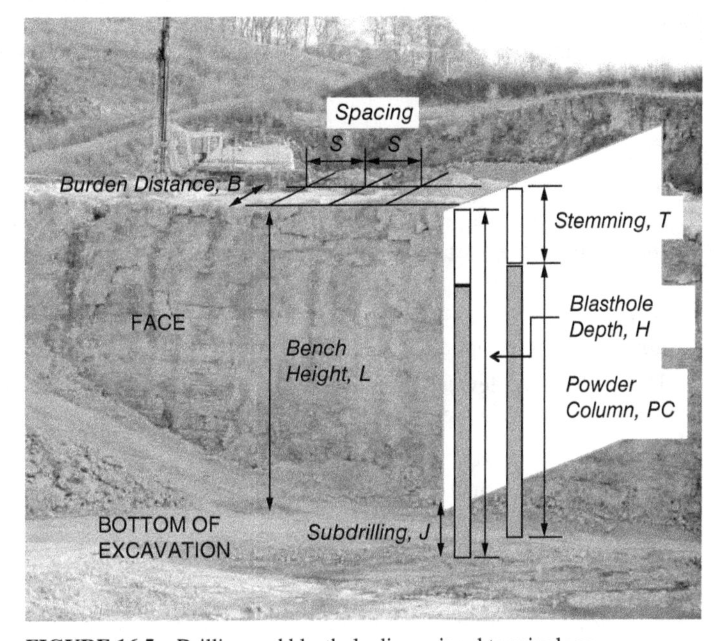

FIGURE 16.5 Drilling and blasthole dimensional terminology.

Glossary of Terms

ANFO An ammonium nitrate and fuel oil mixture used extensively as a construction explosive. This explosive is the cheapest source of explosive energy and must be detonated by special primers, and is much safer than dynamite.

Bench The horizontal ledge of an excavation face along which holes are drilled for blasting. Benching is the process of excavating using ledges in a stepped pattern.

Bench height The vertical distance between the base of an excavation and the ledge above where the blastholes are drilled and shot.

Bit The portion of a drilling tool cutting through the rock or soil by a combination of crushing and shearing actions. There are many types of bits.

Blasthole A hole drilled (borehole) into rock to enable the placing of an explosive.

Blasting (shot) The detonation of an explosive to fracture the rock.

Blasting agent An explosive with ingredients that, by themselves, are nonexplosive and can only be detonated by a high explosive charge, consisting of an oxidizer and a fuel. It is less sensitive to initiation and cannot be detonated with a No. 8 detonator when unconfined. Therefore, blasting agents are covered by different handling regulations than high explosives.

Blasting cap A smaller primary explosive used to detonate a larger secondary explosive. The cap initiates the explosion using electric current, a nonelectric chemical charge, or fuse.

Booster A cap-sensitive explosive not containing a detonator. It is designed to complete the initiation work of the primer in the explosive column.

Burden The horizontal distance from a rock face to the first row of drill holes or the distance between rows of drill holes. When considering all material to be removed, the burden is the horizontal distance from the rock face to the excavation limits.

Burden distance The distance between the rock face and a row of holes or between adjacent rows of holes.

Cuttings These are the disintegrated rock particles caused by the action of the drill bit against the rock.

Deflagration A rapid chemical reaction in which the output of heat is enough to enable the reaction to proceed and be accelerated without input of heat from another source. The effect of a true deflagration under confinement is an explosion.

Density An explosive's density is its specific weight, usually expressed as grams per cubic centimeter (g/cc). In the case of some explosives, density and energy are correlated.

Detonator A device initiating an explosion. Methods for the detonator to initiate the explosives include electric current, nonelectric pyrotechnic charge, or a fuse.

Drills

Abrasion A drill that grinds rock into small particles through the abrasive effect of a bit as it rotates in the hole.

Blasthole A rotary drill consisting of a steel-pipe drill stem, on the bottom of which is a roller bit. As the bit rotates it grinds the rock.

Churn A percussion-type drill consisting of a long steel bit mechanically lifted and dropped to disintegrate the rock. It is used to drill deep holes, usually 6 in. in diameter or larger.

Diamond A rotary abrasive-type drill whose bit consists of a metal matrix embedded with a large number of diamonds. As the drill rotates, the diamonds disintegrate the rock.

Downhole drill or downhole bit The bit and the power system providing rotation and percussion are one unit suspended at the bottom of the drill steel.

Percussion A drill to break rock into small particles by the impact from repeated blows. Compressed air or hydraulic fluids can power percussion drills.

Shot A rotary abrasive-type drill whose bit consists of a section of steel pipe with a roughened surface at the bottom. As the bit is rotated under pressure, chilled-steel shot is supplied under the bit to accomplish the disintegration of the rock.

Heave The displacement of rock as a result of the expansion of gases from firing an explosive.

Lead wires The wires conducting the electric current from its source to the leg wires of the electric detonator.

Leg wires The wires conducting the electric current from lead wires to an electric detonator.

MS delay detonator (millisecond) A detonator with a built-in delay element. These detonators are commonly available in millisecond increments.

Nitroglycerin A powerful explosive liquid obtained by treating glycerol with a mixture of nitric and sulfuric acids. Pure nitroglycerin is a colorless, oily, and somewhat toxic liquid. It was first prepared in 1846 by the Italian chemist Ascanio Sobrero.

PETN An abbreviation for the chemical content (pentaerythritol tetranitrate) of a high explosive having a very high rate of detonation. PETN is commonly used as a primary explosive to initiate the blasting agent.

Prill In the United States, most ammonium nitrate, both agricultural and blasting grade, is produced by the prill tower method. Ammonium nitrate liquor is released as a spray at the top of a prilling tower. Prills of ammonium nitrate congeal in the upcoming steam and air of the tower. The moisture driven out by the dropping process leaves voids within the prills.

Propagation The movement of a detonation wave, either in the borehole or from hole to hole.

Rounds A term including all the blastholes drilled, loaded, and fired at one time.

Sensitiveness A measure of an explosive's cartridge-to-cartridge propagating ability.

Shank or striker bar A short piece of strengthened steel attached to the percussion drill piston for receiving the blow and transferring the energy to the drill steel.

Spacing The distance between adjacent holes in the same row.

Subdrilling The depth to which a blasthole will be drilled below the proposed final grade. This extra depth is necessary to ensure rock breakage will occur completely to the required elevation.

Tamping The process of compacting the stemming material placed in a blasthole.

TNT A high explosive whose chemical content is trinitrotoluene or trinitrotoluol.

The simplest type of drilling pattern is a straight line of vertical holes parallel to a vertical face. The distance from each hole to the face is termed burden, and the distance between holes their spacing. The holes are drilled with slightly more depth than the face so any ridges left between them will not project above the new grade. Blasts tend to overbreak at the top and not shatter completely at the base. As a result, faces tend to slope back. The projection of the bottom beyond the vertical line from the top is called the toe or bottom of the excavation (Fig. 16.5). The extra burden at the toe may be handled by subdrilling, where explosives slightly below the quarry floor simultaneously create uplift and shearing forces.

Holes may be drilled at an angle so they are parallel to the slope of the face. This angle may be from 5 to 30 degrees. Angle or slope drilling keeps the burden at the toe from being greater than at the top, so there is no need for heavy charges or subdrilling at the bottom. Improved bottom breakage reduces the need for drilling below the floor (subdrilling).

Face Height

Face height affects the method of drilling and the size and placement of holes. A rock mass may be taken out in a single layer or in a series of benches. Highway work is usually done in benches, as are some open-pit mining, but quarry operators may take single 100-ft (30.5- or even 61-m) high slices. This is partly because quarry rock is often sound enough to stand at great heights without much danger of collapse.

High faces are usually developed by pushing low or moderate ones back into a hillside, as in Fig. 16.6A. Where low faces are preferred, such a cut may be made in a series of benches, as in Figs. 16.6B and 16.6C. This is the safest method in most formations. The final slope should be based on an angle-of-repose geotechnical analysis. Figure 16.7 illustrates a benched rock excavation.

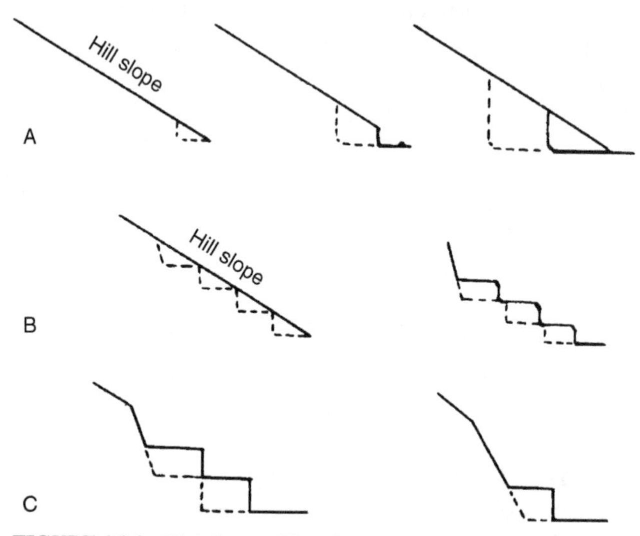

FIGURE 16.6 High face and bench cuts.

FIGURE 16.7 Benched rock excavation on a highway project.

Accessibility to the top surface may be a determining factor in excavating the top layer. Drills can climb and work on slopes over 30 degrees. However, risks and delays are involved in working on very steep or rough ground, and it may be more economical to rip portions of the top surface for steep surface slopes.

Depths over 30 ft (9.1 m) call for blasthole drills of the rotary, downhole, or churn types. These heavy and expensive machines should be kept on fairly even ground. They can dig through any depth of overburden readily, and can start their work from pioneer roads notched into soil slopes by dozers.

Hole Size

Hand drills will produce hole diameters of 1 to 2 in. (2.5 to 5.1 cm), tracked drifter drills 1½ to 5 in. (3.8 to 12.7 cm), and blasthole drills 4 to 12 in. (10.2 to 30.5 cm). Pneumatic drills of all sizes make tapering holes with steel bits, as the hole gets smaller toward the bottom as they wear. Carbide bits and rotary and churn drills produce holes with little or no taper.

Table 16.1 is a loading density chart used to calculate the weight of explosive required for a blasthole. For example, the diameter of the borehole and explosives is 3 in., and the explosive has a specific gravity density of 0.8. Bench face height is 20 ft, and 2 ft is added for subdrilling to shear the bench toe, so the total borehole length is 22 ft. Stemming of angular-faced aggregate at the top of the borehole to contain the blast is 4 ft. Using Table 16.1, the loading density is found to be 2.45 lb/ft of charge. The total powder column length containing explosives is the total hole length minus the stemming, 18 ft in this case (20 ft + 2 ft subdrilling – 4 ft stemming). The total weight of explosive used per blasthole would be 18 ft × 2.45 lb/ft = 44.1 lb.

The explosive in each hole is designed to break out a section of the rock between the line of holes and the face.

In order to drill and load holes accurately, it is necessary to know the face height and the amount of the toe or subdrilling. With low faces, or in casual operations, or in working upper

TABLE 16.1 Explosive Loading Density Chart in Pounds per Foot of Borehole

Column Diam. (in.)	Explosive Specific Gravity							
	0.80	0.90	1.00	1.10	1.20	1.30	1.40	1.50
1	0.27	0.31	0.34	0.37	0.41	0.44	0.48	0.51
1¼	0.43	0.48	0.53	0.59	0.64	0.69	0.74	0.80
1½	0.61	0.69	0.77	0.84	0.92	1.00	1.07	1.15
1¾	0.83	0.94	1.04	1.15	1.25	1.36	1.46	1.56
2	1.09	1.23	1.36	1.50	1.63	1.77	1.91	2.04
2½	1.70	1.92	2.13	2.34	2.55	2.77	2.98	3.19
3	2.45	2.76	3.06	3.37	3.68	3.98	4.29	4.60
3½	3.34	3.75	4.17	4.59	5.01	5.42	5.84	6.26
4	4.36	4.90	5.45	6.00	6.54	7.08	7.63	8.17
4½	5.52	6.21	6.89	7.58	8.27	8.96	9.65	10.34
5	6.81	7.66	8.51	9.36	10.22	11.07	11.92	12.77
5½	8.24	9.27	10.30	11.33	12.36	13.39	14.42	15.45
6	9.81	11.03	12.26	13.48	14.71	15.93	17.16	18.39
6½	11.51	12.95	14.39	15.82	17.26	18.70	20.14	21.58
7	13.35	15.02	16.68	18.35	20.02	21.69	23.36	25.03
8	17.43	19.61	21.79	23.97	26.15	28.33	30.51	32.69
9	22.06	24.82	27.58	30.34	33.10	35.85	38.61	41.37
10	27.24	30.64	34.05	37.46	40.86	44.26	47.67	51.07

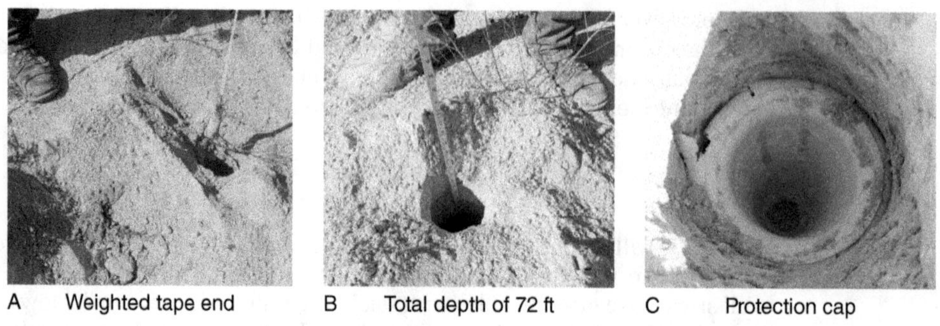

A Weighted tape end B Total depth of 72 ft C Protection cap

FIGURE 16.8 Measuring borehole depth with weighted nylon tape for 70 ft face plus 2 ft subdrilling.

lifts to temporary grades, depths may be estimated. Length of drill steel rods in the borehole, with common section lengths of 10 ft, 15 ft, 20 ft, and up to 40 ft, is a practical estimation method. Laser measurement instruments are a newer technology; however, standard nylon measuring tapes are still used, as in Fig. 16.8 measuring a 72-ft borehole depth. Each drill hole may be marked according to the cut to bottom grade or by the drilling depth desired. The top of the borehole is protected against debris and disturbance with a cap or inverted traffic cone.

COMMERCIAL EXPLOSIVES

Commercial explosives detonate with introduction of a suitable initiation stimulus. On detonation, the ingredients of an explosive compound react at high speed, liberating gas and heat, thus causing very high-pressure, high-temperature gases. Pressures just behind the detonation front are on the order of 150,000 to 4,000,000 psi (10,340 to 275,730 bars), and temperature can range from 3,000 to 7,000°F (1,600 to 3,900°C). High explosives contain at least one high-explosive ingredient. Low explosives contain no ingredients capable of exploding by themselves. Both high and low explosives can be initiated by a single No. 8 detonator. The No. 8 detonator is considered an industry standard for strength and reliability detonating a wide variety of explosives. It contains a sleeved fuse head and primer PETN base charge. PETN (pentaerythritol tetranitrate) is a high explosive with a very high rate of detonation.

Explosives are different in the following ways:

- Strength
- Sensitivity—input energy needed to start the reaction
- Detonation velocity
- Water resistance
- Flammability
- Generation of toxic fumes
- Bulk density

Strength. The term *strength* refers to the energy content of an explosive, which is the measure of the force it can develop and its ability to do work. There are no standard strength measurement methods used by explosives manufacturers. Strength ratings are misleading and do not accurately compare rock fragmentation effectiveness with explosive type.

Sensitivity. The sensitivity of an explosive product is defined by the amount of input energy necessary to cause the product to detonate reliably. Some explosives, such as dynamite, require very little energy to detonate reliably. An electric detonator or blasting cap alone will not reliably initiate bulk-loaded ANFO and some slurries. To detonate these, a primer or booster would have to be used in conjunction with the electric detonator or the cap.

Detonation Velocity. Explosive velocity is the speed at which the detonation wave moves through the column of explosive. Commercially available explosives have extremely high velocities in the range of 5,000 to 25,000 ft/sec. High-velocity explosives should be used in hard rock, whereas lower-velocity explosives give better results in softer rock.

Water Resistance. The water resistance of an explosive can broadly be defined as its ability to detonate after exposure to water. Water resistance is generally expressed as the number of hours a product can be submerged in static water and still be detonated reliably. A standard water resistance test is used primarily for classification of dynamite explosives. Coatings on the dynamite casing are a primary factor in water resistance.

Flammability. The characteristic of an explosive used to describe its ease of initiation from spark, fire, or flame. Explosives that detonate by burning are called low explosives. Black powder, one of the oldest-known explosives, may be ignited or exploded by flame, heat, sparks, or concussion, and requires more careful handling than most explosives. A special hazard is spilled powder igniting on the ground or the magazine floor if stepped on or scuffed.

Generation of Toxic Fumes. The amount of toxic gases produced by an explosive during the detonation process. The inability to fully oxidize the blasting agent may produce toxic fumes, as in the case of ANFO releasing nitrogen oxides, as evident with reddish-colored gases above the blast.

Bulk Density. The density of an explosive is normally expressed in terms of weight per unit volume. The density determines the weight of explosive loaded into a specific borehole diameter (Table 16.1). The specific gravity of an explosive is the ratio of an explosive's weight to the weight of an equal volume of water. For some explosives, density is commonly used as a way to approximate the explosive's strength.

There are four main categories of commercial high explosives: (1) dynamite, (2) ANFO, (3) slurries, and (4) two-component explosives. To be a high explosive, the material must be cap sensitive and react at a speed faster than the speed of sound, and the reaction must be accompanied by a shock wave. The first three categories—(1) dynamite, (2) ANFO, and (3) slurries—are the principal explosives used for blasthole charges. Two-component or binary explosives are normally not classified as explosive until mixed. Therefore, they offer advantages in shipping and storage. This makes them attractive alternatives on small jobs. But their unit price is significantly greater than what other high explosives cost.

Dynamite

Dynamite is the best-known commercial high explosive. Until the development of lower-cost ANFO (ammonium nitrate and fuel oil), dynamite was one of the most widely used explosives. This nitroglycerin-based product is the most sensitive of all the generic classes of explosives in use today. It is available in many grades and sizes to meet the requirements of a particular job. Common applications are demolition of structures and a priming agent for ANFO. Straight dynamite contains no ammonium nitrate and is not appropriate for construction applications because it is very sensitive to shock.

The dynamite name includes several different chemical groups, wrapped and marketed in about the same manner. The "straight" dynamites consist primarily of a mixture of nitroglycerin, sodium nitrate, and combustible absorbents such as wood pulp wrapped in strong paper to make a cylindrical cartridge. Although a wide variety of sizes are available, the most popular are 8 in. (20.3 cm) long and 1⅛ or 1¼ in. (2.86 or 3.18 cm) in diameter. The percentage of nitroglycerin by weight contained in the mixture is used to classify according to strength, ranging from 15% to 60%. Strength does not increase in proportion to the percentage of nitroglycerin because the other ingredients also contribute expansive gas and heat. For example, a 60% dynamite is about 1½ times as strong as a 20% dynamite. Higher percentages are faster and more sensitive. The more common commercial dynamite sticks for construction are 40% nitroglycerin in half-pound sticks, 8 × 1¼ in. A detonator or a Primacord fuse may be used to fire the dynamite. If a detonator is used, one of the cartridges serves as a primer. The detonator is placed within a hole made in this cartridge.

The most widely used product in mining, quarries, and construction applications is "high-density extra dynamite," but individual explosive manufacturers have their own trade names for dynamite products. This product is less sensitive to shock than straight dynamite because some of the nitroglycerin has been replaced with ammonium nitrate. Dynamite with an ammonium nitrate chemical compound is sometimes referred to as ammonia dynamite. PETN is a high explosive with a very high rate of detonation and commonly used as a primer or booster. Pentolite is a mixture of equal parts of TNT and PETN with a detonation velocity of up to 25,000 ft/sec. This product is often called a cast primer or booster to provide an extremely rapid and high detonation pressure to initiate a blasting agent, most commonly with ANFO. Although traditional dynamite has been used extensively for charging blastholes, especially for smaller size holes, the economics and blasting efficiency have favored the use of nitrate-based explosives. Handling and sensitivity to shock are also important factors in this trend.

ANFO

ANFO, an ammonium nitrate and fuel oil mixture, is a nitrogen fertilizer largely replacing dynamite in medium and large borehole blasting. This explosive represents about 80% of all explosives used in the United States, primarily due to lower cost and ease in handling. Because it must be detonated by special primers, it is much safer than dynamite.

Standard ANFO is a mixture of prilled industrial-grade ammonium nitrate and 5.7% No. 2 fuel oil. This is the optimum mixture. The mixture is blended at the jobsite with 116 to 120 lb of ammonium nitrate injected with 1 gal of fuel oil. The detonation efficiency is controlled by the amount of fuel oil. It is less detrimental to have a fuel deficiency, but fuel percentage variances, either surplus or deficiency, affect the blast. With too little fuel, the explosive will not perform properly. With a fuel percentage of only 5%, there will be a 5.3% energy loss because of excess oxygen, and reddish-orange nitrous oxide fumes may be produced. Maximum energy output is also reduced when too much fuel is used. The ANFO prills should not be confused with ammonium nitrate fertilizer prills. By comparison, ANFO is considered a low-grade fertilizer for agricultural purposes. A blasting prill is porous to better absorb and distribute the fuel oil. Keeping the ammonium nitrate and the fuel oil in separate compartments allows transportation with standard hazardous material placards. Ammonium nitrate by itself is so insensitive that it is not rated as an explosive and can be shipped and stored free of special blasting regulations and permits.

Because the ANFO is free flowing, it can be either blown or augured from bulk trucks directly into the blastholes (Fig. 16.9). Augers move prills from a truck compartment into a stream of compressed air, in which they are mixed with oil and blown into the borehole through flexible hose. Loading prills by compressed air may build up charges of static electricity in the borehole and its vicinity. Such charges create the danger of premature discharge of electric caps, and even possible hazards with fuse caps. Minimum precautions are grounding of the pneumatic loader and use of semiconducting hose. Caps should be of types least sensitive to stray currents.

A Blowing ANFO into borehole B Tamping with non-static rod

FIGURE 16.9 Loading bulk ANFO into blastholes.

Filling boreholes is visually monitored as the ANFO flows from an air-blown flexible hose or a bucket. In shallow boreholes, a rod made of wood or nonstatic aluminum gently tamps the ANFO column to eliminate bridging and monitor filling height (Fig 16.9B). The top portion of the borehole is filled with drill cuttings or crushed stone to confine the blast, known as the stemming, so the filling of ANFO must terminate before the bottom of the stemming is reached (Fig. 16.10). Markers on the tamping rod can indicate when the ANFO column is at the bottom of the stemming segment. Excessive ANFO with an insufficient stemming height can result in an air blast and loss in blasting efficiency. If there are cavities in the borehole wall from rock seams or soft disintegrated rock, these will be filled with excessive ANFO. In such cases, ANFO loading is stopped and a stem plug is inserted into the borehole to limit the pounds of explosives in the powder column, as in Fig. 16.11. ANFO in such void spaces can cause an irregular blast and flyrock.

ANFO is detonated by primers consisting of charges of explosive placed at the bottoms of the holes. Sometimes primers are placed at the bottom and at intermediate depths to better control blast movement, a technique known as "double priming." Electric detonators or Primacord can be used to detonate the primer.

The detonation velocity and, therefore the efficiency of ANFO, is dependent on the diameter and density of the power column. Higher detonation velocities can be achieved from more densely compacted ANFO. Air-emplaced ANFO in a 1¼-in.-diameter column

FIGURE 16.10 Stemming added to top of boreholes with crushed aggregate.

FIGURE 16.11 Stemming plug pushed into borehole to control explosives placement in the borehole.

will detonate at velocities of 7,000 to 10,000 ft/sec (fps). If placed in larger-diameter columns, the velocity increases: 3-in.-diameter column, 12,000 to 13,000 fps; 9-in. and 12-in.-diameter column, 14,000 to 15,000 fps. The placement method also affects the velocity. The velocities for poured ANFO are 2-in.-diameter column, 6,000 to 7,000 fps; 3-in.-diameter column, 10,000 to 11,000 fps; 9-in. and 12-in.-diameter column, 14,000 to 15,000 fps.

ANFO is not water-resistant. Detonation will be marginal if ANFO is placed in water, even if the interval between loading and shooting is very short. If it is to be used in wet holes, there is a densified ANFO cartridge. This product has a density greater than water, so it will sink to the bottom of a wet hole. Standard ANFO has a product density of 0.84 g/cc. The poured density of ANFO is generally between 0.78 and 0.85 g/cc. Air-blown ANFO can reach densities of 0.95 g/cc. Cartridges or bulk product sealed in plastic bags will not sink. Another method to exclude the water and to enable the use of bulk ANFO in wet conditions is to pre-line the holes with plastic tubing with a closed bottom. The tubes, whose diameters should be slightly smaller than the holes, are installed in the holes by placing rocks or other weights in their bottoms. To prevent drill cuttings and other debris entering the borehole, the top should be capped with a bag of drill fines or an inverted safety cone until charging the borehole (Fig. 16.8C).

Slurries

Slurries are a generic term for both water gels and emulsions. A slurry explosive is made of ammonium, calcium, or sodium nitrate and a fuel sensitizer along with varying amounts of water.

In comparison to ANFO, slurries have a higher cost per pound and they have less energy. However, in wet conditions they are very competitive with ANFO, because the ANFO is water-sensitive and must be protected in lined holes, or a bagged product has to be used. Both of these measures add to the total cost of the ANFO.

An advantage of slurries over dynamite is separate ingredients can be hauled to the project in bulk and mixed immediately before loading the blastholes. The mixture can be poured directly into the hole. Slurries may be packaged in plastic bags or for placement in the holes, as in Fig. 16.12. Because they are denser than water, they will sink to the bottom of holes containing water.

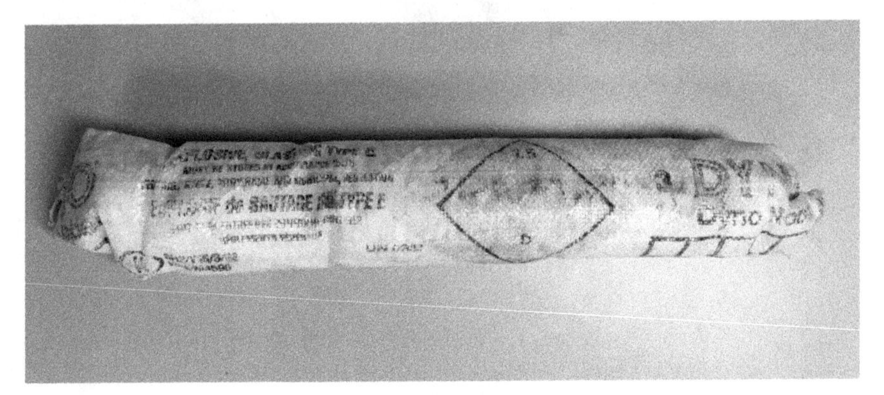

FIGURE 16.12 Slurry explosive packaged in a bag for wet boreholes.

Some slurries may be classified as high explosives, whereas others, if they cannot be initiated by a No. 8 detonator, are classified as blasting agents. This difference in classification is important for *magazine storage*, or a building or chamber for the storage of explosives.

A slurry explosive mixture of oxidizing salts, fuels, and water-resistant sensitizers by the cross-linking of gums or waxes is explicitly referred to as a *water gel*. The primary sensitizing methods are the introduction of air throughout the mixture, the addition of aluminum particles, and the addition of nitrocellulose.

A slurry explosive mixture of oxidizing salts and fuels and containing no chemical sensitizers is made water resistant by an emulsifying agent, and is explicitly referred to as an *emulsion*. Emulsions will have a somewhat higher detonation velocity than water gels. Some emulsions tend to be wet and will adhere to the blasthole, causing bulk loading problems. Emulsions can be loaded into the hole in plastic bags or chubs.

Primers and Boosters

Most ANFO cannot be detonated dependably by blasting caps or regular Primacord. Even when detonated, it may not maintain full speed of explosion, or even any explosion, for the length of the hole. It is therefore necessary to use primers or boosters or Primacord to produce sufficient explosion to detonate the ANFO at high velocity.

A prime charge is an explosive ignited by an initiator (detonator or cap), which, in turn, initiates a noncap-sensitive explosive or blasting agent (Fig. 16.13). The resulting detonation is then transmitted to an equally or less sensitive explosive. Often, primers are cartridges of high-density dynamites, PETN, a pentolite mixture of TNT and PETN, highly sensitized slurries, or emulsions together with detonators or a detonating cord. Good priming improves fragmentation, increases productivity, and lowers overall cost. The primer should never be tamped.

FIGURE 16.13 Placing an electric detonator in a booster and then lowering into the blasthole.

To obtain optimum performance of ANFO, the size of the primer should match the borehole diameter. A mismatch in the size of the primer and the borehole diameter will affect the detonation velocity of ANFO. The detonation side of a downhole primer is facing upright (Fig 16.13, right) and contacting the ANFO for greatest blast efficiency.

There are exceptions to where the primer should be placed in the explosive column, and care should be exercised when there are soft seams or fractures in the rock. A bottom primer should be just off the bottom of the hole and facing upright. Tension in the downhole wiring or detonating cord can ensure the primer is in the vertically upright position, or in cases of sloped boreholes, the primer is parallel to the borehole wall. Maintain good contact between the ANFO column and the primer. Contact between the primer and the explosive column is extremely important for an efficient blast.

Boosters are highly sensitized explosives or blasting agents, used either in bulk form or in packages and of weights (amounts) greater than those used for primers. Boosters are placed within the explosive column where additional breaking energy is required. They are a cap-sensitive explosive but do not contain a detonator and are designed to assist the initiation work of the primer.

It is a good plan to check a deep, large borehole before loading it by inspecting it with a flashlight or sunlight reflected from a mirror, or sounding it with a tamping block, to make sure it is not obstructed. The block can be used to knock obstructing pieces or scale to the bottom. Cartridges may be lowered into shallow or deep holes with the cord (Fig. 16.13, right). If the hole is deep and rough and there is a possibility they may stick partway down, the cartridges should be lowered carefully. The impact of the cartridges on the bottom and the weight of the column above frequently compress charges well enough so tamping is not necessary.

INITIATING SYSTEMS

Initiator is a term used in the explosive industry to describe any device used to initiate a detonation. An initiation system is a combination of explosive devices and component accessories designed to convey a signal and initiate an explosive charge from a safe distance. The signal function may be electric or nonelectric. Fragmentation, backbreak (rock broken beyond the last row of blastholes), vibration, and violence of a blast are all controlled by the firing sequence of the individual blastholes. The order and timing of the detonation of individual holes are regulated by the initiation system. When selecting the proper system, one should consider both blast design and safety. Electric systems are more sensitive to lightning than nonelectric systems, but both are susceptible. The cost of electric detonation is more expensive than nonelectric with the benefit of efficiently programming the blasting sequence and performing system diagnostic checks.

Electric Detonators

The most widely used explosive initiator is the electric detonator (Fig. 16.14). An electric detonator passes an electric current through a wire bridge, similar to an electric lightbulb filament, causing an explosion. The current, approximately 1.5 amps, heats the bridge to incandescence and ignites a heat-sensitive flash compound. The flash compound sets off a primer (initiating charge) that in turn fires a base charge in the detonator (Fig. 16.13, left). This charge detonates with sufficient violence to fire a charge of explosive.

Electric detonators are supplied with two leg wires in lengths varying from 2 to 200 ft. These wires are connected with the wires from other blastholes to form a closed electric circuit for firing purposes. The leg wires of electric detonators are typically made of copper for superior connectivity. For ease in wiring a blast, each leg wire on an electric detonator

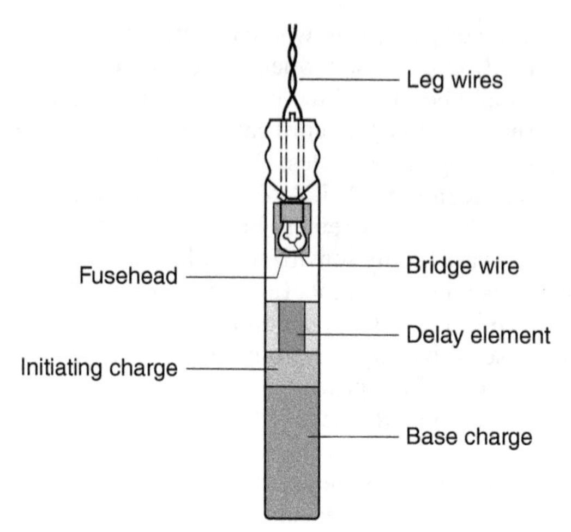

FIGURE 16.14 Electric initiating device.

may be a different color. There are instantaneous, short-period delay, long-period delay, electronic delay type, and seismic (geophysical exploration) electric detonators. Instantaneous electric detonators are made to fire within a few milliseconds after current is applied.

When two or more electric detonators are connected in the same circuit, they must be products of the same manufacturer. This is essential to prevent a misfire, because detonators of different manufacturers do not have the same electrical characteristics. Detonators are extremely sensitive. They must be protected from shock and extreme heat. They are never to be stored or transported with other explosives.

Electric Detonator Safety

These devices are to be kept away from all sources of electrical energy: batteries, outlets, radios, calculators, pagers, and cellular phones. Early electric detonators could be initiated by the batteries, speakers, or even the antennas of such devices. It was common to place warning signs entering the jobsite to prohibit the use of cellular phones near a blasting spread. Developments in technology have largely eliminated the potential for a cellular phone to accidentally trigger a blast sequence. A system diagnostic check of the detonators and wiring can detect the possibility of a misfire.

Nonelectric Detonators

Nonelectrical detonator caps (NONEL) are thin, metal, cylindrical tubes open on one end for the insertion of a safety fuse (Fig 16.15). The safety fuse contains a low explosive enclosed in a suitable covering, and when ignited, it will burn at a predetermined speed (35 to 45 sec/ft). The NONEL detonation tube itself contains two explosives, one layer upon the other. The bottom layer is called the base charge and is usually an insensitive high explosive. The top layer is the initiating charge and is a sensitive explosive. The ignition powder ensures the flame is picked up from the safety fuse. At one time, the standard blasting cap contained 2 g of mercury fulminate and was called a No. 8 cap. A No. 6 cap contained

FIGURE 16.15 Nonelectrical detonator cap and cord.

1 g of fulminate. Other caps with less strength had lower numbers. Many of the current No. 8 caps contain 0.8 g of PETN. These detonators are sensitive to heat, shock, and crushing.

Many of the early NONEL detonators were instantaneous, but technological developments now incorporate a delay element. The burn rate of the detonator is factored into the delayed ignition sequence using the manufacturer's specified rate, such as 5,000 ft/sec. There is "cap scatter" or variability in the burn rate with nonelectrical detonators that must be built into the blast timing. A tolerance in cap scatter of +/−8 ms per borehole is factored into most blasts and thus requires mathematical calculations to ensure the desired delay sequence is achieved. Temporary paint markings on the borehole grid can assist in the layout of nonelectrical detonation.

Initiating Sequence

In construction and surface blasting applications, millisecond-delay electric detonators are frequently used. Millisecond-delay blasting can be used both in shooting a single row and in multiple-row shots. When each charge breaks rock mass from the burden before the next charge detonates, there is a gain in blasting efficiency. Delayed sequencing can minimize ground vibration, air blast, and flyrock by disorienting wave propagation. It also has the benefit of increasing rock fragmentation. The delaying procedure effectively reduces the apparent burden for the holes in the succeeding rows.

Millisecond-delay detonators have individual timing intervals ranging from 25 to about 650 ms. Long-period-delay electric detonators have individual intervals ranging from about 0 ms to over 8 sec. They are primarily used in tunneling, underground mining, and shaft sinking. With electronic detonation systems, a tagging device can program the delay time for individual borehole detonators. The tagger is connected to the downhole detonator wiring at the top of each borehole and programmed. A diagnostic check can also be performed to confirm the electrical connection with the detonator.

Detonating Cord (Primacord)

This is a nonelectric initiation system consisting of a flexible cord with a center core of high explosive, usually PETN. It is used to detonate cap-sensitive explosives. The explosive core is encased within textile fibers and covered with a waterproof sheath for protection. Detonating cord is insensitive to ordinary shock or friction. The detonation rate or velocity of the cord's explosive will depend on the manufacturer; typical rates are between 18,000 and 26,000 ft/sec (5,500 and 7,850 m/sec). Delays can be achieved by attaching in-line delay devices.

When several blastholes are fired in a round, the cord is laid on the surface between the holes as a trunkline (Fig 16.16, left). At each hole, one end of a detonating cord downline is attached to the trunkline, whereas the other end of the downline extends into the blasthole. If it is necessary to use a blasting cap and/or a primer to initiate the blast in the hole, the bottom end of the downline may be cut square and securely inserted into the cap.

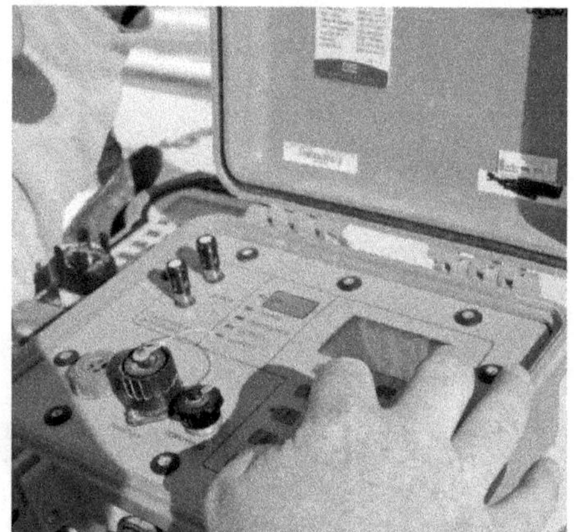

FIGURE 16.16 Electric wiring and firing sequence device.

Sequential Blasting Machines

There are programmable blast timing machines for firing electric detonators (Fig 16.16, right). These machines enable precise millisecond/microsecond accuracy of blast circuit firing intervals. This provides the blaster the option of many delays within a blast. Because many delays are available, the pounds of explosive fired per delay can be reduced to better control noise and vibration.

ROCK FRAGMENTATION

Rock fragmentation is the most important aspect of production blasting because of its direct effect on the costs of drilling and blasting, and on the economics of the subsequent operations of loading, hauling, and crushing (Fig. 16.17). Many variables affect rock fragmentation: rock properties, site geology, in situ fracturing, moisture content, and blasting parameters—known collectively as the blast design.

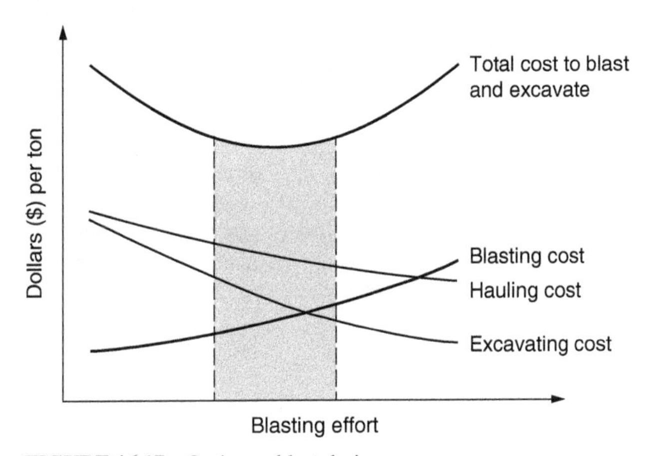

FIGURE 16.17 Optimum blast design.

Blasting is considered by many to be more art than science. Advancements in electronic technology and computer software are leading to a greater understanding of blast engineering. Experience and best practices predict blast fragmentation size distribution resulting from explosive detonation in a blasthole. However, the major mechanisms of rock breakage clearly result from the sustained gas pressure produced in the borehole by the explosion. First, this pressure will cause radial cracking. Such cracking is similar to what happens in the case of frozen water pipes—a longitudinal split occurs parallel to the axis of the pipe. A borehole is analogous to the frozen pipe as a cylindrical pressure vessel. But there is a difference in the rate of loading. A blasthole is pressurized instantaneously. Thus, failure, instead of being at the one weakest seam, is in many seams parallel with the borehole. Burden distance (Figs. 16.5 and 16.18), the direction to the free face, will control the course and extent of the radial crack pattern.

When the radial cracking occurs, the rock mass is transformed into individual rock wedges. If relief is available perpendicular to the axis of the blasthole, the gas pressure pushes against these wedges, putting the opposite sides of the wedges into tension and compression. The exact distribution of such stresses is affected by the location of the charge in the blasthole. In this second breakage mechanism, flexural rupture of the wedge is controlled by the burden distance and bench height. The bench height divided by the burden distance is known as the "stiffness ratio." This is the same mechanism a structural engineer is concerned with when analyzing the length of a column in relation to its thickness.

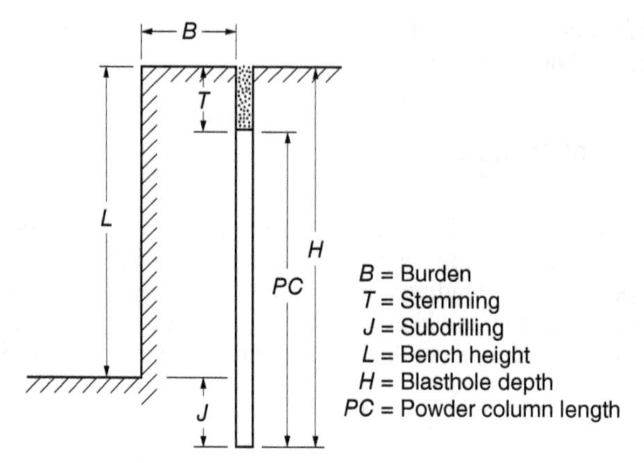

FIGURE 16.18 Blasthole geometry.

There is a greater degree of difficulty in breaking rock when the burden distance is equal to the bench height. As bench height increases compared to burden distance, the rock is more easily broken. If the blast is not designed properly and the burden distance is too great, blast energy relief will not be available by this mechanism. When this happens, either the blasthole will crater or the stemming will blow out.

BLAST DESIGN

To minimize the total costs of rock excavation, one must start with careful planning of the blasts. Every blast must be designed to meet the existing conditions of the rock formation and overburden and to produce the desired final result. Blast design is not an exact science, and there is no single solution to the rock removal problem. Rock is not a homogeneous material. Sedimentary rock may have multiple layers of unique minerals requiring individual extraction. There are fracture planes, seams, and changes in bench height to be considered. Wave propagation is faster in hard rock than in soft rock. Initial blast designs use idealized assumptions. The engineer develops a blast design, realizing material discontinuities exist in the field. Because of these facts, it must always be understood that the theoretical design is only the starting point for blasting operations on the project.

Empirical formulas provide an estimate of the work that can be accomplished by a given explosive. The application of these formulas is provided in the following sections. Their use results in a series of blasting dimensions suitable for trial shots, such as the burden distance, blasthole diameter, top stemming depth, subdrilling depth, and hole pattern and spacing. Adjustments made from investigating the product of the trial shots should result in the optimum blast dimensions.

Burden Distance

Burden distance and borehole diameter are the two most important factors affecting blast performance (Fig. 16.18). Burden is the shortest distance to stress relief—the nearest free face—at the time a blasthole detonates. It is normally the distance to the free face in an excavation, be it a quarry situation, mining operation, or a highway cut. Internal free faces can be created by blastholes fired on an earlier delay within a shot. When the burden distance is insufficient, rock will be thrown for excessive distances from the face, fragmentation may be excessively fine, and air blast levels will be high.

A rule of thumb to ensure the blaster is using the proper burden is $B = 24$ to 30 times the charge (explosive) diameter of the blasthole. When blasting with ANFO or other low-density blasting agents (explosive densities near 53 pcf, 0.85 g/cc) and typical rock (density of roughly 170 pcf, 2.7 g/cc), the burden should be approximately 25 times the charge diameter. When using denser products, such as slurries or dynamites with densities near 75 pcf (1.2 g/cc), the burden should be approximately 30 times the charge diameter, but these are only first approximations or guideline values. For example, for production blastholes with a diameter of 0.5 ft (6 in.), by rule of thumb, the recommended burden distance would be 24×0.5 ft = 12 ft and 30×0.5 ft = 15 ft.

An empirical formula for approximating a burden distance to be used for a first trial shot is:

$$B = \left(\frac{2SG_e}{SG_r} + 1.5 \right) \times D_e \tag{16.1}$$

where B = burden in feet
SG_e = specific gravity of the explosive
SG_r = specific gravity of the rock
D_e = diameter of explosive in inches

The explosive diameter will depend on the manufacturer's packaging container thickness, or it will equal the blasthole diameter if a granular or slurry explosive is poured directly into the hole. If the specific explosive product is known, the exact specific gravity information should be used; but in the case of developing a design before settling on a specific product, an allowance should be made.

Rock density is an indicator of rock strength, which, in turn, establishes the amount of energy required to cause breakage. Approximate specific gravities for various rocks are given in Table 16.2.

TABLE 16.2 Density by Nominal Rock Classifications

Rock Classification	Specific Gravity	Density Broken (ton/cy)
Basalt	2.8–3.0	2.36–2.53
Diabase	2.6–3.0	2.19–2.53
Diorite	2.8–3.0	2.36–2.53
Dolomite	2.8–2.9	2.36–2.44
Gneiss	2.6–2.9	2.19–2.44
Granite	2.6–2.9	2.19–2.28
Gypsum	2.3–2.8	1.94–2.36
Hematite	4.5–5.3	3.79–4.47
Limestone	2.4–2.9	1.94–2.28
Marble	2.1–2.9	2.02–2.28
Quartzite	2.0–2.8	2.19–2.36
Sandstone	2.0–2.8	1.85–2.36
Shale	2.4–2.8	2.02–2.36
Slate	2.5–2.8	2.28–2.36
Trap rock	2.6–3.0	2.36–2.53

Sample Calculation: Bulk ANFO, specific gravity 0.8, opens an excavation in granite rock. The drilling equipment will drill a 3-in. blasthole. The trial burden distance for the first trial shot uses Table 16.2, specific gravity of granite 2.6 to 2.9, use an average value 2.75:

$$B=\left(\frac{2\times 0.8}{2.75}+1.5\right)\times 3=6.2\text{ ft}$$

Compare the calculated burden with the rule of thumb:

$$24\times 0.25\text{ ft}=6\text{ ft}$$

and

$$30\times 0.25\text{ ft}=7.5\text{ ft}$$

Therefore, the burden distance of 6.2 ft is between 6 and 7.5 ft. If the calculated burden were to lie outside this range, consider other factors in the design such as the measured specific gravity of the rock, stiffness ratio, and desired fragmentation.

Relative Bulk Strength

Relative bulk strength is a comparative strength of an explosive to standard ANFO. Standard ANFO is assigned a strength rating of 100. Explosive density is used in Eq. (16.1) because of the proportional relationship between explosive density and strength. There are, however, some explosive emulsions exhibiting differing strengths at equal densities. In such a case, Eq. (16.1) will not be valid. An equation based on relative bulk strength instead of density can be used in such situations.

The relative bulk strength rating of an explosive should be based on test data under specified conditions, but sometimes the rating is based on calculations. Manufacturers will supply specific values for their individual products. The relative energy equation for burden distance is

$$B=0.67\,D_e\sqrt[3]{\frac{St_v}{SG_r}} \tag{16.2}$$

where, St_v = relative bulk strength compared to ANFO = 100.

When one or two rows of blastholes are fired in the same shot, the burden distance between rows would be equal. If more than two rows are to be fired in a single shot, either the burden distance of the rear rows must be adjusted or the millisecond-delay times must be used to allow the face rock from the front rows to move before the back rows are fired.

Field experiments have shown 2 to 2½ ms per foot of effective burden is the minimum considered if any relief is to be obtained for firing successive rows. Consider an example where three successive rows of blastholes will be fired in one shot. The rows are 10 ft apart. It is desired to obtain good relief for firing each row; therefore, the milliseconds of delay between firing the rows: 2 to 2½ ms per foot of effective burden are required. There should be 20 ms (10 × 2 ms) of delay at a minimum, and 25 ms (10 × 2.5 ms) would be better; in some cases where maximum relief is desired 5 to 6 ms per foot is appropriate.

Geologic Variations

Rock is not the homogeneous material assumed by the empirical formulas; therefore, it is often necessary to employ burden correction factors for specific geological conditions. Table 16.3 provides burden distance correction factors for rock deposition, K_d (Fig. 16.19), and rock structure, K_s.

$$B_{\text{corrected}} = B \times K_d \times K_s \tag{16.3}$$

TABLE 16.3 Burden Distance Correction Factors

Rock Deposition	K_d
Bedding steeply dipping into cut	1.18
Bedding steeply dipping into face	0.95
Other cases of deposition	1.00
Rock Structure	**K_s**
Heavily cracked, frequent weak joints, weakly cemented layers	1.30
Thin, well-cemented layers with tight joints	1.10
Massive intact rock	0.95

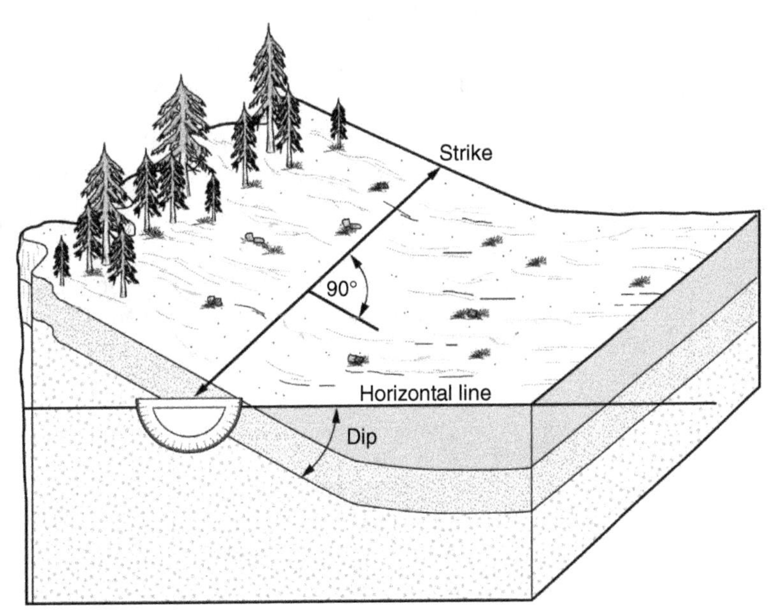

FIGURE 16.19 Rock deposition terminology.

Sample Calculation: A quarry is in a limestone (SG, 2.6) formation with horizontal bedding, with numerous weak joints. From a borehole test-drilling program, it is believed the limestone is highly laminated with many weakly cemented layers. Because of possible wet conditions, a cartridged slurry (relative bulk strength of 140) will be used as the explosive. The 4.5-in. blastholes will be loaded with 4-in.-diameter cartridges. What is the calculated burden distance?

$$B = 0.67 \times 4 \times \sqrt[3]{\frac{140}{2.6}} = 10.12 \text{ ft}$$

Using correction factors from Table 16.3, $K_d = 1$, horizontal bedding, $K_s = 1.3$, and numerous weakly cemented layers:

$$B_{\text{corrected}} = 10.12 \times 1 \times 1.3 = 13.1 \text{ ft}$$

Blasthole Diameter

The diameter of the blasthole will affect rock fragmentation, air blast, flyrock, and ground vibration. Larger blastholes have higher explosive loading densities (lb/ft or kg/m). As a result, blast patterns can be expanded while maintaining the same energy factor within the rock mass. This pattern expansion increases production by reducing the total number of holes required.

However, as a general rule, large-diameter blastholes of 6 to 15 in. (15 to 38 cm) have limited applications on most construction projects because of the fragmentation requirements and depth of cut limits. Construction blasthole diameters usually vary from 3.5 to 5.5 in. (89 to 140 mm), and the normal drilling depth is less than 40 ft (12 m). A rule of thumb is to use blasthole diameters smaller than the bench height divided by 60. Therefore, for a bench height of 13 ft, blasthole diameter should be in the order of 2.6 in. $\left(\frac{13 \text{ ft}}{60} \times 12 \text{ in. per ft}\right)$.

Another consideration in selecting the proper blasthole diameter is the detonation characteristics of the explosive. All commercial explosives are nonideal explosives. The ideal is the expected velocity based on equilibrium chemical thermodynamics. Nonideal explosives, in certain cases, have velocity of only one-third of the ideal detonation velocity.

Nonideal explosives have a critical diameter below which they will not reliably detonate. An ANFO's performance falls off quickly below 3 in. (76 mm). Figure 16.20 presents the results from an empirical model of the effect charge diameter has on detonation velocity. It is based on field detonation velocity measurements. The blasthole diameter should be large enough to permit the optimum detonation of the explosives.

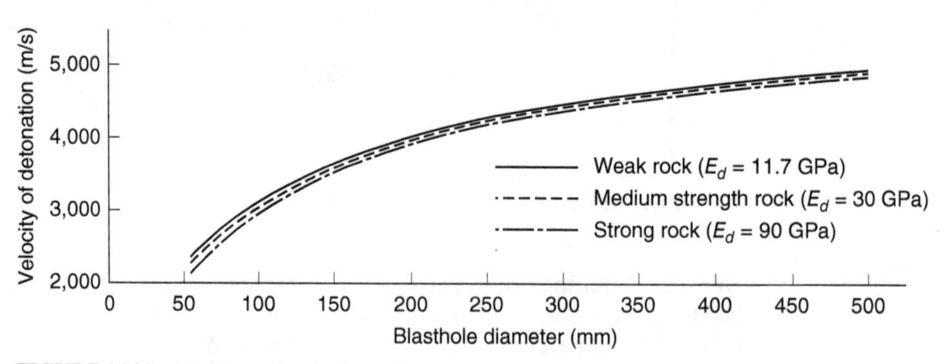

FIGURE 16.20 Model-predicted effect of charge (hole) diameter on a blended explosive.

In some situations, as in a quarry, the blaster can adapt the bench height to optimize the blast, but on a construction project, the existing ground and the specified final project grades set limits on any bench-height modification.

One of the parameters in both Eqs. (16.1) and (16.2) was the diameter of the explosive, D_e. The diameter of the explosive is limited by the diameter of the blasthole. Consider the earlier example: if the contractor had wanted to use a 5-in. bit, the charge would now be 5 in. in diameter.

Using the 5-in. explosive diameter, the burden distance would be

$$B=\left(\frac{2\times 0.8}{2.75}+1.5\right)\times 5=10.4 \text{ ft}$$

By increasing the blasthole diameter, the number of holes that will have to be drilled and loaded has been reduced. However, the question of bench height has not been considered in either example. If the bench height was only 13 ft because of the required excavation depth, which size blasthole should be used? The stiffness ratios (SRs) for the two blasthole diameters under consideration are

$$\text{3-in. blasthole: } \frac{13 \text{ ft height}}{6.2 \text{ ft burden}} = 2.1 \text{ SR}$$

$$\text{5-in. blasthole: } \frac{13 \text{ ft height}}{10.4 \text{ ft burden}} = 1.3 \text{ SR}$$

Table 16.4 gives the relationship between stiffness ratio and the critical blasting factors. The data in Table 16.4 indicate that for the 5-in. blasthole there will be blasting problems. Even the 3-in. blasthole can only be expected to yield fair results, which indicates the shot should be redesigned.

TABLE 16.4 Stiffness Ratio's Effect on Blasting Factors

Stiffness Ratio	1	2	3	4 and Higher*
Fragmentation	Poor	Fair	Good	Excellent
Air blast	Severe	Fair	Good	Excellent
Flyrock	Severe	Fair	Good	Excellent
Ground vibration	Severe	Fair	Good	Excellent

*Stiffness ratios above 4 yield no increase in benefit.

It would be good to have an SR value of at least 3 for the blast to yield good results. Try a 2-in. blasthole and an explosive diameter of 2 in.

$$B=\left(\frac{2\times 0.8}{2.75}+1.5\right)\times 2.0=4.2 \text{ ft}$$

$$\text{2-in. blasthole: } \frac{13 \text{ ft height}}{4.2 \text{ ft burden}} = 3.1 \text{ SR}$$

Of course, a reduction in burden distance will proportionally increase the amount of drilling and blasting. The added cost of drilling and blasting must be balanced with excavation and crushing for the most cost-effective operation (Fig. 16.17).

If the rule-of-thumb blasthole diameter value of 2.6 in., obtained by dividing the height by 60, had been used, the SR would be 2.4. Therefore, rules of thumb are clearly reasonable magnitude checks or first approximations, but a full analysis is necessary to develop a blast design. This means more rows will be required, but better fragmentation will result. Drilling cost will be increased, but secondary blasting and handling cost should be reduced.

Three factors affect burden distance: (1) the specific gravity of the rock, (2) the diameter of the explosives, and (3) either the specific gravity of the explosives or the relative bulk strength of the explosive in the case of an emulsion. The underlying geological conditions of a project are a given, and the blasting engineer must work with the given site environment. There are many different commercial explosives with various strengths, but across the complete range of explosive strengths, the effect on calculated burden distance would be very small, only 2 to 3 ft. If the burden distance must be altered to achieve an effective blast providing good fragmentation and not causing damage, explosive diameter is the parameter to be adjusted.

Top Stemming

The purpose of top stemming (collar distance) is to confine the explosive energy to the blasthole. To function properly, the material used for stemming must lock into the borehole. It is common practice to use drill cuttings as the stemming material. Very fine cuttings or drill cuttings are essentially dust and will not accomplish the desired purpose. To function properly, the stemming material should have an average diameter 0.05 times the diameter of the hole, and it should be angular. It may be necessary in such cases to bring in crushed stone. It is recommended No. 8 stone (½ in.) be used as stemming for holes of less than 4 in. diameter and No. 57 stone (¾ in.) for holes greater than 4 in. diameter. Standard practice is to drop the stone directly on the filled ANFO powder column (Fig. 16.10). Very coarse materials do not make good stemming because they tend to bridge and will be ejected from the hole.

If the stemming depth is too great, there will be poor top breakage from the explosion and backbreak will increase. When the stemming depth is inadequate, the explosion will escape prematurely from the hole (Fig. 16.21).

FIGURE 16.21 Explosive energy escaping from the blastholes.

Under normal conditions with good stemming material, a stemming distance, T, of 0.7 times the burden distance, B, will be satisfactory, but stemming depth can range from 0.7 to 1.3 B. The stemming equation is

$$T = 0.7 \times B \tag{16.4}$$

Another design approach is to use the ratio of stemming depth to blasthole diameter. An appropriate ratio is from 14:1 to 28:1, depending on the velocities of the explosive and rock, the physical condition of the rock, and the type of stemming used. Greater top-stemming depth is required where the velocity of the rock exceeds the detonation velocity of the explosive or where the rock is heavily fractured or low density.

When drilling dust is used for stemming material, the stemming distance will have to be increased by another 30% of the burden distance because the dust will not lock into the hole.

Sometimes it is necessary to deck load a blasthole. Deck loading is the operation of placing inert material (stemming material) in a blasthole to separate several explosive charges in the hole from each other. This is necessary when the blasthole passes through a seam or crack in the rock. In such a situation, if the shot were fired with explosive material completely filling the hole, the force of the blast would blow out through the weak seam. Therefore, when such conditions are encountered, a charge is placed below and above the weak seam, and stemming material is tamped between the charges. A stem plug made of plastic or flexible vinyl is inserted to the desired borehole depth to separate stemming and blasting material (Fig. 16.11).

Subdrilling

A shot will normally not break to the very bottom of the blasthole. This can be understood by remembering the second mechanism of breakage is flexural rupture. To achieve a specified grade by blasting, it is necessary to drill below the desired floor elevation (Fig. 16.5). This portion of the blasthole below the desired final grade is termed "subdrilling." The subdrilling distance, J, required can be approximated by the formula

$$J = 0.3 \times B \tag{16.5}$$

Satisfactory subdrilling depths can range from 0.2 to 0.5 B, as results are dependent on the type of rock and its structure. In many cases, however, the project specifications may limit subdrilling to 10% or less of the bench height. In blasting for foundations for structures where a final grade is specified, subdrilling of the final cut layer is normally severely restricted by specification—in some cases no greater than 6 in. (15 cm)—to minimize drill trace. The final layer removed in structural excavations is usually limited to a maximum depth of between 5 and 10 ft (1.5 to 3 m). Typically, subdrilling is not allowed in a final 5-ft (1.5-m) cut layer and is restricted to 2 ft (0.6 m) for the final 10-ft (3-m) cut layer. This is to prevent damage to the rock upon which the foundation will rest. Additionally, the diameter of the blasthole is often limited to 3.5 in. (90 mm).

Hole Patterns

The three commonly used blasting patterns are (1) the square, (2) the rectangular, and (3) the staggered (Fig. 16.22). The square drill pattern has equal burden and spacing distances. With the rectangular pattern, the spacing distance between holes is greater than the burden distance.

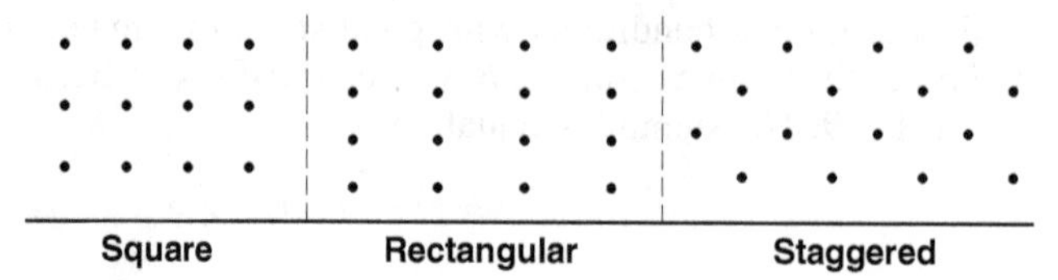

FIGURE 16.22 Common blasting patterns.

Both of these patterns place the holes of each row directly behind the holes in the preceding row. In the staggered pattern, the holes in each row are positioned at the midpoint between the holes in the preceding row. In the staggered pattern, the spacing distance between holes should be greater than the burden distance. Staggered patterns are effective with delayed ignition at an angle to the free face when project conditions warrant this blasting approach, such as removing benches in corners or adjacent to structures. The blasting pattern is always designated as *Burden* × *Spacing*, so the design parameters are never confused; for example, a 10 ft × 12 ft design has a 10-ft burden and 12-ft spacing.

A V-cut firing sequence is used with square and rectangular drilling patterns (Fig. 16.23). The burdens and subsequent rock displacement are at an angle to the original free face when the V-cut firing order is used. The staggered drilling pattern is used for row-by-row firing, where the holes of one row are fired before the holes in the immediately succeeding row. V-cut firing orients the movement of blasted rock toward the center of the free face, not perpendicular, with the goal of depositing the fractured rock closer to the new free face.

FIGURE 16.23 V-pattern firing sequence. Numbers indicate firing order.

Spacing is the distance between adjacent blastholes, measured perpendicular to the burden. Where the rows are blasted one after the other, as when firing a staggered blasthole pattern, the spacing is measured between holes in the same row. Where the blast progresses at an angle to the original free face, as in Fig. 16.23, the spacing is measured at an angle from the original free face (Fig. 16.24). Spacing distance is calculated as a function of the burden.

FIGURE 16.24 A presplit rock face.

When holes are spaced too close and fired instantaneously, there will be venting of the energy resulting in air blast and flyrock. When the spacing is extended, there is a limit beyond which fragmentation will become harsh. Before beginning a spacing analysis, two questions must be answered concerning the shot:

(1) Will the charges be fired instantaneously or will delays be used?

(2) Is the stiffness ratio greater than 4?

An SR of less than 4 is considered a low bench, and a high bench is an SR value of 4 or greater. This means there are four design cases to be considered:

1. Instantaneous initiation, with the SR greater than 1 but less than 4.

$$S = \frac{L+2B}{3} \tag{16.6}$$

where S = Spacing
L = Bench height

2. Instantaneous initiation, with the SR equal to or greater than 4.

$$S = 2B \tag{16.7}$$

3. Delayed initiation, with the SR greater than 1 but less than 4.

$$S = \frac{L+7B}{8} \tag{16.8}$$

4. Delayed initiation, with the SR equal to or greater than 4.

$$S = 1.4B \tag{16.9}$$

The actual spacing utilized in the field should be within ±15% of the calculated value.

Consider a project loading 4-in. diameter blastholes with bulk ANFO. The bench height is 35 ft, and each hole is to be fired on a separate delay. The blasting engineer would like to use an 8 × 8 drill pattern (8-ft burden × 8-ft spacing). Assuming the burden distance is correct, determine whether the 8-ft spacing is acceptable. First, check the stiffness ratio for high or low bench:

$$SR = \frac{L}{B} = \frac{35}{8} = 4.4;\text{ this is >4, therefore high bench.}$$

Delay timing is used and Eq. (16.9):

$$S = 1.4 \times 8 = 11.2 \text{ ft}$$

Range, 11.2 ± 15%: $9.5 \le S \le 12.9$

The proposed spacing of 8 ft does not appear to be sufficient. At a minimum, the pattern should be changed to 8-ft burden × 9.5-ft spacing for the first trial shot in the field.

Blast Design Example

Develop a blast design for a project in granite rock with an average bench height of 20 ft. The contractor's equipment can easily drill 3-in. diameter holes. Loading the borehole with bulk ANFO with a specific gravity of 0.8 has been proposed. Delayed blasting techniques will be utilized. The specific gravity of granite is between 2.6 and 2.9 (Table 16.1), so use the average, 2.75. Using Eq. (16.1), the trial burden is obtained:

$$B = \left(\frac{2 \times 0.8}{2.75} + 1.5\right) \times 3.0 = 6.2 \text{ ft}$$

Use 6 ft for burden distance. Remember the numbers calculated here will be used in the field. Make it easy for those actually doing the work under often-challenging project conditions and round the calculated numbers to the nearest foot or half-foot.

$$\text{The stiffness ratio} = \frac{L}{B} = \frac{20}{6} = 3.3$$

From Table 16.4, the stiffness ratio is good.

The stemming depth, from Eq. (16.4), is

$$T = 0.7 \times 6 = 4.2 \text{ ft}$$

Use 4 ft for stemming.

The subdrilling depth, from Eq. (16.5), is

$$J = 0.3 \times 6 = 1.8 \text{ ft}$$

Use 2 ft for subdrilling.

The spacing for an SR greater than 1 but less than 4 and using delay initiation, from Eq. (16.8), is

$$S = \frac{20 + (7 \times 6)}{8} = 7.75 \text{ ft}$$

7.75 ± 15%: The range for S is 6.6 to 8.9 ft.

As a first trial shot, use a 6-ft burden × 8-ft spacing pattern. Note the blast has been designed to require only integer measurements in the field.

Powder Factor

The amount of explosive required to fracture a ton or cubic yard of in situ rock is a measure of the economy of a blast design, or the *powder factor*. Powder factors for surface blasting can vary from 0.25 to 2.5 lb of explosives per cy rock (0.2 to 1.5 kg/m^3), with 0.5 to 1.0 lb/cy (0.3 to 0.6 kg/m^3) being typical values. Higher-energy explosives, such as those containing greater amounts of aluminum, will break more rock per unit weight than lower-energy explosives.

Many blasters, quarry operators, and miners may choose to measure the powder factor in terms of rock tonnage blasted per pound of explosives. Because the specific gravity and density of rock vary, as well as the specific gravity of explosives, each blaster will establish their unique powder factor. A general powder factor for blasting standard rock with median SG = 2.75 and bulk ANFO is at least 1.0 ton/lb (2.0 Mton/kg), with an ideal factor of 2.0 ton/lb (4.0 Mton/kg).

Table 16.1 is a loading density chart enabling the engineer to calculate the weight of explosive required for a blasthole. In the previous example, the diameter of the explosives was 3 in. and the explosive had a density of 0.8. Using Table 16.1, the loading density is found to be 2.45 lb/ft of charge. The powder column length is the total hole length minus the stemming: 18 ft in this case [20 ft + 2 ft (subdrilling) – 4 ft (stemming)]. Not considering employment of a primer, the total weight of explosive used per blasthole would be 18 ft × 2.45 lb/ft = 44.1 lb.

The amount of rock fractured by one blasthole is the pattern area times the depth to grade. For the 6 ft × 8 ft pattern with a 20-ft depth to grade, each hole would have an affected volume of 35.6 cy:

$$\text{Volume} = \left(\frac{6\ \text{ft} \times 8\ \text{ft} \times 20\ \text{ft}}{27\text{cf/cy}}\right)$$

The powder factor will bc 1.29 lb/cy

$$\text{Powder Factor} = \left(\frac{44.1\ \text{lb}}{35.6\ \text{cy}}\right) = 1.29\ \text{lb/cy}$$

Converting the powder factor in rock tonnage per lb of explosives, 1.87 ton/lb:

$$\text{Powder Factor}\left(\frac{35.6\ \text{cy}}{44.1\ \text{lb}} \times \frac{27\ \text{cf}}{1\ \text{cy}} \times \frac{(2.75)(62.4\ \text{lb})}{1\ \text{cf}} \times \frac{1\ \text{ton}}{2000\ \text{lb}}\right) = 1.87\ \text{ton/lb}$$

A powder factor of 1.87 tons/lb suggests this blast is nearly ideal with a relatively high yield from the standpoint of blasted rock output versus explosives input.

In all the design examples up to this point, it has been assumed that only one explosive was used in a blasthole. This is not typically the case; if a hole is loaded with ANFO, it will require a primer to initiate the explosion. When the bottom of some holes is expected to be wet, an allowance must be made for a water-resistant explosive, such as a slurry. In the case of a powder column 18 ft deep and loaded with ANFO and specific gravity 0.8, a primer will have to be placed at the bottom of the hole. If an 5-in.-long × 2-½-in.-dia. stick of pentolite PETN/TNT primer, specific gravity 1.6, is used, there will be 18 ft minus 5 in. of ANFO and 5 in. of primer.

In this case, assuming all dry holes, the weight of explosives based on a 3-in. explosive diameter would be:

ANFO	2.45 lb/ft × (211 in. ÷ 12 in./ft) =	43.08 lb
Dynamite	3.40 lb/ft × (5 in. ÷ 12 in./ft) =	1.42 lb
		44.50 lb

The total weight of explosive per hole is 44.50 lb for 18 ft of powder column.

To check vibration, it is necessary to know the total explosive weight. The breakout as to the amount of agent and the amount of primer is also important, because the prices of the two are considerably different. Primer can cost over ten times the unit price of blasting agent. Consider the market prices of No. 2 fuel oil and low-grade ammonium nitrate fertilizer. In this case, primer is about 3.2% by weight of the total amount of explosives.

Trench Rock

When excavating a trench in rock, the diameter or width of the structural unit to be placed in the trench, whether a pipe or conduit, is the principal consideration. When considering necessary trench width, the space required for working and backfill placement requirements must be taken into account. Many specifications will specifically address backfill width. Another important consideration is the size of the excavation equipment.

The geology will have a considerable impact on the blast design. Trenches are at the ground surface; usually, they will be extending through soil overburden and weathered unstable rock into solid rock. This nonuniform condition must be recognized. The blaster must check each individual hole to determine the actual depth of rock. Explosives are placed only in rock, not in the overburden.

If only a narrow trench in an interbedded rock mass is required, a single row of blastholes located on the trench centerline is usually adequate. Equation (16.1) or Eq. (16.2) will provide a first trial for hole spacing. The delayed timing of the shot should sequence down the row. The firing of the first hole provides the free face for the progression, which is why the equations are applicable. In the situation where the trench is shallow or there is little overburden, blasting mats laid over the alignment of the trench may be necessary to control flyrock. In the case of a wide trench or when solid rock is encountered, a double row of blastholes is common. Due to the shallow borehole depths and small volume of excavation in a trench, a lesser amount of explosives is necessary for trench blasting than conventional downhole blasting.

BREAKAGE CONTROL TECHNIQUES

Controlled blasting employs the use of reduced explosive quantities loaded into holes generally smaller in diameter and spaced closer than the main blast. The holes are often placed along the periphery of an excavation or individual shots. Breakage control techniques are used to limit overbreak, reduce fractures within remaining rock walls, and reduce ground vibrations.

Overbreak

Rock usually tends to overbreak at the top of the bank, and special drilling or loading may be required to avoid leaving a hard bottom rib at each blast junction. It may be necessary to set back the first row of holes for the next blast for more than the normal burden, as their burden may be partly shattered, so the shot is not confined as well as in the other holes. Drills may not be able to work close to the edge, or may not be able to penetrate shattered material. In such conditions the front row may not pull the bottom unless it is drilled considerably deeper than the others, some horizontal bottom snake holes are drilled, or a denser or faster explosive is used.

Line Drilling

One technique used to control overbreak outside the blasting limits and achieve a definite final excavation surface is line drilling. Line drilling is simply the drilling of a single row of unloaded, closely spaced holes along the perimeter of the excavation, creating a plane of weakness that the primary blast can break. The diameter of these holes ranges from 2 to 3 in. (50 to 75 mm). They are spaced 2 to 4 diameter apart. These holes are left empty. The maximum practical depth to which line drilling can be done is dependent on the accuracy the drilling can achieve. Seldom can the alignment of the holes be held at depths greater than 30 ft (10 m). They create a line of weakness along which the rock should be sheared by the preliminary blast. The back row of primary holes may be closer to the line than the normal between-row interval, and may be more closely spaced and lightly loaded than other rows.

This method is best suited to formations with a minimum of bedding planes, joints, and other weaknesses that may affect breakage more than the line of holes. Weak-bedded rock may respond satisfactorily if its planes are nearly at right angles with the proposed slope, or exactly parallel to it.

In large-area blasting, control of overbreak and underbreak is mostly a matter of getting the most efficient use from equipment and explosives. In road cuts and other work where remaining rock is a permanent part of the finished work, accurate finishing may be required by the contract. In most rock formations, good-to-precise faces may be cut by line drilling, cushion blasting, preshearing, and various combination techniques. However, selection and refinement of method for a particular formation may be difficult.

For success, the rock must be reasonably sound and cohesive, and capable of standing at the slope to which it is trimmed. It is essential the drills and the drillers be able to keep the holes in accurate alignment, both to obtain the desired edge and because the holes often show in the completed work.

Presplitting

Presplitting is a technique for creating an internal free face, usually at the boundary of an excavation, which will contain stress waves from successively detonated holes. Presplit blastholes can be fired either in advance of other operations or just before any adjacent main blastholes (Fig. 16.24). If the presplit holes are fired just before the main blastholes, there should be a minimum of a 200-ms delay between the presplit blast and the firing of the nearest main blastholes.

Usually, the presplit blastholes are 2½ to 3 in. in diameter and are drilled along the desired surface at spacings varying from 18 to 36 in., depending on the characteristics of the rock. When the rock is highly jointed, the hole spacing must be reduced. There have, however, been some presplit operations using holes as large as 12¼ in. in diameter with depths of more than 80 ft. But the maximum depth is limited by the accuracy of the drilling, and this usually deteriorates at about 50 ft (15 m). Depths of 20 to 40 ft (6 to 12 m) are more common. The presplit holes are loaded with one or two sticks of explosive at their bottoms and then have smaller charges, such as 1¼- × 4-in. sticks spaced at 12-in. intervals, to the top of the hole. The sticks may be attached to Primacord with tape, or hollow sticks may be used, which permits the Primacord to pass through the sticks with cardboard tube spacers between the charges. It is important the charges be less than half the presplit hole diameter, and they should not touch the walls of the hole.

Overbreak is a concern with any presplit blast. As a precaution, many project specifications limit the explosive diameter to half the diameter of the presplit borehole diameter. If the presplit material face does not materialize as planned, a series of buffer holes can be drilled about 3 ft (1 m) from the presplit line at a spacing of 3 to 5 ft (1 to 1.5 m). Explosives and cardboard spacers or stem plugs are staggered in the borehole so less than 50% of the powder column is loaded with explosives.

When the explosives in these holes are detonated ahead of the main blastholes, the webs between the holes will fracture, leaving a surface joint serving as a barrier to the shock waves from the main blast. This will essentially eliminate breakage beyond the fractured surface.

Presplit Explosive Load and Spacing

The approximate load of explosive per foot of presplit blasthole is given by

$$d_{ec} \frac{D_h^2}{28} \tag{16.10}$$

where, d_{ec} = explosive load in pounds per foot
D_h = diameter of blasthole in inches

When this formula is used to arrive at an explosive loading, the spacing between blastholes can be determined by the equation

$$S_p = 10\,D_h \tag{16.11}$$

where, S_p = presplit blasthole spacing in inches

Because of the variations in the characteristics of rocks, the final determination of the spacings of the holes and the quantity of explosive per hole should be determined by tests conducted at the project. Many times field trials will allow the constant in Eq. (16.11) to be increased from 10 to as much as 14.

Presplit blastholes are not extended below grade (floor of the excavation) to eliminate drill trace. Instead of subdrilling, a concentrated charge of two to three times d_{ec} should be placed in the bottom of the hole.

Because a presplit blast is only meant to cause fracture, drill cuttings can be used as stemming. The purpose is to only momentarily confine the gases and reduce noise. Stemming of 2 to 5 ft is normal.

Consider the following example of a presplit design. By contract specification, the walls of a highway excavation through rock must be vertically presplit. The contractor will be using drilling equipment capable of a 3-in.-dia. hole. Explosive load and hole spacing must be determined for the first presplit shot on the project. Using Eqs. (16.10) and (16.11),

$$d_{ec} = \frac{3^2}{28} = 0.32 \text{ lb/ft}$$

$$S_p = 10 \times 3 = 30 \text{ in.}$$

The bottom load should be $3 \times 0.32 = 0.96$ lb.

Vibration

Explosive blasts are an instantaneous conversion of blasting solids and liquids to gas. This extremely rapid transformation and release of energy undoubtedly have the potential to produce ground vibration and increased air pressure. Estimates by blasting engineers suggest 95% of the explosive energy dissipates as the rock fractures, while the remaining 5% residual energy is transferred through the ground and air, eventually dispersing as it travels away from the blast epicenter. Understanding vibration is important because it both affects surrounding people and property and affords valuable insight concerning the blasting design parameters. If the vibration is excessive, not only will it disturb the surrounding environment but it also presents an opportunity to modify future blasting designs for greater efficiency.

Rock exhibits the property of elasticity. When explosives are detonated, elastic waves are produced as the rock is deformed and then regains its shape. The two principal factors affecting how this motion is perceived at any discrete point are the mass of the detonated explosive charge and distance to the charge. Though the vibrations diminish in strength with distance from the source, they can achieve audible and physically detectable movement close to the blast site. Rarely do these vibrations reach levels causing damage to structures, but the issue of vibration problems is controversial. The case of old, fragile, or historical buildings is a situation where special care must be exercised in controlling vibrations because there is a danger of significant structural damage. The issue of vibration can cause restraints on blasting operations and lead to additional project cost and time. The determination of "acceptable" vibration levels is, in many cases, very difficult duc to its subjective nature with regard to being a nuisance. Humans and animals are very sensitive to vibration, especially in the low-frequency range (1 to 100 Hz).

It is the unpredictability and unusual nature of a vibration source, rather than the level itself, that is most likely to result in complaints. The effect of intrusion tends to be psychological rather than physiological, and is more of a problem at night, when occupants of buildings expect no unusual disturbance from external sources.

Vibration Strength Levels

When vibration levels from an "unusual source" exceed the human threshold of perception, complaints may occur. Vibration is measured near blast sites by the peak particle velocity (PPV), where humans begin to experience some level of disorientation between 0.008 and 0.012 in./sec. Peak particle velocity can be described as the maximum particle speed oscillating about a point of equilibrium from a passing wave, or the maximum amplitude per unit time. In an urban situation, serious complaints are probable when

PPV exceeds 0.12 in./sec, even though these levels are much less than what would result from a thunderstorm or passing train. Personal tolerance will be improved, provided the origin of the vibrations is known in advance and no damage results. With proactive communication and warning adjacent property owners and tenants of an imminent blast, much of the blast sensation is alleviated. It is important to provide people with a motivation to accept some temporary disturbance. Appropriate practice is to avoid vibration-causing activities at night.

There are published studies comparing the stresses imposed on structures by typical environmental charges and equivalent particle velocities. A 35% change in outside humidity imposes stress equivalent to almost a 5.0 in./sec particle velocity. A 12°F change in inside temperature imposes stress equivalent to about a 3.3 in./sec particle velocity. A 23-mph wind differential imposes a stress equivalent to about a 2.2 in./sec particle velocity. Typical construction blasting creates particle velocities of less than 0.5 in./sec.

Consequently, it must be remembered people can perceive very low levels of vibration, and at the same time they are unaware of the silent environmental forces acting on and causing damage to their homes. So even though construction activities cause movements significantly less than those created by common natural occurrences, the impact perceived by humans can cause problems.

Therefore, because blasting operations may cause actual or alleged damages to buildings, structures, and other properties located in the vicinity of the blasting operations, it is desirable to examine, photograph, and document any structures for which charges of damages may be made following blasting operations. To be of value, the survey must be thorough and accurate. Before firing a blast, seismic (geophone) and audible (microphone) recording instruments can be placed in the vicinity to monitor the magnitudes of vibrations and sound waves caused by the shot (Fig. 16.25). Minimum trip levels for recording the event may be set at 125 decibels for sound and 0.04 in./sec for vibration. Minimum recording levels vary by project site. The persons responsible for the blasting may conduct the monitoring, or if the company responsible for the blasting carries insurance covering this activity, a representative of the insurance carrier may provide this service.

FIGURE 16.25 Microphone and geophone to record blast vibration.

Vibration Mitigation

Delays in the blast initiation sequence will reduce vibrations because the mass of the individual charges is less than the total that would have been fired without the delays. The U.S. Bureau of Mines has proposed a formula to evaluate vibration and as a way to control blasting operations:

$$D_s = \frac{d}{\sqrt{W}} \tag{16.12}$$

where D_s = scaled distance (nondimensional factor)
d = distance from shot to a structure in feet
W = maximum charge weight per delay in pounds

A scaled distance value of 50 or greater indicates a shot is safe with respect to vibration according to the Bureau of Mines. Some regulatory agencies require a value of 60 or greater.

Covered Blasts

Throw can be closely controlled by working downward, using small blasts, and covering them with mats or chained timber. If the cover is large and heavy in proportion to the strength of the explosion, it will prevent any scattering of fragments. If the charge is heavy enough to lift the cover, it will move somewhat less than the average distance of throw to be expected from an uncovered blast and fragments with higher-than-usual velocity will be held in. It is important the cover extend several feet beyond the area being shot, particularly if the charge is heavy enough to lift the mat, as fragments might escape under its edges. When a front shovel or excavator is used to remove the shot rock, it is advantageous to use a woven steel mat, as it is easily handled with chains and provides a quicker and more secure cover than timber. The mat is lowered over the holes or dragged in such a manner that it will not damage the wiring and cause misfires. Timber or railroad ties are used when no mat is available. Two chains should be laid on the ground first, the timber piled, and the chains fastened over them, preferably by wired square knots.

Chaining is important, as unfastened logs may be thrown farther than rocks. Neither mats nor logs should be laid directly over boreholes, as they are liable to be thrown long distances and be severely damaged as well. Blasting mats should be used wherever there is the slight possibility of fragments reaching people or property. Even a small amount of flyrock on adjacent property is an indication loading should be reduced or the technique changed.

SAFETY

An accident involving explosives can easily kill or cause serious injury and property damage. A majority of injuries and fatalities from blasting operations involve failure to properly clear the blast area. The four major causes of blasting-related injuries in surface mining operations are lack of blast area security, flyrock, premature blasts, and misfires. The primary causes of blast area security problems are

- Failure by employees and visitors to evacuate the blast area
- Failure to understand the instructions of the blaster or supervisors
- Inadequate guarding of the access roads leading to the blast area

Blast area security accidents are preventable with good training and communications. Stories of flyrock accidents appear in the news regularly, and the injured or damaged parties are usually innocent.

Improper loading and firing practices contribute to the creation of flyrock. Any irregularity in the geological structure surrounding the borehole can cause an uneven stress field resulting in flyrock. There were 15 reported flyrock incidents in Tennessee during the 1999 to 2003 time frame where the rock traveled more than 500 ft from the blast site. In six of these instances, flyrock travel exceeded 1,000 ft, and in one case, flyrock travel approached half a mile.

In shooting charges of explosives, one or more charges may fail to explode; this is referred to as a "misfire." It is necessary to dispose of this explosive before excavating the loosened rock. The most satisfactory method is to shoot these charges if possible. In cases of ANFO, simple dilution with buckets of water is a remedial measure.

If electric detonators are used, the leading wires should be disconnected from the source of power prior to investigating the cause of the misfire. If the leg wires to the cap are available, test the cap circuit; if the circuit is satisfactory, try again to set off the charge.

When it is necessary to remove the stemming to gain access to a charge in a hole, the stemming should be removed with a wooden tool instead of a metal tool. If water or compressed air is available, either one can be used with a rubber hose to wash the stemming out of the hole. A new primer, set on top of or near the original charge, can be used to fire the charge.

The prevention of accidents depends on careful planning and faithful observation of proper blasting practices. The Mine Safety and Health Administration (MSHA) within the U.S. Department of Labor is the primary regulator of blasting activities. MSHA develops and enforces safety and health rules for all U.S. mines, regardless of size, number of employees, commodity mined, or method of extraction. In addition to jobsite inspections and issuing fines or warnings, MSHA sponsors training courses and assists state and local agencies with assistance.

The Occupational Safety and Health Administration (OSHA) is also active in the regulation of workplace safety and publishes key requirements applicable to blasting found under Standards: General Provisions 1926.900, Section U, Blasting and the Use of Explosives. Important subsections include Blaster Qualifications 1926.901(a), Surface Transportation of Explosives 1926.902, Storage of Explosives and Blasting Agents 1926.904, Firing the Blast 1926.909, Inspection after Blasting 1926.910, and Misfires 1926.911. Each state has specific guidelines for blasting, both within and outside the construction and mining industries, and has requirements for blaster certification training and licensing. Most licenses are renewed on an annual basis by completing a refresher course, which helps to educate on changes in blasting technology and regulations.

Explosive manufacturers will provide safety information on their specific products. An excellent source for material on blasting safety practices is the Institute of Makers of Explosives in Washington, D.C. Blasting manufacturers offer local and regional training programs and annually partner to sponsor the "Quarry Academy" focusing on best practices to improve process efficiency, both in blasting and rock crushing.

Large quantities of explosives should be transported in special vehicles marked and placarded in accordance with state or interstate laws. Smaller quantities may be carried in an ordinary car or truck, with any required warning signs made so they can be removed when not in use. Caps and explosives should be carried in different trips or vehicles unless quantities are small, in which case they may be carried in one vehicle if kept well separated and if permitted by law. ICC regulations are accepted by most of the U.S. states for intrastate transportation, but some have more restrictive laws.

Different classes of explosives should be kept in separate magazines in accordance with regulations. Magazine areas should be as far as practicable from roads or structures, and should be posted with warning signs and fenced if possible. Many regulatory agencies specify minimum distances between the magazine storage and private buildings, anywhere from 100 ft to 2,000 ft, depending upon pounds of stored explosives. Regulations also apply to fireworks magazine storage. Magazines should be constructed of cohesive fire-resistant material, such as sheet iron, or soft material that will tear or crush rather than separate into flying fragments. Ventilation and protection from grass fires and from excessive heat should be provided. Doors must be secured and locked.

CHAPTER 17
AGGREGATE PROCESSING

There are many types and sizes of crushing and screening plants. The capacity of a crusher will vary with the type of stone, size of feed stone, required size of the finished product, and the extent to which the stone is fed uniformly into the crusher. The screening process is based on the simple premise that particle sizes smaller than the screen cloth opening size will pass through the screen and oversized particles will be retained. After stone is crushed and screened, it is necessary to handle it carefully and store it in separate stockpiles to prevent contamination of the particle sizes. Belt conveyors provide an economical method for handling and transporting materials in a continuous flow. They are the primary stone movement equipment at a stone-processing plant.

ROCK CRUSHING

Pit and Quarry Planning

Most pit and quarry operations begin with the removal of soil or rock lying over the deposit to be mined. Overburden may include topsoil, subsoil, sand, gravel, clay, and also rock. The depth of overburden to be removed depends on its character and accessibility, the value of the underlying formation, and the extent to which the spoil can be sold or utilized.

Need for Stripping. Stripping overburden may be a large part of the cost of mining, and a number of factors should be considered before it is undertaken. First, is it necessary to strip it? It may be possible to mix the overburden material with the product or to separate it at lower cost during processing. If the pit has gravel or stone screening, separating, or washing equipment of adequate capacity, then separation as part of the material processing operation may be possible. In this case, thorough clearing is necessary, as sod and brush can clog screens and crushers.

Regrading. Replacement of topsoil and/or replanting may be required by law or by private contract between the pit operator and the land owner. Reclamation involves grading to smooth contours; planting with grass, hardy trees, or other vegetation; and protecting the graded land against erosion until the plants have taken hold. A reclamation plan must usually be filed and bonded before pits or quarries are developed.

Selective Digging

Selective digging may be done to separate at the face two or more materials of value and to remove them for processing. Any or all of the spoil from these operations may be hauled away or side-cast.

Layers. If the different formations lie horizontally, any excavator can move them if they can be cut as separate banks. If they are horizontal and two or more must be removed at once, the excavator should be able to work from the top down. When horizontal layers are separated by a dragline, it should have a boom at least twice as long as the bank is high. The boom angle should be low and the dump cable short to make it possible to pick up the bucket at a distance. Selective digging is commonly required in stripping overburden from gravel and clay pits.

Boulders. A common problem in pits that are dug without blasting is the occurrence of boulders too large for the loading or processing machinery. These are found in glacial and stream deposits, in disintegrated rock, and near steep slopes. In pits selling directly from the bank, there is no convenient way of utilizing boulders or disposing of them. Blasting will reduce them to a size that can be loaded, but the market for coarse rock is so limited that they may have to be sold as second-grade fill or wasted. The pit operator will generally prefer to allow them to accumulate along the bank, or will have holes dug to bury them. Occasionally, an abandoned pit is close and deep enough to permit disposal by pushing them over the edge.

If allowed to remain where they fell out of the bank or pushed into occasional piles, they will present obstacles to orderly pit development. In general, the nuisance value increases with the size of the boulders, relative to the power of the dozer that must handle them.

It is occasionally possible to sell boulders for use in jetties or breakwaters at a price high enough to justify hiring a machine of the size needed to load them.

Hauling

Pit hauling includes the movement of material from the bank to the plant or to storage, and also between the plant and storage in both directions. The principal hauling units for pit use are trucks and conveyor belts.

Conveyor belts may be considered either hauling units or part of the plant itself. They move and elevate material with minimum effort. They may be used instead of haul roads and trucks for delivery of a large volume of heavy material to a single point many miles away.

Trucks are excellent flexible, general-purpose units. They are available in a wide range of standard sizes and can be adapted to different-size loaders or production schedules by varying the number on the haul. Truck hauls may be kept short by adding conveyor belts to the plant. The new belt will dump on the receiving end of the previous belt. Such installations may be quite long and are justified whenever considerable yardage will be handled. Hoppers, which are built so a truck can drive straight across instead of backing to dump, are more expensive to construct but will allow a faster truck cycle. Such hoppers can also be used for scrapers.

PROCESSING PLANTS

The processing of aggregates includes screens, crushers, and washers with their feeding and discharge mechanisms. Four functions are required to accomplish the desired results:

1. Reducing particle size—crushing
2. Separating into particle size ranges—sizing/screening
3. Eliminating undesirable materials—washing
4. Handling and moving the crushed materials—storage and transport

Processing plants are capable of all four functions, ranging in size from smaller portable plants to larger fixed plants.

Portable Plants

The simplest crushing and screening equipment are small portable plants mounted on one or more trailers. Portable or mobile plants typically have crushing, screening, and optional washing equipment mounted on one or more wheeled trailers. Some equipment may be on one trailer in cases of construction demolition (Fig. 17.1) or topsoil screening (Fig. 17.2).

The use of portable units allows flexibility. However, their capability to produce a range of product specifications is limited because the difficulty in packing all needed components on a trailer frame.

They can often be used profitably on remote jobs or for filling special orders, or in a pit or quarry to support a fixed plant. They may also be used at subsidiary pits. Hauling is a

FIGURE 17.1 Portable crushing unit.

FIGURE 17.2 Portable topsoil screening plant.

major cost factor in delivering gravel or crushed rock to a project, and the ability to open and process material near the job can be advantageous. Although not very readily sold at the end of a project, they are more or less standardized, and if found to be of the wrong size and type, can be sold with a far smaller loss than a fixed plant. Also, the use of one or more of these units may permit the pit to be developed until adequate space is excavated for a fixed plant. If the pit or quarry is on a hillside, with steadily increasing bank height, getting the permanent plant positioned will result in cheaper primary hauling over the whole job.

Fixed Plants

A fixed plant should be the right type for the material being processed; it must be large enough for the job and within the capital budget. The first consideration is generally the most important, for if business is good, the plant can be expanded, although at relatively higher cost, and exceeding a budget may be less damaging than trying to use the wrong equipment. Plant manufacturers are ready to supply good engineering advice on every aspect of plant layout. It is a sound plan to get at least general recommendations from two or more companies and to compare their findings with local practice. Even with these precautions, no person without a good working knowledge of the business should make a heavy investment in machinery from catalogs.

Plant Location. A permanent or semi-permanent plant should be as close to the excavation area as it can be without being in danger from blasting and slides. If the pit is wide or includes many sections, the plant should be near the center or the side that produces the greatest yardage. Dig-down pits may supply the plant by means of ramps and trucks, vertical bucket conveyors, clamshells, or elevators. If material supply is vertical, the plant should be located on the pit edge, or nearby on firm footing. This method is ordinarily used only in rock pits, but not all rock will support a plant on the edge of a cliff. If the material is trucked, it is best to locate the plant well back from the pit to allow for ramps, storage, and room for expansion.

Wherever a fixed plant is placed, the cost of hauling to it will increase as the digging progresses, whether laterally, downward, or both. Taking down, moving, and setting up a permanent plant are usually tedious and expensive operations, particularly large plants. Even if it is of prefabricated, knock-down construction, rust, wear, and patching may make it hard to handle, and the foundations are generally left behind. Many mechanics consider it best to salvage only the operating units and to order or build new frames to carry the units. It is usually sound policy to charge the entire cost of such a plant against the material to be handled at its original location. If the pit area is limited by property boundaries, zoning restrictions, or change of ground and the depth of the deposit is known, the yardage can be calculated. It is best to make a liberal allowance for occurrence of unexpected masses of unusable material.

CRUSHING

Complete plants for the processing of blasted rock or pit-run gravel perform the function of taking in a coarse and variable material and turning out one or more classes of graded and uniform product. Equipment includes a hopper into which mined material is dumped, screens to separate the pieces according to size both before and after crushing, from one to three crushers to reduce oversize pieces, and hoppers or chutes and conveyors for individual handling of different sizes separated by the screens. Provision must be made for removal or storage of the products.

Classification

Crushers are often classified according to their stage of crushing, such as primary, secondary, and tertiary (Fig. 17.3). A primary crusher receives the stone directly from the excavation after blasting and performs the initial size reduction. The output of the primary crusher is fed to a secondary crusher for further size reduction. Some of the material may pass through four or more crushers before it is reduced to the desired size.

Crushers are classified by their method of mechanically transmitting energy to fracture the rock. Jaw, gyratory, and roll crushers work by applying compressive force. As the name implies, impact crushers use high-speed impact force to accomplish fracturing. By using units of differing size, crushing chamber configuration, and speed, the same mechanical-type crusher can be employed at different stages in the crushing operation.

Size Reduction

As stone passes through a crusher, its size reduction can be expressed as a reduction ratio of input feed size to output product size, as in Eq. (17.1):

$$\text{Reduction ratio} = \frac{\text{Input feed size}}{\text{Output product size}} \tag{17.1}$$

Crushing plants use step reduction because the degree of size reduction is directly related to applied energy and production. When there is a large difference between the size of the feed material entering the crusher and the resulting crushed product size, a greater

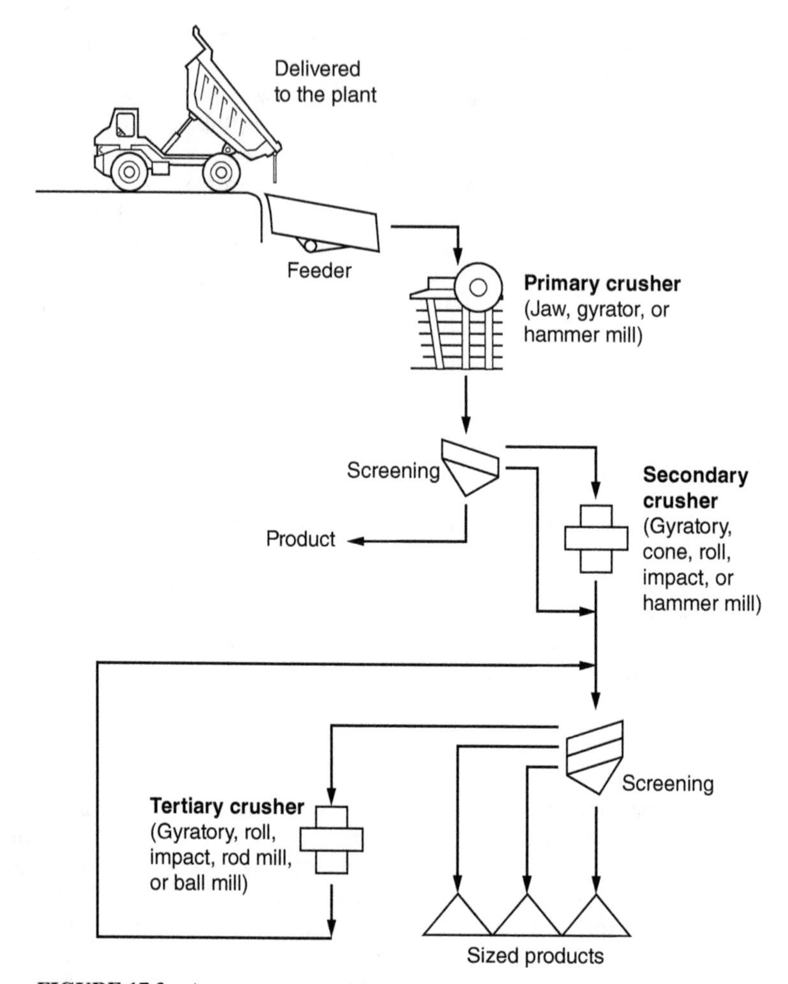

FIGURE 17.3 Aggregate processing steps.

amount of energy is required. Stone will be in the crusher chamber for a longer period, thereby slowing production and increasing cost. If this energy were concentrated in a single-step process, excessive fines would be generated, and normally there is only a limited market for fines. Fines are a nonrevenue waste material at many plants. Therefore, to minimize the quantity of waste material and increase production, the degree of breakage is spread over several stages as a means of closely controlling product size.

Table 17.1 lists the major types of crushers and presents data on attainable material reduction ratios. Primary crushers processing larger aggregates have lower reduction ratios than secondary and tertiary crushers. A multistep reduction process permits better control of particle size. The reduction ratio of a roller crusher could be estimated as the ratio of the dimension of the largest stone the rolls can nip divided by the smallest distance between the faces of the rolls. A reduction of ratio of 5:1 to 8:1 is common on most crushing operations. A more accurate measurement of reduction ratio is to use the ratio of size corresponding to 80% passing for both the feed and the product.

TABLE 17.1 The Major Types of Crushers

Crusher Type	Reduction Ratio Range
Jaw	
Double toggle	
Blake	4:1–9:1
Overhead pivot	4:1–9:1
Single toggle: Overhead eccentric	4:1–9:1
Gyratory	
True	3:1–10:1
Cone	
Standard	4:1–6:1
Attrition	2:1–5:1
Roll	
Compression	
Single roll	Maximum 7:1
Double roll	Maximum 3:1
Impact	
Single rotor	to 15:1
Double rotor	to 15:1
Hammer mill	to 20:1
Specialty crushers	
Rod mill	15:1–20:1
Ball mill	20:1–200:1

Jaw Crushers

Jaw crushers are typically employed as primary units because of their large energy-storing flywheels (Fig. 17.4) and high mechanical advantage. True gyratories are the other crusher type employed as primary crushers. These have recently become the unit of choice for primary crushing. A true gyratory is an excellent primary crusher because it provides continuous crushing and can handle slabby material. Jaw crushers do not handle slabby

A Electric power and flywheel

B Jaw plates

FIGURE 17.4 Portable jaw crusher.

material well. Impact crushers are increasingly used for primary crushing. They crush by hurling the stone against interior steel apron walls. Roller crushers squeeze and fracture the rock with a single- or double-roller system, and impact crushers use velocity to hurl and fracture the rock against internal steel walls. Models of gyratory, roll, and impact crushers can be found in both secondary and tertiary applications.

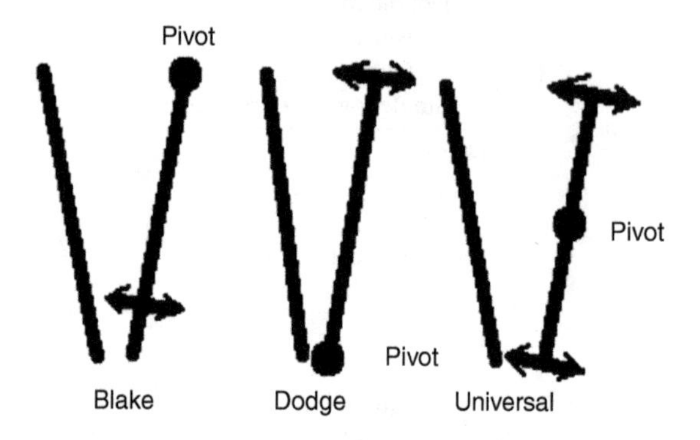

Jaw crushers are classified by the fixed location of the pivoting jaw plate. The most common jaw crusher is the Blake type, named for the inventor, which fixes the swing jaw at the top of the movable plate. Two less common jaw crusher configurations are the universal type, which fixes the movable jaw at the mid position, and the Dodge type, which fixes it at the lower position.

Jaw crushers are also classified by the number of toggles used to transmit flywheel rotational energy to the movable plate. Jaw crushers are usually designed with the toggle as the weakest part. The toggle will break if the crusher encounters an uncrushable object or is subjected to overload. This limits the damage to the crusher and provides safety to personnel.

Double Toggle. The Blake-type, double-toggle jaw crusher has a movable jaw suspended from a shaft mounted on the crusher frame. The rotation of a second shaft, which is eccentric and is located behind the movable jaw, raises and lowers the pitman to actuate two toggles and produce an elliptical crushing action. As the pitman raises the two toggles, a high pressure is exerted near the bottom of the swing jaw and partially closes the opening at the bottom of the two jaws. This operation is repeated as the eccentric shaft is rotated.

Jaw plates are replaceable. The jaw plates may be smooth or corrugated (Fig. 17.4B), in cases where slabby rock needs to be reduced. Corrugations are more common. They have the ability to confine the elongated slabby rock in the plate walls. This is important when specifications limit the percentage of thin and elongated particles, such as a maximum of 5% in concrete and asphalt mixtures. The swing jaw may be straight or slightly curved to reduce the possibility of choking. Elliptical movement of the jaw plates is effective in limiting choking and continuously moving rock through the crusher. Because double-toggle, Blake-type jaw crushers are so large and heavy, they do not lend themselves to portable applications.

Table 17.2 gives representative capacities for various sizes of Blake-type jaw crushers. In selecting a jaw crusher, consideration must be given to the size of the feed stone. The top opening of the jaw should be at least 2 in. wider than the largest stones fed into the crusher.

Capacity tables can be based on the open or closed position of the bottom of the swing jaw; therefore, the table should specify which setting applies. The closed position is commonly used for most crushers and is the basis for the values given in Table 17.2. However, Blake-type jaw crushers are often rated based on the open-size setting. The capacity is

TABLE 17.2 Representative Capacities of Blake-Type Jaw Crushers, in Tons per Hour (Metric Tons per Hour) of Stone*

Size Crusher, in. (mm)†	Maximum rpm	Maximum hp (kW)	Closed Setting of Discharge Opening, in. (mm)													
			1½ (38)	2 (51)	2½ (64)	3 (76)	3½ (89)	4 (102)	5 (127)	6 (152)	7 (178)	8 (203)	9 (229)	10 (254)	11 (279)	12 (305)
32 × 20 (800 × 510)	350	100 (75)	60 (55)	85 (77)	110 (100)	120 (109)	160 (145)	180 (164)	230 (209)	270 (245)	320 (290)					
37 × 23 (930 × 580)	330	125 (90)			135 (123)	155 (141)	195 (177)	220 (200)	275 (250)	330 (300)	385 (350)					
40 × 30 (1000 × 760)	260	150 (110)			140 (127)	165 (150)	205 (186)	235 (213)	290 (263)	345 (313)	375 (340)	460 (417)				
42 × 28 (1060 × 700)	280	150 (110)			160 (145)	190 (172)	230 (209)	260 (236)	320 (290)	380 (345)	440 (400)	500 (454)				
45 × 32 (1150 × 800)	260	175 (132)			180 (163)	200 (181)	255 (231)	280 (254)	335 (304)	400 (363)	460 (417)	520 (472)				
47 × 34 (1200 × 870)	230	220 (160)			195 (177)	215 (195)	280 (254)	310 (281)	375 (340)	445 (404)	510 (463)					
51 × 39 (1300 × 1000)	220	220 (160)						350 (317)	425 (385)	495 (449)	570 (517)	640 (580)	715 (648)	785 (712)		
55 × 47 (1400 × 1200)	220	300 (200)							445 (404)	525 (476)	600 (544)	680 (617)	760 (689)	835 (757)		
63 × 47 (1600 × 1200)	220	350 (250)								570 (517)	655 (594)	740 (671)	825 (748)	905 (821)	990 (898)	1075 (975)
79 × 59 (2000 × 1500)	200	500 (400)									835 (757)	940 (853)	1035 (939)	1145 (1039)	1240 (1125)	1345 (1220)

*Based on the closed position of the bottom swing jaw and stone weighing 100 lb/cf when crushed.
†The first number indicates the width of the feed opening, whereas the second number indicates the width of the jaw plates.

given in tons per hour based on a standard material unit weight of 100 lb/cf when crushed. Production is directly related to crusher size and closed (output) setting, with larger crushers and closed settings having higher production rates. Larger crushers also have slower flywheel rpm.

Single Toggle. When the eccentric shaft of the single-toggle crusher is rotated, the movable jaw shifts in both a vertical and horizontal motion. This type of crusher is used quite frequently in portable rock-crushing plants because of its compact size, lighter weight, and reasonably sturdy construction. A pair of lower spring arms help dampen the plate movement and minimize vibration transfer to the crusher frame. The capacity of a single-toggle crusher is usually rated at the closed size setting and is less than that of a double-toggle, Blake-type unit.

Sizes of Jaw and Roll Crusher Product. Although the setting of the discharge opening of a crusher will determine the maximum-size stone produced, the aggregate sizes will range from slightly greater than the crusher setting to a fine dust. Experience gained in the crushing industry indicates that for any given setting for a jaw or roll crusher, approximately 15% of the total amount of stone passing through the crusher will be larger than the setting. This is due in part to the irregular shapes of crushed aggregate particles, where the smaller dimension passes the closed setting and the larger dimension is later retained on a square mesh screen. If the output from a crusher is set at the same size as the openings of a receiving screen, 15% of the output will not pass through the screen.

Gyratory Crushers

Gyratories are the most efficient of all primary-type crushers. A gyrating mantle mounted within a deep bowl characterizes these crushers. They provide continuous crushing action and are used for both primary and secondary crushing of hard, tough, abrasive rock. To protect the crusher from uncrushable objects and overload, the outer crushing surface can be spring-loaded, or the mantle height may be hydraulically adjustable.

True Gyratory. A section through a gyratory crusher is shown in Fig. 17.5. The crusher unit consists of a heavy cast-iron or steel frame, with an eccentric shaft and driving gears in the lower part of the unit. In the upper part, there is a cone-shaped crushing chamber lined with hardened steel or manganese-steel plates called the *concaves*. The crushing member includes a hardened-steel crushing head mounted on a vertical steel shaft. This shaft and head are suspended from the spider at the top of the frame to allow vertical adjustment of the shaft. The eccentric support at the bottom causes the shaft and the crushing head to gyrate as the shaft rotates, thereby varying the space between the concaves and the head. Rock fed in at the top of the crushing chamber moves downward and undergoes a reduction in size until it passes through the opening at the bottom of the chamber. Aggregate that surges at the top of the shaft indicates an excessive feed rate or an insufficient closed setting. The goal is a balance between the highest allowable feed rate and closed setting that achieves product size.

The size of a gyratory crusher is the width of the receiving opening, measured between the concaves and the crusher head. The setting is the width of the bottom opening and may be the open or closed dimension. When a setting is given, it should be specified whether it is the open or closed dimension. Normally the capacity of a true gyratory crusher is based on an open-size setting. The ratio of reduction for true gyratory crushers usually ranges from 3:1 to 10:1, with an average value around 8:1.

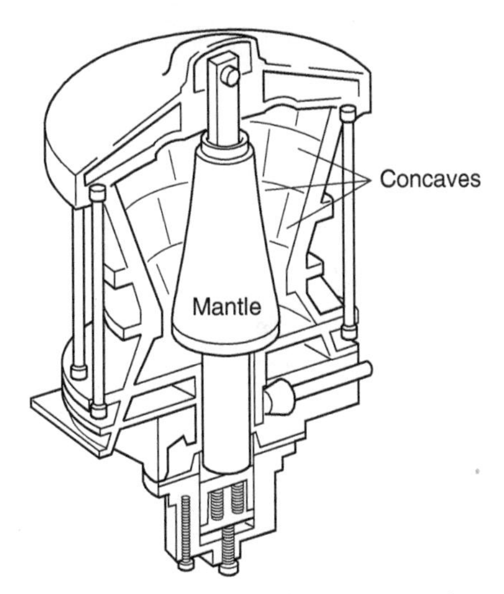

FIGURE 17.5 Section through a gyratory crusher.

When a gyratory crusher is used as a primary crusher, the selected size may be dictated by the size of the rock from the blasting operation, or it may be dictated by a desired capacity. In cases when a gyratory crusher is used as a secondary crusher, increasing its speed within reasonable limits may increase the capacity. Speed of the pinon shaft typically ranges from 300 to 500 rpm. A lubricated differential gear transfers rotation of the pinion shaft to the vertical main shaft supporting the mantle. A differential gear ratio of about 2:1 reduces the main shaft rotation to 100 to 200 rpm. Insufficient speed can stall the crusher, whereas excessive speed may produce a load imbalance and generate heat detrimental to the gear lube.

Table 17.3 gives representative capacities of gyratory crushers, expressed in tons per hour, based on a continuous feed of stone with a unit weight of 100 lb/cf when crushed. Gyratories with straight concaves are commonly used as primary crushers, whereas those with nonchoking concaves are commonly used as secondary crushers.

Cone Crushers

Cone crushers are used as secondary or tertiary crushers. They are capable of producing large quantities of uniformly fine crushed stone. A cone crusher differs from a true gyratory crusher in the following respects:

1. Shorter cone
2. Smaller receiving opening
3. Rotates at a higher speed, about twice the speed of a true gyratory
4. Produces a more uniformly sized product

Standard models have large feed openings for secondary crushing and produce stone in the 1- to 4-in. range. The capacity of a standard model is usually rated based on a closed-size setting.

TABLE 17.3 Representative Capacities of Gyratory Crushers, in Tons per Hour (Metric Tons per Hour) of Stone*

Size of Crusher [in. (cm)]	Approximate Power Required [hp (kw)]	Open-Side Setting of Crusher [in. (mm)]											
		1½ (38)	1¾ (44)	2 (51)	2¼ (57)	2½ (63)	3 (76)	3½ (89)	4 (102)	4½ (114)	5 (127)	5½ (140)	6 (152)
Straight Concaves													
8 (20.0)	15–25 (11–19)	30 (27)	36 (33)	41 (37)	47 (42)								
10 (25.4)	25–40 (19–30)		40 (36)	50 (45)	60 (54)								
13 (33.1)	50–75 (37–56)				85 (77)	100 (90)	133 (120)						
16 (40.7)	60–100 (45–75)						160 (145)	185 (167)	210 (190)				
20 (50.8)	75–125 (56–93)							200 (180)	230 (208)	255 (231)			
30 (76.2)	125–175 (93–130)								310 (281)	350 (317)	390 (353)		
42 (106.7)	200–275 (150–205)										500 (452)	570 (515)	630 (569)
Modified Straight Concaves													
8 (20.0)	15–25 (11–19)	35 (32)	40 (36)	45 (41)									
10 (25.4)	25–40 (19–30)		54 (49)	60 (54)	65 (59)								
13 (33.1)	50–75 (37–56)					95 (86)	130 (117)						
16 (40.7)	60–100 (45–75)						150 (135)	172 (155)	195 (176)				
20 (50.8)	75–125 (56–93)							182 (165)	200 (180)	220 (199)			
30 (76.2)	125–175 (93–130)								340 (308)	370 (335)	400 (362)		
42 (106.7)	200–275 (150–205)										607 (550)	650 (589)	690 (625)
Nonchoking Concaves													
8 (20.0)	15–25 (11–19)	42 (38)	46 (42)										
10 (25.4)	25–40 (19–30)	51 (46)	57 (52)	63 (57)	69 (62)								
13 (33.1)	50–75 (37–56)	79 (71)	87 (79)	95 (86)	103 (93)	111 (100)							
16 (40.7)	60–100 (45–75)			107 (96)	118 (106)	128 (115)	150 (135)						
20 (50.8)	75–125 (56–93)				155 (140)	169 (152)	198 (178)	220 (198)	258 (233)	285 (257)	310 (279)		

*Based on continuous feed and stone weighing 100 lb/cf when crushed.

Attrition models are for producing stone with a maximum size of about ¼ in. The capacity of an attrition model cone crusher may not be related to closed-size setting.

Figure 17.6 shows the difference between a gyratory and a standard cone crusher. The conical head on the cone crusher is usually made of manganese steel. It is mounted on a vertical shaft and serves as one of the crushing surfaces. The other surface is the concave; it is attached to the upper part of the crusher frame. The bottom of the shaft is set in an eccentric bushing to produce the gyratory effect as the shaft rotates.

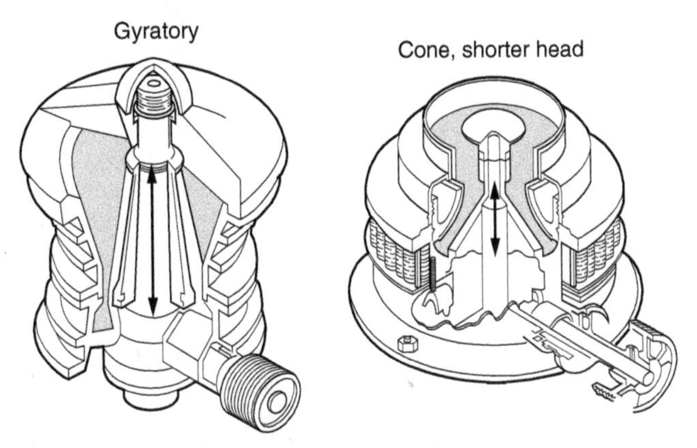

FIGURE 17.6 Sections through a gyratory and a cone crusher.

The maximum diameter of the crusher head can be used to designate the size of a cone crusher. However, the size of the rocks fed to the crusher is limited by the size of the feed opening, the width of the opening at the entrance to the crushing chamber. The magnitude of the eccentric throw and the setting of the discharge opening can be varied within reasonable limits. Because of the high speed of rotation, all particles passing through a cone crusher will be reduced to sizes no larger than the closed-size setting that designates the size of the discharge opening.

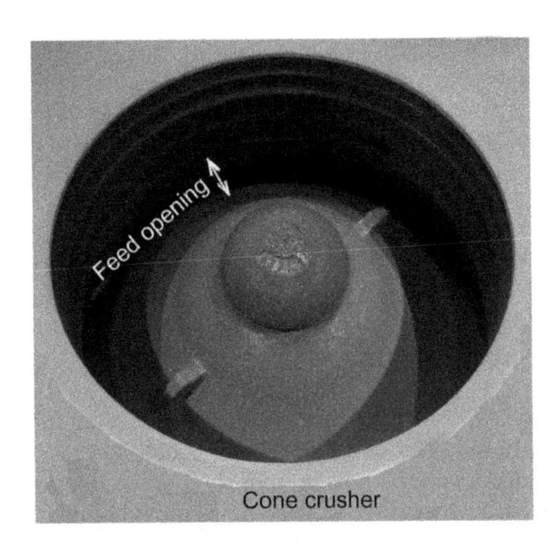

TABLE 17.4 Representative Capacities of Standard Cone Crushers, in Tons per Hour (Metric Tons per Hour) of Stone*

Size of Crusher, ft (m)	hp (kW)	Size of Feed, in. (mm)	Closed Setting of Discharge Opening, in. (mm)										
			¼ (6)	⅜ (9)	½ (13)	⅝ (17)	¾ (19)	⅞ (22)	1 (25)	1¼ (32)	1½ (38)	1¾ (48)	2 (51)
5	125	5½	55	65	75	90	95	100	105	130			
(1.5)	(90)	(141)	(50)	(59)	(68)	(82)	(86)	(91)	(95)	(118)			
6⅜	200	7¼		115	145	175	185	195	205	235	250		
(1.95)	(132)	(185)		(104)	(132)	(159)	(168)	(177)	(186)	(213)	(227)		
7	300	8⅝		140	175	210	230	255	270	290	330	365	
(2.15)	(220)	(220)		(127)	(159)	(180)	(209)	(231)	(245)	(263)	(300)	(331)	
7¼	300	9⅛		140	185	220	240	260	280	315	375	435	
(2.21)	(200)	(233)		(127)	(168)	(200)	(218)	(236)	(254)	(286)	(340)	(395)	
7¾	400	11¾		175	230	280	315	340	370	410	470	535	605
(2.37)	(315)	(300)		(159)	(209)	(254)	(286)	(308)	(336)	(372)	(426)	(485)	(549)
8¼	400	9⅞		200	250	300	330	365	385	410	470	525	
(2.51)	(315)	(252)		(181)	(227)	(272)	(300)	(331)	(349)	(372)	(426)	(476)	
8⅞	500	12¼		240	300	360	395	430	455	490	540	600	
(2.70)	(370)	(312)		(218)	(272)	(326)	(358)	(390)	(412)	(444)	(490)	(545)	
9	500	13⅛		215	285	345	395	430	450	520	580	665	760
(2.73)	(355)	(335)		(195)	(258)	(313)	(358)	(390)	(408)	(472)	(526)	(603)	(689)
10	600	13		285	360	425	475	520	550	590	675	750	
(3.06)	(450)	(330)		(258)	(326)	(385)	(431)	(471)	(499)	(535)	(612)	(680)	
12⅛	800	13⅞		325	415	485	540	590	675	740	795	955	1090
(3.70)	(600)	(353)		(295)	(376)	(440)	(490)	(535)	(612)	(671)	(721)	(866)	(987)

*Based on stone weighing 100 lb/cf when crushed.

Table 17.4 gives representative capacities for standard cone crushers, expressed in tons of stone per hour, for material with a unit weight of 100 lb/cf when crushed.

Roll Crushers

Roll crushers are used to produce further reductions in the sizes of stone after one or more stages of prior crushing. A *roll crusher* consists of a heavy steel frame equipped with either one or more hardened steel rolls, each mounted on a separate horizontal shaft. The surface of the rolls is either smooth or textured with narrow beads; the beads provide friction to grip and pull the rock through the crusher.

Single Roll. With a single-roll crusher, the material is drawn up and forced between a large-diameter roller and an adjustable liner. Because the material is dragged against the liner, these crushers are not economical for crushing highly abrasive materials, but they can handle sticky materials.

Double Roll. Roll crushers with two rollers are manufactured so each roll is driven independently by a flat-belt pulley or a V-belt sheave. A drive pulley, often lined with a tire tread, allows minor slippage in the belt when intermittent high tension occurs when crushing very hard rock. One of the rolls is mounted on a slide frame to permit an adjustment in the width of the discharge opening between the two rolls. The movable roll is spring-loaded to provide safety against damage to the rolls when noncrushable material passes through the machine.

Triple Roll. Triple-roll crushers are manufactured to provide two-stage reduction all in one crusher assembly. A single roll is arranged above a double roll so the stone undergoes both primary reduction and secondary reduction. Stone entering the crusher chamber is first crushed by the single-roll system, then cascades down to the double-roll arrangement for secondary reduction. The triple-roll crusher eliminates extra handling and a conveyor belt between the reduction stages.

Feed Size. The maximum size of material fed to a roll crusher is directly proportional to the diameter of the rolls. If the feed contains stones that are too large, the rolls will not grip the material and pull it through the crusher. The angle of nip (grip), B in Fig. 17.7, which is constant for smooth rolls, has been found to be 16°45′.

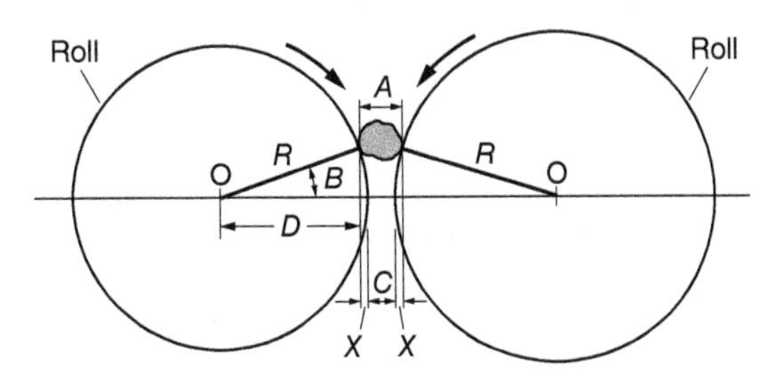

FIGURE 17.7 Crushing rock between two rolls.

The maximum size of stone that can be crushed is determined as follows. Referring to Fig. 17.7, these terms are defined as

R = radius of rolls, in inches
B = angle of nip (16°45′)
$D = R \cos B = 0.9575R$
A = maximum feed size, in inches
C = roll setting (size of finished product), in inches

Then

$$\begin{aligned} X &= R - D \\ &= R - 0.9575R = 0.0425R \\ A &= 2X + C \\ &= 0.085\,R + C \end{aligned} \tag{17.2}$$

Capacity. The capacity of a roll crusher will vary with the stone type, feed size, finished product size, width of rolls, the speed at which the rolls rotate, and the extent to which the stone is fed uniformly into the crusher. Referring to Fig. 17.7, the theoretical volume of a solid ribbon of material passing between the two rolls in 1 min would be the product of the width of the opening times the width of the rolls times the speed of the surface of the rolls. The volume can be expressed in cubic inches per minute or cubic feet per minute (cfm). In actual practice, the ribbon of crushed stone will never be continuous. A more realistic volume should approximate *one-fourth* to *one-third* of the theoretical volume. An equation to use as a guide in estimating the capacity is derived using these terms:

C = distance between rolls, in inches
W = width of rolls, in inches
S = peripheral speed of rolls, in inches per minute
N = speed of rolls, in rpm
R = radius of rolls, in inches
V_1 = theoretical volume, in cubic inches per minute or cfm
V_2 = actual volume, in cubic inches per minute or cfm
Q = probable capacity, in tons per hour

Then

$$V_1 = CWS$$

Assume one-third of the theoretical volume, then

$$V_2 = \frac{V_1}{3}$$

$$= \frac{CWS}{3} \text{ cu in./min}$$

Divide by 1,728 cu in. per cf

$$V_2 = \frac{CWS}{5{,}184} \text{ cfm}$$

Assume the crushed stone has a unit weight of 100 lb/cf.

$$Q = \frac{100 \text{ lb/cf} \times (60 \text{ min/hr}) V_2}{2{,}000 \text{ lb/ton}} = 3 \text{ min/cf} \times V_2 \text{ tph}$$

$$= \frac{CWS}{1{,}728} \text{ tph} \qquad (17.3)$$

S can be expressed in terms of the diameter of the roll and the speed in rpm:

$$S = 2\pi RN$$

Substituting this value of S in Eq. (17.3) gives

$$Q = \frac{CW\pi RN}{864} \text{ tph} \qquad (17.4)$$

TABLE 17.5 Representative Capacities of Double-Roll Crushers, in Tons per Hour (Metric Tons per Hour) of Stone*

Size of Crusher, in. (mm)†	Speed, rpm	HP Electric (HP Diesel)	Maximum Feed Size, in. (mm)	Closed Setting of Discharge Opening, in. (mm)								
				¼ (6)	½ (13)	¾ (19)	1 (25)	1¼ (32)	1½ (38)	2 (51)	2½ (64)	3 (76)
24 × 16	270	50	½–2¾	16	31	47	63	79	94			
(610 × 416)		(75)	(13–70)	(14)	(28)	(43)	(57)	(72)	(85)			
30 × 18	325	100	½–3½	25	50	75	100	125	150	200		
(763 × 456)		(150)	(13–89)	(23)	(45)	(68)	(91)	(113)	(136)	(181)		
30 × 24	325	125	½–3½	33	66	100	133	166	200	266		
(763 × 610)		(175)	(13–89)	(30)	(60)	(91)	(121)	(150)	(181)	(241)		
30 × 30	300	200	½–3½	41	82	125	166	207	276	344	414	
(763 × 763)		(300)	(13–89)	(37)	(74)	(113)	(150)	(188)	(227)	(301)	(375)	
40 × 22	325	150	⅝–5	34	69	103	138	172	207	276	344	414
(1016 × 558)		(200)	(16–127)	(31)	(62)	(93)	(125)	(156)	(188)	(250)	(312)	(375)
40 × 30	310	250	⅝–5	53	106	160	213	266	320	426	532	640
(1016 × 763)		(325)	(16–127)	(48)	(96)	(145)	(193)	(241)	(290)	(386)	(483)	(580)
42 × 40	320	300	⅝–5	70	141	213	284	354	426	568	709	853
(1067 × 763)		(400)	(16–127)	(64)	(128)	(193)	(257)	(321)	(386)	(514)	(644)	(773)
54 × 24	310	250	¾–6	44	87	131	175	228	262	350	437	525
(1374 × 610)		(325)	(19–152)	(40)	(79)	(119)	(159)	(207)	(238)	(317)	(396)	(476)
55 × 36	250	350	¾–6	65	130	195	261	326	390	522	652	782
(1397 × 914)		(475)	(19–152)	(59)	(118)	(177)	(237)	(296)	(354)	(473)	(591)	(709)

*Based on stone weighing 100 lb/cf when crushed.
†The first number indicates the diameter of the rolls, and the second indicates the width of the rolls.

Table 17.5 gives representative capacities for smooth-roll crushers, expressed in tons per hour for stone with a unit weight of 100 lb/cf when crushed. These capacities should be used as a guide only in estimating the probable output of a crusher. Production rates may be more or less than the given values. Adjusting the unit weight from 100 lb/cf will proportionally affect tonnage production. If a roll crusher is producing a finished aggregate, the reduction ratio should not be greater than 4:1. However, if a roll crusher is used to prepare feed for a fine grinder, the reduction may be as high as 7:1.

Impact Crushers

Impact crushers fracture the feed stone by applying high-speed impact forces. A spinning impeller and rebound between the individual stones and against the machine surfaces fully exploits the impact energy. The design of some impact crushers also utilizes shear and compression, in addition to impact action, to fracture the stones. Cubical stone is produced by these internal forces, and both the concrete and asphalt industries are significant markets for impact crushers. The cubical shape is accomplished by forcing the stone between the revolving and stationary parts of the crusher. Speed of rotation is important for effective operation of these crushers, as the energy available for impact varies as the square of the rotational speed. Primary impact crushers have a rotational speed of 400 to 500 rpm, whereas secondary impact crushers have a slightly higher rotational speed.

The rotor orientation classifies these crushers as either a horizontal shaft impactor (HSI) or vertical shaft impactor (VSI). The HSI arrangement is common in primary crushing, whereas VSI is increasingly used for secondary and tertiary crushing. These impact crushers are also referred to as New Holland style or Andreas style crushers. Impact crushers may be further classified by the number of rotors, either a single or double configuration.

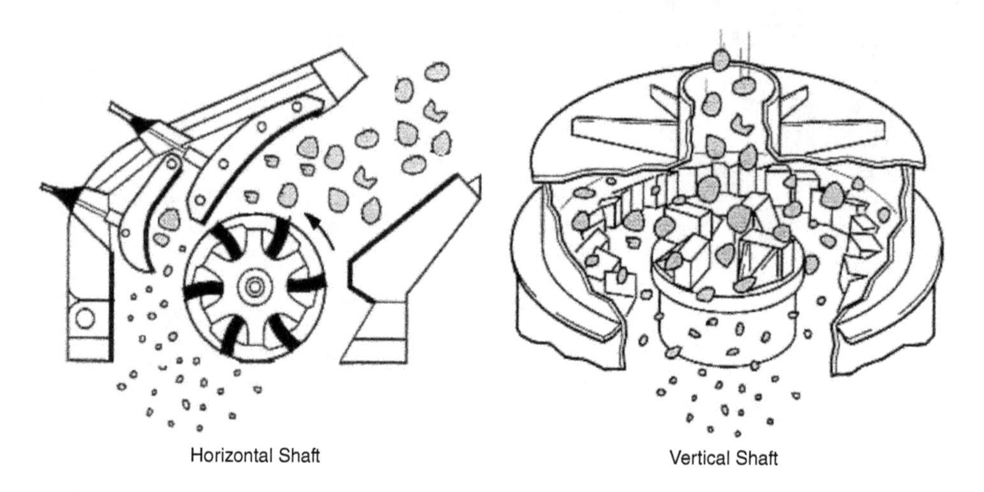

Single Rotor. The single-rotor impact crusher breaks the stone, both by the impact action of the impellers striking the feed material and by the impact that results when the impeller-driven material strikes against the aprons within the crusher unit. These crushers produce a cubical product, but are economical only for low-abrasion feeds.

The crusher's production rate is affected by the rotor speed. The speed also affects the reduction ratio. Therefore, any speed adjustment should be made only after consideration is given to both the production rate and final product size. There are primary and secondary phase models in either HSI or VSI configuration capable of over 800 tph.

Double Rotor. These units are similar to the single-rotor models and accomplish aggregate size reduction by the same mechanical action. Rotation of the rotors can be in the same direction or opposite direction depending on the model setup. They will produce a somewhat higher proportion of fines. With both single- and double-rotor crushers, the impacted material flows freely to the bottom of the units without any further size reduction.

Hammer Mills. The hammer mill, which is the most widely used impact crusher, can be used for primary or secondary crushing. The basic parts of a unit include a housing frame; a horizontal shaft extending through the housing; a number of arms and hammers attached to a spool, which is mounted on the shaft; multiple hardened steel or manganese-steel breaker plates; and a series of grate bars, whose spacing can be adjusted to regulate the width of openings through which the crushed stone flows.

As the stone to be crushed is fed into the mill, the hammers, which revolve at a high rpm, strike the particles, breaking them and driving them against the breaker plates, which further reduces their size. Final size reduction is accomplished by grinding the material against the bottom grate bars. This mechanical action is different from the rotor-type impact crushers, which throw the rock against sidewalls and individual particles.

The size of a hammer mill may be designated by the size of the feed opening. The capacity will vary with the size of the unit, the kind of stone crushed, the size of the material fed into the mill, and the speed of shaft rotation. Hammer mills will produce a high proportion of fines and cannot handle wet or sticky feed material.

Feeders

Effective and efficient crushing requires a constant and adequate supply of feed material. Compression-type crushers (jaw crushers) are designed to use particle interaction in the crushing process. An underfed compression crusher produces a larger percentage of oversize material, as the necessary material is not present to fully develop interparticle crushing. In an impact crusher, efficient use of interparticle collisions is not possible with an underfed machine. Gyratory-type crushers do not need feeders.

The capacity of compression- and impact-type crushers will be increased if the stone feed is at a uniform rate. Surge feeding tends to momentarily overload a crusher followed closely by an insufficient supply of stone. Using a feeder ahead of a crusher eliminates crusher capacity problems caused by surge feeding. The installation of a feeder may increase the capacity of a jaw crusher as much as 15%.

There are several different types of feeders:

1. Apron
2. Vibrating
3. Belt
4. Plate

Apron Feeder. A feeder constructed of overlapping pans forming a continuous belt is referred to as an apron feeder. It provides a continuous positive feed of material. These feeders have the advantage that they can be obtained in considerable lengths. A disadvantage is potential damage from stone entering spaces between overlapping pans, links, and the drive assembly.

Vibrating Feeder. There are both simple vibrating feeders activated by a vibrating unit similar to the ones used on horizontal screens (discussed in the next section) and vibrating grizzly feeders. A vibrating grizzly feeder diverts fines from entering the primary crusher and slowing production (Fig. 17.8). The grizzly allows fines to fall between straight or tapered forks immediately before entering the crusher chamber. A short conveyor belt then collects and moves the finer material to the discharge side of the primary crusher. A vibrating feeder requires less maintenance than an apron feeder.

Belt Feeder. Operating on the same principle as the apron feeder, the belt feeder is used for smaller sizes of material, usually sand or small-diameter aggregates.

Plate Feeder. By means of rotating eccentrics, a plate can be made to move back and forth on a horizontal plane. The motion of the plate will uniformly feed material to a crusher. Larger stationary plants have reciprocating plates to feed larger tonnage into the crusher chamber.

FIGURE 17.8 Vibrating grizzly feeder with tapered forks mounted on a jaw crusher.

Selecting Crushing Equipment

Certain information must be known prior to selecting equipment to support a crushing requirement. At a minimum, this would include the following but other particulars may also be important:

1. Type of stone to be crushed
2. Needed output production (required capacity of the plant)
3. Maximum size of the feed stones with information concerning the size ranges of the feed
4. Method of feeding the crushers
5. Specified size ranges of the product

Plant Layout. The actual layout and erection of the plant are the culminating tasks in the plant design. Plans should initially consider the appropriate configuration of equipment within the plant. Special attention should be given to creating a productive, logical flow of material from the point where the trucks or loader deposit the raw material to the point where the trucks leave the plant with the crushed aggregate product. The physical/environmental requirements of each piece of equipment, such as supporting foundations, water supply requirements, and power demands, should be estimated to ensure they are properly located during erection of the plant. Aerial drone surveys are an effective site planning tool to map ore deposits and evaluate appropriate layouts of plant equipment and stockpiles.

Site drainage is of critical importance in locating the plant. The plant design should include adequate space around the needed individual units. Access must be provided for maintenance personnel to perform repairs and to move cranes and lifting devices for handling the heavy pieces of crushing equipment. There must also be adequate material storage areas. Periodic quarry blasting may also require the use of portable equipment or the temporary removal of plant units away from the blast area and their later repositioning.

Power Source. The power source for crushing plants is usually either electric or diesel. Plant operators favor electric power for its availability, less maintenance than diesel power, and reduced cost. Whether electric or diesel power is chosen, crushing plants demand a considerable amount of energy. An individual crusher unit may be driven by single or dual 100- to 350-hp (75- to 265-kW) motors; there is also the need for 10- to 50-hp motors on conveyors and ancillary equipment such as vibrating feeders, screen decks, log washers, and control panels. Stationary plants will have a dedicated hard-line electric service to the plant. In the case of portable plants, there is a choice of a temporary hard-line electric service or a diesel-powered electric generator. An electrical generator and electrical switch gear housed in separate trailers is a common setup on most portable crushing plants. Even though a 1,000-kW (1,250-amp) generator housed in a portable trailer may provide a steady flow of power, it may only be able to power one crusher plus other ancillary equipment. For this reason, many portable crushing plants may have one crusher solely powered by a diesel engine. Another option is an additional generator, but this complicates wiring diagrams and centralized control.

Surge Piles

A stationary crushing plant may include several types and sizes of crushers, each followed by a set of screens supported by a belt conveyor to transport the stone to the next crushing operation or to storage (Fig. 17.9). A plant may be designed to provide temporary storage for stone between the successive stages of crushing. Such a plan has the advantage of

FIGURE 17.9 A multistage crushing plant with surge piles feeding several types of crushers and screening units.

eliminating or reducing the surge effect that frequently exists when the crushing, screening, and handling operations are sequenced linearly.

Stone in temporary storage ahead of a crusher is referred to as a *surge pile.* Adding one or more surge piles can keep at least a portion of a plant in operation at all times. Within reasonable limits, the use of a surge pile enables the crusher to be fed uniformly at the most satisfactory rate, regardless of variations in the output of other equipment ahead of the crusher. The use of surge piles has enabled some plants to increase their final output production by as much as 20%.

Advantages derived from using surge piles include the following:

1. Enhanced uniform feed and higher crusher efficiency
2. Continuation of other plant crushing units, should the primary crusher break down
3. Repair of primary or secondary plant crushers without a complete production stoppage

Arguments against the use of surge piles include the following:

1. Additional storage area requirement
2. Need to construct storage bins or reclaiming tunnels
3. Increased stone handling, particularly if wheeled loaders must be added to the operation
4. Possible segregation

The decision to use surge piles should be based on an analysis of the advantages and disadvantages for each plant operation. Consideration of the planned production rate, plant layout, and market should be part of the analysis.

Segregation

The different-size pieces within a material being piled or distributed have a tendency to separate from each other, so a disproportionately large amount of coarse pieces will be found in one part of the pile and finer ones in another. This phenomenon is called segregation or separation. Whenever size gradation is important to the use of the material, the system for handling product must be checked to make sure it either will not cause segregation or will include adequate remixing.

Dry. With dry materials, a principal cause of separation is sliding and rolling down slopes. When a pile is built from the top, material falls or settles onto the pile, picking up greater or less momentum from the downward movement. Small particles develop little energy and tend to come to rest almost immediately. Larger ones have enough momentum to slide or roll down the surface, with the biggest tending to reach the ground at the edge of the pile, whereas intermediate sizes stop on the slope.

The extent of the separation increases as the range in sizes becomes greater and when the material has low internal friction. Fine or sticky material tends to pile up to unstable slopes and then fall or slide in masses, minimizing separation. For example, damp sand does not usually separate, whereas dry sand or fine crushed rock separates a little, and gap-graded coarse and fine rock and rounded river gravel may segregate almost completely.

When sliding in chutes or subjected to random movements or strong vibration, large pieces tend to be wedged upward by small ones working under them, with a layering effect opposite to a top-built pile.

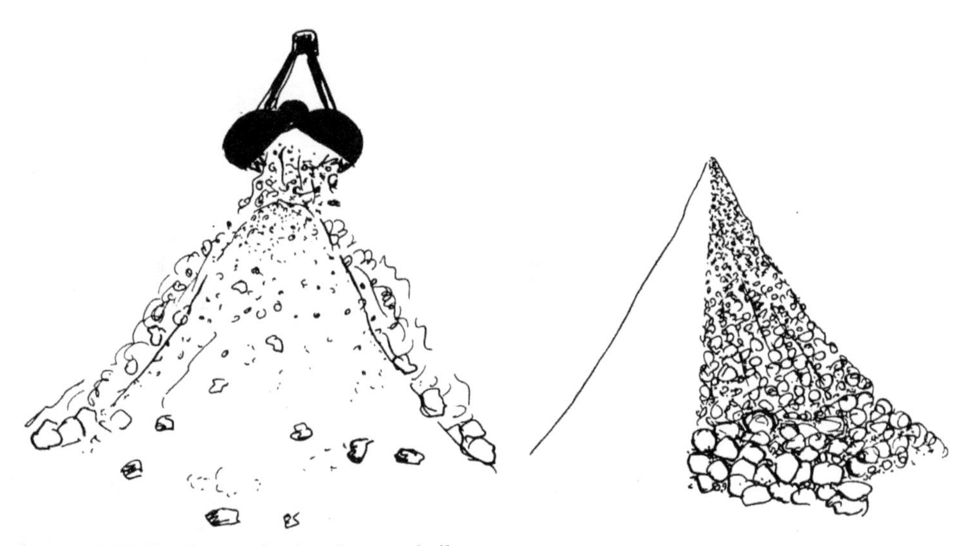

FIGURE 17.10 Segregation in a dry gravel pile.

The segregation caused by building a pile of assorted-size pieces from the top, with a clamshell or a fixed-discharge conveyor, is shown in Fig. 17.10. The pile is always surrounded at ground level by a ring of the coarsest pieces. As the pile expands, its bottom layer is made up of these. Above it is a zone of somewhat smaller particles, with a fairly even gradation to mostly fines at the top. Such separation is almost never complete, as big pieces will get trapped in the top and small ones cascade to the bottom, but it is often sufficient to make the pile unusable until recombined.

A clamshell operator can largely avoid layering by estimating pile area in advance and building it in layers. Each successive layer must be smaller enough than the one below it to prevent sliding over the edges. But this is not possible when conveyors are used and material simply falls off the end of the belt.

Remixing. A dragline working from thc top of a pile, or a loader at the bottom, can make long shallow cuts from bottom to top, to get some of each layer in each bucket load. A clamshell should dig at various layers in turn, trying to provide a good average mix in the hopper or haul unit. Layers are more or less mixed when loading is done from the bottom by a loader. Undermining causes the finer upper part to fall and slide, mixing with the bottom.

One method of reducing stockpile segregation is by distributing the crushed stone with a telescoping conveyor belt (Fig. 17.11). Adopting technology from loading and unloading of bulk materials on ships, the telescoping conveyor belt is capable of extending and retracting to change the target of spooling stone. The conveyor length is shortened and lengthened, while the main axis is oscillated to each side on a swing axle for placing the stone in layered rows. Automated controls limit horizontal swing angle, while depth sensors detect the stockpile height to incrementally incline the conveyor angle. These movements reduce segregation found in traditional cone-shaped stockpiles. Aggregate producers have reduced variability in stockpile gradation using this approach, which in turn has resulted in more uniform aggregate blends in concrete and asphalt mixtures.

FIGURE 17.11 Telescoping conveyor belt to reduce segregation.

PRODUCT SIZING

Scalping Crushed Stone

Scalping removes those stones too large for the crusher opening or that are small enough to be used without further crushing from the main mass of material to be processed/crushed. Removal of fines ahead of a primary crusher improves crusher efficiency. Scalping can be performed ahead of a primary crusher, and it represents good crushing practice to scalp all crushed stone following each successive stage of reduction.

Scalping ahead of a primary crusher serves two purposes. It prevents oversized stones from entering the crusher and blocking the opening, and it can be used to remove dirt, mud, or other debris not acceptable as finished product. This is important in the expansion of rock quarries or pits when soil overburden is removed from above the ore body and remnants intermingle with the stone. If the product of the blasting operation contains oversized stones, it is desirable to remove them ahead of the crusher by scalping. The use of a *grizzly*, which consists of a number of widely spaced parallel bars, can be used to scalp large material ahead of the primary crusher (Fig. 17.12).

The product from the blasting operation can contain an appreciable amount of stone that meets the specified size requirements. In this event, it may be good economy to remove such stone ahead of the primary crusher, thereby reducing the total load on the crusher and increasing the overall capacity of the plant. This also applies to rock pits, where naturally loosened alluvial or glacial gravel deposits in the earth's crust are mined without blasting.

It is usually economical to install a scalper after each stage of reduction to remove specification sizes. This stone may be transported to grading screens where it can be sized and placed in appropriate storage.

FIGURE 17.12 Grizzly used to scalp oversize material.

Screening Aggregate

In all but the most basic crushing operations, the crushed rock particles must be separated into two or more particle size ranges. This separation enables the selection of certain material for additional or special processing, or the diversion of certain material from unnecessary processing. The screening process is based upon the simple premise that particle sizes smaller than the screen cloth opening size will pass through the screen and the oversized particles will be retained.

Screen openings can be described by either of two terms: (1) mesh and (2) clear opening. The term "mesh" refers to the number of openings per linear inch. The number of openings to an inch can be counted by measuring from the center of the screen wire to a point 1-in. distance. Clear opening, or "space," is a term that refers to the distance between the inside edges of two parallel wires.

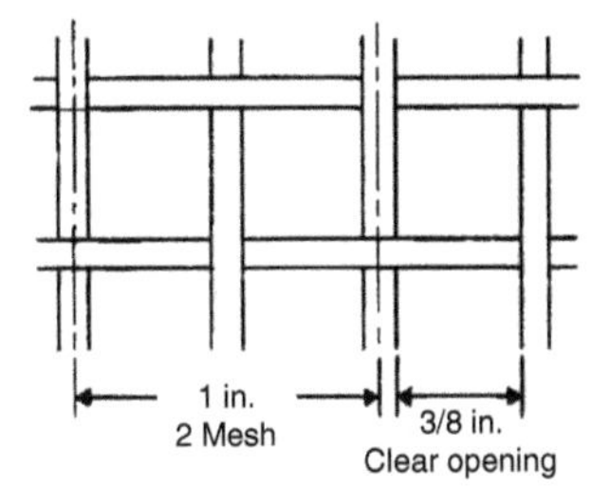

Most specifications covering the use of aggregate stipulate the blending of different sizes to produce a material with a given size distribution. Specification writers realize it is not possible to crush and screen material with complete precision, and accordingly they allow some tolerance in the specified size distribution. The extent of tolerance can be indicated by a statement such as: "The quantity of aggregate passing a 1-in. screen and retained on a ¼-in. screen shall be not less than 30% or more than 40% of the total quantity of aggregate." Integer-level accuracy is commonly specified for passing percentages.

Revolving Screens. Revolving screens have several advantages over other types of screens, especially when they are used to wash and screen both sand and gravel. The operating action is slow and simple, and the maintenance and repair costs are low. If the aggregate to be washed contains silt and clay, a scrubber can be installed near the entrance of a screen to agitate the material in water. At the same time, streams of water can be sprayed on the aggregate as it moves through the screen.

Vibrating Screens. Vibrating screens consist of one or more layers, or *decks,* of open mesh wire cloth mounted one above the other in a rectangular metal box (Fig. 17.13). These are the most widely used aggregate production screens. The vibration is obtained by means of a hydraulically actuated eccentric shaft, a counterweight shaft, or electromagnets attached to the frame or to the screens. Modular screen sections can be easily removed and replaced. A greater occurrence of bent screen wire and damage occurs directly under the conveyor belt discharge.

FIGURE 17.13 Vibrating screen box.

A unit may be horizontal or inclined with slight slope (20 degrees or less) from the receiving to the discharge end. The vibration, 750 to 1,250 strokes per minute, causes the aggregate to flow over the surface of the screen. Screen models with frequencies of over 4,000 vibrations per minute (vpm) are available but are limited to screening very fine aggregate or minerals. Normally, large amplitude and slower speed are necessary for large screen openings, whereas the opposite is effective for small screen openings. In the case of a horizontal screen, the throw of the vibrations must move the material both forward and upward. For that reason, its line of action is 45 degrees relative to the horizontal.

Most of the particles smaller than the openings in a screen will drop through the screen, whereas the oversize particles will flow off the screen at the discharge end. For a multiple-deck unit, the openings size will be progressively smaller for each lower deck.

A screen will not pass all material whose sizes are equal to or less than the dimensions of its openings. Some of this material will be retained on and carried over the discharge end of a screen. A screen deck with a steep incline can reduce the screen effectiveness in allowing the correct size material to pass the mesh; however, they can be operated in a static mode with reduced power requirements and maintenance.

The efficiency of a screen can be defined as the ratio of the amount of material passing through a screen divided by the total amount sufficiently small to pass through. The amounts could be expressed in tons or in tons per hour. The ratio is expressed as a percentage. The material sufficiently small to pass the mesh is sometimes referred to as *undersize* material.

$$\text{Screen Efficiency} = \frac{\text{Undersized materials passing the openings}}{\text{Undersized feed materials}} \times 100\% \tag{17.5}$$

Particle shape is a significant factor affecting screen efficiency. Elongated, flat, or similarly irregular shapes do not pass screen openings easily. Moreover, they hinder the passage of other particles. Attention to crusher choice and setting can improve particle shape and improve screen efficiency. Particle moisture will affect screen efficiency. Cleanliness of the screen decks is important. Decks should be regularly inspected for aperture blockage.

The highest efficiency is obtained with a single-deck screen, usually amounting to 90% to 95%. As additional decks are installed, the efficiencies of these decks will decrease, being above 85% for the second deck and 75% for the third deck. Wet screenings will increase screening efficiency, but additional equipment is necessary for handling the water.

The capacity of a screen is the number of tons of material that 1 sf (square foot) will separate per hour. The capacity of a screen is *not* the total amount of material fed and passed over its surface, but the rate at which it separates desired material from the feed. However, some screen manufacturers designate capacity in tonnage per hour. The capacity will vary with the size of the openings, type of material screened, moisture content of the material screened, and other factors.

Because of the variable factors affecting screen capacity, it will seldom, if ever, be possible to calculate in advance the exact capacity of a screen. If a given number of tons of material must be passed per hour, it is prudent to select a screen whose total calculated capacity is 10% to 25% greater than the required capacity.

The chart in Fig. 17.14 provides guide capacities for dry screening to use in selecting the correct size screen for a given flow of material. The capacities given in the chart should be modified by the application of appropriate correction factors.

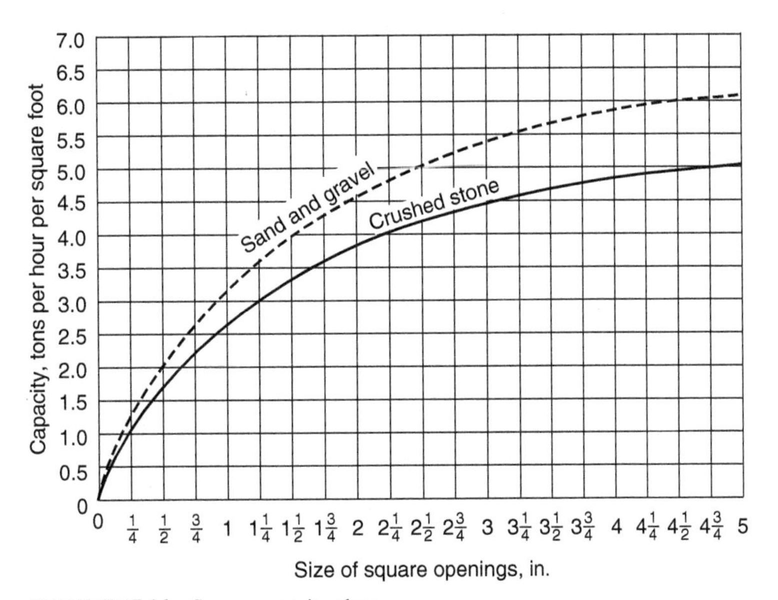

FIGURE 17.14 Screen capacity chart.

Efficiency Factors. If a low screening efficiency is permissible, the actual capacity of a screen will be higher than the values given in Fig. 17.14. Table 17.6 gives factors by which the chart values of capacities in Fig. 17.14 can be multiplied to obtain corrected capacities for given efficiencies.

Deck Factors. This is a factor whose value will vary with the particular deck position for multiple-deck screens. The deck factor values are given in Table 17.7.

TABLE 17.6 Efficiency Factors for Aggregate Screening

Permissible Screen Efficiency (%)	Efficiency Factor
95	1.00
90	1.25
85	1.50
80	1.75
75	2.00

TABLE 17.7 Deck Factors for Aggregate Screening

For Deck Number	Deck Factor
1	1.00
2	0.90
3	0.75
4	0.60

Aggregate-Size Factors. The capacities of screens given in Fig. 17.14 are based on screening dry material that contains particle sizes, such as would be found in the output of a representative crusher. If the material to be screened contains a surplus of small sizes, the capacity of the screen will be increased, whereas if the material contains a surplus of large sizes, the capacity of the screen will be reduced. Table 17.8 gives representative factors that can be applied to the capacity of a screen to correct for the effect of excess fine or coarse particles.

TABLE 17.8 Aggregate-Size Factors for Screening

Percent of Aggregate Less Than ½ the Size of Screen Opening	Aggregate Size Factor
10	0.55
20	0.70
30	0.80
40	1.00
50	1.20
60	1.40
70	1.80
80	2.20
90	3.00

Determining the Screen Area Required. Figure 17.14 provides the theoretical capacity of a screen in tons per hour per square foot based on material weighing 100 lb/cf when crushed. The corrected capacity of a screen is given by the equation

$$Q = ACEDG \tag{17.6}$$

where Q = capacity of screen, in tons per hour
A = area of screen, in square feet

C = theoretical capacity of screen, in tons per hour per square foot
E = efficiency factor
D = deck factor
G = aggregate-size factor

The minimum area of a screen to provide a given capacity is determined from the equation

$$A = \frac{Q}{CEDG} \tag{17.7}$$

Sand Preparation and Classification Machines. When the specifications for sand and other fine aggregates require the materials to meet specific gradation requirements, it is frequently necessary to produce the required sizes by mechanical equipment. Several types of equipment are available for this purpose. There are mechanical and water flow machines that classify sand into a multiple number of individual sizes. Sand and water are fed to these classifiers at one end of the unit's tank. As the water flows to the outlet end of the tank, the sand particles settle to the bottom of the tank, the coarse ones first and the fine ones last. When the depth of a given size reaches a predetermined level, a sensing paddle will actuate a discharge valve at the bottom of the compartment to enable that material to flow into the splitter box, from which it can be removed and stockpiled.

Another machine for handling sand is the screw-type classifier (Fig. 17.15). This unit can be used to produce specification sand. A single or double screw is erected so the material must move up the screw to be discharged. In the case of the single screw, the motor is at the discharge end. Sand and water are fed into the hopper. As the spiral screws rotate, the sand is moved up the tank to the discharge outlet under the motor. Undesirable material is flushed out of the tank with overflowing water.

FIGURE 17.15 Screw classifier for cleaning specification sand.

Conveyors

The belt conveyor is a transporting, elevating, or distributing machine made up of an endless wide belt, which carries a load on its upper surface. It operates between a head and a tail pulley and is supported by idlers, which in turn are supported by a frame or by steel cables.

As independent units, they are well suited for rapid transportation of loose material. They have less mobility and flexibility than trucks and scrapers and are therefore used chiefly where large volumes of material are to be moved along one route. They are particularly applicable where the load must be lifted steeply. They are desirable as feeders for processing plants because they provide an even and continuous flow (Fig. 17.16). However, they are not adapted to hauling big chunks, which clog hoppers, damage the belt, and are likely to fall off in transit.

FIGURE 17.16 Multiple conveyors at an aggregate processing plant.

Their mechanical efficiency is high, as very little dead weight must be moved with the load, friction is at a minimum, and power-consuming starts and stops are rare. The large permanent type of belt conveyor is almost unique among machines—it is usually custom-built, and rules of design and construction are flexible so that it can be tailored to the job.

The belt extends between a head pulley (which may be the drive pulley) and a tail or return pulley, and carries its load on its upper surface, usually toward the head pulley. Its upper strand is supported by idler sets whose three rollers are arranged to shape it into a trough (Fig. 17.17), and the lower strand is supported at wider intervals by flat rollers called return idlers.

Figure 17.18 illustrates four belt-conveyor systems based on the location of the drive pulley. There is a frame that keeps the working parts in position, an engine that turns a drive pulley that moves the belt by friction, a tail pulley to reverse the direction of belt travel, and idlers and return idlers.

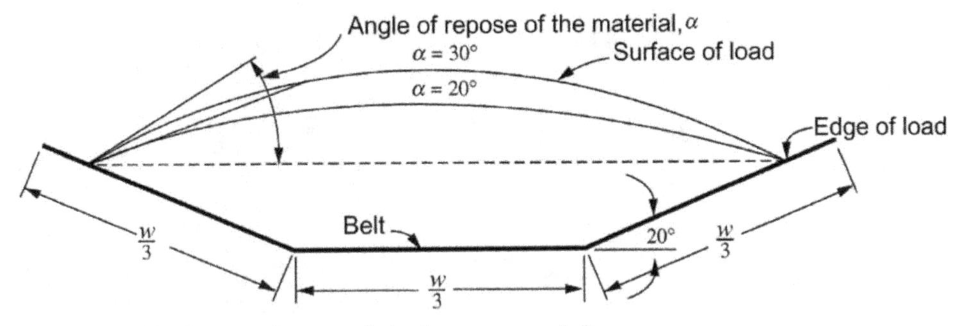

FIGURE 17.17 Cross-section area of a load on a conveyor belt.

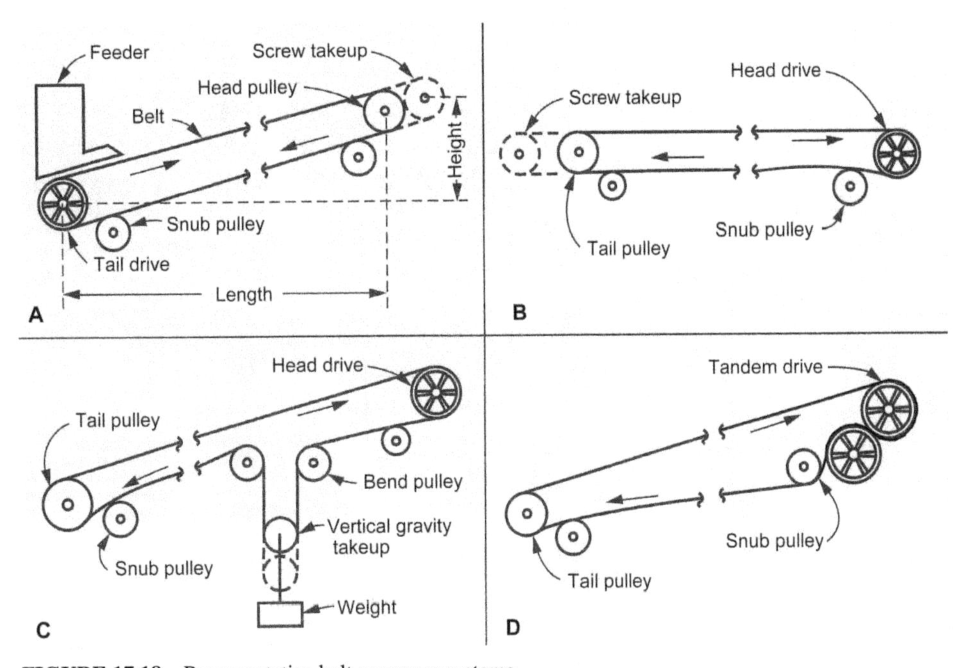

FIGURE 17.18 Representative belt conveyor systems.

Frame. Many frames are of the sectional type. A head and a tail piece must always be used and as many intermediate sections as are required to obtain the desired length. Sections are bolted to each other.

Belt. The belt is an endless flat strip of rubber-covered cotton or rayon fabric laid up in plies. Very long or heavily loaded belts may be reinforced with steel cable. The type of fabric, the number of plies, and any reinforcement that is present determine the strength of the belt. The rubber cover serves only to protect the fabric from abrasion and weather. Its thickness and quality are varied to suit different types of service. Hardness of the rubber is measured using the durometer scale. Several constructions are shown in Fig. 17.19.

STANDARD

Multiple layers of suitable duck of the same thickness and ply across the entire Belt. Standard is suitable for nearly all types of Conveyor service.

SHOCK PAD

Substantially the same as Standard but with a reinforced top *Cover* consisting of an abrasion-resistant tire-tread stock on the top surface, backed by a thick pad of resilient rubber. The pad yields to sudden, extreme impacts and pressures—protecting the cover from puncture or breakage and preventing rupture of the carcass.

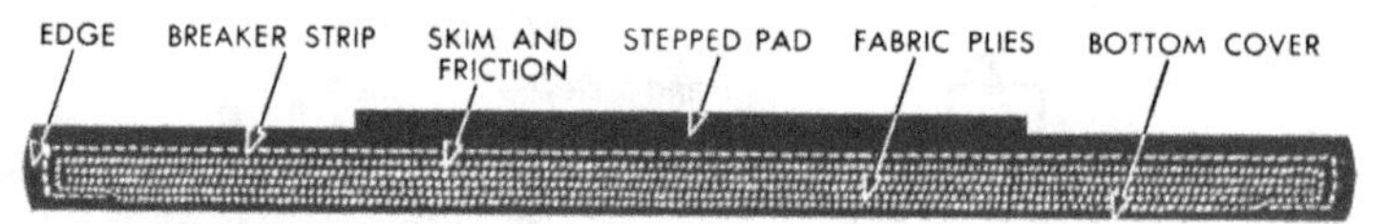

STEPPED PAD . . . Stacker-type Cover

Substantially the same as Standard construction except that it is moulded with the *Cover* having an additional thickness standing out in relief in the center Belt area. It is recommended when the loading of abrasive material is concentrated in the center of the Belt with only slight abrasion at the cover edges. Stepped Pad Belting is not recommended for two-pulley or internal drives.

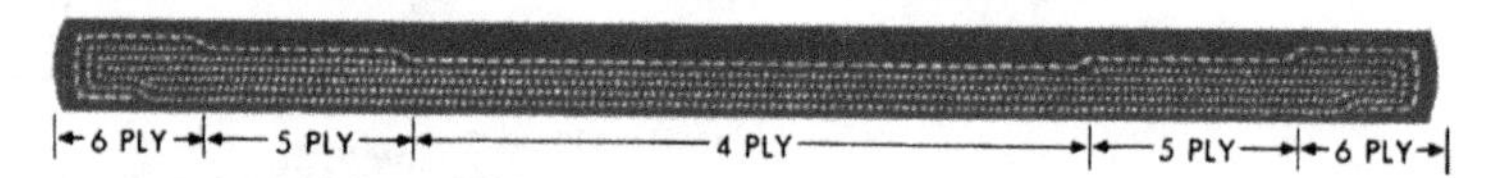

STEPPED PLY

A smooth-top construction having a heavier cover in the center of the Belt than at the edges. This is accomplished by moving the middle portion of one or two of the top fabric plies to the sides and filling in the extra space with cover stock. A Stepped Ply Belt has more crosswise flexibility and troughs more easily. The extra thickness of cover stock gives longer life to the Belt under loading conditions where abrasion is concentrated in the center Belt area.

FIGURE 17.19 Conveyor system belt cross-sections. (*Courtesy of Hewitt-Robins Incorporated.*)

There is no definite limit to the length of a single belt. More friction surface for the drive and stronger belt construction are needed as the unit is made longer, the climb steeper, or the load heavier. Common practice limits the carrying distance to ¼ mile (0.4 km), but belts with a carry of 1 mile (1.6 km) or more have been constructed.

Belts may be from 8 in. (20 cm) to 8 ft (2.4 m) in width, but standard belts range from 12 in. (0.3 m) to 60 in. (1.5 m), with 30 in. (0.76 m) very common.

Drive. Power carries from the drive pulley to the belt by friction. If the resistance of the belt to moving is greater than the friction, the pulley will spin or slip inside the belt, with resultant loss of power, and will wear on both surfaces. The amount of friction or traction is determined by the nature of the surfaces, the slack side tension on the belt, and the area of contact. Speed of the pulley must be balanced with carrying friction.

The load or tension on the carrying strand of the belt, which tends to cause slippage, is made up of the pull of gravity on this strand and its load, friction in idlers, pulleys, the belt, and its load, as well as the inertia of the whole system when starting or accelerating. The drive pulley surface may be bare metal or covered by a smooth face or grooved rubber lagging. Such lagging may be bolted or vulcanized in place. It increases traction, particularly when the belt is wet or frosty, and prevents pulley wear.

The belt is held in full contact with the pulley by its tension on the slack or low-tension side. This tension is normally regulated by some form of gravity takeup, in which a hanging weight exerts a pull on the tail pulley (Fig. 17.18C), or a special takeup pulley, which moves outward if the belt slackens and inward if it tightens. If the incline is steep, the weight of the slack side may maintain sufficient tension. Very short conveyors may have threaded adjustments to move the tail pulley in or out.

The amount of drive traction that may be obtained by increasing tension is limited by sharply increased power requirements and shortened life of a too-tight belt. The area of contact is determined by pulley diameter, the arc of contact, and the number of pulleys. A thicker pulley not only increases the contact area but also reduces flexing strain on the belt. Its disadvantage is the cost of the pulley itself and of changes in frame and layout to accommodate it.

If an existing drive pulley is replaced by a larger one, the belt will then be driven at higher speed with less power, unless the gearing or the motor is changed. Putting on lagging has the same effect.

A belt whose strands are parallel will have 180 degrees of contact with the drive pulley. This contact can be increased by a snubbing pulley on the slack side. Increasing the degree of wrap in this manner is the cheapest way to increase contact area.

Head pulley drives are adequate for short conveyors and for those so steeply inclined that the weight of the return strand maintains a high tension on the slack side. For conditions requiring greater traction, tandem pulleys are used (Fig. 17.18D). Up to 440 degrees of wrap can be obtained in this manner.

A belt that is pulling a load changes shape as it goes over a drive pulley. It is stretched thin where it first develops contact and then fattens as its tension is reduced. The belt moves fastest where it is thinnest, in the same manner of water accelerating through a restricted width in a channel. The change in belt thickness and speed is quite small but requires the second of the tandem pulleys to turn a bit more slowly than the first, if extra stress is to be avoided. The amount of difference varies with the load. Where an electric drive is used, separate motors on the two pulleys can be made to automatically adjust their speeds to each other.

A possible disadvantage of a tandem drive is that one pulley works on the load-carrying surface of the belt, which may be wet and slippery so as to afford a poor grip, or gritty so that excessive wear of both belt and pulley will occur. This may be avoided by using the head and lower pulley for drive and the intermediate one as an idler. A number of drive and takeup arrangements are shown in Fig. 17.18. Short reversible belts may have drive pulleys at both ends, connected by roller chain.

Idlers. Idler rollers support the upper or working surface of the belt. They are usually of the troughing type (Fig. 17.17), in which a horizontal center roll supports the loaded part of the belt, and a pair of outer rolls turn up the edges to create a trough cross-section, which keeps the load from spilling off the sides.

At loading points, shock and wear to both belt and idler can be reduced by using rubber idlers. The idlers turn on ball or antifriction roller bearings.

Adjustment. Short belts commonly have a screw adjustment or takeup on the tail pulley (Fig. 17.18B), which slides in and out on a track. Care should be taken to adjust both sides equally to keep the axle at right angles to the direction of belt travel. Any cocking will make the belt tighter on one side than the other, and it will tend to climb the pulley and run off the tight side.

With long belts, a fixed tension is not satisfactory, as the length of the frame is affected by changes in temperature, and the belt is affected by both temperature and moisture.

Two common types of automatic takeup keep the belt under constant tension. In the horizontal gravity or counterweight-and-rigging method, tension is controlled by a tail pulley in a track, which pulls it outward by means of a weight hanging from a pulley. With a vertical gravity takeup, a fixed tail pulley is used and the weighted pulley hangs between two return idlers, preferably at or near the point of minimum belt tension (Fig. 17.18C). If the belt stretches too far to be adjusted, a piece is cut out, the ends are stapled or vulcanized together, and the adjustments are reset.

Alignment. The belt is sensitive to tiny changes in frame and pulley alignment, which will cause it to wander out of a straight line. Internal changes in belt tension or a splice that is beginning to pull apart may have the same effect. Trouble with wandering can be greatly reduced by making both pulleys and frames wide enough that the belt does not have to run absolutely straight to keep out of trouble. The wider construction is more expensive, of course, but through the years it should more than pay for itself in longer belt life and in reduced checking and adjusting.

If the framework is out of line, the carrying strand may still track well enough, being steered by the load seeking the idler troughs. The return strand, however, will find the shortest path or allow itself to be influenced by sloping idlers so that it will rub against stationary parts. Unfortunately, it is almost standard practice to carry the return strand inside the framework, where it is very difficult to see just what it is doing. It is much better to hang it below the frame members, where irregularities can be readily observed and necessary corrections made before serious damage is done.

Troughed idlers will steer the belt if they are tipped in the direction it is moving. However, only a very slight tilt can be used, because if it is overdone, it will set up a drag against the bottom of the belt, which will wear it rapidly and consume extra power.

Self-aligning or training idlers are mounted on a center swivel and have vertical spools set at each edge. If the belt rubs against a spool, it tilts and presses a lined brake shoe against the adjoining roller, slowing it, swinging the idler, and shifting the belt back toward center. This device creates very little drag, and if placed every 50 ft (15.2 m) in place of regular idlers, will keep a belt in line under any ordinary conditions. Two placed at 30-ft (9.1-m) intervals ahead of the tail pulley will line the belt up properly to go under the loading point.

Holdbacks. An inclined conveyor (Fig. 17.20) tends to run backward when power is cut off, because the weight of the load pulls the upper part of the belt downhill. This tendency can be overcome by means of a brake of sufficient size, but it is often more convenient to use a device that will automatically lock it against turning backward, without interfering with normal movement of the belt.

Hoppers and Loading. Hoppers are of two principal types: the primary or loading hopper and the transfer (Fig. 17.21). The primary usually has enough storage capacity to steadily supply material to the belt, although it is loaded intermittently by bucket or truck loads. It may have a screen or grizzly to keep out oversize pieces.

Loading hoppers must have a restricted chute capacity or a gate or feeding device to regulate the rate of delivery to the belt. In some installations, increased flow is obtained by merely raising the chute so more material can pass under its forward edge. For any rate of

FIGURE 17.20 Long inclined conveyors moving aggregate to a processing plant.

FIGURE 17.21 Excavators feeding two loading hoppers and a transfer hopper feeding the main conveyor.

feed, the load per foot (meter) of belt will be increased by slowing the belt, and decreased by accelerating it.

Various feeding devices are available to ensure even flow from the hopper to the belt. A short feeder conveyor (transition belt) may be used, either for convenience of location or to save the long belt the extra wear of taking a part in measuring out material. Another common device is a pan that moves back and forth under the hopper chute, carrying and pushing a fixed amount of material toward the belt with each move. Other models use shaking or vibrating plates, or rotating rolls or vanes.

A conveyor belt is loaded by a chute from a hopper or another conveyor or by a feeding device. The material always has some vertical drop and may be moving in a different direction and at a different speed than the belt. As a result, there is impact to the belt in stopping its fall and giving it proper speed and direction. Chutes and baffles should be constructed so they convert as much vertical and cross movement as possible in the direction of belt travel and so they deposit their load uniformly in the center of the belt.

Coarse, lumpy materials, particularly when they include sharp edges, require greater precautions at the loading point than fine or malleable ones. Impact damage to the belt can be reduced by using thicker rubber on both the top and bottom surfaces and by using closely spaced rubber-cushioned idlers at the loading point.

A certain amount of belt wear from impact and abrasion occurs each time the belt goes over an idler, as it is a high point with a sag on each side. The slight but steady wear at these points can be reduced by minimizing the sag, which is affected by belt tension and idler spacing. Maximum smoothing effect can be obtained from any given number of idlers by using wide spacing near the head, where tension is greatest, and progressively narrower spacing toward the tail pulley.

Spillage. There is usually very little spillage off a belt that is wide enough for its load. However, many belts are overloaded in bulk, if not in weight, and the bouncing over idlers, movements in the load, and changes in alignment result in pieces rolling and bouncing off the sides. If decking is placed under the full length of the belt or sometimes under the ends and the loading point, this material will be kept from spilling onto the return belt. A clean belt and clean pulleys and idlers increase belt life.

Wet dirt and many other materials stick to belts and will build up to considerable amounts if permitted to do so. A scraper may be placed just after the discharge point to clean the belt and drop the scrapings into the same receptacle as the mainstream. This cleaner may consist of a rubber or steel blade, or a series of them, or be a serrated rubber roll or bristle brush that revolves oppositely to the belt. Proper pressure against the belt and adjustment for wear are provided by a counterweight or springs.

Troughs and Skirts. Most belts carrying loose material are troughed so the center is lower than the sides. The side idlers are usually at a slope of 20 degrees (Fig. 17.17). However, there is a trend toward steeper slopes, and 35 degrees and 45 degrees may now be used to increase load and/or reduce spillage.

Skirts are used to prevent spillage off the sides of a belt at a loading point and to increase the capacity of short belts (Fig. 17.22). They usually consist of vertical or sloped faces of metal or wood, with a lower piece of flexible rubber that makes light contact with the belt. This rubber should be lowered as it wears and replaced whenever necessary to avoid damage from material wedging under it.

Both troughs and skirts may be used at a loading point. The trough tends to keep material from reaching or pressing against the skirt and working under it. Trough slopes as steep as 50 degrees may be used.

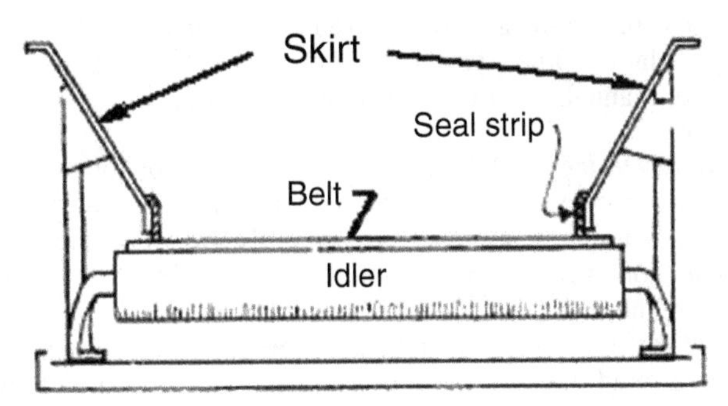

FIGURE 17.22 Skirted belt.

Trippers. A conveyor belt may discharge into any of several bins or piles. For a discharge point not at the end of the belt frame, a tripper may be used to dump the load into a sidecasting chute at any point. As shown in Fig. 17.23, it consists primarily of a pair of straight idler pulleys, one of which turns the belt under and dumps it, and the second returns it to its original direction.

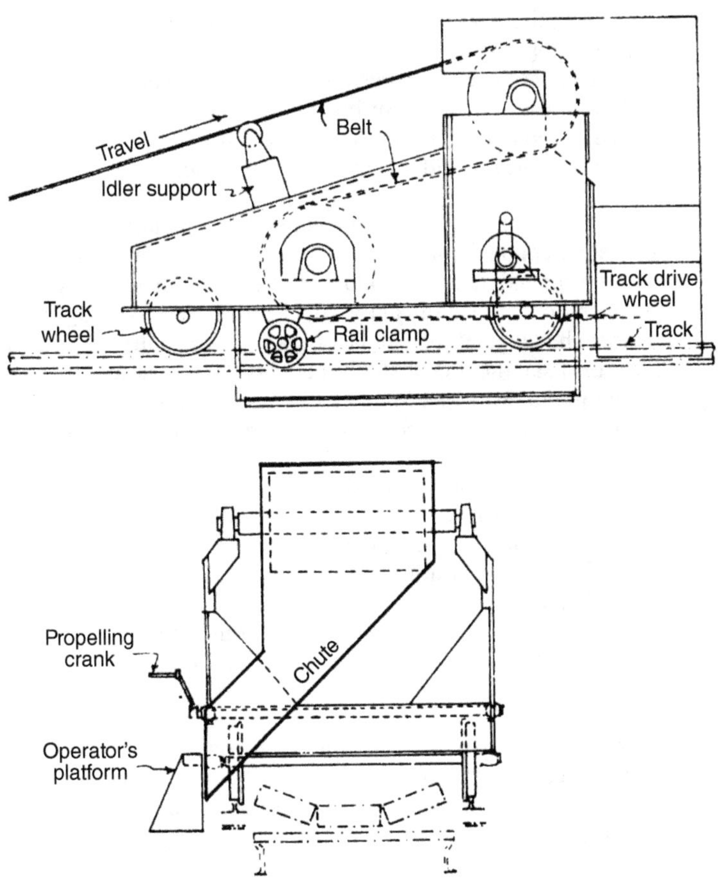

FIGURE 17.23 Tripper.

The model illustrated is moved along tracks by a hand crank, and is anchored in position by a clamp. Other types can be propelled by the belt or by a separate motor, and can be under manual, automatic, or remote control.

Safety Devices. A belt conveyor will run for long periods without attention. However, sudden accidents may happen, which can be very costly if their results are not controlled, and it is impossible to keep a worker always on the alert for events that may not happen for years, or ever. Controls should therefore be installed that will automatically react to emergencies.

If the power fails on an inclined belt, it will tend to run backward, jamming its load into and around the loading chute. This movement can be prevented by a holdback ratchet on the head or drive pulley, which will allow free working motion but immediately lock against backsliding.

If objects jam between the belt and a pulley, if discharged material backs up into the belt, or if the belt runs off its rollers, the power requirement will be sharply increased. If the drive is electric, an automatic overload switch in the line can be set to cut off the power and prevent or limit the resulting damage.

It is also a good plan to use a motor that is not too big for its job. A 200-hp (149-kW) motor can drag a belt into a lot more trouble than a 100-hp (75-kW) can, without increasing current requirement as sharply. The oversize motor will also put greater stress on the belt when starting it with a load.

A moderate rise in power consumption without increase in load indicates increasing friction. It may be in dry or broken idlers, result from the belt rubbing the frame, or be caused by too much tension. A record of current consumption will often reveal such conditions before they would be observed otherwise.

Repairs. Damage to conveyor belts is an operating challenge. Constant tension on the belt fabric coupled with sharp-edged aggregate continually cascading down onto the belt causes wear and imminent failure. Just one tear in a lengthy conveyor run is enough to stop the operation and significantly affect production. Bearing failures on pulley housings are another challenge. If the belt is torn, a stitch is applied by applying tension in the opposing ends, then stapling metal stitches across the tear at uniform spacing. Stitches are perpendicular to the belt edge. Skewed stitches have also been successfully applied but have the potential to irregularly spool the belt across an idler pulley.

Inclines. Conveyors are used for uphill (Fig. 17.20), horizontal, and downhill (decline) transporting. A belt will lift dry sand on an incline of about 15 degrees, and wet sand up to 20 degrees. Some cohesive materials can be carried on grades as steep as 28 degrees on standard belts. The limiting factors are slippage between the belt and the load and sliding of the load on itself.

Rough-surface belts may be used on somewhat steeper inclines if the load is thin. Belts with metal cleats will carry loads on inclines of up to 35 or 45 degrees. Belts or chains with attached buckets can lift at any slope or vertically.

Capacities. Output is usually figured in tons per hour and converted to yards when necessary by means of average weight data. Production is determined by the width of the belt, the speed of the belt, and the height to which material is piled. Power needs are proportioned to tonnage and lift.

Those variables make design of a conveyor system a complex matter. In general, increasing belt width means more expensive construction throughout. An increased speed beyond specifications shortens belt life, and may involve lost power through slippage between the belt and the load. Heavier loading may call for stronger frame and idler construction, as well as heavier belting. In addition, a problem of spilling off the sides may be encountered. If capacity of an existing installation is to be increased, the most economical, although often not the soundest, way is to increase power and speed and to maintain the original load. Recommended speeds for various materials range from 300 to 1,200 ft (91 to 366 m) per minute, with dirt, sand, and gravel in the higher rate range.

GLOSSARY

Adhesion The soil characteristic of sticking to buckets, blades, and other parts of excavators.

A-frame An open structure tapering from a wide base to a load-bearing top.

Air receiver The air storage tank on a compressor.

Air waves Airborne vibrations caused by explosions.

Auger A rotating drill having a screw thread that carries cuttings away from the face.

Auxiliary A helper or standby engine or unit.

Axle, dead A fixed shaft functioning as a hinge pin. A fixed shaft or beam on which a wheel revolves.

Axle, live A revolving horizontal shaft.

Backfill The material and/or process of refilling a ditch or other excavation.

Bailer A hollow cylinder used for removing rock chips and water from churn drill holes.

Ballast Heavy material, such as water, sand, or iron, which has no function in a machine except to increase weight.

Ball joint A connection, consisting of a ball and socket, that limits hinge movement in any direction.

Bank Specifically, a mass of soil rising above a digging or trucking level. Generally, any soil that is to be dug from its natural position.

Bank measure Volume of soil or rock in its original condition in the ground.

Batter Inward slope from bottom to top of the face of a wall.

Bearing A part in which a shaft or pivot revolves.

Bearing, antifriction A bearing consisting of an inner and outer ring, separated by balls or rollers held in position by a cage.

Bed A base for machinery.

Bedding Material used to support pipe laid in the ground.

Bedding plane A separation or weakness between two layers of rock, caused by changes during the building up of the rock-forming material.

Bedrock A solid rock mass, as distinguished from fragmented boulders.

Bell An expanded part at one end of a pipe section into which the next pipe fits.

Benching A trench protection method where level steps are excavated into the slope extending out of a trench. *See also* Sloping.

Benchmark A point of known or assumed elevation used as a reference in determining and recording other elevations.

Bentonite A colloidal clay that forms a slurry when mixed with water. Typically the primary component in an HDD drilling fluid.

Berm An artificial ridge of earth.

Binder Fines that hold gravel together when it is dry. A deposit check that makes a contract valid.

Blasting machine (battery) A hand-operated generator used to supply firing current to blasting circuits.

Bleed To remove unwanted air or fluid from passages.

Block, snatch A sheave in a case with a pull hook or ring.

Block holing Blasting boulders by means of drilled holes.

Blue tops Grade stakes whose tops indicate the finish grade level.

BM Benchmark.

Body The load-carrying part of a truck or scraper.

Body, quarry A dump body with sloped sides.

Bogie A two-axle driving unit in a truck. Also called tandem drive unit or a tandem.

Boom, live A shovel boom that can be lifted and lowered without interrupting the digging cycle.

Boulder Per ASTM D653, a rock fragment with a dimension of 12 in. or more.

Brake, tooth (jaw brake) A brake used to hold a shaft by means of a tooth or teeth engaging with fixed sockets. Not used for slowing or stopping.

Brake horsepower The horsepower output of an engine or mechanical device as measured at the flywheel, driveshaft or belt, usually by some form of mechanical brake.

Bridge In an electric blasting cap, the wire that is heated by electric current so as to ignite the charge. Sometimes the shunt connection between the cap wires.

Bulkhead A wall or partition erected to resist ground or water pressure.

Bull gear A toothed driving wheel that is the largest or strongest in the mechanism.

Butt joint (open joint) In pipe, flat ends that meet but do not overlap.

Cab guard On a dump truck, a heavy metal shield extending up from the front wall of the body and forward over the cab.

Caisson A box or chamber used in construction work underwater. A foundation element.

Cam A rotating or sliding piece, or a projection on a wheel, used to impart exactly timed motion to light parts.

Camber Vertical convex curve in a culvert barrel or a structural member. Outward lean of the front wheels of a motor vehicle.

Cantilever A lever-type beam that is held down at one end, supported near the middle, and supports a load on the free end.

Capillary water Underground water drawn above the water table by capillary attraction.

Carriage A sliding or rolling base or supporting frame.

Carrier A rotating or sliding mounting or case.

Caster A wheel mounted in a swivel frame so that it is steered automatically by movements of its load. In an automotive vehicle, the toe-in of the front wheels.

Catwalk A pathway, usually of wood or metal, that gives access to parts of large machines.

Center of mass In a cut or a fill, a cross-section line that divides its bulk into equal halves.

Centerpin (center pintle) In a revolving shovel, a fixed vertical shaft around which the shovel deck turns.

Check valve Any device allowing fluid or air to pass through it in only one direction.

Chock A block used under and against an object to prevent it from rolling or sliding.

Choker A chain or cable so fastened that it tightens on its load as it is pulled.

Choker hook (round hook) A hook that can slide along a chain.

Chuck The part of a drill that rotates the steel. A device that clamps a rod or shaft.

Cleavage plane Any uniform joint, crack, or change in quality of formation along which rock will break easily when dug or blasted.

Cloth, wire Screen composed of wire or rod woven and crimped into a square or rectangular pattern.

Clutch A device that connects and disconnects two shafts that revolve in line with each other.

Clutch, jaw (positive or denture clutch) A toothed hub and a sliding toothed collar that can be engaged to transmit power between two shafts having the same axis of revolution.

Clutch, lockup A clutch that can be engaged to provide a nonslip mechanical drive through a fluid coupling.

Clutch, overrunning (freewheeling unit) A coupling that transmits rotation in only one direction and disconnects when the torque is reversed.

Clutch, slip (safety clutch) A friction clutch that protects a mechanism by slipping under excessive load.

Cobble Per ASTM D653, a rock fragment with dimensions between 3 and 12 in.

Cofferdam A set of temporary walls designed to keep soil and/or water from entering an excavation.

Collaring Starting a drill hole. When the hole is deep enough to hold the bit from slipping out, it is said to be collared.

Compacted yards Measurement of soil or rock after it has been placed and compacted in a fill.

Concussion Shock or sharp air waves caused by an explosion or heavy blow.

Core A cylindrical piece of an underground formation, cut and raised by a rotary drill, with a hollow bit.

Core barrel A hollow cylinder containing a socket and choker springs for holding a section of drilled rock.

Counterweight A "dead" or nonworking load attached to one end or side of a machine to balance weight carried on the opposite end or side.

Coyote holes Horizontal drilled holes in which explosives are packed for blasting a high rock face.

Crown The elevation of a road center above its sides. The curved roof of a tunnel or pipe.

Curtain drain (intercepting drain) A drain that is placed between the water source and the area to be protected.

Cut and cover A work method involving excavation in the open and placing of a temporary roof over it to carry traffic during further work.

Datum Any level surface taken as a plane of reference from which to measure elevations.

Decking, explosives Separating charges of explosives by inert material, which prevents passing of concussion, and placing a primer in each charge.

Decking, screens Several levels of screening in an aggregate plant.

Detergent A chemical compound that acts to clean surfaces and to keep foreign matter in solution or suspension.

Differential A device that drives two axles and allows them to turn at different speeds to adjust to varying resistance.

Dike A long, low dam. A thin rock formation that cuts across the structure of surrounding rock.

Dimension stone Rock quarried in blocks to predetermined sizes in such a manner as not to weaken or shatter it.

Dipper A digging bucket rigidly attached to a stick or arm.

Dipper trip A device that unlatches the door of a shovel bucket to dump the load.

Drawbar In a tractor, a fixed or hinged bar extending to the rear, used as a fastening for lines and towed machines or loads. In a grader, the connection between the circle and the front of the frame.

Drawbar horsepower A tractor's flywheel horsepower minus friction and slippage losses in the drive mechanism and the tracks or tires.

Drift A small, nearly horizontal tunnel.

Drill pipe The sections of a rotary drilling string connected togther with turning action.

Drill string In rotary drills, all revolving parts below the ground. In churn drills, the tools hanging from the drilling cable.

Drive To dig or make a tunnel. To hammer down piling.

Drum A rotating cylinder with side flanges used for winding in and releasing cable.

Dry well A deep hole, covered and usually lined or filled with rocks, that holds drainage water until it soaks into the ground.

Eccentric A wheel or cam with an off-center axis of revolution.

Erosion Wear caused by moving water or wind.

Fairlead A device that lines up cable so that it will wind smoothly onto a drum.

Faulting In geology, the movement producing relative displacement along a fracture in rock.

Finish grade The final grade required by specification.

Fishing The operation of recovering an object left or dropped in a drill hole.

Flange A ridge that prevents a sliding motion. A rib or rim for strength or for attachments.

Fleet angle The maximum angle between a rope and a line perpendicular to the drum on which it winds.

Flight The screw thread (helix) of an auger.

Foot In tamping rollers, one of a number of projections from a cylindrical drum.

Foot pin The hinge that attaches the boom to a revolving shovel.

Forepole A plank driven ahead of a tunnel face to support the roof or wall during excavation.

Four-part line A single rope or cable reeved around pulleys so that four lines connect the fixed and the movable units.

French drain (rubble or stone drain) Covered ditch or pipes with open joints and covered by a layer of fitted or loose stone or other pervious material.

Frostline The greatest depth to which ground may be expected to freeze.

Gantry An overhead structure that supports machines or operating parts. An upward extension of a shovel's revolving frame that holds the boom line sheaves.

Gauge size The width of a drill bit along the cutting edge.

Gear, idler A gear meshed with two other gears that does not transmit power to its shaft. Used to reverse direction of rotation in a transmission.

Gear, planetary set A gear set consisting of an inner (sun) gear, an outer ring with internal teeth, and two or more small (planet) gears meshed with both the sun and the ring.

Gear, sprocket A gear that meshes with roller or silent chain.

Gooseneck An arched connection, usually between a tractor and a trailer.

Grade stake A stake indicating the amount of cut or fill required to bring the ground to a specified level.

Gradient Slope along a specific route, as of a road surface, channel, or pipe.

Grapple A clamshell-type bucket with three or more leaves.

Grid A set of surveyor's closely spaced reference lines laid out at right angles, with elevations taken at line intersections.

Grizzly A coarse screen used to remove oversize pieces from earth or blasted rock. A gate or closure on a chute.

Ground pressure The weight of a machine divided by the area in square inches or centimeters of the ground directly supporting it.

Grout A cementing or sealing mixture of cement and water to which sand or other fillers may be added.

Grubbing Digging out roots.

Guy A line that steadies a high piece or structure by pull against an off-center load or other guys.

Hardpan Hard, tight soil. A hard layer that may form just below plow depth on cultivated land.

Harrow An agricultural tool that loosens and works the ground surface.

Head Height of water above a specified point. The back pressure against a pump from a high outlet.

Heading In a tunnel, a digging face and its work area.

Holdback An automatic safety device that prevents a conveyor belt from running backward.

Hook, grab A chain hook that will slide over any one link but will not slide along the chain.

Hook, pintle A towing bracket with a fixed lower part and a hinged upper part, which when locked together, make a round opening that can hold a tow ring.

Hopper A storage bin or a funnel that is loaded from the top and discharges through a door or chute in the bottom.

Horizontal directional drilling (HDD) A surface-launched trenchless technology for the installation of pipes, conduits, and cables.

Horsepower, drawbar Horsepower available to move a tractor and its load, after deducting losses in the power train.

Horsepower, rated Theoretical horsepower of an engine based on dimensions and speed. Power of an engine according to a particular standard.

Horsepower, shaft (flywheel or belt horsepower) Actual horsepower produced by the engine, after deducting the drag of accessories.

Hub The strengthened inner part or mounting of a wheel or gear.

Humus Decayed organic matter. A dark, fluffy swamp soil composed chiefly of decayed vegetation; also called peat.

Hydraulic fill Fill moved and placed by running water.

Hydraulic fracture Commonly associated with horizontal directional drilling, a form of inadvertent drilling fluid return in which drilling fluid pressures exceed the strength and confining stress of the soil or rock above the bore. Also called a hydrofracture, fracking, or a "frac-out."

Hydro-excavation Excavation performed using high-pressure water to loosen soil and remove the soil with a suction created by a vacuum pump or fan blower. Commonly used for potholing. *See also* Vacuum excavation.

Idler A wheel or gear changing the direction of shaft rotation, or the direction of movement of a chain or belt. An unpowered pulley or wheel.

Injector In a diesel or gasoline engine, the unit that sprays fuel into the combustion chamber.

Interference, active An electromagnetic signal produced by an energized source such as a traffic loop or overhead electric lines that interferes with the receipt of electromagnetic signal from an HDD transmitter.

Interference, passive A condition that attenuates, dilutes, or weakens the signal received by a receiver from a transmitter, such as the reinforcing steel in structural concrete or rock formations containing iron or other metallic ores.

Invert The inside bottom elevation of a pipe or tunnel.

Jack boom A boom supporting sheaves between the hoist drum and the main boom in a pull shovel or a dredge.

Jacking pipe A sectional pipe with a low-profile bell used in pipe jacking in order to withstand the hydraulic jacking forces transmitted through the pipe while maintaining the structural integrity of the pipe itself.

Jackknife A tractor and trailer assuming such an angle to each other that the tractor cannot move forward.

Jetting Drilling with high-pressure water or air jets.

Jib boom An extension piece hinged to the upper end of a crane boom.

Jumbo A number of drills mounted on a mobile carriage and used in tunnels.

Kelly A square or fluted pipe that is turned by a drill rotary table, while it is free to move up and down in the table. Also called grief stem or Kelly bar.

Key A hard steel strip inserted in matching grooves (keyways) in a shaft and a hub to make them turn as a unit.

Knife The dirt-cutting edge of a digging machine.

Lacing Small boards or patches that prevent dirt from entering an excavation through spaces between sheeting or lagging planks.

Ladder The digging boom assembly in a hydraulic dredge or chain-and-buckets ditcher.

Ladder ditcher A machine that digs ditches by means of buckets in a chain that travels around a boom.

Lag Delay in one action following another. To install lagging, or increase the diameter of a drum.

Lagging The surface or contact area of a drum or flat pulley, especially a detachable surface or one of special composition. In a tunnel, planking placed against the dirt or rock walls and ceiling, outside the ribs. Boards fastened to the back of a shovel for blast protection.

Lay The direction of twist in wires and strands in wire rope.

Lay, lang A wire–rope construction in which the wires are twisted in the strands in the same direction as the strands are twisted in the rope.

Lay, regular A wire–rope construction in which the direction of twist of the wires in the strands is opposite to that of the strands in the rope.

Leg A side post in tunnel timbering. A wire or connector in one side of an electric circuit.

Line oiler An oil reservoir and metering device placed in a compressed-air line to lubricate air tools.

Lip The cutting edge of a bucket. Applied chiefly to edges, including tooth sockets.

Load To place explosives in a hole. To transfer material to a hauling unit or hopper.

Load, deck Charges of dynamite spaced well apart in a borehole and fired by separate primers or by detonating cord.

Loam A soft, easily worked soil containing sand, silt, and clay.

Low bed A machinery trailer with a low deck.

Lug down To slow down an engine by increasing its load beyond its capacity.

Magazine A structure or container in which explosives are stored.

Mast A tower or vertical beam carrying one or more load lines at its top.

Mat A heavy, flexible fabric of woven wire rope or chain used to confine blasts. A wood platform used in sets to support machinery on soft ground.

Millisecond delay (short-period delay) A type of delay cap with a definite but extremely short interval between the passing of current and the explosion.

Misfire Failure of all or part of an explosive charge to go off.

Mixed face In tunneling, digging in both dirt and rock in the same heading at the same time.

Muck Mud rich in humus. Finely blasted rock.

Mudcapping Blasting boulders or other rock by means of explosive laid on the surface and covered with mud.

Nip The seizing or gripping of stone between the jaws or rolls of a crusher.

Nip, angle of In a roll crusher, the angle between tangents to the roll surfaces at the widest point at which they will grip a stone.

One-part line A single strand of rope or cable.

Open cut A method of excavation in which the working area is kept open to the sky. Used to distinguish from cut-and-cover and underground work.

Ore Rock or earth containing workable quantities of a mineral or minerals of commercial value.

Outrigger An outward extension of a frame that is supported by a jack or block. Used to increase stability of a machine.

Overbreak Moving or loosening of rock as a result of a blast beyond the intended line of cut.

Pad (shoe or plate) The part of a crawler-type track that contacts the ground.

Parts of line Separate strands of the same rope or cable used to connect two sets of sheaves.

Pass A working trip or passage of an excavating, grading, or compacting machine.

Peat (humus) A soft, light swamp soil consisting mostly of decayed vegetation.

Perched water table Underground water lying over dry soil and sealed from it by an impervious layer.

Petcock A small drain valve.

Pintle A vertical pin fastened at the bottom that serves as a center of rotation.

Pintle hook A towing device consisting of a fixed lower jaw, a hinged and lockable upper jaw, and a socket between them to hold a tow ring.

Pioneering The first working over of rough or overgrown areas.

Pitch arms (pitch braces, pitch rods) Rods, usually adjustable, that determine the digging angle of a blade or bucket.

Ply One of several layers of fabric or of other strength-contributing material.

Portal A nearly level opening into a tunnel.

Potholing A small excavation used for locating buried utility lines or other subsurface features, typically accomplished using vacuum excavation.

Power takeoff A place in a transmission or engine to which a shaft can be attached so as to drive an outside mechanism.

Prime To provide means to start a process, as to supply sufficient water to a pump to enable it to start pumping. In blasting, to place a detonator in a cartridge or a charge of explosive.

Puddle To compact loose soil by soaking it and allowing it to dry.

Puff blowing (blowing) Blowing chips out of a hole by means of exhaust air from the drill.

Pusher A tractor that pushes a scraper to help it pick up a load.

Quicksand Fine sand that is prevented from settling firmly together by upward movement of groundwater. Any wet, inorganic soil so unsubstantial that it will not support any load.

Radial Lines converging at a single center.

Reamer A cutting device that enlarges or straightens a hole.

Reciprocating Having a straight back-and-forth or up-and-down motion.

Reclaiming Digging from stockpiles. Reprocessing previously rejected material.

Reeving Threading or placement of a working line.

Relief holes Holes drilled closely along a line, which are not loaded and which serve to weaken the rock so that it breaks on that line.

Retaining wall A wall separating two levels.

Reverse bend To bend a line over a drum or a sheave and then in the opposite direction over another sheave.

Revetment A wall sloped back sharply from its base. A masonry or steel facing for a bank.

Rifling Forming a spiral thread on the wall of a drill hole, which makes it difficult to pull out the bit.

Riprap Heavy stones placed on a slope or at water's edge to protect soil from waves or current.

Rocker arm A lever resting on a curved base so that the position of its fulcrum moves as its angle changes. A bell crank with the fulcrum at the bottom.

Roller, swing In a revolving shovel, one of several tapered wheels that roll on a circular turntable and support the upper works.

Roller, track In a crawler machine, the small wheels that rest on the track and carry most of the weight of the machine.

Sandhog An underground worker who works in compressed air.

Scaling Prying loose pieces of rock off a face or roof to avoid danger of their falling unexpectedly.

Scarifier An accessory on a grader, roller, or other machine used chiefly for shallow loosening of road surfaces.

Scour Erosion in a streambed, particularly if caused or increased by channel changes.

Seam A layer of rock, coal, or ore.

Shackle A connecting device for lines and drawbars, which consists of a U-shaped section pierced for a cross-bolt or a pin.

Sheave (pronounced "shiv") A grooved wheel used to support cable or change its direction of travel.

Sheave, traveling A sheave block that slides in a track.

Sheave block A pulley and a case provided with means to anchor it.

Shoring Temporary bracing to hold the sides of an excavation from caving.

Shoulder The graded part of a roadway on each side of the pavement. The side of a horizontal pipe at the level of the centerline.

Shuttle A back-and-forth motion of a machine that continues to face in one direction.

Sidecasting Piling spoil alongside the excavation from which it is taken.

Sidehill cut A long excavation in a slope that has a bank on one side and is near original grade on the other.

Sling A lifting hold consisting of two or more strands of chain or cable.

Sling block A frame in which two sheaves are mounted so as to receive lines from opposite directions.

Sloping A trench protection method where the walls of the trench are excavated in a V-shaped manner to reduce lateral pressure on the soil mass and reduce the likelihood of soil failure and trench collapse.

Sod A rolled or cut layer of grass or turf that is held together by its roots in a thin layer of soil matting.

Solid loading Filling a drill hole with all the explosive that can be crammed into it, except for the top stemming.

Spiraling rifling A drill hole twisting into a spiral around its intended centerline.

Spoil Dirt or rock that has been removed from its original location.

Spotter In truck or scraper use, the person who directs the driver into loading or dumping position.

Spur A rock ridge projecting from a sidewall after inadequate blasting.

Stoper A hand-operated air drill mounted on a column or other support.

Supercharger A blower that increases the intake pressure of an engine.

Surge bin A compartment for temporary storage that will allow converting a variable rate of supply into a steady flow of the same average amount.

Swivel In an HDD operation, it is the device installed between the reamer and product pipe used to isolate the pipe string from the rotating reamer and drill string.

Tagline A line from a crane boom to a clamshell bucket that holds the bucket from spinning out of position.

Tail The rear of a shovel deck.

Tailboard Tailgate.

Tailgate The hinged rear wall of a dump truck body. The hinged or sliding rear wall of a scraper bowl (ejector).

Tailings Second-grade or waste material separated from orebody or pay material during screening or processing.

Tail swing The clearance required by the rear of a revolving shovel.

Talus Loose rock or gravel formed by disintegration of a steep rock slope.

Tandem A double-axle drive unit for a truck or grader. A pair in which one part follows the other.

Tandem drive A multi-axle vehicle having two adjacent driving axles.

Three-part line A single strand of rope or cable doubled back around two sheaves so that three parts of it pull a load together.

Track frame In a crawler mounting, a side frame to which the track roller and idler are attached.

Track roller In a crawler machine, the small wheels that are under the track frame and that rest on the track.

Tread The surface of a tire or track contacting the ground,

Trunnion (walking beam or bar) An oscillating bar that allows changes in angle between a unit fastened to its center and another attached to both ends. A heavy horizontal hinge.

Turntable A base that supports a part and allows it to rotate or swing. In a shovel, the upper part of the travel unit.

Two-part line A single strand of rope or cable doubled back around a sheave allowing the two parts to hoist a load together.

Universal joint ("U" joint) A connection between two revolving shafts that allows them to turn or swivel at an angle.

Vacuum excavation An excavation performed using high-pressure water or air to loosen soil and remove the soil with a suction created by a vacuum pump or fan blower. Commonly used for potholing or pavement surface profiling.

Vein A layer, seam, or narrow irregular body of material that is different from surrounding formations.

Vertical drains Usually, columns of sand used to vent water squeezed out of humus by weight of fill.

Walking dragline A dragline shovel that moves itself along the ground by means of side-mounted shoes.

Waste Digging, hauling, and dumping of valueless material to get it out of the way; or the valueless material itself.

Water table The surface of underground, gravity-controlled water.

Working cycle A complete set of operations. In an excavator, it usually includes loading, moving, dumping, and returning to the loading point.

INDEX

Note: Page numbers followed by *f* denote figures; page numbers followed by *t* denote tables.

Printed in the USA
CPSIA information can be obtained
at www.ICGtesting.com
LVHW082023220923
757462LV00009B/6